PARTICLE DARK MATTER
Observations, Models and Searches

Dark matter is among the most important open problems in modern physics. Aimed at graduate students and researchers, this book describes the theoretical and experimental aspects of the dark matter problem in particle physics, astrophysics and cosmology. Featuring contributions from 46 leading theorists and experimentalists, it presents many aspects, from astrophysical observations to particle physics candidates, and from the prospects for detection at colliders to direct and indirect searches.

The book introduces observational evidence for dark matter along with a detailed discussion of the state-of-the-art of numerical simulations and alternative explanations in terms of modified gravity. It then moves on to the candidates arising from theories beyond the Standard Model of particle physics, and to the prospects for detection at accelerators. It concludes by looking at direct and indirect dark matter searches, and the prospects for detecting the particle nature of dark matter with astrophysical experiments.

GIANFRANCO BERTONE is Coordinator of the Theoretical Physics group at the Institut d'Astrophysique de Paris, and Visiting Professor at the Institute for Theoretical Physics, University of Zurich. His technical work mainly focuses on theoretical and phenomenological aspects of the dark matter problem, and he participates in important experimental collaborations on the elaboration of optimal strategies for the detection of dark matter particles.

PARTICLE DARK MATTER

Observations, Models and Searches

Edited by

GIANFRANCO BERTONE
Institut d'Astrophysique de Paris

CAMBRIDGE
UNIVERSITY PRESS

CAMBRIDGE UNIVERSITY PRESS
Cambridge, New York, Melbourne, Madrid, Cape Town,
Singapore, São Paulo, Delhi, Mexico City

Cambridge University Press
The Edinburgh Building, Cambridge CB2 8RU, UK

Published in the United States of America by Cambridge University Press, New York

www.cambridge.org
Information on this title: www.cambridge.org/9781107653924

First published 2010
First paperback edition 2013

A catalogue record for this publication is available from the British Library

Library of Congress Cataloguing in Publication Data
Particle dark matter : observations, models and searches / edited by Gianfranco Bertone.
p. cm.
ISBN 978-0-521-76368-4 (Hardback)
1. Dark matter (Astronomy)   2. Mass (Physics)   I. Bertone, Gianfranco, 1975–
II. Title.
QB791.3.P365 2009
523.1′126–dc22

2009044256

ISBN 978-0-521-76368-4 Hardback
ISBN 978-1-107-65392-4 Paperback

Contents

Contributors

Elena Aprile
1016 Pupin Hall/Nevis, MC 5231, Box 31, 538 W 120 St, New York, NY 10027, USA

Stephen Asztalos
Lawrence Livermore National Laboratory, High Energy Physics and Accelerator Technology Group, L-50 Livermore, CA 94551, USA

Marco Battaglia
Lawrence Berkeley National Laboratory, Physics Division, 1, Cyclotron Road, Mail Stop 50A2161, Berkeley, CA 94720-8143, USA

Laura Baudis
University of Zurich, Physik Institut, Winterthurerstrasse 190, CH-8057 Zurich, Switzerland

Jacob Bekenstein
Racah Institute of Physics, The Hebrew University of Jerusalem, Givat Ram, Jerusalem 91904, Israel

Pierluigi Belli
Università degli Studi di Roma 'Tor Vergata', Via della ricerca Scientifica 1, 00133 Roma, Italy

Lars Bergström
Department of Physics, Stockholm University, AlbaNova University Centre, S-106 91 Stockholm, Sweden

Rita Bernabei
Università degli Studi di Roma 'Tor Vergata', Via della ricerca Scientifica 1, 00133 Roma, Italy

Gianfranco Bertone
Institut d'Astrophysique de Paris, 98 bis Bd Arago, 75014 Paris, France

Fawzi Boudjema
LAPTH CNRS, Chemin de Bellevue, F-74941 Annecy-le-Vieux, France

James Bullock
Physics and Astronomy Department, 4168 Frederick Reines Hall, University of California, Irvine, CA 92697-4575, USA

David Cerdeño
Departmento de Física Teórica, Universidad Autónoma de Madrid, Módulo C-Xl, 608, Cantoblanco, Madrid 28049, Spain

Juerg Diemand
Department of Astronomy and Astrophysics, University of California, Santa Cruz, CA 95064, USA

Fiorenza Donato
University of Turin, via Giuria 1, 10125, Italy

Joakim Edsjö
Department of Physics, AlbaNova, Stockholm University, SE-106 91 Stockholm, Sweden

John Ellis
CERN, Theory Division, CH-1211 Geneva 23, Switzerland

Jonathan Feng
Physics and Astronomy Department, 4168 Frederick Reines Hall, University of California, Irvine, CA 92697-4575, USA

Nicolao Fornengo
University of Turin, via Giuria 1, 10125 Torino, Italy

Jules Gascon
Institut de Physique Nucleaire, 4 rue Enrico Fermi, 69622 Villeurbanne cedex, France

Graciela Gelmini
UCLA Physics and Astronomy, Box 951547, PAB 7-320, Los Angeles, CA 90095-1547, USA

Gilles Gerbier
Centre d'Etudes de Saclay (CEA-Saclay), Orme des Merisiers, F-91191 Gif-sur-Yvette Cedex, France

Paolo Gondolo
The University of Utah, Department of Physics, 115 S 1400 E, Salt Lake City, UT 84112-0830, USA

Anne Green
The School of Physics and Astronomy, University of Nottingham, University Park, Nottingham NG7 2RD, UK

Francis Halzen
University of Wisconsin-Madison, Department of Physics, Phenomenology Institute, 5293 Chamberlin Hall, Madison, Wisconsin, USA

Dan Hooper
Wilson Hall 6 West, Fermilab, PO Box 500, Batavia, IL 60510, USA

Karsten Jedamzik
Laboratoire Physique Théorique & Astroparticules, Université de Montpellier II, Place Eugène Bataillon, FR 34095 Montpellier Cedex 5, France

Manoj Kaplinghat
Physics and Astronomy Department, 4168 Frederick Reines Hall, University of California, Irvine, CA 92697-4575, USA

Kyoungchul Kong
Fermilab, Theory Department, PO Box 500 MS 106, Batavia, IL 60510, USA

Konstantin Matchev
Department of Physics, University of Florida, PO Box 118440, Gainesville, FL 32611-8440, USA

Yannick Mellier
Institut d'Astrophysique de Paris, 98 bis Bd Arago, 75014 Paris, France

David Merritt
Department of Physics, 85 Lomb Memorial Drive, Rochester Institute of Technology, Rochester, NY 14623-5604, USA

Ben Moore
Institute for Theoretical Physics, University of Zurich, Winterthurerstrasse 190, CH-8057 Zurich, Switzerland

Keith Olive
William I. Fine Theoretical Physics Institute, School of Physics and Astronomy, University of Minnesota, 116 Church Street SE, Minneapolis, MN 55455, USA

Michael Peskin
Theory Group, MS 81, SLAC, Stanford University, 2575 Sand Hill Road, Menlo Park, CA 94025, USA

Tilman Plehn
Institute for Theoretical Physics, Heidelberg University, Philosophenweg 16, 69120 Heidelberg, Germany

Giacomo Polesello
INFN Sezione di Pavia, Via Bassi 6, 27100 Pavia, Italy

Maxim Pospelov
Department of Physics and Astronomy, University of Victoria, Victoria, BC, V8P 1A1, Canada

Stefano Profumo
Department of Physics, ISB 325, University of California, 1156 High Street, Santa Cruz, CA 95064, USA

Pierre Salati
LAPTH, 9 Chemin de Bellevue BP 110, 74941 Annecy le Vieux Cedex, France

Geraldine Servant
CERN, Theory Division, CH-1211 Geneva 23, Switzerland

Mikhail Shaposhnikov
Ecole polytechnique fédérale de Lausanne, Institut de théorie des phénomènes physiques, LPPC, BSP - Dorigny, CH-1015 Lausanne, Switzerland

Pierre Sikivie
Physics Department, University of Florida, Gainesville, FL 32611, USA

Joseph Silk
University of Oxford, Department of Physics, Denys Wilkinson Building, Keble Road, Oxford, UK

Neil Spooner

Room E23, Hicks Building, Dept of Physics and Astronomy, University of Sheffield, Sheffield S3 7RH, UK

Louis Strigari

Kavli Institute (KIPAC), Physics Department, Stanford University, SLAC 2575 Sand Hill Road, M/S 29, Menlo Park, CA 94025, USA

Piero Ullio

SISSA, via Beirut, 2–4, 34014 Trieste TS, Italy

Preface

Dark matter (DM) is one of the pillars of the Standard Cosmological Model, but the nature of this elusive component of the matter budget of the Universe remains unknown, despite the compelling evidence at all astrophysical scales. The possible connection with theories beyond the Standard Model of particle physics makes DM one of the most important open problems in modern cosmology and particle physics, as witnessed by the enormous theoretical and experimental effort that is being put towards its identification.

Many different strategies have been devised to achieve this goal. First, the Large Hadron Collider, which is just starting operations, is expected to provide insights of paramount importance into possible extensions of the Standard Model of particle physics. Whether or not a specific candidate is "observed" at the LHC, any evidence for new physics (or lack thereof) will inevitably change our understanding of physics, and in particular our understanding of DM. If DM candidates are actually found, the question will arise of whether they actually are *the* DM in the Universe.

A convincing identification can probably be obtained only by combining the results of accelerator searches with astrophysical searches, based on the direct or indirect detection of DM particles in the local Universe. Direct DM searches are based on the measurement of the recoil energy of nuclei struck by DM particles in large detectors. This field has evolved dramatically in the past decade, and the different experimental strategies (cryogenic, liquid noble gases, superheated) developed over the years have led to a spectacular improvement of the constraints on DM–nucleon interactions. Alternatively, DM could be detected indirectly, through the detection of its annihilation or decay products. With this aim, many important instruments are currently taking data, in particular in the energy range between 1 GeV and 1 TeV, like the antimatter satellite PAMELA and the gamma-ray satellite Fermi, launched in June 2008.

The material is arranged in five parts: Dark matter in cosmology; Candidates; Accelerator searches; Direct detection; Indirect detection and astrophysical constraints. The interested reader will find an introduction to the DM problem and a detailed overview of the contents of the book in Chapter 1.

In brief, Part I is devoted to the astrophysical and cosmological aspects of the DM problem. The current understanding of the distribution of DM in the Universe, based on numerical simulations and astrophysical observations, is reviewed here, along with the most recent lensing observations that provide a 'direct' proof of the existence of DM, and a discussion of alternative theories that seek to dispense with DM. The chapters on DM distribution are particularly important for the reader interested in the indirect DM searches discussed in Part V.

The particle physics aspects of the DM problem are discussed in Part II. This part contains a discussion of the production mechanisms of DM in the early Universe, including among others thermal production, relevant for the broad class of candidates generically referred to as weakly interacting massive particles (WIMPs). The most widely discussed extension of the Standard Model of particle physics, supersymmetry (SUSY), and the most widely discussed DM candidate, the neutralino, are also introduced here. The discussion is, however, enlarged to include a systematic review of alternative extensions of the Standard Model, and alternative DM candidates, including non-SUSY WIMPs, and non-WIMP candidates.

Parts III, IV and V are dedicated to DM searches. Accelerator searches for new physics are discussed in Part III. Of particular importance for the reader interested in the DM problem are the chapters on strategies to discover and identify extensions of the Standard Model of particle physics at the LHC, and on the techniques that may allow, in case of detection, the identification of possible DM candidates. Part III also contains a chapter on DM 'tools', software that has been developed over the years by many different groups in an effort to allow systematic scans of the parameter space of new theories, along with the determination of physical quantities relevant for cosmology and accelerator, direct and indirect searches in each of the models explored in random scans.

Direct detection – the detection of DM through the measurement of the recoil energy of nuclei struck by DM in low-background detectors – is presented in Part IV. The main avenues in this field of research, cryogenic detectors and detectors using liquid noble gases, are presented here, along with a phenomenological overview and a discussion of 'directional' detectors currently under study.

Finally, Part V contains a discussion of the prospect for detecting DM 'indirectly', that is, through the observation of the products of its annihilation or decay. This is a field that is witnessing an explosion of interest, owing to the new data of the Fermi and PAMELA satellites, and it is anticipated that the excitement will remain high in the upcoming years, when a new generation of experiments will become available. A critical assessment of the possible DM interpretation of existing data is presented in this last part of the book, along with a discussion of the strategies that may provide the long-awaited smoking-gun signature for DM.

This book aims at presenting in a coherent way the state-of-the-art of the relevant aspects of the disciplines involved in the DM problem (astrophysics, cosmology and particle physics), along with the detection strategies, in order to build a common language among the different communities and, we hope, to prepare for the age of discoveries. It is more than a collection of review papers, in the sense that particular care has been taken to ensure a coherent and complete presentation of the DM problem, and to bear in mind as its eventual readers graduate students and researchers who want to obtain a better understanding of the many different aspects of the DM puzzle. The 46 authors who joined the project met the challenge of summarizing in chapters of 15 to 20 pages an entire field of research, in an effort to be at the same time accessible and complete. I thus believe that it can become a tool to increase exchanges among the different communities involved in DM searches, and to pave the road to a truly multidisciplinary approach to DM.

As is the case for all major endeavours, it is by no means certain that the search for DM will succeed. But this is how scientific research proceeds. If all our attempts fail, we will have to perform a radical revision of our understanding of Nature. But if one or more of the strategies so far devised turns out to be successful, the discovery of DM may well be remembered as one of the most exciting adventures in the history of science.

Acknowledgements

This book is the result of the collaborative effort of 46 authors. I am grateful to all of them for the great quality of their contributions, and for keeping up with the tight schedule I have imposed. I am indebted to Lars Bergström, Enzo Branchini, Marco Cirelli, Stephane Colombi, Gilles Esposito Farese, Nicolao Fornengo, Fabio Iocco, Yannick Mellier, Lidia Pieri, Pasquale Serpico, Volker Springel, Marco Taoso and Matteo Viel for reading parts of the book and providing very useful comments. It is a pleasure to thank Simon Capelin of Cambridge University Press, who solicited a book on dark matter on the occasion of the PASCOS'08 conference at the Perimeter Institute, and his assistant Laura Clark, who assisted me during the completion of this book. I also thank the Galileo Galilei Institute of Theoretical Physics in Florence, where the final touches were put to this project. Finally, thanks to Nadia. She knows why.

Symbols and abbreviations

Symbols

Throughout the book, quantities describing the properties of DM particles are denoted with the subscript χ, unless otherwise specified. Here are some of the most frequently used symbols:

m_χ, mass of the DM particle

σ_p^{SI}, spin-independent scattering cross-section off protons

σ_n^{SI}, spin-independent scattering cross-section off neutrons

σ_p^{SD}, spin-dependent scattering cross-section off protons

σ_n^{SD}, spin-dependent scattering cross-section off neutrons

$\langle \sigma v \rangle$, thermal average of the annihilation cross-section

(σv), annihilation cross-section in the non-relativistic limit

M_{GUT}, Grand-unification scale

M_{Pl}, Planck scale

M_{SUSY}, Supersymmetry scale

R_0, Galactocentric radius of the Sun

ρ_{crit}, critical density of the Universe

ρ_χ, DM density

$\rho_\chi(R_0) \equiv \rho_0$, DM density in the solar neighbourhood

Ω_χ, relic abundance of DM (in units of ρ_{crit})

Ω_{M}, relic abundance of matter (same units)

Ω_{b}, relic abundance of baryons (same units)

a_{o}, MOND parameter

r_{vir}, virial radius

M_{vir}, virial mass

c_{vir}, virial concentration

δ_{vir}, virial overdensity

Acronyms and abbreviations

AQUAL, Aquadratic Lagrangian (theory)

BAO, Baryon Acoustic Oscillation

BBN, Big Bang Nucleosynthesis

BSM, (theories) Beyond the Standard Model

cMSSM, constrained MSSM

CDM, Cold Dark Matter

CMB, Cosmic Microwave Background

Crest, Collisionally Regenerated Structure

DM, Dark Matter

DSph, Dwarf Spheroidal Galaxy

EM, Electroweak

FP, Fokker–Planck (equation)

GEM, Gas Electron Multiplier

GC, Galactic centre

GR, General Relativity

IDM, Inert Doublet Model

IMBH, Intermediate Mass Black Hole

KK, Kaluza–Klein

LHC, Large Hadron Collider

LMC, Large Magellanic Cloud

LSP, Lightest supersymmetric particle

LKP, Lightest Kaluza–Klein particle

LTP, Lightest T-odd Particle

LZP, Lightest Z_3 Particle

LTR, Low-Temperature Reheating (models)

MOND, Modified Newtonian Dynamics

MSSM, Minimal Supersymmetric Standard Model

MW, Milky Way

NFW, Navarro, Frenk and White (profile)

NS, Neutron Star

mSUGRA, Minimal Supergravity

NTD, Neutron Transmutation Doped (germanium sensors)

SD, Spin-Dependent (coupling)

SDSS, Sloan Digital Sky Survey

SI, Spin-Independent (coupling)

SM, Standard Model

SMC, Small Magellanic Cloud

SMBH, Supermassive Black Hole

SNR, Signal-to-Noise Ratio
SQUID, Superconducting QUantum Interference Device
SUSY, Supersymmetry
TES, Transition Edge Sensors
TeVeS, Tensor Vector Scalar theory
TPC, Time Projection Chamber
UED, Universal Extra Dimensions
WD, White Dwarf
WDM, Warm Dark Matter
WIMP, Weakly Interacting Massive Particle

Part I

Dark matter in cosmology

1
Particle dark matter

Gianfranco Bertone and Joseph Silk

1.1 Introduction

Dark matter is surely at the heart of modern cosmology. It undoubtedly
pervades the Universe, unless we are being completely misled by diverse
data sets, yet it has not been detected. The possible connection with proposed extensions of the Standard Model of particle physics, currently being
searched for at accelerators, makes the identification of dark matter one of
the highest priority goals in cosmology and particle physics. In this chapter
we provide an introduction to the dark matter situation and an overview of
the material presented in this book.

Dark matter has a venerable history (see e.g. ref. [742] for an historical
account). One could even cite Solar System arguments for dark matter,
including anomalies in the orbit of Uranus and the advance of Mercury's
perihelion. One led to the discovery of a previously dark planet, Neptune,
the other to a new theory of gravitation. Similar parallels may be drawn
today. There are advocates of new theories of gravitation (see Chapter 6),
who seek to dispense with dark matter, and there are observations of large-
scale structure, such as gravitational lensing, the cosmic web and the cosmic
microwave background acoustic fluctuations, that are notoriously difficult
to reproduce in the absence of a dominant component of cold dark matter
(CDM) particles (we refer the interested reader to Chapter 4, which includes
an introduction to gravitational lensing in the context of the CDM paradigm,
and its potential to discriminate CDM from modified-gravity theories). A
strong case for the dominance of dark matter in galaxy clusters was made
as long ago as 1933 [1977]. It is remarkable that our understanding of its
nature has not advanced since then. Of course, modern observations have

Particle Dark Matter: Observations, Models and Searches, ed. Gianfranco Bertone. Published by
Cambridge University Press. © Cambridge University Press 2010.

led to an increasingly sophisticated exploration of the distribution of dark matter, now confirmed to be a dominant component relative to baryonic matter over scales ranging from those of galaxy haloes to that of the particle horizon.

1.2 The baryon budget

It is useful to begin our overview by noting that diffuse baryons cannot account for the dark matter. There are three methods for determining the baryon fraction in the high-redshift Universe. The traditional approach is via primordial nucleosynthesis of ^{4}He, ^{2}H and ^{7}Li, and it provides a unique value of $\Omega_b = 0.04 \pm 0.02$ which is generally consistent with recent data. There are, however, some potential difficulties, such as the tension between ^{2}H and ^{4}He, on the one hand, which in principle are the most sensitive barometers, and ^{7}Li, whose abundance in metal-poor stars may or may not be depleted by stellar convection [498] (see Chapter 28). An independent probe of Ω_b comes from measuring the relative heights of the peaks in the acoustic temperature fluctuations of the cosmic microwave background. With conventional priors, the data yields excellent agreement between the baryon abundance at redshift $z \approx 1000$ and that inferred from nucleosynthesis at $z \approx 10^9$. A third independent measure of Ω_b, this time at $z \approx 3$, comes from modelling the Lyman alpha forest of the intergalactic medium (IGM). This depends on the square root of the ionizing photon flux, which in this redshift range is due predominantly to quasars and is measured. These are high-z measurements. At $z \sim 0$, cold intergalactic gas at the current epoch is sparser, and its detection ideally requires a far ultraviolet telescope such as FUSE. However, the IGM is found to dominate the known baryon fraction today, and amounts to about 30% of the total baryon fraction.

The most reliable present-epoch measure of the primordial baryon fraction comes from galaxy clusters, which are usually considered to be laboratories that have retained their primordial baryon fraction. The observed baryon fraction in massive clusters is about 15%, which is consistent with $\Omega_b = 0.04$ for $\Omega_M = 0.26$, the WMAP5-preferred value being $\Omega_b h^2 = 0.02273 \pm 0.00062$ with Hubble parameter $h = 0.719 \pm 0.027$. However, there is a shortfall: the hot diffuse gas in clusters only accounts for 90% of the baryons expected from primordial nucleosynthesis considerations [987]. There are indications, motivated as much by theory as by observations, that the remaining baryons are in the warm-hot intergalactic medium (WHIM) at a temperature of 10^5–10^7 K [502]. According to the simulations, in fact, 30% or more of the baryons are actually heated by a combination of gravitational clustering and galactic outflows by the present epoch and remain diffuse. Observations seem to confirm the existence of some WHIM, which may contain as much as 50% of all the baryons in the Universe, in particular via detection of redshifted rest-frame UV OVI absorption towards quasars, extended soft X-ray emission near clusters, and OVII/OVIII X-ray absorption along lines of sight to active galactic nuclei (AGN) [1508], although the statistical significance of this detection is still controversial [1628]. Although too few lines of sight have so far been probed to confirm the expected WHIM mass fraction, the baryons are in plausible environments with plausible heating sources, and with every expectation of being present at the anticipated level.

1.3 The case for cold dark matter: good news and bad news

There are some noteworthy success stories for CDM. First and foremost is its success in predicting the initial candidates for structure formation, which culminated in the discovery of the temperature fluctuations in the cosmic microwave background (CMB). The amplitude of the Sachs–Wolfe effect was predicted to within a factor of 2, under the assumption, inspired qualitatively by inflation but quantitatively by the theory of structure formation via gravitational instability in the expanding Universe, of adiabatic scale-invariant initial density fluctuations. The first direct test of this theory came with the detection and mapping of the acoustic peaks. These are the hallmarks of galaxy formation, first predicted some three decades previously, and demonstrate the imprint of the density fluctuation initial conditions on the last scattering surface of the CMB at $z \approx 1000$. The first six peaks have been measured at high resolution by the ACBAR experiment [1635], and fit precisely onto the best-fit WMAP5 power spectrum [729]. This is a remarkable confirmation of the essential correctness of the hypothesis that approximately scale-invariant primordial adiabatic density fluctuations are the origin of structure in the Universe. It is also worth mentioning that the fluctuations seen in the CMB can be extrapolated to the present, to predict the peculiar velocities that distort redshift-space clustering. This extrapolation has been found to be in agreement with direct observations from galaxy surveys, thus providing a remarkable test of the CDM model [1550].

Another dramatic demonstration of the essential validity of CDM has come from the simulations of the large-scale structure of the Universe (see Chapter 2). Once the initial conditions, including gaussianity, are specified, growth occurs by gravitational instability, and the sole requirements on dark matter are that it be weakly interacting and cold. Thus was born CDM, and the CDM scenario works so well that we cannot easily distinguish the artificial universe from the actual Universe mapped by means of redshift surveys [1791]. More to the point, perhaps, is that the simulations are used to generate mock galaxy catalogues and maps that yield the observed correlations and clustering of galaxies, and precise values of the cosmological parameters, in combination with the CMB maps and the other cosmological probes.

Dark-matter-dominated haloes of galaxies are another generic success of CDM, as mapped out by rotation curves. However, the detailed predicted properties of haloes do not seem to be well matched to observations in the inner regions of many disk galaxies [1672]. Dark matter cusps (density $\rho \propto r^{-\alpha}$ with $1 < \alpha < 1.5$) are a robust CDM prediction [1504] and are not found in most low surface brightness dwarfs observed at high resolution [1863] (see Chapter 5 for a discussion of DM distribution at the centre of galaxies). Observational issues, however, can complicate the modelling of the data in well-studied examples [1887]. Nor is the predicted dark matter concentration ($C \equiv r_{200}/r_s \approx 5$–$10$, where r_{200} is the radius at density contrast 200 and r_s is the halo scale length) consistent with the dark matter distribution in barred disk galaxies, possibly including our own Galaxy. Another issue is the predicted number of satellites as well as the frequency of massive galaxies both at present and in the past, cf. [1388]. There might be excess numbers of predicted satellites and too many massive galaxies predicted today, and maybe the converse is the case at $z \approx 2$–3 (see Chapter 3 for an exhaustive discussion). On the baryonic front, an unresolved challenge for theorists is the excessive loss of angular momentum by the contracting and cooling baryons in the dark halo. The resulting spheroids are too massive, and continuing minor mergers make the final disks too thick. Finely tuned merging models, with late gas infall forming a cold disk, may be able to reproduce late-type galaxies with small bulges [1000], although we cannot yet explain the observed frequency of thin disk-dominated galaxies or the distribution of angular momentum. The essential problem is that one simultaneously has to explain disk galaxies as well as account for the dark matter concentrations and the galaxy luminosity function. Hitherto, solving any one of these problems has generally aggravated the other contact points with data.

It is difficult, however, to be definitive about any possible contradiction between theory and observation. For example, reformation of bars by gas infall can avoid the problem of bar spin-down by dynamical friction [1722], gas bulk flows can dynamically soften the dark matter cusps of dwarf galaxies [1395] as well as massive galaxies [1164], and astrophysical processes, including supernova-driven winds, can render the dwarf satellites optically invisible. Many extremely faint dwarfs are indeed being discovered in our halo that could plausibly be identified with the expected faded relics. The observed numbers are within a factor of ~4 of accounting for the missing satellite problem [1763]. At high redshift ($z \gtrsim 4$), the galaxy luminosity function slope is steep at the low luminosity end with $\alpha \approx 1.7$ [420], almost in accord with CDM expectations. Numerical simulations at high resolution of the subhalo mass function yield $\alpha \approx 1.9$ [1790]. Massive galaxies undergo strong AGN feedback that prevents excessive growth at late epochs, and also terminates but possibly accelerates early star formation. Hence astrophysical processes should be able to reconcile CDM reasonably well with the data.

This certainly is the conservative viewpoint, while a more radical view is that the tension between CDM and data motivates a more radical overhaul of the theory of dark matter, or even gravitational theory. For example, changing the weakly interacting nature of the dark matter by increasing the scattering cross-section helps to alleviate several problems, such as cuspiness and clumpiness. However, the resulting dark haloes are found to be too spherical [615]. CDM simulations are in excellent agreement with the data, predicting massive galaxy halo shapes with mean flattenings of 1:2 [64], whereas gravitational weak lensing studies find ellipticities of around 0.3 [1546] for all types and slightly larger values (~0.4) for early types. Another approach modifies the law of gravity to the extent that one may be able to dispense entirely with dark matter. In particular, the MOND scheme or paradigm of Milgrom [1434; 1435; 1436] is singularly successful in explaining empirical facts such as the detailed shapes of rotation curves of galaxies, as well as the so-called Tully–Fisher relation, according to which the total mass in visible stars and gas (baryonic mass) in a disk galaxy is proportional to the fourth power of the asymptotic rotational velocity. Bekenstein has subsequently proposed a tensor–vector scalar theory, or TeVeS [254; 255; 258], which represents an embodiment of MOND into a full relativistic theory. Weak lensing observations currently pose the strongest challenge to these models, since they show that the source of the gravitational potential does not track baryonic matter [96; 433; 562; 827], thus pointing towards the need for DM. The successes and challenges of modified theories of gravity are thoroughly discussed in Chapter 6.

1.4 Portrait of a suspect

Baryons are not the dark matter. Nor are neutrinos. We are then left with the necessity of postulating new particles, arising in theories Beyond the Standard Model (BSM) of particle physics. In fact, the possible connection with BSM physics has prompted an enormous proliferation of dark matter candidates, which are currently being sought in an impressive array of accelerator, direct and indirect detection experiments. However, as our understanding of particle physics and astrophysics improves, we are accumulating information that progressively reduces the allowed regions in the parameter space of DM particles. In practice, a particle can be considered a good DM candidate only if a positive answer can be given to all of the following questions [1839]:

  (i) Does it match the appropriate relic density?
 (ii) Is it cold?
(iii) Is it neutral?
 (iv) Is it consistent with Big Bang nucleosynthesis (BBN)?
  (v) Does it leave stellar evolution unchanged?
 (vi) Is it compatible with constraints on self-interactions?
(vii) Is it consistent with direct DM searches?
(viii) Is it compatible with gamma-ray constraints?
 (ix) Is it compatible with other astrophysical bounds?
  (x) Can it be probed experimentally?

The particle physics aspects of the DM problem are discussed in detail in Part II. In particular, the production mechanisms of DM in the early Universe are discussed in Chapter 7. The first and most studied mechanism is thermal production, positing that DM particles were in chemical and thermodynamical equilibrium with ordinary matter in the early Universe, until the DM annihilation rate dropped below the expansion rate of the Universe. The fact that the annihilation cross-section at the electroweak scale naturally leads to the appropriate relic abundance has made Weakly Interacting Massive Particles an excellent dark matter candidate, as already noticed in the late 1970s [657; 1144; 1317; 1692; 1911]. Among the most important physical processes that can modify the simplest version of thermal freeze-out, we mention co-annihilations with near-degenerate particles [355; 735; 1020]. Examples and references to this scenario, and to alternatives such as models in which the DM is produced gravitationally or through the decay of heavier particles, and scenarios with a non-standard expansion rate of the Universe, are discussed in detail in Chapter 7.

The most widely discussed extension of the Standard Model of particle physics is Supersymmetry (SUSY), and the most widely discussed DM candidate is the supersymmetric neutralino, although other candidates, such as the gravitino, naturally arise in SUSY scenarios (see Chapter 8 for a discussion of SUSY DM). SUSY is aesthetically appealing, and it has been built on a solid theoretical ground, with the aim of addressing some of the most important problems of the Standard Model of particle physics. In fact, SUSY has been used as a benchmark theory by both the ATLAS and CMS collaborations [144; 145; 237], part of the Large Hadron Collider, which started operation at CERN in 2009. Furthermore, numerical codes have been developed to scan the supersymmetric parameter space automatically, determine the properties of the neutralino (such as mass and couplings) and study the consequences for accelerator, direct and indirect searches (see Chapter 16).

The different steps involved in the hypothetical discovery and measurement of supersymmetry at the LHC, and the strategies for constraining the soft breaking parameters, are reviewed in Chapter 13, while a detailed discussion of the reconstruction of the DM properties, with particular attention to the relic density, starting from collider measurements is presented in Chapter 14. It is interesting to stress that there is a large portion of the parameter space where upcoming accelerator measurements cannot constrain the underlying SUSY scenario sufficiently to prove that newly discovered particles are actually *the* DM in the Universe. It is therefore important to perform direct and indirect DM searches that would provide complementary information, as we shall see later.

Despite the good theoretical motivation, there is no experimental evidence for SUSY. Many alternative theories have been proposed, usually abandoning the ambitious attempt to cure the Standard Model, and trying instead to address the DM problem in a phenomenological way. A review of alternative candidates at the electroweak scale can be found in Chapter 9, where the motivations for and properties of Kaluza–Klein, Little Higgs, technicolour, minimal and mirror DM candidates are discussed in detail. Kaluza–Klein DM is particularly appealing, and there are many efforts being made to understand how to discriminate the signatures of theories with (universal) extra dimensions from SUSY signatures with collider experiments (Chapter 15).

It is also possible that the DM is not a WIMP. For instance, we mentioned above the case of the gravitino, which is representative of a class of superweakly interacting particles, or superWIMPs, that also include Kaluza–Klein gravitons, axinos, quintessinos and others; furthermore, the mass scale may not be in the range $100\,\mathrm{GeV}$ to $1\,\mathrm{TeV}$, as in the so-called WIMP-less scenarios (see Chapter 10 for a detailed discussion).

Axions provide another example of perfectly viable non-WIMP DM candidates. The theoretical motivation and possible production mechanism are discussed in Chapter 11. Finally, sterile neutrinos also provide an interesting alternative, as discussed in Chapter 12.

1.5 Observing cold dark matter

As we have seen, accelerator measurements will be invaluable to establish
the theoretical framework chosen by nature. The discovery of new particles
would be per se an extraordinary discovery, but it might not allow a straightforward identification with a successful DM candidate. In this case, progress
will only be achieved by observing cold dark matter directly or indirectly.

Direct detection experiments aim at detecting the fraction of WIMPs, con-
stantly flowing through the Earth with a flux $10^5(100\,\mathrm{GeV}/m_\chi)\,\mathrm{cm}^{-2}\,\mathrm{s}^{-1}$,
that elastically scatter off nuclei in underground detectors, through the mea-
surement of nuclear recoils [988]. The rate, R, and energies, E_R, of nuclear
recoils can then be used to reconstruct the properties of the DM particle (see
Chapter 17 for a review of the theoretical aspects of direct DM detection).

The total cross-section for the scattering of WIMPs off target nuclei can
be separated into a spin-independent and a spin-dependent part: the spin-
independent part arises from scalar and vector couplings to quarks, whereas
the spin-dependent one originates from axial-vector couplings. This is important because different target nuclei are required to probe these two quantities. There are many uncertainties associated with the reconstruction of
the WIMP properties from direct detection experiments, including particle
physics uncertainties in the determination of scattering cross-section, and
especially astrophysical uncertainties, such as the local density and velocity
distribution of WIMPs. The prospects for detecting specific DM candidates,
and for reconstructing their properties starting from direct detection data,
are discussed in Chapter 17.

Scattering of WIMP particles leads to nuclear recoils that can be measured
by at least three different techniques, namely applications of scintillation,
phonons and ionization. The DAMA/LIBRA experiment, which makes use
of NaI (Tl) (i.e. sodium iodide doped with thallium) scintillators (see Chapters 18 and 19) is the only experiment claiming the detection of a signal
consistent with a DM interpretation, i.e. an annual modulation of the event
rate, possibly due to the effect of the motion of the Earth around the Sun,
that was predicted long ago as a peculiar signature of halo WIMPs [719; 881].
The evidence for the signal is strong, but an actual interpretation in terms
of existing candidates and standard assumptions for the astrophysical properties of DM has to face the fact that no other experiments have observed a
signal. It is however true that a comparison between different experiments
can only be performed under specific assumptions, perhaps affected by theoretical prejudices. Possible ways to reconcile the DAMA/Libra signal with
other experiments are discussed in Chapter 19.

Cryogenic detectors, i.e. solid state or superfluid ^{3}He detectors operated at temperatures lower than 77 K, encompass a large category of DM experiments, such as CDMS, CRESST, EDELWEISS and ROSEBUD (see Chapter 20). Historically, the first detectors of this type aimed to detect particles in a crystal by measuring the increase of temperature induced by the energy deposition. However, in order to reject the overwhelming background induced by gamma-rays and X-rays from radioactive contaminants, 'double parameter measurement' bolometers have been built, which are able to measure charge and phonon signals in ionization detectors, and light and phonon signals in scintillating bolometers (Chapter 20).

Noble liquid gases such as xenon, argon and neon appear today as extraordinarily promising targets for direct DM detection. They are in fact excellent scintillators (and, for Xe and Ar, also very good ionizers), and they can be scaled up to a large mass thanks to their modest cost. In practice, these experiments consist of time projection chambers (TPCs) with two phases, liquid and gas, of noble elements (liquid xenon for XENON, ZEPLIN and LUX, and liquid argon for ArDM and WArP), or a single phase of noble elements (XMASS and DEAP/CLEAN). For two-phased TPCs, ionization electrons and prompt scintillation photons produced by scattering events are simultaneously detected, and the ionization electrons are extracted from the liquid to the gas where they emit proportional scintillation photons. Photomultiplier tubes are then used to detect the prompt and delayed scintillation signals, thus providing information that allows one to separate WIMP-like events from electron recoil events produced by background beta- and gamma-rays. A complete discussion of liquid noble gas experiments is presented in Chapter 21.

Complementary information on the nature of DM may come from indirect detection searches, based on the search for secondary products (photons, neutrinos or antimatter) produced in the annihilation or decay of DM particles [1759].

Among these secondary products, gamma-rays appear particularly interesting, since they travel in straight lines and are practically unabsorbed in the local Universe. The observational strategies are different for the cases of annihilation and decay. In fact, while in the first case the galactic centre (central source, or surrounding diffuse emission) and possibly a few bright substructures represent the optimal targets, in the latter it is best to focus on the diffuse emission at high galactic latitudes, where the astrophysical backgrounds are less intense. The great success of the new generation of air Cherenkov telescopes, and the excitement about the upcoming data from the Fermi satellite, make this field very lively, as discussed in Chapter 24.

As we have seen, WIMPs are expected to have small but finite scattering cross-sections. As they pass through the Solar System, they may occasionally scatter off nuclei in the Sun or in the Earth, and lose enough momentum to become gravitationally bound and sink to the centre of the Sun or Earth. Among their annihilation products, the only ones that can escape without interacting with the dense surrounding matter are neutrinos. The prospects for detecting neutrinos from DM annihilation with the next generation of neutrino experiments, in the framework of different DM candidates, are reviewed in Chapter 25.

Among indirect searches, the most discussed as we write are antimatter searches (see Chapter 26). In fact, a feature in the positron fraction has been detected above 10 GeV, initially by balloon experiments and now by the PAMELA satellite [24], that cannot easily be attributed to cosmic-ray secondary production of e^+. The ATIC experiment [515] also reports a rise in the positron flux, followed by a sharp drop just below 1 TeV. This can be explained either by a nearby positron source or by dark matter annihilation, although the DM explanation requires particle physics and astrophysical parameters that theory struggles to explain.

It is worth noting that several astrophysical and cosmological probes of the DM nature exist. In the last two chapters of this book, we will see how the physics of Big Bang nucleosynthesis and the structure and evolution of stars provide useful information on, and possibly observational evidence for, DM particles (see Chapters 28 and 29).

1.6 The future

Advances on the particle physics front seem likely. As we have seen, there are strong motivations to believe that new physics beyond the Standard Model should exist. Whatever the result of the measurements at the LHC, we will need either to understand the structure of the new theory or to abandon (at least the most naive incarnations of) the WIMP paradigm.

Whether or not new particles will soon be found in accelerators, indirect and direct searches may provide crucial information, and it is likely that to find a full solution to the DM problem we will need to correlate complementary signals and corroborate astrophysical detections with accelerator evidence, and vice versa. These carefully laid plans are useless if the DM particle is not a WIMP. In that case, alternative candidates such as the axion, or alternative theories of gravity, would gain popularity.

Let us stay optimistic, however. This book presents a series of studies of the various ways that particle dark matter may be detected. The hunt will be long and multifaceted, combining direct detection, indirect detection and accelerator experiments. It may be that we will learn more from the pursuit than we can envisage now, about the physics of dark matter, about its connections to high-energy physics and astrophysics, and even about our dark halo and other nearby dark matter concentrations. But the journey will surely spawn new generations of enthusiastic particle astrophysicists who will lead us on to an even more challenging future.

2

Simulations of cold dark matter haloes

Ben Moore and Juerg Diemand

Numerical studies of the formation of cold dark matter haloes have produced several robust results that allow unique tests of the hierarchical clustering paradigm. Universal properties of haloes, including their mass profiles and substructure properties, are being tested against observational data from the scales of dwarf galaxies to galaxy clusters. Resolving the fine-grained structure of haloes has enabled us to make predictions for ongoing and planned direct and indirect dark matter detection experiments taking us beyond the smooth spherical isotropic model for the Galactic halo.

2.1 From cold collapse to hierarchical clustering – a brief history

N-body simulations of the gravitational collapse of a collisionless system of particles pre-date the CDM model. Early simulations in the 1960s studied the formation of elliptical galaxies from the collapse of a cold top-hat perturbation of stars [1089; 1556; 1889]. The resulting virialization process gave rise to equilibrium structures with de Vaucouleurs [633] or Einasto [741; 743] type density profiles. Profiles of the same form but with higher concentrations are widely used to describe the light distribution of elliptical galaxies. It is remarkable that the end state of almost any gravitational collapse, independent of the small-scale structure and hierarchical merging pattern, leads to a similar global structure of the final equilibrium system [1143; 1426; 1474].

Computer simulations in the 1970s attempted to follow the expansion and collapse of a spherical overdensity to relate to the observed properties of virialized structures such as galaxy clusters [1930]. Using a random distribution of particles with a Poisson power spectrum led to the initial formation of many bound clumps, but it was observed that these bound structures were

Particle Dark Matter: Observations, Models and Searches, ed. Gianfranco Bertone. Published by Cambridge University Press. © Cambridge University Press 2010.

destroyed as the final system formed, resulting in a smooth distribution of matter. This overmerging problem persisted for over two decades and motivated the development of semi-analytical models for galaxy formation [1931].

We now have a 'standard cosmological model' in which the initial conditions for structure formation are known to a reasonable precision [729; 1559], allowing computer simulations to follow the non-linear evolution of perturbations. The final quasi-equilibrium structures are the dark matter haloes that are observed to surround all galactic and galaxy cluster systems. Numerical simulations provide a reliable method of calculating this evolution to make predictions for the clustering of the dark matter component, and more recently, but less reliably, for the combined growth of dark matter and baryonic fluctuations. However, there is still a lot to learn; the detailed interaction between the dark matter and baryons is poorly understood, and many comparisons with observations and predictions for experimental searches rely on extrapolations well below the current resolution scale.

During the 1980s, the first simulations of the hierarchical cold dark matter model were carried out. Large cubes of the Universe were simulated in an attempt to match the large-scale clustering of galaxies. Some of the most basic properties of collapsed structures were discovered – the distribution of halo shapes, spin parameters etc. [888; 1618]. It was not until the simulations of Dubinski and Carlberg that individual objects were simulated at sufficiently high resolution to resolve their inner structure on scales that could be compared with observations [721]. Using a million-particle simulation of a cluster mass halo run on a single workstation for an entire year, these authors found central cusps and density profiles with a continuously varying slope as a function of radius. They fitted Hernquist profiles to their initial simulations but a Navarro, Frenk and White (NFW) profile provides an equally good fit (see Figure 2.1). Most probably because of a large softening length, the final virialized structure was almost completely smooth.

Navarro *et al.* published results of simulations of halo density profiles from scales of galaxies to galaxy clusters. They demonstrated that all dynamically relaxed haloes could be reasonably well fitted by a simple function with a concentration parameter that was related to the halo mass [1501]. With only $\sim 10^4$ particles they could only resolve the halo structure to about 5–10% of its virial radius. Shortly afterwards, simulations with 10^6 particles showed cusps slightly steeper than r^{-1} down to the innermost resolved point. These simulations also resolved the overmerging problem [1472]: the resolution was sufficient to resolve cusps in the progenitor haloes, enabling the structures to survive the merging hierarchy [940; 1250; 1471]. The final surviving substructure population is a relic of the entire merger history of a given CDM halo.

B. Moore and J. Diemand

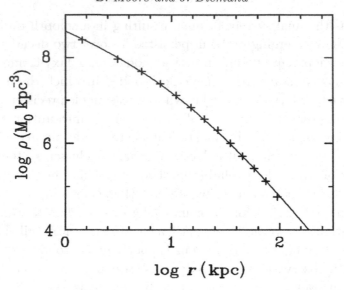

Fig. 2.1. Density profile of the million-particle dark matter halo simulation of Dubinski and Carlberg [721] (crosses). The solid line shows the best fit NFW profile to the original data. This figure was adapted from [1090] by John Dubinski and is reproduced here with his permission.

Algorithmic and hardware development have increased the mass and spatial resolution by orders of magnitude (parallel computing, special purpose hardware, graphics pipelines etc.). The first simulations used just a few hundred particles with length resolution that was a large fraction of the final structure. Today we can simulate individual collapsed structures, in a full cosmological context with up to 10^9 particles and spatial resolution that is better than 0.1% of the virialized region, and $\sim 10^5$ substructure haloes can be resolved (see Figures 2.2 and 2.3, and refs. [660; 1790; 1794]).

The final virialized structures can be described by various universal empirical relations, including the density profiles, the phase-space density profile [1846], the velocity anisotropy and distribution functions [1071; 1072]. Understanding the origin of these relations is a challenging and fascinating problem for dynamicists. Most aspects of CDM halo structure resulting from gravitational collapse are well resolved, and different numerical techniques and published studies all agree, although some confusion still exists in the literature because of different interpretation, different cosmologies and claimed resolution limits. Our detailed understanding and predictive power is now limited by the complex interplay between the dark matter and baryonic components, which is an area of intense current activity.

Whilst discrepancies between observations and simulations have been the subject of much debate in the literature, the details of the galaxy formation process needs to be resolved in order to test the CDM paradigm fully. Whatever the true nature of the dark matter particle, it must not be too different from a cold neutralino-like particle to maintain all the successes of the model in matching large-scale structure data and the global properties of haloes, which are mostly in good agreement with observations.

2.2 Results from collisionless simulations

Before we summarize the main simulation results on the properties of dark matter haloes we introduce definitions of the mass and radius of dark matter haloes. Deciding which material belongs to a halo and what lies beyond it is a non-trivial question, which has not received much attention until quite recently [598; 661; 664; 1596].

The usual halo definitions are still based on the simple spherical collapse picture [992; 1039]. These models assume that there is no internal kinetic energy in an overdense sphere of matter as it turns around and that this sphere virializes, rather than falling back radially into one point. The virial theorem predicts that the final halo radius is 0.5 of its turnaround radius and that this 'virial radius' would enclose an overdensity of $178\rho_{crit}$ in an Einstein–deSitter cosmology ($\Omega_M = 1.0$). For the now favoured Lambda cold dark matter (ΛCDM) cosmology this overdensity is close to 100 times ρ_{crit} or roughly 300 times ρ_M. However, CDM haloes form very differently than assumed in the classical spherical collapse model: material ending up near this so-called 'virial radius' does not undergo a collapse by anywhere near a factor of two [664]. The ratio of final radius to turnaround radius is much larger in the outer halo, and typical orbits extend out to 90% of the turnaround radius [661], well beyond the formal virial radius. All the conventional overdensity definitions, by which the halo radius encloses about 200 times ρ_{crit} or ρ_M, are therefore too small, and they underestimate the extent and mass of dark matter haloes significantly [1596]. However, spherical overdensities are easy to measure in simulations and we adopt r_{200}, defined to enclose 200 times ρ_M, in the following. As a proxy for halo mass we will use $M_{200} = M(< r_{200})$.

One has to keep in mind that practical, ad hoc halo definitions such as r_{200} and M_{200} are rather different from the concept of halo masses defined in theoretical models of structure formation. This can be confusing and has affected many attempts at comparisons of theory with simulations, for example halo mass functions (see Section 2.2.1). Halo mergers and merger trees are well

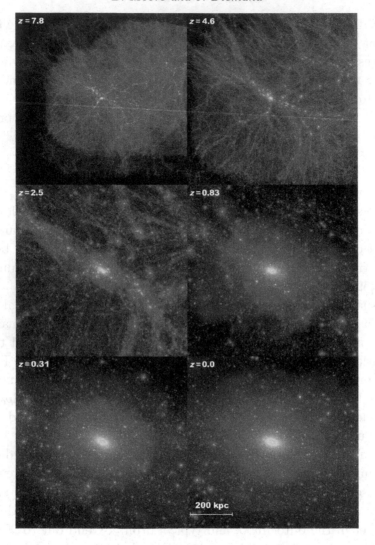

Fig. 2.2. Dark matter density maps from the 'Via Lactea II' (VL-II) simulation [660]. Cubes of 800 proper kpc are shown at different redshifts, always centred on the main progenitor halo. VL-II has a mass resolution of $4100 M_\odot$ and a force resolution of 40 pc. Initial conditions were generated with a modified, parallel version of GRAFIC2 [346]. The high-resolution region (some of its border is visible in the upper left panel) is embedded within a large periodic box (40 comoving Mpc) to account for the large-scale tidal forces. More images, movies and data are available at http://www.ucolick.org/~diemand/vl/

defined in theory, but not in practice since the M_{200} of the remnant may be smaller than the sum of its progenitors M_{200} and because some (sub-)haloes do not stay within r_{200}, i.e. they de-merge [661; 664; 946; 1467]. By convention we will also refer only to haloes within the virial radius of a larger halo

Fig. 2.3. Dark matter density map within the inner 200 kpc of the 'GHALO' simulation [1794]. This galaxy-scale halo is resolved with over one billion particles of $1000 M_\odot$ each. It contains over 100,000 subhaloes, the largest of which are visible as bright spots in this image.

as 'subhaloes', but this actually mis-classifies many satellite haloes beyond this ad hoc scale as isolated haloes, resulting in environmental dependencies in 'field' haloes [664]. Furthermore, the halo mass $M_{200}(z)$ and radius $r_{200}(z)$ grow non-stop by definition, even when no mass is accreted [664].

2.2.1 Mass function of haloes

For a range of practical halo mass definitions, e.g. based on a given spherical overdensity, the abundance of field haloes as a function of mass can be accurately measured in cosmological simulations of large, representative, periodic volumes [1854].

The classic approach by Press and Schechter [1601] combines the statistics of a hierarchical, Gaussian random field with the spherical, radial top-hat collapse model, and it allows one to predict the abundance of collapsed objects as a function of mass. The predictions of this idealized model are surprisingly close to the halo abundances measured in simulations, but the model predicts too many small haloes ($M < M_*$, where M_* is the mass of a typical, one-sigma halo forming now). This difference is usually interpreted as being caused by the assumption of a spherical collapse. Allowing for an ellipsoidal collapse introduces free parameters, which allows for a more accurate fit to the measured mass functions [1743; 1854].

Alternatively, one can argue that the main difference between the model and the simulation is the collapse (or lack thereof): the infalling mass hardly loses any energy, but typically orbits back out to 90% of its turnaround radius [661], significantly beyond the 'virial' radius. In other words, one cannot expect to find the collapsed mass predicted by the Press–Schechter approach inside the overdensity radii commonly used to define halo masses in simulations. A recently suggested improvement is to use the stationary mass, i.e. the mass within the largest radius with zero mean radial velocity [598; 1596]. The static mass exceeds the 'virial' mass substantially in haloes below M_* [1596] and at $z = 0$ the Press–Schechter mass function fits the abundance of haloes as a function of their static mass very well [598]. At $z = 1$ and $z = 2$ the agreement is less impressive; the Press–Schechter mass function lies above simulated abundances [598].

Further improvements in these directions seem possible, since Press–Schechter predicts the collapsed mass, not the static mass. Above M_* accretion typically sets in near r_{vir}, i.e. static and virial radius are similar [1596]. But significant amounts of previously collapsed material are still found beyond the virial and beyond the static radius [661; 664; 946; 1467].

2.2.2 Global halo properties

Here we briefly summarize some basic, global properties of CDM haloes: concentration and density profiles; shapes, spin and velocity distribution.

Halo formation, density profiles and concentrations

Self-similar infall models predict scale-free, nearly isothermal ($\rho \propto r^{-2}$) profiles [344; 853; 992; 1039]. Simulated profiles are found to be steeper than $\rho \propto r^{-2}$ in the outer parts and shallower in the inner regions [721], i.e. their circular velocity profiles $V_{\mathrm{c}}(r) = \sqrt{GM(< r)/r}$ have a well-defined peak, which serves as a natural halo size scale V_{max} and scale radius $r_{V\mathrm{max}}$.

Self-similar spherical infall maintains a distinct relation between the initial conditions and the final structure, while violent relaxation assumes that information on the initial conditions is chaotically lost. Simulations show that despite their non-isothermal density profiles, CDM haloes still do have much in common with idealized spherical infall models. (i) Particles that collapse in rare, early peaks (e.g. formation sites of first stars and old globular clusters) are located closer to the potential minima of the entire turnaround region because of peaks biasing and those particles do end up closer to the centre of the final halo [665; 1466]. (ii) The typical particle apocentre distances are close to the turnaround radii [661]. Modified infall models are indeed able to reproduce some of the features of halo density profiles found in cosmological simulations [135].

Over a mass range spanning 20 decades, from micro-haloes to galaxy clusters, the spherically averaged CDM halo density profile can be approximated with the same universal form (NFW [1501]):

$$\rho(r) = \frac{\rho_s}{(r/r_s)^\gamma (1 + r/r_s)^{3-\gamma}}, \tag{2.1}$$

where $\gamma = 1$. The scale radius r_s is related to the peak circular velocity scale by $r_{V\max} = 2.163 r_s$ and it is used to define the halo concentration $c_{\mathrm{vir}} = r_{\mathrm{vir}}/r_s$, where the 'virial' radius r_{vir} is defined following one of the ad hoc overdensity criteria described above. Halo concentrations c_{vir} (and equivalently scale densities ρ_s) are related to the halo formation time: early forming haloes tend to have higher c_{vir} and ρ_s at $z = 0$. The mass variance $\sigma(M)$ in CDM decreases from dwarf to galaxy cluster scales, i.e. smaller haloes form earlier on average than larger ones and they end up having higher median concentrations [465; 1286; 1363; 1364; 1501]. At a given mass, the concentrations of individual haloes have a large scatter: the variance in log c_{vir} is 0.18. On subsolar mass scales the CDM power spectrum approaches $P(k) \propto k^{-3}$, i.e. $\sigma(M)$ approaches a constant, which leads to very similar halo formation times and halo concentrations over a wide range of masses [567; 668; 1013]. Even Earth-mass micro-haloes, the first and smallest systems in the CDM hierarchy, have NFW-like density profiles. A systematic study of their typical concentrations is still lacking, but values found in the small sample of ref. [668] ($c_{\mathrm{vir}}(z = 0) \simeq 80$) seem consistent with the predictions of the Bullock *et al.* model [465].

A simpler and more general measure of halo concentrations is the mean density within $r_{V\max}$. It is well defined both for isolated haloes and for subhaloes and is independent of assumptions on their 'virial' radius or their density profile [664]:

$$c_V \equiv \frac{\bar{\rho}(< r_{V\max})}{\rho_{\mathrm{crit},0}} = 2\left(\frac{V_{\max}}{H_0 r_{V\max}}\right)^2 = \left(\frac{V_{\max}}{r_{V\max}}\right)^2 \frac{3}{4\pi G \rho_{\mathrm{crit},0}}, \tag{2.2}$$

where $\rho_{\mathrm{crit},0} = 1.48 \times 10^{-7}\ \mathrm{M}_\odot\,\mathrm{pc}^{-3}$. For the NFW profile it is easy to convert from c_V to c_{vir} [664]. Since the NFW form is not a very good fit to most CDM haloes, the measured c_{vir} depends somewhat on the details of the fitting procedure [465; 1286; 1363; 1364]. These complications could be avoided by using c_V, which is a robust concentration measure as long as V_{max} and $r_{V max}$ are resolved in the simulation.

Inner density profiles

As larger simulations started to resolve scales around $0.03 r_{V max}$, it became clear that most haloes are significantly denser than the best-fit NFW profile on these scales [666; 901; 902; 903; 939; 940; 1247; 1471; 1503; 1842]. Samples of high resolution, relaxed, isolated CDM haloes revealed significant halo-to-halo scatter [666; 1503]: some haloes follow the NFW form quite well, while others are better approximated with a steeper profile as suggested by ref. [1471], but most haloes lie somewhere in between and their average deviations from both fitting functions are larger than 10% (more than 20% at some radii) [666]. Improved fits require an additional free parameter to account for the substantial halo-to-halo scatter. Good fits are obtained by letting the inner slope γ of the NFW function (2.1) become a free parameter, instead of forcing $\gamma = 1$. Another option is to use the Einasto profile [741; 743; 1503]

$$\rho(r) = \rho_s e^{-\frac{2}{\alpha}[(r/r_s)^\alpha - 1]}, \tag{2.3}$$

where α is the additional free parameter. Down to $0.03 r_{V max}$, both forms are very similar and both fit simulated CDM density profiles quite well [666; 1002; 1426; 1503]. The two functions only differ significantly at very small radii (below about 0.6% of $r_{V max}$, i.e. below about 500 pc in a galaxy halo). Only very few (if any) simulations are currently able to resolve such small scales. The haloes in refs. [660; 670] are denser than their Einasto fits in the inner parts and they are well approximated by $\gamma \simeq 1.2$ cusps. The same is true for the billion-particle GHALO down to 400 pc [1794]. However, the higher mass resolution of GHALO might allow significantly smaller scales to be resolved, and a new functional form for the fitting function of the density profile has been proposed [1794]:

$$\rho(r) = \rho_0 e^{-\lambda[\ln(1+r/R_\lambda)]^2}. \tag{2.4}$$

This function has a constant logarithmic slope down to a scale R_λ, beyond which it approaches the central maximum density ρ_0 as $r \rightarrow 0$. If one makes a plot of $d \ln \rho / d \ln(1 + r/R_\lambda)$ versus $\ln(1 + r/R_\lambda)$, then this profile forms an exact straight line with slope -2λ. Good fits are obtained with $\lambda \simeq 0.10$ [1794].

However, to demonstrate convergence on such extremely small scales, several issues have to be addressed first. One necessary condition is to resolve the short dynamical times in these inner, very high density regions. The widely used, but unphysical, time step criterion $\Delta t \propto \sqrt{\epsilon/|a|}$, where ϵ is the force softening and $|a|$ the local acceleration, can lead to artificially low central densities [670]. The VL-II and GHALO simulations use an improved criterion based on the local dynamical time [1961]. Another caveat is that the material ending up in the inner 500 pc of a galaxy halo comes mostly from very early forming, dense progenitor haloes (4σ-peaks and higher [665]). At the starting redshifts of these simulations the variation on the smallest resolved mass scales $\sigma(M_{\min})$ is in the range 0.1 to 0.2. Four-sigma peaks already have the (mildly) non-linear density contrast of 0.4 to 0.8, which artificially lowers their formation redshift and scale density.

Beyond 0.6% of $r_{V\max}$, there is a general consensus that haloes are denser than NFW and that they are well fitted by functions that are steeper and cuspier than NFW at these radii, like Eq. (2.1) with $\gamma \simeq 1.2$, or the Einasto form (2.3) with $\gamma \simeq 0.17$. Resolving the remaining uncertainties in the density profiles of pure CDM haloes below 0.6% of $r_{V\max}$ is of some numerical and theoretical interest, but it does not affect the observational predictions from CDM significantly:

- In galaxies, groups and clusters these central regions are dominated by the baryons. Predictions on the dark matter and total mass distribution require a realistic treatment of the baryons and their dynamical interactions with the dark matter (Section 2.2.4).
- The predicted kinematics of dark-matter-dominated dwarf galaxies (3) are indistinguishable for these two density profiles, even if large numbers of accurate proper motion measurements become available [1813].
- The annihilation properties of small dark-matter-dominated haloes (Section 2.3.1) do not change significantly: the resulting annihilation luminosities of both fits and of the measured profile in Figure 2.4 all lie within 5% and the half-light radii are within 10% of each other. The best NFW fit to the VL-II density profile, however, underestimates the luminosity of the main halo by a factor of 1.4, and it overestimates the half-light radius by 1.3.

Shapes, spin and velocity distributions

Another main result from cosmological collisionless simulations is that CDM haloes are rather elongated, significantly more than, for example, elliptical galaxies [721]. The mean minor-to-major axis ratio is well described by $<s> = 0.54(M_{\mathrm{vir}}/M_*)^{-0.050}$, where s is measured at $0.3r_{\mathrm{vir}}$. The scatter around the mean is about $\sigma = 0.1$ [64]. The length of the intermediate axis is usually closer to the minor axis, i.e. most haloes are prolate. Towards the centre, CDM haloes are even more elongated, i.e. the axis ratios become

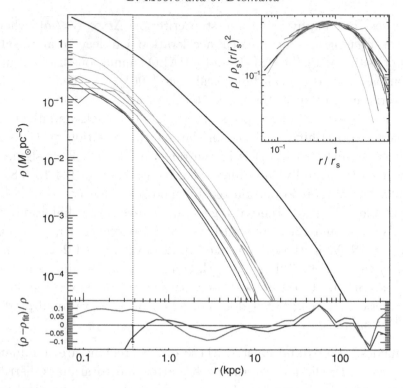

Fig. 2.4. Density profiles of the VL-II main halo and several of its subhaloes. Main panel: Profile of the main cold dark matter halo (thick line) and of eight large subhaloes (thin lines). The lower panel gives the relative differences between the simulated main halo profile and the Einasto fitting formula (Eq. (2.3)), with best-fit parameters $\alpha = 0.170$, $r_s = 21.5\,\mathrm{kpc}$, $\rho_s = 1.73 \times 10^{-3}\,M_\odot\,\mathrm{pc}^{-3}$ (grey curve), and function (2.1) with a best-fit inner slope of $\gamma = 1.24$, $r_s = 28.1\,\mathrm{kpc}$, $\rho_s = 3.50 \times 10^{-3}\,M_\odot\,\mathrm{pc}^{-3}$ (black curve). The vertical dotted line indicates the estimated convergence radius of 380 pc: simulated local densities are only lower limits inside 380 pc and they should be correct to within 10% outside this region. The cuspy profile is a good fit to the inner halo, while the Einasto profile has too shallow a slope in the inner few kiloparsecs, causing it to overestimate densities around 4 kpc and to underestimate them at all radii smaller than 1 kpc. The same behaviour in the inner few kiloparsecs is also found at higher redshifts, while the large residuals in the outer haloes on the other hand are transient features. Inset: Rescaled host (thick line) and subhalo (thin lines) density profiles multiplied by radius squared to reduce the vertical range of the figure. Reproduced with permission from ref. [660].

smaller while their orientations remain fairly well aligned. At a growing radius like $0.3r_{\mathrm{vir}}$, halo samples become rounder with time [64; 1173], while individual haloes have quite stable shapes at some fixed, inner radius (except during major mergers) [1284].

CDM haloes are supported by nearly isotropic velocity dispersions, not by rotation. The tangential and radial velocity dispersions measured in spherical bins define the anisotropy parameter $\beta = 1 - 0.5\sigma_t^2/\sigma_r^2$. The typical, relaxed CDM halo has $\beta(r) \simeq 0.35(r/r_{\mathrm{vir}})^{0.35}$, i.e. β changes from nearly isotropic near the centre to mildly radial in the outer halo.

The velocity distributions of CDM haloes are Gaussian only near the scale radius, i.e. where the density profile is roughly isothermal ($\rho \propto r^{-2}$). Analytical calculations predict more peaked distributions for the inner halo, and broader distributions beyond the scale radius [1090; 1219; 1943; 1944] in agreement with the measured distributions in CDM simulations [805; 1072; 1943; 1944].

Halo shapes are supported by velocity dispersion ellipsoids, which are elongated in the same direction as the mass distribution, both globally [64] and also locally [1960]. Fairly symmetric velocity distributions in the tangential directions are found throughout the halo [1960], i.e. there is about as much negative as positive angular momentum material relative to any given reference axis. Small asymmetries lead to a residual median spin $\lambda = |\mathbf{J}||E|^{1/2}/(GM^{5/2})$ of only 0.04 with a lognormal scatter of $\sigma \simeq 0.56$, both for the dark matter and for adiabatic gas. However, the two spins are often poorly aligned, with a median misalignment angle of 30 degrees [1892]. The dark matter angular momentum tends to align roughly with the minor axis of the halo shape, with a mean misalignment of 25 degrees [172], while disk galaxy orientation and halo shape beyond $0.1r_{\mathrm{vir}}$ are completely uncorrelated [171]. It is often assumed that the net halo angular momentum correlates with disk galaxy size and orientation, but only some selected fraction of the halo material with its wide variety of angular momenta can be incorporated into a realistic disk [1892]. In our Galaxy, for example, a significant fraction of the total available baryonic angular momentum is not in the disk but in the polar orbit of the Magellanic clouds, owing to their large distances and proper motions [1196]. How exactly disk galaxies form out of the angular momentum distributions available in ΛCDM haloes remains an open question despite much recent progress [999; 1404].

2.2.3 Substructure

A major contribution of N-body simulations to our understanding of structure formation was to demonstrate how hierarchical merging gives rise to a vast amount of surviving substructure, both gravitationally bound (subhaloes) and unbound (streams).

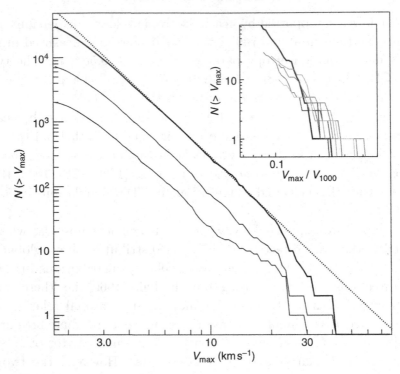

Fig. 2.5. Subhalo and sub-subhalo abundances in the VL-II halo. Number of sub-haloes above $V_{\max}$ within $r_{200} = 402$ kpc (thick solid lines) and within 100 and 50 kpc of the galactic centre (thin solid lines). The dotted line is $N(> V_{\max}) = 0.036 \, (V_{\max}/V_{\max,\mathrm{host}})^{-3}$, where $V_{\max,\mathrm{host}} = 201 \, \mathrm{km \, s^{-1}}$ (at $r_{V\max,\mathrm{host}} = 60$ kpc). It fits the subhalo abundance above $V_{\max} \simeq 3.5 \, \mathrm{km \, s^{-1}}$. The number of smaller subhaloes is artificially reduced by numerical limitations. The inset shows the sub-subhalo abundance within r_{1000} (enclosing 1000 times the mean matter density) of the centres of eight (same ones as in Fig. 2.4) large subhaloes (thin solid lines). The radius r_{1000} is well inside the tidal radius for these systems. The thick solid line shows the subhalo abundance of the host halo inside its $r_{1000} = 213$ kpc. The (sub-)subhalo $V_{\max}$ values are given in units of $V_{1000} = \sqrt{GM(< r_{1000})/r_{1000}}$ of the corresponding host (sub)halo. Lines stop at $V_{\max} = 2 \, \mathrm{km \, s^{-1}}$. The mean sub-substructure abundance is consistent with the scaled-down version of main halo, and both the mean abundance and the scatter are similar to the values found in refs. [1158; 1632] for distinct field haloes. This figure is from ref. [660].

Subhalo abundance: velocity and mass functions

The abundance of subhaloes within $r_{200} = 402$ kpc of the Galactic centre in the VL-II simulation (Figure 2.5) is well approximated by

$$N(> V_{\max}) = 0.036 \, (V_{\max}/V_{\max,\mathrm{host}})^{-3} \, , \tag{2.5}$$

where $V_{\mathrm{max,host}} = 201\,\mathrm{km\,s}^{-1}$. It is close to the median abundance found in a large sample of haloes simulated using similar cosmological parameters [1158]. In earlier halo samples using a higher normalization of the power spectrum ($\sigma_8 = 1.0$) a higher normalization of 0.042 for the median abundance was found [1632]. This difference is consistent with the expected cosmology dependence [1962]. Both samples demonstrate significant halo-to-halo variance of about a factor of two. Some of the variance comes from the scatter in halo concentration, which introduces scatter in $V_{\mathrm{max,host}}$ at a given host halo mass. Normalizing the subhalo function to the $V_c(r_{200})$ (i.e. to M_{200}) reduces the variance to about a factor of 1.3 [1158]. Without normalizing to the host halo size, the differential V_{max}-distribution function of subhaloes in dwarf galaxies to cluster haloes scatters around

$$\mathrm{d}n/\mathrm{d}V_{\mathrm{max}} = 1.5 \times 10^8 V_{\mathrm{max}}^{-4.5}\ (h^3\mathrm{Mpc}^{-3}\,\mathrm{km\,s}^{-1}), \qquad (2.6)$$

again with a halo-to-halo variance of about a factor of two [1632]. The median subhalo abundance given by the normalized (2.5) or non-normalized velocity function (2.6) seems to be approximately self-similar, i.e. independent of host halo mass and redshift [1632]. Even tiny host haloes ($<M_\odot$) show similar abundances, perhaps surprisingly so given that the density contrasts between subhaloes and the host are much smaller on these scales owing to their similar formation times caused by the nearly flat $\sigma(M)$ [662; 746].

While cosmological simulations are able to resolve the subhalo content of a given dark matter halo accurately, the exact abundance of substructure around a given galaxy, for example the Milky Way, remains uncertain:

- The $V_{\mathrm{max,host}}$ of the CDM halo in which a galaxy with a rotation speed of $220\,\mathrm{km\,s}^{-1}$ would form could lie anywhere between 160 and $220\,\mathrm{km\,s}^{-1}$ [731; 1249; 1772]. The resulting uncertainty in $N(>V_{\mathrm{max}})$ spans about a factor of $(220/160)^3 = 2.6$.
- At a given $V_{\mathrm{max,host}}$, and also at a given M_{200}, the subhalo abundance within r_{200} has a substantial halo-to-halo variance [1158; 1632].

These theoretical uncertainties have to be considered when CDM subhaloes are compared to the subhaloes around the Milky Way satellite galaxies (see Chapter 3).

Cumulative mass functions of subhaloes can be approximated by power laws of the form $M^{-\alpha}$, with $\alpha = 1.9$ to 1.0 and normalized so that the mass in subhaloes larger than $10^{-6}M_{\mathrm{host}}$ is between about 5 and 15% [629; 662; 663; 664; 667; 916; 939; 940; 1250; 1279; 1369; 1468; 1469; 1470; 1471; 1790]. The steep slope of the mass function means that there is a significant amount of mass in small subhaloes, which are still unresolved in current simulations.

Numerical convergence studies show that about four hundred particles per subhalo are required to resolve the mass function within r_{200}, and more than that to reach convergence in the inner regions of the host and in the velocity functions [667; 916; 1790]. Including smaller, under-resolved mass scales [1790] artificially lowers the slope α of the best fitting power law.

Subhalo evolution and their final spatial distribution

The abundance of field haloes of a given (moderate) mass is proportional to the dark matter density of the environment. This proportionality is altered as matter and haloes fall into a host halo and tidal forces reduce the mass of systems, especially those orbiting close to the host halo centre. Subhaloes selected by their final remaining mass are therefore more extended than the matter distribution (see Figure 2.6) [569; 660; 663; 667; 809; 916; 940; 1281; 1790]. The number density of subhaloes is independent of the mass threshold and roughly proportional to the dark density times radius: $n_{\mathrm{sub},M0}(r) \propto \rho_{\mathrm{DM}}(r) \times r$ [664; 667; 916; 1790]. The value of V_{max} is less affected by tides [664; 1280] and the spatial distribution of V_{max} selected subhaloes differs less from the dark matter distribution.

During tidal mass loss $r_{V\mathrm{max}}$ becomes smaller and the enclosed mean subhalo density c_V increases. Subhaloes near the centre of the host halo end up having much larger concentrations c_V than field haloes (see Figure 2.6) [660; 664]. This increase partially compensates for the reduced V_{max} so that the annihilation luminosity $L \propto V_{\mathrm{max}}^3 \sqrt{c_V}$ (see Section 2.3.1) resulting from the subhalo sample closely traces the dark matter distribution, except in the very inner regions (about 5% of r_{200}, or 20 kpc for the VL-II halo). In other words subhalo luminosity closely traces the dark matter distribution, because L is not significantly affected by tidal stripping for most subhaloes. This is not surprising, given that half the annihilation originates from the rather small radius of about $0.07r_{V\mathrm{max}}$. Smaller haloes than those resolved in current simulations have higher typical concentrations, i.e. higher densities inside $r_{V\mathrm{max}}$ and $0.07r_{V\mathrm{max}}$, and their annihilation luminosity is expected to resist tidal losses even better.

Tidal stripping removes mass from the outer, loosely bound regions of subhaloes [663; 940; 1280]. Stripping is well approximated by removing the mass beyond the tidal radius over some timescale [663; 1844; 1962]. The tidal radius is defined so that the host halo density is equal (or similar) to the subhalo density at the subhalo's tidal radius. High-density parts of subhaloes are therefore able to survive intact even close to the centre of the host. This explains why practically all (97% since $z = 1$) subhaloes survive until the present time, despite substantial mass loss in some cases [663]. Subhaloes

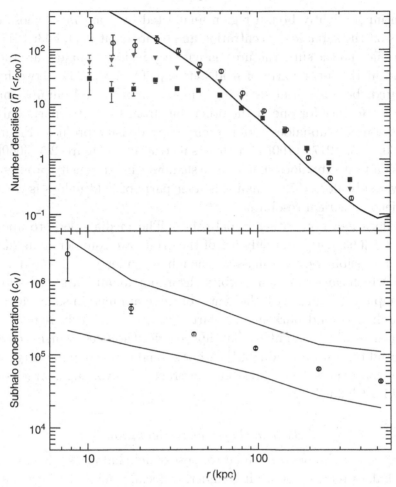

Fig. 2.6. Abundance and concentrations of subhaloes vs. distance from the galactic centre in the VL-II simulation. Top: The number density profile of subhaloes (symbols) is more extended than the dark matter density profile $\rho(r)$ (thick line). The difference relative to the matter distribution is largest for mass-selected subhalo samples: squares show a subhalo sample with a mass above $8 \times 10^5 M_\odot$. Selecting subhaloes by their peak circular velocity V_{max} reduces the difference: triangles are for subhaloes with $V_{max} > 3\,\mathrm{km\,s^{-1}}$. Selecting by subhalo annihilation luminosity ($L \propto V_{max}^3 \sqrt{c_V} > (5\,\mathrm{km\,s^{-1}})^4/(0.15\,\mathrm{kpc})$) eliminates the difference throughout most of the halo (circles). Subhalo luminosity closely traces the dark matter distribution, except in the very inner regions (about 5% of r_{200}, or 20 kpc for the VL-II halo). Bottom: Subhalo concentrations c_V (median and 68% range are shown) increase towards the centre, where the stronger tidal force removes more of the outer, low-density parts from the subhaloes. To make sure their c_V are resolved, only subhaloes larger than $V_{max} = 5\,\mathrm{km\,s^{-1}}$ are used. The error bars indicate the statistical uncertainties in both panels. This figure was adapted from [660].

have an inner, tightly bound region unaffected by mass loss, whose extent depends on the subhalo concentration and on its orbit [544; 663; 1217]. The mass profile and the substructure content deep inside subhaloes are the same as found in the inner parts of field haloes [660; 1790]. These two findings are related, because tidal stripping removes smooth and clumpy material without preference for one or the other, but from the outer parts only. The inner regions of subhaloes retain their cuspy density profiles (Figure 2.4) [544; 660; 663; 1217; 1790] and their substructure (Figure 2.5) [660]. The impression that subhaloes should have shallower inner density profiles [1812] and fewer subhaloes [1790] than the inner parts of field haloes is caused by insufficient numerical resolution.

Subhaloes move on rather radial orbits. The median peri- to apocentre ratio is 1:6 [663; 940], and only 5% of the orbits are rounder than 2:3 [663]. Only a few subhaloes are massive enough to suffer significant dynamical friction, which causes decaying orbits, disproportionally large mass loss and even complete merging with the centre of the main halo in some cases [663]. Most subhaloes, and dark matter particles, move on fairly regular orbits: they reach nearly constant median apocentric distances, which lie close to their initial turnaround radii [661]. A few subhaloes even gain energy during their pericentre passage in three-body interactions involving a larger subhalo and the host halo [1671].

Subhalo shapes and orientations

The shapes of subhaloes are similar to those of field haloes (see Section 2.2.2), but subhaloes tend to be a bit rounder, especially near the host halo centre [1284]. Tidal interactions make individual subhaloes rounder over time [1473], and they also tend to align their major axis towards the centre of the host halo [810; 1251; 1284; 1565]. The alignment is often maintained over most of the subhaloes' orbits, except during pericentre passages [1284]. A similar radial alignment has been found for red galaxies in SDSS groups [811]. The major axes of the Milky Way dwarf satellites might also be preferentially aligned radially, i.e. roughly towards us, which might bias spherical mass estimates (see Chapter 3) towards higher values.

Other halo substructure: caustics, streams and voids

Besides the gravitationally bound, dense subhaloes discussed above, there exists additional structure in the phase space structure of cold dark matter haloes. Current simulations are now starting to resolve some of this structure (see Figure 2.7), although finite mass resolution and artificial numerical

Fig. 2.7. Local phase-space densities calculated with EnBiD [1740] in a 40 kpc cube in the centre of the VL-II halo at $z = 0$ (compare to the last panel in Figure 2.2). Subhaloes have very high central phase-space densities ($>10^{-5}\,M_\odot\,\mathrm{pc}^{-3}\,\mathrm{km}^{-3}\,\mathrm{s}^3$) because of their steep inner density cusps and their relatively small internal velocity dispersions. The most clearly visible streams still have quite low densities (about 100 times lower than the local density) but owing to their low velocity dispersion (about 10 times smaller than that of background particles) they just barely manage to stand out in local phase-space density (these streams have about $10^{-9}\,M_\odot\,\mathrm{pc}^{-3}\,\mathrm{km}^{-3}\,\mathrm{s}^3$).

heating [357; 669] still severely limit our ability to detect and resolve fine-grained phase space structure. The coherent elongated features in Figure 2.7 are dark matter streams which form out of material removed from accreted and disrupted subhaloes. In cases where the disrupted subhalo hosted a luminous satellite galaxy, the resulting streams would contain not only dark matter but also stars and would produce detectable features in galactic stellar haloes (see Chapter 3).

Cosmological infall into a halo does sometimes lead to the characteristic patterns in the $f(v_r, r)$ plane [661], which are predicted in the classic secondary infall models [853; 344]. However, in real space significant density enhancements (infall caustics) do not occur, since large random motions among the clumpy, infalling material prevent their formation [661]. Truly cold infall (at the very low WIMP thermal velocity dispersion) is expected to occur in the first and smallest haloes forming at the bottom of the CDM hierarchy. The resulting caustics on these very small scales might propagate into larger haloes at lower redshifts [25].

2.2.4 Effects of baryons

There is a lot of work remaining to be done to quantify the effects that baryons can have in modifying the distribution of dark matter. Simulations become more complex and expensive and we do not as yet have a clear understanding of how galaxies form.

Dark matter density profiles can steepen in response to the dissipation and contraction in the baryonic component [370; 966; 1723; 1954]. The strength of this effect depends on the baryonic fraction that slowly dissipates via radiative cooling. However, accretion of baryons via cold flows may dominate the growth of many galaxies [1222], so it is not yet clear how strongly this changes the inner distribution of dark matter in galaxies. For a halo that cools the cosmologically available baryons into a disk component, the dark matter density at a few per cent of the virial radius increases by about a factor of two and the final density profile can resemble an isothermal sphere, comparable to observed constraints on elliptical galaxies [966]. For isolated galaxies less massive than the Milky Way, the baryon fraction decreases rapidly, $M_{\mathrm{baryon}} \propto V_{\mathrm{vir}}^4$, such that the smallest galaxies have captured and cooled fewer than 10% of the available baryons [1405].

The growth of supermassive black holes or central nuclei can make the central dark matter cusp steeper or shallower, depending on whether these structures grow adiabatically or through mergers of smaller objects. Gondolo and Silk [985] explored the effects of slow central black hole formation on the CDM cusp in the context of an enhanced annihilation signal. This mechanism can create isothermal cusps on parsec scales, with a boost factor of several orders of magnitude. On the other hand, if supermassive black holes form via mergers, then dark matter particles can be ejected from the central halo region via three-body encounters, resulting in a constant density region of dark matter with a mass deficit roughly proportional to the mass of the sinking binary objects (see Chapter 5). Similar behaviour would result from the formation of central stellar nuclei in galaxies. Dissipative growth would increase the central dark matter density, but formation via merging of existing star clusters would lead to an inner core [970]. A similar mechanism was studied in the context of cluster haloes, whereby energy transfer to the dark matter background from dynamical friction acting on massive satellite galaxies gave rise to a constant density inner region [745]. All of these processes have yet to be studied in a realistic cosmological context.

Feedback from the star formation process has frequently been invoked to flatten cusps, especially in dwarf galaxies which have challenged the CDM paradigm through observations of rotation curves, stellar velocities and star-cluster kinematics. A single violent event, which somewhat unrealistically ejects a cosmological baryon fraction from the inner region, can redistribute the dark matter through a central revirialization. However, the most careful study of this process shows the effect to be modest, with a reduction in the central halo density by at most a factor of two to six [967]. More realistic smoothed particle hydrodynamics (SPH) simulations in a cosmological context show that supernovae-driven turbulent gas motions can impart sufficient energy to the dark matter to create a core as large as 400 parsecs in a Fornax-sized galaxy [1395]. This effect requires both that there should be a significant early central baryon fraction and that the Jeans mass should be accurately followed, since bulk motions are driven by starbursts in giant molecular clouds. It will be interesting to compare these experiments with higher-resolution adaptive mesh techniques including the effects of reionization.

Over half of disk galaxies have stellar bars which can transfer angular momentum to dark matter particles through orbital resonances and dynamical friction. The magnitude of this process has been debated in the literature [1248; 1722; 1924], but even when a rigid perturber mimicking a bar was placed at the centre of a CDM halo, it only affected the dark matter particles within $\sim 0.1\% R_{200} \approx 300$ pc in our Galaxy. The most recent highest-resolution study of this process demonstrates that the effect of bars on the central dark matter distribution is negligible [720].

The shapes of dark matter haloes can be dramatically modified as particles on box orbits supporting the triaxial configurations are destroyed by the central potential [637; 1209; 1218]. Since particles move on eccentric orbits with a typical apocentric to pericentric distance ratio of 6:1, haloes can be visibly affected out to half the virial radius, and become almost spherical close to the galaxy. The change in shape depends on the central baryonic fraction, which is highest for elliptical galaxies and lowest for galaxy clusters and dwarf galaxies whose haloes should barely be affected. The detailed modification of particle orbits within the disk region has yet to be explored but could also change the detailed predictions for direct detection experiments.

Galaxy formation also leads to the accretion of gas, stars and dark matter from satellites into the disk: systems on roughly co-planar orbits suffer dynamical friction against the disk, which brings them into the disk plane where they are disrupted [1433; 1630]. This process produces a dark matter disk, which could contribute a significant fraction of the local dark matter density [1630].

On the smallest scales, substructures orbiting through the Galactic disk will lose significant amounts of mass as they suffer disk shocking or heating from individual stars [968; 1967]. The disruption timescale for inner substructures due to disk shocking is an order of magnitude higher than for globular clusters (the disruption timescale scales as the square of the mean radius of the system and CDM subhaloes are much larger than globular clusters). Thus the inner 20 kpc of the Galactic halo could be smooth in configuration space, but rich in phase space. For the smallest substructures with sizes smaller than a few hundred parsecs, impulsive collisional heating due to encounters with disk stars dominates their mass loss. Over a Hubble time, most of their particles will be lost into small streams, although an inner dense and bound core is expected to survive.

The importance of many of these processes remain to be quantified, and this will be an area of activity whilst simulators attempt to create realistic galaxies from cosmological initial conditions. They will also play an important role in precision cosmology in which future observational missions plan to measure the power spectrum, or mass distribution on large scales to precision levels of a few per cent, which requires a detailed knowledge of how baryons affect the global properties of their haloes.

2.3 Predictions for direct and indirect detection experiments

Here we summarize the main predictions from cosmological N-body simulations for dark matter detection, and refer readers to the later chapters for detailed discussion of detection experiments.

2.3.1 γ-rays from dark matter annihilation

For the NFW density profile and its variants discussed above, the total luminosity from a dark matter annihilation in a halo scales as

$$L \propto \int \rho^2 \mathrm{d}V \propto c_V^2 r_{\mathrm{Vmax}}^3 \propto V_{\max}^4 / r_{\mathrm{Vmax}} \propto V_{\max}^3 \sqrt{c_V} \,. \tag{2.7}$$

The half-light radius is $r_{1/2} \simeq 0.07 r_{\text{Vmax}}$ for the best-fit profiles in Figure 2.4, and $0.09 r_{\text{Vmax}}$ for the shallower NFW profile. Combining Eq. (2.7) with the steep velocity functions found for field haloes [1632] and subhaloes [660; 663; 940; 1632], $N(> V_{\text{max}}) \propto dN/d\log V_{\text{max}} \propto V_{\text{max}}^{-3}$, implies that small field haloes and subhaloes emit more gamma-rays per decade in halo size (V_{max} or mass) than larger ones. In the latest galaxy halo simulations, all resolved subhaloes together are about as luminous as the main halo [660; 663; 1790], i.e. the boost factor $B = L_{\text{total}}/L_{\text{mainhalo}}$ is at least two for CDM galaxy haloes. Extrapolations down to the smallest CDM subhaloes increase the boost to $B = 4$ to 16 [660], where we assume small mass cutoffs between 10^{-12} and $1.0 M_\odot$ and we take the nearly constant halo concentrations on the smallest scales into account (see Section 2.2.2). These boost factors imply that small-scale structure ($<10^6 M_\odot$) dominates the diffuse extragalactic dark matter annihilation signal. The flux from small-scale clumps is proportional to the dark matter density averaged over larger scales (Figure 2.6).

The spatial and spectral distribution of diffuse gamma-rays was measured by EGRET [1142] and it is consistent with a superposition of (poorly constrained) astrophysical diffuse components [1823]. Even under the very optimistic assumption of a perfect subtraction of astrophysical diffuse foregrounds, the DM detection window of the recently launched Fermi satellite (formerly known as GLAST) in the diffuse galactic component is quite small [180]. However, the significantly improved spatial resolution of Fermi relative to EGRET might allow it to detect gamma-rays from dark matter annihilation in subhaloes, both in dwarf galaxies and in smaller, dark subhaloes [180; 1285]. Small-scale clumpiness within subhaloes also increases their signal slightly, which makes a somewhat larger number of subhaloes detectable [1285]. A thorough discussion of DM indirect detection can be found in Part V of this book.

2.3.2 Nearby dark matter distribution and charged particles from dark matter annihilation

Besides gamma-rays, dark matter annihilation would produce charged particles and antiparticles that, owing to magnetic field entanglement, propagate over much smaller distances within the Galaxy. Using the local subhalo abundance from Figure 2.6 and extrapolating down to micro-subhalo scales, one finds that nearby subhaloes produce a total flux of 40% of the local smooth halo signal [660]. In other words, the local boost factor is 1.4, and the uncertainty from the extrapolation is about ±0.2. Explaining the positron excess measured by HEAT [242] and PAMELA [24] with local dark matter annihilation requires significantly larger enhancements [292; 1311]. When a relatively large subhalo happens to lie within 1 kpc, the local boost factor increases, but much larger values are unlikely: only 5.2% of all random realizations have a boost factor of 3 or larger (caused by a clump with $V_{max} \geq 3.4\,km\,s^{-1}$ within 1 kpc). In only 1.0% of the cases the boost factor reaches 10 or higher because of a nearby, large subhalo with $V_{max} \geq 5.6\,km\,s^{-1}$ [660].

2.3.3 Local dark matter distribution, direct detection and capture in the Sun

Most of the local dark matter is in a smooth component [660; 1790], and therefore the probability that the Solar System is currently passing through a subhalo is quite small, even when the smallest micro-subhaloes are taken into account [668; 1201]. The large number of overlapping streams in the inner halo also leads to rather smooth local velocity distribution functions [1470; 1085; 1960]. Even the most prominent streams apparent in Figure 2.7 account for less than 1% of the local dark matter density, i.e. even if we happened to be located within such a stream today, the bulk of detected WIMPs would still come from the 'hot' background. Whilst current cosmological simulations are able to probe the local density and velocity distributions on kiloparsec scales [1960], alternative methods are needed to study finer structures [813; 1808]. Further studies are needed to quantify or exclude the relevance for dark matter detection experiments of possible very fine-grained features in the local six-dimensional dark matter distribution.

At 8 kpc from the centres of galaxy-scale pure CDM haloes, velocity distributions are peaked (positive kurtosis), because of the shallower than isothermal potential of pure CDM haloes (see Section 2.2.2). The resulting excess of slow and fast WIMPs relative to the Gaussian standard halo model, however, is too small to change the interpretation of direct detection results significantly [805]. Galaxy formation is likely to have changed the local velocity distribution significantly (cf. Section 2.2.4): for example, the almost constant Milky Way rotation curve implies a nearly isothermal potential, and close to Gaussian velocity distributions for the dark matter.

The shape of the local velocity ellipsoid correlates with the shape of the halo: it is radially anisotropic on the major axis and tangentially on the minor axis [1960]. However, only the inner parts of pure CDM haloes are very elongated [64; 1284]; the shape of the local Milky Way halo is expected to be fairly round [637; 1209; 1218]. The dark matter disk (cf. Section 2.2.4) is probably the most marked deviation from the standard halo model, and it may well have significant implications for dark matter detection [452; 1630].

The time-averaged WIMP capture rate relevant for neutrino production in the centre of the Sun (and Earth) smears out the small variations due to local clumps and streams. In pure CDM simulations the rate is close to the rate obtained from the standard halo, just slightly higher owing to the small excess of low-velocity WIMPs relative to a Gaussian [805]. As for direct detection, the presence of accreted dark matter in the Milky Way's disk [452; 1630] is probably the most relevant deviation from the standard halo model, and further studies are required to better understand the properties of dark disks. A detailed discussion of the theoretical and experimental aspects of direct DM searches can be found in Part IV of this book.

Acknowledgements

We thank John Dubinski for preparing Figure 2.1 and for his permission to use it here. J.D. acknowledges support from NASA through a Hubble Fellowship grant HST-HF-01194.01, and from KITP through National Science Foundation Grant No. PHY05-51164.

3

Milky Way satellites

James Bullock, Manoj Kaplinghat and Louis Strigari

3.1 Satellite galaxies

3.1.1 Historical review

Since Hubble first resolved stars in external galaxies and confirmed that
these 'island universes' were beyond the realm of our own Milky Way Galaxy,
astronomers have sought to understand the properties of galaxies over many
orders of magnitude in luminosity and distance from the Milky Way. To
deal with the various morphologies of observed galaxies, Hubble proposed a
classification in which galaxies are broadly identified as variations of spirals,
ellipticals and irregulars. Although galaxies are observed with widely varying
morphologies, the mass of most of them appears to be dominated by an
unseen dark matter component, as was shown by Vera Rubin, Ken Freeman
and others using measurements of gas clouds in spiral galaxies in the 1970s.
Since these early studies, observations of the mass distributions of many
galaxies have been studied, with results showing that the ratio of dark matter
to luminous matter varies from galaxy to galaxy; the largest clusters of
galaxies and the smallest known dwarf galaxies have the highest ratio of
dark to luminous matter.

It is now known that even though galaxies with brightness similar to that
of the Milky Way dominate the luminosity distribution of galaxies, by far
the most numerous galaxies in the Universe are dwarf galaxies, which fall
under Hubble's irregular category. The first recorded discovery of a dwarf
galaxy came perhaps as early as the tenth century in the Persian astronomer
Al-Sufi's *Book of Fixed Stars*. This galaxy is known today as the Large
Magellanic Cloud (LMC), named after Magellan, who viewed it along with
its neighbour in the sky, the Small Magellanic Cloud (SMC), during his

Particle Dark Matter: Observations, Models and Searches, ed. Gianfranco Bertone. Published by
Cambridge University Press. © Cambridge University Press 2010.

Table 3.1. *Discovery dates, distances and visual*
magnitudes of Milky Way satellites.

Satellite	Discovery date	Distance (kpc)	Magnitude
Large Magellanic Cloud	1519	48	−18.5
Small Magellanic Cloud	1519	55	−17.1
Sculptor	1938	80	−9.8
Fornax	1938	138	−13.1
Leo I	1950	250	−11.9
Leo II	1950	205	−10.1
Draco	1954	80	−9.4
Ursa Minor	1954	66	−8.9
Carina	1977	101	−9.4
Sextans	1990	86	−9.5
Sagittarius	1994	25	−15
Willman 1	2005	38	−2.7
Ursa Major I	2005	106	−5.5
Ursa Major II	2006	32	−4.2
Hercules	2006	138	−6.6
Leo IV	2006	158	−5.0
Canes Venatici I	2006	224	−8.6
Canes Venatici II	2006	151	−4.9
Coma Berenices	2006	44	−4.1
Segue 1	2006	25	−1.5
Leo T	2004	417	−8.0
Bootes I	2004	66	−6.3
Bootes II	2007	60	−2.7
Leo V	2008	170	−4.3

voyage in 1519. These galaxies are now known to lie at distances of about 48 and 55 kpc (LMC and SMC, respectively), and represent the first discovery of a satellite galaxy of the Milky Way. It was not until 1938 that another neighbouring galaxy of the Milky Way was discovered in the constellation Sculptor by Shapley [1733]. Described by Shapley as 'A Stellar System of a New Type', Sculptor was observed to have a scale size similar to globular clusters and smaller spheroidal galaxies.

Between 1938 and 2004, eight more satellite galaxies of the Milky Way were discovered. With the exception of the LMC and SMC, which are classified as dwarf irregulars, all of these galaxies are classified as dwarf spheroidals (dSphs). The discovery dates, along with some basic properties of these satellites, are shown in Table 3.1. As is seen, up until 1994 and the discovery of the Sagittarius dwarf spheroidal at a distance of about 25 kpc from the Sun, the nearest known galaxy to the Milky Way was the LMC.

3.1.2 The Milky Way in the era of the Sloan Digital Sky Survey

At the turn of the century, it was not known if galaxies or star clusters with surface brightnesses below that of the known population of dwarf satellites even existed. At this time, only galaxies and star clusters with surface brightnesses greater than about 27 mag arcsec^{-2} were detectable in photometric surveys. This limit arises primarily because galaxies of this low surface brightness are only detectable through their population of resolved stars, revealing their presence as statistically significant overdensities above the Galactic foreground.

Beginning in 2004, however, the census of ultra-faint stellar systems and Galactic satellites changed dramatically, with the advent of the Sloan Digital Sky Survey (SDSS) and its ability to detect galaxies via stellar overdensities down to surface brightness of about 29 mag arcsec^{-2}. The first new object to be detected in the SDSS data was SDSSJ1049+5103 by B. Willman and collaborators, an object which has now become widely known as Willman 1 [1937]. Willman 1 was detected as an overdensity of resolved old, blue and metal-poor stars at a distance of 45 ± 10 kpc with a half-light radius of 23 ± 10 pc. The discovery of Willman 1 was closely followed up by the detection of Ursa Major I [1938], an object with a magnitude $M_V = -5.6$ at a distance of $\sim$100 kpc and with a half light radius of $\sim$250 pc. Since these discoveries, more than a dozen new satellites have been discovered, with magnitudes that vary by over three orders of magnitude, distances from the Sun ranging anywhere from $\sim$25 to 400 kpc and half-light radii in the range $\sim$20–500 pc [272; 1975; 1976]. The observed properties of all of these new satellite galaxies are summarized in Table 3.1. Interestingly, all of these satellites have luminosities less than the least luminous of the previously known satellites, implying that the census of satellites at the high-luminosity end may be largely complete.

3.1.3 Classification of Milky Way satellites

What is the nature of the newly discovered SDSS satellites? Are they more akin to dwarf spheroidal galaxies or globular clusters? Answering this question requires an understanding of the sometimes ambiguous discriminants between these two types of low-luminosity stellar systems.

Perhaps the most straightforward classical distinction between globular clusters and dwarf galaxies results from an analysis of their magnitudes and half-light radii, as shown in Figure 3.1. At a fixed magnitude, globular clusters are less extended than dwarf galaxies; the characteristic half-light radius for a globular cluster is less than about 10 pc, while for a dwarf galaxy the characteristic half-light radius is much greater than 10 pc. While many of the new SDSS satellites lie firmly in the dwarf galaxy region of the figure, several objects, in particular Willman 1, Segue 1, Coma Berenices,

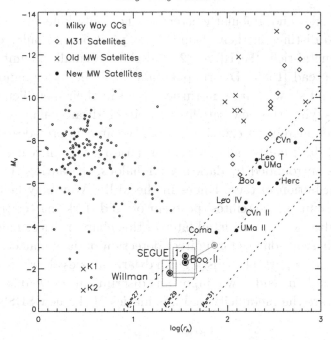

Fig. 3.1. The magnitude versus half-light radius for globular clusters (GCs) and dwarf galaxies of the Milky Way and M31. The objects highlighted with squares towards the bottom centre are several recently discovered low-luminosity systems. Circled points are objects with half-light radii intermediate between GCs and dwarf galaxies. Figure courtesy of Shane Walsh, private communication. Data for GCs from www.physics.mcmaster.ca/~harris/mwgc.dat.

Bootes II and Ursa Major II, populate a region of this figure at low luminosity and intermediate half-light radius that was not previously inhabited by a known dwarf galaxy or globular cluster (GC). Thus, uncovering the true nature of the aforementioned extremely low-luminosity objects requires a more detailed investigation beyond the simple interpretation of Fig. 3.1.

In addition to having different half-light radii, dwarf galaxies and globular clusters are distinct objects in terms of both their kinematics and chemical abundances. Kinematically, both the most luminous dwarf galaxies and globular clusters have velocity dispersions in the range ~5–15 km s^{-1}. For a typical high-luminosity globular cluster (~$10^6 L_\odot$), the half-light radius and the velocity dispersion imply that the dynamical mass within the stellar distributions is dominated by the stars. On the other hand, the more extended dwarf galaxies with this velocity dispersion have inferred dynamical total mass-to-light ratios of ~100. Thus dwarf galaxies all appear to be strongly dominated by dark matter, while GCs are not, at least within the stellar tidal radii.

Dwarf galaxies and globular clusters are not only kinematically distinct, they are also distinct in their chemical composition. Globular clusters are on average metal-rich, [Fe/H] > -2, and show a small amount of internal metallicity spread [1606]. Dwarf spheroidals, on the other hand, are more metal-poor and show a larger intrinsic spread in their metallicities [1398].

Thus classifying the new satellites as either dwarf galaxies or globular clusters is not merely an exercise in bookkeeping; it provides clues to the properties of dark matter and galaxy formation on small scales. If the new satellites are surrounded by dark matter haloes, measuring the properties and the distribution of these haloes in the Milky Way may help in understanding the 'missing satellites' problem of cold dark matter [1250; 1468]. Also, as we discuss in more detail later in this chapter, these haloes may be great astrophysical sources for indirect detection of dark matter. Because of these issues it is important to distinguish dwarf galaxies clearly from globular clusters at the faint end. Two important discriminators we discuss in more detail below are the metallicity and kinematics of the new SDSS satellites.

3.1.4 Dark matter in satellite galaxies

Determining the dark matter content of dwarf spheroidals requires spectroscopy of their stars. Aaronson first measured the velocity dispersion of Draco and Ursa Minor from stellar spectroscopy, deriving a mass-to-light ratio nearly an order of magnitude larger than that of GCs [6]. The velocity dispersions for several other dSphs were then measured by Mateo *et al.* in 1993 [1397]. Thanks to high-resolution spectra, the past few years have seen a great increase in the number of line-of-sight velocities for stars in all classical, pre-SDSS dSphs [1915], and the first measurements of the velocity dispersions of the new SDSS satellites [1763].

Given the wealth of new data available, it is important to derive accurate models of the dynamical mass of dSphs. When modelling the stellar distribution of a dwarf galaxy in close proximity to the Milky Way (MW), it is important to consider the tidal effects of the Milky Way. We provide simple estimates here. The internal gravitational force on stars in the outer parts of the dwarf galaxy is of order σ_r^2/R_s, where σ_r is the radial velocity dispersion and R_s is the half-light radius of the dwarf. The external tidal force from the MW potential is of order $(\sim 220 \text{ km s}^{-1})^2 \, R_s/D^2$, where D is the distance to the dwarf from the centre of the MW, and 220 km s^{-1} is the approximate rotation speed of the MW in the outer regions of the halo where the dwarfs are located. The most luminous of the MW dwarfs have half-light radii of $R_s \approx 400 \text{ pc}$, and the least luminous have half-light radii of $R_s \approx 10\text{–}100 \text{ pc}$.

The velocity dispersions vary in the range $\sigma \approx 5$–$15\,\mathrm{km\ s^{-1}}$. Comparing the internal and external forces on dwarfs in the observed distance range of ~ 20–$250\,\mathrm{kpc}$, we find that the internal gravitational forces are typically larger by a factor of ~ 100. Note that this estimate does not exclude the possibility that the dwarfs have been tidally stripped in the past; it does, however, allow us to proceed with the assumption that the surviving stellar distributions trace the local potential.

Given that the timescale between stellar encounters in dSphs is large, and the relaxation times are larger than the age of the Universe, it is appropriate to use the collisionless Jeans equation to determine the mass distributions, which assumes that the galaxies are spherical and supported entirely by their velocity dispersions. For the high-luminosity dSphs, the central light distributions are approximately spherically symmetric; including stars even from the outer regions of the galaxies near the tidal radii shows that the total ellipticities, defined as $1 - b/a$ where b is the minor axis and a is the major axis, vary in the range from ~ 0.1 to 0.6 [1156]. The faint dSphs are markedly elliptical [1390] and require non-spherical Jeans analysis. However, the small number of measured radial velocities do not yet warrant detailed non-spherical analysis – the errors on the measured dynamical masses are currently quite large [1816]. To date, there has been no significant detection of rotation in any dSphs; the one possible exception is Sculptor [228], although given the extent of this galaxy this rotational signal may be mimicked by the gradient in the line-of-sight velocities [1208; 1914].

The velocity dispersion profiles of the highest-luminosity dSphs that were discovered before SDSS all remain constant out to, and in some cases beyond, the stellar tidal radius of the galaxy [1915]. For several dSphs, there are over a thousand line-of-sight velocities measured. As a result mass-follows-light models, which assume that all of the mass distribution is accounted for by the stars, are conclusively ruled out. Thus a large dark matter component is required to explain the dynamics of these systems, and to obtain a more accurate assessment of the mass distribution of the satellites, two-component models are required which separately account for the distribution of stars and dark matter. For the new satellites discovered by SDSS, anywhere from tens to hundreds of line-of-sight velocities have been measured, and even with these small data sets it is possible to conclude that dark matter dominates the dynamics of many of these systems [1763]. At present, the only new satellites without line-of-sight velocities are Bootes II and Leo V.

Given the data sets now available, it is important to ask, in the context of a two-component model, which physical quantities are best constrained by the line-of-sight velocity data. Answering this question first requires defining a set of model parameters. The cold dark matter haloes typically are described by several free parameters: these include a scale density, a scale radius, an asymptotic inner and outer slope, and a transition between these latter two slopes. An example of a widely considered halo profile along these lines is given by $\rho(r) = \frac{\rho_0}{(r/r_0)^a[1+(r/r_0)^b]^{(c-a)/b}}$. Additionally the anisotropy in the velocity dispersion of the stars is unknown, and this quantity could change with physical radius in the galaxy.

Although none of the above shape parameters, including the velocity dispersion anisotropy, is well determined by the present data, the integrated mass within the approximate half-light radius can be measured [1815]. The fact that the mass is well constrained at the half-light radius allows us to make a comparison between the observed satellite population and the population of subhaloes in numerical simulations of cold dark matter [660; 1790]. As an example, the mass of the pre-SDSS satellites within their inner 600 pc is constrained to be $\sim [1-6] \times 10^7 \; M_\odot$ [1815; 1915], accounting for model uncertainties. Comparing the observed mass function in this mass range to the mass function of dark matter subhaloes, one finds that the population of MW satellites is well described by either the earliest forming subhaloes or the largest before accretion onto the Milky Way [1815].

The radius of 600 pc represents an average of the half-light radii of the satellite population known before SDSS. However, the half-light radii of the SDSS satellites are in some cases smaller by about an order of magnitude, and their stellar tidal radii in these extreme cases only extend to ~ 100 pc. Further, the mass within 600 pc may be ill defined for several of the extremely low-luminosity satellites, as this radius is beyond the present value of the Jacobi radius [922]. Thus to make the same comparison as above to numerical simulations, only now including the SDSS satellites, the mass within a smaller radius must be measured. We choose 300 pc to display results (see Fig. 3.3), because it also corresponds to the minimum resolvable radius in CDM simulations at present. When calculating the masses within 300 pc, the spread in masses is observed to become much narrower; in fact all of the satellites have a mass very close to $10^7 \; M_\odot$ within 300 pc, in spite of the spread of nearly five orders of magnitude in the luminosities of the satellites [1816].

The fact that all of the dwarf satellites share a common mass scale of $10^7 \; M_\odot$ within 300 pc provides clues to either the formation of galaxies at small scales, or the nature of the particle dark matter, or perhaps both. Is this result of a common mass scale consistent with CDM simulations? Although no currently published numerical simulations are able to resolve the mass of subhaloes within 300 pc from their centres, initial results that

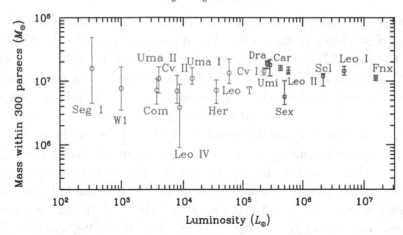

Fig. 3.2. The mass within 300 pc for the population of Milky Way satellites with measured velocity dispersions. Square points are well-known satellites, circles are newly discovered SDSS satellites. The error bars reflect the region within which the likelihood is greater than 10% of its peak value. Figure taken from [1816].

extrapolate the subhalo mass distributions from larger radii show that this large luminosity scatter at fixed subhalo mass can be explained by understanding the distribution of halo masses upon accretion into the Milky Way [1365].

Thus while the luminosities and in some cases the half-light radii of the ultra-faint satellites are similar to those of globular clusters, spectroscopic studies firmly suggest that the majority of these objects are dark-matter-dominated dwarf galaxies. Further, even though the luminosities of the Milky Way satellite population vary by over four orders of magnitude, their dynamical masses within a fixed radius are observed to be very close to constant. As discussed below, future imaging surveys of stars in the Milky Way will provide a more complete census of low-luminosity Milky Way satellites, with the prospects of determining whether astrophysics or fundamental dark matter physics is responsible for setting the common mass scale. The masses for the faintest dwarf galaxies will become more strongly constrained with more line-of-sight velocity data. This will sharpen the observational picture of galaxy formation on these small scales and provide data around which theories of galaxy formation may be built.

3.1.5 Metallicity, gas content and star formation in dwarfs

The mean stellar metallicity of a galaxy is believed to reflect the metallicity conditions of the intergalactic medium at the formation time of the galaxy, and a wide metallicity spread in the stellar populations is indicative of multiple episodes of star formation throughout the history of the galaxy. The dwarf satellites of the Milky Way discovered prior to the SDSS were known to follow a linear relationship between their luminosity and metallicity, [Fe/H] [1006; 1398]. In contrast, GCs tend to exhibit a larger scatter in their [Fe/H] over the range of luminosities, and the metallicity spreads are much lower than those observed in dwarf galaxies.

Do the metallicities and luminosities of the SDSS satellites follow the same log-linear relationship as the higher-luminosity dSphs? Answering this question is also important in establishing them as either dwarf galaxies or globular clusters. Recent measurements do indeed seem to indicate that the metallicity–luminosity trend established for the high-luminosity satellites continues down through the luminosity range of the SDSS satellites [1240]. These results further strengthen the case that the new satellites more resemble dwarf galaxies, and may also have important implications for understanding the assembly history of the Milky Way stellar halo.

In addition to the above measurements of metallicity, there are strong observational upper limits on the HI gas content in Milky Way dSphs [1616]. The only satellites with known detections of HI are the LMC, SMC and Leo T. The latter is the only dSph type of object with observed HI, and it is located at a distance of ~420 kpc from the Milky Way centre.

3.2 Streams

3.2.1 Historical review

The origin of the diffuse, low-density stellar halo around the Milky Way is a long-standing astronomical problem. The poles of the debate are defined by the monolithic collapse model of Eggen, Lynden-Bell and Sandage in 1962 [737] and the chaotic accretion model of Searle and Zinn in 1978 [1716]. The Searle and Zinn scenario bears a strong resemblance to the hierarchical galaxy formation scenario characteristic of CDM cosmological models, where the accretion of smaller dark matter haloes into larger ones is common [1807]. The stellar debris from recent accretion events can remain in relatively distinct spatial structures for many gigayears [1175] and therefore their presence provides a means to constrain the small-scale accretion history of the Galaxy. Lambda-CDM-based simulations predict a significant amount of recognizable halo substructure, and the discovery and characterization of this substructure promises to determine whether structure formation is hierarchical on the smallest scales [466; 467; 630].

In the past decade, much more direct evidence for a Searle and Zinn or LCDM picture of lumpy build-up of the Galaxy has emerged in the form of clumps and streams in the stellar distribution. The most striking example in the latter category is the discovery of the tidal tail of the Sagittarius dwarf galaxy, which has been traced around the entire Galaxy [1148; 1375; 1951]. More importantly for dark matter models, the SDSS has revealed a significant number of unrelated stellar streams and overdensities that seem to span the Galactic sky [267; 271; 724; 1022; 1185; 1375; 1506], suggesting that low-mass accretions have been frequent. The Andromeda galaxy, M31, has at least as much of this type of structure as the Milky Way [520; 1147]. Overall, the observed structure is quite encouraging for the standard LCDM model. Specifically, both the radially averaged density structure and the degree of clumpiness are in line with LCDM expectations [267].

3.2.2 Cold streams as a probe of substructure

While the count and character of stellar streams around the Milky Way are in line with LCDM expectations, these streams only provide a direct probe of dark matter halo properties on scales that are massive enough to form stars, $M > 10^8 M_\odot$ [1615]. Searches for less massive substructures must rely on gravitational effects. The streams of debris from the destruction of Galactic satellites and globular clusters provide cold structures that could be scattered by low-mass substructure in the Milky Way halo potential [1148; 1177; 1563; 1751]. These streams tend to align along a single orbit [1176] and hence would individually have lower cross-section to interactions than the Galactic disk. However, the Galactic disk is relatively hot compared with thin Galactic streams, and is only sensitive to the most massive substructures which host a visible galaxy, such as the LMC halo [1216].

Most recent simulations have concluded that the uncertainties associated with the shape of the host Milky Way halo and the orbit of the stream progenitor are large enough that they do not allow current data to place interesting constraints on the presence of small-scale substructure [1751]. However, substructure can act to shift the location of sections of debris, and this may provide smoking gun signatures as larger surveys cover more sky. Cold globular cluster streams hold particular promise for constraining the number of small haloes [1148], although these constraints will have to wait for future large-sky surveys (for globular-stream discoveries) and essential proper motion information from astrometric satellite missions such as ESA's GAIA (for local streams) and NASA's SIM Planetquest (for more distant streams). The role of SIM Planetquest is essential for this purpose because the outer halo is predicted to have a much higher abundance of substructure than the inner halo probed by GAIA.

3.3 Central profiles and the nature of dark matter

Measuring the central density profiles of dark matter haloes has important ramifications for the particle properties of dark matter. From the perspective of particle dark matter, if, for example, either the Milky Way's dark matter halo or the haloes of Local Group satellites had steep central cusps, this would lead to a higher annihilation rate of dark matter into Standard Model particles, providing better prospects for detection in high-energy particle observatories. On the other hand, if the central density profiles of dark matter haloes approach constant density cores over observable scales, this may imply a low phase-space density for the dark matter, and provide a measurement of the temperature of dark matter.

As dwarf satellites are the least massive, most dark-matter-dominated galaxies known, they provide unique testbeds for understanding how the dark matter behaves in the central regions of haloes. Precisely because these objects are dominated by dark matter, their central densities are a direct reflection of dark matter physics, whereas in higher-mass galaxies baryonic processes become increasingly important in the central regions of the haloes. Further, from an observational perspective the interpretation of the density profiles of rotationally supported low-mass galaxies is somewhat ambiguous: some haloes seem to prefer cores, some prefer cusps and others are well fitted by either [1293; 1762].

From an empirical perspective, however, determining the central density slope of the most dark-matter-dominated galaxies has been difficult. Several indirect lines of evidence suggest that a couple of dSphs are consistent with cored dark matter profiles. For example, a substructure clump with low velocity dispersion has been observed in the Ursa Minor dSph [1246], which may be a gravitationally unbound star. If this were the case, then the tidal forces in Ursa Minor must be weak, possibly implying the existence of a constant density core [952]. Additionally, the Fornax dSph is observed to have five surviving globular clusters. Given the projected distribution of these globular clusters, it is likely that their orbits would have decayed if the dark matter density profile had a steep central cusp. The estimated size of the core to stabilize the globular clusters over a Hubble time is a few hundred parsecs [969].

Unfortunately, direct dynamical evidence does not substantiate the claim that cored dark matter profiles are preferred in dSphs. Only models with dark matter cores ~ 1 kpc are ruled out by present data sets [1814]. If the dSphs are in dynamical equilibrium, which we showed to be a good description in the previous section, the stars act as test particles that trace the gravitational potential. Dark matter profiles of dSphs are directly inferred through the measured line-of-sight velocity distribution of the stars. Although many thousands of line-of-sight velocities are now available from several dSphs, even with the large data sets equilibrium model solutions to the Jeans equation are degenerate: even though the stellar kinematics strongly constrains the integrated mass within a fixed radius, it does not measure well enough the shape of the density profile of these haloes at radii necessary to probe the phase-space cores. More specifically, there is a strong degeneracy between the logarithmic slope of the density profile of the dark matter halo at a given radius and the velocity anisotropy of the stellar distribution. This degeneracy can be explicitly seen by considering the Jeans equations for velocities projected along the line of sight (los), and for the two directions perpendicular to the line of sight, in the plane of the sky (R and t):

$$\sigma_{\mathrm{los}}^2(R) = \frac{2}{I_\star(R)} \int_R^\infty \left(1 - \beta \frac{R^2}{r^2}\right) \frac{\nu_\star \sigma_{\mathrm{r}}^2 r \mathrm{d}r}{\sqrt{r^2 - R^2}}, \tag{3.1}$$

$$\sigma_R^2(R) = \frac{2}{I_\star(R)} \int_R^\infty \left(1 - \beta + \beta \frac{R^2}{r^2}\right) \frac{\nu_\star \sigma_{\mathrm{r}}^2 r \mathrm{d}r}{\sqrt{r^2 - R^2}}, \tag{3.2}$$

$$\sigma_t^2(R) = \frac{2}{I_\star(R)} \int_R^\infty (1 - \beta) \frac{\nu_\star \sigma_{\mathrm{r}}^2 r \mathrm{d}r}{\sqrt{r^2 - R^2}}. \tag{3.3}$$

Here β is the stellar velocity anisotropy, $I_\star$ is the surface density, $\nu_\star$ is the three-dimensional surface density and σ_{r} is the solution to the three-dimensional Jeans equation. It is clear that, from just the projection along the line of sight, solutions to the Jeans equation are degenerate with β. This situation is shown more clearly in Fig. 3.3, which shows the projected errors on the log-slope of Draco, both for the case where the proper motions of stars are measured (i.e. the R and t components as well as line-of-sight velocities) and for line-of-sight velocities only. As is seen, utilizing proper motions reduces the error at radii closer to the centre of the galaxy. Although Figure 3.3 shows the scenario where β is constant as a function of radius, for simplicity, the above results hold even for more complicated, and radially variable, anisotropy functions.

3.3.1 Constraining models of particle dark matter

There are many theoretically well-motivated dark matter candidates that predict dramatically different structure formation on dwarf galaxy scales compared with CDM. More precise observations and better theoretical modelling will allow dSph observations to put constraints on the particle

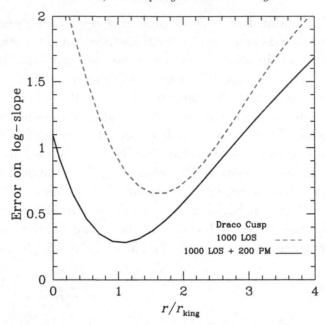

Fig. 3.3. The projected errors on the log-slope of the dark matter density profile of Draco as a function of radius (r_{king} is the king core radius), assuming a fiducial NFW profile that is consistent with the stellar dynamics of Draco. This NFW profile has a log-slope of ~ 1.2 over range ~ 100–300 pc. The dashed curve uses only the line-of-sight velocities of 1000 stars, while the black curve uses the same number of line-of-sight velocities plus the proper motions (PM) of 200 stars. As is seen, including proper motions implies a reduced error on the log-slope at radii closer to the centre of the galaxy.

properties of dark matter. A large class of alternative models, such as thermal warm dark matter (WDM), could have phase-space limits that imply central density cores that would be observable on the scales probed by stellar kinematics – a well-known example is a sterile neutrino [690; 904]. In addition to thermal WDM there is a large class of models in which the dark matter is created from decays that also predict cores [501; 1207].

The difference between different dark matter models arises from at least two separate effects. The dark matter density fluctuations are smoothed out on scales smaller than the free-streaming and damping lengths [345; 1114; 1207; 1348; 1611]. This smoothing scale depends on the microphysical properties of the dark matter particle (including its mass and its scattering cross-section off Standard Model fermions) and it sets the minimum dark matter halo mass. For WIMPs this minimum mass could be a few orders of magnitude larger or smaller than the mass of the Earth ($\sim 10^{-6}~M_{\odot}$). In contrast, for dark matter from decays or a sterile neutrino the smallest halo mass could be of the order of the mass of the haloes of the Milky Way's satellite galaxies ($\sim 10^8$ $M_\odot$). A census of dwarf galaxies is thus essential in discriminating between different dark matter models.

The primordial phase-space density (Q) of dark matter gives an indication of how cold the dark matter particle is. The fact that dark matter is almost collisionless implies that Q sets the maximum density that dark matter haloes can have in their centres [1116; 1207; 1864]. WIMPs and axions are cold particles and have very high Q values (e.g. $\sim 10^{15} M_\odot$ pc^{-3}(km/s)$^{-3}$ for some WIMPs) compared with average phase-space densities in galaxies. Haloes formed of WIMPs or axions would show a sharp rise in density towards the centre. On the other hand, warmer particles such as sterile neutrinos have much lower Q values (as low as $\sim 10^{-5} M_\odot$pc^{-3}(km/s)$^{-3}$) and can produce observable flat density cores in the dark matter haloes of small galaxies. Definitive observation of these cores would disfavour axion and WIMP dark matter. The galaxies with the highest measured values of Q are the lowest-mass galaxies known, the dSphs: the ratio of the core size to the halo virial radius is largest in these low-mass haloes.

It is possible to obtain upper limits on the coarse-grained Q-values from dSphs [1763]. These limits are strong enough to rule out several varieties of dark matter that have fine-grained phase-space densities well below these measured Q values. Further, combining these results with limits on the dark matter free-streaming length provides an interesting constraint on dark matter properties [1814]. Future measurements of both the small-scale power spectrum and the phase-space density in dSphs thus promise to be a unique window into the nature of dark matter.

3.4 Indirect detection of dark matter from Milky Way satellites

In a wide variety of models, including well-motivated models from supersymmetry and extra dimensions, dark matter particles will annihilate into Standard Model particles that may be visible in current and future high-energy particle detectors. Of particular interest are the models in which gamma-ray photons are among the particles produced in the annihilations. To detect these gamma-rays from dark matter annihilation, there are a variety of astrophysical targets to consider: for example the Galactic centre, the diffuse Galactic and extragalactic emission, and dwarf galaxies. We will focus here on using the dark-matter-dominated dSphs to constrain the properties of dark matter via annihilation radiation.

Because of their high mass-to-light ratios and their relative proximity to the Milky Way, dwarf spheroidals provide independent targets that have been widely considered in the literature [178; 299; 803; 876; 1610; 1817; 1877]. Their status as bona fide targets for indirect detection has become even more interesting recently, given that the known number of satellites has more than doubled in the past few years, coupled with the discovery that the least luminous of these satellites are the most dark-matter-dominated objects known [1816]. The fact that the mean central dark matter densities for all satellites are similar largely implies that the best targets are simply those objects that are the nearest to the Earth. Although, in general, the actual gamma-ray flux from the dSphs is smaller than the flux from the Galactic centre, the astrophysical gamma-ray backgrounds tend to be reduced in the direction of these objects, particularly if we account for the fact that many of these objects are located at high Galactic latitudes. As an additional benefit, dSphs may have low intrinsic emission from astrophysical gamma-ray sources, as all of them with the exception of the distant satellite Leo T have strong upper limits on the HI gas content [1616]. Dwarf galaxies thus play an important complementary role to the Galactic centre when it comes to indirect detection; optimistically, a signal from both the Galactic centre and the satellites would be a convincing demonstration of dark matter annihilations.

Assuming that the haloes of the dSphs are smooth (we will consider the alternative case of clumpy haloes due to the presence of substructure below), the dark matter annihilation flux from a dSph is given by:

$$\frac{\mathrm{d}N}{\mathrm{d}A\mathrm{d}t} = L\mathcal{P} \tag{3.4}$$

$$L = \frac{1}{4\pi} \int_{\mathrm{los}} \rho^2(r)\mathrm{d}l \tag{3.5}$$

$$\mathcal{P} = \int_{E_{\mathrm{th}}}^{M_\chi} \sum_i \frac{\langle \sigma v \rangle}{M_\chi^2} \frac{\mathrm{d}N_{\gamma_i}}{\mathrm{d}E} \mathrm{d}E \tag{3.6}$$

where E is energy, E_{th} is threshold energy, A is area, $\langle \sigma v \rangle$ is the annihilation cross-section and N_{γ_i} is the number of photons. The 'L' component of the flux depends only on the astrophysics of the dark matter distribution in the dSph haloes. Thus in order to gain constraints on the properties of the dark matter particle (the '$\mathcal{P}$' component) we must fully understand the uncertainty in the mass modelling of the dark matter haloes.

Numerical simulations of CDM show that the density profiles of subhaloes are fitted well by a relation of the form $\rho(r) = \frac{\rho_s}{(r/r_s)^\gamma [1+r/r_s]^{\delta-\gamma}}$, where ρ_s and r_s are the scale density and the scale radius, respectively. Recent simulations suggest that $\delta \simeq 3$, while γ has a range of value of ~ 0.7–1.2. From a kinematic standpoint alone, there is a wide range of ρ_s and r_s values that fit the velocity dispersion data in dSphs. However, this degenerate range of $\rho_s - r_s$ parameter space can be significantly reduced by appealing to the relation between mass and concentration in CDM haloes (here, concentration is defined as the halo outer radius divided by the inner scale radius).

In principle, a major source of uncertainty in the gamma-ray luminosity of a dSph is the slope of the central density profile. Indeed, the uncertainty in gamma-rays from the Galactic centre implies an uncertainty in flux of about three orders of magnitude. However, this is not the case for dSphs, and the simple reason for this is because of the kinematics. For a profile similar to the one measured in CDM simulations, with $\gamma \simeq 1$, ~90% of the gamma-ray emission comes from within r_s. For a typical dSph, this corresponds to a physical radius of a couple of hundred parsecs. In order to maintain consistency with the velocity dispersion data, for changes in the inner slope over the allowed region, the scale density changes roughly in such a manner that the L values remain nearly constant. Thus, assuming a smooth halo, the astrophysical uncertainty is much smaller than the corresponding uncertainty in the dark matter halo profile in the Galactic centre.

Figure 3.4 shows the best fitting values for the gamma-ray flux from Segue 1, assuming three different energy thresholds. Because of its proximity, 25 kpc from the Sun, Segue 1 is the most promising source of dark matter annihilation amongst all of the dSphs. Because of the low flux values, which are, at best, near the sensitivity limits of current gamma-ray detectors, extraction of the signal from a dSph will depend strongly on the understanding and subtraction of astrophysical backgrounds. Most dSphs, including Segue 1, are at high Galactic latitudes, so astrophysical backgrounds will be much smaller, and probably less spatially variable, than the backgrounds from the Galactic centre. The dominant astrophysical background is expected to come from diffuse extragalactic emission, whose energy spectrum is well fitted by a power law, $dN/dE \sim E^{-2.7}$. Because of this fairly well-characterized and uniform background, as well as their well-understood dark matter distributions, dSphs make a strong target for future gamma-ray detectors.

3.4.1 Boost to annihilation signal from subhaloes within subhaloes

Not only is the Milky Way's dark matter halo expected to contain sub-structures, some of which correspond to the observed dwarf galaxies, but these substructures also contain substructure. Thus substructures should also be present in the dark matter haloes of the dSphs, and they may have a significant effect on the dark matter annihilation signal. On the one hand,

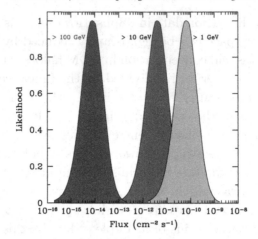

Fig. 3.4. The maximum likelihood values for the gamma-ray flux from Segue 1. The three distributions assume the three different energy thresholds that are shown. For the dark matter haloes we assume central slopes that are within the range indicated by present high-resolution numerical simulations.

it may be easier for substructure to survive in the haloes of dSphs, since these objects are dark matter dominated (and under the assumption that they have always been dark matter dominated). On the other hand, there are some effects that may reduce the number of substructures within substructures; for example, the haloes of dwarf satellites do not get replenished by accreted field haloes to the extent that a Milky Way mass halo would. Also, substructures within dSph haloes are older, giving them more time to assimilate into the smooth component of the dark matter haloes.

If these sub-substructures survive as bound objects within the dSph host dark matter haloes, they will increase the annihilation rate from the object. The important quantity is the increase in the signal relative to the smooth component. Thus, the gamma-ray flux (L) is the combination of the flux due to the smooth host halo ($\tilde{L}$) plus flux due to its subhaloes. The flux enhancement from halo substructure may be parameterized by the 'boost' factor $B(M, m_{\min})$, which we define as

$$L(M) = [1 + B(M, m_{\min})]\,\tilde{L}(M). \qquad (3.7)$$

Here $\tilde{L}(M)$ is defined as annihilation flux from the smooth density distribution for a halo of fixed mass M. The minimum mass halo here is defined as $m_{\min}$, which is derived from the cutoff scale in the CDM power spectrum [1348; 1611]. The boost factor depends not only on enhancement due to the subhaloes of the host halo, but also on enhancement due to the substructure of the subhaloes themselves, and on the enhancements due to substructure of the sub-subhaloes, and so on. For some CDM models, this requires the knowledge of the subhalo mass population over ~ 13 orders of magnitude, well beyond the limitations of our current numerical simulations.

Relatively little is known about the boost from halo substructures at the mass scales of the smallest CDM halo. The primary reason for this is the uncertain relation between the concentration and mass at these low mass scales. The concentration may continue to rise as a power law to the smallest masses, or flatten out because the power spectrum flattens out on small scales. To better understand the origin of this uncertainty, we write the boost as

$$\tilde{L}(M)B(M, m_{\min}) = AM^\alpha \int_{m_{\min}}^{qM} [1 + B(m, m_{\min})] \, \tilde{L}(m)m^{-1-\alpha} dm \, . (3.8)$$

The last step was achieved with an assumed substructure halo mass function that scaled as $A(M/m)^\alpha$ for $m < qM$ and self-similar substructure. This would lead to an overestimate in the boost. Even so, ref. [1817] showed that the boost is no larger than about 100, typical numbers being about 10 or smaller for $m_{\min} \approx 10^{-6}M_\odot$ with A, q and α set to match simulations by ref. [664]. The exact value is very sensitive to the extrapolation to small masses and scales roughly as $(M/m_{\min})^{\alpha-\gamma'}$ where $\gamma' \equiv d\ln \tilde{L}(M)/d\ln M$, and we have the scaling $\tilde{L}(M) \sim Mc^{2.2} \sim M^{0.9}$, where c is dark matter halo concentration. It is clear that, since $\alpha \approx 1$, it is essential to estimate the slopes γ' and α correctly.

3.5 Future prospects

Is the census of Milky Way satellites and streams complete? The likely answer to this question is no, starting from the simple fact that the SDSS surveyed only about one-fifth of the sky near the North Galactic Cap. In fact, cosmologically motivated corrections to the luminosity function of Milky Way satellites in combination with SDSS incompleteness limits [1270; 1917] suggest that there may be hundreds of satellites with luminosity greater than the least luminous satellites detected so far [1858]. Future, deep wide-field surveys are expected to deliver a complete census of ultra-faint dwarf satellites out to the Milky Way virial radius, offer new limits on the free-streaming scale of dark matter and provide unprecedented constraints on the low-luminosity threshold of galaxy formation. If the dark matter particle has annihilation cross-sections of the order of the weak interaction cross-sections (as predicted by many particle physics models that will be tested at the Large Hadron Collider), then the Milky Way satellites will play a pivotal role in establishing the nature of the dark matter.

4

Gravitational lensing and dark matter

Yannick Mellier

4.1 Introduction

Gravitational lensing effects arise from the modification of space-time metric produced by mass concentrations. Following an early prediction of General Relativity, gravitational fields deflect the light path of photons and modify the apparent flux and shape of astronomical sources. By observing gravitationally lensed images, cosmologists can probe dark matter almost 'directly'. They can in principle examine it without the need to speculate on the distribution of matter inside the gravitational potential responsible for the light deflection, nor on its dynamical state and on its thermodynamical properties.

A most attractive application of gravitational lensing concerns the quest for dark matter candidates and for the properties of dark matter particles. Successful examples are the microlensing experiments (see [798] and references therein) carried out over the past decade to detect sub-stellar dark compact objects in the Galaxy. Similar searches for invisible cosmologically distributed compact objects have been carried out using microlensing on quasars [609]. These experiments set limits on compact dark objects (CDO) in the mass range $10^{-6} M_\odot < m < 10^6 M_\odot$ [1918] and prove they represent a fraction $\Omega_M(\text{CDO}) < 0.1$ of invisible mass in the Universe, if any.

Beside microlensing, other extragalactic gravitational lensing effects on quasars and extended sources (galaxies) can also probe dark matter. Although they are less frequent and more complex phenomena than microlensing, these lensing configurations provide much more detail on the lens and on the amount and distribution of matter inside. Observations of multiply imaged quasars, Einstein rings or gravitational arcs probe deflecting masses in the range

Particle Dark Matter: Observations, Models and Searches, ed. Gianfranco Bertone. Published by Cambridge University Press. © Cambridge University Press 2010.

$m > 10^6 M_\odot$ and, eventually, can explore the properties of non-baryonic dark matter particles from their effect on structure formation in the Universe and on the dynamics of gravitational systems. Instead of using the amplitude and the duration of transient magnifications, as for microlensing, gravitational optics is used to unveil dark matter from the description of the image multiplicity, positions, shapes and flux ratios of lensed sources.

The aim of this chapter is to present applications of such gravitational lensing effects to explore dark matter. We focus on gravitational lensing by galaxies, clusters of galaxies and large-scale structure of the Universe and on applications that may help to track down its nature. Gravitational lensing is primarily sensitive to the amount of matter. Understanding the nature of the matter from it is not trivial and is usually derived from indirect arguments. Observations of lenses and lens models of the dark matter distribution are compared to theoretical predictions, such as the dark matter power spectrum or the properties of dark haloes at non-linear scales. The gravitational lensing signal is primarily compared with numerical simulations and predictions of the most popular scenarios, such as cold dark matter universes. However, it can also reveal tiny and subtle features like those produced by neutrino components or those expected from alternative views to dark matter, namely modified gravity models.

The chapter is organized as follows. In Section 4.2, elements of gravitational lensing theory are presented. Section 4.3 summarises the observational evidence of dark matter from strong gravitational lenses. Section 4.4 is focused on collisionless cold dark matter and the gravitational lensing evidence supporting the ΛCDM cosmological model. This section also includes an introduction to weak gravitational lensing theory and to weak lensing by large-scale structure of the Universe (i.e. cosmic shear). Section 4.5 presents the results on neutrino mass derived from cosmic shear surveys. The comparison of gravitational lens observations to predictions of modified gravity models is discussed in Section 4.6. Conclusions and outlook are given in the final section.

4.2 Gravitational lensing theory

This section introduces the foundations and the concepts of gravitational lensing theory that are relevant for our review. We only focus on the description of stationary lenses in the weak field limit and the small deflection angle regimes (see below for justifications). More comprehensive reviews of gravitational lens theory can be found in [225; 1707] or in [1709].

4.2.1 Deflection angle and lens equation

The mass range of gravitational lenses discussed in this review is between 10^{11} and 10^{15} $M_\odot$ and corresponds to velocity dispersion of $\sigma_v \approx 100$–3000 km s^{-1}. The amplitude of the deflecting gravitational potential field is then always small, i.e. $\phi \propto \sigma_v^2 \ll c^2$, and the weak field limit applies. In this regime, the metric of space-time is described by a locally perturbed Minkowski space-time,

$$\mathrm{d}s^2 = \left(1 + 2\frac{\phi}{c^2}\right) c^2 \mathrm{d}t^2 - \left(1 - 2\frac{\phi}{c^2}\right) \mathrm{d}l^2, \tag{4.1}$$

where ϕ is the Newtonian gravitational potential of the deflector. In the weak field limit, and using the Fermat principle to compute the travel time of photons from the emitting source point, S, to the observer, O, one can show that the deflection angle is

$$\boldsymbol{\alpha}(\boldsymbol{x}) = -\frac{2}{c^2} \int_S^O \nabla_\perp \phi \, \mathrm{d}l, \tag{4.2}$$

where $\nabla_\perp$ is the gradient perpendicular to the line of sight. The deflection angle at position $\boldsymbol{x}$, $\boldsymbol{\alpha}(\boldsymbol{x})$, produced by a mass concentration is then written as

$$\boldsymbol{\alpha}(\boldsymbol{x}) = \frac{4G}{c^2} \int \mathrm{d}^2 x' \int \rho(\boldsymbol{x}', z) \frac{\boldsymbol{x} - \boldsymbol{x}'}{|\boldsymbol{x} - \boldsymbol{x}'|^2} \mathrm{d}z, \tag{4.3}$$

where z is the radial direction along the line of sight and ρ is the mass density of the lens.

Assume a gravitational lensing configuration composed of an observer, O, a gravitational lens, L, and a source located behind the lens, S (see Fig. 4.1). In the case where the size of the lens is much smaller than D_{OL} and D_{LS}, the thin lens approximation applies and the deflection angle is

$$\boldsymbol{\alpha}(\boldsymbol{x}) = \frac{4G}{c^2} \int \Sigma(\boldsymbol{x}') \frac{\boldsymbol{x} - \boldsymbol{x}'}{|\boldsymbol{x} - \boldsymbol{x}'|^2} \mathrm{d}^2 x', \tag{4.4}$$

where $\Sigma(\boldsymbol{x}') = \int \rho(\boldsymbol{x}', z) \mathrm{d}z$ is the mass density projected along the line of sight. If the lens is a point of mass M, the mass density reduces to a Dirac distribution and the deflection angle is

$$\alpha = \frac{4GM}{c^2 \, x}, \tag{4.5}$$

where x is the impact parameter, the radial distance between the lens position and the photon in the lens plane. The small deflection angle approximation applies if $\alpha \ll 2GM/(c^2 R_{\mathrm{S}})$, where R_{S} is the Schwarzschild radius

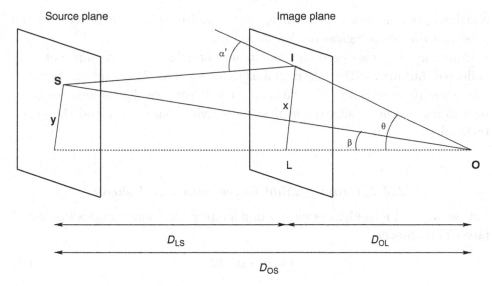

Fig. 4.1. Lensing configuration.

of the lens. That is, $R_S \ll x$, which is verified for all lenses we consider in this review.

The point mass approximation is always valid for microlensing experiments on compact objects but does not strictly apply to extended lenses. Nevertheless, it provides an acceptable indication of deflection angles of these lenses as well. Applying Eq. (4.5) to galaxies and clusters of galaxies, it is then easy to show that, for typical impact parameters between 1 and 100 kpc, $1'' < \alpha < 1'$ for masses between 10^{12} $M_\odot$ and 10^{15} $M_\odot$. These values demonstrate that the small angle approximation is valid and therefore gravitational lens optics based on the paraxial approximation is sufficient to describe the lens configurations.

The amplitude of the gravitational deflection does not only depend on the mass inside the lens, but also on the geometric configuration of the 'optical bench'. It is described by the angular distances D_{OL}, D_{LS} and D_{OS}, that is, the redshifts of the lens and the source and the underlying cosmological model. The *lens equation* relates the true position of the source in the source plane to the deflection angle and to the apparent position of images in the lens plane:

$$y = \frac{D_{OS}}{D_{OL}} x - D_{LS}\alpha(x). \tag{4.6}$$

It completely describes the lensing configuration and how the deflecting Newtonian potential can be recovered, if the deflection angle (or at least the relative apparent positions of images), the redshifts of the source and the lens, and the cosmological model are known.

Equation (4.6) also predicts that images of a lensed source are not only deflected, but are multiplied, magnified and distorted. The relations between the image properties and the (dark) matter density inside the lens are given by two important lensing quantities: the gravitational shear and the gravitational convergence.

4.2.2 Gravitational convergence and shear

Let us first write the lens equation in a simpler way using angles instead of physical distances

$$\boldsymbol{\beta} = \boldsymbol{\theta} - \boldsymbol{\alpha}'(\boldsymbol{\theta}), \tag{4.7}$$

with $\boldsymbol{\beta} = \boldsymbol{y}/D_{\mathrm{OS}}$, $\boldsymbol{\theta} = \boldsymbol{x}/D_{\mathrm{OL}}$ and $\boldsymbol{\alpha}' = D_{\mathrm{LS}}/D_{\mathrm{OS}}\,\boldsymbol{\alpha}(\boldsymbol{x})$. From the angular lens equation one can easily derive the expression of the scaled deflection angle, $\boldsymbol{\alpha}'$,

$$\boldsymbol{\alpha}'(\boldsymbol{\theta}) = \frac{4G}{c^2}\frac{D_{\mathrm{OL}}D_{\mathrm{LS}}}{D_{\mathrm{OS}}}\int \Sigma\left(D_{\mathrm{OL}}\boldsymbol{\theta}'\right)\frac{\boldsymbol{\theta}-\boldsymbol{\theta}'}{|\boldsymbol{\theta}-\boldsymbol{\theta}'|^2}\mathrm{d}^2\theta'$$

$$= \frac{1}{\pi}\int \frac{\boldsymbol{\theta}-\boldsymbol{\theta}'}{|\boldsymbol{\theta}-\boldsymbol{\theta}'|^2}\kappa\left(\boldsymbol{\theta}'\right)\,\mathrm{d}^2\theta', \tag{4.8}$$

where

$$\kappa = \frac{\Sigma\left(D_{OL}\boldsymbol{\theta}\right)}{\Sigma_{\mathrm{crit}}} \quad \text{and} \quad \Sigma_{\mathrm{crit}} = \frac{c^2}{4\pi G}\frac{D_{\mathrm{OS}}}{D_{\mathrm{OL}}D_{\mathrm{LS}}} \tag{4.9}$$

are the gravitational convergence and the critical projected mass density.

The gravitational convergence κ is the dimensionless projected mass density of the lens. It characterizes the strength of a gravitational lensing effect. The strong lensing regime produces image multiplications, strong magnifications and strong distortions (rings and arcs), and corresponds to $\kappa > 1$. Since Σ_{crit} depends on angular distances, the strength of a lens also depends on the relative distances to the lens of the source and the observer as well as on the underlying cosmology. The weak lensing regime corresponds to $\kappa \ll 1$ and will be discussed later.

In the weak field limit, κ is related to the projected Newtonian gravitational potential, $\psi(\boldsymbol{\theta})$, via the projected Poisson equation,

$$\nabla^2\psi = 2\kappa, \tag{4.10}$$

so that

$$\psi\left(\boldsymbol{\theta}\right) = \frac{1}{\pi} \int \kappa\left(\boldsymbol{\theta}'\right) \ln|\boldsymbol{\theta} - \boldsymbol{\theta}'| \, \mathrm{d}^2\theta' \quad \text{and} \quad \boldsymbol{\alpha}'\left(\boldsymbol{\theta}\right) = \frac{1}{\pi} \int \kappa\left(\boldsymbol{\theta}'\right) \frac{\boldsymbol{\theta} - \boldsymbol{\theta}'}{|\boldsymbol{\theta} - \boldsymbol{\theta}'|^2} \, \mathrm{d}^2\theta'.$$

$$(4.11)$$

By differentiating the lensing equation over $\boldsymbol{\theta}$,

$$\frac{\partial \boldsymbol{\beta}}{\partial \boldsymbol{\theta}} = \delta_{ij} - \frac{\partial^2 \psi\left(\boldsymbol{\theta}\right)}{\partial \theta_i \partial \theta_j} = \delta_{ij} - \partial_i \partial_j \psi, \tag{4.12}$$

where δ_{ij} is the Kronecker delta. Equation (4.12) describes the gravitational lens mapping, i.e. how a deflected point of the source plane is related to its image(s) in the image plane via the Newtonian gravitational field ψ. It contains the convergence

$$\kappa = \frac{1}{2} \left(\partial_1 \partial_1 + \partial_2 \partial_2\right) \psi, \tag{4.13}$$

as well as two other terms,

$$\gamma_1 = \frac{1}{2} \left(\partial_1 \partial_1 - \partial_2 \partial_2\right) \psi \quad \text{and} \quad \gamma_2 = \partial_1 \partial_2 \psi. \tag{4.14}$$

The mapping is fully described by the magnification matrix, $M\left(\boldsymbol{\theta}\right)$,

$$M^{-1}\left(\boldsymbol{\theta}\right) = \begin{pmatrix} 1 - \kappa - \gamma_1 & -\gamma_2 \\ -\gamma_2 & 1 - \kappa + \gamma_1 \end{pmatrix}$$

$$= (1 - \kappa) \begin{pmatrix} 1 & 0 \\ 0 & 1 \end{pmatrix} + \begin{pmatrix} -\gamma_1 & -\gamma_2 \\ -\gamma_2 & +\gamma_1 \end{pmatrix}. \tag{4.15}$$

The right-hand side decomposition shows that gravitational lensing modifies lensed images in two ways. It produces an isotropic magnification of the lensed source, given by $(1 - \kappa)$, and a distortion of images, described by γ_1 and γ_2. $\gamma = (\gamma_1; \gamma_2)$ is the gravitational shear. The gravitational shear term is responsible for the transformation of lensed galaxies into rings or giant arcs.

From an observational point of view, the magnification is the ratio of the total intensity of the observed image, $I^{\mathrm{I}}\left(\boldsymbol{\theta}\right) |\mathrm{d}\Omega^{\mathrm{I}}|$, to the total intensity of the source, $I^{\mathrm{S}}\left(\boldsymbol{\theta}\right) |\mathrm{d}\Omega^{\mathrm{S}}|$, where the $\mathrm{d}\Omega$s are the solid angles subtended by the image and the source. As consequences of the Liouville theorem, the conservation of photons and the achromaticity of gravitational deflection, the surface brightness of a lensed source is conserved,

$$I^{\mathrm{I}}\left(\boldsymbol{\theta}\right) = I^{\mathrm{S}}\left[\boldsymbol{\beta}\left(\boldsymbol{\theta}\right)\right]. \tag{4.16}$$

For any region of the image plane much smaller than the typical scale of variation of the magnification matrix, the magnification is almost constant and one can use the linearized lens equation

$$\boldsymbol{\beta}\left(\boldsymbol{\theta}\right) = \boldsymbol{\beta}\left(\boldsymbol{\theta}_0\right) + M\left(\boldsymbol{\theta}_0\right)^{-1}\left(\boldsymbol{\theta} - \boldsymbol{\theta}_0\right). \tag{4.17}$$

Therefore

$$I^{\mathrm{I}}\left(\boldsymbol{\theta}\right) = I^{\mathrm{S}}\left[\boldsymbol{\beta}\left(\boldsymbol{\theta}_0\right) + M\left(\boldsymbol{\theta}_0\right)^{-1}\left(\boldsymbol{\theta} - \boldsymbol{\theta}_0\right)\right] \tag{4.18}$$

which means that the linear lens mapping transforms a circular source into an ellipse.

From the gravitational lensing point of view, the magnification is the result of the convergence and the shear effects. The global magnification, $|\mu\left(\theta_0\right)|$, is then given by the determinant of the magnification matrix

$$|\mu\left(\theta_0\right)| = \mathrm{Det}M = \frac{1}{\left(1 - \kappa\right)^2 - |\gamma|^2}. \tag{4.19}$$

There are several important properties of the magnification matrix that are relevant for this review.

- The matrix M^{-1} usually has two eigenvalues, $1 - \kappa \pm |\gamma|$, that provide the axis ratio of the lensed image, $b/a = \frac{1-\kappa+|\gamma|}{1-\kappa-|\gamma|}$.

- When $(1 - \kappa)^2 - |\gamma|^2 = 0$, the magnification diverges. These extreme lens configurations produce multiple images, giant arc and Einstein ring images with extremely high magnifications. The loci of the image plane where the magnification is infinite are the critical lines. The corresponding loci of the source plane are the caustic lines. The shapes of caustic and critical lines are set by the lensing configuration which depends on the distances between the observer, the lens and the source, and on the geometry and the mass distribution inside the lens. For a given configuration, the properties of the lensed images depend on the position of the source with respect to the caustic lines and on the intrinsic morphology of the source. In practice, the strongly lensed images are good tracers of critical lines and therefore reveal many properties of the lens configuration. A perfect circular ring image corresponds to a source aligned with the lens and the observer, and a circular mass distribution. The critical line is a perfect circle drawn by the ring (see Figure (4.2)). The corresponding caustic line is a single point of the source plane.

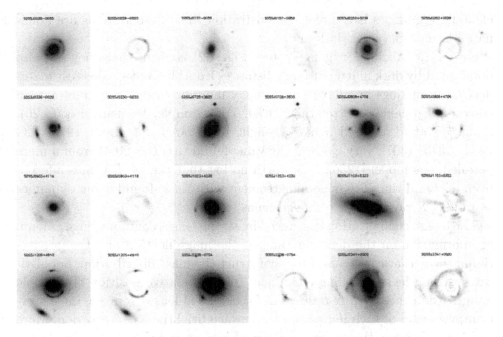

Fig. 4.2. A set of gravitational arcs and rings from the SLACS survey. All lenses are galaxies. Each panel is composed of a pair of images: an original image (left) and an image after subtraction of the model for the lens (right). The morphology of each lensed image reveals the lens configuration. From [920].

4.3 Evidence for dark matter from strong lenses

The strong lensing regime defines lenses with $\kappa \geq 0.5$ and $|\gamma| \geq 0.5$. The most spectacular cases of gravitational arcs, rings or multiple images of lensed quasars can easily be identified in the sky and correspond to first discoveries of gravitational lensing in the Universe [1092; 1356; 1780; 1916].

For Einstein rings and some spectacular giant arcs, the total mass can be easily estimated assuming the projected mass density inside a circle drawn by the arc or the ring (θ_{ring}) is about the critical mass density:

$$M\left(\theta < \theta_{\mathrm{ring}}\right) = \pi \theta_{\mathrm{ring}}^2 D_{\mathrm{OL}}^2 \Sigma_{\mathrm{crit}}. \qquad (4.20)$$

Applying Eq. (4.20) to the giant arc discovered in Abell 370 [1780; 1781] with the redshifts of the lens and the source ($z_{\mathrm{cluster}} = 0.374$, $z_{\mathrm{arc}} = 0.724$) [1781] showed that the mass-to-light ratio of this cluster is about 300, close to the values found later [1253] with a more sophisticated lens model. Similar simplistic mass estimates have been applied to many clusters of galaxies and lead to similar values (see [1417]). The accuracy of the estimate is poor, about ± 20–40%, but sufficient to claim that the discovery of giant arcs in clusters of galaxies is a direct demonstration that these systems are dominated by non-baryonic dark matter.

Gravitational rings in galaxies (see Fig. 4.2) confirm that they are also dominated by dark matter at large distances from the centre. Mass estimates derived from rings in galaxies are usually more accurate than giant arcs in clusters of galaxies (better than 10%). For example, by using Eq. (4.20), Langston *et al.* derived a mass-to-light ratio, M/L, of about 19 in MG 1654+1346 [1307], very close to the value found ($M/L = 20.4$) from a more detailed lens model [1256]. It is worth noting that the M/L values inferred from arcs and rings are in good agreement with those found using dynamics as well as X-ray analysis of hot plasmas.

The image configurations produced by strong lenses contain a huge amount of information on the deflection angles, the magnification and the gravitational shear that can be used to reconstruct the mass distribution. Sophisticated strong lens modelling techniques use the relative position of multiple images, their relative flux ratio, and, for extended lensed sources, the distortion properties of each image and the surface brightness conservation. More precisely, they use the following pieces of information:

- The angular lens equation sets the relation between each image position, $\theta_i^I, \theta_j^I, \theta_k^I, \ldots$, and the unique position back in the source plane:

$$\theta^S = \theta_i^I - \alpha\left(\theta_i^I\right) = \theta_j^I - \alpha\left(\theta_j^I\right) = \theta_k^I - \alpha\left(\theta_k^I\right) = \ldots \qquad (4.21)$$

- The flux ratio between images is an estimate of their magnification ratio:

$$\frac{F_i^I}{F_j^I} = \frac{|\mu_i^I|}{|\mu_j^I|}; \quad \frac{F_i^I}{F_k^I} = \frac{|\mu_i^I|}{|\mu_k^I|}; \quad \ldots \qquad (4.22)$$

- When sent back to the source plane, the morphology of each unlensed image should be identical.

To date, several tens of rings around galaxies and arcs inside clusters of galaxies have been discovered and modelled [378; 379; 483; 572; 821; 1088; 1282; 1464]. In some cases, the observational data (image resolution, image positions, flux ratios, image distortion and image parity, as well as redshifts of the lens and the source) are exceptional and the deflecting gravitational potentials have been modelled with high precision, including details of substructures inside the lens and predictions of extra images.

Despite the huge diversity of mass reconstruction techniques and lens models explored in the literature, strong lens studies of galaxies and clusters of galaxies have converged to a series of important results about dark matter.

- Galaxies and clusters of galaxies are dominated by a dark matter component, if one assumes General Relativity to be valid. The mass-to-light ratios derived from the total mass inside rings and arcs are high, about 10–20 for galaxies and 100–300 for clusters of galaxies. Therefore, clusters of galaxies are dominated by a non-baryonic dark matter component.
- The mass-to-light ratio of galaxies increases with mass as well as with radius [848; 849; 920]. More massive galaxies seem dark-matter-dominated. However, most galaxies seem to be baryon-dominated in the innermost regions.
- There is no evidence of misalignment between the light and the matter distributions, for either the galaxy or galaxy cluster population [605; 848; 1220; 1269; 1417].
- The (dark+baryon) mass profiles of galaxies are very well modelled by an isothermal profile ($\rho(r) \propto r^{-2}$). For clusters of galaxies, conclusions are harder to draw without using weak gravitational lensing. However, the discovery of radial arcs in the innermost regions of several clusters of galaxies is difficult to reconcile with an internal shallow mass density profile.

These exceptional cases of gravitational lensing benefit from their spectacular visibility on astronomical images. Since rings and arcs only provide the total projected mass inside the cones subtended by the critical line, projection effects may produce overestimates of the mass-to-light ratios of galaxies and clusters. The strong lens samples may then be biased toward the uppermost mass-to-light ratio limits.

4.4 Cold dark matter confronted to gravitational lenses

Since strong lens models are only based on gravitational optics theory, the details of lens configurations are sufficient to get a robust and reliable mass estimate, (almost) independent of the dynamical state of the lens. Unfortunately, strong lenses only probe matter inside a small region, located inside the critical lines, and cannot tell us much about areas beyond this region. Weak gravitational lensing studies take advantage of the high galaxy number density of deep astronomical images to use all distant galaxies as a sampling population of the distortion grid generated by the shear field of a foreground gravitational lens. In contrast with strong gravitational lensing, which only concerns a few lensed sources, weak gravitational lensing does modify the shape of all sources. It results in a very small change of the ellipticity of lensed galaxies that can be analysed statistically using the correlated signal induced in all of them. The gravitational distortion field can then be sampled over the whole lens, even at very large distance from the centre. Its feasibility has been demonstrated on real data [382; 804; 1878].

When used together, strong and weak lensing probe the structural and dynamical properties of dark haloes from the internal regions to the outermost periphery. Specific predictions of cosmological models of structure formation and, indirectly, the nature and the properties of dark matter can then be tested. In particular, several predictions of collisionless cold dark matter models, such as the radial mass profile or the ellipticity and the lumpiness of haloes, can be compared with observations. In this section we present some important results relevant for the cosmological CDM scenario. We will also briefly introduce the theory of weak gravitational lensing and its extension to large-scale structure of the Universe.

4.4.1 *Weak gravitational lensing*

Weak lensing corresponds to $\kappa \ll 1$ and $|\gamma| \ll 1$. This is roughly the regime experienced by all background sources that are not close to caustic lines, which is the case for the majority of sources. All lensed sources are still magnified and distorted by the gravitational shear, but the amplitude of the distortion is very weak compared with the intrinsic ellipticity of each source, and is undetectable on individual galaxies. The lensing signal is derived from the averaged ellipticity of large galaxy samples and after a difficult deconvolution process [441; 1093].

Fortunately, in the weak lensing regime the relation between the gravitational shear and the galaxy ellipticity is simple. If one defines the shape of galaxies on astronomical images by their weighted centroid

$$\boldsymbol{\theta}_0 = \frac{\int \boldsymbol{\theta} \, W \left(I \left(\boldsymbol{\theta} \right) \right) \mathrm{d}^2 \theta}{\int W \left(I \left(\boldsymbol{\theta} \right) \right) \mathrm{d}^2 \theta} \tag{4.23}$$

and their second brightness moment components

$$Q_{ij} = \frac{\int \left(\theta_i - \theta_{0i} \right) \left(\theta_j - \theta_{0j} \right) W \left(I \left(\boldsymbol{\theta} \right) \right) \mathrm{d}^2 \theta}{\int W \left(I \left(\boldsymbol{\theta} \right) \right) \mathrm{d}^2 \theta} \tag{4.24}$$

where $i, j = 1, 2$ and W is a weight function, then the complex ellipticity of a galaxy, ε, can be defined as

$$\varepsilon = \frac{Q_{11} - Q_{22} + 2iQ_{12}}{Q_{11} + Q_{22} + 2\left(Q_{11}Q_{22} - Q_{12}^2\right)^2} \quad \text{with} \quad |\varepsilon| = \frac{1 - \frac{b}{a}}{1 + \frac{b}{a}}, \quad (4.25)$$

where b/a is the axis ratio of the elliptical galaxy isophotes.

The Q_{ij} can be formally defined for the lensed images and for the source as well. They are related by the angular lens equation and the surface brightness conservation. It is then possible to express how the gravitational distortion transforms the (complex) ellipticity of a lensed source, ε^S, into the observed (complex) ellipticity of its image, ε^I ([1719]; see also Schneider's lecture in [1709]):

$$\varepsilon^S = \begin{cases} \dfrac{\varepsilon^I - g}{1 - g^*\varepsilon^I} & \text{if } |g| \leq 1 \\[3mm] \dfrac{1 - g\varepsilon^{*I}}{\varepsilon^{*I} - g^*} & \text{if } |g| > 1 \end{cases} \quad (4.26)$$

where g is the (complex) reduced shear:

$$g\left(\boldsymbol{\theta}\right) = \frac{\gamma\left(\boldsymbol{\theta}\right)}{1 - \kappa\left(\boldsymbol{\theta}\right)}. \quad (4.27)$$

In the weak lensing regime, Eq. (4.26) simplifies, and therefore intrinsic and shear-induced ellipticities just add up as follows

$$\varepsilon^I \simeq \varepsilon^S + \gamma. \quad (4.28)$$

If one assumes that the intrinsic orientation of unlensed galaxies is randomly distributed, then $\langle \varepsilon^S \rangle = 0$ and

$$\langle \varepsilon^I \rangle = \gamma. \quad (4.29)$$

Therefore, the local gravitational shear can be directly measured from galaxy ellipticities. This result applies at all scales and can be used to probe shear fields around galaxies, clusters of galaxies or large-scale structure of the Universe.

To relate the estimate of the shear to the projected mass density inside the lens, we use Eq. (4.13) and Eq. (4.14). The relation between the shear and the convergence is written as

$$\gamma\left(\boldsymbol{\theta}\right) = \frac{1}{\pi} \int \mathcal{D}\left(\boldsymbol{\theta} - \boldsymbol{\theta}'\right) \kappa\left(\boldsymbol{\theta}'\right) \mathrm{d}^2\theta' \quad (4.30)$$

where

$$\mathcal{D} = \frac{\theta_2^2 - \theta_1^2 - 2i\theta_1\theta_2}{|\boldsymbol{\theta}|^4}. \tag{4.31}$$

This operation is formally a convolution, so it can be inverted. It is then possible to reconstruct the projected mass density (κ) from the shear-induced ellipticity of galaxies ($\gamma = \langle \varepsilon^{\mathrm{I}} \rangle$):

$$\kappa(\boldsymbol{\theta}) = \frac{1}{\pi} \int \mathcal{D}^* \left(\boldsymbol{\theta} - \boldsymbol{\theta}'\right) \gamma \left(\boldsymbol{\theta}'\right) \mathrm{d}^2\theta'. \tag{4.32}$$

The details and the performances of mass reconstruction techniques are beyond the scope of this review (see for example [224; 1190; 1720], and reviews from [1417] or [225]). Mass reconstruction from weak lensing is now a mature technique. Its performance is more limited by the accuracy of the measurement of galaxy ellipticities, the number density of lensed galaxies that samples the shear field and the knowledge of redshifts of lensed sources than by the reconstruction methods themselves.

Mass reconstruction from weak lensing is optimal for clusters of galaxies located within the redshift range $0.1 < z < 0.8$. For these clusters the scaling factor $D_{\mathrm{LS}}/D_{\mathrm{OS}}D_{\mathrm{OL}}$ is large enough, and the galaxy number density of lensed sources samples the cluster field with enough points (about 10–50 galaxies per arcmin^2, with a cluster angular size of about $30 \times 30\,\mathrm{arcmin}^2$).

An important application of weak lensing to probe galaxy haloes is the galaxy–galaxy lensing. This is the weak gravitational distortion of background galaxies produced by foreground galaxies (see, for example, the early seminal work of ref. [436], or, more recently, [1379]).

Beyond supercluster scales the weak lensing signal drops below the detection limit of present-day instruments. Weak lensing no longer applies on individual lensing haloes, but statistically, using all galaxies over the whole sky. It will be discussed in Section 4.4.4.

The literature on weak lensing and its applications to galaxies, groups, clusters, superclusters of galaxies and large-scale structure of the Universe is abundant and flourishing. In the following sections we focus only on results relevant for dark matter. Other applications and more detailed results are presented elsewhere [225; 1111; 1417; 1490; 1633; 1893].

4.4.2 The collisionless nature of dark matter

The most spectacular results obtained about the nature of dark matter from weak gravitational lensing are those on the 'Bullet cluster' 1E0657-56 [434; 562; 563] and more recently on the lensing cluster MACS J0025.4-1222 [433].

In both clusters the joined weak and strong lensing mass reconstructions show two massive substructures that are offset with respect to the baryon distribution observed in X-rays by Chandra (see Fig. 4.3). In contrast, the cluster galaxy population follows the dark matter distribution. These two clusters are strong cases in favour of the existence of dark matter and seem difficult challenges for alternative views, where the mass distribution must coincide with the baryon distribution.

The offsets between the hot gas distributions and the gravitational potentials result from violent merging, experienced recently by both clusters. Remarkably, the mass-to-light ratios found in the dark matter haloes, as well as the gas-to-total-mass ratios, are consistent with other massive clusters of galaxies. In the merging scenario, the clear decoupling of the dark matter and the hot baryonic plasma components implies that dark matter should be collisionless. Otherwise it should also experience ram pressure [907; 1389]. Using theoretical models and hydrodynamical numerical simulations of merging processes, Randall *et al.* [1627] found a conservative upper limit to the self-interaction cross-section per unit mass of dark matter particles in 1E0657−56

$$\frac{\sigma}{m} < 1.25 \text{ cm}^2 \text{ g}^{-1}, \tag{4.33}$$

and Bradač *et al.* [433] found a limit of

$$\frac{\sigma}{m} < 4 \text{ cm}^2 \text{ g}^{-1} \tag{4.34}$$

in MACS J0025.4−1222.

These remarkable results are based on joint weak+strong lensing studies, so the loci of substructures and the amount of dark matter inside can be assessed without ambiguity from the properties of arcs as well as the weak distortion fields.

The limits on the self-interaction cross-section are based on the modelling of the merging history and on the accuracy of the mass-to-light ratio estimates in each dark halo. They may still be uncertain and the cross-section values should be taken with caution.

These results on the nature of dark matter need to be more securely established from a larger sample of clusters. Indeed, a similar study done by Mahdavi *et al.* [1372] in the merging cluster Abell 520 led to different conclusions. In contrast with 1E0657-56 and MACS J0025.4-1222, the weak lensing reconstruction of dark matter in Abell 520 reveals a massive dark core that coincides with the hot gas. However, the core does not contain any galaxies inside. From arguments similar to those for 1E0657-56 and MACS

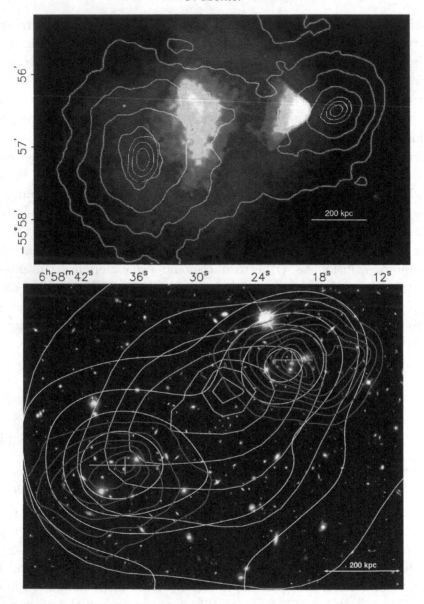

Fig. 4.3. Lensing mass and Chandra X-ray emissivity maps of 1E0657-56 (top; [562]) and MACS J0025.4-1222 (bottom; [433]) showing the offsets between the gravitational convergence field and the distribution of hot baryons. Top: the X-ray emissivity image obtained with Chandra of 1E0657-56. The contours overlaid on the image show the convergence map, reconstructed using weak and strong lensing data. Bottom: Hubble Space Telescope image of the lensing cluster MACS J0025.4-1222 with the weak+strong convergence map overlaid in grey contours and the Chandra X-ray emissivity map overlaid in white. A second set of white contours represents the galaxy light distribution in I-band. From [562] and [433].

J0025.4-1222, it may imply that dark matter is self-interacting. Mahdavi *et al.* [1372] concluded that the self-interaction cross-section of dark matter particles is

$$\frac{\sigma}{m} = 3.8 \pm 1.1 \ \text{cm}^2 \, \text{g}^{-1}. \tag{4.35}$$

Another interesting consequence of self-interacting dark matter particles concerns the innermost regions of dark matter haloes in clusters of galaxies. Self-interaction produces spherical dark haloes and shallow inner profiles, and damp (lower contrast) substructures. If dark matter is self-interacting, observation of inner elliptical haloes should be rare, and the number of giant tangential and radial arcs should be significantly lower than for collisionless dark matter particles. Radial arcs in particular should be extremely rare events.

Constraints on the self-interaction cross-section from the inner ellipticity of dark haloes were first explored by Miralda-Escudé using the lensing cluster MS2137-23 [1453]. This cluster shows a unique lensing configuration of multiple radial and tangential arcs that provide enough clues to reconstruct the gravitational potential from a strong lensing model. In particular, the shape and positions of multiple arcs set tight constraints on the ellipticity of the gravitational potential [878; 1066; 1418; 1452]. Assuming the rotation of the cluster is negligible, the dark matter velocity dispersion is $1000 \, \text{km s}^{-1}$ and its cross-section is constant, Miralda-Escudé concluded that

$$\frac{\sigma}{m} < 0.02 \ \text{cm}^2 \, \text{g}^{-1}. \tag{4.36}$$

Meneghetti *et al.* drew similar conclusions based on the occurrence of tangential and radial arcs expected from cosmological simulations with self-interacting dark matter [1419]. Although the fraction of giant tangential arcs already seems underestimated in cosmological simulations with collisionless dark matter, introducing self-interacting dark matter particles would be even worse. It reduces this fraction by one order of magnitude. Furthermore, self-interacting dark matter cannot easily produce radial arcs. From these considerations, they estimated that

$$\frac{\sigma}{m} < 0.1 \ \text{cm}^2 \, \text{g}^{-1}. \tag{4.37}$$

The studies carried out in refs. [1453] and [1419] are still debated. They are based on very few lensing clusters, and a generalization to the whole cluster population is premature. Cosmological interpretation of statistics of arcs in clusters of galaxies also depends on observational biases in the detections of strong lenses, on the intrinsic size of sources and on the image quality of astronomical data. It is, however, interesting that the arc statistics provide similar conclusions to the lensing analysis of merging clusters.

Natarajan *et al.* set limits on the dark matter cross-section by comparing the size of galaxy haloes in the lensing cluster Abell 2218 to predictions of collisional and collisionless dark matter particles [1498]. While the cluster galaxy truncation radius in a collisionless dark matter cluster is first controlled by tidal interaction, it is mainly controlled by ram pressure if dark matter is collisional. In that case the truncation radii of cluster galaxies should be significantly smaller [907]. Natarajan *et al.* derived the truncation radii of 25 galaxies using a strong+weak lensing model of Abell 2218. They found values that are compatible with tidal radii predicted for collisionless dark matter and derived

$$\frac{\sigma}{m} < 42 \ \text{cm}^2 \, \text{g}^{-1}. \tag{4.38}$$

Their upper limit is significantly higher than those discussed before. However, it is derived from a totally different approach that does not suffer from selection effects like the arc statistics. It seems rather easy to apply to many lensing clusters but demands a careful control of degeneracies between the main cluster potential and the cluster galaxy haloes in order to derive reliable values for the galaxy truncation radii.

4.4.3 The CDM haloes compared with lensing observations

Besides the investigation of dark matter particle cross-section, one can also confront predictions of the most popular cosmological model to the constraints on dark halo properties derived from gravitational lensing. Indirect assessments of the validity of the collisionless cold dark matter model can then be derived.

An obvious test is the comparison of the radial mass profile derived from gravitational lensing reconstructions of galaxies and clusters of galaxies to current predictions of ΛCDM hierarchical models of structure formation. Numerical simulations predict that virialized CDM haloes have a 'universal' mass density profile, common to all haloes regardless of their mass, size and concentration. In particular, it must be the same for galaxies, groups and clusters of galaxies. Several best-fitting analytical mass density profiles have been proposed. The Navarro, Frenk and White (NFW) mass profile [1502],

$$\rho\left(r\right) = \frac{200}{3} \frac{c_{200}^3}{\ln\left(1 + c_{200}\right) - c_{200}/\left(1 + c_{200}\right)} \frac{\rho_{\text{c}}\left(z\right)}{\left(\dfrac{r}{r_{\text{s}}}\right)\left(1 + \dfrac{r}{r_{\text{s}}}\right)^2}, \tag{4.39}$$

is the most popular and the most debated. Here, $\rho_c(z)$ is the critical density of the Universe at redshift z, $c_{200} = r_{200}/r_s$ is the concentration parameter, r_s is a characteristic scale radius and r_{200} is the radius within which the mean mass density of the halo is $200 \times \rho_c(z)$. Another parameterization consists in using $c_{vir} = r_{vir}/r_s$, where r_{vir} is the virial radius within which the mean mass density of the halo is $178 \times \rho_c(z)$.

The inner slope of the NFW mass density profile has $\rho(r) \propto r^{-1}$, while $\rho(r) \propto r^{-3}$ for the outer regions. The profile is similar for all dark haloes, but numerical simulations predict that the concentration parameter should weakly decrease with scales. Typical expectation values are $c = 6\text{--}8$ for galaxies and a rather small value $c \approx 3\text{--}4$ for clusters of galaxies.

These predictions can in principle easily be tested using gravitational lensing, in particular on clusters of galaxies, where giant arcs and weak lensing can probe the dark matter profile from ~ 50 kpc to ~ 10 Mpc. They have indeed been measured on many clusters of galaxies and compared to NFW lensing predictions (see [223] for an analytic derivation of NFW lensing quantities). There is an abundant literature on this issue (see for example refs [199, 608, 1380, 1381]). The results are, however, unclear and often contradictory, sometimes on the same clusters.

A recent investigation [572] and two comprehensive analyses of clusters of galaxies in the SDSS [1174; 1381] show good examples of the confusion. Strong and weak lensing studies and compilation of 100 clusters with $0.18 < z < 0.8$ by Comerford and Natarajan led to a concentration value [572]

$$c_{vir} = 14.8 \pm 6.1 \, (1+z)^{-1} \left(\frac{M_{vir}}{1.3 \times 10^{13} \, M_\odot} \right)^{-0.14 \pm 0.12}, \tag{4.40}$$

significantly higher than the ΛCDM prediction. In contrast, the values found by Johnston *et al.* [1174] and by Mandelbaum *et al.* [1381] on SDSS samples are

$$c_{200} = 4.1 \pm 1.5 \left(\frac{M_{200}}{1.0 \times 10^{14} \, M_\odot} \right)^{-0.12 \pm 0.04} \tag{4.41}$$

and

$$c_{200} = 4.6 \pm 0.7 \left(\frac{M_{200}}{1.0 \times 10^{14} \, M_\odot} \right)^{-0.13 \pm 0.07}, \tag{4.42}$$

respectively, in very good agreement with ΛCDM models.

The 'good news' is that the slopes found in the three studies for the evolution of the concentration parameter as function of the halo mass agree with ΛCDM predictions, which confirms that the concentration weakly decreases with mass scale. However, the concentration parameters are very different.

Furthermore, similar discrepancies arise from a careful investigation of individual cases analysed from ground-based or space data. In general, detailed strong and weak lensing studies of clusters of galaxies lead to high concentration value (see [448] and references therein; see however the cluster sample in ref. [608], or [1335] on Abell 1703).

The puzzling discrepancy in the concentration parameter may result from observational biases. The samples discussed above in particular are very different. Comerford and Natarajan's compilations mix weak and strong lensing analyses and put together clusters without a well-defined selection function. It is possible that the selection of X-ray and strong lensing clusters biases their sample toward a high mean concentration value. Their sample may contain several 'super-lenses' [1523], for example. However, their fraction of clusters with high concentration seems to exceed predictions from numerical simulations [1087]. In contrast, the SDSS sample is more homogeneous and likely to be less biased. It is interesting to notice that an earlier study [1380] using galaxy–galaxy lensing in SDSS also led to a rather small value, and that the analysis of the RCS survey by Hoekstra et al. [1112] also found concentration parameters that agree very well with CDM predictions.

Strong and weak lensing mass reconstruction have also been carried out to probe the slope of the mass density profile. Although the results also show a huge scatter from cluster to cluster, there is no conclusive evidence that the slope significantly deviates from an NFW profile in clusters of galaxies. Kneib et al. used wide-field HST observations of Cl0024+1654 to demonstrate that the lensing mass reconstruction definitely rules out an isothermal model and perfectly fits an NFW profile [1252]. However, many clusters are equally well fitted by non-singular isothermal or Sérsic profiles [1180]. Furthermore, the internal complex structure of clusters of galaxies as well as the role of baryons in the innermost regions make the exploration of the inner slope difficult and its interpretation in a cosmological context questionable.

Similar conclusions have been drawn in ref. [920] on the SLACS early-type galaxy sample. Strong and weak lensing mass reconstructions of these galaxies show that their dark matter profile is well modelled by an NFW profile. However, the strange 'conspiracy' between the stellar and dark component to form in combination an isothermal mass profile is embarrassing. Neither the stellar nor the dark matter profiles can be fitted in isolation by an isothermal component, whereas the final result is perfectly fitted by a r^{-2} profile over two decades.

Other probes of CDM models, like the triaxiality of haloes or the abundance of substructures, have also been explored using gravitational lensing. Multiply imaged lensed quasars are excellent strong lens configurations for

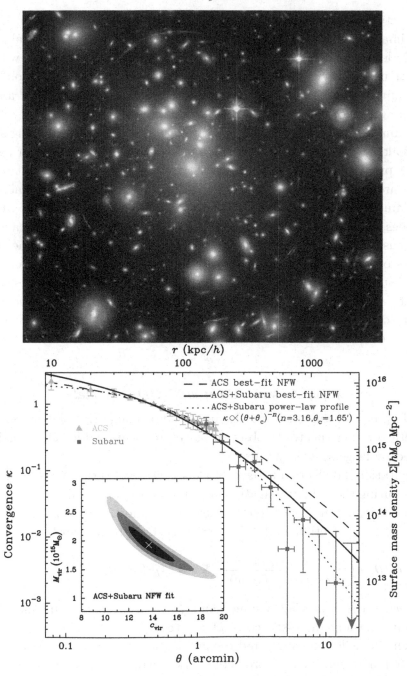

Fig. 4.4. The lensing cluster Abell 1689. Top: A close-up of the HST/ACS deep image, showing the cluster centre and the strongly lensed arcs and weakly lensed galaxies. The distortion field is visible from the distribution of lensed galaxies. Image credit: NASA, Benitez, Broadhurst, Ford, Clampin, Hartig, Illingworth, the ACS science team and ESA. Bottom: Mass profile reconstruction of Abell 1689 using HST/ACS and Subaru images. The curves show some best fits. This figure shows the complementarity of space and ground data to probe both the innermost regions and the periphery of the cluster. The plot inside shows the best-fit value for $r_{\rm vir}$. It is clearly higher than CDM predictions. Note the power law mass profile

substructures. In contrast to extended lensed galaxies, the flux of each magnified image of quasars can be measured very accurately. For several exceptionally lensed quasars, the image flux ratios significantly disagree with the predictions of gravitational lens models using one single big halo. These cases cannot be reconciled unless one assumes there is a substantial amount of dark matter concentrated in invisible subhaloes [1257].

In summary, the exploration of the properties of dark haloes using strong and weak lensing has not yet provided indisputable evidence for or against ΛCDM predictions. In many cases the agreement is excellent, but several results are still puzzling, in particular for the concentration parameters of some famous lensing clusters observed with HST. Regarding NFW at cluster scales, the reconstruction of radial mass profiles is not easy, because the weak lensing signal is noisy, strong lens models still have degeneracies and clusters of galaxies are not simple spherical systems but show many massive substructures. It is, however, clear that gravitational lensing as seen in present-day observations of dark matter haloes at galaxy, group or galaxy cluster scales cannot rule out that the matter content of the Universe is dominated by collisionless cold dark matter particles.

4.4.4 Cosmic shear and CDM models

Cold dark matter universes can also be tested by comparing the properties of theoretical power spectra to observations of gravitational lensing of galaxies induced by large-scale structures.

Gravitational deflection by large-scale structure of the Universe is derived by computing the light propagation in an inhomogeneous universe. Assuming that the Friedmann–Robertson–Walker metric and the Born approximation are valid, the deflection angle of a perturbed light beam can be written as

$$\boldsymbol{\alpha}\left(\boldsymbol{\theta}, w\right) = \frac{2}{c^2} \int_0^w \frac{f_K\left(w - w'\right)}{f_K\left(w\right)} \boldsymbol{\nabla}_\perp \Phi \left(f_K\left(w'\right) \boldsymbol{\theta}, w'\right) \, \mathrm{d}w', \qquad (4.43)$$

where w is the comoving distance and f_K is the angular distance (more precisely, the angular distance is $D = a(z)f_K(w(z))$. Using the relations between the deflection angle and the gravitational convergence (that is, the projected mass density), the projected dark matter power spectrum resulting from all deflections along a line of sight, $P_\kappa\left(l\right)$, is related to the three-dimensional power spectrum, $P_\delta\left(k\right)$, as follows

$$P_\kappa\left(l\right) = \int G^2\left(w\right) P_\delta \left(\frac{l}{f_K\left(w\right)}\right) \, \mathrm{d}w, \qquad (4.44)$$

where

$$G(w) = \frac{3}{2} \left(\frac{H_0}{c}\right)^2 \Omega_{\mathrm{M}} \int_w^{w_{\mathrm{lim}}} \frac{f_K(w'-w)}{f_K(w')} p(w') \mathrm{d}w' \qquad (4.45)$$

and $p(w')$ is the redshift distribution of sources, so that $p(w)\,\mathrm{d}w = p(z)\,\mathrm{d}z$. A remarkable consequence of the relations between κ, γ and the projected Newtonian potential is

$$P_\kappa(l) = P_\gamma(l). \qquad (4.46)$$

Hence, the projected dark matter power spectrum can be derived from the gravitational distortion field produced by foreground structures (or cosmic shear), as for single gravitational lenses.

Equations (4.44) and (4.45) show that the lensing signal depends on the angular distances of lenses and sources as well as on the growth history of cosmic structure, from $P_\delta(k)$. Assuming that the dark matter power spectrum is a power law, $P_\delta(k) \propto \sigma_8^2 k^n$, and the mean redshift of sources is z_{s}, then [329]

$$P_\kappa(\theta) \simeq 10^{-4} \, \sigma_8^2 \, \Omega_{\mathrm{M}}^{1.5} \, z_{\mathrm{s}}^{1.5} \, \theta^{-(n+2)}. \qquad (4.47)$$

The shape of the projected dark matter power spectrum as a function of angular scale, $P_\kappa(\theta)$, is measured from the galaxy ellipticity correlation function, that is, the shear power spectrum. The signal is then compared to numerical simulations of cold dark matter universes.

Early observations of weak gravitational lensing by large-scale structure [151; 1191; 1384; 1894; 1942] ruled out extreme cold dark matter models with $\Omega_{\mathrm{M}} = 1$, and clearly showed that the lensing signal follows predictions of hierarchical CDM-dominated models of structure formation with $\Omega_{\mathrm{M}} > 0.1$. These conclusions were confirmed by all surveys done later (see refs. [1111; 1490; 1633; 1893] for reviews). The most recent results [275; 892; 1396] demonstrate that the hierarchical ΛCDM scenario agrees very well with lensing observations, with $\Omega_{\mathrm{M}} \approx 0.25$ and $\sigma_8 \approx 0.8$.

4.5 Hot dark matter: limits on neutrino masses from lensing

Even in a CDM-dominated universe, the presence of a small fraction of hot dark matter particles modifies the expansion rate of the universe and the history of structure formation. It results in a different evolution of the dark matter power spectrum with look-back time as compared with a universe only composed of baryons and cold dark matter particles. If the mass fraction of neutrinos is high enough, its contribution to the convergence power

Table 4.1. *Constraints on neutrino mass from recent cosmic shear surveys together with SNIa, CMB and redshift surveys.*

Reference	Cosmic shear data	Neutrino mass
[1331]	[275]	$\sum m_\nu < 0.47$ eV (95% C.L.)
[1848]	[892]	$\sum m_\nu < 0.54$ eV (95% C.L.)
[986]	[275]	$\sum m_\nu < 0.80$ eV (2σ)
[1149]	[892]	$\sum m_\nu < 0.54$ eV (95% C.L.)

spectrum may be detectable with enough accuracy to set upper limits on the total mass of neutrino particles. The expected deviation from the standard cosmological model is weak and (partly) degenerate with other parameters, like dark energy. Nevertheless the precision of the most recent cosmological lensing observations is now good enough to explore subtle effects produced by neutrinos on the distortion field.

In a recent series of papers, several teams used the data sets of [275] or [892] to constrain the total mass of neutrinos, $\sum m_\nu$. Assuming the ΛCDM model is valid, all these studies used weak lensing surveys together with other non-lensing data (CMB, baryon acoustic oscillations, Type Ia supernovae) to explore jointly the equation of state of dark energy, w, and the total neutrinos mass parameters.

The results are summarized in Table 4.1. There is good agreement between these works, despite different weak lensing and non-weak lensing data set, and despite the use of different methods to explore the parameter space. Tereno *et al.* used extra SNIa data from the SNLS sample and indeed derived tighter limits [1848], setting a very first tentative mass range of 0.03 eV $< \sum m_\nu < 0.54$ eV.

Degeneracies and weak lensing systematics hamper weak lensing studies from better probing effects from $\sum m_\nu$ with present-day weak lensing data. Although dark energy and neutrinos do not change the power spectrum the same way and can be decoupled, lensing systematics is a more serious barrier than degeneracies. The promise of weak lensing to derive information on neutrino mass will depend on the reliability and the accuracy of the ellipticity measurements on galaxies as well as on better control of astrophysical systematics (redshift distribution of galaxies, intrinsic galaxy alignment and so on).

4.6 Dark matter or modified gravity?

In the previous sections, we presented direct and indirect evidence that gravitational lensing provides in favour of a Universe dominated by cold collisionless dark matter particles. However, there is an alternative to the cold dark matter paradigm, if one assumes, instead, a modified theory of gravitation with no extra dark matter. Modified gravity theories, in particular TeVeS, will be presented in Chapter 6. In this section we discuss this option in the context of gravitational lensing.

An interesting consequence of a theory of gravitation like TeVeS [254] is the subtle difference between the gravitational convergence and the projected mass density. In contrast with General Relativity, in TeVeS a non-zero convergence does not mean a non-zero projected mass density. Using this remarkable property, Angus and co-workers [95; 96] argued that they could solve the apparent contradiction between the convergence and the projected mass density field, and they successfully modelled the Bullet cluster without using extra non-baryonic cold dark matter. So, the Bullet cluster does not prove the existence of dark matter (see also Chapter 6), although a significant amount of collisionless particles is still needed in TeVeS to model the cluster. This extra component could be a substantial contribution of massive neutrinos ($\Omega_\nu > 0.1$), provided their mass is about 2 eV. If so, the KATRIN experiment will challenge this model soon, and may eventually rule out a pure collisionless CDM-dominated universe.

There are several issues that are still unclear with the TeVeS alternative view. First, even for the Bullet cluster the TeVeS+neutrino model fails to predict the observed strong lensing features (arcs) around the cluster. Second, modified Newtonian dynamics (MOND) and TeVeS models produce circular gravitational potentials, which seems hard to reconcile with the galaxy–galaxy lensing results. Observations of galaxy haloes in the RCS sample [1112] and the CFHTLS samples [1546] show the dark matter distribution is elliptical at galaxy scale. Third, [1852] found that the scaling relation between the lensing signal and the stellar mass follows a power law with a significantly steeper slope than the MOND prediction. These results put modified gravity models in an uncomfortable position. However, none of these results, taken individually, definitely rule them out. A possible way out is to assume that the elliptical shape is not intrinsic but results from the presence of multiple external fields that superimpose on each galaxy halo. These fields may contaminate the galactic lensing signal with an anisotropic term.

A more difficult issue is the contradiction between the upper limit on neutrino masses set by cosmological weak lensing presented in the previous section and the need for a 2 eV neutrino to model the Bullet cluster with TeVeS. The difference is significant and these results are hard to reconcile unless the data sets of [275] and [892] have underestimated errors or systematics residuals. It seems, however, that the 2 eV neutrino mass is also not supported by the steep slope of the gravitational shear profile measured in lensing clusters [1832], nor by the phase space densities of clusters derived from lensing [1499]. Further, it has been argued that the clumpy structure oberved in 1E0657-56 cannot be explained by neutrinos with mass less than 4.5 eV [563]. In order to agree with present-day neutrino mass constraints from cosmic shear, this lower limit implies that Ω_ν is small. If these results are confirmed, then to preserve the TeVeS alternative one has to assume that the 'missing mass' is composed either of cold baryons [94] or of non-standard neutrinos.

The underlying key questions are whether General Relativity is valid and, from our point of view, whether observational tests using gravitational lensing may rule out this theory or exclude some families of gravitation theories. Since one historical validation of Einsteinian gravitation results from gravitational deflection of light behind the Sun in 1919, it would be ironic if the gravitational lensing effect were eventually to kill it almost 100 years later.

Weak gravitational lensing by large-scale structure can indeed challenge General Relativity against modified gravity. Since the growth rate of structures is driven by gravity, the gravitational convergence power spectrum is a direct imprint of gravity in action. Deviations from General Relativity are expected to be weak and hard to describe in detail (see [1702] and references therein), but present-day cosmic shear data can already falsify predictions of many families of models. Several teams used the cosmological weak lensing signal to test gravitation on scales between ~ 10 kpc and ~ 50 Mpc. White and Kochanek [1929] used the VIRMOS-DESCART data from [1895], Wang *et al.* [1922] used the CTIO weak lensing survey [1165], and Doré *et al.* [708] and Thomas *et al.* [1850] used the [892] CFHTLS data. They explored different phenomenological models of modified theory of gravity using the weak lensing data together with priors and other constraints from CMB, SNIa or redshift surveys. Remarkably, they all find that the present-day data do not show deviations from General Relativity. The families of modified gravity models are in all cases disfavoured, at least over the range of parameter values examined in these works.

4.7 Conclusion and outlook

The present-day observations of strong and weak lensing provide almost direct probes of matter and gravity over physical scales between ~ 1 kpc and ~ 100 Mpc. Despite the diversity of tests, and the broad range of size, mass, mass density contrast and gravitational lensing regimes that have been examined so far, gravitational lensing is not really able to reveal easily the deep nature of dark matter particles, if any. However, the mass-to-light ratios and the mass profiles derived from lensing models, giant tangential and radial arcs as well as weak lensing by galaxies, clusters of galaxies and large-scale structure do agree with predictions of hierarchical cosmological models dominated by collisionless cold dark matter, with $\Omega_M \approx 0.2$–0.3, and none of these observations show any deviations from predictions of general relativity. Indeed, the most recent cosmic shear analyses seem to strengthen the standard flat ΛCDM model with $\Omega_M = 0.25$ and $\Omega_b = 0.04$ [1229].

The upper limits on the dark matter particle cross-section and on the neutrino mass open the route for more direct explorations of dark matter from gravitational lensing. If the upper limits on neutrino mass derived from cosmological weak lensing are confirmed, it will put the present version of the MOND/Bekenstein modified gravitation theory in serious trouble, and would strengthen the dark matter interpretation of refs. [563] and [433]. This controversy shows that the mass maps of 1E0657-56 and MACS J0025.4-1222 are among the most impressive results obtained on dark matter from gravitational lensing (together with X-ray) since the discovery of gravitational arcs and rings.

In contrast to this optimistic view, it is important to keep in mind that many results presented in this review are still debated. Most conclusions drawn depend on the quality of observations, on the reliability of the shear measurement or on the lens model, and several statistical samples may be biased [1382]. It is therefore important to interpret observations with caution.

These caveats set the grounds of future challenges and provide useful insights into the potential of next-generation instruments. The James Webb Space Telescope seems a most promising telescope for strong and weak lensing observations at all scales and in particular in clusters of galaxies. Ultra-deep observations will increase the galaxy number density of lensed galaxies by an order of magnitude. They will provide a much better sampling of the dark matter mass profile, of the substructures, as well as of the innermost regions, where radial arc systems and de-magnified odd images are located (see [919] for a discussion of this issue). The degeneracy between NFW and isothermal models could then be definitely broken (but baryons must be taken into account). Application of more sophisticated lensing techniques using flexions (see [150], [1708] and references therein) will produce mass density maps of clusters with sufficient details to obtain accurate size of cluster galaxy haloes in order to challenge tidal against ram pressure models. One can also expect to go one step further and perhaps pin down decaying relic axions, as was tentatively done by Grin *et al.* [1024], or a fermionic dark matter component, as tested by Nakajima and Morikawa [1492], by using hundreds of clusters of galaxies instead of very few and biased samples.

Regarding tests on gravity and neutrino mass, next-generation cosmic shear surveys look very promising. Present-day data are still too few and suffer from non-optimal optical designs of telescopes for these kinds of extreme observations. Technical issues regarding image deconvolution and shear measurements, intrinsic galaxy alignments or the redshifts of lenses and sources have been overlooked on purpose in this review, but are the real challenges of next-generation gravitational lensing surveys. The greatest promise lies in dedicated projects. Surveys like those planned with the Large Synoptic Survey Telescope, or with Joint Dark Energy Mission (JDEM)- and/or EUCLID-type space missions, will be far better suited. They will be sufficiently accurate to describe the power spectrum of the convergence field at different redshifts and the history of the growth of structure, with matter and gravity in action over cosmological timescales, much more precisely than the early tentative interpretation of COSMOS data [1396]. They may improve the constraints on the physical properties of dark matter haloes, on dark matter particles, on neutrino mass or on modified gravity models by two orders of magnitude.

Acknowledgements

I thank T. Broadhurst, D. Clowe, R. Gavazzi, M. Kilbinger, J.-P. Kneib, M. Limousin and J.-P. Uzan for their comments and careful reading of the manuscript, and M. Bradač, T. Broadhurst, D. Clowe and R. Gavazzi for the figures.

5

Dark matter at the centres of galaxies

David Merritt

Dark matter haloes formed in ΛCDM cosmologies exhibit a characteristic dependence of density on distance from the centre (Chapter 2). Early studies [721; 1501] established $\rho_{\mathrm{DM}} \sim r^{-3}$ or r^{-4} at large radii and $\rho_{\mathrm{DM}} \sim r^{-1}$ inside the virial radius. On still smaller scales, the form of $\rho_{\mathrm{DM}}(r)$ was little more than an ansatz since the relevant scales were barely resolved in the N-body simulations. A debate ensued as to whether the profiles were indeed universal and, if so, what power of the radius described the dark matter density in the limit $r \to 0$. Subsequent studies found central profiles both steeper [666; 940; 1247; 1471] and shallower [1431; 1503; 1504] than r^{-1}.

The focus of this chapter is the dark matter distribution on sub-parsec scales. At these radii, the gravitational force in many galaxies is known to be dominated by the observed baryonic components (stellar bulge, nuclear star cluster) and by the supermassive black hole. Dark matter densities at these radii are barely constrained observationally; however, they could plausibly be orders of magnitude higher than the local value at the solar circle ($\sim 10^{-2} M_{\odot}$ pc^{-3}), owing both to the special location at the centre of the halo and to interactions between dark matter and baryons during and after formation of the galaxy. High dark matter densities make the centres of galaxies preferred targets for indirect detection studies, in which secondary particles and photons from the annihilation or decay of supersymmetric dark matter particles are detected on the Earth (Chapter 24).

5.1 Phenomenology of galactic nuclei

The distribution of baryonic matter at the centres of galaxies is relevant to the dark matter problem for several reasons.

Particle Dark Matter: Observations, Models and Searches, ed. Gianfranco Bertone. Published by Cambridge University Press. © Cambridge University Press 2010.

- Many dynamical processes affect the dark and luminous components in similar ways. The distribution of stars at the centre of a galaxy can tell us something about the distribution of dark matter.
- If the nuclear relaxation time (Eq. (5.5)) is shorter than the age of the Universe, stars will exchange kinetic energy with dark matter particles, causing the dark matter distribution to evolve in predictable ways.
- Supermassive black holes (SMBHs) appear to be ubiquitous components of galactic nuclei. Depending on its mode of growth, an SMBH can greatly increase, or decrease, the density of dark matter in its vicinity.

Galactic nuclei are the innermost regions of stellar spheroids: either elliptical galaxies, or the bulges of spiral galaxies. Most galaxies are too distant for individual stars to be resolved, and descriptions of their structure are generally based on their luminosity profiles, the surface brightness of starlight as a function of distance from the centre. Luminosity profiles of galactic spheroids are well fitted at most radii by Sérsic's law [1728],

$$\ln I = \ln I_0 - bR^{1/n}, \tag{5.1}$$

where I is the surface brightness at projected radius R and n is the Sérsic index; $n = 4$ is the de Vaucouleurs law [633]. Sérsic's law predicts a space density that increases as $\rho_\star \sim r^{-(n-1)/n}$ toward the centre, or $\rho_\star \sim r^{-3/4}$ in the case of de Vaucouleurs' law. However, in the best-resolved galaxies – which include both the nearest, and the intrinsically largest, galaxies – deviations from Sérsic's law often appear near the centre. Bright spheroids exhibit deficits (known as cores) with respect to Sérsic's law. Faint spheroids exhibit excesses, or nuclear star clusters (NSCs), with sizes in the range 1–100 pc and luminosities in the range 10^6–$10^8 L_\odot$ [377; 584]. The transition from deficit to excess occurs at spheroid luminosities of roughly $10^{10} L_\odot$ [583]. While NSCs are generally unresolved (a notable exception [1711] being the NSC at the centre of the Milky Way), cores in luminous elliptical galaxies can extend hundreds of parsecs.

Galactic nuclei also contain supermassive black holes. In a handful of galaxies, the presence of the SMBH is indicated by a clear Keplerian rise of stellar or gas velocities inside a radius $\sim r_h$, the gravitational influence radius:

$$r_h = GM_\bullet/\sigma^2. \tag{5.2}$$

Here $M_\bullet$ is the black hole mass and σ is the one-dimensional, root mean square (rms) velocity of stars in the spheroid. In other galaxies, indications are seen of a central rise in velocity but the implied SMBH mass is very uncertain [1425]. Among the dozen or so galaxies with well-determined

SMBH masses, there is a remarkably tight correlation between $M_\bullet$ and σ, the $M_\bullet - \sigma$ relation [845]:

$$M_\bullet/10^8 M_\odot \approx 1.66(\sigma/200 \text{ km s}^{-1})^\alpha, \qquad \alpha \approx 4.86. \tag{5.3}$$

Combining Eqs. (5.2) and (5.3),

$$r_{\rm h} \approx 18 \text{ pc } (\sigma/200 \text{ km s}^{-1})^{2.86} \approx 13 \text{ pc } (M_\bullet/10^8 M_\odot)^{0.59}. \tag{5.4}$$

The $M_\bullet - \sigma$ relation extends at least down to $M_\bullet \approx 10^{6.6} M_\odot$, the mass of the Milky Way SMBH [938; 947]. Indirect, but contested, evidence exists for the existence of intermediate-mass black holes in some low-luminosity spheroids, active galaxies and star clusters [1448].

The connection between SMBHs and nuclear structure is circumstantial but reasonably compelling. Observed core radii are a few $r_{\rm h}$ in the brightest elliptical galaxies, consistent with a model in which the cores were created when stars were displaced by a pre-existing binary SMBH [1423]. At the other extreme in spheroid luminosity, NSCs sometimes appear to co-exist with SMBHs, but there is only a handful of galaxies in which the presence of both components can unambiguously be established [1731]; thus there is no clear evidence that SMBHs are associated with an excess of (luminous) mass at the centres of galaxies.

The nuclear relaxation time

$$t_{\rm r} = \frac{0.34\sigma^3}{G^2 \rho_\star m_\star \ln \Lambda} \tag{5.5}$$

measures the timescale over which gravitational encounters redistribute energy between stars; $\ln \Lambda \approx 12$ is the Coulomb logarithm [1786]. Relaxation times greatly exceed 10 Gyr at all radii in spheroids more massive than $\sim 10^{10} M_\odot$ [1428]; in these 'collisionless' systems, star–star and star–dark matter interactions occur too rarely to alter the distribution of either component significantly over the lifetime of the galaxy. In fainter spheroids, and particularly those containing dense NSCs, central relaxation times can be shorter [1424]; for instance, at the Galactic centre, $t_{\rm r}$ falls below 10 Gyr inside $r_{\rm h}$ [1710]. In these 'collisional' nuclei, the distribution of stars around an SMBH is expected to evolve, in a time $\sim t_{\rm r}$, to the quasi-steady-state form

$$\rho_\star(r) \propto r^{-7/4} \tag{5.6}$$

at $r \lesssim r_{\rm h}$: a Bahcall–Wolf cusp [166]. If multiple mass groups are present, Eq. (5.6) describes the central behaviour of the most massive component, while the lowest-mass component (e.g. dark matter particles) obeys

$$\rho(r) \propto r^{-3/2} \tag{5.7}$$

D. Merritt

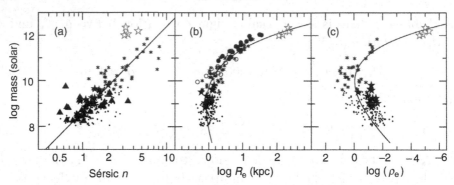

Fig. 5.1. Relations between the parameters that define the mass, size and density
of galaxy spheroids and N-body dark matter haloes. Because both types of system
are well described by the same empirical density law, they can be plotted together
on the same axes. Here n = Sérsic index, R_e = effective (projected half-mass or
half-light) radius, ρ_e = mass density at $r = R_e$. The filled symbols are dwarf
and giant elliptical galaxies; structural parameters for these objects were derived
by fitting Sérsic laws, Eq. (5.1), to their observed luminosity profiles. Luminosity
densities were converted to mass densities using an assumed value for the stellar
mass-to-light ratio, neglecting dark matter. The open stars are a set of simulated,
galaxy-sized dark matter haloes [666]. Mass density profiles for these objects were
fitted with de-projected Sérsic laws and total mass was defined as mass within
the virial radius. The solid lines are semi-empirical fitting relations. Adapted from
[1003].

[167]. Equation (5.6) approximately describes the distribution of luminous
stars at the Galactic centre [1710], but no other galaxy containing an SMBH
is near enough that a Bahcall–Wolf cusp could be resolved even if present.

Scaling relations between spheroid luminosities and masses or velocity
dispersions are continuous over many decades in mass, from giant elliptical galaxies down to globular clusters (Fig. 5.1). Only the class of dwarf
spheroidal galaxies (dSphs) departs systematically from these relations, in
the sense of having too large an inferred (dynamical) mass compared with
their luminosities: these systems appear to be dark-matter-dominated even
at their centres (Section 5.6).

5.2 Dark matter models

5.2.1 Cusps versus cores

Traditionally there have been two approaches to estimating the density of
dark matter at the centres of galaxies. Unfortunately, they often lead to
different conclusions.

In the first method, N-body simulations of gravitational clustering follow the growth of dark matter haloes as they evolve via mergers in an expanding, cold-dark-matter universe. Halo density profiles in these simulations are well determined on scales $10^{-2} \lesssim r/r_{\rm vir} \lesssim 10^0$, where the virial radius $r_{\rm vir}$ is of order 10^2 kpc for a galaxy like the Milky Way; hence inferences about the dark matter density on parsec or sub-parsec scales require an extrapolation from the N-body results. A standard parameterization of $\rho_{\rm DM}$ in these simulated haloes is [1501]

$$\rho_{\rm DM}(r) = \rho_0 \xi^{-1} (1+\xi)^{-2}, \tag{5.8}$$

the NFW profile, where $\xi = r/r_{\rm s}$ and $r_{\rm s}$ is a scale length of order $r_{\rm vir}$. In the Milky Way, $r_{\rm vir} \gg R_\odot$ (the radius of the Solar circle), so Eq. (5.8) is essentially a power law at $r < R_\odot$, and the implied dark matter density is

$$\rho_{\rm DM}(r) \approx 10^2 M_\odot {\rm pc}^{-3} \left(\frac{\rho_\odot}{10^{-2} M_\odot {\rm pc}^{-3}} \right) \left(\frac{R_\odot}{8\ {\rm kpc}} \right) \left(\frac{r}{1\ {\rm pc}} \right)^{-1} \tag{5.9}$$

where $\rho_\odot \equiv \rho_{\rm DM}(R_\odot)$ and $\rho_\odot \approx 8 \times 10^{-3} M_\odot {\rm pc}^{-3}$ (from the Galactic rotation curve).

In contrast, rotation-curve studies of low-surface-brightness spiral galaxies are generally interpreted as implying much lower central dark matter densities [476; 624; 625; 933; 1673]. While there are caveats to this interpretation – systematic biases in long-slit observations [1782], non-circular motions [1762], gas pressure [1887], etc. – these effects do not seem capable of fully explaining the discrepancies between rotation curve data and expressions like Eq. (5.8) [623; 933]. A model for $\rho_{\rm DM}(r)$ that is often fitted to rotation curve data is

$$\rho_{\rm DM}(r) = \rho_{\rm c} (1+\xi)^{-1} (1+\xi^2)^{-1}, \tag{5.10}$$

the Burkert profile [476], where $\xi \equiv r/r_{\rm c}$ and $r_{\rm c}$ is the core radius. Inferred core radii are $\sim 10^2$–10^3 pc and inferred central densities are $10^{-2} \lesssim \rho_{\rm c} \lesssim 10^0 M_\odot$ pc^{-3}.

Since the N-body haloes are not resolved on the scales ($\sim 10^2$ pc) where rotation curves are measured, the mismatch between theory and observation may be due in part to a poor choice of empirical function used to describe the N-body models. An alternative parameterization,

$$\rho_{\rm DM}(r) = \rho_0 \exp\left[-(r/r_0)^{1/n} \right], \tag{5.11}$$

the Einasto profile [740], has recently been shown to describe N-body haloes even better than Eq. (5.8) [1423; 1503; 1598]. The low central density of the

Einasto model alleviates some, but not all, of the disagreement with rotation curve studies [1003].

Remarkably, Eq. (5.11) has the same functional form as Sérsic's law (5.1) that describes the projected density profiles of galactic spheroids. In fact, the two descriptions are roughly equivalent if $n_{\text{Einasto}} \approx n_{\text{Sérsic}} + 1$ [1431], showing that luminous spheroids and simulated dark-matter haloes are essentially rescaled versions of each other (Fig. 5.1), at least over the range in radii that is resolvable by the N-body simulations.

Several resolutions have been suggested for the persistent conflict between predicted and measured central dark matter densities [1841; 1681; 1758] (see also Chapter 6), but none is universally agreed upon.

5.2.2 Effects of baryonic dissipation

N-body simulations of dark matter clustering typically ignore the influence of the baryons (stars, gas) even though these components may dominate the gravitational force in the inner kiloparsec or so. One simple, though idealized, way to account for the effect of the baryons on the dark matter is via adiabatic contraction models, which posit that the baryons contracted quasi-statically and symmetrically within the pre-existing dark matter halo, pulling in the dark matter and increasing its density in the process [370]. When applied to a dark matter halo with the density law (5.8), i.e. $\rho_{\text{DM}} \sim r^{-1}$, the result is the more steeply rising $\rho \sim r^{-\gamma_c}$, $\gamma_c \approx 1.5$ [966; 1378; 1597].

5.3 Dark matter in collisionless nuclei

The dark matter annihilation signal from a region of volume V is proportional to $\langle \rho_{\text{DM}}^2 \rangle V$. If the dark matter density rises steeply toward the centre of a galaxy, the annihilation flux can be dominated by dark matter within the central parsec or so. Neither N-body simulations nor rotation curve studies are a reliable guide to ρ_{DM} on these small scales. In addition, in many galaxies, the total gravitational force in the inner parsecs is dominated by the SMBH.

We consider first 'collisionless' nuclei, in which central relaxation times exceed ~ 10 Gyr; this is the case in spheroids more massive than $\sim 10^{10} M_\odot$ [1428]. In these systems, the distribution of stars and dark matter near the galaxy centre has probably remained essentially unchanged since the era at which the nucleus and the SMBH were created.

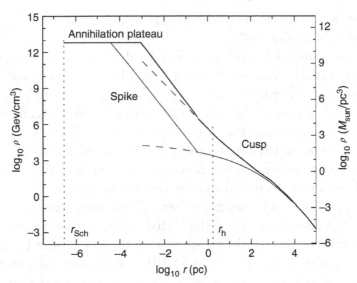

Fig. 5.2. Possible models for the dark matter distribution at the centre of a galaxy like the Milky Way. The thin curve shows an Einasto density profile, Eq. (5.11), and the thick curve labelled 'cusp' is the same model after 'adiabatic compression' by the baryons (stars and gas). Lines labelled 'spike' show the additional increase in density that would result from spherically symmetric growth of the SMBH. The 'annihilation plateau' is the density that satisfies $\rho = m_\chi/\langle \sigma v \rangle t$; this density was computed assuming $m_\chi = 200$ GeV, $\langle \sigma v \rangle = 10^{-28}$ cm^3 s^{-1} and $t = 10^{10}$ yr. Dashed vertical lines indicate the SMBH's Schwarzschild radius (left) and gravitational influence radius (right). Effects of the dynamical processes discussed in Section 5.4 (scattering of dark matter off stars, loss of dark matter into the SMBH) are excluded from this plot; in nuclei (like that of the Milky Way) with relaxation times less than ∼10 Gyr, these processes would generally act to decrease the dark matter density below what is shown here, particularly in the models with a spike (cf. Figs. 5.4 and 5.5). Adapted from [338].

5.3.1 Black hole adiabatic growth models

If the SMBH grew to its final size in the simplest possible way – via spherically symmetric infall of gas – the density of matter around it would increase [1557; 1954], in the same way that contracting baryons steepen the dark matter density profile on somewhat larger scales (Section 5.2.2). In the limit that the growth timescale of the SMBH is long compared with orbital periods, this scenario predicts a final density (of stars or dark matter) near the SMBH of

$$\rho_f(r) \approx \rho_i(r_f)(r/r_f)^{-\gamma_f}, \qquad \gamma_f = 2 + 1/(4 - \gamma_i), \qquad (5.12)$$

where $\rho_i \propto r^{-\gamma_i}$ is the pre-existing density profile, and $r_f \approx 0.2r_h$. Even for $\gamma_i \approx 0$, Eq. (5.12) predicts $\gamma_f > 2$ – a density spike (Fig. 5.2). Such a steep dark matter density profile near the SMBH would imply very high rates of dark matter annihilation [985].

Stars would respond in the same way as dark matter particles to the growth of an SMBH. A $\rho \sim r^{-2}$ density cusp in the stars is not observed near the centre of any galaxy, however, even those close enough that a spike could be resolved if present. In the case of the most luminous galaxies, this is an expected consequence of core formation by binary SMBHs, as discussed in the next section. In low-luminosity spheroids like the bulge of the Milky Way, relaxation times are short enough to convert a stellar spike into a shallower, Bahcall–Wolf cusp in roughly one relaxation time.

It is also possible that spikes never form. Even small (compared with r_h) and temporary displacements of the SMBH from its central location are sufficient to inhibit the growth of a spike or to destroy it after it has formed [1883]. Most models for the growth of SMBHs invoke strong departures from spherical symmetry during galaxy mergers in order to remove excess angular momentum from the infalling gas [1747].

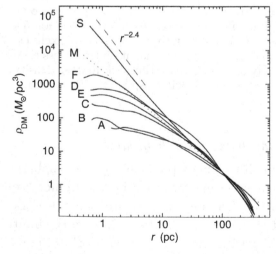

Model	M_1/M_2	ρ_1/ρ_2	$M/M_\bullet$
A	1	1	0.01
B	1	1	0.03
C	3	1	0.03
D	3	1/3	0.03
E	5	1	0.03
F	10	1	0.03

Fig. 5.3. The effect of mergers, including SMBHs, on the central densities of dark matter haloes [1401]. The curves labelled M and S are the density profiles of the larger of the two haloes before and after an SMBH was grown adiabatically at its centre. The other curves show the final density profile of the merged halo, for various choices of the initial halo parameters, as given in the table. M_1 (M_2) is the mass of the large (small) dark matter halo and ρ is the central halo density before growth of the SMBH. Mass and length scalings were based on the Milky Way.

5.3.2 Binary black holes and core creation

Most spheroids are believed to have experienced at least one major merger (defined as a merger with mass ratio ~3:1 or less) since the epoch at which the SMBHs formed [1210]. If two merging galaxies each contain an SMBH, a massive binary forms [249], displacing stars and dark matter as the two holes spiral in to the centre. The process can be understood as a kind of dynamical friction, with the 'heavy particles' (the SMBHs) transferring their kinetic energy to the 'light particles' (stars, dark matter). However, most of the energy transfer takes place after the two SMBHs have come within each other's spheres of influence, and in this regime the interaction with the background is dominated by another mechanism, the gravitational slingshot [1691]. The massive binary ejects passing stars or dark matter particles at high velocity, removing them from the nucleus and simultaneously increasing its binding energy [1617].

This process stops, or at least slows, when the two SMBHs reach a separation $r_{\text{stall}} \approx q/(1 + q)^2 r_{\text{h}}$, the stalling radius; here $q \equiv m_2/m_1 \leq 1$ is the binary mass ratio. At this separation, the binary has already removed essentially all material on intersecting orbits and the inspiral stops; or, it continues at a much lower rate that is limited by how fast the depleted orbits can be repopulated [1428]. The size of the low-density core that is produced by inspiral from $r \approx r_{\text{h}}$ to $r \approx r_{\text{stall}}$ is around a few times r_{h}, quite consistent with the sizes of the stellar cores observed in many galaxies [1001; 1423]. Dark matter cores would presumably be of similar size, or even larger if multiple mergers occurred [1423; 1910] or if the pre-binary dark matter distribution was characterized by a core as in the Burkert model described above. However, this mechanism probably cannot explain the kiloparsec-scale dark matter cores inferred in many spiral galaxies: the mergers that formed the bulges of these systems would have resulted in much smaller, parsec-scale cores.

5.3.3 Gravitational-wave recoil

If the two SMBHs at the centre of a merged galaxy manage to overcome the 'final-parsec problem' and coalesce, another mechanism comes into play that can affect the central density of stars and dark matter. Emission of gravitational waves during the final plunge is generically anisotropic, resulting in a transfer of linear momentum to the coalesced SMBH [1631]. The resultant 'kick' can be as large as ~4000 km s^{-1} if the two holes have equal mass and optimal spins (i.e. maximal amplitude, oppositely aligned, and parallel to the binary orbital plane) [486]. While such extreme kicks are probably rare, even a mass ratio of 0.1 can result in kicks of ~ 1000 km s^{-1} if spins are optimal; and if the spins are maximal but oriented parallel to the orbital angular momentum, the kick velocity V_{kick} peaks at ~ 600 km s^{-1} for $m_1 = m_2$. By comparison, kicks large enough to remove SMBHs from galaxy cores range from ~ 90 km s^{-1} for spheroid masses of $3 \times 10^9 M_\odot$ to ~ 750 km s^{-1} for $M_{\text{sph}} = 3 \times 10^{11} M_\odot$ and to ~ 1000 km s^{-1} for $M_{\text{sph}} = 3 \times 10^{12} M_\odot$ [1429].

Sudden removal of the SMBH from the galaxy centre impulsively reduces the force that binds stars and dark matter to the centre [431; 1429]. If V_{kick} is less than the galaxy central escape velocity, still more energy is injected into the core by the kicked SMBH as it passes repeatedly through the centre before finally coming to rest. Cores enlarged in this way can be several times larger than r_{h}, and indeed a few of the brightest elliptical galaxies have such over-sized cores [1310]; dark matter cores are presumably of comparable size in these galaxies.

5.4 Dark matter in collisional nuclei

Nuclear relaxation times fall below 10 Gyr in spheroids fainter than $\sim 10^{10} L_\odot$, roughly the luminosity at which NSCs first appear [1428]. As discussed above, at least some of these galaxies (including the Milky Way) also contain SMBHs. In these collisional nuclei, a Bahcall–Wolf cusp in the stars can re-form even if it had been previously destroyed by a binary SMBH [1432]. Both the Milky Way and the nearby dwarf elliptical galaxy M32 exhibit steeply rising stellar density profiles within the influence radii ($r \lesssim 1$ pc) of their SMBHs [1309; 1711], probably the result of collisional evolution.

Dark matter particles in these galaxies are still collisionless: their individual masses are so small that gravitational encounters between them are negligible. But even massless particles can scatter off stars, and the associated timescale is of the same order of magnitude as the star–star relaxation time, Eq. (5.5) [966; 1150; 1422].

Naively, one would expect the stars to act like a heat source, transferring kinetic energy to the dark matter particles and lowering their density. This does occur; but in addition, the phase-space density of dark matter particles is driven toward a constant value as a function of orbital energy, $f(E) \approx f_0$. A constant phase-space density with respect to E implies a configuration-space density that rises as $\rho \sim r^{-3/2}$ in the $1/r$ potential of an SMBH. The term 'crest', for 'collisionally regenerated structure', has been coined to describe the result of this process [1427].

Figure 5.4 illustrates the joint evolution of the stellar and dark-matter densities near an SMBH at the centre of a galaxy in which the density

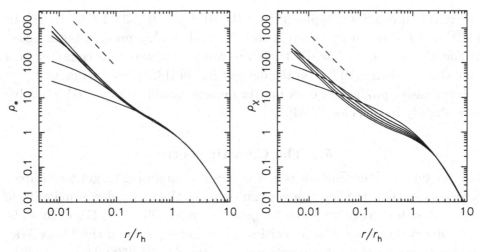

Fig. 5.4. Joint evolution of the stellar (left) and dark matter (right) densities around an SMBH due to star–star and star–dark matter gravitational encounters [1427]. Length unit r_h is the SMBH influence radius (roughly 1 pc at the Galactic centre). Density is in units of its initial value at r_h. Curves show density profiles at times $(0, 0.2, 0.4, 0.6, 0.8, 1.0)$ in units of the initial relaxation time (Eq. 5.5) at $r = r_\mathrm{h}$. Line thickness indicates elapsed time (thicker line, later time). Dashed lines are the 'steady-state' solutions, Eqs. (5.6) and (5.7).

of both components had previously been lowered by a binary SMBH. The stars are seen to attain the $\rho \sim r^{-7/4}$, Bahcall–Wolf form in approximately one relaxation time. Simultaneously, the dark matter particles evolve to the shallower $\rho \sim r^{-3/2}$ profile, increasing their density near the SMBH. The normalization of the dark matter density continuously drops as the stars transfer kinetic energy to the dark matter. Simulations like these suggest that the presence of a Bahcall–Wolf (collisional) cusp in the stars at the centre of a galaxy should always be associated with a shallower, $\sim r^{-3/2}$ 'crest' in the dark matter, regardless of how the nucleus and the SMBH formed [1427]. Note, however, that this argument cannot be used to infer the normalization of the dark matter density.

Conditions for the formation of crests are relaxed somewhat if there is a top-heavy spectrum of stellar masses, since the dark matter scattering time scales as $\tilde{m}_\star^{-1}$ where $\tilde{m}_\star = \langle m_\star^2 \rangle / \langle m_\star \rangle$ is the second moment of the stellar mass function [1422]. The stellar cusp can also evolve more quickly in this case [235]. However, owing to their gradual dissolution, dark matter crests might only be present with significant amplitudes in galactic nuclei having a fairly narrow range of properties: older than about one relaxation time but younger than a few relaxation times. The corresponding range in spheroid luminosities is approximately $3 \times 10^8 L_\odot \lesssim L \lesssim 3 \times 10^9 L_\odot$ [1427]. In addition, low-luminosity spheroids may not all contain massive black holes. In the absence of an SMBH, the stars would undergo core collapse, in a time shorter than $\lesssim 10^{10}$ yr in the densest nuclei [1424], producing an $r^{-2.25}$ stellar density profile; the dark matter density would be expected to evolve only slightly in this case [1234].

5.5 The Galactic centre

The proximity of the Galactic centre makes it a promising target for indirect detection studies: predicted fluxes can be more than an order of magnitude higher than for any other potential galactic source [303; 341; 1798]. In addition, observations of stellar velocities in the inner parsec of the Milky Way yield a highly precise value for the mass in the SMBH [938; 947], as well as (somewhat less precise) estimates of the distributed mass [1711; 1865]. In principle, dark matter might be detected by observing its effects on the stellar orbits [1060; 1956], but uncertainties about the masses associated with other 'dark' components – neutron stars, stellar mass black holes, etc. – probably render this approach unfeasible for the foreseeable future.

Given a detector with angular acceptance $\Delta\Omega$ sr, the observed flux of photons produced by annihilation of dark matter particles is [303]

$$\Phi(\Delta\Omega, E) \approx 1.9 \times 10^{-12} \frac{dN}{dE} \frac{\langle \sigma v \rangle}{10^{-26}\,\mathrm{cm}^{-3}\,\mathrm{s}^{-1}} \left(\frac{1\,\mathrm{TeV}}{m_\chi}\right)^2 \overline{J}_{\Delta\Omega}\, \Delta\Omega\ \mathrm{cm}^{-2}\,\mathrm{s}^{-1} \tag{5.13}$$

where dN/dE is the spectrum of secondary photons per annihilation, m_χ is the particle mass, $\langle \sigma v \rangle$ is the velocity-averaged self-annihilation cross-section and $\overline{J}_{\Delta\Omega}$ contains the information about the dark matter density:

$$\overline{J}_{\Delta\Omega} = K \Delta\Omega^{-1} \int_{\Delta\Omega} d\psi \int_\psi \rho_{\mathrm{DM}}^2 dl. \tag{5.14}$$

Here, dl is an element of length along the line of sight and ψ is the angle with respect to the Galactic centre. The normalizing factor K is typically set to $K^{-1} = (8.5\,\mathrm{kpc})(0.3\,\mathrm{GeV/cm^3})^2$, which is the product of the distance to the Galactic centre, and the squared, local value of the dark matter density, the latter derived from the measured rotation curve assuming an NFW halo. Henceforth we write $\overline{J}_{\Delta\Omega=10^{-5}} \equiv \overline{J}_5$: $\Delta\Omega = 10^{-5}$ sr (~ 10 arcminutes) is the approximate angular resolution of atmospheric Cerenkov telescopes like H.E.S.S. [1115] and of Fermi [1463].

Extrapolation of a halo model like that of Eq. (5.8) into the Galactic centre region gives $\overline{J}_5 \approx 10^3$, large enough to produce observable signals

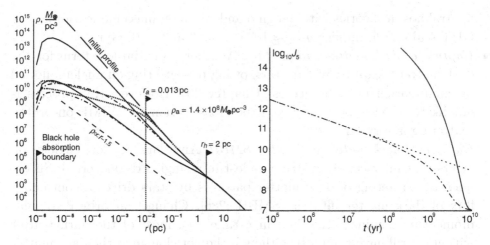

Fig. 5.5. Left: Time evolution of the dark matter density at the centre of the Milky Way, including the effects of scattering off stars, self-annihilations and absorption into the SMBH. The initial density profile, shown by the dashed line, represents an adiabatically compressed NFW profile with $\rho \sim r^{-2.3}$. The solid, dotted and dash-dotted lines show evolved density profiles assuming scattering only; annihilation only; and both annihilation and scattering, respectively. The top curve in each set is for $t = 2.5$ Gyr and the bottom curve is for $t = 10$ Gyr. Right: Evolution of the form factor $\bar{J}_5$ for the models in the left panel. In the absence of self-annihilations (solid line), the central density remains very high until ~1 relaxation time has elapsed, yielding large J_5 values. When annihilations are included (dotted, dash-dotted lines), the dark matter density near the SMBH drops rapidly and J_5 is much smaller at early times. The almost equal values of J_5 at 10 Gyr in the two models that include scattering are partly a coincidence due to the assumed value of the annihilation cross-section. Adapted from [1897].

for many interesting choices of $\langle \sigma v \rangle$ and m [303]. On the other hand, the detections by the Whipple and H.E.S.S. collaborations of gamma-rays from the Galactic centre with energies up to 10 TeV [37; 1272] would require very large values of $\langle \sigma v \rangle \bar{J}$ [1119; 1131; 1607], motivating the exploration of models in which the central dark matter density is enhanced with respect to standard models – for instance, via the collisionless 'spikes' discussed above.

In models with such high initial densities, ρ_{DM} evolves rapidly near the SMBH, owing both to scattering off stars and to a number of other processes.

- *Self-annihilations.* The same annihilations that produce observable radiation also cause the dark matter density to decay. Self-annihilations limit the density to $\rho_{\mathrm{ann}} \approx m/\langle \sigma v \rangle t$, with $t \approx 10$ Gyr the time since formation of the spike [288]. The result is a weak, $\sim r^{-0.5}$ density plateau near the SMBH [1896]. Assuming a 'maximal' $\langle \sigma v \rangle \approx 3 \times 10^{-26}$ cm^3 s^{-1}, appropriate for a thermal relic, and a 'minimal' $m_\chi \approx 50$ GeV, appropriate to neutralinos in theories with gaugino and sfermion mass unification at the GUT scale [739], implies $\rho_{\text{ann}} \approx 10^6 M_\odot \text{ pc}^{-3}$ at $t = 10$ Gyr.

- *Capture of dark matter within stars.* Another potential loss term for the dark matter is capture within stars, owing to scattering off nuclei followed by annihilation in stellar cores (Chapter 29). However, this process is not likely to be important unless the cross-section for WIMP-on-proton scattering is very large.

- *Capture of dark matter within the SMBH.* Any dark matter particles on orbits that intersect the SMBH are lost in a single orbital period. Subsequently, scattering of dark matter particles by stars drives a continuous flux of dark matter into the SMBH [286]. Changes in orbital angular momentum dominate the flux; in a time $\sim t_r$, most of the dark matter within r_h will have been lost, although the net change in the dark matter density profile will be more modest than this suggests since more particles are continuously being scattered onto depleted orbits [1422].

A strict inner cutoff to the dark matter density is set by the SMBH's event horizon, $r_{\text{Sch}} = 2GM_\bullet/c^2 \approx 3 \times 10^{-7}$ pc, although for reasonable values of m_χ and $\langle \sigma v \rangle$, the density is limited by self-annihilations well outside r_{Sch}.

These various effects can be modelled in a time-dependent way via the orbit-averaged Fokker–Planck (FP) equation [1422]. In its simplest, energy-dependent form, the FP equation can be written

$$\frac{\partial f}{\partial t} = -\frac{1}{4\pi^2 p} \frac{\partial F_E}{\partial E} - f(E)\nu_{\text{coll}}(E) - f(E)\nu_{\text{lc}}(E). \qquad (5.15)$$

Here $f(E)$ is the phase-space mass density of dark matter, $E \equiv -v^2/2 + \phi(r)$ is the energy per unit mass of a dark-matter particle, P is a phase-space volume element and $\phi(r)$ is the gravitational potential generated by the stars and the SMBH. F_E is the flux of particles in energy space and depends on the stellar density profile and the stellar masses. The two loss terms, ν_{coll} and ν_{lc}, represent decay of dark matter due to self-annihilations and capture within stars; and loss of particles into the SMBH, respectively. A similar equation describes the evolution of the stellar distribution [1427].

Self-annihilations add a second timescale to the problem that depends on the particle physics parameters, the annihilation time t_{ann}:

$$t_{\text{ann}} \equiv \left(\rho_{\text{DM}} \frac{\langle \sigma v \rangle}{m_\chi} \right)^{-1}$$

$$\approx 0.8 \text{ Gyr} \left(\frac{m_\chi}{100 \text{ GeV}} \right) \left(\frac{\rho_{\text{DM}}}{10^{10} M_\odot \text{ pc}^{-3}} \right)^{-1} \left(\frac{\langle \sigma v \rangle}{10^{-26} \text{ cm}^3 \text{ s}^{-1}} \right)^{-1}.$$

$$(5.16)$$

Even assuming the 'maximal' annihilation model defined above, t_{ann} becomes comparable to t_{r} only for dark matter densities greater than $\sim 10^8 M_\odot$ pc^{-3}, corresponding to $r \lesssim 10^{-2}$ pc in models with a 'spike'. Figure 5.5 illustrates this. At early times, annihilations dominate the changes in $\overline{J}$, but after ~ 1 Gyr, heating of dark matter particles by stars tends to repopulate orbits near the SMBH, tending toward a $\rho_{\mathrm{DM}} \sim r^{-3/2}$ 'crest' density profile at $r \lesssim r_{\mathrm{h}}$.

The predicted spectrum of annihilation products depends separately on m_χ (shape) and $\langle \sigma v \rangle \overline{J}$ (amplitude) [331], while for a given initial dark matter model, the final distribution of mass in the evolutionary models depends on $t_{\mathrm{r}}/t_{\mathrm{ann}} \propto \langle \sigma v \rangle / m_\chi$. Assuming a dark matter origin for the TeV gamma-rays observed by H.E.S.S. [37], the spectrum implies $10\,\mathrm{TeV} \lesssim m_\chi \lesssim 20\,\mathrm{TeV}$ and $\langle \sigma v \rangle \overline{J}_5 \approx 10^{-18}$ cm^3 s^{-1} [1607]. The latter value requires either $\overline{J}_5 \gtrsim 10^6$ or a substantial enhancement in the dark matter relic abundance compared with expectations for thermal freeze-out. Figure 5.5 suggests that a dark matter 'crest' can yield sufficiently high values of $\overline{J}$ if the initial density profile is sufficiently steep.

5.6 Dwarf spheroidal galaxies

As noted above, dwarf spheroidal (dSph) galaxies depart systematically from the scaling relations obeyed by other 'hot' stellar systems between size, mass and luminosity [872]. Dwarf spheroidals have total luminosities and internal velocity dispersions comparable with those of globular star clusters, but are much larger, implying very high ratios of (dynamical) mass to (stellar) light, roughly $10-100$ times the solar value [1398]. Since the stars in these systems contribute a negligible fraction of the total mass, dSphs are ideal test-beds for theories of dark matter: in principle, $\rho_{\mathrm{DM}}(r)$ can be mapped directly given sufficiently large samples of radial velocities [1421], without the need to correct for baryon 'contamination'.

Modelling of this sort has been carried out now for roughly a dozen dSphs [e.g. 1245; 1394; 1915]. In practice, the kinematical data are not copious enough for a fully non-parametric approach and the inferred densities are still somewhat model-dependent. Interestingly, the kinematical data seem rarely if ever to demand cusped dark matter density profiles like that of Eq. (5.8). Halo models with low-density cores, e.g. Eq. (5.10), are sometimes preferred [969; 1246]. Assuming that $\rho_{\mathrm{DM}}(r)$ follows an NFW profile, inferred mean densities within 10 pc are $\sim 10 M_\odot$ pc^{-3}, compared with $\sim 0.1 M_\odot$ pc^{-3} if the inner density profile is flat. Inferred total masses depend less strongly on the assumed profile shape [952]. Thus, while the existence of large amounts of dark matter is clearly established in the dSph galaxies, as expected from the N-body models, the detailed distribution of mass within these systems appears to be at odds with the N-body predictions. Dwarf spheroidal galaxies are similar to low-surface-brightness spiral galaxies in this regard.

The number of known dSph satellites of the Local Group has roughly doubled during the past decade [272] and will probably continue to rise; at last count the Milky Way halo contained at least 18 dSphs. Their proximity, combined with their large masses, makes them good candidates for indirect detection studies, although the predicted fluxes are interestingly large only if the dark matter distribution is cusped [178; 1817]. Angular sizes are small enough that a large fraction of their dark matter could be imaged in a single pointing of a telescope like Fermi. The predicted annihilation flux then scales simply as $\sim \rho_{\rm s}^2 r_{\rm s}^3$, where $\rho_{\rm s}$ is the dark matter density at the scale radius $r_{\rm s}$. If the dark matter is clumped, fluxes could be boosted by up to two orders of magnitude [1818]. In addition, the low baryonic densities imply that dSphs should be relatively free of intrinsic gamma-rays from other astrophysical sources, making the interpretation of a signal much more straightforward than in the case of the Galactic centre. A particularly attractive object is the recently discovered object Willman 1 with a luminosity of only $\sim 10^3 L_\odot$ [1936]. At a distance of 38 kpc, this object is so close that it could be marginally resolved by FERMI, in principle allowing a determination of the dark matter distribution [1818].

Acknowledgements

I thank A. Graham and E. Vasiliev for supplying modified versions of figures from their published work.

6

Modified gravity as an alternative to dark matter

Jacob D. Bekenstein

6.1 Missing mass in galaxies and clusters of galaxies

A look at the other papers in this volume will show the present one to be singular. Dark matter is a prevalent paradigm. So why do we need to discuss alternatives? While observations seem to suggest that disk galaxies are embedded in giant haloes of dark matter, this is just an inference from accepted Newtonian gravitational theory. Thus if we are missing understanding about gravity on galactic scales, this inference may be deeply flawed. And then we must remember that, aside from some reports which always seem to contradict established bounds, DM is not seen directly. Finally, were we to put all our hope on the DM paradigm, we would be ignoring a great lesson from the history of science: accepted understanding of a phenomenon has usually come through confrontation of rather contrasting paradigms.

To construct a competing paradigm to DM, it is best to bear in mind concrete empirical facts. Newtonian gravity with the visible matter as source of the Poisson equation properly describes all observed systems from asteroid scale up to the scale of globular clusters of stars ($\sim 10^5$ stars bound together in a ball the size of a few tens of light years). But as we move up to galaxies, ours or external ones, troubles appear. In essence, the way that disk-like galaxies rotate is incompatible with the Newtonian gravitational force generated by only the visible stars, gas and dust. From the centrally concentrated light distribution of the typical disk galaxy, we would expect a rotation linear velocity which first rises with galactocentric radius r, and then drops asymptotically as $r^{-1/2}$. But as is clear from Fig. 6.1, most disk galaxy rotation curves become flat and stay so to the outermost measured point, which generally lies well beyond the edge of the optical galaxy. And

Particle Dark Matter: Observations, Models and Searches, ed. Gianfranco Bertone. Published by Cambridge University Press. © Cambridge University Press 2010.

J. Bekenstein

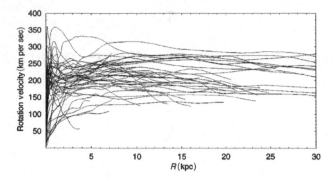

Fig. 6.1. Examples of rotation curves of nearby spiral galaxies from Sofue and Rubin [1778]. These resulted from combining Doppler data from CO molecular lines for the central regions, optical lines for the disks and the HI 21 cm line for the outer gaseous regions. The galactocentric radius R is in kpc (1 kpc ≈ 3000 light years). Reprinted, with permission, from the *Annual Review of Astronomy and Astrophysics* **31**. © 2001 *Annual Reviews* www.annualreviews.org.

typically the mass lying out to the last measured point of the rotation curve, as calculated *à la* Newton, is at least an order of magnitude larger than the baryonic mass actually seen. These are empirical facts begging explanation. Dark matter, if appropriately distributed in each case, can explain the shape and scale of rotation curves. But the required mass distributions are not always reasonable from the point of view of galactogenesis: for example, the predicted central cusps in density are not observed.

Another fact to be explained is the Tully–Fisher law for disk galaxies. Originally discovered as a correlation between blue luminosity and the peak rotation velocity of a disk, it has metamorphosed into McGaugh's baryonic Tully–Fisher law: the total mass in visible stars and gas (baryonic mass) in a disk galaxy is accurately proportional to the fourth power of the asymptotic (terminal) rotational velocity of that galaxy. This law extends over six orders of magnitude in mass, and is a tight correlation, as can be appreciated from Fig. 6.2. In the low-surface-brightness galaxies (LSBs), those with low central brightness per unit area (typically below 21.65 magnitudes per square arcsec), the rotation curve is still on the rise at the last measured point. Yet LSBs, which can be luminous and massive or small dim galaxies, all fall on the same Tully–Fisher relation as more conventional disks [1410].

Within the DM paradigm the Tully–Fisher law must arise from galaxy formation since it connects luminosity of baryonic matter with a dynamical property, rotation, which is seen as dominated by the DM halo. But it has not been easy to derive Tully–Fisher from any natural connection between the two components. And as Robert Sanders is wont to point out, the messiness

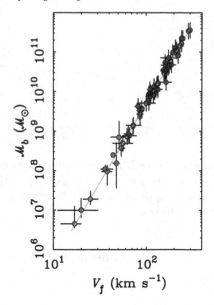

Fig. 6.2. McGaugh's baryonic Tully–Fisher correlation [1410] for disk galaxies. $\mathcal{M}_b$ is the total baryonic (stars plus gas) mass; V_f is the asymptotic rotation velocity. For galaxies from Sanders and McGaugh [1681] (shaded circles), the mass in stars comes from a fit of the shape of the rotation curve with MOND. For the eight dwarf spirals (unshaded circles, all in lower left) the mass in stars (relatively small) is inferred directly from the luminosity. The solid line, with slope 4, is MOND's prediction. Reproduced by permission of the American Astronomical Society.

of galaxy formation is hardly the natural backdrop for such a sharp correlation between galaxy properties. The sharpness needs a dynamical reason as opposed to an evolutionary one.

Turn from galaxies to clusters of galaxies. In these systems, containing sometimes hundreds of galaxies and much hot intergalactic gas, the Newtonian virial theorem can be used to estimate the cluster's mass from the velocities of galaxies in the cluster. The determined masses are very large compared with the mass seen directly as galaxies and hot gas. The mass discrepancy also shows up when the overall cluster mass is determined *à la* Newton from the assumption that the hot gas is in a hydrostatic state, or when gravitational lensing by a cluster is analysed in the framework of General Relativity (GR). The conventional solution is to assume that the typical cluster contains DM to the tune of about five times the visible mass.

There are other aspects of the missing mass problem, but the above will furnish enough background for the ensuing discussion here. The questions before us are two. Are there other scenarios, apart from DM, that can account for all the mentioned facts? And can one single out a particular

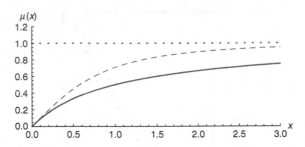

Fig. 6.3. Two widespread choices for Milgrom's interpolation function: the 'simple' function $\tilde{\mu}(x) = x/(1+x)$ (solid) and the 'standard' function $\tilde{\mu}(x) = x/\sqrt{1+x^2}$ (dashed). The dotted line corresponds to strict Newtonian behaviour, obtainable also as the limit $a_0 \to 0$ of MOND.

alternative scenario as especially promising, both in terms of physical basis and explanatory success?

6.2 The MOND scheme

6.2.1 The paradigm

The MOND scheme or paradigm of Milgrom [1434; 1435; 1436] serves first and foremost as an effective summary of much extragalactic phenomenology. Milgrom postulated a preferred scale of acceleration, $a_0 \approx 10^{-8}\,\mathrm{cm\,s^{-2}}$, of the order of the centripetal accelerations of stars and gas clouds in the outskirts of disk galaxies. In terms of it MOND relates the acceleration $\mathbf{a}$ of a test particle to the ambient Newtonian gravitational field $-\nabla\Phi_{\mathrm{N}}$ generated by the baryonic mass density alone by

$$\tilde{\mu}(|\mathbf{a}|/a_0)\,\mathbf{a} = -\nabla\Phi_{\mathrm{N}}. \tag{6.1}$$

Milgrom assumes that the positive smooth monotonic function $\tilde{\mu}$ approximately equals its argument when this is small compared with unity (deep MOND limit), but tends to unity when that argument is large compared with unity (Fig. 6.3).

The stars or gas clouds orbiting in the disk of a spiral galaxy of baryonic mass M at radius r from its centre do so in nearly circular motion with velocity $V(r)$. Obviously $|\mathbf{a}| = V(r)^2/r$. Sufficiently outside the main mass distribution we may estimate $|\nabla\Phi_{\mathrm{N}}| \approx GM/r^2$. And at sufficiently large r, $|\mathbf{a}|$ will drop below a_0 and we should be able to approximate $\tilde{\mu}(x) \approx x$. Putting all this together gives $V(r)^4/r^2 \approx GMa_0/r^2$. It follows, first, that well outside the main mass distribution $V(r)$ must become independent of r, that is, the rotation curve must flatten at some value V_f, in agreement with observations. Second, from the coefficients it follows that

$$M = (Ga_0)^{-1}V_f{}^4. \tag{6.2}$$

But this is just McGaugh's baryonic Tully–Fisher law (Fig. 6.2). The MOND-predicted proportionality coefficient $(Ga_0)^{-1}$ agrees well with the measured coefficient of the baryonic Tully–Fisher law, namely $50\,M_\odot\,\mathrm{km}^{-4}\,\mathrm{s}^{-4}$. Thus, MOND's single formula unifies two central facts of spiral galaxy phenomenology.

6.2.2 Several roles of a_0

Of course a devil's advocate could claim that MOND works well for the above phenomena because the low-argument form of $\tilde{\mu}(x)$ and the value of a_0 have been rigged to obtain these results: after all, both flat rotation curves and early forms of the Tully–Fisher law were already known before MOND was formulated. Note, however, that the supposed prearrangement need not guarantee that the parameter a_0 needed to recover the observed Tully–Fisher law should have anything to do with the detailed shape of rotation curves interior to the flat regions. Yet MOND is singularly successful in explaining the detailed shapes of rotation curves, as made clear by Fig. 6.4. The observed points are radio-determined velocities. The MOND predictions were made using the 'simple' form of $\tilde{\mu}(x)$. Once the value of a_0 is adopted, the only parameter that need be adjusted to fit the velocities using the observed photometry is the stellar mass-to-luminosity ratio Υ of each disk. The required Υs turn out, in most cases, to be reasonable in view of stellar evolutionary models [1681]. So MOND is a successful and consistent one-parameter-per-galaxy paradigm. By contrast DM halo models usually require three fitting parameters, e.g. Υ, length scale and central velocity dispersion, and yet they do not do a better job than MOND. For disk galaxies MOND is more economical, and more falsifiable, than the DM paradigm.

One cannot emphasize enough that, a priori, the a_0 that enters in the Tully–Fisher law is, in principle, a different parameter from that which determines the shape of rotation curves. The value $a_0 = 1.2 \times 10^{-8}\,\mathrm{cm\,s}^{-2}$ that needs to be adopted for good fits to over 100 galaxies [250] agrees very well with the a_0 determined from the baryonic Tully–Fisher law. It is only in the MOND paradigm that these two (and other) roles of a_0 have a common origin, and observations agree that there is only one a_0.

Yet a third role for a_0 results from Milgrom's observation that $\Sigma_m = a_0/G$ sets a special scale for mass surface density. Wherever in a system the actual

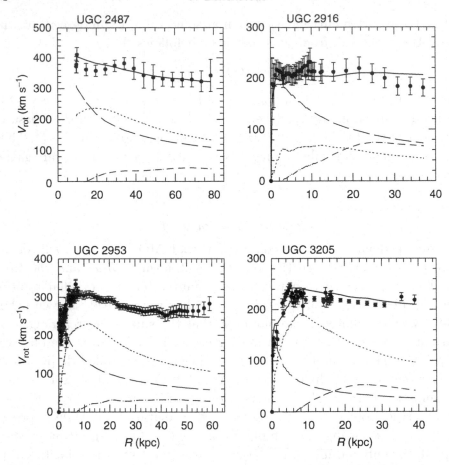

Fig. 6.4. MOND fits to the measured rotation curves of four disk galaxies from the Uppsala catalogue, modelled by the method of Sanders and Noordermeer [1682]. Each solid curve is the fit (based on the simple function $\tilde{\mu}$ of Fig. 6.3) generated by the distribution of stars and neutral hydrogen. Long-dashed, dotted and short-dashed curves give the Newtonian rotation curves generated separately by the galaxy's bulge, disk and gas, respectively. Figure courtesy of R. H. Sanders.

surface mass density drops below Σ_m, Newtonian gravitational behaviour gives way to MOND dynamics [1435]. In galaxies like ours this occurs some way out in the disk; that is why the rotation curves may exhibit a brief drop (attempting to act *à la* Newton) before becoming flat asymptotically as befits MOND. Of course, if at the centre of a disk the surface mass density is already below Σ_m, then MOND holds sway all over the disk. Thus Milgrom predicted that were disk galaxies to exist with surface mass densities everywhere below Σ_m, they should show especially large acceleration discrepancies coming from the fact that $\tilde{\mu} \ll 1$ everywhere in them. Likewise he noted that the shapes of the rotation curves should be independent of the precise way $\tilde{\mu}(x)$ changes from a linear function to a constant, since the non-linear range of μ does not come into play. A population of such galaxies became known in the late 1980s [250]; these are none other than the LSBs. Much subsequent work has confirmed both facets of Milgrom's prediction [1411]. In particular, it is easy to understand why in these galaxies the rotation curves are still on the rise at the outermost observed point.

Use of the MOND analogue to the virial theorem [1440] to estimate the masses of galaxy clusters softens, but does not fully resolve, the acceleration discrepancy. Clusters still seem to contain a factor of 2–3 more matter than is actually observed in all known forms [1681]. One should keep in mind that clusters may contain much invisible matter of rather prosaic nature, either baryons in a form which is hard to detect optically [1443], or massive neutrinos [1581; 1676; 1679] so that, while troubling, the above lingering discrepancy should not stymie further investigation of MOND.

Excellent reviews exist of other phenomenological successes of MOND [1409; 1444; 1681; 1698]. So I forgo further review of MOND's implications, and turn to ask the obvious question.

6.3 Modified gravity theory for MOND

6.3.1 What is behind MOND?

There is no doubt that MOND is a useful paradigm for summarizing extragalactic data. But what is the physical basis of its success? The most conservative answer is that MOND is merely a happy summary of how much DM there is in galaxies, and how it is spatially distributed in the various types of objects. Surprisingly, this minimal interpretation, which would surely excite no opposition from conventional astrophysicists, does not hold out well. First, such interpretation should explain how the scale a_0 enters into spiral galaxy properties through some regularity in the formation process of DM haloes. Yet detailed arguments have shown that the haloes cannot have an intrinsic scale (e.g. ref. [1680]). In addition, it is hard to see how such an interpretation could give an account of the multiple roles of a_0 alluded to above [1441].

Another interpretation of MOND [1435; 1439] is that it reflects modified inertia, e.g. Eq. (6.1) represents a modification of the Newtonian law $\mathbf{f} = m\mathbf{a}$, whether the force $\mathbf{f}$ is purely gravitational or includes other forces. In evaluating this alternative it is important to realize that the MOND formula cannot be exact. For a binary system with unequal masses, m_1 and m_2, the time derivative of $m_1\mathbf{v}_1 + m_2\mathbf{v}_2$, as calculated from Eq. (6.1), does not vanish in general since the inertial factors $\tilde{\mu}$ values will generally be unequal. In light of this failure of momentum conservation, Milgrom suggested that Eq. (6.1) is meaningful only for test particle motion in a background gravity field. While this restriction is not onerous if all one wants is to analyse the rotation curve of a galaxy, it would put problems such as the dynamics of a binary galaxy system out of the reach of MOND.

Even as a theory of test particle motion, the modified inertia interpretation of Eq. (6.1) is problematic. Milgrom [1439] showed rigorously that no Lagrangian exists for it which (i) is Galilean invariant, (ii) generates Newtonian dynamics for $a_0 \to 0$ and (c) generates deep MOND dynamics when $a_0 \to \infty$. If one instead demands relativistic invariance, the desired Lagrangian is even more remote [697]. To break this *impasse* Milgrom suggested giving up the locality requirement and instead deriving the modified inertia from a non-local action functional, one which is not an integral over a Lagrangian density. More recently he has suggested, as an alternative, giving up Galilean invariance in order to make a Lagrangian possible [1442]. Clearly the modified inertia viewpoint of MOND is a difficult one to formalize.

6.3.2 *MOND as modified gravity – AQUAL theory*

When introducing MOND Milgrom also suggested that it may instead represent a modification of the Newtonian law of gravitation. To see this, just replace the right-hand side of $m\mathbf{a} = \mathbf{f}$ with a non-Newtonian gravitational force that depends non-linearly on its sources in such a way that the content of this equation coincides that of Eq. (6.1). Of course, the resulting theory will also flout the conservation of momentum for the same reason as above. This defect is easily repaired by deriving the requisite theory from a suitable Lagrangian [252].

First one stipulates that a particle's acceleration $\mathbf{a}$ is given by $-\nabla\Phi$ where Φ is some potential distinct from Newton's; in this way the motion is conservative. As the Lagrangian for Φ one selects the simplest generalization of the Newtonian field Lagrangian which is still rotationally and Galilean invariant, but depends on a scale of acceleration a_0 which should be identified with MOND's:

$$L = -\int \left[\frac{a_0{}^2}{8\pi G} F\left(\frac{|\nabla\Phi|^2}{a_0{}^2}\right) + \rho\Phi \right] \mathrm{d}^3 x. \tag{6.3}$$

Here F is some positive function and ρ is the visible matter's mass density. This aquadratic Lagrangian theory (AQUAL) reduces to Newton's for the choice $F(X^2) = X^2$, when it readily leads to Poisson's equation for Φ. Generically it yields the equation

$$\nabla \cdot [\tilde{\mu}(|\nabla\Phi|/a_0)\nabla\Phi] = 4\pi G\rho, \qquad (6.4)$$

$$\tilde{\mu}(X) \equiv dF(X^2)/d(X^2). \qquad (6.5)$$

The $\tilde{\mu}$ function here should be identified with Milgrom's $\tilde{\mu}$ for the following reason.

Comparison of the AQUAL equation, Eq. (6.4), with Poisson's for the same ρ allows one to determine the former's first integral,

$$\tilde{\mu}(|\nabla\Phi|/a_0)\nabla\Phi = \nabla\Phi_N + \nabla \times \mathbf{h}, \qquad (6.6)$$

where $\mathbf{h}$ is some calculable vector field (it may be required to be solenoidal). Now in situations with spherical or cylindrical symmetry $\nabla \times \mathbf{h}$ must vanish by symmetry, and it is plain to see that Eq. (6.6) reduces to the MOND equation because $\mathbf{a} = -\nabla\Phi$. Thus, at least for highly symmetric situations, the MOND equation can be recovered exactly; in more realistic geometries it acquires some 'corrections'. AQUAL has the correct limits if we stipulate that $F(X^2) \to X^2$ for $|\nabla\Phi| \gg a_0$ so that $\tilde{\mu}$ is constant (Newtonian regime), and $F(X^2) \to \frac{2}{3}X^{3/2}$ for $|\nabla\Phi| \ll a_0$ so that $\tilde{\mu}(|\nabla\Phi|/a_0) \to |\nabla\Phi|/a_0$ in this deep MOND regime.

At present AQUAL is the best available embodiment of the MOND idea in the non-relativistic regime. Being Lagrangian-based, AQUAL respects the energy, momentum and angular momentum conservation laws. It removes an ambiguity inherent in the MOND formula: how is a composite object, such as a star, to move if its constituents (ions) are subject to strong collision-related accelerations ($\gg a_0$), but the composite moves as a whole in a weak gravitational field ($\ll a_0$), such as the far field of the Galaxy? AQUAL makes it clear that, regardless of the internal makeup, the composite's centre of mass moves in the ambient field according to the MOND formula [252]. This is in harmony with the fact that all sorts of stars, gas clouds and so on delineate one and the same rotation curve for a galaxy. AQUAL has permitted the calculation of the MOND force between two galaxies in a binary from first principles [1437; 1440]. It establishes the existence of the conjectured 'external field effect', namely, the partial suppression of MOND effects in a system with internal accelerations weak on scale a_0 which happens to be immersed in a gravitational field strong on scale a_0 [252]. This may well explain the total absence of missing mass effects in the open galactic clusters, systems with weak internal accelerations [1434].

6.3.3 N-body simulations à la MOND

Newtonian cold disk instabilities originally motivated the idea that DM haloes surround disk galaxies. By dint of their dominant gravity, the haloes were thought to moderate dangerous instabilities. Using a mixture of MOND formula and AQUAL equation arguments, Milgrom [1438] showed analytically that MOND enhances local stability of disks against perturbations, as compared with the Newtonian situation. And AQUAL has opened the door to numerical simulations of stellar systems within the MOND paradigm [1437] which help to explore stability and other issues. Thus Brada and Milgrom [432] combined an N-body code with a numerical solver for the AQUAL equation to verify that MOND stabilizes galaxy disks locally, and to demonstrate that it enhances global stability as well. The degree of stabilization saturates in the deep MOND regime, so no disk is absolutely stable in MOND. This agrees with the fact that a fraction of disk galaxies have bars which must have arisen from instabilities.

More recently the Paris group of Tiret and Combes studied evolution of bars in spiral disks using their own AQUAL based N-body code. They find that bars actually form more rapidly in MOND than in DM halo Newtonian models, but that MOND bars then weaken as compared with those in Newtonian models. The long-term distribution of bar strengths predicted by MOND is in better agreement with the observed distribution than that obtained from Newtonian halo models [1855; 1856]. Another N-body simulator based on AQUAL has been developed by the Bologna school of Ciotti, Nipoti and Londrillo, and used in a number of investigations. In studying dissipationless collapse according to AQUAL, they find that the profiles of the ensuing objects look like those of real elliptical galaxies, and when the resulting dynamics are interpreted Newtonially, the objects seem to contain little DM, as indeed found in real life [1512]. In their study of merging galaxies with AQUAL they find that merging is slower than that in Newtonian theory (because of the lack of DM-generated dynamical friction). This may be a problem for MOND because of widespread evidence that galaxies merge fast. However, the outcome of a merger is indistinguishable from the merger product in Newtonian theory with DM [1513]. In joint work with Binney [1511] the Bologna group have investigated dynamical friction with the AQUAL-based code. They confirm the conclusion of an earlier analytic study by Ciotti and Binney [555] that MOND shortens the dynamical friction timescale as compared with that of a system with the same phase space distribution in Newtonian theory with DM. The situation with regard to slowing of bars and mergers is satisfactory; however, a persistent problem remains with regard to the timescale for digestion of globular clusters in the dwarf spheroidal galaxies, which is too brief [555; 1674].

As could be expected, AQUAL, being nothing but a reformulation of the MOND equation, does not resolve the acceleration discrepancy in clusters of galaxies. We shall return to this at the end. For a more comprehensive review of the AQUAL representation of MOND, the reader is referred elsewhere [256].

6.4 TeVeS and other relativistic MOND theories

6.4.1 Early paths to relativistic MOND

Neither the MOND formula nor AQUAL pretends to encompass relativistic phenomena. Yet we need to deal with gravitational lenses, gravitational waves and cosmology, all relativistic phenomena connected with DM. How to make the MOND idea relativistic? Since AQUAL has had significant success, we should start with that.

The obvious way to make AQUAL into a covariant theory is to let the AQUAL potential Φ metamorphose into a scalar field ϕ whose Lagrangian is the covariant version of Eq. (6.3). Einstein's equations of GR are retained so that the desired theory will approximate standard gravity theory, and so partake of the latter's many successes, for example in the Solar System. To achieve the requisite deviation from exact GR, which after all cannot reduce to MOND, one couples matter in the new theory to ϕ by the simple technique of writing the matter Lagrangian entirely with the new metric $\tilde{g}_{\alpha\beta} = e^{2\phi} g_{\alpha\beta}$, where henceforth we set $c = 1$. This programme was carried out by Bekenstein and Milgrom [252], but the resulting relativistic AQUAL turns out to have superluminal propagation (of ϕ waves). Even more disconcerting is its failure to deal with the observational fact that whatever it is that enhances the pull of galaxies and clusters on their components also significantly enhances their gravitational lensing power [253]; AQUAL does not do this in any significant way. The root of the problem is the conformal relation between the two metrics: it is well known that conformally related metrics are equivalent insofar as light propagation is concerned.

6.4.2 The vector field

It took a while to gather together the elements for a viable relativistic embodiment of MOND. One of these is a timelike 4-vector field with unit norm $\mathcal{U}_\alpha$ first proposed by Sanders [1675]. It is needed to break the above-mentioned conformal relation, which Sanders replaced by

$$\tilde{g}_{\alpha\beta} = e^{-2\phi} g_{\alpha\beta} - (e^{2\phi} - e^{-2\phi}) \mathcal{U}_\alpha \mathcal{U}_\beta. \tag{6.7}$$

The space-time described by $\tilde{g}_{\alpha\beta}$ is related to that of $g_{\alpha\beta}$ by a stretching in the direction of $\mathcal{U}^\alpha \equiv g^{\alpha\beta}\mathcal{U}_\beta$ and a compression orthogonal to it. Sanders regarded $\mathcal{U}^\alpha$ as established by the cosmos, and as pointing in the time direction. With this proviso, the problem with the lensing can be cured.

However, the prescription that a 4-vector should point in the time direction is not a covariant one. Instead one should endow $\mathcal{U}_\alpha$ with covariant dynamics of its own, for we do want a covariant theory. In the tensor–vector–scalar theory, or TeVeS [254; 255; 258], these dynamics are derived from a covariant action of the general form (K and $\bar{K}$ are dimensionless coupling constants)

$$S_v = -\frac{1}{32\pi G} \int \left[K g^{\alpha\beta} g^{\mu\nu} \mathcal{U}_{[\alpha,\mu]} \mathcal{U}_{[\beta,\nu]} + \bar{K}(g^{\alpha\beta}\mathcal{U}_{\alpha;\beta})^2 \right. \tag{6.8}$$

$$\left. - 2\lambda(g^{\mu\nu}\mathcal{U}_\mu\mathcal{U}_\nu + 1) \right] (-g)^{1/2} \mathrm{d}^4 x. \tag{6.9}$$

The first term in the integrand is a Maxwell-like one; it takes care of approximately aligning $\mathcal{U}^\alpha$ with the 4-velocity of matter in the region in question, this in lieu of the non-covariant requirement that $\mathcal{U}^\alpha$ point in the time direction. But, as remarked by Contaldi *et al.* [573], the 'gauge freedom' inherent in this term causes the integral lines of $\mathcal{U}_\alpha$ to form caustics, thus compromising the predictivity of the theory. Thus in later work the second term in the integrand is introduced to break the gauge freedom. Finally, the last term is a constraint (λ being a Lagrange multiplier function) which forces $\mathcal{U}_\alpha$ to have unit negative norm with respect to metric $g_{\alpha\beta}$, and so always to be timelike. Zlosnik *et al.* [1973] have shown how to reformulate TeVeS solely in terms of the metric $\tilde{g}_{\alpha\beta}$ (which turns out to satisfy Einstein-like equations) and the vector field; the ϕ is eliminated with the help of the above constraint. In this form TeVeS might figuratively be described as GR with a DM (the vector field). However, the second form is very intricate. We continue to discuss TeVeS in its original form.

6.4.3 The scalar field

The dynamics of ϕ in relativistic AQUAL must be modified to obviate the superluminal propagation [254]. Here we use the notation of later work [1667] to write the scalar action as

$$S_{\mathrm{s}} = -\frac{1}{2k^2\ell^2 G} \int \mathcal{F}\left(k\ell^2 h^{\alpha\beta}\phi_{,\alpha}\phi_{,\beta} \right) \sqrt{-g}\,\mathrm{d}^4 x, \tag{6.10}$$

where $\mathcal{F}$ is a positive function, $h^{\alpha\beta} \equiv g^{\alpha\beta} - \mathcal{U}^\alpha \mathcal{U}^\beta$, k is another dimensionless coupling constant and ℓ is a constant scale of length. Each choice of the function $\mathcal{F}(y)$ defines a separate TeVeS theory; $\mu(y) = \mathrm{d}\mathcal{F}(y)/\mathrm{d}y$ functions somewhat like the $\tilde{\mu}(x)$ in MOND. We need only consider functions such that $\mathcal{F} > 0$ and $\mu > 0$ for either positive or negative arguments. The correct MOND behaviour will emerge if $\mu(y) \to 1$ for $y \to \infty$ and $\mu(y) \approx D\sqrt{y}$ for $0 < y \ll 1$ with D a dimensionless positive constant. The action of TeVeS comprises S_v, S_s as well as the customary Einstein–Hilbert action for $g_{\alpha\beta}$ and the matter action written in the usual fashion entirely in terms of $\tilde{g}_{\alpha\beta}$ of Eq. (6.7), which I term the physical metric.

6.4.4 Properties of TeVeS

Consequences of TeVeS with the above sort of $\mathcal{F}$ have been investigated mostly for the case $\bar{K} = 0$. TeVeS reduces to GR in the parameter regime $k \to 0$, $K \propto k$ and $\ell \propto k^{-3/2}$ [254]. As a consequence there is a non-relativistic quasi-stationary regime in which TeVeS is Newtonian, and which serves to describe the overall features of Earth and Solar System gravity. In contrast to many gravity theories with a scalar sector, TeVeS evidences no cosmological evolution of the Newtonian 'constant' G_N [257]. My original choice of $\mathcal{F}(y)$ involves a singular passage into the cosmological domain; more successful choices have been propounded by Zhao and Famaey [1969] and by Sanders [1678].

The post-Newtonian behaviour of TeVeS in terms of the celebrated PPN coefficients has been investigated [254; 258; 944; 1678; 1834]. The β and γ parameters agree exactly with those of GR. The parameters $\zeta_1, \zeta_2, \zeta_3, \zeta_4$ and α_3 are all expected to vanish here, just as in GR, because TeVeS is derived from an action principle and is, like GR, a conservative theory (this conclusion extends to the $\bar{K} \neq 0$ case). Eva Sagi (unpublished PhD work) has shown that the preferred location parameter ξ vanishes too. There are strong indications that the preferred frame coefficients α_1 and α_2 are non-vanishing, but extant calculations of them lack consistency. Reliable calculation of these two coefficients is a high priority because experimental bounds on them are strong.

Out of the parameters of TeVeS one can construct the MOND scale of acceleration. Omitting a factor close to unity and reinstating c, this is

$$a_0 = \frac{c^2 \sqrt{k}}{4\pi\ell}. \tag{6.11}$$

The Newtonian regime mentioned above sets in when the naive gravitational field is well above a_0. For weaker ones which are not far from spherically symmetric, the gravitational field obeys an equation like Eq. (6.4) with $\tilde{\mu}$ directly related to the TeVeS μ function [254; 257]. Our stipulation of the small argument limit of $\mathcal{F}$ leads automatically to the extreme MOND limit for fields $\ll a_0$. With a suitable form of $\mathcal{F}$ at intermediate arguments, TeVeS thus reproduces the essence of AQUAL and MOND, and inherits their success in the theatre of galaxies. To conclude, one should mention that when worked out in detail, the scale a_0 is found to have a rather weak dependence on epoch of the universal expansion [257].

6.4.5 Other theories

Sanders has proposed a variant of TeVeS, a bi-scalar tensor vector theory (BSTV) with three free functions and a free parameter [1677]. BSTV is a more appropriate frame for generating cosmological evolution of a_0. Since numerically $a_0 \sim cH_0$, it is often argued that this scale must be determined by cosmology, and should thus vary on a Hubble timescale [1434]. As mentioned, in TeVeS a_0 evolves very slowly, but its change is faster in BSTV. Discrimination between the two behaviours may be possible with good rotation curves of disk galaxies at redshifts $z = 2$–5. Such curves are just now coming into range [859; 934].

Zlosnik *et al.* have also proposed a variant of TeVeS, a tensor-vector theory in which the timelike vector is normalized with respect to $\tilde{g}_{\alpha\beta}$ [1974]. The vector's action is taken to be a function F of $\mathcal{K}$, the quadratic form in the derivatives of the vector field from Einstein–Aether theories. The theory has four parameters: a length scale and three dimensionless parameters. The form of $F(\mathcal{K})$ can be deduced approximately from the requirement that MOND arise in the non-relativistic quasi-static limit, and from the stipulation that a cosmology built on this theory shall have an early inflationary period and an accelerated expansion at late times.

6.5 Gravitational lenses and cosmology in TeVeS
6.5.1 The background

How does TeVeS measure up to the task of describing gravitational lensing? Because all matter and field actions, apart from those for $g_{\alpha\beta}, \phi$ and $\mathcal{U}_\alpha$, are constructed with the physical metric $\tilde{g}_{\alpha\beta}$, light rays in TeVeS are null geodesics of $\tilde{g}_{\alpha\beta}$. Recall that this is built out of $\mathcal{U}_\alpha, \phi$ and $g_{\alpha\beta}$. Since the lensing by galaxies and clusters of galaxies spans cosmological distances, one must first understand a little about TeVeS cosmology. In its isotropic cosmological models the $\mathcal{U}_\alpha$ is pointed precisely in the time direction, basically as a consequence of their spatial isotropy. Consequently, models with baryonic matter content alone tend to be very similar to the corresponding GR models because the scalar ϕ's energy density stays small [254; 540; 1966] while the vector's is nil. According to Chiu *et al.* [540], these models give a reasonable relation between redshift and angular distances, and provide just as good a scaffolding for the analysis of cosmologically distant gravitational lenses as do the GR Robertson–Walker models.

Just as in isotropic cosmological models, in static situations like the environment of a galaxy, the vector $\mathcal{U}^\alpha$ is pointed precisely in the time direction (this is already a result of computation and may not be the unique solution [944]). To compute the light ray deflection in linearized theory, one also needs the scalar field ϕ and the metric $g_{\alpha\beta}$, both to first order in the Newtonian gravitational potential Φ_N. The line element takes the form [254; 257]

$$\mathrm{d}\tilde{s}^2 = -(1 + 2\Phi)\mathrm{d}t^2 + (1 - 2\Phi)(\mathrm{d}x^2 + \mathrm{d}y^2 + \mathrm{d}z^2), \qquad (6.12)$$

where $\Phi = \Phi_N + \phi$; here we again omit the mentioned factor of nearly unity mentioned in connection with Eq. (6.11). (In linearized GR one gets the same form of metric, but with Φ_N in place of Φ.) Since the same potential Φ appears in both terms of this isotropic form of the line element, the light ray bending, which leans on both to equal degree, measures the same gravitational potential as do the non-relativistic dynamics, which are themselves sensitive only to the temporal part of the line element. Thus in TeVeS an extragalactic system lenses light, or radio waves, just as would GR, were the latter supplemented by DM in the amount and with the distribution necessary to reproduce the observed galactic dynamics [254].

In GR Φ_N is all there is, and its Laplacian, as determined from the lensing observations or from the dynamics, will give the total mass distribution. But DM is not visible directly, so its mass distribution (inferred by subtracting the observed baryonic component) cannot be checked; its status is better described with the terminology of physical plausibility. By contrast in TeVeS the observationally determined Φ is to be broken up into two parts: $\Phi = \Phi_N + \phi$. The Φ_N part is given by Poisson's equation, the ϕ part by a non-linear AQUAL-type equation (compare Eq. (6.10) with Eq. (6.3)); the observed baryonic mass density ρ is the source of both. Evidently the TeVeS scheme is falsifiable – by comparison of the calculated potential with that inferred from the lensing – to a larger extent than is the DM paradigm for which any discrepancy can be tucked away into the invisible component.

6.5.2 Strong lensing

Some features of TeVeS gravitational lensing by a pointlike mass M are worked out by Chiu *et al.* [540]. They note that the deflection angle in the deep MOND regime (impact parameter $b \gg b_0 \equiv (k/4\pi)(GM/a_0)^{1/2}$) approaches a constant, as might have been expected from naive arguments, but is less predictable in the intermediate MOND regime $b \sim b_0$. This may serve as a warning against doing lensing analysis with a mixture of GR and MOND concepts (e.g. ref. [1482]). In analogy with GR, Chiu *et al.* work out the lens equation for TeVeS (which controls the amplifications of the various images in strong lensing). For two images they find that the difference in amplifications is no longer unity, as in GR, and may depend on the masses. They also investigate the gravitational time delay in TeVeS which is important for interpreting differential time delays in doubly imaged variable quasars.

The baryon distribution in a galaxy is better represented by the Hernquist model than by a point source model. Zhao *et al.* [1966] use both to compare TeVeS predictions with a large sample of quasars doubly imaged by intervening galaxies. The lensing galaxy masses are estimated by comparing observations both with predicted image positions and with predicted amplification ratios; the two methods are found to give consistent results, themselves well correlated with the luminosities of the galaxies. The corresponding mass-to-light ratios are found to be in the normal range for stellar populations, with some exceptions. This result clashes with the claim by Ferreras *et al.* [850] that lensing by galaxies from the very same sample can only be explained in MOND by including a lot of DM apart from neutrinos. But the last authors use a mixture of MOND and GR instead of TeVeS.

What should be the probability distribution by angular separation of the two images in a sample of lensed quasars? This important question has proved troublesome for the DM paradigm. In TeVeS it has been investigated by Chen and Zhao [530] and lately by Chen [529]. Again modelling the shapes of the mostly elliptical galaxies with Hernquist profiles, and describing their space distribution with the Fontana stellar-mass function, these workers compare predictions of both TeVeS for a purely baryonic universe with cosmological constant and of GR with DM and baryons with the CLASS/JVAS quasar survey. After the preliminary work the later paper reports that TeVeS comes out on top. All the above is accomplished with spherical mass models of the galaxies; a step towards the modelling of asymmetric lenses within TeVeS has been taken by Shan *et al.* [1732].

6.5.3 Weak lensing

When it comes to weak lensing (distorted but unsplit images) by clusters of galaxies, a pure MOND account is less than satisfactory. The case of spherically symmetric clusters is fairly summarized by Takahashi and Chiba [1832]. In spherical symmetry the Poisson and AQUAL equations of weak field TeVeS for Φ_N and ϕ are easily solved, and the total potential $\Phi = \Phi_N + \phi$ is found to be related to Φ_N by a MOND relation with some complicated $\tilde{\mu}$ function. These authors use several such interpolation functions to predict the shear and convergence of the light lensed by a large number of quasi-spherical clusters in terms of the visible baryonic matter, but fail to get a fit with observations unless they add a neutrino component *à la* Sanders [1676; 1679]; the required neutrino mass is unrealistically large, so it seems that a DM component is needed to buttress the MOND effect. Similar conclusions are reached by Natarajan and Zhao [1499].

Non-spherical cluster systems are also problematic. In the massive colliding clusters systems MACSJ0025.4-1222 [433] and 1E0657-56 [563], the galaxy components have been rudely separated from the hot gas concentrations. Weak lensing mapping using background galaxies shows the gravitating mass to be preponderantly located in the regions containing the galaxies, rather than in the gas which accounts for the bulk of the visible baryonic mass [433; 562]. Collisionless DM would indeed be expected to move together with the galaxies and get separated from the collisionless gas; hence the widespread belief that these systems are decisive proof of the existence of DM. However, this view conflicts with the finding by Mahdavi *et al.* [1372] that in the merging clusters A520, the lensing centre is in the hot gas which is separate from the galaxy concentration. Angus *et al.* [95] considered it possible to explain the lensing seen in 1E0657-56 by TeVeS with a reasonable purely baryonic matter distribution, but later concluded [96] that a collisionless component is needed after all, with neutrinos just barely supplying a resolution. This conclusion is confirmed by a careful study by Feix *et al.* [827] who devised a Fourier solver for the AQUAL equation in TeVeS, and conclude that the source of gravity in 1E0657-56 must include an invisible component. This study also shows that non-linearity of the AQUAL equation cannot prevent the lensing from tracking the baryonic matter.

The weak lensing by cluster Cl0024+17 provides another relevant case study. Jee *et al.* [1171] find its deduced mass surface density to exhibit a ring which does not coincide either with the galaxy distribution or with the hot gas. Again this has been hailed as graphic proof of the existence of DM. But

Milgrom and Sanders [1445] argue that such a feature is actually expected in MOND, lying as it does at the transition between the Newtonian and the MOND regime. Famaey *et al.* [812] conclude that the lensing in Cl0024+17 can be modelled in MOND by including 2 eV neutrinos. A truly TeVeS model of Cl0024+17 is still outstanding.

6.5.4 TeVeS cosmology

Let us turn now to cosmology. Critics of MOND used to argue that the complex power spectrum of cosmological perturbations of the background radiation, which is said to be well fitted by the 'concordance' DM model of the Universe, proves that DM is essential to any rational picture of the cosmos. They tacitly assumed that the MOND paradigm could never measure up to the same test. With TeVeS on the scene, this could be checked for the first time.

In a massive work, Skordis [1768] has provided the full covariant formalism for evolution of cosmological perturbations in TeVeS, the analogue of that in GR. Using it Skordis *et al.* [1769] have shown that, without invoking DM, TeVeS can largely be made consistent with the observed spectrum of the spatial distribution of galaxies, and of the cosmic microwave radiation, if one allows for contributions to the energy density from massive neutrinos, and for a cosmological constant. The role of DM in GR cosmology is taken over by a feedback mechanism involving perturbations of the scalar field ϕ. Dodelson and Liguori [689] have independently calculated perturbation growth, and claimed that it is rather the vector field in TeVeS which is responsible for growth of large-scale structure without needing DM for this. Thus, although the elimination of galaxy-bound DM was the original motivation for MOND, and thus for TeVeS, the latter may potentially provide a way to eliminate cosmological (homogeneously distributed) DM.

Apart from the DM mystery, cosmology furnishes us with a 'dark energy' mystery. Dark energy is the agent responsible for the observed acceleration of the Hubble expansion in the context of GR cosmological models. An interesting question is whether modified gravity can supplant the dark energy. Here we touch only upon TeVeS-related work in this direction. Diaz-Rivera *et al.* [656] find an exact deSitter solution of TeVeS cosmology which can represent either early time inflation epochs or the late time acceleration era. Hao and Akhoury [1074] conclude that with a suitable choice of the TeVeS function $\mathcal{F}$, the scalar field can play the role of dark energy. According to Zhao [1968], the choice of $\mathcal{F}$ implicit in the work of Zhao and Famaey [1969] leads to cosmological models that evolve at early times like those of standard cold DM cosmology, and display late time acceleration with the correct present Hubble scale, all this without needing DM or dark energy. Likewise, in the related Einstein–Aether theory, Zlosnik *et al.* [1974] find that with suitable choice of their theory's function $\mathcal{F}$, the vector field can both drive early inflation and double for dark energy at late times.

General relativity and TeVeS differ also in the cosmological arena, in the relation they stipulate between the matter overdensity and the local depth of the gravitational potential. The advent of large lensing surveys may open a way to distinguish between these two theories, as well as between GR and other modified gravities, by exposing correlations between galaxy number density and weak lensing shear [1705; 1965]. The effect of the dark energy on the expansion can be separated out by comparing cosmological models with the same expansion history in two theories. And the ultimate confrontation between GR and TeVeS cosmology may be accomplished by cross-correlating galaxy number density with CMB antenna temperature [1706].

6.5.5 Epilogue

As we have seen, both gravitational lensing and cosmology provide innumerable tests of TeVeS. The increasing sophistication of the measurements in both fields should eventually clearly distinguish between the various modified gravity theories, and between each of these and GR. It will also provide an improved basis for the modifications of TeVeS that already seem to be necessitated by the extant data on weak gravitational lensing. Space limitations have here prevented coverage of possibilities for testing TeVeS in the Solar System, another rich field of inquiry with many opportunities in our space age.

Part II
Candidates

7

DM production mechanisms

Graciela Gelmini and Paolo Gondolo

7.1 Dark matter particles: relics from the pre-BBN era

A general class of candidates for non-baryonic cold dark matter is weakly interacting massive particles (WIMPs). The interest in WIMPs as dark matter candidates stems from the fact that WIMPs in chemical equilibrium in the early Universe naturally have the right abundance to be cold dark matter. Moreover, the same interactions that give the right WIMP density make the detection of WIMPs possible. The latter aspect is important as it provides a means to test the WIMP hypothesis.

The argument showing that WIMPs are good dark matter candidates is old [657; 1144; 1317; 1692; 1911]. The density per comoving volume of non-relativistic particles in equilibrium in the early Universe decreases exponentially with decreasing temperature, owing to the Boltzmann factor, until the reactions which change the particle number become ineffective. At this point, when the annihilation rate becomes smaller than the Hubble expansion rate, the WIMP number per comoving volume becomes constant. This moment of chemical decoupling or freeze-out happens later, i.e. at smaller WIMP densities, as the WIMP annihilation cross-section σ_{ann} is larger. If there is no subsequent change of entropy in matter plus radiation, the present relic density of WIMPs is approximately

$$\Omega h^2 \approx \frac{3 \times 10^{-27}\,\mathrm{cm}^3\,\mathrm{s}^{-1}}{\langle \sigma_{\mathrm{ann}} v \rangle}. \tag{7.1}$$

For weak cross-sections this gives the right order of magnitude of the DM density (and a temperature $T_{\mathrm{fo}} \simeq m/20$ at freeze-out for a WIMP of mass m). This is a ballpark argument. A more precise derivation will be presented in Section 7.2.

Particle Dark Matter: Observations, Models and Searches, ed. Gianfranco Bertone. Published by Cambridge University Press. © Cambridge University Press 2010.

It is important to realize that the determination of the WIMP relic density depends on the history of the Universe before Big Bang nucleosynthesis (BBN), an epoch from which we have no data. BBN (200 s after the Big Bang, $T \simeq 0.8$ MeV) is the earliest episode from which we have a trace, namely the abundance of light elements D, ^{4}He and ^{7}Li. The next observable in time is the cosmic microwave background radiation (produced 3.8×10^4 yr after the Big Bang, at $T \simeq$ eV) and the next one is the large-scale structure of the Universe. WIMPs have their number fixed at $T_{\rm fo} \simeq m/20$, so WIMPs with $m \gtrsim 100$ MeV would freeze out before BBN and would thus be the earliest remnants. If discovered, they would for the first time give information on the pre-BBN phase of the Universe.

As things stand now, to compute the WIMP relic density we must make assumptions about the pre-BBN epoch. The standard computation of the relic density relies on the assumptions that the entropy of matter and radiation were conserved, that WIMPs were produced thermally, i.e. via interactions with the particles in the plasma, that they decoupled while the expansion of the Universe was dominated by radiation and that they were in kinetic and chemical equilibrium before they decoupled. These are just assumptions, which do not hold in all cosmological models. In particular, in order for BBN and all the subsequent history of the Universe to proceed as usual, it is enough that the earliest and highest temperature during the radiation dominated period, the so-called reheating temperature $T_{\rm RH}$, is larger than 4 MeV [1067]. At temperatures higher than 4 MeV, when the WIMP freeze-out is expected to occur, the content and expansion history of the Universe may differ from the standard assumptions. In non-standard cosmological models, the WIMP relic abundance may be higher or lower than the standard abundance. The density may be decreased by reducing the rate of thermal production (through a low $T_{\rm RH} < T_{\rm fo}$) or by producing radiation after freeze-out (entropy dilution). The density may also be increased by creating WIMPs from decays of particles or extended objects (non-thermal production) or by increasing the expansion rate of the Universe at the time of freeze-out.

Non-thermal production mechanisms may also be at work within standard cosmological scenarios. For example, WIMPs may be produced in the out-of-equilibrium decay of other particles whose density may be fixed by thermal processes. A particular type of heavy WIMP, WIMPZILLAs, could be formed during the reheating phase at the end of an inflationary period through gravitational interactions. Another production mechanism involves quantum-mechanical oscillations: for example a sterile neutrino may be produced in the early Universe by the oscillation of active (interacting) neutrinos into sterile neutrinos (for the latter, see Chapter 12).

In the rest of this chapter we review the standard production mechanism and some of the non-standard scenarios.

7.2 Thermal production in the standard cosmology

7.2.1 The standard production mechanism

In the standard scenario, it is assumed that in the early Universe WIMPs were produced in collisions between particles of the thermal plasma during the radiation-dominated era. Important reactions were the production and annihilation of WIMP pairs in particle–antiparticle collisions, such as

$$\chi\bar{\chi} \leftrightarrow e^+e^-, \mu^+\mu^-, q\bar{q}, W^+W^-, ZZ, HH, \ldots \qquad (7.2)$$

At temperatures much higher than the WIMP mass, $T \gg m_\chi$, the colliding particle–antiparticle pairs in the plasma had enough energy to create WIMP pairs efficiently. Also, the inverse reactions that convert WIMPs into Standard Model particles (annihilation) were initially in equilibrium with the WIMP-producing processes. Their common rate was given by

$$\Gamma_{\text{ann}} = \langle \sigma_{\text{ann}} v \rangle n_{\text{eq}}, \qquad (7.3)$$

where σ_{ann} is the WIMP annihilation cross-section, v is the relative velocity of the annihilating WIMPs, n_{eq} is the WIMP number density in chemical equilibrium and the angle brackets denote an average over the WIMP thermal distribution.

As the Universe expanded, the temperature of the plasma became smaller than the WIMP mass. While annihilation and production reactions remained in equilibrium, the number of WIMPs produced decreased exponentially as $e^{-m_\chi/T}$ (the Boltzmann factor), since only particle–antiparticle collisions with kinetic energy in the tail of the Boltzmann distribution had enough energy to produce WIMP pairs. At the same time, the expansion of the Universe decreased the number density of particles n, and with it the production and annihilation rates, which are proportional to n. When the WIMP annihilation rate Γ_{ann} became smaller than the expansion rate of the Universe H, or equivalently the mean free path for WIMP-producing collisions became longer than the Hubble radius, production of WIMPs ceased (chemical decoupling). After this, the number of WIMPs in a comoving volume remained approximately constant (or in other words, their number density decreased inversely with volume).

In many of the current theories, WIMPs are their own antiparticles. For WIMPs of this kind (e.g. neutralinos and Majorana neutrinos), the WIMP density is necessarily equal to the antiWIMP density. In the following we restrict our discussion to this case. We refer the reader interested in cosmological WIMP–antiWIMP asymmetries, as might apply for example to a Dirac neutrino, to ref. [1019].

The current density of WIMPs can be computed by means of the rate equation for the WIMP number density n and the law of entropy conservation:

$$\frac{dn}{dt} = -3Hn - \langle \sigma_{\mathrm{ann}} v \rangle \left(n^2 - n_{\mathrm{eq}}^2 \right), \tag{7.4}$$

$$\frac{ds}{dt} = -3Hs. \tag{7.5}$$

Here t is time, s is the entropy density, H is the Hubble parameter and as before n_{eq} and $\langle \sigma_{\mathrm{ann}} v \rangle$ are the WIMP equilibrium number density and the thermally averaged total annihilation cross-section. The first and second terms on the right-hand side of Eq. (7.4) take into account the expansion of the Universe and the change in number density due to annihilations and inverse annihilations, respectively.

It is customary (see e.g. [735; 982; 1261; 1793]) to combine Eqs. (7.4) and (7.5) into a single one for $Y = n/s$, and to use $x = m/T$, with T the photon temperature, as the independent variable instead of time. This gives:

$$\frac{dY}{dx} = \frac{1}{3H} \frac{ds}{dx} \langle \sigma v \rangle \left(Y^2 - Y_{\mathrm{eq}}^2 \right). \tag{7.6}$$

Here and in the rest of the chapter we will simply write $\langle \sigma v \rangle$ for $\langle \sigma_{\mathrm{ann}} v \rangle$ when no ambiguity can arise.

According to the Friedman equation, the Hubble parameter is determined by the mass-energy density ρ as

$$H^2 = \frac{8\pi}{3M_{\mathrm{P}}^2} \rho, \tag{7.7}$$

where $M_{\mathrm{P}} = 1.22 \times 10^{19}$ GeV is the Planck mass. The energy and entropy densities are related to the photon temperature by the equations

$$\rho = \frac{\pi^2}{30} g_{\mathrm{eff}}(T) T^4, \quad s = \frac{2\pi^2}{45} h_{\mathrm{eff}}(T) T^3, \tag{7.8}$$

where $g_{\mathrm{eff}}(T)$ and $h_{\mathrm{eff}}(T)$ are effective degrees of freedom for the energy density and entropy density respectively. Recent computations of $g_{\mathrm{eff}}(T)$ and $h_{\mathrm{eff}}(T)$ that include quantum chromodynamic (QCD) effects can be found in ref. [1099]. If the degrees of freedom parameter $g_*^{1/2}$ is defined as

$$g_*^{1/2} = \frac{h_{\mathrm{eff}}}{g_{\mathrm{eff}}^{1/2}} \left(1 + \frac{1}{3} \frac{T}{h_{\mathrm{eff}}} \frac{dh_{\mathrm{eff}}}{dT} \right), \tag{7.9}$$

then Eq. (7.6) can be written in the following way,

$$\frac{dY}{dx} = -\left(\frac{45}{\pi M_P^2}\right)^{-1/2} \frac{g_*^{1/2} m}{x^2} \langle \sigma v \rangle \left(Y^2 - Y_{eq}^2\right). \tag{7.10}$$

This single equation is then numerically solved with the initial condition $Y = Y_{eq}$ at $x \simeq 1$ to obtain the present WIMP abundance Y_0. From it, the WIMP relic density can be computed as

$$\Omega_\chi h^2 = \frac{\rho_\chi^0 h^2}{\rho_c^0} = \frac{m_\chi s_0 Y_0 h^2}{\rho_c^0} = 2.755 \times 10^8 \, Y_0 m_\chi / \text{GeV}, \tag{7.11}$$

where ρ_c^0 and s_0 are the present critical density and entropy density respectively. In obtaining the numerical value in Eq. (7.11) we used $T_0 = 2.726\,\text{K}$ for the present background radiation temperature and $h_{eff}(T_0) = 3.91$ corresponding to photons and three species of neutrinos.

The numerical solution of Eq. (7.10) (see Fig. 7.1 for an illustration) shows that at high temperatures Y closely tracks its equilibrium value Y_{eq}. In fact, the interaction rate of WIMPs is strong enough to keep them in thermal and chemical equilibrium with the plasma. But as the temperature decreases, Y_{eq} becomes exponentially suppressed and Y is no longer able to track its equilibrium value. At the freeze-out temperature (T_{fo}), when the WIMP annihilation rate becomes of the order of the Hubble expansion rate, WIMP production becomes negligible and the WIMP abundance per comoving volume reaches its final value. In the standard cosmological scenario, the WIMP

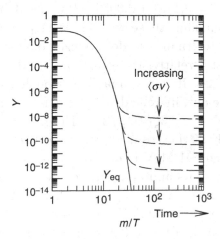

Fig. 7.1. Typical evolution of the WIMP number density in the early Universe during the epoch of WIMP chemical decoupling (freeze-out).

freeze-out temperature is about $T_{\text{fo}} \simeq m_\chi/20$, which corresponds to a typical WIMP speed at freeze-out of $v_{\text{fo}} = (3T_{\text{fo}}/2m_\chi)^{1/2} \simeq 0.27c$.

An important property that Fig. 7.1 illustrates is that smaller annihilation cross-sections lead to larger relic densities ('The weakest wins'). This can be understood from the fact that WIMPs with stronger interactions remain in chemical equilibrium for a longer time, and hence decouple when the Universe is colder, wherefore their density is further suppressed by a smaller Boltzmann factor. This leads to the inverse relation between $\Omega_\chi h^2$ and $\langle\sigma_{\text{ann}}v\rangle$ in Eq. (7.1).

From this discussion follows that the freeze-out temperature plays a prominent role in determining the WIMP relic density. In general, however, the freeze-out temperature depends not only on the mass and interactions of the WIMP but also, through the Hubble parameter, on the content of the Universe. Some examples of how modifications of the Hubble parameter affect the WIMP density are discussed in Section 7.4 below.

7.2.2 Annihilations and co-annihilations

A pedagogical example of the dependence of the relic density on the WIMP mass is provided by a thermally produced fourth-generation Dirac neutrino ν with Standard Model interactions and no lepton asymmetry, although it is excluded as a cold dark matter candidate by a combination of LEP and direct detection limits [39; 91; 1021]. Figure 7.2 summarizes its relic density $\Omega_\nu h^2$ as a function of its mass m_ν. The narrow band between the horizontal lines is the current cosmological measurement of the cold dark matter density $\Omega_{\text{cdm}}h^2 = 0.1131 \pm 0.0034$ [1100]. Neutrinos with $\Omega_\chi > \Omega_{\text{cdm}}$ are said to be overabundant, those with $\Omega_\chi < \Omega_{\text{cdm}}$ are called underabundant. A neutrino lighter than ~ 1 MeV freezes out while relativistic. If it is so light as to be still relativistic today ($m_\nu \lesssim 0.1$ meV), its relic density is $\rho_\nu = 7\pi^2 T_\nu^4/120$. If it was massive enough to have become non-relativistic after freeze-out, its relic density is determined by its equilibrium number density as $\rho_\nu = m_\nu 3\zeta(3)T_\nu^3/2\pi^2$. Here $T_\nu = (3/11)^{1/3}T_\gamma$, where $T_\gamma = 2.725 \pm 0.002$ K is the cosmic microwave background temperature. A neutrino heavier than ~ 1 MeV freezes out while non-relativistic. Its relic density is determined by its annihilation cross-section, as in Eq. (7.1). The shape of the relic density curve above ~ 1 MeV in Figure 7.2 is a reflection of the behaviour of the annihilation cross-section into lepton–antilepton and quark–antiquark pairs $f\bar{f}$: the Z-boson resonance at $m_\nu \simeq m_Z/2$ gives rise to the characteristic V shape in the Ωh^2 curve. Above $m_\nu \approx 100$ GeV, new annihilation channels into Z- or W-boson pairs open up (thresholds at

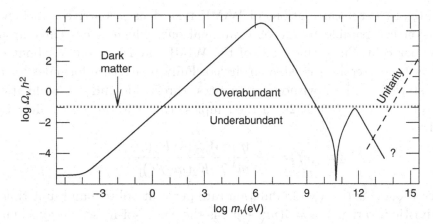

Fig. 7.2. Relic density $\Omega_\nu h^2$ of a thermal Dirac neutrino with Standard Model interactions as a function of the neutrino mass m_ν (solid line). The very close horizontal dashed lines enclose the current 1σ band for the cold dark matter density [1100].

$m_n u = m_W$ and $m_\nu = m_Z$, respectively). When the new channel annihilation cross-sections dominate, the relic density decreases. Soon, however, the perturbative expansion of the cross-section in powers of the (Yukawa) coupling constant becomes untrustworthy (the question mark in Figure 7.2). A general unitarity argument [1018] limits the relic density to the dashed curve on the right,

$$\Omega_\nu h^2 \simeq 3.4 \times 10^{-6} \sqrt{\frac{m_\nu}{T_{\text{fo}}}} \left(\frac{m_\nu}{1\,\text{TeV}}\right)^2. \tag{7.12}$$

The relic density of other WIMP candidates exhibits features similar to that of the Dirac neutrino just discussed. In general, because of the presence of resonances and thresholds in the annihilation cross-section, one should not rely on a Taylor expansion of $\sigma_{\text{ann}}v$ in powers of v, because it would lead to unphysical negative cross-sections. Let us remark that resonant and threshold annihilation, including the co-annihilation thresholds discussed below, are ubiquitous for neutralino dark matter (see Chapter 8). In the non-relativistic limit $v \to 0$, the product $\sigma_{\text{ann}}v$ tends to a constant, because of the exoenergetic character of the annihilation process that makes the annihilation cross-section σ_{ann} diverge as $1/v$ as $v \to 0$. One can safely Taylor-expand $\sigma_{\text{ann}}v$ in powers of v^2 if $\sigma_{\text{ann}}v$ varies slowly with v, $\sigma_{\text{ann}}v = a + bv^2 + \cdots$. Then the thermal average is $\langle\sigma_{\text{ann}}v\rangle = a + b\frac{3T}{2m} + \cdots$. Close to resonances and thresholds, however, σ_{ann} varies rapidly with v and more sophisticated procedures (described in the following) should be used [982; 1020].

State-of-the-art calculations of WIMP relic densities strive to achieve a precision comparable to the observational one, which is currently around a few per cent. Since the speed of the WIMPs at freeze-out is about $c/3$, relativistic corrections must be included. Fully relativistic formulas for any cross-section, with or without resonances, thresholds and co-annihilations, were obtained in [735; 982]. In the simplest case without co-annihilations, one has

$$\langle \sigma_{\mathrm{ann}} v \rangle = \frac{\int_0^\infty \mathrm{d}p \, p^2 \, W_{\chi\chi}(s) \, K_1(\sqrt{s}/T)}{m_\chi^4 \, T \, [K_2(m_\chi/T)]^2}, \qquad (7.13)$$

where $W_{\chi\chi}(s)$ is the $\chi\chi$ annihilation rate per unit volume and unit time (a relativistic invariant), $s = 4(m_\chi^2 + p^2)$ is the centre-of-mass energy squared, and $K_1(x)$, $K_2(x)$ are modified Bessel functions. The Lorentz-invariant annihilation rate $W_{ij}(p)$ for the collision of two particles of 4-momenta p_i and p_j is related to the annihilation cross-section σ_{ij} through

$$W_{ij}(s) = \sigma_{ij} F_{ij}, \qquad (7.14)$$

where

$$F_{ij} = 4\sqrt{(p_i \cdot p_j)^2 - m_i^2 m_j^2} \qquad (7.15)$$

is the Lorentz-invariant flux factor. Co-annihilations are an essential ingredient in the calculation of the WIMP relic density. They are processes that deplete the number of WIMPs through a chain of reactions, and occur when another particle is close in mass to the dark matter WIMP (mass difference $\Delta m \sim$ temperature T). In this case, scattering of the WIMP off a particle in the thermal 'soup' can convert the WIMP into the slightly heavier particle, since the energy barrier that would otherwise prevent it (i.e. the mass difference) is easily overcome. The particle participating in the co-annihilation may then decay and/or react with other particles and eventually effect the disappearance of WIMPs. We give two examples in the context of the minimal supersymmetric Standard Model. Neutralino co-annihilation with charginos $\tilde{\chi}^\pm$ may proceed, for instance, through

$$\tilde{\chi}_1^0 e^- \rightarrow \tilde{\chi}_2^- \nu_e, \qquad \tilde{\chi}_2^- \rightarrow \tilde{\chi}_2^0 \mathrm{d}\bar{u}, \qquad \tilde{\chi}_2^0 \tilde{\chi}_1^0 \rightarrow W^+ W^-. \qquad (7.16)$$

Neutralino co-annihilation with tau sleptons $\tilde{\tau}$ may instead involve the processes

$$\tilde{\chi}_1^0 \tau \rightarrow \tilde{\tau} \gamma, \qquad \tilde{\tau} \tilde{\chi}_1^+ \rightarrow \tau W^+. \qquad (7.17)$$

Co-annihilations were first included in the study of near-degenerate heavy neutrinos in [355] and were brought to general attention in [1020]. The relativistic treatment was formulated in [735]. Under the conditions described below, which are reasonable during WIMP freeze-out, one replaces $\langle \sigma_{\mathrm{ann}} v \rangle$ in Eq. (7.4) with

$$\langle \sigma_{\mathrm{eff}} v \rangle = \frac{\int_0^\infty \mathrm{d}p_{\mathrm{eff}}\, p_{\mathrm{eff}}^2\, W_{\mathrm{eff}}(s)\, K_1(\sqrt{s}/T)}{m_\chi^4\, T \left[\sum_{i=1}^N \frac{g_i}{g_\chi} \frac{m_i^2}{m_\chi^2} K_2(m_i/T) \right]^2}, \tag{7.18}$$

where $s = 4p_{\mathrm{eff}} + 4m_\chi^2$, g_i is the number of internal degrees of freedom (statistical weight factor) for the ith co-annihilating particle and

$$W_{\mathrm{eff}}(s) = \sum_{ij} \frac{F_{ij}}{F_{\chi\chi}} \frac{g_i g_j}{g_\chi^2} W_{ij}(s). \tag{7.19}$$

The sums extend over all the N co-annihilating particles, including the χ, and $m_1 = m_\chi$, $g_1 = g_\chi$. The assumptions underlying Eq. (7.18) are: (1) all co-annihilating particles decay into the lightest one, which is stable, and their decay rate is much faster than the expansion rate of the Universe – so the final WIMP abundance is simply described by the sum of the density of all co-annihilating particles; (2) the scattering cross-sections of co-annihilating particles off the thermal background are of the same order of magnitude as their annihilation cross-sections – since the relativistic background particle density is much larger than each of the non-relativistic co-annihilating particle densities, the scattering rate is much faster and the momentum distributions of the co-annihilating particles remain in thermal equilibrium; (3) all co-annihilating particles are semi-relativistic, so the Fermi–Dirac and Bose–Einstein thermal distributions can be replaced by a Maxwell–Boltzmann distribution $f_i = \mathrm{e}^{-E_i/T}$.

An important aspect of the effective annihilation rate in Eq. (7.18) is that co-annihilations appear as thresholds at a value of $\sqrt{s}$ equal to the sum of the masses of the co-annihilating particles. As an example of this, Figure 7.3, taken from ref. [735], shows that co-annihilation thresholds and regular final state thresholds appear on the same footing in the invariant annihilation rate W_{eff}.

Computations of WIMP relic densities can become quite involved, especially in the presence of co-annihilations. There exists publicly available software [262; 980] that can handle these calculations for generic WIMPs (see Chapter 16).

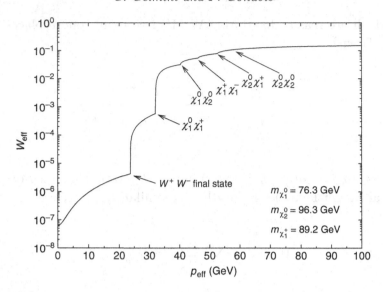

Fig. 7.3. The effective invariant annihilation rate $W_{\rm eff}$ as a function of $p_{\rm eff}$ for a particular supersymmetric model examined in [735]. The final state threshold for annihilation into W^+W^- and the co-annihilation thresholds appearing in Eq. (7.19) are indicated. The $\chi_2^0\chi_2^0$ co-annihilation threshold is too small to be seen. Adapted from ref. [735].

7.3 Non-thermal production in the standard cosmology

7.3.1 Gravitational mechanisms

WIMPZILLAs illustrate a fascinating idea for generating matter in the expanding Universe: the gravitational creation of matter in an accelerated expansion [548; 549; 550; 551; 1294; 1295]. This mechanism is analogous to the production of Hawking radiation around a black hole, and of Unruh radiation in an accelerated reference frame.

WIMPZILLAs are very massive relics from the Big Bang: they can be the dark matter in the Universe if their mass is $\sim 10^{13}$ GeV. They might be produced at the end of inflation through a variety of possible mechanisms: gravitationally, during preheating, during reheating, in bubble collisions. It is possible that their relic abundance does not depend on their interaction strength but only on their mass, giving great freedom in their phenomenology. To be the dark matter today, they are assumed to be stable or to have a lifetime of the order of the age of the Universe. In the latter case, their decay products may give rise to the highest energy cosmic rays.

Gravitational production of particles is an important phenomenon that is worth describing here. Consider a scalar field (particle) X of mass M_X in the expanding Universe. Let η be the conformal time and $a(\eta)$ the time dependence of the expansion scale factor. Assume for simplicity that the Universe is flat. The scalar field X can be expanded in spatial Fourier modes as

$$X(\mathbf{x}, \eta) = \int \frac{\mathrm{d}^3 k}{(2\pi)^{3/2} a(\eta)} \left[a_k h_k(\eta) \mathrm{e}^{i\mathbf{k}\cdot\mathbf{x}} + a_k^\dagger h_k^*(\eta) \mathrm{e}^{-i\mathbf{k}\cdot\mathbf{x}} \right]. \qquad (7.20)$$

Here a_k and $a_k^\dagger$ are creation and annihilation operators, and $h_k(\eta)$ are mode functions that satisfy (i) the normalization condition $h_k h_k'^* - h_k' h_k^* = i$ (a prime indicates a derivative with respect to conformal time), and (ii) the mode equation

$$h_k''(\eta) + \omega_k^2(\eta)\, h_k(\eta) = 0, \qquad (7.21)$$

where

$$\omega_k^2(\eta) = k^2 + M_X^2 a^2 + (6\xi - 1) \frac{a''}{a}. \qquad (7.22)$$

The parameter ξ is $\xi = 0$ for a minimally coupled field and $\xi = \frac{1}{6}$ for a conformally coupled field. The mode equation, Eq. (7.21), is formally the same as the equation of motion of a harmonic oscillator with time-varying frequency $\omega_k(\eta)$. For a given complete set of positive-frequency solutions $h_k(\eta)$, the vacuum $|0_h\rangle$ of the field X, i.e. the state with no X particles, is defined as the state that satisfies $a_k|0_h\rangle = 0$ for all k. Since Eq. (7.21) is a second-order equation and the frequency depends on time, the normalization condition is in general not sufficient to specify the positive-frequency modes uniquely, contrary to the case of constant frequency ω_0 for which $h_k^0(\eta) = \mathrm{e}^{-i\omega_0\eta}/(2\omega_0)^{1/2}$. Different boundary conditions for the solutions $h_k(\eta)$ define in general different creation and annihilation operators a_k and $a_k^\dagger$, and thus in general different vacua.[1] For example, solutions which satisfy the condition of having only positive frequencies in the distant past,

$$h(\eta) \sim \mathrm{e}^{-i\omega_k^-\eta} \quad \text{for } \eta \to -\infty, \qquad (7.23)$$

contain both positive and negative frequencies in the distant future,

$$h(\eta) \sim \alpha_k \mathrm{e}^{-i\omega_k^+\eta} + \beta_k \mathrm{e}^{+i\omega_k^+\eta} \quad \text{for } \eta \to +\infty. \qquad (7.24)$$

Here $\omega_k^\pm = \lim_{\eta\to\pm\infty} \omega_k(\eta)$. As a consequence, an initial vacuum state is no longer a vacuum state at later times, i.e. particles are created. The

---

[1] The precise definition of a vacuum in a curved space-time is still subject to some ambiguities. We refer the interested reader to [363; 905; 906; 1913] and to the discussion in [552] and references therein.

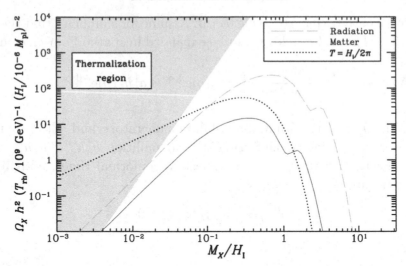

Fig. 7.4. Relic density of gravitationally produced WIMPZILLAs as a function of their mass M_X. H_I is the Hubble parameter at the end of inflation, T_{rh} is the reheating temperature and $M_{pl} \approx 3 \times 10^{19}$ GeV is the Planck mass. The dashed and solid lines correspond to inflationary models that smoothly end in a radiation- or matter-dominated epoch, respectively. The dotted line is a thermal distribution at the Gibbons–Hawking temperature $T = H_I/2\pi$. Outside the 'thermalization region' WIMPZILLAs cannot reach thermal equilibrium. Adapted from [551].

number density of particles is given in terms of the Bogolubov coefficient β_k in Eq. (7.24) by

$$n_X = \frac{1}{(2\pi a)^3} \int d^3k |\beta_k|^2. \qquad (7.25)$$

These ideas have been applied to gravitational particle creation at the end of inflation by [551] and [1294]. Particles with masses M_X of the order of the Hubble parameter at the end of inflation, $H_I \approx 10^{-6} M_{Pl} \approx 10^{13}$ GeV, may have been created with a density which today may be comparable to the critical density. Figure 7.4 shows the relic density $\Omega_X h^2$ of these WIMPZILLAs as a function of their mass M_X in units of H_I. Curves are shown for inflation models that have a smooth transition to a radiation-dominated epoch (dashed line) and a matter-dominated epoch (solid line). The third curve (dotted line) shows the thermal particle density at temperature $T = H_I/2\pi$. Also shown in the figure is the region where WIMPZILLAs are thermal relics. It is clear that it is possible for dark matter to be in the form of heavy WIMPZILLAs generated gravitationally at the end of inflation.

7.3.2 Decays

Dark matter may be produced in the decay of other particles. If the DM particles are non-interacting when the decay occurs, then they inherit (except for some entropy dilution factor) the density of the parent particle P

$$\Omega_{\rm DM} \, h^2 \simeq \frac{m_{\rm DM}}{m_{\rm P}} \Omega_{\rm P} \, h^2 \, . \tag{7.26}$$

This is the case for superWIMPs (see Chapter 10), extremely weakly interacting particles produced in the late decays of WIMPs (e.g. axinos or gravitinos from the decay of neutralinos or sleptons) which practically only interact gravitationally and cannot be directly detected. In some models the superWIMP may produce WIMPs through its decay. This is the case, for example, for gravitinos decaying to produce Winos (which otherwise would have a very low thermal relic density) with the right DM abundance [889; 936].

7.4 Thermal and non-thermal production in non-standard cosmologies

The relic density (and also the velocity distribution before structure formation) of WIMPs, and other DM candidates such as heavy sterile neutrinos and axions, depends on the characteristics of the Universe (expansion rate, composition, etc.) immediately before BBN, i.e. at temperatures $T \gtrsim 4 {\rm MeV}$. The standard computation of relic densities relies on the assumption that radiation domination began before the main epoch of production of the relics and that the entropy of matter and radiation has been conserved during and after this epoch. Any modification of these assumptions would lead to different relic density values. Any extra contribution to the energy density of the Universe would increase the Hubble expansion rate H and lead to larger relic densities (since the decreasing interaction rate Γ becomes smaller than H earlier, when densities are larger). This can happen in the Brans–Dicke–Jordan [1203] cosmological model, models with anisotropic expansion [222; 1203; 1612], scalar–tensor [494; 495; 611; 1685] or kination [1612; 1669] models and other models [107; 202; 203]. In some scalar–tensor models H may be decreased, leading to smaller relic densities [495]. These models alter the thermal evolution of the Universe without an extra entropy production.

Not only the value of H but also the dependence of the temperature T on the scale factor of the Universe could be different, if entropy in matter and radiation is produced. This is the case if a scalar field ϕ oscillating around its true minimum while decaying is the dominant component of the Universe just before BBN. This field may be an inflaton or another late decaying field, such as a modulus in supersymmetric models. Models of this type include some with moduli fields, either the Polonyi field [1215; 1481] or others [1480], or an Affleck–Dine field and Q-ball decay [893; 894], and thermal inflation [1360]. Moduli fields correspond to flat directions in the supersymmetric potential, which are lifted by the same mechanisms that give mass to the supersymmetric particles of the order of a few to tens of TeV, and they usually have interactions of gravitational strength. The decays of the ϕ field finally reheat the Universe to a low reheating temperature T_{RH}, which could be not much larger than 5 MeV. In these low temperature reheating (LTR) models there can be direct production of DM relics in the decay of ϕ which increases the relic density, and there is entropy generation, through the decay of ϕ into radiation, which suppresses the relic abundance.

Thus, in non-standard cosmological scenarios, the relic density of WIMPs Ω_χ may be larger or smaller than that in standard cosmologies Ω_{std}. The density may be decreased by reducing the rate of thermal production (through a low $T_{\mathrm{RH}} < T_{\mathrm{fo}}$), by reducing the expansion rate of the Universe at freeze-out or by producing radiation after freeze-out (entropy dilution). The density may be increased by creating WIMPs from particle (or extended objects) decay (non-thermal production) or by increasing the expansion rate of the Universe at freeze-out. Usually these scenarios contain additional parameters that can be adjusted to modify the WIMP relic density. However, these are due to physics that does not manifest itself in accelerator or detection experiments.

Let us comment that not only the relic density of WIMPs but also their characteristic speed before structure formation in the Universe can differ in standard and non-standard pre-BBN cosmological models. If kinetic decoupling (the moment when the exchange of momentum between WIMPs and radiation ceases to be effective) happens during the reheating phase of LTR models, WIMPs can have much smaller characteristic speeds, i.e. be much 'colder' [931], with free-streaming lengths several orders of magnitude smaller than in the standard scenario. Much smaller DM structures could thus be formed, a fraction of which may persist as mini-haloes within our Galaxy and be detected in indirect DM searches. The signature would be a much larger boost factor of the annihilation signal than expected in standard cosmologies for a particular WIMP candidate. WIMPs may instead be much 'hotter' than in standard cosmologies; they may even be warm DM instead of cold, which would leave an imprint on the large-scale structure spectrum [929; 1102; 1339].

7.4.1 Low temperature reheating (LTR) models

Let us consider a late decaying scalar field ϕ of mass m_ϕ and decay width Γ_ϕ which dominates the energy density of the Universe while oscillating about the minimum of its potential and decays reheating the Universe to a low reheating temperature T_{RH}, with 5 MeV $\lesssim T_{RH} \lesssim T_{fo}$, so BBN is not affected. The usual choice for the parameter T_{RH} is the temperature the Universe would attain under the assumption that the ϕ decay and subsequent thermalization are instantaneous,

$$\Gamma_\phi = H_{decay} = \sqrt{\left(\frac{8\pi}{3}\right)\rho_R} = \sqrt{\frac{8}{90}\pi^3 g_\star}\,\frac{T_{RH}^2}{M_P}. \tag{7.27}$$

Here, Γ_ϕ is the decay width of the ϕ field, $\Gamma_\phi \simeq m_\phi^3/\Lambda_{eff}^2$. If ϕ has non-suppressed gravitational couplings, as is usually the case for moduli fields, the effective energy scale $\Lambda_{eff} \simeq M_P$ (but Λ_{eff} could be smaller [1225]). Thus, with $g_\star \simeq 10$,

$$T_{RH} \simeq 10 \text{ MeV} \left(\frac{m_\phi}{100 \text{ TeV}}\right)^{3/2} \left(\frac{M_P}{\Lambda_{eff}}\right). \tag{7.28}$$

Numerical calculations in which the approximation of instantaneous decay is not made show that the parameter T_{RH} provides a good estimate of the first temperature of the radiation-dominated epoch (see Figure 7.5).

Both thermal and non-thermal production mechanisms in LTR models have been discussed [51; 52; 550; 710; 711; 792; 877; 926; 930; 957; 1203; 1215; 1225; 1407; 1480; 1481; 1543], mostly in supersymmetric models where the WIMP is the neutralino. The decay of ϕ into radiation increases the entropy, diluting the WIMP number density. The decay of ϕ into WIMPs increases the WIMP number density. In supersymmetric models ϕ decays into supersymmetric particles, which eventually decay into the lightest such particles (the LSP, typically a neutralino). Call b the net number of WIMPs produced on average per ϕ decay, which is a highly model-dependent parameter [51; 52; 930; 1480].

A combination of T_{RH} and the ratio b/m_ϕ can bring the relic WIMP density to the desired value Ω_{cdm} [930]. The equations which describe the evolution of the Universe depend only on the combination b/m_ϕ and not on b and m_ϕ separately. They are

$$\dot\rho = -3H(\rho + p) + \Gamma_\phi\rho_\phi \tag{7.29}$$

$$\dot{n}_\chi = -3Hn_\chi - \langle \sigma v \rangle \left(n_\chi^2 - n_{\chi,\mathrm{eq}}^2 \right) + \frac{b}{m_\phi} \Gamma_\phi \rho_\phi \tag{7.30}$$

$$\dot{\rho}_\phi = -3H\rho_\phi - \Gamma_\phi \rho_\phi \tag{7.31}$$

$$H^2 = \frac{8\pi}{3M_P^2}(\rho + \rho_\phi). \tag{7.32}$$

In Eqs. (7.29–7.32), a dot indicates a time derivative, ρ_ϕ is the energy density in the ϕ field, which is assumed to behave like non-relativistic matter; ρ and p are the total energy density and pressure of matter and radiation at temperature T, which are assumed to be in kinetic but not necessarily chemical equilibrium; n_χ is the number density of WIMPs, and $n_{\chi,\mathrm{eq}}$ is its value in chemical equilibrium; finally, $H = \dot{a}/a$ is the Hubble parameter, with a the scale factor. The first principle of thermodynamics in the form $\mathrm{d}(\rho a^3) + \mathrm{d}(\rho_\phi a^3) + p\mathrm{d}a^3 = T\mathrm{d}(sa^3)$ can be used to rewrite Eq. (7.29) as

$$\dot{s} = -3Hs + \frac{\Gamma_\phi \rho_\phi}{T}, \tag{7.33}$$

where $s = (\rho + p - m_\chi n_\chi)/T$ is the entropy density of the matter and radiation. For $\rho_\phi \to 0$ these equations reduce to the standard scenario.

During the ϕ-oscillation-dominated epoch, $H \propto T^4$ [1407]. This can be seen using Eq. (7.29) while the matter content is negligible. In Eq. (7.29) with $p = \rho/3$, substitute $\rho \simeq T^4$ and $\rho_\phi \simeq M_P^2 H^2$. Then use $H \sim t^{-1}$, write $T \propto t^\alpha$, where α is a constant, match the powers of t in all terms, and determine that $\alpha = -(1/4)$. Hence, $H \propto t^{-1} \propto T^4$ (and $\rho_\phi \propto H^2 \propto T^8$). Since H equals T_{RH}^2/M_P at $T = T_{\mathrm{RH}}$, it is $H \simeq T^4/(T_{\mathrm{RH}}^2 M_P)$.

The initial conditions are specified through the value H_I of the Hubble parameter at the beginning of the ϕ-oscillation-dominated epoch. This amounts to giving the initial energy density $\rho_{\phi,\mathrm{I}}$ in the ϕ field at the beginning of the reheating phase, or equivalently the maximum temperature of the radiation T_max. Indeed, one has $H_I \simeq \rho_{\phi,\mathrm{I}}^{1/2}/M_P \simeq T_\mathrm{max}^4/(T_{\mathrm{RH}}^2 M_P)$. The latter relation can be derived from $\rho_\phi \simeq T^8/T_{\mathrm{RH}}^4$ and the consideration that the maximum energy in the radiation equals the initial (maximum) energy $\rho_{\phi,\mathrm{I}}$. As the ϕ begins to decay, the temperature of the radiation bath rises sharply to T_max [550], decreases slowly as a function of the scale factor a during the ϕ-oscillation-dominated phase, as $T \sim a^{-3/8}$ until it reaches T_{RH}, when the radiation-dominated phase starts and $T \sim a^{-1}$.

Figure 7.5a shows how the WIMP density $\Omega_\chi h^2$ depends on T_{RH} for illustrative values of the parameter $\eta = b(100\,\mathrm{TeV}/m_\phi)$, both for WIMPs that

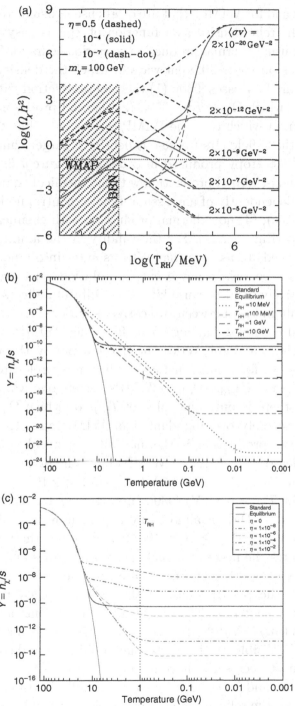

Fig. 7.5. a, WIMP density $\Omega_\chi h^2$ as a function of the reheating temperature T_{RH} for illustrative values of the ratio $\eta = b(100\,\mathrm{TeV}/m_\phi)$ [930]. b, Evolution of the neutralino χ abundance for different values of T_{RH} and $\eta = 0$ in an mSUGRA model with $M_{1/2} = m_0 = 600\,\mathrm{GeV}$, $A_0 = 0$, $\tan\beta = 10$, $\mu > 0$, $m_\chi = 246\,\mathrm{GeV}$ and standard relic density $\Omega_{\mathrm{std}} h^2 \simeq 3.6$ [926]. The short vertical lines indicate T_{RH} [926]. c, Same as b, but for $T_{\mathrm{RH}} = 1\,\mathrm{GeV}$ and several values of η.

are underdense and for WIMPs that are overdense in usual cosmologies. The behaviour of the relic density as a function of T_{RH} is easy to understand physically. The usual thermal production scenario occurs for $T_{\mathrm{RH}} > T_{\mathrm{fo}}$. For $T_{\mathrm{RH}} < T_{\mathrm{f.o.}}$, there are four different ways in which the density Ωh^2 depends on T_{RH}. There are four cases [930]: (1) thermal production without chemical equilibrium, for which $\Omega_\chi \sim T_{\mathrm{RH}}^7$ [550]; (2) thermal production with chemical equilibrium, in which case the WIMP freezes out while the Universe is dominated by the ϕ field. Its freeze-out density is larger than usual, but it is diluted by the entropy produced in ϕ decays (Figure 7.5b). In this case $\Omega_\chi \propto T_{\mathrm{RH}}^4$; (3) non-thermal production without chemical equilibrium, where $\Omega_\chi \propto \eta T_{\mathrm{RH}}$ (independently of any assumption on neutralino kinetic equilibrium) (Figure 7.5c); (4) non-thermal production with chemical equilibrium, where $\Omega \propto T_{\mathrm{RH}}^{-1}$ (Figure 7.5c). For the validity of the annihilation term in Eq. (7.29) one needs to assume that WIMPs enter into kinetic equilibrium before production ceases. In any event the solutions just presented should remain qualitatively valid because kinetic equilibrium affects only solutions which interpolate in T_{RH} between two correct solutions, namely the solution of the standard cosmology at high T_{RH} for which WIMPs are initially in kinetic equilibrium, and the WIMP production purely through the scalar field decay (case 3), for which kinetic equilibrium is irrelevant.

For all overdense ($\Omega_{\mathrm{std}} > \Omega_{\mathrm{cdm}}$) WIMPs, given one value of $\eta \lesssim 10^{-4}$ $(100\,\mathrm{GeV}/m_\chi)$ there is only one value of T_{RH} for which $\Omega_\chi = \Omega_{\mathrm{cdm}}$. The exception is a severely overabundant light WIMP with $\Omega_{\mathrm{std}} \gtrsim 10^{12}$ $(m_\chi/100\,\mathrm{GeV})^4$ (if the production is thermal with chemical equilibrium as is usual). Underdense ($\Omega_{\mathrm{std}} < \Omega_{\mathrm{cdm}}$) WIMPs have one or two solutions $\Omega_\chi = \Omega_{\mathrm{cdm}}$ per η, if $\Omega_{\mathrm{std}} \gtrsim 10^{-5}(100\,\mathrm{GeV}/m_\chi)$ and $\eta \gtrsim 10^{-7}$ $(100\,\mathrm{GeV}/m_\chi)^2$ $(\Omega_{\mathrm{cdm}}/\Omega_{\mathrm{std}})$ (for $T_{\mathrm{RH}} > 5$ MeV) [930]. In particular the neutralino density can be that of cold DM in almost any supersymmetric model, provided $10^{12}(m_\chi/100\,\mathrm{GeV})^4 \gtrsim \Omega_{\mathrm{std}} \gtrsim 10^{-5}(100\,\mathrm{GeV}/m_\chi)$ and the high-energy theory accommodates the necessary combinations of values of b/m_ϕ and T_{RH}.

Let us comment on other DM candidates. Sterile neutrinos ν_{s} would also be remnants of the pre-BBN era. If they are produced through oscillations with active neutrinos ν_{a} their production rate has a sharp peak at $T_{\mathrm{max}} \simeq 13\,\mathrm{MeV}(m_{\mathrm{s}}/1\,\mathrm{eV})^{1/3}$ [192; 690; 796; 797], which for $m_{\mathrm{s}} > 10^{-3}$ eV is above 1 MeV. 'Visible' ν_{s} (i.e. those that could be found soon in neutrino experiments) must necessarily have mixings $\sin(\theta)$ with ν_{a} large enough to be overabundant, and thus be rejected, in standard cosmologies. In LTR with $T_{\mathrm{RH}} < T_{\mathrm{max}}$, the relic abundance of visible ν_{s} could be reduced enough for them to be cosmologically acceptable, both if they are lighter and heavier than 1 MeV [927; 928; 958; 1947]. For example, for ν_{s} lighter than 1 MeV

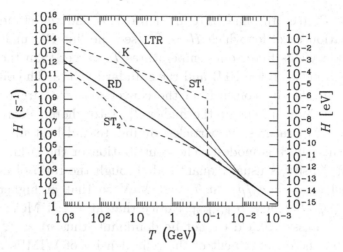

Fig. 7.6. The Hubble parameter H as a function of the photon temperature T before primordial nucleosynthesis for several cosmological models.

produced through oscillations, the ratio of sterile and active neutrino number abundance is [928]

$$n_{\rm s}/n_{\rm a} \simeq 10 \sin^2 2\theta \, (T_{\rm RH}/5 \text{ MeV})^3 \qquad (7.34)$$

(this ratio is within one order of magnitude of numerically derived results [1947]). Thus $n_{\rm s}$ is small for low $T_{\rm RH}$, even if the active–sterile neutrino mixing $\sin\theta$ is large. Another example is that of thermally produced axions, whose abundance can be strongly suppressed if $T_{\rm RH}$ is smaller than their freeze-out temperature ~ 50 MeV in standard cosmologies [957; 1025]. Also superWIMPs may be produced in LTR models [1527].

Finally, let us remark that LTR scenarios are more complicated than the standard cosmology and no consistent all-encompassing scenario exists yet. In particular, baryogenesis should happen during the reheating epoch too, possibly through the Affleck–Dine mechanism [694; 893; 894; 1481].

7.4.2 Models that only change the pre-BBN Hubble parameter

We will consider two of these models, in which the change in WIMP relic density is more modest than in LTR: kination and scalar–tensor gravitational models. An homogeneous field ϕ, e.g. a candidate for quintessence, has an energy density $\rho_\phi = \dot{\phi}^2/2 + V(\phi)$. Kination is an epoch in which the kinetic term dominates over the potential $V(\phi)$ so $\rho_{\rm total} \simeq \dot{\phi}^2/2 \sim a^{-6}$. No entropy is produced in this period, so $T \sim a^{-1}$ as usual. Thus $H \sim \sqrt{\rho_{\rm total}} \sim T^3$ (see line K in Figure 7.6). This case is intermediate between LTR, for which $H \sim T^4$ (see the line LTR in Figure 7.6) and the standard radiation domination case, for which $H \sim T^3$ (see the line RD in Figure 7.6). Thus kination yields freeze-out temperatures $T_{\rm fo}$ larger than the standard, somewhere in between the LTR and the standard values. The only entropy dilution of the density comes from the conversion of a larger number of degrees of freedom present at the higher $T_{\rm fo}$ into photon degrees of freedom at low temperatures, as particles annihilate and heat up the photon bath, and this effect is modest. The contribution of the ϕ kinetic energy to the total density is usually quantified through the ratio of ϕ-to-photon energy density, $\eta_\phi = \rho_\phi/\rho_\gamma$, at $T \simeq 1$ MeV so that at higher temperatures $H \simeq \sqrt{\eta_\phi}(T/1{\rm MeV})H_{\rm standard}$. Notice that at $T \simeq 1$ MeV, i.e. during BBN, the quintessence field cannot be dominant, thus $\eta_\phi < 1$. According to ref. [1669], the enhancement of the relic density of WIMPs in kination models is

$$\Omega_{\rm kination}/\Omega_{\rm std} \simeq \sqrt{\eta_\phi}10^3(m_\chi/100\,{\rm GeV}). \qquad (7.35)$$

Thus, WIMPs that are underdense in the standard cosmology could account for the whole of the dark matter.

Scalar–tensor theories of gravity [494; 495; 611; 1685] incorporate a scalar field coupled only through the metric tensor to the matter fields. In many of these models the expansion of the Universe drives the scalar field towards a state where the theory is indistinguishable from General Relativity, but the effect of the scalar field changes the expansion rate of the Universe at earlier times, either increasing or decreasing it. Theories with a single matter sector typically predict an enhancement of H before BBN. In ref. [494] the H is enhanced by a factor A, which is $A \simeq 2.19 \times 10^{14}(T_0/T)$ (T_0 is the present temperature of the Universe) for large temperatures $T > T_\phi$. At T_ϕ, A drops sharply to values close to 1 before BBN sets in (see the line ST$_1$ in Figure 7.6). WIMPs freeze out at $T > T_\phi$ while $H \sim T^{1.2}$, but at the transition temperature T_ϕ, H drops sharply to the standard value, and becomes smaller than the WIMP reaction rate. The already frozen WIMPs are still abundant enough at T_ϕ to start annihilating again. This is a post-freeze-out 'reannihilation phase' peculiar to these models. The WIMP relic abundance is reduced in this phase, but nonetheless remains much larger than in the standard case. The amount of increase in the WIMP relic abundance was found in ref. [494] to be between 10 and 10^3. With more than one matter sector, of which only one is 'visible' and the other 'hidden', scalar–tensor models may also produce a reduction of H by as much as 0.05 of the standard value (see line ST_2 in Figure 7.6) before the transition temperature T_ϕ at which H increases sharply to the standard value before BBN [495]. The maximum reduction of the WIMP relic abundance is larger for larger WIMP masses, ranging from a factor of 0.8–0.9 for masses close to 10 GeV to 0.1–0.2 for those close to 500 GeV [494].

8

Supersymmetric dark matter candidates

John Ellis and Keith A. Olive

8.1 Motivations

Supersymmetry is one of the best-motivated proposals for physics beyond the Standard Model. There are many idealistic motivations for believing in supersymmetry, such as its intrinsic elegance, its ability to link matter particles and force carriers, its ability to link gravity to the other fundamental interactions, and its essential role in string theory. However, none of these aesthetic motivations gives any hint as to the energy scale at which supersymmetry might appear. The following are the principal utilitarian reasons to think that supersymmetry might appear at some energy accessible to forthcoming experiments.

The first and primary of these was the observation that supersymmetry could help stabilize the mass scale of electroweak symmetry breaking, by cancelling the quadratic divergences in the radiative corrections to the mass-squared of the Higgs boson [1374; 1829; 1940], and by extension to the masses of other Standard Model particles. This motivation suggests that sparticles weigh less than about 1 TeV, but the exact mass scale depends on the amount of fine-tuning that one is prepared to tolerate.

Historically, the second motivation for low-scale supersymmetry, and the one that interests us most here, was the observation that the lightest supersymmetric particle (LSP) in models with conserved R-parity, being heavy and naturally neutral and stable, would be an excellent candidate for dark matter [760; 973]. This motivation requires that the lightest supersymmetric particle should weigh less than about 1 TeV, if it had once been in thermal equilibrium in the early Universe. This would have been the case for a neutralino χ or a sneutrino $\tilde{\nu}$ LSP, and the argument can be extended

Particle Dark Matter: Observations, Models and Searches, ed. Gianfranco Bertone. Published by Cambridge University Press. © Cambridge University Press 2010.

to a gravitino LSP because it may be produced in the decays of heavier, equilibrated sparticles.

The third reason that emerged for thinking that supersymmetry may be accessible to experiment was the observation that including sparticles in the renormalization-group equations (RGEs) for the gauge couplings of the Standard Model would permit them to unify [79; 763; 764; 961; 1306], whereas unification would not occur if only the Standard Model particles were included in the RGEs. However, this argument does not constrain the supersymmetric mass scale very precisely: scales up to about 10 TeV or perhaps more could be compatible with grand unification.

The fourth motivation is the fact that the Higgs boson is (presumably) relatively light, according to the precision electroweak data – an argument reinforced by the negative results (so far) of searches for the Higgs boson at the Fermilab Tevatron collider. It has been known for some 20 years that the lightest supersymmetric Higgs boson should weigh no more than about 140 GeV, at least in simple models [786; 787; 1046; 1528; 1529]. Since the early 1990s, the precision electroweak noose has been tightening, and the best indication now (incorporating the negative results of searches at LEP and the Tevatron) is that the Higgs boson probably weighs less than about 140 GeV [1030; 1032], in perfect agreement with the supersymmetric prediction.

Fifthly, if the Higgs boson is indeed so light, the present electroweak vacuum would be destabilized by radiative corrections due to the top quark, unless the Standard Model is supplemented by additional scalar particles [788]. This would be automatic in supersymmetry, and one can extend the argument to 'prove' that any mechanism to stabilize the electroweak vacuum must look very much like supersymmetry.

There is a sixth argument that is still controversial, namely the anomalous magnetic moment of the muon, $g_\mu - 2$. As is well known, the experimental measurement of this quantity [277] disagrees with the Standard Model prediction [1547], if this is calculated using low-energy e^+e^- annihilation data. On the other hand, the discrepancy with the Standard Model is greatly reduced if one uses τ decay data to estimate the Standard Model contribution to $g_\mu - 2$. Normally, one would prefer to use e^+e^- data, since they are related more directly to $g_\mu - 2$, with no need to worry about isospin violations, etc. Measurements by the BABAR collaboration using the radiative-return method [616] yield a result intermediate between the previous e^+e^- data and τ decay data. Until the discrepancy between these data sets has been ironed out, one should take $g_\mu - 2$ *cum grano salis*.

8.2 The MSSM and R-parity

We refer to [825; 1509] for the general structure of supersymmetric theories. We restrict ourselves here to theories with a single supersymmetry charge, called simple or $N = 1$ supersymmetry, as these are the only ones able to accommodate chiral fermions and hence the violation of parity and charge conjugation. We recall that the basic building blocks of $N = 1$ supersymmetric models are so-called chiral supermultiplets, each consisting of a Weyl fermion and a complex scalar, and gauge supermultiplets, each consisting of a gauge field and a gaugino fermion. The renormalizable interactions between the chiral supermultiplets are characterized by a superpotential that couples the chiral supermultiplets in bilinear and trilinear combinations that yield masses and Yukawa interactions, and by gauge interactions. In this framework, bosons and fermions must appear in pairs with identical internal quantum numbers. Since the known particles do not pair up in this way, it is necessary to postulate unseen particles to partner those known in the Standard Model.

In order to construct the minimal supersymmetric extension of the Standard Model (MSSM) [822; 823; 824], one starts with the complete set of chiral fermions needed in the Standard Model, and adds a complex scalar superpartner to each Weyl fermion, so that each matter field in the Standard Model is extended to a chiral supermultiplet. These are denoted by L^i, Q^i, e^c, d^c and u^c, where i, j are $SU(2)_L$ doublet indices and generation indices have been suppressed as were colour indices for the quarks. In order to avoid a triangle anomaly, Higgs supermultiplets must appear in pairs with opposite hypercharges, and the minimal possibility is a single pair H_1^i, H_2^i. One must also add a gaugino for each of the gauge bosons in the Standard Model so as to complete the gauge supermultiplets. The minimal supersymmetric Standard Model (MSSM) [1048] is defined by this minimal field content and the minimal superpotential necessary to account for the necessary Yukawa couplings and mass terms, namely:

$$W = \epsilon_{ij}\left(y_e H_1^j L^i e^c + y_d H_1^j Q^i d^c + y_u H_2^j Q^j u^c\right) + W_\mu, \qquad (8.1)$$

where

$$W_\mu = \epsilon_{ij}\mu H_1^i H_2^j. \qquad (8.2)$$

In Eq. (8.1), the Yukawa couplings, y, are all 3×3 matrices in generation space, with no generation indices for the Higgs multiplets. A second reason for requiring two Higgs doublets in the MSSM is that the superpotential must be a holomorphic function of the chiral superfields. This implies that there would be no way to account for all of the Yukawa terms for both up- and down-type quarks, as well as charged leptons, with a single Higgs doublet. The physical Higgs spectrum then contains five states: two charged Higgs bosons $H^{\pm}$, two scalar neutral Higgs bosons h, H, and a pseudoscalar Higgs boson A. The bilinear mixing term W_{μ} must be added to the super-potential, in order to avoid a massless Higgs state.

The MSSM must be coupled to gravity, which requires the introduction of a graviton supermultiplet containing a spin-3/2 gravitino as well as the spin-2 graviton itself, which may or not be coupled minimally to the MSSM. The consistency of supergravity at the quantum level requires the breaking of supersymmetry to be spontaneous, with the gravitino mass acting as an order parameter [588; 589]. The mechanism whereby supersymmetry is broken is unknown, as is how this feeds into the MSSM. We adopt here a phenomenological approach, parameterizing the results of this mechanism in terms of differing amounts of explicit supersymmetry breaking in the masses and couplings of the unseen supersymmetric partners of Standard Model particles [120; 193; 1509].

In order to preserve the hierarchy between the electroweak and GUT or Planck scales, it is necessary that this explicit breaking of supersymmetry be 'soft', i.e. in such a way that the theory remains free of quadratic divergences, which is possible with the insertion of weak scale mass terms in the Lagrangian [954]. The possible forms for such terms are

$$\mathcal{L}_{\text{soft}} = -\frac{1}{2} M^a \lambda^a \lambda^a - \frac{1}{2} (m^2)^{\alpha}_{\beta} \phi_{\alpha} \phi^{\beta *}$$
$$- \frac{1}{2} (BM)^{\alpha\beta} \phi_{\alpha} \phi_{\beta} - \frac{1}{6} (Ay)^{\alpha\beta\gamma} \phi_{\alpha} \phi_{\beta} \phi_{\gamma} + h.c., \qquad (8.3)$$

where the M^a are masses for the gauginos λ^a, m^2 is a matrix of soft scalar masses-squared that carries two field indices, α, β, for scalars ϕ_{α}, A is a trilinear coupling term with three field indices and B is a bilinear supersymmetry breaking term associated with a superpotential bilinear mass term such as μ in Eq. (8.2). Masses for the gauge bosons are, as usual, induced by the spontaneous breaking of gauge invariance, and the masses for chiral fermions are induced by the Yukawa superpotential terms when the electroweak gauge symmetry is broken. For a more complete discussion of supersymmetry and the construction of the MSSM see [747; 1391; 1530; 1568].

In defining the MSSM, we have limited the model to contain a minimal field content: the only new fields are those which are required by supersymmetry. Consequently, apart from superpartners, only the Higgs sector was enlarged from one doublet to two. Moreover, in writing the superpotential (8.1), we have also made a minimal choice regarding interactions. We have limited the types of interactions to include only the minimal set required in the Standard Model and its supersymmetric generalization.

However, even with the minimal field content, there are several other superpotential terms that one could envisage adding to (8.1) which are consistent with all of the gauge symmetries of the theory. Specifically, one could consider adding any or all of the following terms that violate R-parity:

$$W_R = \frac{1}{2}\lambda\epsilon_{ij}L^iL^je^c + \lambda'\epsilon_{ij}L^iQ^jd^c + \frac{1}{2}\lambda''u^cd^cd^c + \mu'L^iH_2^i. \qquad (8.4)$$

Each of the terms in (8.4) has one or more suppressed generation indices. We note that the terms proportional to λ, λ' and μ' both violate lepton number by one unit, whereas the term proportional to λ'' violates baryon number by one unit.

Each of the terms in (8.4) predicts new particle interactions and can be to some extent constrained by the lack of observed exotic phenomena. In particular, any combination of terms which violate both baryon and lepton number would be unacceptable, unless the product of coefficients was extremely small. For example, consider the possibility that both λ' and λ'' were non-zero. This would lead to the following proton decay processes: $p \rightarrow e^+\pi^0, \mu^+\pi^0, \nu\pi^+, \nu K^+$, etc. The rate of proton decay due to this process would have no suppression by any superheavy masses, since there is no GUT- or Planck-scale physics involved: this is a purely (supersymmetric) Standard Model interaction involving only the electroweak scale. The (inverse) rate can be easily estimated to be

$$\Gamma_p^{-1} \sim \frac{\tilde{m}^4}{m_p^5} \sim 10^8 \text{GeV}^{-1}, \qquad (8.5)$$

assuming a supersymmetry breaking scale of $\tilde{m}$ of order 100 GeV. This should be compared with current limits to the proton lifetime of $\gtrsim 10^{63}$ GeV^{-1}. Clearly the product of λ' and λ'' must be very small, if not exactly zero.

It is possible to eliminate the unwanted superpotential terms by imposing a discrete symmetry on the theory called R-parity [816]. This can be represented as

$$R = (-1)^{3B+L+2s}, \qquad (8.6)$$

where B, L and s are the baryon number, lepton number and spin, respectively. It is easy to see that, with the definition (8.6), all the known Standard Model particles have R-parity $+1$. For example, the electron has $B = 0$, $L = -1$ and $s = 1/2$, and the photon has $B = L = 0$ and $s = 1$, so in both cases $R = 1$. Similarly, it is clear that all superpartners of the known

Standard Model particles have $R = -1$, since they must have the same value of B and L as their conventional partners, but differ by $1/2$ unit of spin. If R-parity is exactly conserved, then all four superpotential terms in (8.4) must be absent.

The additive conservation of the quantum numbers B, L and s implies that R-parity must be conserved multiplicatively. A first important corollary is that the collisions of conventional particles must always produce supersymmetric particles in pairs, and a second corollary is that heavier supersymmetric particles can decay only into lighter supersymmetric particles. For our purposes here, an even more important corollary of R-parity conservation is the prediction that the lightest supersymmetric particle (LSP) must be stable, because it has no legal decay mode. In much the same way that baryon number conservation predicts proton stability, R-parity predicts that the lightest $R = -1$ state is stable. This makes supersymmetry an extremely interesting theory from the astrophysical point of view, as the LSP naturally becomes a viable dark matter candidate [760; 973].

8.3 Possible supersymmetric dark matter candidates

What options are available in the MSSM for the stable LSP? Any electrically charged LSP would bind to conventional matter, and be detectable as an anomalous heavy nucleus, since the 'Bohr radius' for the LSP 'atom' would be less than the nuclear radius. Similarly, strongly interacting LSPs would also form anomalous heavy nuclei. However, experiments searching for such objects [1086; 1639; 1773] have excluded their presence on Earth down to an abundance far lower than the expected abundance for the LSP (see below for more details how this is calculated). Therefore, the stable LSP is presumably electrically neutral and can have only weak interactions. For this reason, the commonly expected signature of supersymmetric particle production at colliders is missing energy carried away by undetected LSPs.

This still leaves us with several possible dark matter candidates in the MSSM, specifically the sneutrino with spin zero, the neutralino with spin $1/2$ and the gravitino with spin $3/2$. However, a sneutrino LSP would have relatively large coherent interactions with heavy nuclei, and experiments searching directly for the scattering of massive dark matter particles on nuclei exclude a stable sneutrino weighing between a few GeV and several TeV [807]. The possible loophole of a very light sneutrino was excluded by measurements of the invisible Z-boson decay rate at LEP [570].

The LSP candidate that is considered most often is the lightest neutralino. In the MSSM there are four neutralinos, each of which is a linear combination of the following $R = -1$ neutral fermions [760]: the neutral wino $\tilde{W}^3$ (the partner of the third component of the $SU(2)_L$ triplet of weak gauge bosons); the $U(1)$ bino $\tilde{B}$; and two neutral Higgsinos $\tilde{H}_1$ and $\tilde{H}_2$ (the supersymmetric partners of the neutral components of the two Higgs doublets).

The composition of the LSP χ can be expressed as a linear combination of these fields:

$$\chi = \alpha\tilde{B} + \beta\tilde{W}^3 + \gamma\tilde{H}_1 + \delta\tilde{H}_2, \tag{8.7}$$

whose mass and composition are determined by the $SU(2)_L$ and $U(1)$ gaugino masses, $M_{2,1}$, the Higgs mixing parameter μ, and $\tan\beta$, the ratio of the vacuum expectation values $v_{1,2} \equiv < 0|H_{1,2}|0 >$ of the two neutral Higgs fields $\tan\beta \equiv v_2/v_1$. The mass of the LSP χ and the mixing coefficients α, β, γ and δ in (8.7) for the neutralino components that compose the LSP can be found by diagonalizing the mass matrix

$$(\tilde{W}^3, \tilde{B}, \tilde{H}_1^0, \tilde{H}_2^0) \begin{pmatrix} M_2 & 0 & \frac{-g_2 v_1}{\sqrt{2}} & \frac{g_2 v_2}{\sqrt{2}} \\ 0 & M_1 & \frac{g_1 v_1}{\sqrt{2}} & \frac{-g_1 v_2}{\sqrt{2}} \\ \frac{-g_2 v_1}{\sqrt{2}} & \frac{g_1 v_1}{\sqrt{2}} & 0 & -\mu \\ \frac{g_2 v_2}{\sqrt{2}} & \frac{-g_1 v_2}{\sqrt{2}} & -\mu & 0 \end{pmatrix} \begin{pmatrix} \tilde{W}^3 \\ \tilde{B} \\ \tilde{H}_1^0 \\ \tilde{H}_2^0 \end{pmatrix}. \tag{8.8}$$

In different regions of the supersymmetric parameter space, the LSP may be more bino-like, wino-like or Higgsino-like, depending on the relative magnitudes of the coefficients α, β, γ and δ.

The relic abundance of an LSP candidate such as the lightest neutralino is calculated by solving the Boltzmann equation for the LSP number density in an expanding Universe:

$$\frac{dn}{dt} = -3\frac{\dot{R}}{R}n - \langle\sigma v\rangle(n^2 - n_0^2), \tag{8.9}$$

where n_0 is the equilibrium number density of neutralinos. Defining the quantity $f \equiv n/T^3$, we can rewrite this equation in terms of the reduced temperature $x \equiv T/m_\chi$:

$$\frac{df}{dx} = m_\chi \left(\frac{8\pi^3}{90}G_N N\right)^{-1/2} \langle\sigma v\rangle(f^2 - f_0^2), \tag{8.10}$$

where G_N is Newton's constant and N is the number of relativistic degrees of freedom at a given temperature. The solution to this equation at late times and low temperatures, and hence small x, yields a constant value of f, so that $n \propto T^3$.

The technique [1793] used to determine the neutralino relic density is similar to that used previously for computing the relic abundance of massive neutrinos [1144; 1317; 1911], with the substitution of the appropriate annihilation cross-section. This and hence the relic density depend on additional parameters in the MSSM beyond M_1, M_2, μ and $\tan\beta$, which include the sfermion masses, $m_{\tilde{f}}$, and mass of the pseudoscalar Higgs boson, m_A. In much of the parameter space of interest, the LSP is a bino and the annihilation proceeds mainly through crossed t-channel sfermion exchange. The exception is if the sum of two neutralino masses happens to lie near a direct-channel pole, such as $m_\chi \simeq m_Z/2$ or $m_h/2$, in which case there are large contributions to the annihilation through direct s-channel resonance exchange. Since the neutralino is a Majorana fermion, away from such a resonance the s-wave part of the annihilation cross-section is generally suppressed by the outgoing fermion masses, and the annihilation occurs mainly through the p wave, which is also suppressed because the annihilating LSPs are non-relativistic at low temperatures (small x). This means that one can approximate the annihilation cross-section including p-wave corrections by incorporating a term proportional to the temperature if neutralinos are in thermal equilibrium: $\sigma v = a + bx + \cdots$, where the expansion coefficients a, b are model-dependent.

Annihilations in the early Universe continue until the annihilation rate $\Gamma \simeq \sigma v n_\chi$ drops below the expansion rate, after which it is a good first approximation to assume that annihilations are negligible – the freeze-out approximation. The final neutralino relic density, expressed as a fraction Ω_χ of the critical energy density and denoting the present-day Hubble expansion rate as h in units of $100\,\mathrm{km\ s^{-1}\ Mpc^{-1}}$, can be written as [760]

$$\Omega_\chi h^2 \simeq 1.9 \times 10^{-11} \left(\frac{T_\chi}{T_\gamma}\right)^3 N_{\mathrm{f}}^{1/2} \left(\frac{\mathrm{GeV}}{ax_{\mathrm{f}} + \frac{1}{2}bx_{\mathrm{f}}^2}\right), \qquad (8.11)$$

where $(T_\chi/T_\gamma)^3$ accounts for the subsequent reheating of the photon temperature with respect to χ, due to the annihilations of particles with mass $m < x_{\mathrm{f}} m_\chi$ [1531; 1804], and $x_{\mathrm{f}} = T_{\mathrm{f}}/m_\chi$ is proportional to the freeze-out temperature T_{f}. Equation (8.11) yields a very good approximation to the relic density except near direct s-channel annihilation poles or thresholds, and in regions where the LSP is nearly degenerate with the next lightest supersymmetric particle [1020].

When there are several particle species i that are nearly degenerate in mass, co-annihilations between the different species become important. In this case [1020], the rate equation (8.9) still applies, provided n is interpreted as the total number density,

$$n \equiv \sum_i n_i, \tag{8.12}$$

n_0 is interpreted as the total equilibrium number density,

$$n_0 \equiv \sum_i n_{0,i}, \tag{8.13}$$

and the effective annihilation cross-section as

$$\langle \sigma_{\rm eff} v_{\rm rel} \rangle \equiv \sum_{ij} \frac{n_{0,i} n_{0,j}}{n_0^2} \langle \sigma_{ij} v_{\rm rel} \rangle. \tag{8.14}$$

In Eq. (8.10), m_χ is now understood to be the mass of the lightest sparticle under consideration.

We turn finally to the third LSP candidate within the MSSM, namely the gravitino. Since it has only gravitational-strength interactions, it is not expected to have been in thermal equilibrium in the early Universe. However, it could have been produced in high-energy particle collisions in the early Universe, or in the decays of heavier supersymmetric particles. The fact that the gravitino has only gravitational-strength interactions implies that only decays of the next-to-lightest supersymmetric particle (NLSP) would be significant sources of gravitinos, and the NLSP would be metastable. As we discuss in more detail later, there are important cosmological and astrophysical constraints on the possible mass and lifetime of the NLSP, derived principally from the agreement between astrophysical observations and Big Bang nucleosynthesis calculations of light-element abundances.

What might be the nature of the NLSP in such a gravitino LSP scenario? One option is the lighter of the two supersymmetric partners of the τ lepton, denoted by $\tilde{\tau}_1$. Being a metastable charged particle, it would have a distinctive experimental signature at the LHC or other colliders. Studies within such a scenario have shown that the mass of the $\tilde{\tau}_1$ could be measured very accurately, and that one could easily reconstruct heavier sparticles that decay into the $\tilde{\tau}_1$ [785].

Alternatively, the NLSP might be the lighter supersymmetric partner of the top quark, denoted by $\tilde{t}_1$ [655; 1260; 1689], which would have even more distinctive signatures at the LHC. Immediately after production, it would become confined inside a charged or neutral hadron. As it moves through an LHC detector, it would have a high probability of changing its charge as it interacts with the material in the detector. This combined with its non-relativistic velocity would provide a truly distinctive signature.

Yet another possibility is that the NLSP might be some flavour of sneutrino [774], in which case the characteristic signature would be missing energy carried away by the metastable sneutrino. This could nevertheless be distinguished from the conventional missing-energy signature of a neutralino LSP (or NLSP), because the final states would be more likely to include the charged lepton with the same flavour as the sneutrino NLSP, either e, μ or τ.

These are just a few examples of the possible alternatives to the conventional missing-energy signature of supersymmetry. Studies have shown that the LHC would also have good prospects for detecting such signatures.

8.4 Renormalization-group equations and electroweak symmetry breaking

The fact that measurements of the strengths of the Standard Model gauge interactions measured at low energies are in excellent agreement with the predictions of a supersymmetric gauge theory [79; 763; 764; 961; 1306] was already cited as an important motivation for low-energy supersymmetry. It can also be regarded as a motivation for thinking that other parameters of the effective low-energy theory, for example, the soft supersymmetry breaking parameters, can also be calculated and related using renormalization-group equations (RGEs) below the grand unification scale. For example, the one-loop RGEs for the gaugino masses are:

$$\frac{\mathrm{d}M_i}{\mathrm{d}t} = -b_i \alpha_i M_i / 4\pi. \tag{8.15}$$

If the gaugino masses have a common value $m_{1/2}$ at the grand unification scale, these equations can be used to relate the physical low-energy, on-shell values of the gaugino masses to the corresponding gauge coupling strengths α_i:

$$M_i(t) = \frac{\alpha_i(t)}{\alpha_i(M_{\mathrm{GUT}})} m_{1/2}, \tag{8.16}$$

which implies that

$$\frac{M_1}{g_1^2} = \frac{M_2}{g_2^2} = \frac{M_3}{g_3^2} \tag{8.17}$$

at the one-loop level. When applying this relation within a specific grand unified theory, one must remember to incorporate the difference of the normalization of the U(1) factor from that in the Standard Model, so that we have $M_1 = \frac{5}{3} \frac{\alpha_1}{\alpha_2} M_2$ for the one-loop relation between the bino and wino

masses. Also, the simple relations (8.17) are modified by threshold corrections at the electroweak scale, and by two-loop effects in the RGEs. The soft
supersymmetry-breaking scalar masses-squared m^2 and the trilinear couplings A are renormalized analogously to (8.15), with the difference that
Yukawa interactions contribute as well as gauge interactions. However, the
Yukawa contributions are small, except for the supersymmetric partners of
third-generation fermions.

   As described above, the MSSM has over 100 undetermined parameters,
which are mainly associated with the breaking of supersymmetry. It is often
assumed that the soft supersymmetry-breaking parameters M^a, m^2 and A
have some universality properties. There are phenomenological arguments,
based on the success of the Standard Model in describing the observed suppression of flavour-changing interactions, that, at some input scale (often
assumed to be that of grand unification), the parameters m^2 and A must
be universal for supersymmetric particles with the same gauge quantum
numbers, e.g. the supersymmetric partners of the e, μ and τ. There is no
strong argument why these parameters should be universal for supersymmetric particles with different quantum numbers, e.g. d, u and e, although
this may occur in some grand unified theories, as may unification of the gaugino masses M^a. The simplified version of the MSSM in which universality
at the grand unification scale is assumed for each of M^a, m^2 and A is called
the constrained MSSM (CMSSM)[122; 154; 156; 157; 158; 159; 208; 209; 524;
525; 679; 715; 748; 749; 750; 752; 753; 772; 776; 1301; 1302; 1303; 1489; 1655].

   Once one has chosen a set of boundary conditions at the grand unification
scale and run the RGEs down to the electroweak scale, one must check the
properties of the electroweak vacuum, which are characterized by specifying the mass of the Z boson, M_Z, and the ratio of the two Higgs vacuum
expectation values, $\tan\beta$. These electroweak symmetry-breaking conditions
should be used as consistency conditions on the solutions to the RGEs,
e.g. of the CMSSM. They are frequently used to fix, as functions of the
input values of the common gaugino mass $m_{1/2}$, m, A and $\tan\beta$, the magnitudes of the Higgs mixing mass parameter, μ, and of the bilinear coupling, B, which determines the pseudoscalar Higgs mass, m_A. The sign of μ
remains free.

   An example of the running of the mass parameters in the CMSSM as
functions of the renormalization scale is shown in Fig. 8.1, using as inputs
the choices $m_{1/2} = 250$ GeV, $m_0 = 100$ GeV, $\tan\beta = 3$, $A_0 = 0$ and
$\mu < 0$. We notice in the figure several characteristic features of the sparticle spectrum. For example, the coloured sparticles are typically the heaviest, because of the large positive corrections to their masses arising from

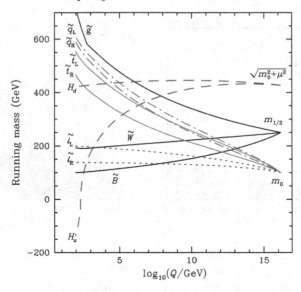

Fig. 8.1. The renormalization-group evolution of the mass parameters in the CMSSM, assuming $m_{1/2} = 250\,\mathrm{GeV}$, $m_0 = 100\,\mathrm{GeV}$, $\tan\beta - 3$, $A_0 = 0$ and $\mu < 0$. We thank Toby Falk for providing this figure.

α_3-dependent terms in the RGEs. Also, one finds that the bino, $\widetilde{B}$, is typically the lightest sparticle. Most importantly, we notice that one of the Higgs masses squared goes negative, triggering electroweak symmetry breaking [73; 761; 767; 1145; 1146]. (The negative sign in the figure refers to the sign of the mass squared, even though it is the mass of the sparticles which is depicted.)

8.5 The CMSSM

For given values of $\tan\beta$, A_0 and $sgn(\mu)$, the regions of the CMSSM parameter space that yield an acceptable relic density and satisfy the other phenomenological constraints may conveniently be displayed in the $(m_{1/2}, m_0)$ plane. Figure 8.2 displays, for $\tan\beta = 10$ (a) and 50 (b), the impacts of the most relevant constraints. These include the LEP lower limits on the chargino mass: $m_{\chi^\pm} > 104$ GeV [1028], on the selectron mass: $m_{\tilde{e}} > 99$ GeV [1029] and on the Higgs mass: $m_h > 114$ GeV [187; 870]. The former two constrain $m_{1/2}$ and m_0 directly via the sparticle masses, and the latter indirectly via the sensitivity of radiative corrections to the Higgs mass to the sparticle masses, principally $m_{\tilde{t},\tilde{b}}$. Here the code FeynHiggs [1082; 1081] is used for the calculation of m_h. It would be prudent to assign an uncertainty of 3 GeV to this calculation. Nevertheless, the Higgs limit imposes

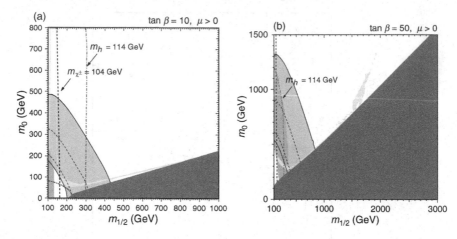

Fig. 8.2. The $(m_{1/2}, m_0)$ planes for (a) $\tan\beta = 10$ and (b) $\tan\beta = 50$, assuming $\mu > 0$, $A_0 = 0$, $m_t = 175\,\mathrm{GeV}$ and $m_b(m_b)^{\overline{MS}}_{SM} = 4.25$ GeV. The near-vertical dot-dashed lines are the contours for $m_h = 114$ GeV, and the near-vertical dashed line is the contour $m_{\chi^\pm} = 104$ GeV. Also shown by the dot-dashed curve in the lower left is the region excluded by the LEP bound $m_{\tilde{e}} > 99$ GeV. The medium shaded region is excluded by $b \to s\gamma$, and the light shaded area is the cosmologically preferred region. In the dark shaded region, the LSP is the charged $\tilde{\tau}_1$. The region allowed by the E821 measurement of a_μ at the 2σ level is shaded and bounded by solid black lines, with dashed lines indicating the 1σ ranges.

important constraints, principally on $m_{1/2}$ and particularly at low $\tan\beta$. Another constraint is the requirement that the branching ratio for $b \to s\gamma$ be consistent with the experimental measurements [190]. These measurements agree with the Standard Model, and therefore provide bounds on MSSM particles [640], such as the chargino and charged Higgs bosons, in particular. Typically, the $b \to s\gamma$ constraint is more important for $\mu < 0$, but it is also relevant for $\mu > 0$, particularly when $\tan\beta$ is large. The constraint imposed by measurements of $b \to s\gamma$ also exclude small values of $m_{1/2}$. Finally, there are regions of the $(m_{1/2}, m_0)$ plane that are favoured by the Brookhaven National Laboratory measurement [277] of $g_\mu - 2$. Here we assume the Standard Model calculation [1547] of $g_\mu - 2$ using e^+e^- data, and indicate by dashed and solid lines the contours of 1σ and 2σ level deviations induced by supersymmetry.

The most precise constraint on supersymmetry may be that provided by the density of cold dark matter, as determined from astrophysical and cosmological measurements by WMAP and other experiments [729]:

$$\Omega_{\mathrm{CDM}} = 0.1099 \pm 0.0062. \tag{8.18}$$

Applied straightforwardly to the relic LSP density $\Omega_{\mathrm{LSP}}h^2$, this would give a very tight relation between supersymmetric model parameters, fixing some combination of them at the per cent level, which would essentially reduce the dimensionality of the supersymmetric parameter space by one unit. Let us assume for now that the LSP is the lightest neutralino χ, whose density is usually thought to be fixed by freeze-out from thermal equilibrium in the early Universe, as discussed previously. In this case, respecting the constraint (8.18) would force the CMSSM into one of the narrow WMAP 'strips' in planar projections of the parameters [776], as illustrated by the narrow light shaded regions in Fig. 8.2. However, caution should be exercised before jumping to this conclusion.

Supersymmetry might not be the only contribution to the cold dark matter, in which case (8.18) should be interpreted as an upper limit on $\Omega_{\mathrm{LSP}}h^2$. However, most of the supersymmetric parameter space in simple models gives a supersymmetric relic density that exceeds the WMAP range (8.18), e.g. above the WMAP 'strip' in Fig. 8.2, and the regions with lower density generally correspond to lower values of the sparticle masses, i.e. below the WMAP 'strip' in Fig. 8.2.

However, even if one takes them seriously, the locations of these WMAP 'strips' do vary significantly with the choices of other supersymmetric parameters, as can be seen by comparing the cases of $\tan\beta = 10, 50$ in Fig. 8.2 (a, b). As one varies $\tan\beta$, the WMAP 'strips' cover much of the $(m_{1/2}, m_0)$ plane.

Several different regions of the WMAP 'strips' in the CMSSM $(m_{1/2}, m_0)$ plane can be distinguished, in which different dynamical processes are dominant. At low values of $m_{1/2}$ and m_0, simple $\chi - \chi$ annihilations via crossed-channel sfermion exchange are dominant, but this 'bulk' region is now largely excluded by the LEP lower limit on the Higgs mass, m_h. At larger $m_{1/2}$, but relatively small m_0, close to the boundary of the region where the lighter stau is lighter than the lightest neutralino: $m_{\tilde{\tau}_1} < m_\chi$, co-annihilation between the χ and sleptons is important in suppressing the relic χ density into the WMAP range (8.18), as seen in Fig. 8.2. At larger $m_{1/2}, m_0$ and $\tan\beta$, the relic χ density may be reduced by rapid annihilation through direct-channel H, A Higgs bosons, as seen in Fig. 8.2(b). Finally, the relic density can again be brought down into the WMAP range (8.18) at large m_0 (not shown in Fig. 8.2), in the 'focus-point' region close to the boundary where electroweak symmetry breaking ceases to be possible and the lightest neutralino χ acquires a significant Higgsino component [832].

As seen in Fig. 8.2, the relic density constraint is compatible with relatively large values of $m_{1/2}$ and m_0, and it is interesting to look for any indication of where the supersymmetric mass scale might lie within this range, using the available phenomenological and cosmological constraints. A global likelihood analysis enables one to pin down the available parameter space in the CMSSM and the related models discussed later. One can avoid the dependence on priors by performing a pure likelihood analysis as in [778], or a purely χ^2-based fit as done in [455; 762]. Here we present results from one such analysis [456], which used a Markov-chain Monte Carlo (MCMC) technique to explore efficiently the likelihood function in the parameter space of the CMSSM. A full list of the observables and the values assumed for them in this global analysis are given in [455], as updated in [456].

The 68% and 95% confidence-level (C.L.) regions in the $(m_{1/2}, m_0)$ plane of the CMSSM are shown in Fig. 8.3 [456]. Also shown for comparison are the physics reaches of ATLAS and CMS with 1/fb of integrated luminosity [142; 238]. (MET stands for missing transverse energy, SS stands for same-sign dilepton pairs, and the sensitivity for finding the lightest Higgs boson in cascade decays of supersymmetric particles is calculated for 2/fb of data.) The likelihood analysis assumed $\mu > 0$, as motivated by the sign of the apparent discrepancy in $g_\mu - 2$, but sampled all values of $\tan \beta$ and A_0: the experimental sensitivities were estimated assuming $\tan \beta = 10$ and $A_0 = 0$, but are probably not very sensitive to these assumptions. The global maximum of the likelihood function (indicated by the black dot) is at $m_{1/2} = 310$ GeV, $m_0 = 60$ GeV, $A_0 = 240$ GeV, $\tan \beta = 11$ and $\chi^2/N_{\mathrm{dof}} = 20.4/19$ (37% probability). It is encouraging that the best-fit points lie well within the LHC discovery range, as do the 68% and most of the 95% C.L. regions. It is also encouraging that the two best-fit points have similar values of $m_{1/2}, m_0$ and $\tan \beta$, the most important parameters for the sparticle spectrum, indicating that the likelihood analysis is relatively insensitive to the theoretical model assumptions.

In contrast to this neutralino LSP scenario, the gravitino dark matter (GDM) scenario in the CMSSM is tightly constrained by the astrophysical constraints on the cosmological abundances of light elements, as seen in Fig. 8.4 [603]. However, such a scenario might have some advantages, e.g. by enabling the cosmological prediction for the abundance of ^{7}Li [604] to be improved, as also shown in Fig. 8.4(b).

Recently, new attention has been focused on the regions in which a metastable stau is the next-to-lightest sparticle (NSP) in a GDM scenario, due to its ability to form bound states (primarily with ^{4}He). When such bound states occur, they catalyse certain nuclear reactions such as ^{4}He(D, γ)^{6}Li, which is normally highly suppressed because of the production of a low-energy γ, whereas the bound-state reaction is not [359; 1062; 1591].

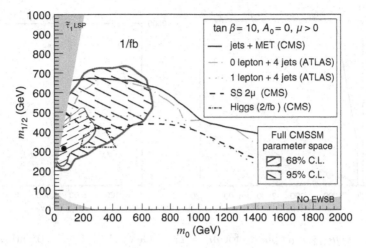

Fig. 8.3. The $(m_0, m_{1/2})$ plane in the CMSSM showing the regions favoured in a likelihood analysis at the 68% and 95% confidence levels [456]. The best-fit point is shown as the black point. Also shown are the discovery contours in different channels for the LHC with 1/fb (2/fb for the Higgs search in cascade decays of sparticles) [142; 238].

In Figure 8.4(a), the $(m_{1/2}, m_0)$ plane is displayed showing explicit element abundance contours [603] when the gravitino mass is $m_{3/2} = 0.2m_0$ in the absence of stau bound-state effects. To the left of the solid black line the gravitino is not the LSP. The diagonal dotted line corresponds to the boundary between a neutralino and stau NSP: above it, the neutralino is the NSP, and below it, the stau is the NSP. Very close to this boundary, there is a diagonal solid line. Above this line, the relic density of gravitinos from NSP decay is too high, i.e.

$$\frac{m_{3/2}}{m_{\text{NSP}}} \Omega_{\text{NSP}} h^2 > 0.12. \tag{8.19}$$

Thus we should restrict our attention to the area below this line.

The thick line labelled $^7\text{Li} = 4.3$ corresponds to the contour where $^7\text{Li}/\text{H} = 4.3 \times 10^{-10}$, a value very close to the standard BBN result for $^7\text{Li}/\text{H}$. It forms a 'V' shape, whose right edge runs along the neutralino-stau NSP border. Below the V, the abundance of ^7Li is smaller than the standard BBN result. However, for relatively small values of $m_{1/2}$, the ^7Li abundance does not differ much from the standard BBN result: it is only when $m_{1/2} \gtrsim 3000$ GeV that ^7Li begins to drop significantly. The stau lifetime drops with increasing

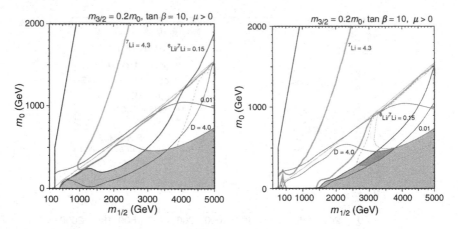

Fig. 8.4. The $(m_{1/2}, m_0)$ planes for $m_t = 172.7$ GeV, $A_0 = 0$, $\mu > 0$ and $\tan \beta = 10$ with $m_{3/2} = 0.2m_0$ without (a) and with (b) the effects of metastable stau bound states included. The regions to the left of the solid black lines are not considered, since there the gravitino is not the LSP. In the light shaded regions, the differences between the calculated and observed light-element abundances are no greater than in standard BBN without late particle decays. In the dark shaded region in (b), the abundances lie within the ranges favoured by observation. The significances of the other lines and contours are explained in the text.

$m_{1/2}$, and when $\tau \sim 1000$ s, at $m_{1/2} \sim 4000$ GeV, the ^{7}Li abundance has been reduced to an observation-friendly value close to 2×10^{-10} as reported in [1166; 1167] and shown by the (unlabelled) thin dashed contours.

The region where the ^{6}Li/^{7}Li ratio lies between 0.01 and 0.15 forms a band which moves from lower left to upper right. As one can see in the light shading, there is a large region where the lithium isotopic ratio can be made acceptable. However, if we restrict to D/H $< 4.0 \times 10^{-5}$, we see that this ratio is interesting only when ^{7}Li is at or slightly below the standard BBN result.

Turning now to Figure 8.4(b), we show the analogous results when the bound-state effects are included in the calculation. The abundance contours are identical to those in panel (a) above the diagonal dotted line, where the NSP is a neutralino and bound states do not form. We also note that the bound-state effects on D and ^{3}He are quite minimal, so that these element abundances are very similar to those in Figure 8.4(a). However, comparing panels (a) and (b), one sees dramatic bound-state effects on the lithium abundances. Everywhere to the left of the solid lines labelled 0.15 is excluded. In the stau NSP region, this means that $m_{1/2} \gtrsim 1500$ GeV. Moreover, in the stau region to the right of the ^{6}Li/^{7}Li $= 0.15$ contour, the

^{7}Li abundance drops below 9×10^{-11} (as shown by the thin dotted curve). In this case, not only do the bound-state effects increase the ^{6}Li abundance when $m_{1/2}$ is small (i.e. at relatively long stau lifetimes), but they also decrease the ^{7}Li abundance when the lifetime of the stau is about 1500 s. Thus, at $(m_{1/2}, m_0) \simeq (3200, 400)$ GeV, we find that ^{6}Li/^{7}Li $\simeq$ 0.04, ^{7}Li/H $\simeq 1.2 \times 10^{-10}$ and D/H $\simeq 3.8 \times 10^{-5}$. Indeed, when $m_{1/2}$ is between 3000 and 4000 GeV, the bound-state effects cut the ^{7}Li abundance roughly in half. In the darker shaded region, the lithium abundances match the observational plateau values, with the properties ^{6}Li/^{7}Li > 0.01 and $0.9 \times 10^{-10} < {}^7$Li/H $< 2.0 \times 10^{-10}$. This example demonstrates that it is possible to resolve the ^{6}Li/^{7}Li by postulating GDM with a stau NSP.

8.6 mSUGRA

Minimal supergravity (mSUGRA) is often used as a basis for phenomenological studies [193; 442; 1509]. The framework termed above the CMSSM is occasionally referred to as the mSUGRA. However, models based strictly on minimal supergravity should employ two additional constraints [775; 780]. One is a relation between the soft supersymmetry-breaking bilinear and trilinear parameters: $B_0 = A_0 - m_0$, and the other is a relation between the gravitino and input scalar masses: $m_{3/2} = m_0$. In the simplest version of mSUGRA [193; 442; 1509; 1583], where supersymmetry is broken by a single field in a hidden sector, the universal trilinear soft supersymmetry-breaking terms are $A_0 = (3 - \sqrt{3})m_0$ and the bilinear soft supersymmetry-breaking term is $B_0 = (2 - \sqrt{3})m_0$, which is a special case of the general relation $B_0 = A_0 - m_0$.

Given such a relation between B_0 and A_0, one can no longer use the standard CMSSM boundary conditions, in which $m_{1/2}$, m_0, A_0, $\tan\beta$ and $sgn(\mu)$ are input at the GUT scale, and then μ and B are determined by the electroweak symmetry-breaking conditions. In this case, it is natural to use B_0 as an input and calculate $\tan\beta$ from the minimization of the Higgs potential [775; 780].

Phenomenologically distinct planes may be determined by specifying a choice for A_0/m_0, with the above-mentioned simplest hidden sector being one example. In Figure 8.5, two such planes are shown assuming $m_t = 172.7$ GeV with (a) $A_0/m_0 = 3 - \sqrt{3}$, as predicted in the simplest model of supersymmetry breaking [1583], and with (b) $A_0/m_0 = 2.0$. We show in these $(m_{1/2}, m_0)$ planes the contours of $\tan\beta$ as solid lines. Also shown are the contours where $m_{\chi^\pm} > 104$ GeV (near-vertical black dashed lines)

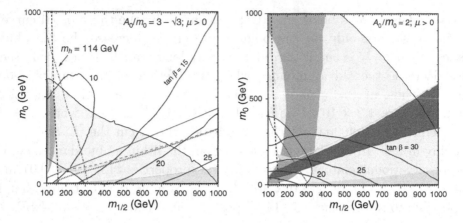

Fig. 8.5. Examples of mSUGRA $(m_{1/2}, m_0)$ planes with contours of $\tan\beta$ super-posed, for $\mu > 0$ and (a) the simplest Polonyi model with $A_0/m_0 = 3 - \sqrt{3}$, and (b) $A_0/m_0 = 2.0$, all with $B_0 = A_0 - m_0$. In each panel, we show the regions excluded by the LEP lower limits on MSSM particles and those ruled out by $b \to s\gamma$ decay (medium shading): the regions favoured by $g_\mu - 2$ are very light shaded, bordered by a thin (black) line. The dark solid lines separate the neutralino and gravitino LSP regions. The regions favoured by WMAP in the neutralino LSP case have light shading. The dashed line corresponds to the maximum relic density for the grav-itino LSP, and regions allowed by BBN constraint neglecting the effects of bound states on NSP decay have light shading.

and $m_{\rm h} > 114$ GeV (diagonal dash-dotted lines). The regions excluded by $b \to s\gamma$ have medium shading, those where the relic density of neutralinos lies within the WMAP range have light shading, and the region suggested by $g_\mu - 2$ at 2σ has very light shading, as in the CMSSM planes shown previously. As one can see, relatively low values of $\tilde{b}$ are obtained in most of the visible planes.

Another difference between the CMSSM and models based on mSUGRA concerns the mass of the gravitino. In the CMSSM, it is not specified and can be taken suitably large so that the neutralino or the lighter stau is the LSP. In mSUGRA, the scalar masses at the GUT scale, m_0, are determined by (and equal to) the gravitino mass. In Figure 8.5, the gravitino LSP and the neutralino LSP regions are separated by dark solid lines. Above these lines, the neutralino (or stau) is the LSP, whilst below them the gravitino is the LSP [777; 837; 841]. As one can see by comparing the two panels, the potential for neutralino dark matter in mSUGRA models is dependent on A_0/m_0. In panel (a), the only areas where the neutralino density is not too large occur where the Higgs mass is far too small or, at higher m_0, the chargino mass is too small. At larger A_0/m_0, the co-annihilation strip rises above the neutralino–gravitino LSP boundary. In panel (b), we see the familiar co-annihilation strip. It should be noted that the focus-point region is not realized in mSUGRA models as the value of μ does not decrease with increasing m_0 when A_0/m_0 is fixed and $B_0 = A_0 - m_0$. There are also no funnel regions, as $\tilde{b}$ is never sufficiently high.

In the gravitino LSP regions, the NSP may be either the neutralino or stau, which are now unstable. (Note that in panel (b) there is also a region which is excluded because the stau is the LSP.) The relic density of gravitinos is acceptably low only below the dashed line. This excludes a supplementary domain of the $(m_{1/2}, m_0)$ plane in panel (a) which has a neutralino NSP (the dotted curve in panel (a) separates the neutralino and stau NSP regions). However, the strongest constraint is provided by the effect of neutralino or stau decays on Big Bang nucleosynthesis [507; 602; 784; 843; 1259; 1801]. Outside the light shaded region, the decays spoil the success of BBN.

8.7 Other possibilities

These cosmologically preferred regions move around in the $(m_{1/2}, m_0)$ plane if one abandons the universality assumptions of the CMSSM. For example, if one allows the supersymmetry-breaking contributions to the Higgs masses to be non-universal (NUHM), the rapid-annihilation WMAP 'strip' can appear at different values of $\tan\beta$ and $m_{1/2}$, as seen in Fig. 8.6 [751; 773]. Rapid annihilation through the direct-channel H, A poles suppresses the relic density between the two parallel vertical WMAP strips at smaller values of $m_{1/2}$, and the relic density is suppressed in the rightmost strip because the neutralino LSP has a significant Higgsino component. A complete exploration of the parameter space of the NUHM, which has two additional parameters compared to the CMSSM, lies beyond the scope of this review.

The appearance of the $(m_{1/2}, m_0)$ plane is also changed significantly if one assumes that the universality of soft super symmetry-breaking masses in the CMSSM occurs not at the GUT scale, but at some lower renormalization scale [769; 770; 771], M_{in}, as occurs in some 'mirage unification' models [545; 546; 808]. In this case, the sparticle masses are generally closer together. As a consequence, the bulk, co-annihilation, rapid-annihilation and focus-point regions approach each other and eventually merge as the mirage unification scale is reduced, as illustrated in Fig. 8.7, where they form an 'atoll'. At smaller values of the mirage unification scale, the atoll contracts and eventually disappears, and there is no WMAP-compatible within the displayed portion of the $(m_{1/2}, m_0)$ plane. In such 'GUTless'

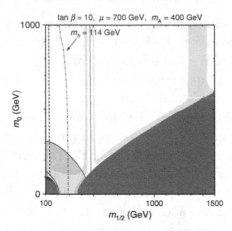

Fig. 8.6. The $(m_{1/2}, m_0)$ plane in the NUHM for $\tan\beta = 10$, $\mu = 700$ GeV and $m_A = 400$ GeV [751]. The styles of the shadings and contours are the same as in Fig. 8.2.

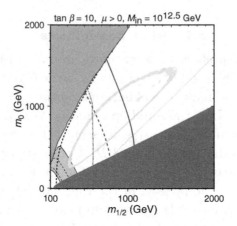

Fig. 8.7. The $(m_{1/2}, m_0)$ plane in a GUTless model with $\tan\beta = 10$, $\mu > 0$ and $A_0 = 0$, assuming universality at $M_{\rm in} = 10^{12.5}$ GeV [770]. The colours of the shadings and contours are the same as in Fig. 8.2.

models, $\Omega_{\rm LSP}h^2$ falls below the WMAP range (8.18) in larger regions of the $(m_{1/2}, m_0)$ plane than in the conventional CMSSM with unification at the GUT scale.

8.8 Summary

As we have discussed above, there are many sound theoretical and phenomenological reasons to favour supersymmetric extensions of the Standard Model. In particular, supersymmetry predicts the existence of cold dark matter in a very natural way, and there are several plausible candidates for the lightest supersymmetric particle that would be present as a relic from the Big Bang. The most prominent candidate is the lightest neutralino, and we have described how its relic density may be calculated, and the regions of supersymmetric parameter space in which its density falls within the range favoured by astrophysics and cosmology. However, other candidates for the cold dark matter are also possible, such as the gravitino. In that case, the next-to-lightest supersymmetric particle would be metastable, and comparisons between the observed light-element abundances and those predicted by Big Bang nucleosynthesis calculations impose important constraints on the parameter space. We have given examples of neutralino and gravitino dark matter scenarios in the minimal supersymmetric extension of the Standard Model, under various different theoretical assumptions. It will be for collider and dark matter detection experiments to determine which, if any, of these options has been adopted by Nature.

9

Dark matter at the electroweak scale: non-supersymmetric candidates

Géraldine Servant

Despite the impressive successes of the Standard Model (SM) in describing nearly all experimental data collected so far in particle physics, it is not viewed as a fundamental theory but as an effective field theory valid on scales less than a few TeV, at most. The problem lies in the difficulty of understanding the relatively low values of the Higgs mass parameter $|m_H^2| \sim (100 \text{ GeV})^2$ in a framework in which the SM is valid up to some ultra-high scale, for instance of the order of the Planck scale. This is because the Higgs boson mass parameter receives radiative corrections (dominantly from the top loop, the W and Z gauge bosons and from the Higgs itself) that are quadratically divergent, and therefore proportional to Λ^2 where Λ is the maximum mass scale that the theory describes:

$$\delta m_H^2 = \frac{3\Lambda^2}{8\pi^2 v^2} \left(2m_W^2 + m_Z^2 + m_H^2 - 4m_t^2\right) \sim -(0.23\Lambda)^2. \qquad (9.1)$$

For large values of Λ, tree-level and radiative contributions to the Higgs mass parameter must cancel. For the SM to be valid up to 5 TeV a cancellation by two orders of magnitude is already required and to reach the Planck scale requires an adjustment finely tuned to 32 orders of magnitude. This is the so-called hierarchy problem. Therefore, a theory with a light Higgs is not a satisfactory effective description since it does not incorporate the dynamics at work in the cancellation of quadratic divergences.

Over the past two decades, this hierarchy problem has been the main reason to think that the SM should be overthrown at the electroweak (EW) scale. Theories that solve this naturalness problem, i.e in which the ratio between the EW scale and the Planck scale can be understood dynamically without recourse to fine-tunings, have been proposed, starting with the early

Particle Dark Matter: Observations, Models and Searches, ed. Gianfranco Bertone. Published by Cambridge University Press. © Cambridge University Press 2010.

proposals of supersymmetry and technicolour through to the more recent ideas of large and warped dimensions, and the Little Higgs. In all these models, there is a whole set of new states arising at the TeV scale, responsible for cancelling the Higgs divergences. They are the main focus of potential discoveries at the LHC.

Supersymmetry has long appeared to be the most realistic extension of the Standard Model with Higgs mass stabilized under quantum corrections. However, after the LEP2 results, the lack of discovery of a Higgs boson below 114 GeV or of any supersymmetric particles has forced minimal supersymmetry into fine-tuning territory, partially undermining its original motivation. In recent years, models of dynamical EW symmetry breaking have received a revival of interest, triggered by the realization that extra-dimensional models may provide a weakly coupled dual description to technicolour type theories. The connection between strongly interacting gauge theories and gravity on warped geometries has led the way to the construction of new models of EW symmetry breaking (Holographic Higgs, Higgsless, Little Higgs).

More recently, 'unnatural' directions have also been opened up, as a result of both experimental observations and theoretical speculations. On the one side, the evidence for dark energy has reopened the question of the cosmological constant, which has satisfactory anthropic justification, but no successful explanations based on symmetry or dynamics. Also, as just mentioned, the negative LEP searches for new physics have created some conflict in essentially all known models that can naturally explain the weak scale. The scale of new physics Λ must be close to 1 TeV to keep the Higgs light. However, the new physics should be rather special as its effects should not be visible in the EW precision tests. This is the 'LEP paradox' [198]. On the theoretical side, the formulation of the string landscape together with an inflationary picture has given a more solid justification for a multiverse description, where some of the properties of our Universe are described by environmental selection.

Independent of one's opinion on the relevance of the naturalness problem and even ignoring all other theoretical puzzles unanswered by the SM (gauge coupling unification, origin of the three families of quarks and leptons, hierarchical spectrum of fermion masses, neutrino masses, dark energy . . .), the dark matter (DM) puzzle and the baryon asymmetry are the two major hints of physics beyond the SM. Interestingly, they can be solved by adding a few extra ingredients to the theory at the EW scale, most probably the same ingredients that are tied up to the underlying mechanism of EW symmetry breaking. Therefore, in addition to addressing the major question of EW symmetry breaking in theoretical physics, the next generation of colliders will potentially also answer two very fundamental questions related to the cosmological evolution of our Universe.

A weakly interacting massive particle as the explanation for DM is very attractive for two reasons. First, its production mechanism is generic and independent from the initial conditions in the history of the Universe. The only requirement is that the reheating temperature of the Universe should be larger than a few tens of GeV, which is compulsory anyway for baryogenesis. Second, the existence of WIMPs is directly related to the presence of new particles at the electroweak scale and is easily justified and motivated by the naturalness problem exposed above. A symmetry is required to guarantee their cosmological stability, otherwise they would rapidly decay into SM particles. The most studied candidate is the LSP (the lightest supersymmetric particle). In supersymmetry, there is a discrete symmetry called R-parity guaranteeing the stability of the LSP, the neutralino, which acts as a natural dark matter candidate. This scenario has been thoroughly investigated in anticipation of the LHC's starting up, and gives tight constraints on the minimal supersymmetric Standard Model. However, some alternatives to neutralinos and other LSPs do exist, as most of the models introduced to address the naturalness problem also turn out to contain WIMP candidates in their spectrum and therefore naturally provide a solution to the DM puzzle. Besides, a more practical and minimal approach to model building has recently been gaining momentum, where the hierarchy problem is ignored, and dark matter is the sole motivation for considering new physics. These models are typically not theoretically motivated but they lead to interesting observable signatures. In this chapter, we will review models reflecting both approaches but we will put more emphasis on the first category.

9.1 New symmetries at the TeV scale and dark matter

The new physics around the TeV scale that is invoked to cutoff the quadratically divergent quantum corrections to the Higgs mass induces higher-dimensional operators involving the SM particles. This results in a tension with precision tests of the SM, in both the EW and the flavour sector. To be consistent with the EW precision tests, flavour-preserving operators generated by new physics typically require the scale of new physics to be larger than a few TeV [197] and are difficult to suppress by any known (approximate) symmetries of the SM.[1] This tension is called the post-LEP 'little hierarchy' problem.

---

[1] Exceptions include custodial isospin for the T parameter [1569].

Table 9.1. *Comparison between some WIMP DM candidates discussed in this chapter.*

	LKP	LZP	LSP	LTP
Set-up	UED	Warped GUTs	Supersymmetry	Little Higgs
Particle	Gauge boson	Dirac fermion	Majorana fermion	Gauge boson
Symmetry	KK parity: $(-1)^n$, consequence of geometry	Z_3: $B - \frac{n_c - \bar{n}_c}{3}$, imposed for proton stability	R-parity: $(-1)^{3(B-L)+2S}$, imposed for proton stability	T-parity imposed for EW precision data
Typical mass range	500–1500 GeV	20 GeV–few TeV	50–1000 GeV	80–500 GeV

A new symmetry at the TeV scale can ameliorate some of these constraints if at least the lightest new physics (NP) states, which a priori give the largest electroweak corrections, are charged under this symmetry while the SM particles are neutral [535; 1945]. In such a case, the charged NP states do not contribute at tree level to the operators constrained by the precision tests since couplings of a single charged state to SM particles are forbidden. New physics contributions arise at loop level only. This makes sub-TeV NP states consistent with EW precision data. These NP states may then play the role of cutting off the Higgs mass divergence without any fine-tuning, thus avoiding the little hierarchy problem. As a spin-off, the new symmetry implies the existence of a new stable particle that can be a dark matter candidate if it is electrically neutral and weakly interacting.

The simplest possibility of a new symmetry at the TeV scale is a discrete Z_2 parity. The classic example is R-parity in supersymmetry. In Little Higgs, a similar role is played by T-parity [535; 536] under which the new gauge bosons are charged. Yet another example is Kaluza–Klein (KK) parity [537] in universal extra dimensions (UED) [100]; see Table 9.1.

In Sections 9.2 to 9.5, we will present DM candidates in scenarios motivated by the (little) hierarchy problem. In Section 9.6, we will review minimal approaches where the DM puzzle is addressed without worrying at all about the hierarchy problem.

9.2 Dark matter from extra dimensions: Kaluza–Klein DM

In the past decade, alternative models for physics beyond the Standard Model have been studied that make use of extra dimensions rather than supersymmetry to solve the gauge hierarchy problem. They also turn out to offer new solutions to the dark matter puzzle. Physics in extra dimensions was initiated at the beginning of the twentieth century through the Kaluza–Klein theory [1197; 1244]. It was ignored for half a century, and finally resuscitated by purveyors of supergravity and superstrings in the 1970s and 1980s. However, it is only recently that the phenomenology of extra dimensions started to be explored. The renaissance of extra dimensions began in 1998 with the Arkani-Hamed, Dimopoulos and Dvali (ADD) proposal [99; 111] to lower the scale of quantum gravity, M_*, to a TeV by localizing the SM to a $3 + 1$-dimensional surface or brane in a higher-dimensional space-time. The n extra dimensions are compactified into a large volume R^n that effectively dilutes the strength of gravity from the fundamental scale (the TeV scale) to the Planck scale: $M_{\mathrm{Pl}}^2 = M_*^{2+n} R^n$. This proposal initiated a whole new field of investigation. The literature on extra dimensions inflated. Very different types of models arose depending on the geometry and scale of compactification, on which fields propagate in extra dimensions, on the type of boundary conditions, etc. The most promising one, the so-called Randall–Sundrum set-up [1625; 1626], is based on a slice of 5D anti de Sitter space.

The basic idea of the Kaluza–Klein picture is that every multidimensional field corresponds to a Kaluza–Klein tower of four-dimensional particles with increasing masses. At low energies, only massless (on the scale $1/R$) particles can be produced, whereas at $E \sim 1/R$ extra dimensions will show up. In the ADD scenario, only gravity propagates in extra dimensions, in which case the constraint on the size of R was advocated to be surprisingly weak ($R < 1$ mm $\sim$ meV^{-1}). On the other hand, in models where ordinary particles can propagate in extra dimensions, the energy scale $1/R$ must be at least in the range of a few hundred GeV since the KK partners of ordinary particles have not been observed, so the size of extra dimensions must be microscopic, $R < 10^{-17}$ cm.

Therefore, among the new ingredients of extra-dimensional theories are the KK excitations of the graviton as well as the radion, the scalar degree of freedom related to the size of extra dimensions. If the extra-dimensional model contains branes, there are also possibly branons, which are associated to brane fluctuations. If the SM lives in the extra dimensions, there are KK excitations for all SM particles. All of them look like potential candidates for dark matter provided that they are stable.

For a particle to be stable, either it has large couplings to SM particles and there must be a symmetry to guarantee its stability, or it interacts so weakly that its lifetime is longer than the age of the Universe, as is the

Table 9.2. *DM candidates in three main classes of extra-dimensional models.*

	Bulk content	DM candidates
ADD models $R \sim$ meV^{-1} (flat extra dimension)	Only gravity is in the bulk, the SM is localized on a 3-brane	• Radion DM, $m \sim$ meV • KK graviton DM, $m \sim$ meV (both finely tuned) • Branon DM (not in the original ADD model, the hierarchy problem remains)
Flat TeV^{-1} **dimensions** $R \sim$ TeV^{-1}	Matter fields may be localized at the boundaries of the extra dimension but if all SM fields are in the bulk ('universal extra dimensions'):	• Radion DM, $m \sim$ meV (finely-tuned); KK graviton is unstable • KK DM, $m \sim$ TeV ⬜ **WIMP or SuperWIMP**
Warped dimensions $R \sim M_{\rm Pl}^{-1}$ $M_{\rm KK} \sim$ TeV	Randall–Sundrum AdS$_5$ geometry. If GUT in the bulk:	• Radion is unstable • KK neutrino DM: ⬜ **WIMP** $m \sim$ few GeV–few TeV

case of light particles with only gravitational couplings. We go through the various possibilities in the next subsections. The situation is summarized in Table 9.2.

Let us start with KK particles. In most extra-dimensional models, KK states are not stable as they are all able to decay into SM particles. Nevertheless, we may wonder whether there are new symmetries available in extra-dimensional contexts which could make a KK mode stable. Indeed, a new dimension potentially means a new conserved momentum along the extra dimension. This leads to the so-called KK parity, a discrete symmetry which remains unbroken in some specific class of extra-dimensional models named 'Universal extra dimensions' [100]. As a result, the lightest KK particle (LKP) is stable. For TeV^{-1} sized extra dimensions, if the LKP is electrically neutral and has interactions of the same size as electroweak interactions, it can act as a WIMP. Some particular examples follow.

• **The KK 'photon'.** One of the most studied examples and the subject of the subsection 9.2.1 is when the LKP is identified with the first KK excitation of the photon. It is actually not really a KK photon because the Weinberg angle for KK modes is very small [538]. It is essentially the KK hypercharge gauge boson: B^1. Relic density [478; 1193; 1194; 1195; 1730; 1268], direct [128; 533; 1729] and indirect detection [183; 221; 293; 294; 340; 360; 443; 533; 1122; 1123] studies of this candidate have been carried out in the past few years. Constraints on these models from radion cosmology have also been discussed [1262].

- **The KK 'Z' and KK Higgs.** The nature of the LKP is model-dependent. The KK mass spectrum depends in particular on the mass terms localized at the boundaries of the extra dimension as well as on the Higgs mass. In some scenarios, Z^1 or H^1 can be the LKP [128] and play the role of dark matter.

- **The KK neutrino.** The possibility that the LKP is the KK excitation of the left-handed neutrino ν_L rather than a KK photon in UED was also studied in refs. [1729; 1730]. This case is excluded experimentally by direct detection experiments because of the large coupling of $\nu_L^{(1)}$ to the Z gauge boson, leading to much too large elastic scattering of the KK neutrino with nucleons.

It could also be that the LKP is the KK excitation of a right-handed neutrino ν_R. To behave as a WIMP, such a particle should interact with TeV KK gauge bosons like in left–right gauge theories such as Pati–Salam or $SO(10)$. This possibility was investigated in UED [1135] as well as in warped grand unified theories (GUTs) [32; 33]. In this latter case, the symmetry which guarantees the stability of the LSP is not the KK parity but a Z_3 discrete symmetry that is primarily postulated to get rid of the proton decay problem. The proton decay problem also arises in extra-dimensional theories, specifically if the cutoff scale is near the TeV scale. It is interesting to investigate whether the symmetry one assumes to get rid of the proton decay can lead to a stable particle, like in supersymmetry. We will indeed present such a solution in Section 9.2.6 where the DM particle is called the LZP. Before presenting the LKP and LZP candidates in more detail, let us review other (non-WIMP) possibilities which have been mentioned in the literature.

- **KK graviton.** In the ADD scenario, only the graviton probes the full bulk space. There is a Kaluza–Klein tower of graviton modes, where the massless mode is the standard 4D graviton, and the other KK modes have mass $m_{KK} \sim R^{-1}$. For instance, if $n = 2$, $m_{KK} \sim M_*^2/M_{Pl}$, which is of the order of meV.

The KK graviton of ADD, with a mass of meV, is stable on cosmological scales (each KK graviton couples only with $1/M_{Pl}$, so it has a decay rate of order $\Gamma \sim T_*^3/M_{Pl}^2$ where T_* is the reheat temperature in the ADD scenario, typically constrained to be below 1 GeV) and could be a DM candidate. It would not be a WIMP and the correct relic density cannot be obtained via the standard thermal calculation. To get the correct relic density requires fine-tuning either in initial conditions for inflation or in the reheat temperature of the Universe, otherwise KK gravitons would overclose the Universe. In addition, there are strong astrophysical constraints on the ADD scenario [493; 1069].

The situation is different in UED models where the right relic abundance can be obtained naturally. The idea is that the standard cold relic abundance is obtained for the next lightest KK particle (NLKP), which is a WIMP (a KK hypercharge gauge boson in UED with mass of the order of TeV) and the NLKP later decays into the LKP which is the KK graviton. That way, the KK graviton, which has a TeV mass and only a gravitational coupling, can still acquire the right abundance as given by the standard thermal relic calculation. This scenario has been intensively studied by Feng *et al.* [836; 837; 838] and is reported in Chapter 10. Finally, let us mention that in Randall–Sundrum models [1625; 1626], KK gravitons have a TeV mass and interact strongly so they cannot play the role of dark matter.

- **Radion.** The radion in ADD has typically the same mass and same coupling as the KK graviton and also suffers from an overclosure problem. As for models with TeV extra dimensions, there is also typically an overclosure problem. Solving it requires modifying the assumptions on the compactification scheme. Details of radion cosmology have been studied in ref. [1262]. The radion overclosure problem does not apply in Randall–Sundrum models where the radion has large interactions and large mass so that it decays fast.

- **Branons.** Branons correspond to brane fluctuations. They control the coordinate position of our brane in the extra dimensions. Those fields can be understood as the Goldstone boson arising from spontaneous breaking of translational invariance by the presence of the brane. They acquire mass by the explicit breaking of the symmetry. The possibility that branons could be dark matter has been investigated in refs. [499; 500]. In this context, the SM lives on a 3D brane embedded in a higher-dimensional $(D = 4+n)$ space-time where the fundamental scale of gravity M_D is lower than the Planck scale. In the original ADD proposal, M_D was the TeV scale. The authors of branon dark matter work in a general brane world scenario with arbitrary fundamental scale (larger than the TeV scale). The branon degree of freedom cannot be neglected when the brane tension scale f is much smaller than M_D, which means that we live on a non-rigid brane. Branon interactions with particles living on the brane can be computed as a function of f, N and the branon mass M. Couplings of KK modes to the fields confined on the brane are exponentially suppressed by the fluctuation of the brane [186]. As f is very small, the KK mode contributions become invisible from our world and the only remaining degrees of freedom are the branons. The gravitational interaction on the brane conserves parity and terms in the effective Lagrangian with an odd number of branons are forbidden. As a consequence, branons are stable. Constraints in the region of parameters made by N, M_D, M and f have been derived. We refer the reader to refs. [499; 500] for details and references.

To summarize, KK particles can arise as stable viable WIMPs in several frameworks: in flat extra dimensions (whether 'universal' with KK parity or non-universal with a new Z_2 symmetry, as we shall see) and in some warped geometries *à la* Randall–Sundrum. We will now discuss some of these possibilities in more detail.

9.2.1 *Universal extra dimensions*

In models with universal extra dimensions (UED) [100], all SM fields propagate in flat toroidal extra dimensions, unlike models in which the SM is localized on a brane. Translation invariance along an extra dimension is only broken by the orbifold projection imposed to recover a chiral SM spectrum (see Section 15.1 for more details). Still, there is a remnant discrete symmetry called KK parity, $(-1)^n$, where n is the KK number. This symmetry ensures that interaction vertices cannot involve an odd number of odd-KK states and, therefore, a vertex with two SM particles (with $n = 0$) and one KK state (with $n = 1$) is forbidden. As a result, the lightest KK particle (LKP) with $n = 1$ cannot decay into SM particles and is stable. Note that KK parity is a reflection about the midpoint of the extra dimension combined with the orbifold projection. For it to be an exact symmetry, one has to assume that the boundary Lagrangians at the two orbifold fixed points are symmetric.

In contrast with supersymmetry where the mass spectrum is largely spread so that at most a few additional particles participate in co-annihilation processes with the LSP, in minimal UED (MUED) the mass spectrum of KK particles is rather degenerate and there are many co-annihilation processes.

The KK mass splittings are essentially due to radiative corrections [538]. The spectrum of KK masses depends also on the values of boundary terms at the cutoff scale, which are not fixed by known SM physics. In this sense, the values of the KK masses can be taken to be arbitrary and the UED scenario has a multitude of parameters. Assuming vanishing boundary terms (this is the so-called MUED hypothesis) and a light Higgs ($m_h = 120$ GeV), the LKP is the KK hypercharge gauge boson B^1 [538].

The viability and relic density of the B^1 LKP were first analysed in [1730] with some simplifying assumptions about the KK spectrum (only one co-annihilation channel involving the KK right-handed electron was considered) and it was shown that including co-annihilation effects can reduce the mass for the DM particle. This is to be contrasted with the supersymmetric case where co-annihilation effects tend to push the prediction for the mass of the neutralino to higher values. The reason is that the annihilation cross-section of the KK photon is not helicity-suppressed and co-annihilation channels do not greatly modify the total effective annihilation cross-section. Therefore, because more degrees of freedom are included to start with, there is no need to compensate with a large mass. In refs. [478; 1268], all co-annihilation channels with KK fermions and KK gauge bosons were included. The net result is that even if the new co-annihilations are Boltzmann suppressed, their effect is still significant because the cross-sections are mediated by weak or strong interactions, while the cross-sections studied so far were purely hypercharge-mediated processes. The conclusion is that in MUED, the LKP mass should be within 500–600 GeV, while in non-minimal UED models, freedom in the KK mass spectrum allows the LKP to be as heavy as 2 TeV. For an analysis taking into account the effects of second-level KK modes, see refs. [1193; 1194]. The effect of co-annihilation with the KK Higgs was studied in [1400]. For a low Higgs mass, precision EW observables set a strong constraint on the KK scale of MUED models ($1/R > 600$ GeV). However, values as low as $1/R \sim 300$ GeV are allowed for heavy Higgs [971]. Previous bounds on $1/R$ from EW precision tests were derived in [100; 101], from direct non-detection and from $b \to s\gamma$ in [29] and from flavour-changing neutral currents in [468; 469]. The collider phenomenology of UED will be presented in Chapter 15. We now move to direct detection constraints.

Direct and indirect detection

As shown in Fig. 9.1, elastic scattering of the B^1 LKP and target nuclei arises from KK quark exchange and Higgs exchange [128; 533; 1729]. Direct

G. Servant

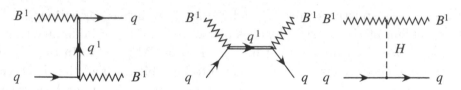

Fig. 9.1. Leading Feynman graphs for effective B^1-quark scattering through the exchange of a KK quark (both q_L^1 and q_R^1) and through the exchange of a zero-mode Higgs boson. The diagrams for a Z^1 LKP are similar.

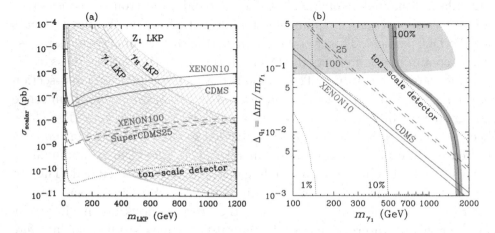

Fig. 9.2. (a) Theoretical predictions for spin-independent LKP-nucleon cross-sections for $m_h = 120$ GeV and $\Delta = (m_{q^1} - m_{LKP})/m_{LKP}$ between 1%, which is the upper boundary of the respective shaded area, and 50%, which is the lower boundary. Limits from CDMS and XENON10 as well as expected sensitivities from future experiments are displayed. (b) The black solid line accounts for all the dark matter in the Universe while the two black dotted lines show the bounds assuming that the LKP would contribute only 1% or 10% to the total amount of dark matter. The darker shaded region represents the preferred WMAP region. The paler shaded region should be covered by the LHC. Both plots are from ref. [128].

detection does not appear the most promising way to probe B^1 LKP dark matter, as the sensitivity of near future experiments does not allow one to probe realistic mass splittings. The situation is summarized in Fig. 9.2(b) where all mass splittings below the respective limit curves are excluded.

Indirect detection through gamma-rays [183; 293; 294; 340; 533], neutrinos and synchrotron flux [340], positrons [533; 1123], antiprotons [221] or antideuterons [162] has also been considered. The neutrino spectrum from LKP annihilation in the Sun was investigated in [1123]. An interesting feature of KK dark matter is, in contrast with the neutralino, that annihilation into fermions is not helicity-suppressed and there can be a direct annihilation into e^+e^- leading to a very valuable positron signal from LKP annihilation into the Galactic halo [533].

9.2.2 'Spinless photon' in 6D UED

Theories with two compact universal extra dimensions also contain a KK parity. Four-dimensional particles are labelled by two positive integers (n, m). These particles are odd under KK parity when $n + m$ is odd. In 6D, there are two spin-0 fields transforming in the adjoint representation of the gauge group. One linear combination is eaten to become the longitudinal degree of freedom of the spin-1 KK particle. The other linear combination remains a physical spin-0 particle [684].

Neglecting effects from cutoff scale physics localized at the corners of the square compactification, the (1,0) mode of the spinless adjoint of the hypercharge gauge group turns out to be the lightest one, the so-called 'spinless photon', and is stable because of Kaluza–Klein parity.

The main difference with the KK photon of 5D UED is that the spinless photon annihilation into fermions is helicity-suppressed and it mainly annihilates into W and Z gauge bosons and Higgs. The mass prediction is rather low (< 500 GeV) and depends on the Higgs mass. This dark matter candidate resembles a neutralino more than a KK photon. Indirect detection with gamma-rays and antimatter is difficult because of the helicity-suppression of the annihilation cross-section into fermions. It has no spin-dependent scattering cross-section and its spin-independent scattering cross-section is also helicity-suppressed. This makes indirect detection in neutrino telescopes (due to scattering in the Sun) almost hopeless and direct detection in underground detectors very challenging.

9.2.3 Flat non-universal extra dimensions

The 5D gauge–Higgs unification model of ref. [1545] provides an example of KK gauge boson dark matter outside of the UED context. This construction addresses EW symmetry breaking (in contrast to UED) in the sense that the Higgs field is the internal component of a 5D gauge field. Gauge–Higgs unification models in flat space generally suffer from problems such as too small top and Higgs masses. To resolve this, the model [1545] has Lorentz symmetry breaking in the bulk, and an extra Z_2 symmetry is imposed to improve naturalness of the model. A consequence of this Z_2 symmetry is the stability of the lightest Z_2 odd particle which could play the role of dark matter.

In contrast with UED, the two localized Lagrangians at the orbifold fixed points are different and there is no KK parity. Consequently, EW precision constraints impose a significantly stronger bound on the LKP mass. For this heavy (in comparison to UED) LKP to have the correct relic density, it is then crucial that co-annihilation effects involving strongly interacting particles take place.

The LKP is the first KK mode of an antiperiodic gauge field, A^1, roughly aligned with $U(1)_Y$. The relic density depends on essentially three parameters: $1/R$, the mass splitting between the LKP and the NLKP, and the amount of bulk Lorentz violation. The NLKP–LKP mass splitting has to be below 7% and one has to tune the Lorentz-violating parameters appropriately to keep the mass splittings small. Nevertheless, this represents a viable example of a cold dark matter candidate that is heavier than standard WIMPs (>2.3 TeV) and whose prospects for detection at colliders or in usual dark matter experiments are unfortunately very low. The generalization of this model to warped geometries is presented in Section 9.2.7.

9.2.4 *Radion constraints in flat extra dimensions*

We close this discussion of flat extra dimensions by mentioning the overclosure problem due to the radion. This typically arises in models with flat TeV extra dimensions, as reviewed in [1262]. For the radion not to overclose the Universe, one should make sure that there is a radion-stabilizing mechanism which leads to a heavy radion. Note that this problem disappears in non-toroidal compactifications or in warped Randall–Sundrum geometries.

9.2.5 *Warped extra dimensions*

The interest in the phenomenology of extra dimensions over recent years has been motivated by the goal of understanding the weak scale. The extra-dimensional geometry which really addresses the hierarchy problem is the Randall–Sundrum geometry [1625], based on a slice of 5D anti de Sitter space, AdS_5 (see Fig. 15.5 of Chapter 15). Owing to the warped geometry, the 4D (or zero-mode) graviton is localized near the UV brane which has a Planckian fundamental scale, whereas the Higgs sector can be localized near the IR brane where the cutoff is of order TeV. Based on the AdS/conformal field theory correspondence [1377; 1941], this model is conjectured to be dual to 4D composite Higgs models [117; 574; 1629]. Particle physics model building in this framework has been flourishing and a favourite class of models has emerged: that where all SM fields propagate in the bulk of AdS$_5$, except for the Higgs (or alternative physics responsible for electroweak symmetry breaking), which is localized on the IR brane. In addition, the electroweak gauge group should be extended to SU(2)$_L$ × SU(2)$_R$ × U(1) to guarantee the custodial symmetry needed to avoid too large deviations with respect to electroweak observables. More details on these constructions are given in Chapter 15.

9.2.6 Warped GUT models with Z_3

When these models are embedded into a GUT (see refs. [32; 33]), a stable KK fermion can arise as a consequence of imposing proton stability in a way reminiscent of R-parity stabilizing the LSP in supersymmetric models. The symmetry is called Z_3 and is a linear combination of baryon number and SU(3) colour. As soon as baryon number is promoted to be a conserved quantum number, the following transformation becomes a symmetry:

$$\Phi \rightarrow e^{2\pi i\left(B - \frac{n_c - \bar{n}_c}{3}\right)}\Phi, \tag{9.2}$$

where B is baryon number of a given field Φ (proton has baryon number $+1$) and n_c ($\bar{n}_c$) is its number of colours (anticolours). SM particles are not charged under this symmetry since only coloured particles carry baryon number in the SM. In higher-dimensional GUTs, baryon number can be assigned in such a way that there exist exotic KK states that have the gauge quantum numbers of a lepton but that carry baryon number as well as KK quarks with non-standard baryon number. These particles carry a non-zero Z_3 charge. The lightest of these, called the lightest Z_3 particle (LZP), is stable since it cannot decay into SM particles [32; 33].

The only massive elementary Dirac fermion which could be a viable WIMP candidate is the neutrino. If such a neutrino had the same coupling to the Z as in the SM, however, it would be excluded by direct detection experiments. Its coupling to the Z, therefore, must be suppressed.[2] Thus, we are left with the possibility of a KK right-handed neutrino. In models where the electroweak gauge group is extended to SU(2)$_L$ × SU(2)$_R$ × U(1), the RH neutrino has gauge interactions in particular with the additional Z'.

---

[2] Note that in supersymmetry, such constraints are weaker because of the Majorana nature of the neutralino.

Nevertheless, its interactions with ordinary matter are feeble because they involve the additional gauge bosons which have a large mass ($M_{KK} \gtrsim 3$ TeV). This opens the possibility of a weakly interacting Dirac RH neutrino. In principle, the LZP is not necessarily the lightest KK particle. There might be KK modes which are lighter but which are unstable because they are not charged under Z_3. There is a large freedom in the choice of parameters determining the KK spectrum. However, the RH neutrino LZP is assumed to be the lightest KK particle to satisfy phenomenological constraints.

In summary, the LZP is a Kaluza–Klein fermion, which is a four-component spinor and vector-like object. As explained in great detail in ref. [33], it can be naturally very light, much lighter than the KK scale of Randall–Sundrum models, namely $M_{KK} \gtrsim 3$ TeV. This is because the RH chirality is localized near the TeV brane while the left-handed one is near the Planck brane. The overlap of wavefunctions is small, resulting in a small Dirac KK mass. Its lightness is related to the top quark's heaviness but not entirely fixed by it, so that LZPs in the mass range of approximately 20 GeV to a few TeV can be considered (see Fig. 15.6). Only the RH chirality has significant interactions and the other chirality decouples. The LZP has the same gauge quantum numbers as the RH neutrino of $SO(10)$ or Pati–Salam. As a result, we refer to it as a 'Dirac RH neutrino'.

Via the AdS/CFT correspondence, the Randall–Sundrum scenario is dual to a 4D composite Higgs scenario, in which the unification of gauge couplings has been studied [27]. In this case, the LZP maps to some low-lying hadron at the composite scale. We also point out that in refs. [32; 33], the strong coupling scale is close to the curvature scale so that $\mathcal{O}(1)$ variations in calculations are expected.

Relic density

An interesting feature of warped GUT models is that GUT states such as X, Y gauge bosons appear at the TeV scale (via the KK excitations). In $SO(10)$, there are also the X', Y', X_s, Z' etc. that the LZP can couple to. In addition, when electroweak symmetry is broken, $Z-Z'$ mixing induces a coupling of the RH neutrino to the SM Z gauge boson. This coupling being suppressed by $(M_Z/M_{Z'})^2$, if $M_{Z'}$ is a few TeV (the mass of KK gauge bosons is set by M_{KK}), the size of this coupling will typically be ideal for a WIMP.

For LZPs lighter than approximately 100 GeV, LZP annihilations proceed dominantly via s-channel Z-exchange and annihilations to light quarks, neutrinos and charged leptons are important. For larger masses, annihilation via the t-channel exchange of X_s into top quarks or via s-channel Z' exchange into $t\bar{t}$, $b\bar{b}$, W^+W^- and Zh dominates. LZPs can generate the observed quantity of dark matter thermally in several mass ranges: near the Z, Z' and Higgs resonances and for considerably heavier masses ($m_{\text{LZP}} \gtrsim$ several hundred GeV) [32; 33].

While annihilations are strongly mass-dependent, the elastic scattering cross-section of Dirac RH neutrino dark matter is solely determined by the LZP–Z coupling.

Direct and indirect detection

Concerning elastic scattering, as is well known for a Dirac neutrino, the spin-independent elastic scattering cross-section via a t-channel Z exchange has the form

$$\sigma_{\text{SI}} \propto \left[Z(1 - 4\sin^2\theta_W) - (A - Z) \right]^2. \tag{9.3}$$

Since $4\sin^2\theta_W \approx 1$, the coupling to protons is suppressed. Nevertheless, scattering off target nuclei puts the strongest constraints on the M_{KK} scale and all the interesting region of parameter space in this model will be probed by direct detection experiments in the near future [32; 33]. Direct detection and relic density constraints were combined in [265]. Indirect detection prospects for the LZP have been studied through three channels in [1127]. First, the prospects for detecting high-energy neutrinos produced through annihilations of LZPs in the Sun are very encouraging. Annihilations of light LZPs in the Galactic halo also generate positrons very efficiently. Finally, LZP annihilations near the Galactic centre may provide an observable flux of gamma-rays not very different from that in the case of annihilating neutralinos. Barrau *et al.* [221] also studied the production of antiprotons from LZP annihilations, and Baer and Profumo [162] looked at antideuteron fluxes.

9.2.7 Warped models with Z_2

In ref. [1544], a procedure to endow any given warped model with a discrete exchange Z_2 symmetry was suggested via a doubling of part of the field content. The idea is that a given field ϕ is replaced by a pair of field ϕ_1 and ϕ_2 and the symmetry $\phi_1 \leftrightarrow \phi_2$ is imposed. The even linear combination is identified with the original field while the orthogonal combination is odd and can play the role of dark matter.[3]

Special attention was given to gauge–Higgs unification models that also address dynamical electroweak symmetry breaking. The Z_2 odd sector that gives rise to dark matter also plays a key role in electroweak symmetry breaking through its contribution to the Higgs potential. The model introduces two $U(1)_{X_1}$ and $U(1)_{X_2}$ factors in the bulk and an extra exchange symmetry $X_1 \leftrightarrow X_2$. The even linear combination can be identified with hypercharge or $U(1)_{B-L}$. The odd combination constitutes the dark matter. In addition, for the annihilation of this DM particle to be efficient enough, some Standard Model fermions have to be charged under these $U(1)_X$, and a subset are also being doubled. The DM candidate (a spin-1 particle) has a mass in the 300–500 GeV range. An interesting implication is the existence of new light vector-like quarks that can be copiously produced at LHC [575; 647].

9.2.8 Warped models with KK parity

Agashe *et al.* [31] attempted to extend Kaluza–Klein parity to non-toroidal compactifications. The initial motivation was to implement the good features of UED, namely KK masses below a TeV and a dark matter candidate, in a warped background – so that the Planck–weak hierarchy is addressed, without giving up some of the virtues of warped extra dimensions such as fermion and Higgs localization (see the discussion of Section 15.2 in Chapter 15).

The first point to address is that, for a single slice of AdS_5, the warp factor is clearly not symmetric under reflection about the midpoint of the extra dimension. Therefore, to implement an analogue of KK-parity of UED in a warped extra dimension requires that two physically distinct slices of AdS_5 are glued and the symmetry interchanging the two AdS_5 slices is imposed. In such a construction, the mass eigenstates can be divided into two classes with different symmetry properties. For any given level n in the KK decomposition, there are KK-even modes $(n+)$, whose profiles are symmetric under reflection around the midpoint of the extra dimension, and KK-odd modes $(n-)$ with antisymmetric profiles. KK-odd modes can only couple in pairs to the KK-even modes and the low-energy, four-dimensional effective theory has KK parity. The SM particles belong to the even towers in this KK decomposition. The odd modes cannot have single couplings to the SM, therefore they are allowed to be lighter than a TeV without contradicting the precision EW constraints. On the other hand, the even modes must be heavier than a few TeV since KK parity by itself is not enough to satisfy EW

---

[3] This trick was also applied in Little Higgs models in [169].

precision tests. A sizable hierarchy, at least a factor of a few, between the lightest KK-even mode and the lightest KK-odd mode can only come from very large IR brane kinetic terms. The dark matter particle can be identified with the lightest KK partner of the Z boson (the KK photon would not lead to the correct abundance since its couplings to the SM are different from the UED case) and the predicted relic abundance is in the correct range. However, large IR brane kinetic terms create a certain tension with perturbativity and the regime where the 5D theory remains weakly coupled is rather narrow. Another problem is that for light fermions localized close to the UV brane the constraints from electroweak precision tests are still quite severe.

To obtain a sizable splitting between the lightest KK-odd and KK-even modes and avoid the strong coupling problem in 5D at the same time, one may need to move the UV brane to an intermediate scale below the Planck scale. This may be a drawback compared with the traditional RS models, but certainly is an improvement over the UED in which the hierarchy problem is simply not addressed at all.

9.2.9 Sterile neutrino DM in warped extra dimensions

Kadota [1188] presents a natural set-up for sterile neutrino dark matter in the context of warped extra dimensions that seems consistent with all experimental and observational constraints. The inclusion of sterile right-handed neutrinos in the bulk of an extra dimension leads to light 4D mass eigenstates that are cosmologically stable without the need for an additional symmetry and can be produced with the correct abundance from the radion decay. The cosmological stability and the keV–MeV mass range can be obtained from wavefunction localization in the bulk, with only a moderate tuning on the 5D Yukawa coupling thanks to the warped geometry. The radion with a mass $\mathcal{O}(100)$ GeV is a generic feature in Randall–Sundrum models. It has large couplings to Standard Model degrees of freedom and therefore inherits a thermal abundance at temperatures of order 100 GeV. It necessarily couples to all degrees of freedom and therefore to the sterile neutrino which is produced from the radion decay. These findings open a new generic possibility for dark matter in Randall–Sundrum models.

9.2.10 Hidden sector string-inspired dark matter

Harling and Hebeker [1076] showed that hidden sector dark matter is a generic feature of type IIB string theory landscape.[4] Flux compactifications

---

[4] See also [531; 814] for other studies of string-inspired hidden dark matter models.

generically contain many strongly warped regions, or 'throats'. KK modes localized in the throat can be very weakly coupled to the Standard Model if the Standard Model lives in the unwarped part of the compact manifold, resulting in very long lifetimes. There is a dual 4D description in terms of a strongly coupled gauge theory with a large number of colours known as the 'conformally sequestered hidden sector'. Scalar glueballs or their spin-1/2 superpartners were considered as possible dark matter candidates. Their mass is given by the IR scale of the throat and is typically larger than 10 TeV to have the correct abundance. A potential observable signature is photons produced by the very late decays and could be detected with Fermi or HESS.

For this class of models, the connection between dark matter and the EW scale is less straightforward, unless the IR scale of the throat is at the TeV scale as expected if these throat configurations are what solve the hierarchy problem. And obviously, testability remains an issue.

9.3 Little Higgs dark matter

In Little Higgs models [1704], the Higgs is a pseudo-Nambu–Goldstone boson associated with spontaneous global symmetry breaking in an extended electroweak sector, which occurs at a scale $\sim$1 TeV. An explicit breaking of the global symmetries by gauge and Yukawa interactions generates a Higgs mass, but in a special collective manner ensuring that the Higgs is light. New particles must be introduced to ensure that the global symmetries are not broken too severely, and these are the states that cutoff the quadratically divergent top, gauge and Higgs loops. Consistency with electroweak precision data requires the introduction of T-parity, which in turn leads to the stability of the lightest T-odd particle, LTP [535; 536]. This particle is typically the heavy partner of the hypercharge gauge boson and can play the role of dark matter [361]. Detailed studies have been performed in the 'littlest Higgs model' based on the SU(5)/SO(5) global symmetry breaking pattern. This heavy 'photon' annihilates dominantly into W, Z or top pairs via s-channel Higgs exchange. Its mass is in the 80–500 GeV range. Direct detection is quite difficult (same diagrams as the KK 'photon' in UED). Concerning indirect searches, the secondary gamma-rays produced in WIMP annihilations could be observed at Fermi but no signal is predicted in neutrino telescopes [1566]. More promising are the LHC signatures of Little Higgs models [148], in particular the potential for discovering the heavy ($\sim$1 TeV) partners of the top quark which are responsible for the cancellation of the main quadratically divergent contribution to the Higgs mass parameter [1399].

9.4 Dark matter in technicolour and composite Higgs theories

Technicolour is a new strongly coupled gauge dynamics which breaks the electroweak symmetry dynamically and solves the hierarchy problem without an elementary Higgs through the mechanism of dimensional transmutation. Given a UV scale, an exponentially smaller scale is generated via a new asymptotically free gauge interaction, $\Lambda_{IR} \sim \Lambda_{UV} \exp[-8\pi^2/b_0 g^2(\Lambda_{UV})]$. The new strongly interacting sector is characterized by the new scale Λ_{IR} associated with the mass of the lowest composite resonance.

These models may feature quasi-stable technibaryons analogous to the proton in QCD that can possess the required properties to play the role of dark matter. There is a new global U(1) symmetry under which technibaryons are charged. That leads to a new conserved quantum number, called the technibaryon number in analogy to the baryon number in QCD. The lightest technibaryon is stable and can be promoted to a dark matter candidate. Technicolour cosmology was investigated in the late 1980s and early 1990s [163; 219; 541; 890; 1521].

The main difficulty in constructing such extensions of the Standard Model is the limited knowledge about strongly interacting theories. Recently, the AdS/CFT correspondence has inspired a new effort in modelling strong dynamics based on using weakly coupled 5D models. In these Holographic Higgs models, EW symmetry breaking is triggered by a light composite Higgs boson, which emerges as a pseudo-Goldstone boson. In this framework, dark matter candidates were suggested as either fermionic composite states made stable through a new conserved quantum number [654] or as skyrmion configurations [1585]. In parallel with the efforts inspired by the AdS/CFT correspondence, there is a different line of investigation reflecting a revival of activity in technicolour model building: a new class of models has been investigated in which technifermions do not belong to the fundamental representation but to larger representations. The resulting dynamics is very different from the one featured by a scaled-up version of QCD. These theories appear to pass EW precision tests and possess various dark matter candidates [1035; 1036; 1189].

9.5 Mirror dark matter

A different direction to address the hierarchy problem involves adding to the SM particles a mirror sector: the entire SM is replicated in a mirror world and a Z_2 symmetry interchanges our world with the mirror world, ensuring identical particles and interactions.[5] An interesting feature of the Higgs sector in these models is that it can alleviate the hierarchy problem because there is a limit in which the Higgs is a pseudo-Goldstone boson of an approximate SU(4) symmetry [195; 514]. The Mirror Model predicts some interesting Higgs physics. There will be two physical neutral Higgs boson mass eigenstates, each with a 50% invisible width, offering a remarkable way to discover mirror particles [195].

The ordinary and mirror particle sectors can interact through gravitation. Besides this, non-gravitational interactions can be induced through the mixing of colourless and neutral particles with their mirror counterparts. Neutrinos, the photon, the Z boson and the physical neutral Higgs boson can mix with the corresponding mirror states. Coloured and/or electrically charged particles are prevented from mixing with their mirror analogues by colour and electric charge conservation laws.

In this context, mirror matter (in particular mirror baryons) seems a natural candidate for dark matter. Various models and experimental signatures have been discussed in the literature [195; 282; 283; 860; 863].

9.6 'Minimal' approaches

As discussed in the introduction to this chapter, in the past few years dark matter model building has been flourishing outside of theories which solve the hierarchy problem. A few years ago, theorists rarely dared to follow such a path. The recent change of attitude is due in part to the emergence of the little hierarchy problem in the post-LEP era [197]: all known Beyond-the-Standard-Model (BSM) theories suffer from a mild fine-tuning problem which, although much less severe than the original hierarchy problem, makes them not fully natural. This and the appearance of the dark energy puzzle both led some to question the relevance of the hierarchy problem as a guide in BSM physics, especially after string theory has provided a natural setting for statistical rather than dynamical explanations of parameter determination [419; 709; 1826]. Moreover, the advent of the LHC certainly motivates a more phenomenological approach.

9.6.1 *Singlet scalar models*

Standard Model extensions that have attracted attention lately are models with additional scalar fields that are singlets under the SM gauge group.

---

[5] The idea that every SM particle is paired with a parity partner was first mentioned by Lee and Yang in their seminal paper on parity violation [1321].

The only renormalizable interaction of such scalars with the SM occurs via the Higgs sector which thus serves as a portal to the hidden sector [421; 799; 800; 1548; 1699].

The simplest model of dark matter one can think of is to add to the Standard Model just a real scalar singlet [471; 1760],[6] with an unbroken Z_2 symmetry that prevents its decay. Its only direct coupling to the SM is to the Higgs boson. The Lagrangian of this model has three new parameters: the mass of the singlet, its self-coupling and its coupling to the Higgs

$$\mathcal{L} = \mathcal{L}_{\text{SM}} + \frac{1}{2}\partial_\mu S \partial^\mu S - \frac{m_0^2}{2}S^2 - \frac{\lambda_S}{4}S^4 - \lambda S^2 H^\dagger H. \qquad (9.4)$$

The scalar singlet S annihilates via Higgs s-channel exchange and scatters elastically off nuclei through a t-channel Higgs. A wide range of singlet masses leads to the correct dark matter relic density but direct detection constraints require it to be larger than 50 GeV. Any Higgs mass can essentially lead to a viable dark matter candidate and typically $\lambda \sim O(0.1)$ (away from the Higgs resonance). Yaguna [1948] argues that the gamma-ray flux from the Galactic centre is within the sensitivity of Fermi. A singlet scalar can also have a significant impact for Higgs boson searches at the LHC. This was investigated with and without the Z_2 symmetry in [206; 1522; 1699]. The Higgs boson can decay to two singlets, thereby reducing the likelihood of discovering the Higgs boson in traditional search modes.

In ref. [281], it was shown that the scalar singlet dark matter provides a simple example for the realization of the idea of self-interacting non-dissipative cold dark matter to overcome the difficulties of the CDM model on galactic scales. And finally, the utility of a scalar singlet for models of sterile neutrino dark matter was discussed in ref. [1291].

9.6.2 *Inert (scalar) doublet model*

The inert doublet model (IDM) is just the addition of a second scalar SU(2) doublet in the Standard Model as well as a Z_2 parity under which this doublet is odd. Recently, the IDM has attracted interest as a model of 'Minimal dark matter'. Despite its simplicity, this model has rich phenomenology. It was already studied back in the 1970s [652] and more recently as a model of light neutrino mass generation: when adding three Majorana fermions odd under Z_2, this model offers the possibility of radiatively generating (purely Majorana) neutrino masses [1361; 1362]. It was actually also studied as a

---

[6] The case of a complex scalar field was considered even earlier in [1408].

model of improved naturalness [196; 490] and electroweak symmetry breaking [1065]. Several analyses have been done over the past two years on the relic density and detection aspects of the neutral scalar dark matter particle [196; 1043; 1352; 1361; 1362]. Direct detection does not put any important constraints on the IDM parameter space. A recent paper [1355] investigated LEP II limits on the IDM and excluded a significant part of the parameter space. The IDM was also compared with recent data on antimatter in cosmic rays in [1507].

9.6.3 Minimal (fermionic) dark matter

As discussed in [196], the IDM with a heavy Higgs ($\gtrsim 400$ GeV) yields a perturbative and natural description of EW physics at all energies up to 1.5 TeV. In the same spirit, it was shown in [791] that adding just a heavy vector lepton doublet as well as a heavy neutral Majorana state to the Standard Model is compatible with EW precision tests if the Higgs is heavy and actually improves naturalness. These extra states are odd under a discrete Z_2 symmetry while the SM states are even. This makes the lightest new state (a Majorana neutrino) stable and a good dark matter candidate if its mass is in the 50–170 GeV range. The same field content was actually introduced in [1371] where the motivation was to study a minimal model for dark matter and gauge coupling unification (see also [649]).

In another way (in the sense that naturalness issues are completely ignored), the so-called 'Minimal dark matter' model of [557] does not need to introduce any new discrete symmetry. It consists of adding to the SM just one multiplet with weak interactions and extrapolating this model up to the Planck scale. To avoid the running of g_2 to hit a Landau pole below the Planck scale, the number of its $SU(2)_L$ components has to satisfy $n \leq 5$ (8) for fermions (scalars). To fulfil direct detection constraints, its hypercharge should be zero, therefore n has to be odd. The fermionic 3-plet and 5-plet with $Y = 0$ are too short-lived to be viable dark matter candidates, unless some ad hoc symmetry is added to the model. Given the SM particle content and interactions, two possibilities emerge: an $n = 5$ fermion or an $n = 7$ scalar are both cosmologically stable, for the same type of reason that the proton is stable in the SM. The authors of [557] focused on the fermionic $SU(2)_L$ 5-plet. Quantum corrections due to a loop of gauge bosons generate a small mass splitting between the components of the multiplet. The lightest component turns out to be the neutral one and the charged partners are 166 MeV heavier. The only free parameter is the tree-level mass and this is fixed by the relic density calculation to be $M = (9.6 \pm 0.2)$ TeV. The corresponding prediction for the elastic scattering cross-section on nucleons is a few 10^{-9} pb, which will be probed by Xenon 1 ton and SuperCDMS experiments in the near future. Indirect searches have been discussed in [558; 560; 561].

9.6.4 Heavy neutrinos

A straightforward possibility is to contemplate the fourth-generation neutrino to serve as cold DM [795]. This is ruled out by direct detection experiments, but there are mechanisms to avoid these bounds, for instance by splitting the neutrino eigenstates with a small Majorana mass. Interestingly, a recent technicolour model [1684] originally introduced as an alternative theory of EW symmetry breaking predicts the existence of a heavy neutrino with normal weak interactions that could play the role of DM [1189]. Compatibility with EW precision tests constrains the relative mass splitting of the charged lepton and the associated neutrino.

A generic extension of the SM containing a stable heavy Dirac neutrino ν' singlet under $SU(2)_L$ as well as an additional Z' gauge boson was studied in [265]. For a mass between 10 and 500 GeV, the main requirement is that the $\nu'-Z$ coupling (potentially induced via $Z-Z'$ mixing after EW symmetry breaking) should be at least a hundred times smaller than the SM neutrino–Z coupling in order to satisfy the direct detection constraint. Once this is satisfied, there is a large range of ν' and Z' masses as well as $\nu'-Z'$ couplings that lead to the correct thermal abundance. The annihilation via Z is the dominant mechanism for ν' masses below 100 GeV. Near $M_Z/2$, the annihilation mechanism is even too efficient. If ν' has a large coupling to the Higgs, $M_{\nu'} \sim M_H/2$ can lead to the correct relic density. Finally, $\nu'-Z'$ couplings open a large spectrum of $M_{\nu'}$ possibilities in the several hundred GeV range up to the Z' mass.

The addition of new fermions charged under a new hidden abelian-gauged sector was also studied in [1593; 1595], the so-called 'secluded' dark matter model. An interesting case is when the new fermions are at the TeV scale while the hidden mediators of the $U(1)'$ gauge group are very light (MeV scale). This allows the annihilation rate into leptons in the Galactic halo to be enhanced through the formation of WIMPonium states, as discussed in the next section.

Another Dirac fermion DM candidate was presented in an effective field-theory approach where it couples to the SM leptons through dimension-6 four-fermion interactions suppressed by a scale of order 1 TeV [1077]. A natural such candidate is a Dirac bino in supersymmetry.

9.7 WIMPonium

We have reviewed simple models of dark matter. However, the true dark sector may turn out to be non-minimal and involve inter-WIMP dynamics. Constraints on the interactions between the WIMPs themselves are quite weak. Indeed, the relic density suggests a size for the interaction between WIMPs and ordinary SM particles, but it says little about the interactions between the WIMPs themselves. On the other hand, inter-WIMP dynamics may prove important in interpreting signals from astrophysical searches for dark matter or even at future colliders.

The interesting possibility of WIMPs forming metastable bound states – WIMPonium, similar to positronium or quarkonium – has been studied lately [1103; 1105; 1387; 1593; 1742]. In some theories of TeV-scale dark matter this can lead to a substantial velocity-dependent amplification of the dark matter annihilation cross-section from threshold resonances (Sommerfeld enhancement), thus providing a dynamical source of 'boost factor' for indirect detection signals [1308; 1386; 1387; 1593]. Interestingly, if WIMPonium is bound by a light mediator (<1 GeV) [113; 1593], this appears to be of sufficient magnitude to boost the observed positron fraction in the multi-GeV range to a level that would be observed by PAMELA [24] as an excess above background without the need for large astrophysical boost factors, thus potentially reconciling observations by PAMELA [24], ATIC [515] and PPB-BETS [1859] with a thermal relic WIMP.[7]

Effective field theories to describe WIMP interactions with the SM were constructed in ref. [1742] where the possibility that WIMPonium states could be produced at future colliders was also considered.

9.8 Connecting dark matter and the baryon asymmetry

Most of the model-building related to dark matter assumes that dark matter consists of a single species. It also assumes thermal freeze-out, which appealingly points to an annihilation cross-section of electroweak size and therefore strongly suggests a DM mass at the energy scale where we expect new physics. On the other hand, in this case, there is no reason to expect a density of dark matter which is similar to the density of baryons, i.e. $\Omega_{DM} \sim 5\Omega_b$. The proximity of the two values is left as an unexplained coincidence. The situation is changed if instead one considers a dark matter candidate and its antiparticle. Indeed, if there is an asymmetry between the density of dark matter particles and dark antimatter particles in the

---

[7] See however refs. [291; 335; 1202]. At the time of writing this review, this is still a hot topic of debate.

early Universe which is related to the baryon–antibaryon asymmetry, the dark matter relic density can be naturally of the same order of magnitude as the baryon density [1521]. The idea is to consider a dark matter candidate which carries a quantum number related to baryon number and transfers its particle–antiparticle asymmetry to the baryons through the electroweak $B + L$ anomaly [219; 1205]. Several implementations have been proposed [33; 219; 582; 817; 818; 1124; 1205; 1241; 1242; 1296; 1851]. In this context, the calculation of the relic density of dark matter is quite different from the usual freeze-out scenario. It does not depend directly on the dark matter annihilation cross-section but on the CP-violating out-of-equilibrium physics producing the matter asymmetry in the dark sector. One calculates the ratio of dark matter to baryon asymmetries at the temperature where the exchange between the two sectors switches off. In the end the crucial point is that independent of whatever dynamics produces an asymmetry in either the baryon sector, or in the dark matter sector, the $(B+L)$-anomaly-induced interactions together with interactions between the dark and the visible sectors automatically distributes the asymmetry between the baryons and the dark matter states, with a predictable ratio.

For example, in the model of ref. [33] discussed in Section 9.2.6, where the (colourless) dark matter particle carries baryon number, dark matter could store the overall negative baryonic charge which is missing in the visible quark sector and therefore be at the origin of the apparent matter–antimatter asymmetry of the Universe. In other words, in a universe where baryon number is a good symmetry, the negative baryonic charge would be carried by dark matter, while the equal and opposite baryonic charge would be carried by ordinary SM quarks (see also [817] for another model where dark matter carries baryon number). A comparable scenario was discussed in [1241] where DM does not carry baryon number but is produced from the decay of a messenger particle that carries baryon number.

9.9 Conclusion

This chapter has aimed at illustrating the plethora of possibilities as far as DM candidates at the EW scale are concerned, but is in no way exhaustive. There are lots of other interesting attempts that we were unable to account for. A complete report of models beyond the Standard Model is clearly beyond the scope of this review. However, it is clear that viable alternatives to supersymmetric dark matter exist. Let us hope that the LHC, together with an impressive array of direct and indirect detection experiments, will soon enable us to reduce significantly the allowed regions in the DM candidates parameter space.

10

Non-WIMP candidates

Jonathan L. Feng

10.1 Motivations

There are many non-WIMP dark matter candidates. Two prominent and highly motivated examples are axions and sterile neutrinos, which are reviewed in Chapters 11 and 12, respectively. In addition, there are candidates motivated by minimality, particles motivated by experimental anomalies, and exotic possibilities motivated primarily by the desire of their inventors to highlight how truly ignorant we are about the nature of dark matter.

In this brief chapter, we focus on dark matter candidates that are not WIMPs but nevertheless share the most important virtues of WIMPs. As discussed in Chapters 7, 8 and 9, WIMPs have several nice properties:

- They exist in well-motivated particle theories.
- They are naturally produced with the correct thermal relic density (the 'WIMP miracle').
- They predict signals that may be seen in current and near-future experiments.

The candidates that we discuss also have all three of these properties. They fall naturally into two classes: superWIMP candidates, which inherit the correct relic density through decays, and WIMPless candidates, which have neither weak-scale masses nor weak interactions, but which nevertheless have the correct thermal relic density. These possibilities appear in the same particle physics frameworks as WIMPs, but they imply very different cosmological histories for our Universe, as well as qualitatively new dark matter signals for both astrophysical observatories and particle physics experiments.

Particle Dark Matter: Observations, Models and Searches, ed. Gianfranco Bertone. Published by Cambridge University Press. © Cambridge University Press 2010.

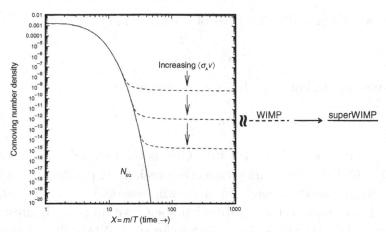

Fig. 10.1. In superWIMP scenarios, WIMPs freeze out as usual, but then decay to superWIMPs, superweakly interacting massive particles that form dark matter.

10.2 SuperWIMP dark matter

10.2.1 Candidates and relic densities

In the superWIMP framework for dark matter, WIMPs freeze out as usual in the early Universe, but later decay to superWIMPs, superweakly interacting massive particles that form the dark matter that exists today. Because superWIMPs are very weakly interacting, they have no impact on WIMP freeze-out in the early Universe, and the WIMPs decouple, as usual, with a thermal relic density Ω_{WIMP} that is naturally near the required density $\Omega_{\text{DM}} \approx 0.23$. Assuming that each WIMP decay produces one superWIMP, the relic density of superWIMPs is

$$\Omega_{\text{SWIMP}} = \frac{m_{\text{SWIMP}}}{m_{\text{WIMP}}} \Omega_{\text{WIMP}}. \qquad (10.1)$$

SuperWIMPs therefore inherit their relic density from WIMPs, and for $m_{\text{SWIMP}} \sim m_{\text{WIMP}}$, they are also naturally produced in the desired amount to be much or all of dark matter. The evolution of number densities is shown in Fig. 10.1.

The superWIMP scenario is realized in many particle physics models. The prototypical example is gravitinos, which exist in all supersymmetric theories [459; 777; 779; 836; 837; 838; 839; 841; 843; 1654; 1920]. In the simplest supersymmetric models, supersymmetry is transmitted to Standard Model superpartners through gravitational interactions, and supersymmetry is broken at a high scale. The mass of the gravitino $\tilde{G}$ is

$$m_{\tilde{G}} = \frac{F}{\sqrt{3}M_*}, \qquad (10.2)$$

and the masses of Standard Model superpartners are

$$\tilde{m} \sim \frac{F}{M_*}, \qquad (10.3)$$

where $M_* = (8\pi G_N)^{-1/2} \simeq 2.4 \times 10^{18}$ GeV is the reduced Planck scale and $F \sim (10^{11} \text{ GeV})^2$ is the supersymmetry-breaking scale squared. The precise ordering of masses depends on unknown, presumably $\mathcal{O}(1)$, constants in (10.3). It is, then, perfectly possible that the gravitino is the lightest supersymmetric particle (LSP) and a candidate for superWIMP dark matter. The role of the decaying WIMP is played by the next-to-lightest supersymmetric particle (NLSP), typically a slepton, sneutrino or neutralino. As required, the gravitino couples very weakly, with interactions suppressed by M_*, and so it is irrelevant during the WIMP's thermal freeze-out.

The gravitino superWIMP scenario differs markedly from other gravitino dark matter scenarios. In previous frameworks [380; 759; 765; 766; 1186; 1227; 1278; 1479; 1493; 1538; 1928], gravitinos were expected to be produced either thermally, with $\Omega_{\tilde{G}} \sim 0.1$ obtained by requiring $m_{\tilde{G}} \sim$ keV, or through reheating, with $\Omega_{\tilde{G}} \sim 0.1$ obtained by tuning the reheat temperature to $T_{\text{RH}} \sim 10^{10}$ GeV. In the superWIMP scenario, the desired amount of dark matter is obtained without relying on the introduction of new, fine-tuned energy scales.

Other examples of superWIMPs include Kaluza–Klein gravitons in scenarios with universal extra dimensions [836; 837; 838], axinos [585; 586; 1624] and quintessinos [351; 352] in supersymmetric theories, and many other scenarios in which a metastable particle decays to the true dark matter particle through highly suppressed interactions, with lifetimes ranging from fractions of a second to beyond the age of the Universe.

10.2.2 Astrophysical signals

Because superWIMPs are very weakly interacting, they are impossible to detect in conventional direct and indirect dark matter search experiments. At the same time, the extraordinarily weak couplings of superWIMPs imply that the decays of WIMPs to superWIMPs may be very late and have an observable impact on, for example, Big Bang nucleosynthesis (BBN), the Planckian spectrum of the cosmic microwave background (CMB), small-scale structure, the diffuse photon flux and cosmic-ray experiments.

In the prototypical case of a slepton decaying to a gravitino superWIMP, the decay width is

$$\Gamma(\tilde{l} \to l\tilde{G}) = \frac{1}{48\pi M_*^2} \frac{m_{\tilde{l}}^5}{m_{\tilde{G}}^2} \left(1 - \frac{m_{\tilde{G}}^2}{m_{\tilde{l}}^2}\right)^4, \tag{10.4}$$

assuming the lepton mass is negligible. This decay width depends on only the slepton mass, the gravitino mass and the Planck mass. For $m_{\tilde{G}}/m_{\tilde{l}} \approx 1$, the slepton decay lifetime is

$$\tau(\tilde{l} \to l\tilde{G}) \simeq 3.6 \times 10^8 \text{ s} \left(\frac{100 \text{ GeV}}{m_{\tilde{l}} - m_{\tilde{G}}}\right)^4 \left(\frac{m_{\tilde{G}}}{\text{TeV}}\right). \tag{10.5}$$

This expression is valid only when the gravitino and slepton are nearly degenerate, but usefully illustrates that decay lifetimes of the order of days or months are perfectly natural. Similar expressions hold for the decay of a neutralino NLSP to a gravitino.

BBN and CMB

Signals in BBN and the CMB are determined primarily by the WIMP lifetime and the energy released in visible decay products when the WIMP decays. This energy release destroys and creates light elements, distorting the predictions of standard BBN. In addition, the injection of electromagnetic energy may also distort the frequency dependence of the CMB away from its ideal black body spectrum. For the decay times of interest with redshifts $z \sim 10^5$ to 10^7, the resulting photons interact efficiently through $\gamma e^- \to \gamma e^-$ and $eX \to eX\gamma$, where X is an ion, but photon number is conserved, since double Compton scattering $\gamma e^- \to \gamma\gamma e^-$ is inefficient. The spectrum therefore relaxes to statistical but not thermodynamic equilibrium, resulting in a Bose–Einstein distribution function

$$f_\gamma(E) = \frac{1}{e^{E/(kT)+\mu} - 1}, \tag{10.6}$$

with chemical potential $\mu \neq 0$.

The energy release is conveniently expressed in terms of

$$\xi_{\text{EM}} \equiv \epsilon_{\text{EM}} B_{\text{EM}} Y_{\text{NLSP}} \tag{10.7}$$

for electromagnetic energy, with a similar expression for hadronic energy. Here ϵ_{EM} is the initial EM energy released in NLSP decay, and B_{EM} is the branching fraction of NLSP decay into EM components. $Y_{\text{NLSP}} \equiv n_{\text{NLSP}}/n_\gamma$ is the NLSP number density just before NLSP decay, normalized to the

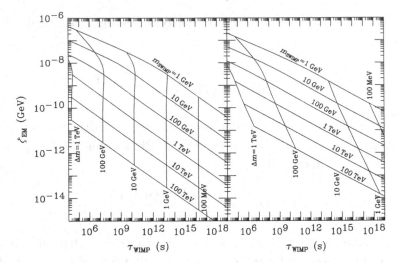

Fig. 10.2. Predicted values of WIMP lifetime τ and electromagnetic energy release $\zeta_{EM} \equiv \varepsilon_{EM} Y_{WIMP}$ in the $\tilde{B}$ (left) and $\tilde{\tau}$ (right) NLSP scenarios for $m_{SWIMP} = 1$ GeV, 10 GeV, ..., 100 TeV (top to bottom) and $\Delta m \equiv m_{WIMP} - m_{SWIMP} = 1$ TeV, 100 GeV, ..., 100 MeV (left to right). For the $\tilde{\tau}$ NLSP scenario, we assume $\varepsilon_{EM} = \frac{1}{2} E_\tau$. Reproduced from ref. [838].

background photon number density $n_\gamma = 2\zeta(3)T^3/\pi^2$. It can be expressed in terms of the superWIMP abundance:

$$Y_{NLSP} \simeq 3.0 \times 10^{-12} \left(\frac{TeV}{m_{\tilde{G}}} \right) \left(\frac{\Omega_{\tilde{G}}}{0.23} \right). \tag{10.8}$$

Once an NLSP candidate is specified, and assuming superWIMPs make up all of the dark matter, with $\Omega_{\tilde{G}} = \Omega_{DM} = 0.23$, both the lifetime and energy release are determined by only two parameters: $m_{\tilde{G}}$ and m_{NLSP}. The results for slepton and neutralino NLSPs are given in Fig. 10.2.

In Fig. 10.3, these predictions are compared with BBN and CMB constraints. The shaded regions are excluded by an analysis of BBN constraints on EM energy release [602]. This analysis has been strengthened by including hadronic constraints and updated and refined in many ways in recent years, as described in Chapter 28. Although the excluded region has shifted around, the basic features remain: some of the gravitino superWIMP parameter space is excluded, and some remains. In addition, late decays to superWIMPs may in fact improve the current disagreement of standard BBN predictions with the observed ^{7}Li and ^{6}Li abundances [173; 599].

Figure 10.3 also includes contours of the chemical potential μ, as determined by updating the analysis of ref. [1136]. The current bound is $\mu < 9 \times 10^{-5}$ [739; 857]. Although there are at present no indications of deviations

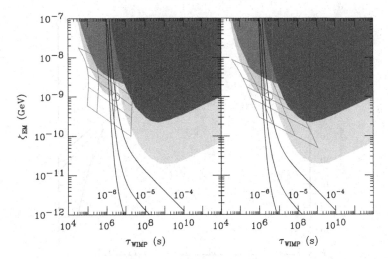

Fig. 10.3. The grid gives predicted values of WIMP lifetime τ and electromagnetic energy release $\zeta_{EM} \equiv \varepsilon_{EM} Y_{WIMP}$ in the $\tilde{B}$ (left) and $\tilde{\tau}$ (right) WIMP scenarios for $m_{SWIMP} = 100$ GeV, 300 GeV, 500 GeV, 1 TeV and 3 TeV (top to bottom) and $\Delta m \equiv m_{WIMP} - m_{SWIMP} = 600$ GeV, 400 GeV, 200 GeV and 100 GeV (left to right). For the $\tilde{\tau}$ WIMP scenario, we assume $\varepsilon_{EM} = \frac{1}{2} E_\tau$. The shaded regions are excluded in one analysis of BBN constraints [602]; the circle gives a region in which ^{7}Li is reduced to observed levels. The contours are for μ, which parameterizes the distortion of the CMB from a Planckian spectrum. From ref. [838].

from black body, current limits are already sensitive to the superWIMP scenario, and future improvements will further probe superWIMP parameter space.

Small-scale structure

In contrast to WIMPs, superWIMPs are produced with large velocities at late times. This has two effects. First, the velocity dispersion reduces the phase space density, smoothing out cusps in DM haloes. Second, such particles damp the linear power spectrum, reducing power on small scales [385; 501; 1207].

Depending on the particular decay time and decay kinematics, super-WIMPs may be cold or warm. As seen in Fig. 10.4, superWIMPs may suppress small-scale structure as effectively as a 1 keV sterile neutrino. Some superWIMP scenarios may therefore be differentiated from standard cold DM scenarios by studies of halo profiles, and may even be favoured by indications that cold DM predicts haloes that are too cuspy, as discussed in Chapter 3.

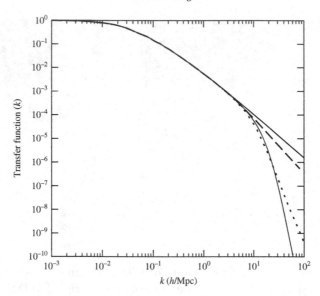

Fig. 10.4. The power spectrum for scenarios in which dark matter is completely composed of WIMPs (solid), half WIMPs and half superWIMPs (dashed), and completely composed of superWIMPs (dotted). For comparison, the solid curve is the transfer function for a 1 keV warm DM model. From ref. [1207].

10.2.3 Astroparticle and collider signals

The possibility of long-lived charged particles in superWIMP scenarios also has many implications for astroparticle and particle physics experiments.

Cosmic rays

In superWIMP (and other similar) scenarios, long-lived charged particles may be produced by cosmic rays, resulting in exotic signals in cosmic-ray and cosmic-neutrino experiments [38; 46; 47; 48; 86; 352; 487]. As an example, ultra-high-energy neutrinos may produce events with two long-lived sleptons through $\nu q \to \tilde{l}\tilde{q}'$ followed by the decay $\tilde{q}' \to \tilde{l}$. The sleptons are metastable and propagate to neutrino telescopes [1138], where they have a typical transverse separation of hundreds of metres. They may therefore be detected above background as events with two upward-going, extremely high-energy charged tracks in experiments such as IceCube.

Colliders

As evident in Eq. (10.5), in supersymmetric superWIMP scenarios, the NLSP decays to the gravitino with lifetimes that may be of the order of

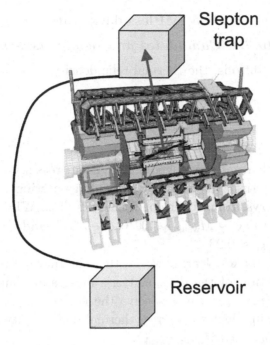

Fig. 10.5. Configuration for slepton trapping in gravitino superWIMP scenarios. From ref. [840].

seconds to months. Such particles are effectively stable in collider experiments, and this scenario therefore implies that each supersymmetric event will be characterized not by missing energy, but by two charged, heavy metastable particles. This is a spectacular signal that cannot escape notice at the LHC [718; 834; 842; 1514]. In addition, given the possibility of long lifetimes, it suggests that decays to gravitinos may be observed by capturing slepton NLSPs and detecting their decays.

The sleptons may be captured in water tanks placed outside collider detectors [840], in the detectors themselves [1063] or by mining LHC cavern walls for sleptons [632]. In the first case, shown in Fig. 10.5, the water tanks may be drained periodically to underground reservoirs where slepton decays may be observed in quiet environments. As many as 10^4 sleptons per year may be stopped in 1 metre thick water tanks, opening up the possibility of a precise measurement of slepton lifetime and the first study of a gravitational process at high-energy colliders, along with many other implications [839].

10.3 WIMPless dark matter

10.3.1 Candidates and relic densities

Under general conditions, the thermal relic density of a particle X is [539; 1701; 1802; 1959]

$$\Omega_X \propto \frac{1}{\langle \sigma v \rangle} \sim \frac{m_X^2}{g_X^4}, \tag{10.9}$$

where $\langle \sigma v \rangle$ is its thermally averaged annihilation cross-section, and m_X and g_X are the characteristic mass scale and coupling entering this cross-section. The last step follows from dimensional analysis. The WIMP miracle is the statement that, for $m_X \sim M_{\text{weak}} \sim 100$ GeV-1 TeV and $g_X \sim g_{\text{weak}} \simeq 0.65$, Ω_X is roughly $\Omega_{\text{DM}} \approx 0.23$.

Equation (10.9) makes clear, however, that the thermal relic density fixes only one combination of the dark matter's mass and coupling, and other combinations of (m_X, g_X) can also give the correct Ω_X. WIMPless models [829] are those in which the correct thermal relic density is achieved with parameters $(m_X, g_X) \neq (M_{\text{weak}}, g_{\text{weak}})$.

Because WIMPless dark matter does not have weak interactions, and existing constraints effectively exclude electromagnetic and strong interactions, WIMPless dark matter is necessarily hidden dark matter, that is, dark matter that has no Standard Model gauge interactions. Hidden sectors have a long history, and hidden sector dark matter has been discussed for decades, beginning with work on mirror matter and related ideas. For a general discussion and references, see Chapter 9. Here we note only that, counter to conventional wisdom, existing constraints place only weak bounds on hidden sectors. For example, light degrees of freedom change the expansion rate of the Universe and thereby have an impact on BBN. The constraint from BBN is highly sensitive to the temperature of the hidden sector, however. Current bounds from BBN on the number of light and heavy degrees of freedom are given in Fig. 10.6. For hidden sector temperatures within a factor of 2 of the observable sector, hundreds of degrees of freedom, equivalent to several copies of the Standard Model or the minimal supersymmetric Standard Model (MSSM), may be accommodated.

Of course, WIMPless dark matter requires hidden sectors with additional structure to guarantee that the hidden sector's dark matter has the desired thermal relic density. Remarkably, this structure may be found in well-motivated models that have been explored previously for many other reasons [829]. As an example, consider supersymmetric models with gauge-mediated supersymmetry breaking (GMSB) [72; 673; 676; 677; 678; 1494].

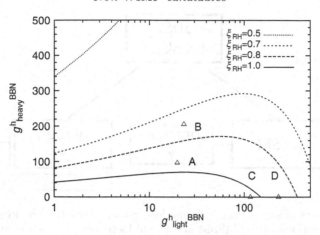

Fig. 10.6. Bounds from BBN in the $(g_{\text{light}}^{\text{h BBN}}, g_{\text{heavy}}^{\text{h BBN}})$ plane, where $g_{\text{light}}^{\text{h BBN}}$ and $g_{\text{heavy}}^{\text{h BBN}}$ are the hidden degrees of freedom with masses $m < T_{\text{BBN}}^{\text{h}}$ and $T_{\text{BBN}}^{\text{h}} < m < T_{\text{RH}}^{\text{h}}$, respectively, for hidden to observable sector reheat temperature ratios $T_{\text{RH}}^{\text{h}}/T_{\text{RH}} = 0.5, 0.7, 0.8, 1.0$ (from top to bottom). The regions above the contours are excluded. The values of $(g_{\text{light}}^{\text{h BBN}}, g_{\text{heavy}}^{\text{h BBN}})$ are marked for four example hidden sectors: (A) one-generation and (B) three-generation flavour-free versions of the MSSM with $T_{\text{BBN}}^{\text{h}} < m_X < T_{\text{RH}}^{\text{h}}$, and (C) one-generation and (D) three-generation flavour-free versions of the MSSM with $m_X < T_{\text{BBN}}^{\text{h}}/2$. From ref. [844].

These models necessarily have several sectors, as shown in Fig. 10.7. The supersymmetry-breaking sector includes the fields that break supersymmetry dynamically and mediate this breaking to other sectors through gauge interactions. The MSSM sector includes the fields of the minimal supersymmetric Standard Model. In addition, supersymmetry breaking may also be mediated to one or more hidden sectors. The hidden sectors are not strictly necessary but, given the discussion above, there is no reason to prevent them, and hidden sectors are ubiquitous in such models originating in string theory.

As is well known, GMSB models generate superpartner masses proportional to gauge couplings squared. Slightly more precisely, the MSSM superpartner masses are

$$m \sim \frac{g^2}{16\pi^2} \frac{F}{M},\qquad(10.10)$$

where g is the largest relevant Standard Model gauge coupling, and F and M are the vacuum expectation values of the supersymmetry-breaking sector's chiral field S, with $\langle S \rangle = M + \theta^2 F$. With analogous couplings of the hidden

J. L. Feng

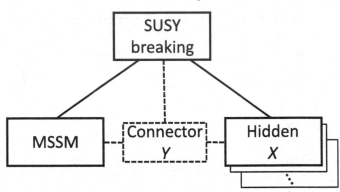

Fig. 10.7. Sectors of supersymmetric models. Supersymmetry breaking is mediated by gauge interactions to the MSSM and the hidden sector, which contains the dark matter particle X. An optional connector sector contains fields Y, charged under both MSSM and hidden sector gauge groups, which induce signals in direct and indirect searches and at colliders. There may also be other hidden sectors, leading to multicomponent dark matter. From ref. [829].

sector fields to hidden messengers, the hidden sector superpartner masses are

$$m_X \sim \frac{g_X^2}{16\pi^2} \frac{F}{M},\qquad(10.11)$$

where g_X is the relevant hidden sector gauge coupling. As a result,

$$\frac{m_X}{g_X^2} \sim \frac{m}{g^2} \sim \frac{F}{16\pi^2 M};\qquad(10.12)$$

that is, m_X/g_X^2 is determined solely by the supersymmetry-breaking sector. As this is exactly the combination of parameters that determines the thermal relic density of (10.9), the hidden sector automatically includes a dark matter candidate that has the desired thermal relic density, irrespective of its mass.

The freeze-out of hidden sector dark matter in such GMSB models has been studied numerically. As an example, in ref. [844] the hidden sector was assumed to be a copy of the MSSM, but with a free superpartner mass scale m_X and all Yukawa couplings $\sim\mathcal{O}(1)$, so that the only light hidden sector particles are the hidden gluon, photon and neutrinos. The results are given in Fig. 10.8. The criterion that the Standard Model weak scale be between 100 GeV and 1 TeV requires values of (m_X, g_X) within the band. The solid curves, where the thermal relic density of hidden dark matter is consistent with dark matter, are seen to lie within this band, confirming the scaling arguments and rough estimates described above.

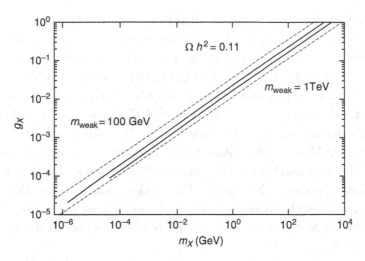

Fig. 10.8. Contours of $\Omega_X h^2 = 0.11$ in the (m_X, g_X) plane for hidden to observable reheat temperature ratios $T^h_{\mathrm{RH}}/T_{\mathrm{RH}} = 0.8$ (upper solid) and 0.3 (lower solid), where the hidden sector is a one-generation flavour-free version of the MSSM. Also plotted are lines of $m_{\mathrm{weak}} \equiv (m_X/g_X^2)g'^2 = 100$ GeV (upper dashed) and 1 TeV (lower dashed). From ref. [844].

In summary, well-known frameworks for hidden sectors include models in which the hidden sector contains a particle whose thermal relic density is automatically in the desired range to be dark matter, even when the particle's mass is not at the weak scale. This property relies on the relation $m_X \propto g_X^2$, which is common to other frameworks for new physics that avoid flavour-changing problems, such as anomaly-mediated supersymmetry breaking. The 'coincidence' required for WIMPless dark matter may also be found in other settings; see, for example, ref. [1130].

10.3.2 Direct and indirect detection signals

The decoupling of the WIMP miracle from WIMPs has many possible implications and observable consequences. In the case that the dark matter is truly hidden, it implies that there are no prospects for direct or indirect detection. Signals must be found in astrophysical observations, as in the case of superWIMPs. Alternatively, there may be connector sectors containing particles that mediate interactions between the Standard Model and the hidden sector through non-gauge (Yukawa) interactions (see Fig. 10.7). Such connectors may generate many signals with energies and rates typically unavailable to WIMPs.

As an example, first consider direct detection. The signal from the DAMA experiment, interpreted as spin-independent, elastic scattering, has conventionally favoured a region in the mass cross-section plane with $(m_X, \sigma_{SI}) \sim (20-200 \text{ GeV}, 10^{-5} \text{ pb})$ [270]. This is now excluded, most stringently by XENON10 [90] and CDMS (Ge) [39], which require $\sigma_{SI} < 10^{-7}$ pb throughout this range of m_X. Gondolo and Gelmini have noted, however, that an alternative region with $(m_X, \sigma_{SI}) \sim (1-10 \text{ GeV}, 10^{-3} \text{ pb})$ may explain the DAMA results without violating other known bounds [983]. DAMA's relative sensitivity to this region follows from its low energy threshold and the lightness of Na nuclei. This region may be extended to lower masses and cross-sections by the effects of channelling [327; 401; 805; 1572; 1695] and may also be broadened by dark matter streams in the solar neighbourhood [983].

The acceptable DAMA-favoured region with $m_X \sim 5$ GeV has masses that are low for WIMPs, given that their masses are expected to be around $M_{\text{weak}} \sim 100$ GeV to 1 TeV. However, in WIMPless models, where the thermal relic density is achieved for a variety of dark matter masses, such masses are perfectly natural. A WIMPless particle X may couple to the Standard Model through Yukawa interactions

$$\mathcal{L} = \lambda_f X \bar{Y}_L f_L + \lambda_f X \bar{Y}_R f_R, \qquad (10.13)$$

where Y is a vector-like connector fermion, and f is a Standard Model fermion. Taking f to be the b quark, and the Y mass to be 400 GeV, consistent with current bounds, these couplings generate spin-independent scattering cross-sections given in Fig. 10.9. We see that WIMPless dark matter may explain the DAMA results without difficulty.

WIMPless dark matter also provides new target signals for indirect detection. For WIMPs, annihilation cross-sections determine both the thermal relic density and indirect detection signals. The thermal relic density therefore constrains the rates of indirect detection signals. In the WIMPless case, however, this connection is weakened, since the thermal relic density is governed by hidden sector annihilation and gauge interactions, while the indirect detection signals are governed by the interactions of Eq. (10.13).

This provides a wealth of new opportunities for indirect detection. As an example, WIMPless dark matter may be detected through its annihilation to neutrinos in the Sun by experiments such as Super-Kamiokande (Super-K). Although such rates depend on the competing cross-sections for capture and annihilation, the Sun has almost certainly reached its equilibrium state, and the annihilation rate is determined by the scattering cross-section [651]. The prospects for Super-Kamiokande may therefore be compared to direct

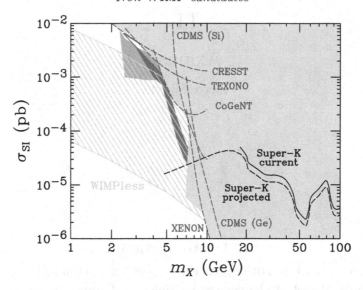

Fig. 10.9. Direct detection cross-sections for spin-independent X-nucleon scattering as a function of dark matter mass m_X. The dark shaded region is DAMA-favoured given channelling and no streams [1572], and the medium shaded region is DAMA-favoured at 3σ given streams and no channelling [983]. The light shaded region is excluded by the direct detection experiments indicated. The cross-hatched region is the parameter space of WIMPless models with connector mass $m_Y = 400$ GeV and $0.3 < \lambda_b < 1.0$. The black solid line is the published Super-K exclusion limit [651], and the dashed line is a projection of future Super-K sensitivity. Adapted from ref. [830].

detection rates [651; 830; 1125]. The results are given in Fig. 10.9. In the near future, Super-Kamiokande may be able to probe the low mass regions corresponding to the DAMA signal.

WIMPless dark matter also provides additional targets for indirect detection experiments looking for photons, positrons and other annihilation products. The connectors may also play an interesting role in collider experiments. Further details may be found in refs. [830; 831; 1125].

11

Axions

Pierre Sikivie

11.1 Introduction

Axions have a double motivation: they solve the 'strong CP problem' of the Standard Model of elementary particles and they are a candidate for the dark matter of the Universe. In this section, we discuss the strong CP problem and its solution through the existence of an axion, axion properties, and the limits on the axion from laboratory searches and from astrophysics.

11.1.1 The strong CP problem

Consider the Lagrangian of QCD:

$$
\mathcal{L}_{\text{QCD}} = -\frac{1}{4} G^a_{\mu\nu} G^{a\mu\nu} + \sum_{j=1}^{n} \left[\bar{q}_j \gamma^\mu i D_\mu q_j \right.
$$

$$
\left. - (m_j q^\dagger_{\text{L}j} q_{\text{R}j} + \text{h.c.}) \right] + \frac{\theta g^2}{32\pi^2} G^a_{\mu\nu} \tilde{G}^{a\mu\nu}. \qquad (11.1)
$$

The last term is a 4-divergence and hence does not contribute in perturbation theory. That term does, however, contribute through non-perturbative effects [485; 1161; 1827; 1828] associated with QCD instantons [266]. Using the Adler–Bell–Jackiw anomaly [21; 268], one can show that θ dependence must be present if none of the current quark masses vanishes. Using the anomaly, one can further show that QCD depends upon θ only through the combination of parameters:

$$
\bar{\theta} = \theta - \arg(m_1, m_2, \ldots, m_n). \qquad (11.2)
$$

If $\bar{\theta} \neq 0$, QCD violates P and CP. The absence of P and CP violations in the strong interactions therefore places an upper limit upon $\bar{\theta}$. The best

Particle Dark Matter: Observations, Models and Searches, ed. Gianfranco Bertone. Published by Cambridge University Press. © Cambridge University Press 2010.

constraint follows from the experimental bound [175] on the neutron electric dipole moment which yields: $\overline{\theta} < 10^{-9}$.

The question then is: why is $\overline{\theta}$ so small? The quark masses originate in the electroweak sector which violates P and CP. In the Standard Model, there is no reason why the overall phase of the quark mass matrix should exactly match the value of θ from the QCD sector to yield $\overline{\theta} < 10^{-9}$. In particular, if CP violation is introduced in the manner of Kobayashi and Maskawa [1255], the Yukawa couplings that give masses to the quarks are arbitrary complex numbers and hence arg det m_q and $\overline{\theta}$ are expected to be of order one.

Peccei and Quinn [1554; 1555] proposed to solve the strong CP problem by postulating the existence of a global $U_{PQ}(1)$ quasi-symmetry. $U_{PQ}(1)$ must be a symmetry of the theory at the classical (i.e. at the Lagrangian) level, it must be broken explicitly by those non-perturbative effects that make the physics of QCD depend upon θ, and finally it must be spontaneously broken. The axion [1926; 1934] is the quasi-Nambu–Goldstone boson associated with the spontaneous breakdown of $U_{PQ}(1)$. One can show that, if a $U_{PQ}(1)$ quasi-symmetry is present, then

$$\overline{\theta} = \theta - \arg(m_1 \ldots m_n) - \frac{a(x)}{f_a}, \tag{11.3}$$

where $a(x)$ is the axion field and $f_a = v_a/N$ is called the axion decay constant; v_a is the vacuum expectation value which spontaneously breaks $U_{PQ}(1)$ and N is an integer which expresses the colour anomaly of $U_{PQ}(1)$. The non-perturbative effects that make QCD depend upon $\overline{\theta}$ produce an effective potential $V(\overline{\theta})$ whose minimum is at $\overline{\theta} = 0$ [1886]. Thus, by postulating an axion, $\overline{\theta}$ is allowed to relax to zero dynamically and the strong CP problem is solved.

11.1.2 Axion properties

The properties of the axion can be derived using the methods of current algebra [201; 707; 758; 1926]. The axion mass is given in terms of f_a by

$$m_a \simeq 0.6 \; eV \; \frac{10^7 GeV}{f_a}. \tag{11.4}$$

All the axion couplings are inversely proportional to f_a. Of particular interest for axion dark matter searches is the axion coupling to two photons:

$$\mathcal{L}_{a\gamma\gamma} = -g_\gamma \frac{\alpha}{\pi} \frac{a(x)}{f_a} \mathbf{E} \cdot \mathbf{B}, \tag{11.5}$$

where **E** and **B** are the electric and magnetic fields, α is the fine structure constant and g_γ is a model-dependent coefficient of order one. In the DFSZ model, $g_\gamma = 0.36$ [675; 1970], whereas $g_\gamma = -0.97$ in the KSVZ model [1232; 1745]. The coupling of the axion to a spin-1/2 fermion f has the form:

$$\mathcal{L}_{a\bar{f}f} = ig_f \frac{m_f}{v} a\bar{f}\gamma_5 f, \qquad (11.6)$$

where g_f is a model-dependent coefficient of order one. In the KSVZ model the coupling to electrons is zero at tree level. Models with this property are called 'hadronic'.

11.1.3 Constraints from laboratory searches and astrophysics

The searches for the axion in high-energy and nuclear physics experiments are discussed in the reviews by Kim [1233] and by Peccei [1552]. If the axion is heavier than 1 MeV and decays quickly into e^+e^- (lifetime of order 10^{-11} s or less), then it is ruled out by negative searches for rare particle decays such as $\pi^+ \to a(e^+e^-)e^+\nu_e$ [738]. The rate of this reaction follows simply from the mixing of a and π^0 and the known decay $\pi^+ \to \pi^0 e^+\nu_e$. Alternatively, the axion is long-lived (10^{-11} s or more). In this case it is severely constrained by negative searches in beam dumps. In a beam dump, axions are produced by many different processes involving independent couplings, such as $a - \pi^0$, $a - \eta$ and $a - \eta'$ mixing, the axion couplings to two photons and to two gluons, and the couplings to quarks and gluons. It is difficult to calculate these processes with precision at all energies. On the other hand, the many processes add up incoherently and they cannot all vanish. Thus one can give a reliable estimate for the total production ($p + N \to a + X$, or $e + N \to a + X$) and interaction ($a + N \to X$) cross-sections. Many such searches have been carried out. They rule out axions with mass larger than about 50 keV.

The astrophysical constraints on the axion are described in detail in the reviews by Turner [1875] and Raffelt [1620]. Axions are emitted by stars in a variety of processes such as Compton-like scattering ($\gamma + e \to a + e$), axion bremsstrahlung ($e + N \to N + e + a$) and the Primakoff process ($\gamma + N \to N + a$). The rate at which a star burns its nuclear fuel is limited by the rate at which it can lose the energy produced. Emission of light weakly coupled bosons, such as the axion, allows a star to radiate energy efficiently because such particles can escape the star at once (without rescattering) whereas photons are emitted only from the stellar surface [1693]. Thus the existence of an axion accelerates stellar evolution, and this may be inconsistent with observation. The longevity of red giants rules out the mass range 200 keV

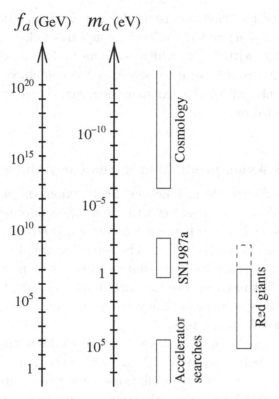

Fig. 11.1. Ranges of axion mass m_a, or equivalently axion decay constant f_a, which have been ruled out by accelerator searches, the evolution of red giants, the supernova SN1987a, and finally the axion cosmological energy density.

$\gtrsim m_a \gtrsim 0.5$ eV [658; 659; 1623] for hadronic axions. Above 200 keV the axion is too heavy to be copiously emitted in the thermal processes taking place in red giants, whereas below 0.5 eV it is too weakly coupled. For axions with a large coupling to electrons ($g_e = 0(1)$ in Eq. (11.6)) the range ruled out can be extended to 200 keV $\gtrsim m_a \gtrsim 10^{-2}$ eV because axion emission through the Compton-like process $\gamma + e \rightarrow a + e$ cools the helium core to such an extent as to prevent the onset of helium burning [636].

Finally the range 2 eV $\gtrsim m_a \gtrsim 3 \times 10^{-3}$ eV is ruled out by Supernova 1987a [768; 1163; 1221; 1619; 1874]. The constraint follows from the fact that the duration of the associated neutrino events in the large underground proton decay detectors [437; 1101] is consistent with theoretical expectations based on the premise that the collapsed supernova core cools by emission of neutrinos. If the axion mass is in the above-mentioned range, the core cools instead by axion emission and the neutrino burst is excessively shortened. The supernova constraint is quite axion model-independent because the axions are emitted by axion bremsstrahlung in nucleon–nucleon scattering $(N + N \rightarrow N + N + a)$ and the relevant couplings follow simply from the mixing of the axion with the π^0 which is a general feature of axion models.

When the limits from laboratory searches are combined with the astrophysical constraints, all of the axion mass range down to approximately 3×10^{-3} eV is ruled out.

11.2 Axion production in the early Universe

In this section we discuss the production of cold axions in the early Universe. In addition to cold axions, there is a thermal axion population [1265] produced by processes such as $q + g \rightarrow q + a$ where q is a quark and g a gluon, or $\pi + \pi \rightarrow \pi + a$ where π is a pion. The temperature of thermal axions is of the same order of magnitude as that of photons or neutrinos. Thermal axions move too fast to constitute the dark matter observed in the haloes of galaxies and clusters of galaxies. They are a form of 'hot dark matter' and will not be discussed further here.

Cold axions are produced when the axion field settles to the CP-conserving minimum of its effective potential. Because the axion field is very weakly coupled, the energy stored in its oscillations cannot dissipate efficiently into other forms. We will first discuss the general evolution of the axion field between the PQ phase transition, in which the $U_{PQ}(1)$ gets spontaneously broken, and the QCD phase transition when the axion acquires a mass. We will then discuss the production of cold axions by the mechanisms of vacuum realignment, string decay and wall decay.

11.2.1 Axion field evolution

The $U_{PQ}(1)$ symmetry gets spontaneously broken at a critical temperature $T_{PQ} \sim v_a$, where v_a is the vacuum expectation value of a complex field $\phi(x)$. The action density for this field is of the form

$$\mathcal{L}_\phi = \frac{1}{2}\partial_\mu \phi^\dagger \partial^\mu \phi - \frac{\lambda}{4}(\phi^\dagger \phi - v_a^2)^2 + \cdots, \tag{11.7}$$

where the dots represent interactions with other fields in the theory. At $T > T_{PQ}$, the free energy has its minimum at $\phi = 0$. At $T < T_{PQ}$, the minimum is a circle, whose radius quickly approaches v_a as T decreases. Afterwards

$$\langle \phi(x) \rangle = v_a\, e^{ia(x)/v_a}, \tag{11.8}$$

where $a(x)$ is the axion field before mixing with the π^0 and η mesons, and has random initial conditions. In particular, at two points outside each other's causal horizon the values of a are completely uncorrelated.

It is well known that the size of the causal horizon is hugely modified during cosmological inflation. Without inflation, the size of the causal horizon is of order the age t of the Universe. Instead, during an inflationary epoch, the causal horizon grows exponentially fast and becomes enormous compared with t. There are two cases to consider. Case 1: inflation occurs with reheat temperature smaller than T_{PQ}, and the axion field is homogenized over enormous distances. The subsequent evolution of this zero momentum mode is relatively simple. Case 2: inflation occurs with reheat temperature larger than T_{PQ}. In case 2, in addition to the zero mode, the axion field has non-zero modes, and carries strings and domain walls as topological defects.

The early Universe is assumed to be homogeneous and isotropic. Its curvature is negligible. The space-time metric can therefore be written in the Robertson–Walker form:

$$-\mathrm{d}s^2 = \mathrm{d}t^2 - R(t)^2 \mathrm{d}\mathbf{x} \cdot \mathrm{d}\mathbf{x}, \qquad (11.9)$$

where $\mathbf{x}$ are comoving spatial coordinates and $R(t)$ is the scale factor. The equation of motion for $a(x)$ in this space-time is:

$$D_\mu \partial^\mu a(x) + V'_a(a(x)) = \left(\partial_t^2 + 3\frac{\dot{R}}{R}\partial_t - \frac{1}{R^2}\nabla_x^2 \right) a(x) + V'_a(a(x)) = 0, \qquad (11.10)$$

where V_a is the effective potential for the axion field, and prime indicates a derivative with respect to a. The potential V_a results from non-perturbative QCD effects associated with instantons [1827; 1828]. They break $U_{PQ}(1)$ symmetry down to a $Z(N)$ discrete subgroup [1753]. The potential $V_a(a)$ is therefore periodic with period $\Delta a = 2\pi v_a/N = 2\pi f_a$. We may write such a potential qualitatively as

$$V_a = f_a^2 m_a^2(t) \left[1 - \cos\left(\frac{a}{f_a}\right) \right], \qquad (11.11)$$

where the axion mass $m_a(t) = m_a(T(t))$ is a function of temperature and hence of time. Equation (11.8) implies that the axion field has range $a \in [0, 2\pi v_a]$. Hence there are N degenerate vacua. The discrete degeneracy implies the existence of domain walls which end up dominating the energy density and cause a cosmological disaster unless $N = 1$. Henceforth we assume $N = 1$.

Substituting Eq. (11.11) into Eq. (11.10), the equation of motion becomes

$$\left(\partial_t^2 + 3\frac{\dot{R}}{R}\partial_t - \frac{1}{R^2}\nabla_x^2\right) a(x) + m_a^2(t)f_a \sin\left(\frac{a(x)}{f_a}\right) = 0. \qquad (11.12)$$

The non-perturbative QCD effects associated with instantons have amplitudes proportional to

$$e^{-\frac{2\pi}{\alpha_c(T)}} \simeq \left(\frac{\Lambda_{\mathrm{QCD}}}{T}\right)^{11 - \frac{2}{3}N_f}, \qquad (11.13)$$

where N_f is the number of quark flavours with mass less than T, and α_c is the fine structure constant for QCD. Equation (11.13) implies that the axion mass is strongly suppressed at temperatures large compared with the QCD scale, but turns on rather abruptly when the temperature approaches Λ_{QCD}.

Because the first three terms in Eq. (11.12) are proportional to t^{-2}, the axion mass is unimportant in the evolution of the axion field until $m_a(t)$ becomes of order $\frac{1}{t}$. Let us define a critical time t_1:

$$m_a(t_1)t_1 = 1. \qquad (11.14)$$

The axion mass effectively turns on at t_1. The value of $m_a(T)$ was obtained [13; 674; 1600] from a calculation of the effects of QCD instantons at high temperature [1026]. The result is:

$$m_a(T) \simeq 4 \times 10^{-9}\,\mathrm{eV}\left(\frac{10^{12}\,\mathrm{GeV}}{f_a}\right)\left(\frac{\mathrm{GeV}}{T}\right)^4 \qquad (11.15)$$

when T is near 1 GeV. The total effective number $\mathcal{N}$ of thermal spin degrees of freedom changes near 1 GeV temperature from a value near 60, valid above the quark–hadron phase transition, to a value of order 30 below that transition. Using $\mathcal{N} \simeq 60$, one has

$$m_a(t) \simeq 0.7 \times 10^{20}\,\frac{1}{\mathrm{sec}}\left(\frac{t}{\mathrm{sec}}\right)^2\left(\frac{10^{12}\,\mathrm{GeV}}{f_a}\right), \qquad (11.16)$$

which implies:

$$t_1 \simeq 2 \times 10^{-7}\mathrm{sec}\left(\frac{f_a}{10^{12}\,\mathrm{GeV}}\right)^{1/3}. \qquad (11.17)$$

The corresponding temperature is:

$$T_1 \simeq 1\,\mathrm{GeV}\left(\frac{10^{12}\,\mathrm{GeV}}{f_a}\right)^{1/6}. \qquad (11.18)$$

11.2.2 Vacuum realignment

Zero mode

In case 1, where inflation occurs after the PQ phase transition, the axion field is homogenized over enormous distances. Equation (11.12) becomes [13; 674; 1600]

$$\left(\frac{\mathrm{d}^2}{\mathrm{d}t^2} + \frac{3}{2t}\frac{\mathrm{d}}{\mathrm{d}t}\right) a(t) + m_a^2(t)f_a \sin\left(\frac{a(t)}{f_a}\right) = 0, \qquad (11.19)$$

where we used $R(t) \propto \sqrt{t}$. For $t \ll t_1$, we may neglect m_a. The solution is then

$$a(t) = a_0 + a_{\frac{1}{2}}t^{-\frac{1}{2}}, \qquad (11.20)$$

where a_0 and $a_{\frac{1}{2}}$ are constants. Equation (11.20) implies that the expansion of the Universe slows the axion field down to a constant value.

When t approaches t_1, the axion field starts oscillating in response to the turn on of the axion mass. We will assume that the initial value of a is sufficiently small that $f_a \sin(a/f_a) \simeq a$. Let us define ψ:

$$a(t) \equiv t^{-\frac{3}{4}}\psi(t). \qquad (11.21)$$

The equation for $\psi(t)$ is

$$\left(\frac{\mathrm{d}^2}{\mathrm{d}t^2} + \omega^2(t)\right)\psi(t) = 0, \qquad (11.22)$$

where

$$\omega^2(t) = m_a^2(t) + \frac{3}{16t^2}. \qquad (11.23)$$

For $t > t_1$, we have $\mathrm{d}(\ln\omega)/\mathrm{d}t \ll \omega \simeq m_a$. That regime is characterized by the adiabatic invariant $\psi_0^2(t)\omega(t)$, where $\psi_0(t)$ is the changing oscillation amplitude of $\psi(t)$. We have therefore

$$\psi(t) \simeq \frac{C}{\sqrt{m_a(t)}}\cos\left(\int^t \mathrm{d}t'\omega(t')\right), \qquad (11.24)$$

where C is a constant. Hence

$$a(t) = A(t)\cos\left(\int^t \mathrm{d}t'\omega(t')\right), \qquad (11.25)$$

with

$$A(t) = \frac{C}{\sqrt{m_a(t)}}\frac{1}{t^{\frac{3}{4}}}. \qquad (11.26)$$

Hence, during the adiabatic regime,

$$A^2(t)m_a(t) \propto t^{-\frac{3}{2}} \propto R(t)^{-3}. \qquad (11.27)$$

The zero-momentum mode of the axion field has energy density $\rho_a = \frac{1}{2}m_a^2 A^2$, and describes a coherent state of axions at rest with number density $n_a = \frac{1}{2}m_a A^2$. Equation (11.27) states therefore that the number of zero-momentum axions per comoving volume is conserved. The result holds as long as the changes in the axion mass are adiabatic.

We estimate the number density of axions in the zero momentum mode at late times by saying that the axion field has a random initial value $a(t_1) = f_a \alpha_1$ and evolves according to Eqs. (11.25, 11.26) for $t > t_1$. α_1 is called the 'initial misalignment angle'. Since the effective potential for a is periodic with period $2\pi f_a$, the relevant range of α_1 values is $-\pi$ to $+\pi$. The number density of zero momentum axions at time t_1 is then [13; 674; 1600]

$$n_a^{\text{vac},0}(t_1) \sim \frac{1}{2}m_a(t_1)(a(t_1))^2 = \frac{f_a^2}{2t_1}(\alpha_1)^2, \qquad (11.28)$$

where we have used Eq. (11.14). The number of axions is an adiabatic invariant after t_1. Hence the axion density from the vacuum realignment zero mode at a later time t

$$n_a^{\text{vac},0}(t) \sim \frac{f_a^2}{2t_1}(\alpha_1)^2\left(\frac{R_1}{R}\right)^3 \qquad \text{for case 1,} \qquad (11.29)$$

where R_1/R is the ratio of scale factors between t_1 and t. Thus we find

$$\rho_a^{\text{vac},0}(t_0) \sim \frac{m_a f_a^2}{2t_1}(\alpha_1)^2\left(\frac{R_1}{R_0}\right)^3 \qquad \text{for case 1} \qquad (11.30)$$

for the axion energy density today.

In case 2, the initial misalignment angle α_1 is different from one QCD horizon to the next. Since the average of $(\alpha_1)^2$ over many QCD horizons is of order one, we have

$$\rho_a^{\text{vac},0}(t_0) \sim \frac{m_a f_a^2}{2t_1}\left(\frac{R_1}{R_0}\right)^3 \qquad \text{for case 2.} \qquad (11.31)$$

However, in case 2 there are additional contributions.

Non-zero modes

In case 2, where there is no inflation after the PQ phase transition, the axion field is spatially varying. Axion strings are present as topological defects, and non-zero momentum modes of the axion field are excited. In this subsection, we consider a region of the Universe which happens to be free of strings. Strings will be added in the next subsection.

The axion field satisfies Eq. (11.12). We neglect the axion mass until $t \sim t_1$. The solution of Eq. (11.12) is a linear superposition of eigenmodes with definite comoving wavevector $\mathbf{k}$:

$$a(\mathbf{x}, t) = \int d^3 k \; a(\mathbf{k}, t) \, e^{i\mathbf{k} \cdot \mathbf{x}}, \qquad (11.32)$$

where the $a(\mathbf{k}, t)$ satisfy:

$$\left(\partial_t^2 + \frac{3}{2t} \partial_t + \frac{k^2}{R^2} \right) a(\mathbf{k}, t) = 0. \qquad (11.33)$$

Equations (11.9) and (11.32) imply that the wavelength $\lambda(t) = \frac{2\pi}{k} R(t)$ of each mode is stretched by the Hubble expansion. There are two qualitatively different regimes in the evolution of a mode, depending on whether its wavelength is outside ($\lambda(t) > t$) or inside ($\lambda(t) < t$) the horizon.

For $\lambda(t) \gg t$, only the first two terms in Eq. (11.33) are important and the most general solution is:

$$a(\mathbf{k}, t) = a_0(\mathbf{k}) + a_{-1/2}(\mathbf{k}) t^{-1/2}. \qquad (11.34)$$

Thus, for wavelengths larger than the horizon, each mode goes to a constant; the axion field is said to be 'frozen by causality'.

For $\lambda(t) \ll t$, let $a(\mathbf{k}, t) = R^{-\frac{3}{2}}(t) \psi(\mathbf{k}, t)$. Equation (11.33) becomes

$$\left(\partial_t^2 + \omega^2(t) \right) \psi(\mathbf{k}, t) = 0, \qquad (11.35)$$

where

$$\omega^2(t) = \frac{k^2}{R^2(t)} + \frac{3}{16t^2} \simeq \frac{k^2}{R^2(t)}. \qquad (11.36)$$

Since $d(\ln \omega)/dt \ll \omega$, this regime is again characterized by the adiabatic invariant $\psi_0^2(\mathbf{k}, t)\omega(t)$, where $\psi_0(\mathbf{k}, t)$ is the oscillation amplitude of $\psi(\mathbf{k}, t)$. Hence the most general solution is:

$$a(\mathbf{k}, t) = \frac{C}{R(t)} \cos \left(\int^t dt' \omega(t') \right), \qquad (11.37)$$

where C is a constant. The energy density and the number density behave respectively as $\rho_{a,\mathbf{k}} \sim \frac{C^2 w^2}{R^2(t)} \propto \frac{1}{R^4(t)}$ and $n_{a,\mathbf{k}} \sim \frac{1}{\omega}\rho_{a,\mathbf{k}} \propto \frac{1}{R^3(t)}$, indicating that the number of axions in each mode is conserved. This is as expected because the expansion of the Universe is adiabatic for modes with $\lambda(t)t \ll 1$.

Let us call $\frac{\mathrm{d}n_a}{\mathrm{d}w}(\omega, t)$ the number density, in physical and frequency space, of axions with wavelength $\lambda = \frac{2\pi}{\omega}$, for $\omega > t^{-1}$. The axion number density in physical space is thus:

$$n_a(t) = \int_{t^{-1}} \mathrm{d}\omega \, \frac{\mathrm{d}n_a}{\mathrm{d}w}(\omega, t), \qquad (11.38)$$

whereas the axion energy density is:

$$\rho_a(t) = \int_{t^{-1}} \mathrm{d}\omega \frac{\mathrm{d}n_a}{\mathrm{d}w}(\omega, t)\omega. \qquad (11.39)$$

Under the Hubble expansion axion energies redshift according to $\omega' = \omega \frac{R}{R'}$, and volume elements expand according to $\Delta V' = \Delta V \left(\frac{R'}{R}\right)^3$, whereas the number of axions is conserved mode by mode. Hence

$$\frac{\mathrm{d}n_a}{\mathrm{d}w}(\omega, t) = \left(\frac{R'}{R}\right)^2 \frac{\mathrm{d}n_a}{\mathrm{d}w}\left(\omega \frac{R}{R'}, t'\right). \qquad (11.40)$$

Moreover, the order of magnitude of $\frac{\mathrm{d}n_a}{\mathrm{d}w}$ for $\omega \sim t^{-1}$ is determined by the fact that the axion field typically varies by $v_a = f_a$ from one horizon to the next. Thus:

$$\omega \frac{\mathrm{d}n_a}{\mathrm{d}w}(\omega, t)\Delta\omega \bigg|_{\omega \sim \Delta\omega \sim \frac{1}{t}} \sim \frac{\mathrm{d}n_a}{\mathrm{d}w}\left(\frac{1}{t}, t\right)\left(\frac{1}{t}\right)^2 \sim \frac{1}{2}(\nabla a)^2 \sim \frac{1}{2}\frac{f_a^2}{t^2}. \qquad (11.41)$$

From Eqs. (11.40) and (11.41), and $R \propto \sqrt{t}$, we have [516]

$$\frac{\mathrm{d}n_a}{\mathrm{d}w}(\omega, t) \sim \frac{f_a^2}{2t^2\omega^2}. \qquad (11.42)$$

Equation (11.42) holds until the moment the axion acquires mass during the QCD phase transition.

Integrating over $\omega > t_1^{-1}$, we find the contribution from vacuum realignment involving higher momentum modes

$$n_a^{\mathrm{vac},1}(t_1) \sim \frac{f_a^2}{2t_1}. \qquad (11.43)$$

Almost all these axions are non-relativistic after t_1. Hence

$$\rho_a^{\mathrm{vac},1}(t_0) \sim \frac{m_a f_a^2}{2t_1}\left(\frac{R_1}{R_0}\right)^3. \qquad (11.44)$$

Note that $\rho_a^{\mathrm{vac},1}$ and $\rho_a^{\mathrm{vac},0}$ are of the same order of magnitude.

11.2.3 String decay

In case 2, axion strings are present as topological defects in the axion field from the PQ to the QCD phase transitions [1904]. The energy per unit length of an axion string is

$$\mu = \pi v_a^2 \ln(v_a L). \tag{11.45}$$

Here, L is an infrared cutoff, which in practice equals the distance to the nearest neighbour string. Because they are strongly coupled to the axion field, the strings decay very efficiently into axions. We will see that practically all axions produced by string decay are non-relativistic after t_1. Because each such axion contributes m_a to the present energy density, it is important to evaluate the number density of axions emitted in string decay.

At a given time t, there is at least on the order of one string per horizon. Indeed, the axion field is completely uncorrelated over distances larger than t. Hence there is non-zero probability that the random values of $a(\mathbf{x}, t)$ wander from zero to $2\pi v_a$ along a closed path in physical space if that closed path has size larger than t. When this is the case, a string perforates the surface subtended by the closed path.

At first, the strings are stuck in the primordial plasma and are stretched by the Hubble expansion. During that time, because $R(t) \propto \sqrt{t}$, the density of strings grows to be much larger than one per horizon. However, expansion dilutes the plasma and at some point the strings become unstuck. The temperature at which strings start to move freely is of order [1075]

$$T_* \sim 2 \times 10^7 \, \mathrm{GeV} \left(\frac{f_a}{10^{12} \, \mathrm{GeV}} \right)^2. \tag{11.46}$$

Below T_*, there is a network of axion strings moving at relativistic speeds. Axions are radiated very efficiently by collapsing string loops and by oscillating wiggles on long strings. By definition, long strings stretch across the horizon. They move and intersect one another. When strings intersect, there is a high probability of reconnection, i.e. of rerouting of the topological flux [1741]. Because of such 'intercommuting', long strings produce loops which then collapse freely. In view of this efficient decay mechanism, the average density of long strings is expected to be of order the minimum consistent with causality, namely one long string per horizon. Hence the energy density in long strings is

$$\rho_{\mathrm{str}}(t) = \xi \frac{\tau}{t^2} \simeq \xi \frac{\pi f_a^2}{t^2} \ln(v_a t), \tag{11.47}$$

where ξ is a parameter of order one.

The equations governing the number density $n_a^{\text{str}}(t)$ of axions radiated by axion strings are [1075]

$$\frac{\text{d}\rho_{\text{str}}}{\text{d}t} = -2H\rho_{\text{str}} - \frac{\text{d}\rho_{\text{str}\to a}}{\text{d}t} \tag{11.48}$$

and

$$\frac{\text{d}n_a^{\text{str}}}{\text{d}t} = -3Hn_a^{\text{str}} + \frac{1}{\omega(t)}\frac{\text{d}\rho_{\text{str}\to a}}{\text{d}t}, \tag{11.49}$$

where $\omega(t)$ is defined by:

$$\frac{1}{\omega(t)} = \frac{1}{\frac{\text{d}\rho_{\text{str}\to a}}{\text{d}t}} \int \frac{\text{d}k}{k}\frac{\text{d}\rho_{\text{str}\to a}}{\text{d}t\,\text{d}k}. \tag{11.50}$$

Here, k is axion momentum magnitude, $\frac{\text{d}\rho_{\text{str}\to a}}{\text{d}t}(t)$ is the rate at which energy density gets converted from strings to axions at time t, and $\frac{\text{d}\rho_{\text{str}\to a}}{\text{d}t\,\text{d}k}(t,k)$ is the spectrum of the axions produced. Therefore $\omega(t)$ is the average energy of axions radiated in string decay processes at time t. The term $-2H\rho_{\text{str}} = +H\rho_{\text{str}} - 3H\rho_{\text{str}}$ in Eq. (11.48) takes account of the fact that the Hubble expansion both stretches ($+H\rho_{\text{str}}$) and dilutes ($-3H\rho_{\text{str}}$) long strings. Integrating Eqs. (11.47) to (11.49), setting $H = \frac{1}{2t}$ and neglecting terms of order one versus terms of order $\ln(v_a t)$, one obtains

$$n_a^{\text{str}}(t) \simeq \frac{\xi\pi f_a^2}{t^{\frac{3}{2}}} \int_{t_{\text{PQ}}}^t \text{d}t' \frac{\ln(v_a t')}{t'^{\frac{3}{2}}\omega(t')}, \tag{11.51}$$

where t_{PQ} is the time of the PQ transition.

To obtain $n_a^{\text{str}}(t)$ we need to know $\omega(t)$, the average energy of axions radiated at time t. If $\omega(t)$ is large, the number of radiated axions is small, and vice versa. Axions are radiated by wiggles on long strings and by collapsing string loops. Consider a process which starts at t_{in} and ends at t_{fin}, and which converts an amount of energy E from string to axions; t_{in} and t_{fin} are both taken to be of order t. It is useful to define the quantity [1052]

$$N_{\text{ax}}(t) \equiv \int \text{d}k \frac{\text{d}E}{\text{d}k}(t)\frac{1}{k}, \tag{11.52}$$

where k is wavevector, and $\frac{\text{d}E}{\text{d}k}(t)$ is the wavevector spectrum of the a field. At the start ($t = t_{\text{in}}$), only string contributes to the integral in Eq. (11.52). At the end ($t = t_{\text{fin}}$), only axions contribute. In between, both axions and string contribute. The number of axions radiated is $N_a = N_{\text{ax}}(t_{\text{fin}})$, and their average energy is $\omega = E/N_a$. The energy stored in string has spectrum $\frac{\text{d}E}{\text{d}k} \propto \frac{1}{k}$ for $k_{\text{min}} < k < k_{\text{max}}$ where k_{max} is of order v_a and k_{min} of order $\frac{2\pi}{L} \sim \frac{2\pi}{t}$. If $\ell \equiv E/\mu$ is the length of string converted to axions, we have

$$N_{\text{ax}}(t_{\text{in}}) = \frac{E}{\ln(tv_a)k_{\text{min}}}. \tag{11.53}$$

Hence

$$\frac{1}{\omega} = \frac{r}{\ln(v_a t)k_{\text{min}}}, \tag{11.54}$$

where r is the relative change in $N_{\text{ax}}(t)$ during the process in question:

$$r \equiv \frac{N_{\text{ax}}(t_{\text{fin}})}{N_{\text{ax}}(t_{\text{in}})}. \tag{11.55}$$

The wavevector k_{min} is of order $2\pi/L$ where L is the loop size in the case of collapsing loops, and the wiggle wavelength in the case of bent strings; L is at most of order t but may be substantially smaller than that if the string network has a lot of small-scale structure. To parameterize our ignorance in this matter, we define χ such that the suitably averaged $k_{\text{min}} \equiv \chi 2\pi/t$. Combining Eqs. (11.51) and (11.54) we find:

$$n_a^{\text{str}}(t) \simeq \frac{\xi \bar{r}}{\chi} \frac{f_a^2}{t}, \tag{11.56}$$

where $\bar{r}$ is the weighted average of r over the various processes that convert string to axions. One can show [1051] that the population of axions that were radiated between t_{PQ} and t have spectrum $\frac{dn_a}{dk} \propto \frac{1}{k^2}$ for $\frac{1}{t} \lesssim k \lesssim \frac{1}{\sqrt{tt_{\text{PQ}}}}$, irrespective of the shape of $\frac{d\rho_{\text{str}\to a}}{dt\,dk}$, provided that $t \gg t_{\text{PQ}}$.

At time t_1, each string becomes the edge of a domain wall, and the process of axion radiation by strings stops. Since their momenta are of order t_1^{-1} at time t_1, the axions radiated by strings become non-relativistic soon after they acquire mass. Therefore, the string decay contribution to the axion energy density today is

$$\rho_a^{\text{str}}(t_0) = m_a n_a^{\text{str}}(t_1)\left(\frac{R_1}{R_0}\right)^3 \simeq m_a \frac{\xi \bar{r}}{\chi} \frac{f_a^2}{t_1}\left(\frac{R_1}{R_0}\right)^3. \tag{11.57}$$

We now discuss the factors on the right-hand side of Eq. (11.57) which are specific to the string decay contribution.

ξ: This parameter determines the density of the string network, Eq. (11.47), with $\xi = 1$ corresponding to a density of one long string per horizon. In ref. [1052] it was argued that $\xi \simeq 1$ because global strings can decay efficiently into axions, and therefore the number density of long strings should be close to the minimum consistent with causality. In numerical simulations of global string networks in an expanding Universe [1949], it was found that indeed $\xi \simeq 1$. So there appears to be good ground for setting $\xi \simeq 1$.

χ: This parameter defines the low wavevector edge of the $\frac{dn_a}{dk} \propto \frac{1}{k^2}$ spectrum through $k_{\min} \equiv \chi \frac{2\pi}{t}$. The parameters χ and ξ are related since the average interstring distance controls both. On dimensional grounds, $\chi \propto \sqrt{\xi}$. So the effect of small-scale structure in the axion string network partially cancels out in the right-hand side of Eq. (11.57). χ is expected to be of order one, but the uncertainty on this is at least a factor two.

$\bar{r}$: This parameter defines the average energy of the axions emitted in string decay, through Eqs. (11.54, 11.55). It is the unknown on which most of the debate has focused in the past. Two basic scenarios have been put forth, which we call A and B. The question is: what is the spectrum of axions radiated by strings? The main source is closed loops of size $L \sim t$. Scenario A postulates that a bent string or closed loop oscillates many times, with period of order L, before it has released its excess energy and that the spectrum of radiated axions is concentrated near $2\pi/L$. In that case one has $\bar{r} \sim \ln(v_a t_1) \simeq 67$. Scenario B postulates that the bent string or closed loop releases its excess energy very quickly and that the spectrum of radiated axions is $dE/dk \propto 1/k$ with a high frequency cutoff of order $2\pi v_a$ and a low frequency cutoff of order $2\pi/L$. In scenario B, the initial and final spectra dE/dk of the energy stored in the axion field are qualitatively the same and hence $\bar{r} \sim 1$. In scenario A, the string decay contribution dominates over the vacuum realignment contribution by the factor $\ln(v_a t_1)$, whereas in scenario B the contributions from string decay and vacuum realignment have the same order of magnitude.

Many authors [233; 234; 607; 617; 618; 619; 1906] have argued in favour of scenario A, adopting the point of view that global strings are similar to local strings and that their coupling to the axion field can be treated perturbatively. My collaborators and I [1052; 1075] have argued in support of scenario B, emphasizing that the dynamics of global strings is dominated by the energy stored in the axion field and that there is no reason to believe that this energy would behave in the same way as the energy stored in the string core. The numerical simulations of the motion and decay of axion strings in refs. [1051; 1052] give strong support to scenario B. These simulations are of oscillating strings with ends held fixed, of collapsing circular loops and of collapsing non-circular closed loops with angular momentum. Over the range of $\ln(v_a L)$ accessible with present technology ($2.5 \lesssim \ln(v_a L) \lesssim 5.0$), it was found that $r \simeq 0.8$ for closed loops and $r \simeq 1.07$ for oscillating strings with ends held fixed. No dependence of r on $\ln(v_a L)$ was found for closed loops, and for bent strings with ends held fixed r was found to decrease slightly with increasing $\ln(v_a L)$, whereas scenario A predicts r to be proportional to $\ln(v_a L)$.

11.2.4 Domain wall decay

The final contribution to the cold axion cosmological energy density in case 2 is from the decay into non-relativistic axions of axion walls bounded by string. We assume here that $N = 1$ in which case each string becomes the edge of one domain wall when the axion mass turns on during the QCD phase transition. The wall energy per unit surface is

$$\sigma \simeq 8 m_a f_a^2 \simeq 5 \times 10^{10} \text{ GeV}^3 \left(\frac{f_a}{10^{12} \text{ GeV}} \right). \tag{11.58}$$

Let t_3 be the time when the decay effectively takes place and $\gamma \equiv \omega'/(m_a(t_3))$ the average Lorentz γ factor of the axions produced; ω' is their average energy. The density of walls at time t_1 was estimated to be of order 0.7 per horizon volume [516]. Hence the average energy density in walls is

$$\rho_{\text{dw}}(t) \sim 0.7 \, \frac{\sigma(t)}{t_1} \left(\frac{R_1}{R} \right)^3 \sim (0.7)(8) m_a(t) \frac{f_a^2}{t_1} \left(\frac{R_1}{R} \right)^3 \tag{11.59}$$

between t_1 and t_3. We assumed that the energy in walls simply scales as $\sigma(t)$. After time t_3, the number density of axions produced in the decay of walls bounded by strings is of order

$$n_a^{\text{dw}}(t) \sim \frac{\rho_{\text{dw}}(t_3)}{\omega'} \left(\frac{R_3}{R} \right)^3 \sim \frac{6 \, f_a^2}{\gamma \, t_1} \left(\frac{R_1}{R} \right)^3. \tag{11.60}$$

Note that the dependence on t_3 drops out of our estimate of n_a^{dw}. In the simulations of the motion and decay of walls bounded by string in ref. [516] it was found that $\gamma \simeq 7$ for $\ln(v_a/m_a) \sim 4.6$ but that γ increases approximately linearly with $\ln(\sqrt{\lambda} v_a/m_a)$. If this behaviour is extrapolated all the way to $\ln(\sqrt{\lambda} v_a/m_a) \simeq 60$, which is the value in axion models of interest, then $\gamma \simeq 60$. In that case the contribution from wall decay is subdominant relative to those from vacuum realignment and string decay.

11.3 Relic density and primordial velocity dispersion

In this section we sum up the various contributions to the cosmological energy density of cold axions and discuss their velocity dispersion.

11.3.1 The cold axion cosmological energy density

To estimate the cosmological energy density of cold axions in case 2, we neglect the contribution from wall decay and assume that scenario B is correct for the string contribution. By adding the right-hand sides of Eqs. (11.31), (11.44) and (11.57) with $N = \bar{r} = \xi = \chi = 1$, we find

$$\rho_a(t_0) \sim 2 \, \frac{f_a^2}{t_1} \left(\frac{R_1}{R_0}\right)^3 m_a \qquad \text{for case 2.} \qquad (11.61)$$

Equation (11.30) gives the cold axion cosmological energy density in case 1. To determine the ratio of scale factors R_1/R_0, we assume conservation of entropy from time t_1 till the present. The number $\mathcal{N}_1$ of effective thermal degrees of freedom at time t_1 is of order 60. Keeping in mind that neutrinos decouple before electron–positron annihilation, one finds

$$\left(\frac{R_1}{R_0}\right)^3 \simeq 0.063 \left(\frac{T_{\gamma,0}}{T_1}\right)^3. \qquad (11.62)$$

Combining Eqs. (11.17), (11.61), (11.30) and (11.62), and dividing by the critical density $\rho_c = 3H_0^2/(8\pi G)$, we find

$$\Omega_a \sim 0.15 \left(\frac{f_a}{10^{12}\,\text{GeV}}\right)^{7/6} \left(\frac{0.7}{h}\right)^2 \alpha_1^2 \qquad \text{for case 1,}$$

$$\qquad \sim 0.7 \left(\frac{f_a}{10^{12}\,\text{GeV}}\right)^{7/6} \left(\frac{0.7}{h}\right)^2 \qquad \text{for case 2,} \qquad (11.63)$$

where h is defined as usual by $H_0 = h \times 100\,\text{km}\,\text{s}^{-1}\,\text{Mpc}^{-1}$.

Equations (11.63) are subject to many sources of uncertainty, aside from the uncertainty about the contribution from string decay. The axion energy density may be diluted by the entropy release from heavy particles which decouple before the QCD epoch but decay afterwards [1313; 1314; 1806], or by the entropy release associated with a first-order QCD phase transition. On the other hand, if the QCD phase transition is first-order [639; 1098; 1873; 1884], an abrupt change of the axion mass at the transition may increase Ω_a. A model has been put forth [1206] in which the axion decay constant f_a is time-dependent, the value $f_a(t_1)$ during the QCD phase transition being much smaller than the value f_a today. This yields a suppression of the axion cosmological energy density by a factor $[f_a(t_1)/f_a]^2$ compared with the usual case. Finally, it has been proposed that the axion density is diluted by 'coherent deexcitation', i.e. adiabatic level crossing of $m_a(t)$ with the mass of some other pseudo-Nambu–Goldstone boson which mixes with the axion [1095].

11.3.2 Velocity dispersions

The axions produced by vacuum realignment, string decay and wall decay all have extremely small velocity dispersion today. In case 1, where the axions are produced in a zero-momentum state, the velocity dispersion is zero. (This ignores the small quantum mechanical fluctuations created during the inflationary epoch, which will be discussed in Section 11.5.)

In case 2, we distinguish two subpopulations of cold axions: Pop. I and Pop. II, with the second kind having velocity dispersion typically a factor 10^3 to 10^4 larger than the first. The Pop. I axions are those produced by vacuum realignment or string decay and which escaped being hit by moving domain walls. They have typical momentum $p_I(t_1) \sim \frac{1}{t_1}$ at time t_1 because they are associated with axion field configurations which are inhomogeneous on the horizon scale at that time. Their velocity dispersion is of order:

$$\beta_I(t) \sim \frac{1}{m_a t_1}\left(\frac{R_1}{R}\right) \simeq 3 \times 10^{-17}\left(\frac{10^{-5}\text{eV}}{m_a}\right)^{5/6}\frac{R_0}{R}. \tag{11.64}$$

The corresponding effective temperature is of order 0.5×10^{-34} K $\left(\frac{10^{-5}\text{eV}}{m_a}\right)^{\frac{2}{3}}$ today. This is very cold indeed!

Population II are axions produced in the decay of domain walls and axions that were hit by moving domain walls. Axions produced in the decay of domain walls have typical momentum $p_{II}(t_3) \sim \gamma m_a(t_3)$ at time t_3 when the walls effectively decay. Their velocity dispersion is therefore of order:

$$\beta_{II}(t) \sim \gamma \frac{m_a(t_3)}{m_a}\frac{R_3}{R} \simeq 10^{-13}\, q\, \left(\frac{10^{-5}\text{eV}}{m_a}\right)^{1/6}\frac{R_0}{R}, \tag{11.65}$$

where $q \equiv \gamma \frac{m_a(t_3)}{m_a}\frac{R_3}{R_1}$ parameterizes our ignorance of the wall decay process. We expect q to be of order one but with very large uncertainties. There is, however, a lower bound on q which follows from the fact that the time t_3 when the walls effectively decay must be after t_2 when the energy density in walls starts to exceed the energy density in strings. This implies

$$q = \frac{\gamma m_a(t_3)}{m_a}\frac{R_3}{R_1} > \frac{\gamma m_a(t_2)}{m_a}\frac{R_2}{R_1} \simeq \frac{\gamma}{130}\left(\frac{10^{-5}\text{eV}}{m_a}\right)^{2/3}. \tag{11.66}$$

Since computer simulations suggest that γ is of order 60, Pop. II axions have much larger velocity dispersion than Pop. I axions, by a factor of 10^3 or more. Whereas Pop. II axions are relativistic or near-relativistic at the end of the QCD phase transition, Pop. I axions are definitely non-relativistic at that time since $m_a >> 1/t_1$. The axions which were produced by vacuum realignment or string decay but were hit by relativistically moving walls at some time between t_1 and t_3 should be included in Pop. II since they are relativistic just after getting hit. The next section will highlight the differences in the behaviours of the two populations of cold axions.

11.4 Axion miniclusters

If there is no inflation after the PQ phase transition (case 2), the initial misalignment angle α_1 changes by $\mathcal{O}(1)$ from one QCD time horizon to the next. Hence, the fluid of cold axions produced by vacuum realignment is inhomogeneous with $\delta\rho_a/\rho_a = \mathcal{O}(1)$ at the time of the QCD phase transition. As will be shown shortly, the streaming length of Pop. I axions is too short for these inhomogeneities to be erased by free streaming before the time t_{eq} of equality between matter and radiation, when density perturbations start to grow in earnest by gravitational instability. At time t_{eq}, the $\delta\rho_a/\rho_a = \mathcal{O}(1)$ inhomogeneities in the axion fluid promptly form gravitationally bound objects, called axion miniclusters [516; 1117; 1263; 1264]. The properties of axion miniclusters are of concern to experimentalists attempting the direct detection of dark matter axions on Earth. Indeed, those experiments would become even more challenging than they are already if most of the cold axions condense into miniclusters and the miniclusters withstand tidal disruption afterwards. Of course, these issues only arise in case 2. There are no axion miniclusters in case 1. As described above, there are two populations of cold axions, Pop. I and Pop. II, with velocity dispersions given by Eqs. (11.64) and (11.65) respectively. Both populations are inhomogeneous at the time of the QCD phase transition. The free streaming length from time t_1 to t_{eq} is:

$$\ell_f = R(t_{eq}) \int_{t_1}^{t_{eq}} dt \frac{\beta(t)}{R(t)} \simeq \beta(t_1)\sqrt{t_1 t_{eq}} \ln\left(\frac{t_{eq}}{t_1}\right). \tag{11.67}$$

The time of equality and the corresponding temperature are respectively $t_{eq} \simeq 2.3 \times 10^{12}\,\text{s}$ and $T_{eq} \simeq 0.77$ eV. The free streaming length should be compared with the size

$$\ell_{mc} \sim t_1 \frac{R_{eq}}{R_1} \simeq \sqrt{t_1 t_{eq}} \simeq 2 \times 10^{13}\text{cm} \left(\frac{10^{-5}\text{eV}}{m_a}\right)^{1/6} \tag{11.68}$$

of axion inhomogeneities at t_{eq}. Using Eq. (11.64) we find for Pop. I:

$$\frac{\ell_{f,I}}{\ell_{mc}} \simeq \frac{1}{t_1 m_a} \ln\left(\frac{t_{eq}}{t_1}\right) \simeq 2 \times 10^{-2} \left(\frac{10^{-5}\text{eV}}{m_a}\right)^{2/3}. \tag{11.69}$$

Hence, in the axion mass range of interest, Pop. I axions do not homogenize. At t_{eq} most Pop. I axions condense into miniclusters. The typical size of axion miniclusters is ℓ_{mc} and their typical mass is [516; 1264]

$$M_{mc} \sim \eta\, \rho_a(t_{eq})\ell_{mc}^3 \sim \eta\, 5 \times 10^{-13} M_\odot \left(\frac{10^{-5}\text{eV}}{m_a}\right)^{5/3}, \tag{11.70}$$

where η is the fraction of cold axions that are Pop. I. We assumed that all Pop. I axions condense into miniclusters, and used Eq. (11.63) (case 2) to estimate $\rho_a(t_{eq})$.

Using Eq. (11.65), we find for Pop. II:

$$\frac{\ell_{f,II}}{\ell_{mc}} \sim q \ln\left(\frac{t_{eq}}{t_3}\right) \simeq 42\,q. \qquad (11.71)$$

Using Eq. (11.66) and assuming the range $\gamma \sim 7$ to 60, suggested by the numerical simulations of ref. [516], we conclude that Pop. II axions do homogenize and hence that the axion energy density has a smooth component at t_{eq}.

However, Pop. II axions may become gravitationally bound to miniclusters later on. It seems rather difficult to model this process reliably. A discussion is given in ref. [516]. It is concluded there that the accretion of Pop. II axions results in miniclusters which have an inner core of Pop. I axions with density of order 10^{-18} g cm^{-3} and a fluffy envelope of Pop. II axions with density of order 10^{-25} g cm^{-3}.

When a minicluster falls onto a galaxy, tidal forces are apt to destroy it. If a minicluster falls through the inner parts of our Galaxy ($r < 10$ kpc) where the density is of order 10^{-24} g cm^{-3}, its fluffy envelope of Pop. II axions is likely to be pulled off immediately. This is helpful for direct searches of dark matter axions on Earth since it implies that a smooth component of dark matter axions with density of order the halo density permeates us whether or not there is inflation after the Peccei–Quinn phase transition. Even the central cores of Pop. I axions may eventually be destroyed. When a minicluster passes by an object of mass M with impact parameter b and velocity v, the internal energy per unit mass ΔE given to the minicluster by the tidal gravitational forces from that object is of order [1117]

$$\Delta E \sim \frac{G^2 M^2 \ell_{mc}^2}{b^4 \beta^2}, \qquad (11.72)$$

whereas the binding energy per unit mass of the minicluster $E \sim G\,\rho_{mc}\ell_{mc}^2$. If the minicluster travels a length $L = \beta t$ through a region where objects of mass M have density n, the relative increase in internal energy is:

$$\frac{\Delta E}{E} \sim \frac{G\rho_M^2 t^2}{\rho_{mc}}, \qquad (11.73)$$

where $\rho_M = Mn$. Equation (11.73) follows from the fact that ΔE is dominated by the closest encounter and the latter has impact parameter b_{min}: $\pi b_{min}^2 n L = 0(1)$. Note that $\Delta E/E$ is independent of M. A minicluster inner core which has spent most of its life in the central part of our Galaxy only barely survives, since $\Delta E/E \sim 10^{-2}$ in that case.

The direct encounter of a minicluster with Earth would be quite rare, happening only every 10^4 years or so. The encounter would last for about 3 days during which the local axion density would increase by a factor of order 10^6.

11.5 Axion isocurvature perturbations

In this section we describe the isocurvature perturbations [147; 1341; 1357; 1717; 1806; 1876] produced if inflation occurs after the Peccei–Quinn phase transition, and derive the constraints on axion parameters from the absence of isocurvature fluctuations in CMB radiation observations.

If the reheat temperature after inflation is less than the temperature T_{PQ} at which $U_{PQ}(1)$ is restored (case 1), the axion field is present during inflation and is subject to quantum mechanical fluctuations, just like the inflaton. In fact, since the axion field is massless and weakly coupled like the inflaton, it has the same fluctuation spectrum [363; 1340; 1796; 1905]

$$P_a(k) \equiv \int \frac{d^3x}{(2\pi)^3} < \delta a(\mathbf{x},t)\delta a(\mathbf{x}',t) > e^{-i\mathbf{k}\cdot(\mathbf{x}-\mathbf{x}')} = \left(\frac{H_I}{2\pi}\right)^2 \frac{2\pi^2}{k^3}, \quad (11.74)$$

where H_I is the expansion rate during inflation. As before, $\mathbf{x}$ are comoving spatial coordinates. The axion fluctuations described by Eq. (11.74) are commonly written in shorthand notation as $\delta a = H_I/2\pi$. The fluctuation in each axion field mode is 'frozen in' after $R(t)/k$ exceeds the horizon length H_I^{-1}.

We do not consider here the possibility of fluctuations in the axion decay constant f_a during inflation. Such fluctuations are discussed in refs. [1342; 1358; 1359].

At the start of the QCD phase transition, the local value of the axion field $a(\mathbf{x},t)$ determines the local number density of cold axions produced by the vacuum realignment mechanism (see Eq. (11.28)):

$$n_a(\mathbf{x},t_1) = \frac{f_a^2}{2t_1}\alpha(\mathbf{x},t_1)^2, \quad (11.75)$$

where $\alpha(\mathbf{x},t_1) = a(\mathbf{x},t_1)/f_a$ is the misalignment angle. The fluctuations in the axion field produce perturbations in the cold axion density

$$\frac{\delta n_a^{iso}}{n_a} = \frac{2\delta a}{a_1} = \frac{H_I}{\pi f_a \alpha_1}, \quad (11.76)$$

where $a_1 = a(t_1) = f_a \alpha_1$ is the value of the axion field at the start of the QCD phase transition, common to our entire visible Universe. These perturbations obey

$$\delta\rho_a^{\text{iso}}(t_1) = -\delta\rho_{\text{rad}}^{\text{iso}}(t_1) \tag{11.77}$$

since the vacuum realignment mechanism converts energy stored in the quark–gluon plasma into axion rest mass energy. In contrast, the density perturbations produced by the fluctuations in the inflaton field [200; 1044; 1080; 1487; 1796] satisfy

$$\frac{\delta\rho_{\text{matter}}}{\rho_{\text{matter}}} = \frac{3}{4}\frac{\delta\rho_{\text{rad}}}{\rho_{\text{rad}}}. \tag{11.78}$$

Density perturbations that satisfy Eq. (11.78) are called 'adiabatic', whereas density perturbations that do not satisfy Eq. (11.78) are called 'isocurvature'. Isocurvature perturbations, such as the density perturbations of Eq. (11.77), make a different imprint on the cosmic microwave background than do adiabatic ones. The CMBR observations are consistent with pure adiabatic perturbations. This places a constraint on axion models if the Peccei–Quinn phase transition occurs before inflation.

Before we derive this constraint, two comments are in order. The first is that, if the Peccei–Quinn transition occurs after inflation, axion models still predict isocurvature perturbations but not on length scales relevant to CMBR observations. Indeed we saw in the previous section that in this case (case 2) the axion field fluctuates by order f_a from one QCD horizon to the next. Those fluctuations produce isocurvature perturbations on the scale of the QCD horizon, which is much smaller than the length scales observed in the CMBR. Their main phenomenological implication is the axion miniclusters which were discussed in Section 11.4. The second comment is that, if the Peccei–Quinn phase transition occurs before inflation (case 1), the density perturbations in the cold axion fluid have both adiabatic and isocurvature components. The adiabatic perturbations $(\frac{\delta\rho_a^{\text{ad}}}{3\rho_a} = \frac{\delta\rho_{\text{rad}}^{\text{ad}}}{4\rho_{\text{rad}}} = \frac{\delta T}{T})$ are produced by the quantum mechanical fluctuations of the inflaton field during inflation, whereas the isocurvature perturbations are produced by the quantum mechanical fluctuations of the axion field during that same epoch. The adiabatic and axion isocurvature components are uncorrelated.

The upper bound from CMBR observations and large-scale structure data on the fraction of CDM perturbations which are isocurvature perturbations is of order 30% in amplitude (10% in the power spectrum) [241; 273; 591;

1562; 1866; 1888]. Allowing for the possibility that only part of the cold dark matter is axions, the bound on isocurvature perturbations implies

$$\frac{\delta\rho_a^{\text{iso}}}{\rho_{\text{CDM}}} = \frac{\delta\rho_a^{\text{iso}}}{\rho_a} \cdot \frac{\rho_a}{\rho_{\text{CDM}}} = \frac{H_I}{\pi f_a \alpha_1} \frac{\Omega_a}{\Omega_{\text{CDM}}} < 0.3 \frac{\delta\rho_m}{\rho_m}, \tag{11.79}$$

where we have used Eq. (11.76). The term $\frac{\delta\rho_m}{\rho_m}$ is the amplitude of the primordial spectrum of matter perturbations. It is related to the amplitude of low multipole CMBR anisotropies through the Sachs–Wolfe effect [14; 1558; 1666]. The observations imply $\frac{\delta\rho_m}{\rho_m} \simeq 4.6 \times 10^{-5}$ [687].

In terms of α_1, the cold axion energy density is given by Eq. (11.63). We rewrite that equation here, assuming $h \simeq 0.7$:

$$\Omega_a \simeq 0.15 \left(\frac{f_a}{10^{12}\,\text{GeV}}\right)^{\frac{7}{6}} \alpha_1^2. \tag{11.80}$$

It has been remarked by many authors, starting with Pi [1575], that it is possible for f_a to be much larger than $10^{12}\,\text{GeV}$ because α_1 may be accidentally small in our visible Universe. The requirement that $\Omega_a < \Omega_{\text{CDM}} = 0.22$ implies

$$\left|\frac{\alpha_1}{\pi}\right| < 0.4 \left(\frac{10^{12}\,\text{GeV}}{f_a}\right)^{\frac{7}{12}}. \tag{11.81}$$

Since $-\pi < \alpha_1 < +\pi$ is the a-priori range of α_1 values and no particular value is preferred over any other, $\left|\frac{\alpha_1}{\pi}\right|$ may be taken to be the 'probability' that the initial misalignment angle has magnitude less than $|\alpha_1|$. (Strictly speaking, the word probability is not appropriate here since there is only one Universe in which α_1 may be measured.) If $\left|\frac{\alpha_1}{\pi}\right| = 2 \times 10^{-3}$, for example, f_a may be as large as $10^{16}\,\text{GeV}$, which is often thought to be the 'grand unification scale'.

The presence of isocurvature perturbations constrains the small α_1 scenario in two ways [1876]. First, it makes it impossible to have α_1 arbitrarily small. Using Eq. (11.74), one can show that the fluctuations in the axion field cause the latter to perform a random walk [1343] characterized by the property

$$\frac{1}{V}\int_V \mathrm{d}^3x < (\delta a(\mathbf{x},t) - \delta a(\mathbf{0},t))^2 >= 4\pi H_I^2 \ln(Rk_{\text{max}}), \tag{11.82}$$

where the integral is over a sphere of volume $V = \frac{4\pi}{3}R^3$ centred at $\mathbf{x} = 0$, and k_{max} is a cutoff on the wavevector spectrum. The value of a_1^2 cannot be smaller than the right-hand side of Eq. (11.82) with R equal to the size of the present Universe and k_{max} equal to the Hubble rate at QCD time, redshifted down to the present. Since $\Omega_a < 0.22$, this implies a bound on H_{I}. Translated to a bound on the scale of inflation Λ_{I}, defined by $H_{\mathrm{I}}^2 = \frac{8\pi}{3} G \Lambda_{\mathrm{I}}^4$, it is

$$\Lambda_{\mathrm{I}} < 5 \times 10^{14}\,\mathrm{GeV} \left(\frac{f_a}{10^{12}\,\mathrm{GeV}} \right)^{\frac{5}{24}}. \tag{11.83}$$

Second, one must require axion isocurvature perturbations to be consistent with CMBR observations. Combining Eqs. (11.79) and (11.80), and setting $\Omega_{\mathrm{CDM}} = 0.22$, $\frac{\delta \rho_m}{\rho_m} = 4.6 \times 10^{-5}$, one obtains

$$\Lambda_I < 10^{13}\,\mathrm{GeV}\ \Omega_a^{-\frac{1}{4}} \left(\frac{f_a}{10^{12}\,\mathrm{GeV}} \right)^{\frac{5}{24}}. \tag{11.84}$$

Let us keep in mind that the bounds (11.83) and (11.84) pertain only if the reheat temperature $T_{\mathrm{RH}} < T_{\mathrm{PQ}}$. One may, for example, have $\Omega_a = 0.22$, $f_a \simeq 10^{12}\,\mathrm{GeV}$ and $\Lambda_{\mathrm{I}} \simeq 10^{16}\,\mathrm{GeV}$, provided $T_{\mathrm{RH}} \gtrsim 10^{12}\,\mathrm{GeV}$, which is possible if reheating is efficient enough.

12

Sterile neutrinos

Mikhail Shaposhnikov

12.1 Particle physics motivation

In the Standard Model (SM) of particle physics interactions, as it was formulated in [964; 1668; 1925], neutrinos are exactly massless as a consequence of the requirements of gauge invariance and renormalizability of the theory. Moreover, three different leptonic numbers, associated with e, μ and τ flavours, are separately conserved. At the same time, it is an experimental fact that neutrinos have tiny but non-zero masses and that neutrinos of different flavours mix with each other (for a review see [1825]). This provides solid experimental laboratory evidence (and the only evidence at present) in favour of physics beyond the SM. Is this a sign of the existence of a new energy scale in particle physics, related to Grand Unification? Is it an indication that the SM has to be replaced by a new renormalizable low-energy theory?

To elucidate the origin of neutrino masses, one can use the effective field theory approach [1927; 1935]. Indeed, the low-energy electroweak Lagrangian can contain all sorts of higher-dimensional $SU(3) \times SU(2) \times U(1)$ invariant operators O_n, containing the SM fields and suppressed by some unknown scale Λ:

$$L = L_{\text{SM}} + \sum_{n=5}^{\infty} \frac{O_n}{\Lambda^{n-4}}. \tag{12.1}$$

The Majorana neutrino masses come from the five-dimensional operator [1927] $O_5 = A_{\alpha\beta}(\bar{L}_\alpha \tilde{\Phi})(\Phi^\dagger L_\beta^c)$, where L_α are the leptonic doublets ($\alpha = e, \mu, \tau$), Φ is the Higgs field, $\tilde{\Phi}_i = \epsilon_{ij}\Phi_j^*$, $A_{\alpha\beta}$ is a complex symmetric dimensionless matrix and c is the sign of charge conjugation. The neutrino mass matrix is

Particle Dark Matter: Observations, Models and Searches, ed. Gianfranco Bertone. Published by Cambridge University Press. © Cambridge University Press 2010.

$$m_\nu = A_{\alpha\beta} \frac{v^2}{\Lambda},\qquad(12.2)$$

where $v \simeq 174$ GeV is the vacuum expectation value of the Higgs field. Equation (12.2) can describe all available confirmed neutrino experiments if A is chosen in some particular way [1825].

An effective field theory approach thus allows us to 'solve' phenomenological aspects of the problem of neutrino masses and oscillations. However, it leaves unanswered the following fundamental questions:

(i) What is the physics behind the non-renormalizable terms?
(ii) What is the value of Λ?

As a first natural possibility one can test a hypothesis that the origin of neutrino masses is the Planck scale physics, related to quantum gravity, so that $\Lambda = M_{\rm Pl} \simeq 10^{19}$ GeV. If true, the prediction of neutrino masses is $m_\nu \sim v^2/M_{\rm Pl} \simeq 10^{-6}$ eV. This number is far away from experimental observations, making this hypothesis unlikely.

Another possibility is to assume that the origin of neutrino masses is the existence of new unseen particles and that the complete theory is a renormalizable extension of the Standard Model. From the SM quantum numbers of active neutrinos one can identify two possible sources for neutrino masses. If no new fermionic degrees of freedom are introduced, one needs to have a Higgs triplet with weak hypercharge 2. Another option is introduction of singlet (with respect to the SM gauge group) Majorana fermions N_I (other names for them are sterile neutrinos or heavy neutral leptons). Here we will consider only the latter possibility. The five-dimensional operator O_5 then appears because of an exchange of the singlet fermions.

This extension of the SM is associated with Lagrangian

$$L = L_{\rm SM} + \bar{N}_I i \partial_\mu \gamma^\mu N_I - F_{\alpha I} \bar{L}_\alpha N_I \tilde{\Phi} - \frac{M_I}{2}\, \bar{N}_I^c N_I + h.c.,\qquad(12.3)$$

where $L_{\rm SM}$ is the Lagrangian of the SM. Since N_I are SU(3)$\times$SU(2)$\times$U(1) singlets, Majorana mass terms for them are consistent with the symmetries of the SM. The number of singlet fermions cannot be deduced from symmetry principles; we take it to be three in analogy with the number of generations of quarks and leptons. This theory also provides the electric charge quantization, coming from the requirement of cancellation of gauge and gravitational anomalies [868].

For $M^{\rm D} \ll M_I$, where $M^{\rm D} = F_{\alpha I} \langle \Phi \rangle$ are Dirac neutrino masses, the see-saw formula of [924; 1450; 1461; 1950] works, leading to active neutrino mass matrix

$$m_\nu = -M^{\mathrm{D}} \frac{1}{M_I} [M^{\mathrm{D}}]^T. \qquad (12.4)$$

This relation tells us nothing about the masses of new particles – singlet fermions. We will discuss below two possible choices. The first is the traditional see-saw, in which M_I are related to the scale of Grand Unification [924; 1450; 1461; 1950]. In the second it is assumed that M_I is of the order of the electroweak (EW) scale [129; 132]. Yet another option is to assume that M_I can be as small as eV. This is discussed in [627; 628] and will not be considered here.

The GUT see-saw

The assumption that Yukawa coupling of N_I to the Higgs and left-handed lepton doublets is similar to those in the quark or charged lepton sector (say, $F_{\alpha I} \sim 1$, as we have for the top quark) allows us to estimate the mass scale of the singlet leptons: $M_I \simeq v^2/m_{\mathrm{atm}} \simeq 6 \times 10^{14}$ GeV, where $m_{\mathrm{atm}} = \sqrt{\Delta m_{\mathrm{atm}}^2} \simeq 0.05$ eV is the atmospheric neutrino mass difference. This scale is rather close to that of Grand Unification, providing an argument that the gauge coupling unification and neutrino masses might be related. In addition, the decays of superheavy N_I in the early Universe may lead to baryogenesis through leptogenesis [900] and rapid anomalous fermion number non-conservation at high temperatures [1297].

The attractive features of the GUT see-saw are the similarities of Yukawa couplings in lepton and quark sector, closeness of the see-saw and GUT scales, and a possibility of explaining the baryon asymmetry of the Universe. At the same time, there are several problems in the GUT see-saw hypothesis, listed below.

(i) Hierarchy problem. Since M_I is much larger than the EW scale, one has to understand not only why $M_W \ll M_{\mathrm{Pl}}$, but also why $M_W \ll M_I$ and why $M_I \ll P_{\mathrm{Pl}}$. In other words, there are three fine-tunings, instead of one [1907], existing in the SM.

(ii) Identification of M_I with the GUT scale requires gauge coupling unification, which does not happen in the SM. The situation is changed if low-energy supersymmetry (SUSY) is introduced. SUSY is an argument for supergravity (SUGRA). The SUGRA theories suffer from copious gravitino production, which can only be avoided if the Universe reheating temperature is small enough, $T_{\mathrm{reh}} \lesssim 10^{10}$ GeV (see [1664] for a recent discussion). This may lead to problems with thermal leptogenesis (for a review see [457]) and to an additional energy scale $M_I \sim 10^{10}$ GeV $\ll M_{\mathrm{GUT}} \sim 10^{16}$ GeV, creating an extra (fourth) hierarchy problem.

Unfortunately, no direct experimental tests of the GUT see-saw are possible because of the large masses of singlet fermions.

The EW see-saw

Let us assume now that Majorana masses of singlet fermions are roughly of the same order as the EW scale [129; 132]. Then, to keep the neutrino masses in the right place, the Yukawa couplings must be much smaller than those in the quark and charged lepton sector:

$$F_{\alpha I} \sim \frac{\sqrt{m_{\text{atm}} M_I}}{v} \sim (10^{-6} - 10^{-13}). \qquad (12.5)$$

We will use the name νMSM (neutrino minimal Standard Model) for the model defined by Eq. (12.3) with this choice of parameters [129; 132]. The advantages of the νMSM are listed below.

(i) No extra energy scale introduced. Therefore, no new hierarchy or fine-tuning problem in comparison with the SM appears. Moreover, M_I and M_W may have a common origin [1737; 1738; 1739]. Note that the singlet fermion masses are automatically protected from large quantum corrections because of conservation of lepton number in the limit $M_I \rightarrow 0$.

(ii) Alternative to SUSY approach to hierarchy problem, based on quantum scale-invariance [1738].

(iii) Baryogenesis due to singlet fermion coherent oscillations right above the electroweak scale [42; 132; 1736].

(iv) Natural dark matter candidate: sterile neutrino with the mass in $O(10)$ keV range.

(v) Common mechanism for production of dark and baryonic matter in the Universe if the mass of the heavier pair of singlet fermions is in the $O(2)$ GeV region [1305; 1736].

The simultaneous explanation of neutrino oscillations, dark matter and baryon asymmetry of the Universe in the framework of the νMSM requires a number of fine-tunings, which may be considered as drawbacks of the model or as a source of predictions of the properties of new singlet fermions.

(i) The Yukawa couplings must be much smaller than those in the quark and charged lepton sector. It is interesting, however, that this requirement not only comes from the smallness of neutrino masses, but is also essential for baryogenesis and dark matter production.

(ii) Degeneracy between a pair of singlet fermions, which is required for baryogenesis.

At the same time, a peculiar choice of parameters of the νMSM, needed for phenomenology, could be a consequence of some underlying symmetry. For example, the existence of a particular U(1) leptonic symmetry could make these fine-tunings automatic [1734].

The aim of this chapter is to discuss the properties of the lightest sterile neutrino as a dark matter candidate. A possibility of direct experimental search for a pair of almost degenerate heavier singlet fermions, responsible for the baryon asymmetry of the Universe in the νMSM, has been discussed in refs. [990; 1735]. The predictions of the νMSM for the neutrinoless double beta decay can be found in [347]. For most of the discussion below, the presence of the heavier singlet states is not essential, so that we can omit N_2 and N_3 from the Lagrangian (12.3). The lightest sterile neutrino, denoted by N_1, will play the role of the dark matter particle.

12.2 Cosmological and astrophysical constraints on sterile neutrino dark matter

12.2.1 Search for DM sterile neutrino in X-rays

The non-zero vacuum expectation value of the Higgs field induces mixing between the active neutrino states and the sterile neutrino N_1. This can be characterized by three angles

$$\theta_{\alpha 1} = \frac{F_{\alpha 1} v}{M_1}, \tag{12.6}$$

giving the admixture of the sterile neutrino in every active neutrino flavour. So, in any reaction where the active neutrino of type α is absorbed or created, the sterile neutrino will disappear or appear with the amplitude suppressed by an extra factor $\theta_{\alpha 1}$. Therefore, the sterile neutrino is unstable. The decay rate of N_1 to three active neutrinos and antineutrinos (main channel) is given by

$$\Gamma_{3\nu} = \frac{G_{\mathrm{F}}^2 M_1^5 \theta^2}{96\,\pi^3} = 10^{14}\ \text{years} \left(\frac{10\ \text{keV}}{M_1}\right)^5 \left(\frac{10^{-8}}{\theta_1^2}\right), \tag{12.7}$$

$$\theta_1 = \frac{m_1}{M_1}, \quad m_1^2 = \sum_{\alpha=e,\mu,\tau} |M^{\mathrm{D}}{}_{\alpha 1}|^2,$$

where G_{F} is the Fermi constant. For example, a choice of $m_1 \sim O(1)$ eV and of $M_1 \sim O(1)$ keV leads to a sterile neutrino lifetime of $\sim 10^{17}$ years [693]. In other words, the νMSM may contain a sterile neutrino with a lifetime exceeding the age of the Universe, and therefore this particle could

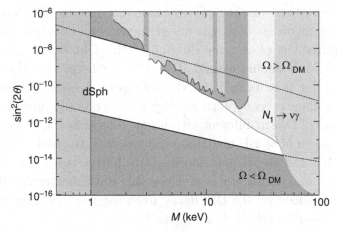

Fig. 12.1. Constraints on the mixing angle of the DM sterile neutrino, coming from X-ray observations (right upper corner), from analysis of dwarf spheroidal galaxies, (left shaded part) and from different theoretical considerations, explained in the text (lower shaded region, and above upper straight line).

be the dark matter. The required stability of DM sterile neutrino favours, in general, small values of its mass and small values of the mixing angle θ_1:

$$\left(\frac{M_1}{10 \text{ keV}}\right)^5 \left(\frac{\theta_1^2}{10^{-4}}\right) \lesssim 1. \tag{12.8}$$

Much stronger constraints than those of Eq. (12.8) can be derived from the analysis of X-ray observations of different objects in the Universe [8; 693]. In fact, dark matter made of sterile neutrinos is not completely dark, since there is a subdominant radiative decay channel $N_1 \to \nu\gamma$ with the width [210; 1540]

$$\Gamma_{\text{rad}} = \frac{9\,\alpha_{\text{EM}}\,G_F^2}{256 \cdot 4\pi^4}\,\sin^2(2\theta_1)\,M_1^5, \tag{12.9}$$

where α_{EM} is the fine structure constant. These decays produce a narrow photon line (its width is determined by the Doppler effect due to the virial motion of DM particles) with energy $E_\gamma = M_1/2$. Until now, no candidate for DM sterile neutrino decay line has been seen. The non-observation of the decay line puts stringent constraints on the mixing angle of DM sterile neutrino,

$$\theta^2 < F(M_1), \tag{12.10}$$

as shown in Fig. 12.1. These were derived from analysis of the observational data from the XMM-Newton and HEAO-1 satellites of different astrophysical objects, such as the Milky Way, Large Magellanic Clouds and Andromeda galaxy, amongst others (for reviews see [423; 425; 429; 1659]). Roughly, these constraints are six orders of magnitude stronger than those coming from the stability requirement (12.8): the lifetime of the DM sterile neutrino with mass $M_1 < 1$ MeV should exceed $10^6 t_U$ [429], where t_U is the age of the Universe, in order that its radiative decays have not been seen by now. The X-ray constraints are very robust as they are based only on Eq. (12.9) and on the presence of DM in different astronomical objects. They do not depend on the mechanism of sterile neutrino production in the early Universe.

The fact that the sterile neutrino has a radiative two-body decay mode is crucial for the experimental search for this dark matter candidate. The energy flux produced by the DM decay from a given direction into a sufficiently narrow solid angle $\Omega \ll 1$ is given by

$$F = \frac{\Gamma_{\rm rad}\Omega}{8\pi} \int\limits_{\rm line\ of\ sight} \rho_{\rm DM}(r){\rm d}r, \qquad (12.11)$$

where $\rho_{\rm DM}(r)$ is the DM density profile. So, to optimize the search one should find the astrophysical objects for which the value of the integral (12.11) is maximal whereas the X-ray background is minimal. It is quite an amazing empirical fact [428] that the signal is roughly the same for many astrophysical objects, from clusters to dwarf galaxies: the Milky Way halo signal is comparable with that of clusters like Coma or Virgo, and the DM flux from Draco or Ursa Minor dwarf spheroidals is only three times as strong as that of the Milky Way (MW) halo. At the same time, the background strongly depends on the astrophysical object. Indeed, for clusters of galaxies (e.g. Coma or Virgo), the temperature of the intracluster medium is in the keV range, leading to strong X-ray emission contributing to both continuous and discrete (atomic lines) background spectra. The continuum X-ray emission from the Milky Way is about two orders weaker than that of a cluster, whereas dwarf satellites of the Milky Way are really dark from the point of view of X-ray background. Therefore, the best objects to look at to find the DM sterile neutrino are the Milky Way [428; 1645] and dwarf satellite galaxies [428]; X-ray quiet outer parts of clusters can be used as well [426].

Unfortunately, the new data from Chandra and XMM-Newton can hardly improve the constraints by more than a factor of 10 because these instruments have an energy resolution greatly exceeding the expected width of the DM line. To go much further, one would need an improvement of spectral resolution up to the natural line width ($\Delta E/E \sim 10^{-3}$), a reasonably wide field of view $\sim 1°$ (size of a dSph) and a wide energy scan, from $O(100)$ eV to $O(100)$ keV. A discussion of the sensitivity of different existing and future space missions can be found in [422]. The laboratory constraints on the mixing angle of DM sterile neutrinos are much weaker than those coming from X-ray observations. They are discussed in [348], together with a proposal of the laboratory search for the DM sterile neutrino, based on precise study of the kinematics of β-decay of tritium.

12.2.2 Lower mass bounds from dwarf spheroidals

The X-ray bounds discussed in the previous subsection tell us nothing about the absolute value of mass of the dark matter sterile neutrinos, if the mixing angle is not fixed. Since this angle is unknown and may be zero, one should try to limit the mass of DM sterile neutrinos from other types of considerations.

A robust and model-independent limit comes from the analysis of rotational curves associated with dark matter distributions of compact astrophysical objects – dwarf spheroidal satellites (dSph) of the Milky Way (for a review see [430]). If dark matter particles are fermions, their phase-space density (PSD) cannot exceed that of the degenerate Fermi gas. For a spherically symmetric distribution of a DM-dominated object with mass M and radius R, the requirement that the maximal Fermi velocity does not exceed the escape velocity $v = (2G_N M/R)^{1/2}$ leads to the lower bound on the mass of the DM particle

$$M_1 > \left(\frac{9\pi}{4\sqrt{2}M^{1/2}R^{3/2}G_N^{3/2}} \right)^{1/4}, \tag{12.12}$$

where G_N is Newton's constant. An analysis of new astrophysical data on dwarf spheroidals has been carried out recently [430; see also 989] and led to a relatively weak but very conservative constraint, coming from Leo IV, CVnII and Coma Berenices, of $M_1 > 0.4$ keV. The limit (12.12) does not depend on the initial dispersion of the velocity of DM particles and can be strengthened, if the primordial momentum distribution of DM particles is fixed. Clearly, this is model-dependent and is determined by a specific mechanism of DM production in the early Universe.

The original idea of such an improvement [1864] is based on the observation that if the dynamics of DM particles can be considered as collisionless, the maximum value of the phase-space density $f_{\max}$ does not change during the evolution. Then, requiring that the observed coarse-grained PSD does not exceed the initial maximum value, one gets another, stronger limit on the mass of a DM particle. Note that this type of limit is not based on the Pauli exclusion principle and is valid for bosonic particles as well.

As a model of the final distribution one can take an isothermal sphere with a core radius r_c and a one-dimensional velocity dispersion σ [1864], leading to

$$f_{\max} = \frac{9\sigma^2}{4\pi G_N (2\pi\sigma^2)^{3/2} r_c^2}.$$ (12.13)

This can be compared with different primordial distributions. We will give them at temperatures corresponding to decoupling of active neutrinos, $T_\nu \simeq 4$ MeV. For example, fermionic dark matter which was in thermal equilibrium and decoupled while being relativistic is described by a Fermi–Dirac distribution

$$f_{\mathrm{FD}} = \frac{g}{e^{p/T_{\mathrm{FD}}} + 1},$$ (12.14)

where g is the number of spin states and $T_{\mathrm{FD}} < T_\nu$ is the temperature accounting for entropy production after decoupling, to be fixed by the requirement that (12.14) gives the correct DM abundance. Then the lower limit is called the Tremaine–Gunn mass bound[1] and reads

$$M_1 > \left[\frac{9(2\pi)^{1/2}}{g G_N \sigma r_c^2} \right]^{1/4}.$$ (12.15)

A conservative analysis of the new data on dSphs, carried out in [430] and making fewer assumptions than [1864], leads numerically to $M_1 > 0.5$ keV.

Constraints are stronger if the primordial distribution of DM sterile neutrinos is given by

$$f = \frac{g\chi}{e^{p/T_\nu} + 1},$$ (12.16)

where $\chi \ll 1$ is determined by observed dark matter abundance (a distribution similar to (12.16) appears when the DM sterile neutrinos are produced through non-resonant active–sterile mixing; see below). Analysis of the same dSphs gives [430] $M_1 > 1.7$ keV. More examples, together with discussion of previous literature, can be found in [430]; see also [610; 1116]. In general, the constraints are in the $O(1)$ keV region.

12.2.3 Lower mass bounds from Lyman-α forest data

If the sterile neutrino mass is in the keV region, it may play a role of warm dark matter (WDM) [7; 690]. Sterile neutrino free-streaming length λ_{FS} at

---

[1] In fact, the consideration of [1864] was applied first to active neutrinos and served as a strong argument for why they cannot play the role of DM: the lower limit strongly exceeded the upper limit on the mass of active neutrinos that arises from the requirement that they do not overclose the Universe [935].

matter–radiation equality is of the order of (see [424] and references therein for a detailed discussion of characteristic scales)

$$\lambda_{FS} \sim 1 \text{ Mpc} \left(\frac{1 \text{ keV}}{M_1}\right) \left(\frac{\langle p_s/T \rangle}{3.15}\right) \qquad (12.17)$$

and the mass inside λ_{FS} is

$$M_{FS} \sim 3 \times 10^7 M_\odot \left(\frac{10 \text{ keV}}{M_1}\right)^3 \left(\frac{\langle p_s/T \rangle}{3.15}\right)^3,$$

where $\langle p_s \rangle$ is an average momentum of the sterile neutrino at temperature T_ν, and $M_\odot$ is the solar mass. The presence of free streaming means that primordial perturbations with size smaller than λ_{FS} are erased, and that structures with mass smaller than M_{FS} are unlikely to be formed. On a more quantitative level, the power spectrum P_{WDM} of WDM is suppressed with respect to that of CDM (P_{CDM}) for large momenta k,

$$T^2(k) = \frac{P_{WDM}(k)}{P_{CDM}(k)} < 1. \qquad (12.18)$$

For the distribution (12.14) the following fit of the numerical simulations was derived [1070]:

$$T(k) = [1 + (\alpha k)^{2\nu}]^{-5/\nu}, \qquad (12.19)$$

where k is the wavenumber in units h Mpc^{-1}, $\nu = 1.12$, and α is given by

$$\alpha = A \left(\frac{\Omega_{DM}}{0.3}\right)^b \left(\frac{h}{0.65}\right)^c \left(\frac{M_1}{0.5 \text{ keV}}\right)^d. \qquad (12.20)$$

Here $A = 1.07, b = 0.11, c = 1.20$ and $d = -1.11$. For large k, $T(k) \propto k^{-10}$. If the primordial distribution of DM particles is different from that of (12.14), the transfer function has to be recomputed, as was done, for example, in [1721] for a distribution close to (12.16).

Clearly, too small a value of M_1 would lead to the well-known difficulties of the hot dark matter scenario. To constrain the mass better, one can try to extract the matter power spectrum at the smallest possible scales with the use of available astrophysical data and compare it with theory predictions, like those in (12.19). In particular, the study of the shift of Lyman-α emission lines in quasar spectra on passing through matter clouds can be used as a probe of the matter power spectrum at comoving scales $(1 - 40)h^{-1}$ Mpc [1070; 1902]. The corresponding analysis is quite complicated and is a subject of different observational and theoretical uncertainties (see [424] for a recent discussion). The published Bayesian credible intervals at 95% C.L. on the mass of a DM particle for the distribution (12.14) are: $M_1 > 2.4$ keV [1721], $M_1 > 2$ keV [1903] and $M_1 > 4$ keV [1901] (the Sloan Digital Sky Survey Lyman-α flux power spectrum is included in the analysis). For the distribution (12.16) the constraints on the mass are stronger (since the free-streaming length for this distribution is larger for a given mass): $M_1 > 14.4$ keV [1721], $M_1 > 10$ keV [1903] and $M_1 > 28$ keV [1901]. For sufficiently smooth primordial distributions of sterile neutrinos, the limit on the mass scales roughly as an average momentum $\langle p_s \rangle$ of the sterile neutrino at $T = T_\nu$, $m_{\min} \propto \langle p_s \rangle$ [1070].

Interestingly, a combination of the X-ray constraints on the mixing between active and sterile neutrinos together with the lower bounds on their mass coming from dSphs or Lyman-α allows one to fix the absolute value of active neutrino masses in the νMSM [427]. One can show [129] that given these constraints, the dark matter sterile neutrino does not contribute to the see-saw formula, so that one of the active neutrinos must be light, $m_\nu < M_1 \sin^2\theta < 10^{-5}$ eV.

12.2.4 Warm versus cold dark matter

For distance scales exceeding λ_{FS} given in (12.17), the predictions of the WDM scenario are the same as for CDM. The difference appears for distances $l < \lambda_{FS}$. If, for example, one takes the distribution (12.16) and fixes the mass to 10 keV (a typical value which appears in Lyman-α constraints) then the cosmological formation of the objects with mass larger than a few times $10^7 M_\odot$ is the same in both WDM and CDM. In other words, the study of scales below or of the order of the typical size of dSphs should eventually allow us to decide whether the dark matter is cold or warm. Note also that the formation of the first stars goes differently in CDM and WDM scenarios [915].

It is intriguing that the standard CDM may indeed experience some problems in describing the astronomical data at small scales. Numerical N-body simulations of pure dark matter reveal that the density profiles of dark matter haloes are universal, singular at the origin, and exist for all scales [1502]. This leads to a number of conclusions which are in tension with existing data. The CDM simulations produce more satellite galaxies than are actually observed [1250; 1469]. The cuspy CDM profiles may be in conflict with astronomical observations of dSphs [951] and with observed dynamics of stellar motion in some dSphs [969]. Observations show [1816] that there exists a common mass scale $\sim 10^7 M_\odot$ for satellite galaxies of the Milky Way, which would be absent for a featureless Navarro–Frenk–White (NFW) profile. The dense galactic cores appearing in CDM would lead eventually to disks with angular momentum an order of magnitude too small [1779].

Perhaps some or all of these problems may have their astrophysical solution in the CDM model (see the discussion in [372]). Interestingly, the WDM scenario does provide an ultraviolet cutoff of the power spectrum, which may cure the CDM model at small distance scales [372].

12.2.5 Astrophysical applications of DM sterile neutrino

The existence of a relatively light particle with very weak interactions may potentially have a number of astrophysical applications. The mean free path of a sterile neutrino inside a collapsing star is much larger than that of an ordinary neutrino, leading to an effective cooling mechanism. However, the constraints on the mixing angle of the sterile neutrino coming from these considerations [7; 693] happen to be weaker than the X-ray bounds discussed above. At the same time, DM sterile neutrinos may play an important role in dynamics of stellar collapse [7; 1094], if their mass and mixing angle satisfy $1 \text{ keV} < M_1 < 5 \text{ keV}$ and $10^{-10} < \sin^2(2\theta_1) < 10^{-8}$, respectively. In addition, according to [1291; 1571], the asymmetric emission of sterile neutrinos from a supernova explosion may lead to an explanation of pulsar kick velocities [211; 904; 1292]. The X-ray photons from early decays of these particles can influence the formation of molecular hydrogen and can be important for the early star formation and re-ionization [353; 1385; 1647; 1648; 1797]. See also [354; 1642] for discussion of formation of supermassive black holes and degenerate heavy neutrino stars.

The effects of sterile neutrinos in astrophysics may be enhanced if the dark matter is multicomponent so that sterile neutrinos constitute only a fraction $z < 1$ of it. Indeed, in this case the X-ray bound on the mixing angle in Fig. 12.1 is weaker by the same factor, $\theta^2 < F(M_1)/z$, where $F(M_1)$ is defined in (12.10). Moreover, the Lyman-α constraints also get diluted for $z < 1$ [424; 1541], if the rest of the dark matter is cold. The analysis of [424] shows that the DM sterile neutrino with the spectrum (12.16) and mass $M_1 > 5 \text{ keV}$ for $z < 0.6$ is compatible with X-ray constraints and Lyman-α data at 99.7% C.L. The use of larger mixing angles and smaller mass makes the influence of sterile neutrinos on supernovae explosions stronger.

12.3 Sterile neutrino production in the early Universe
12.3.1 Active–sterile mixing

Let us now discuss cosmological production of sterile neutrinos. We take first the most conservative point of view and assume that the νMSM is a good effective theory all the way up to the Planck scale. This can be the case if the

Higgs mass lies in the interval $M_H \in [126, 175]$ GeV with some uncertainties in the lower and upper limits coming from experimental errors in the top quark mass, strong coupling constant and higher loop corrections. Above the upper limit the theory is not consistent because there is a Landau pole in the scalar self-coupling (for a review see [1064]), whereas below the lower limit the EW symmetry breaking vacuum is not stable (for a review see [491]). In this case inflation occurs because of the Higgs boson of the Standard Model [349; 1739], and the concentration of DM sterile neutrinos after inflation is negligibly small. Therefore, the only source of DM sterile neutrinos is associated with active–sterile mixing, existing in Lagrangian (12.3).

Qualitatively, the rate of DM sterile neutrino production Γ_N at temperatures below the electroweak scale can be written as [191]

$$\Gamma_N \sim \Gamma_\nu \theta_{\mathrm{M}}(T)^2, \tag{12.21}$$

where $\Gamma_\nu \sim G_{\mathrm{F}}^2 T^5$ is the rate of active neutrino production (G_{F} is the Fermi constant), and $\theta_{\mathrm{M}}(T)$ is the temperature-dependent mixing angle

$$\theta_1 \to \theta_{\mathrm{M}} \simeq \frac{\theta_1}{1 + 2.4(T/200 \text{ MeV})^6(\text{keV}/\mathrm{M}_1)^2}. \tag{12.22}$$

The rate Γ_N peaks roughly at [690]

$$T_{\mathrm{peak}} \sim 130 \left(\frac{M_1}{1 \text{ keV}}\right)^{1/3} \text{ MeV}, \tag{12.23}$$

corresponding to the temperature of the QCD cross-over for keV-scale sterile neutrinos and is strongly suppressed at $T > 100$ MeV, $\Gamma_N \propto T^{-7}$. At temperatures above the electroweak scale the rate is determined mainly by the decays and inverse decays of the Higgs boson to a sterile neutrino and a lepton, and is of the order

$$\Gamma_N \sim \frac{\theta_1^2 M_1^2 T}{v^2}. \tag{12.24}$$

It is easy to see from Eqs. (12.21, 12.22 and 12.24) and Fig. 12.1 that in the region of the parameter space (θ_1, M_1), admitted by X-ray observations, together with Lyman-α or dSph constraints, the DM sterile neutrinos were never in thermal equilibrium in the early Universe. Therefore, to compute the abundance of sterile neutrinos one has to integrate the rate of their production over the whole history of the Universe. Since the maximal production occurs near the temperature of the QCD cross-over, where the quark–gluon plasma is strongly coupled and the dilute quark or hadron gas pictures are not valid, an exact computation of the number of produced sterile neutrinos is a difficult task. At the same time, the fact that DM sterile neutrinos were never in thermal equilibrium allows us to express their abundance through certain equilibrium finite-temperature Green's functions, to be computed in the Standard Model (without inclusion of right-handed fermions) [130; 131; 1305].

Since other particles of the SM equilibrate with high rates, the problem of determining the present number of DM sterile neutrinos can be formulated as follows. Find

$$\mathrm{Tr}\, N\hat{\rho}(t) \tag{12.25}$$

(N is an operator that counts the DM sterile neutrinos), where the density matrix $\hat{\rho}(t)$ satisfies:

$$i\frac{\mathrm{d}\hat{\rho}(t)}{\mathrm{d}t} = [\hat{H}, \hat{\rho}(t)]. \tag{12.26}$$

Here $\hat{H}$ is the total Hamiltonian. The initial condition is fixed as:

$$\hat{\rho}(0) = \hat{\rho}_{\mathrm{SM}} \otimes |0\rangle\langle0|, \tag{12.27}$$

where

$$\hat{\rho}_{\mathrm{SM}} = Z_{\mathrm{SM}}^{-1} \exp\left(-\beta\left[\hat{H}_{\mathrm{SM}} + \mu_\alpha\left(L_\alpha - \frac{1}{3}B\right)\right]\right), \quad \beta \equiv 1/T, \tag{12.28}$$

is the equilibrium SM density matrix at a temperature T, μ_i are the chemical potentials for fermionic numbers exactly conserved in the SM (L_α is the leptonic number of generation α and B is the baryonic number) and $|0\rangle$ is the vacuum state for sterile neutrinos.

Now, since the DM sterile neutrinos are never in thermal equilibrium, the kinetic equations are not necessary and one can use a straightforward perturbation theory with respect to (small) Yukawa couplings.

In the lowest order of perturbation theory one gets for the distribution function of dark matter sterile neutrinos [130]:

$$\frac{\mathrm{d}N_1(x, \mathbf{q})}{\mathrm{d}^4x\, \mathrm{d}^3\mathbf{q}} = \frac{2n_{\mathrm{F}}(q^0)}{(2\pi)^3 2q^0} \sum_{\alpha=1}^{3} |M_{\mathrm{D}}|_{\alpha 1}^2 \mathrm{Tr}\left\{\slashed{Q}\, a_{\mathrm{L}}\left[\rho_{\alpha\alpha}(-Q) + \rho_{\alpha\alpha}(Q)\right]a_{\mathrm{R}}\right\}, \tag{12.29}$$

where $a_{\mathrm{L,R}} = \frac{1}{2}(1 \pm \gamma_5)$ are the chirality projectors and $n_{\mathrm{F}}(q^0)$ is the Fermi distribution. The spectral function is given by

$$\rho_{\alpha\beta}(Q) \equiv \int \mathrm{dt}\, \mathrm{d}^3\mathbf{x}\, \mathrm{e}^{iQ\cdot x}\left\langle \frac{1}{2}\left\{\hat{\nu}_\alpha(x), \hat{\bar{\nu}}_\beta(0)\right\}\right\rangle, \tag{12.30}$$

where $\hat{\nu}_\alpha(x)$ are the active neutrino fields. The use of Eq. (12.29) allows us to compute the abundance of dark matter as a function of the parameters of the model and of the state of the Universe at the time of sterile neutrino production.

The analysis of Eq. (12.29) reveals that the sterile neutrinos are produced most intensively at temperatures (12.23), where QCD interactions are strong and cannot be treated perturbatively. Therefore, a precise computation is only possible if one knows the QCD equation of state at temperatures 10 MeV to 1 GeV and real time correlators of vector and axial vector hadronic currents in this temperature range. Unfortunately, this is impossible with the current state of non-perturbative lattice simulations. Therefore, some semi-phenomenological models have to be used for computations.

As for the QCD equation of state, it is natural to use a gas of hadronic resonances at low temperatures, the most advanced (up to re-summed 4-loop level weak coupling) results at high temperatures [1192], and an interpolation thereof at intermediate temperatures. The result of this procedure can be found in [131; 1304].

The hadronic correlators in (12.30) can be determined as follows: for $T \gg \Lambda_{\text{QCD}}$ one can use quarks for computation, for $T \ll \Lambda_{\text{QCD}}$ one can use hadrons, and interpolate in between. The phenomenological recipe for interpolation reads: use the free quarks at all temperatures and replace the number of colours at the end of the computation by

$$N_{\text{c}} \rightarrow N_{\text{c}} \frac{h_{\text{eff}}^{\text{QCD}}(T)}{58}, \qquad (12.31)$$

where $h_{\text{eff}}^{\text{QCD}}$ is determined through the entropy density as

$$s(T) \equiv \frac{2\pi^2 T^3}{45} h_{\text{eff}}(T). \qquad (12.32)$$

The conservative upper bound on hadronic contribution is derived if one uses free quarks at all temperatures, whereas the conservative lower bound on hadronic contribution is simply to put $N_{\text{c}} = 0$.

Let us discuss now the results of the computations. Besides the SM Hamiltonian, the density matrix $\hat{\rho}_{\text{SM}}$ depends in general on the leptonic asymmetries,

$$\Delta_L^\alpha = \frac{(n_L^\alpha - \bar{n}_L^\alpha)}{(n_L^\alpha + \bar{n}_L^\alpha)}, \quad \Delta_L = \sum \Delta_L^\alpha = 244 \frac{n_{\nu e}}{s(T)}, \qquad (12.33)$$

which may exist during the QCD epoch (here $n_{\nu e}$ is the asymmetry in electron neutrino flavour). It has been shown in [1736] that a substantial lepton

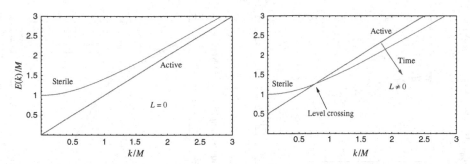

Fig. 12.2. Left: Non-resonant transitions $\nu \rightarrow N_1$, $\Delta_L = 0$. Right: Resonant transitions $\nu \rightarrow N_1$, $\Delta_L \neq 0$.

asymmetry can be created in reactions with heavier singlet fermions of the νMSM below the temperature of the sphaleron freeze-out. It can be as large as $\Delta_L \simeq 0.2$. Note that the constraints on lepton asymmetry coming from Big Bang nucleosynthesis are weaker than that, allowing it to be > 0.2 [1761].

The parameter Δ_L plays an important role in the theory of sterile dark matter production. In Fig. 12.2 we show the dispersion relations (energy of active and sterile neutrino excitations in plasma as a function of momenta) for zero (left) and non-zero (right) leptonic asymmetries. For the case of non-zero asymmetries the dispersion relations for active and sterile neutrinos cross each other at some point, leading to resonance production of sterile neutrinos [1305; 1744].

As a result, the number of created sterile neutrinos and their spectrum depend essentially on three parameters: their mass, the mixing angle θ_1 and the lepton asymmetry. Many plots which show different quantities can be found in [1305]. Here we just summarize the most important results of the analysis.

An example of computation of dark matter abundance and of the average momentum of created sterile neutrinos is shown in Fig. 12.3 for the non-resonant case (zero leptonic asymmetry).

If non-zero leptonic asymmetry is present, it is transferred to the population of dark matter neutrinos in a resonant way (see Fig. 12.4). In this case both baryon asymmetry of the Universe and dark matter are produced by essentially one and the same mechanism, providing an explanation of similar abundances of dark and baryonic matter in the Universe. The right panel in Fig. 12.4 gives the values of parameters necessary to reproduce the observed abundance of dark matter.

The summary of computations is presented in Fig. 12.1. Owing to (resonant and non-resonant) reactions $l\bar{l} \rightarrow \nu N_1$, $q\bar{q} \rightarrow \nu N_1$ etc., sterile neutrinos are created in the early Universe. Their abundance must correctly reproduce

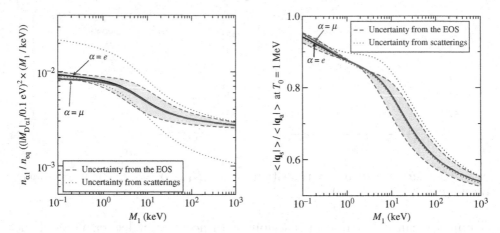

Fig. 12.3. Left: The sterile neutrino abundance normalized to its equilibrium abundance, $n_{\alpha 1}/n_{\rm eq}$, at $T_0 = 1$ MeV. Shown are the two sources of hadronic uncertainties: from the equation of state (EOS) and from hadronic scatterings. The combination $(|M_{\rm D}|_{\alpha 1}/0.1\ {\rm eV})^2$ has been factored out and, for better visibility, the results have been multiplied by $M_1/{\rm keV}$. Right: The average sterile neutrino momentum compared with the active neutrino equilibrium value.

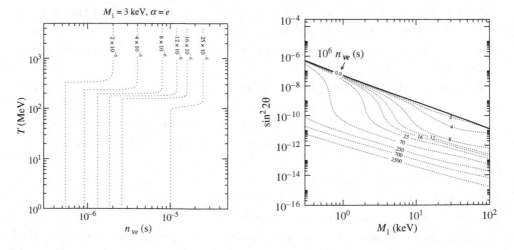

Fig. 12.4. Left: Evolution of lepton asymmetry as a function of temperature. The depletion of Δ_L is due to the resonant transfer of leptonic excess to dark matter sterile neutrinos. Right: A combination of the sterile neutrino mass, mixing angle and of lepton asymmetry, parameterized by $n_{\nu e}/s(T)$, which leads to correct dark matter abundance.

the measured density of DM. Depending on leptonic asymmetry, the admitted region lies between the two thick black lines in Fig. 12.1. The upper line corresponds to zero leptonic asymmetry, the lower one to the maximal possible asymmetry that can be generated in the νMSM. The right upper corner in Fig. 12.1 corresponds to X-ray constraints, discussed in Section 12.2.1. The region to the left of the vertical line corresponds to the part of the parameter space excluded by dSph observations and PSD arguments (see Section 12.2.2).[2]

An interesting feature of Fig. 12.1 is that the admitted region is surrounded by different constraints in all directions, telling us that the hypothesis of sterile neutrino as a dark matter candidate is experimentally testable. Moreover, the $O(10)$ keV scale for the mass of DM is singled out by these considerations.

12.3.2 Particle decays

Up to now we have been using the very conservative point of view that the only interactions the DM sterile neutrinos have are those with ordinary neutrinos through Yukawa couplings, i.e. we assumed that the correct low-energy theory is the νMSM. Giving up this assumption opens other possibilities for primordial creation of DM sterile neutrinos. Here we consider the mechanism [1737] related to the possible coupling of sterile neutrinos to a singlet scalar field χ,

$$L_\chi = \frac{f_1}{2}\, \bar{N}_1{}^c N_1 \chi. \tag{12.34}$$

If $M_\chi > 2M_1$ then this scalar can decay to sterile neutrinos. If these decays occurred at the time when χ was in thermal equilibrium due to the interactions with other particles, the dark matter sterile neutrino abundance can be readily found and is a function of f_1, M_χ (we will assume that $M_\chi \gg M_1$) and the entropy production S which may occur after generation of DM sterile neutrinos.

Sterile neutrinos are produced in χ decays mainly at $T \simeq M_\chi$, and their distribution function $n(p,t)$ (p is the momentum of the sterile neutrino and t is time) can be found from the solution of the kinetic equation

$$\frac{\partial n}{\partial t} - Hp\frac{\partial n}{\partial p} = \frac{2M_\chi \Gamma}{p^2} \int_{p+M_\chi^2/4p}^{\infty} n_\chi(E)\mathrm{d}E, \tag{12.35}$$

where the inverse decays $\chi \leftarrow N_1 N_1$ are neglected, H is the Hubble constant, E is the energy of scalar χ, $n_\chi(E)$ is the χ momentum distribution and $\Gamma = f_1^2 M_\chi/16\pi$ is the partial width of χ for $\chi \to N_1 N_1$ decay. For the case

---

[2] The analysis of [424] shows that for any mass M_1 above the PSD limit $M_1 \geq 1$ keV one can find a leptonic asymmetry for which the DM sterile neutrino spectrum is allowed by the Lyman-α constraints.

when the effective number of degrees of freedom is time-independent, an asymptotic $(t \to \infty)$ analytic solution to (12.35) can be easily found:

$$n(x) = \frac{16\Gamma M_0}{3M_\chi^2}x^2 \int_1^\infty \frac{(y-1)^{3/2}\mathrm{d}y}{\mathrm{e}^{xy}-1}, \qquad (12.36)$$

where $x = p/T$, and $M_0 \approx M_{\mathrm{Pl}}/1.66\sqrt{g^*}$, leading to the number density

$$N_0 = \int \frac{\mathrm{d}^3 p}{(2\pi)^3} n(p) = \frac{3\Gamma M_0 \zeta(5)}{2\pi M_\chi^2} T^3 \qquad (12.37)$$

and to an average momentum of created sterile neutrinos $\langle p \rangle = \pi^6/(378\zeta(5))$ $T = 2.45T$, which is about 20% smaller than that for the equilibrium thermal distribution, $p_T = 3.15T$. Formally, the solution (12.36) diverges at $x \to 0$ and is not valid for very small momenta $x \lesssim \left(10\Gamma M_0/M_\chi^2\right)^2$. This is a result of several approximations that were done in a computation. We have neglected the Fermi blocking factors in the final state, did not take into account the inverse decays and neglected the mass of N_1 in comparison with the mass of χ. However, the number of sterile neutrinos with these momenta is small in comparison with the total number (12.37) and the approximation works well.

If the number of massless degrees of freedom changes during the decays of χ (for example, if χ decays at temperatures of the order of QCD scale ~ 100 MeV), no analytical solution of Eq. (12.35) can be found, but numerical integration poses no problem. The abundance of dark matter sterile neutrinos can be further diluted by a factor S, which accounts for the entropy production after DM sterile neutrinos are produced, $N_0 \to N_0/S$. The same process cools down the spectrum of sterile neutrinos by a factor $S^{\frac{1}{3}}$: $\langle p \rangle \to \langle p \rangle/S^{\frac{1}{3}}$. The entropy production, for example, can occur because of the decays of heavier sterile neutrinos in the νMSM [133], if DM is produced after the QCD cross-over, or because of a decrease of the number of effectively massless degrees of freedom of the Standard Model, if χ decays before the QCD epoch [1291].

A sterile neutrino produced by this mechanism is perfectly consistent with all cosmological and astrophysical observations. As for the bounds on mass versus active–sterile mixing coming from X-ray observations of our Galaxy and its dwarf satellites, they are easily satisfied since the production mechanism of sterile neutrinos discussed above has nothing to do with the active–sterile neutrino mixing leading to the radiative mode of sterile neutrino decay. As for the limits coming from Lyman-α forest considerations for $S = 1$, one gets $m_s > 11.5$ keV (if the limit of [1721] is taken) and an even smaller value if there is some entropy production.

The nature of the scalar field was not essential for the consideration presented above. The only requirement was that this field is in thermal equilibrium owing to other interactions at temperatures of the order of its mass. This is true both for the inflaton of refs. [97; 1737] and for a scalar singlet [1291; 1571].

12.4 Conclusions

New physics, responsible for neutrino masses and mixings, for dark matter and for the baryon asymmetry of the Universe, may hide itself below the EW scale. This possibility can be offered by the νMSM – a minimal model, explaining simultaneously all well-established observational drawbacks of the Standard Model. Of course, there are many other problems in astrophysics and cosmology which cannot be explained by new particles present in the νMSM. They include the origin of high-energy cosmic rays, the existence of the 0.511 MeV annihilation line in the direction of the Galactic centre, etc. However, these phenomena do not provide a 'smoking gun' signature for 'Beyond the SM' physics, as several standard physics explanations have been proposed to deal with them. There are also several anomalies in particle physics experiments, such as the discrepancy between experiment and the theoretical prediction of the anomalous magnetic moment of the muon, the LSND anomaly, excess of low-energy events at the MiniBooNE, evidence of the neutrinoless double decay presented by part of the Heidelberg group, DAMA annual modulations of the count rate, etc. Most probably, singlet fermions of the νMSM have nothing to do with that. However, none of these anomalies has been confirmed by other experiments. All the other SM problems are those of theoretical fine-tuning: 'dark energy' or the cosmological constant problem, the 'gauge hierarchy problem', the strong CP problem, etc.

This new physics (a pair of new neutral leptons, creating the baryon asymmetry of the Universe) can be searched for in dedicated experiments with the use of existing intensive proton beams at CERN, FNAL and planned neutrino facilities in Japan (J-PARC) [990]. Indirect evidence in favour of this proposal will be given by LHC, if it discovers the Higgs boson within the mass interval 126–175 GeV and nothing else. Moreover, the νMSM gives a hint on how and where to search for new physics in this case. In particular, it tells us that in order to uncover new phenomena in particle physics one should go towards high-intensity proton beams or very high-intensity charm or B-factories rather than towards high-energy electron–positron accelerators.

The dark matter candidate in this model is the lightest singlet fermion. Depending on its mass and the creation mechanism, it can play the role of

M. Shaposhnikov

cold or warm dark matter. In the latter case it can help to resolve the potential problems of the CDM scenario at small length scales. To search for a dark matter sterile neutrino in the Universe one needs an X-ray spectrometer in space with good energy resolution $\delta E/E \sim 10^{-3}$–10^{-4} [422] getting signals from our Galaxy and its dwarf satellites. The laboratory search for this particle would require an extremely challenging detailed analysis of kinematics of β-decays of different isotopes [348].

Acknowledgements

This work was supported in part by the Swiss National Science Foundation. It is a pleasure to thank T. Asaka, F. Bezrukov, S. Blanchet, A. Boyarsky, D. Gorbunov, A. Kusenko, M. Laine, A. Neronov, O. Ruchayskiy and I. Tkachev for collaboration on the topics described in this chapter.

Part III

Collider searches

13

SUSY searches at the LHC

Tilman Plehn and Giacomo Polesello

The LHC is a proton–proton collider with centre of mass energy of 14 TeV, which started data-taking at CERN in 2009. Its purpose is to understand the nature of electroweak symmetry breaking and search for physics beyond the Standard Model. Two large general-purpose experiments are installed on the LHC, ATLAS [1] and CMS [22]. In these two experiments, much effort has been devoted to designing detectors sensitive to the broadest possible range of signatures of new physics, and to developing general and efficient search strategies.

While the Standard Model from a theory point of view is a complete renormalizable quantum field theory which does not require an ultraviolet completion, it is clearly experimentally incomplete: adding new stable particles at the weak scale with a typical weak coupling can, for example, explain the observed dark matter density in the Universe. Going up in energy, flavour physics in the quark sector is merely parameterized in the Standard Model, without any hint of what the underlying structure might be. Lepton flavour and in particular neutrino masses can be described by a see-saw mechanism that requires a heavy right-handed neutrino and an effective dimension-five operator at high energy scales. The measured almost-perfect gauge coupling unification in the Standard Model is in contradiction with proton decay searches, but it could easily be rescued by new physics at the weak scale. And last but not least, a proper quantum theory of gravity might involve some kind of string theory (or might be unnecessary owing to a fixed point of gravity in the ultraviolet).

Following this list of experimental arguments, we should consider the Standard Model an effective field theory, with theoretical features like

Particle Dark Matter: Observations, Models and Searches, ed. Gianfranco Bertone. Published by Cambridge University Press. © Cambridge University Press 2010.

renormalizability which we would hope extend to its ultraviolet completion. If we indeed follow the experimental hints which suggest that the Standard Model breaks down at some energy scale, the consistency of the Higgs mechanism at the weak scale is challenged: while the Higgs mechanism can for example only unitarize WW scattering if the Higgs mass is below the TeV scale, and electroweak precision data certainly confirms this mass range, quantum corrections drive the Higgs mass towards the cutoff of the effective Standard Model. This tension in the renormalization of a fundamental Higgs scalar is called the hierarchy problem, and its solution should be considered part of the LHC's main target of understanding electroweak symmetry breaking.

Supersymmetry (SUSY) is for many reasons the most promising candidate for physics beyond the Standard Model. It naturally addresses all the experimental challenges listed above while at the same time maintaining the greatest strength of the Standard Model — being a renormalizable perturbative field theory which ensures maximum predictability. It has been used as a benchmark theory by both the ATLAS and the CMS collaborations, as testified by the very detailed work documented in refs. [144; 145; 237]. A choice of supersymmetric models offering different phenomenologies have been studied in detail through a broad palette of final state signatures, thus allowing a thorough test of the performance of the LHC detectors for discovering new physics at the TeV scale. By far the largest amount of work has been devoted to the development of a generic strategy for R-parity conserving supersymmetry, which provides a weakly interacting stable lightest supersymmetric particle (LSP). This particle (more or less) naturally explains the observed relic density in the Universe. While in principle this neutral weakly interacting particle could also, for example, be a sneutrino, most studies follow the direct-detection constraints and focus on the lightest neutralino.

In this contribution we will give a brief account of the different steps involved in the discovery and measurement of supersymmetry at the LHC, and review the present status of strategies for constraining the soft breaking parameters. These should in turn predict observables of cosmological interest and pave the way for a combined analysis. We particularly emphasize that these strategies can, of course, be extended to any other TeV-scale new physics model, as long as it contains a dark matter agent leading to missing energy signatures.

13.1 Discovery channels

Since squarks and gluinos are strongly interacting, their production cross-sections are completely determined by their masses and the strong coupling.

If *R*-parity is conserved, they each cascade to the colour-neutral lightest supersymmetric particle which escapes the detector. On the way they need to shed their colour charge, so the resulting events are characterized by high jet multiplicity and large missing transverse energy. Moreover, squark and gluino events are expected to be spherical, as is typical for the production and decays of heavy particles. Because of the non-trivial structure of the supersymmetric weak sector, leptons will typically be radiated in the decays of the charginos and neutralinos in the later stages of the cascade.

ATLAS and CMS have studied the reach of inclusive signatures by scanning the parameter space of different supersymmetric models. For each model they establish the statistical significance of the signal over the expected Standard Model backgrounds after appropriate selection criteria.

To present the experimental reach in a simple manner, we perform the scan for specific models of supersymmetry breaking, where the whole phenomenology can be expressed as a function of a limited number of parameters. The baseline (toy) model in both ATLAS and CMS is a model where supersymmetry breaking is communicated through gravitational interactions, commonly denoted as mSUGRA or cMSSM. It assumes that at some scale all SUSY breaking parameters are determined by the mass dimension of the operator in the Lagrangian. In the squark–gluino sector the two-dimensional plane of universal scalar (m_0) and gaugino ($m_{1/2}$) masses can be easily re-rotated into the squark–gluino mass plane.

In Fig. 13.1 we show the 5-sigma reach of LHC searches as computed by ATLAS [145] for various inclusive signatures: E_T^{miss}+jets with no leptons, and with one lepton, two leptons and three leptons. The corresponding plot by the CMS Collaboration [237] shows similar features. The reach is dominated by the E_T^{miss}+jets search but the production of leptons in the cascade decays will provide additional inclusive signatures over most of the parameter space, thus adding robustness to the discovery potential. The approximate discovery reach is 1.3 TeV in squark/gluino masses for the assumed luminosity of 1 fb^{-1}, which corresponds to approximately one month of data taking at $10^{33}\,\mathrm{cm}^{-2}\,\mathrm{s}^{-1}$. The results include systematic errors on the background evaluation for the same luminosity. One year of design luminosity ($10^{34}\,\mathrm{cm}^{-2}\,\mathrm{s}^{-1}$) or 100 fb^{-1} increases the mass reach to approximately 2.5–3 TeV, limited by the signal cross-section dropping sharply for high masses. While the precise results shown in Fig. 13.1 apply only to the mSUGRA toy model, a similar reach in gluino and squark masses should apply to any model in which these states decay into an invisible and relatively light dark matter candidate [77].

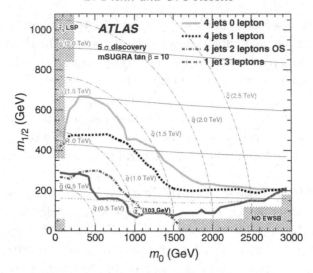

Fig. 13.1. ATLAS 5σ reach in various inclusive channels involving $E_{\mathrm{T}}^{\mathrm{miss}}$ (missing transverse energy) presented in the mSUGRA (toy) mass plane. (OS, opposite sign.) We assume a low integrated luminosity of 1 fb^{-1} [145].

The above discussion shows that the discovery of R-parity conserving supersymmetry should be reasonably straightforward at the LHC, and that a supersymmetric mass scale around 1 TeV should affect the very first data. However, such a SUSY signal buried in a significant Standard Model background will not appear with foolproof distinguishing features. Any discovery will be limited by the achievable experimental understanding of the backgrounds. The main selection criterion is missing transverse energy, and the two main sources of the corresponding backgrounds have been studied: the first source is balanced events, typically from QCD, where the momentum imbalance is generated by a mis-measurement in the detector. Controlling such backgrounds will require careful studies of the detector performance based on collision data and a minimum of several months of work. The second source is events where the missing energy comes from neutrinos in Standard Model processes involving W and Z bosons, like for instance $(Z \to \nu\nu)$+jets, or $\bar{t}t$ production where the latter even has a high jet multiplicity. The key to understand these backgrounds is well-identified and fully reconstructed signals, for example leptonic Z decays and fully reconstructed $\bar{t}t$ events, which can model the production properties. Because we will need sufficient background statistics to accurately predict missing energy events in the Standard Model, the time needed to claim the discovery of supersymmetry will most likely be dominated by the collection of these control samples and not by signal statistics. Therefore, we should re-phrase the requirement of 1 fb^{-1} integrated luminosity as '1 fb^{-1} of well-understood data'.

13.2 LHC measurements

Detailed measurements of supersymmetric parameters hinge on the identification and isolation of complete new-physics decay chains. Only then can we measure kinematic properties of the decay products and extract masses and couplings of the sparticles. Both LHC collaborations have analysed in detail (hopefully representative) benchmark points in parameter space [36; 59; 144]. For specific models, the relevant phenomenology is defined in terms of few parameters. However, benchmark models are often influenced by the theoretical prejudices of the proponents, so we emphasize that the experimental analyses and their results are of course not relying on the motivation of such benchmark points.

The universal starting point of detailed LHC analyses is the identification of decays involving leptons or bottom jets, which allow for a separation of the signal from the overwhelming QCD backgrounds. The classical example is the decay of the second-lightest neutralino. A purely gaugino-type $\tilde{\chi}_2^0$ will be abundantly produced in left-handed squark decays and has a significant branching ratio into pairs of leptons or bottom jets with opposite sign. Over a wide range of model parameters, decays into virtual or real Z bosons, sleptons or Higgs bosons lead to opposite-sign, same-flavour (OS-SF) combinations. This allows us to isolate the $\tilde{\chi}_2^0$ decay chain by selecting a distinctive final state signature, perform a Monte Carlo analysis of this decay and assess the viability of model-independent techniques developed to measure masses and spins of new particles.

13.2.1 Masses

The basic feature of R-parity conserving supersymmetry is the production of two particles, each decaying through chains of variable length into an invisible particle. Thus, each event has two 'legs', each corresponding to the decay chain of one sparticle produced in the proton–proton collision. Each step in the chain produces one or more visible Standard Model particles, leptons or jets, with 4-momenta which we can measure in the detector. Another measurable quantity in the events is the missing transverse energy, the vector sum of the transverse momenta of the two invisible particles.

The problem is how to constrain the unknown masses of the new particles involved, based on these measured quantities and under various assumptions on the length of the decay chains. This is a generic problem of relativistic kinematics and (at leading order) does not require any knowledge of QCD or supersymmetry. The developed methods therefore have no dependence on a specific model and can be applied to any appropriate kinematic situation. A large amount of effort has been devoted to this topic, starting from the pioneering work of ref. [1097]. The proposals to solve this problem can be broadly classified into three classes:

- **Endpoint methods**, where for a single leg we extract mass constraints from lower (threshold) and upper (edge) kinematic endpoints of invariant mass distributions of visible decay products [62; 149; 962; 963; 1096; 1097; 1326; 1446]. Extensions of this method use the kinematic limit of invariant mass distributions of visible particles from both sides of the event (hidden threshold techniques), and correlations between the distributions from each leg (wedgebox techniques) [364; 1137; 1224].

- **Mass relation method**, where we completely reconstruct the kinematics event-by-event, assuming that all of the decaying particles are on-shell. For each event this assumption provides kinematic constraints corresponding to the number of sparticles in the identified decay chain. While for each event the number of free parameters can be larger than the number of measured observables, adding several signal events increases the number of measurements while keeping the number of unknowns constant. At some point the system of equations will solve, provided that all events are really signal events. This method, first proposed by ref. [1515], has been further developed [1212] to address single legs. Using both legs allows us to consider two decays at a time and include the measured missing energy [1517]. This has been successfully applied in the context of a hybrid method combining on-shell constraints and endpoints, and has been further developed in refs. [532; 534].

- **MT2 methods** are based on the global variable m_{T2}, first proposed in [1327], and further developed in [214]. This method generalizes the concept of a transverse mass (for example of a leptonically decaying W boson or top quark) to the pair production of new particles decaying into an invisible final state. With the information from both legs in an event we can construct a variable combining information from the visible part of the events and the missing transverse energy for a given value of the mass of the invisible particle. For the correct value of the LSP mass the m_{T2} distribution has an edge at the mass of the parent particle. It is particular to this method, or of recent variations thereof, that in favourable cases it may allow the measurement of both the parent particle and the LSP

Table 13.1. *LHC measurements in SPS1a [1923]. We show the nominal values, statistical errors, systematic errors from the lepton (LES) and jet energy scales (JES) and theoretical errors. All values in GeV.*

Measurement		Nominal	Statistical	LES	JES	Theoretical
m_h		108.99	0.01	0.25		2.0
m_t		171.40	0.01		1.0	
$m_{\tilde{l}_L} - m_{\chi_1^0}$		102.45	2.3	0.1		2.2
$m_{\tilde{g}} - m_{\chi_1^0}$		511.57	2.3		6.0	18.3
$m_{\tilde{q}_R} - m_{\chi_1^0}$		446.62	10.0		4.3	16.3
$m_{\tilde{g}} - m_{\tilde{b}_1}$		88.94	1.5		1.0	24.0
$m_{\tilde{g}} - m_{\tilde{b}_2}$		62.96	2.5		0.7	24.5
m_{ll}^{max}:	3-particle edge ($\chi_2^0,\tilde{l}_R,\chi_1^0$)	80.94	0.042	0.08		2.4
m_{llq}^{max}:	3-particle edge ($\tilde{q}_L,\chi_2^0,\chi_1^0$)	449.32	1.4		4.3	15.2
m_{lq}^{low}:	3-particle edge ($\tilde{q}_L,\chi_2^0,\tilde{l}_R$)	326.72	1.3		3.0	13.2
$m_{ll}^{max}(\chi_4^0)$:	3-particle edge ($\chi_4^0,\tilde{l}_R,\chi_1^0$)	254.29	3.3	0.3		4.1
$m_{\tau\tau}^{max}$:	3-particle edge ($\chi_2^0,\tilde{\tau}_1,\chi_1^0$)	83.27	5.0		0.8	2.1
m_{lq}^{high}:	4-particle edge ($\tilde{q}_L,\chi_2^0,\tilde{l}_R,\chi_1^0$)	390.28	1.4		3.8	13.9
m_{llq}^{thres}:	threshold ($\tilde{q}_L,\chi_2^0,\tilde{l}_R,\chi_1^0$)	216.22	2.3		2.0	8.7
m_{llb}^{thres}:	threshold ($\tilde{b}_1,\chi_2^0,\tilde{l}_R,\chi_1^0$)	198.63	5.1		1.8	8.0

based on a single-step decay chain [217; 479; 480; 542; 543; 1325; 1518; 1519; 1861].

A detailed discussion of these methods is beyond the scope of this review. A generic feature is that it is reasonably easy to constrain the differences of the squared masses of the sparticles in the chain with good precision, while the absolute scale is much more difficult to fix. This is due to the parametric form of the endpoint formulas, the two invisible sparticles at the end of the chain, and the correlated jet and lepton energy errors.

The endpoint method is the only one which has yet been applied to a complete detector-level analysis [1923]. This analysis assumes a specific parameter point, dubbed SPS1a [59], which is given by $m_0 = 100$ GeV, $m_{1/2} = 200$ GeV, $A_0 = -100$ GeV, $\tan\beta = 10$, $\mu > 0$. It illustrates how we would measure the masses of the maximum number of supersymmetric particles. These measurements we can then use as the input for the model reconstruction. In view of the topic of this volume we will review the key passages of this analysis and give some background information on the measurements listed in Table 13.1.

The first aim of the analysis is the identification of cascades of two-body decays and the measurement of the intermediate sparticle masses. A well-known example is the squark decay [149]

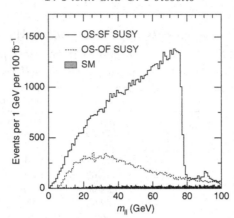

Fig. 13.2. Invariant mass of two leptons after selection cuts for the SPS1a model: SUSY signal Opposite Sign Same Flavour (OS-SF): full line; SUSY signal Opposite Sign Opposite Flavour (OS-OF): dotted line; Standard Model background: grey.

$$\tilde{q}_{\mathrm{L}} \to \tilde{\chi}_2^0 q \to \tilde{\ell}_{\mathrm{R}}^{\pm} \ell^{\mp} q \to \tilde{\chi}_1^0 \ell^+ \ell^- q. \tag{13.1}$$

We can isolate this chain by requiring two isolated leptons in addition to multiple hard jets and large missing energy. The dominant Standard Model background is top pairs. Like any other background involving two independent decays it cancels in the combination $e^+ e^- + \mu^+ \mu^- - e^{\pm} \mu^{\mp}$. Notably, this is also true for two independent chargino decays which is the dominant SUSY background for example in mSUGRA scenarios. We focus on the SPS1a results where the mass scale of the coloured particles is $\sim$600 GeV, the total production cross-section is $\sim$50 pb, and the branching ratio for the decay chain shown in Eq. (13.1) is $\sim$4%.

In Fig. 13.2 we see the characteristic triangular shape of the invariant mass of the opposite-sign lepton pair. It corresponds to a sequence of two-body decays with a scalar intermediate state. The position of the very sharp edge is given by

$$m_{\ell\ell}^2 < \frac{(m_{\tilde{\chi}_2^0}^2 - m_{\tilde{\ell}}^2)(m_{\tilde{\ell}}^2 - m_{\tilde{\chi}_1^0}^2)}{m_{\tilde{\ell}}^2}. \tag{13.2}$$

From this formula we can immediately guess that mass differences are the appropriate observables, simply on the fundamental insight that $A^2 - B^2 = (A + B)(A - B)$. In complete analogy we can measure the edge as well as the non-zero threshold of the $\ell^+ \ell^- q$ distribution (Fig. 13.3), and the edges of the two $\ell^{\pm} q$ invariant mass combinations [144; 149]. All thresholds and edges are known functions of the sparticle masses [62; 962], which means

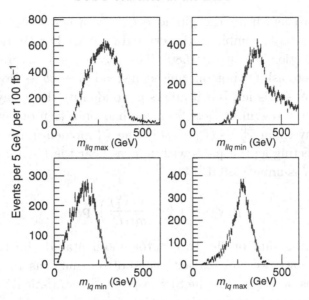

Fig. 13.3. Dilepton and jet invariant mass distributions for the SPS1a parameter point. From ref. [1582].

that given enough measurements we can solve the system for the sparticle masses without any model assumption. As a result all cascade masses are determined accurately as functions of $m_{\tilde{\chi}_1^0}$. For an integrated luminosity of 300 fb^{-1}, the $\tilde{\chi}_1^0$ mass itself we can measure to about 10%. For the same luminosity and model points with a new-physics mass scale up to $\sim$1 TeV, corresponding to supersymmetric cross-sections in the few picobarn range, the statistical error on the determination of the lepton–lepton edge is at the per mil (0.1%) level while for endpoints involving jets it increases to the per cent level. Therefore, we can expect the precision of mass measurements to be affected by systematic effects, such as the modelling of the edge shape [1446] or the knowledge of the energy scales in the detector.

We can use exactly the same method for the decay $\tilde{q}_L \rightarrow \tilde{\chi}_2^0 q \rightarrow \tilde{\tau}_1^{\pm} \tau^{\mp} q \rightarrow \tilde{\chi}_1^0 \tau^+ \tau^- q$. The tau lepton we can detect by tagging the narrow jet from its hadronic decay. The invariant mass distribution of the two τ-jets has a measurable edge, but the sharp triangular shape from the electron and muon case is lost with the momenta of the neutrinos in the tau decay. This way the lighter stau mass can be determined at the LHC.

If the gluino is heavier than the squark it is obvious that all we have to do is extend the squark chain by one step: $\tilde{g} \rightarrow q\tilde{q}$. With one additional jet we can study a large number of invariant mass combinations and measure the mass of the gluino [963]. This measurement is complicated if (like in SPS1a) one of the two jets from the gluino decay is rather soft. Such a jet will be buried in a large combinatorial error due to QCD radiation in massive particle production [75; 1579; 1580]. The way around this probably lethal QCD effect is to ask for bottom quarks instead: $\tilde{g} \to b\tilde{b} \to bb\tilde{\chi}_2^0$. This decay chain has lower statistics, but it allows us to identify the decay jets through the experimental tagging of jets from the hadronization of bottom quarks.

In the decay chain $\tilde{\chi}_2^0 \to \ell\tilde{\ell}_R \to \ell\ell\tilde{\chi}_1^0$ the $\tilde{\chi}_2^0$ momentum can be approximated by selecting lepton pairs with masses near the kinematic edge, even though the $\tilde{\chi}_1^0$ is unmeasured [144]:

$$\mathbf{p}(\tilde{\chi}_2^0) = \left(1 - \frac{m(\tilde{\chi}_1^0)}{m(\ell\ell)}\right) \mathbf{p}_{\ell\ell}, \qquad (13.3)$$

where $\mathbf{p}_{\ell\ell}$ is the sum of the two lepton momenta. If the two neutralino masses are known we can extract the sbottom mass as $m(\tilde{\chi}_2^0 b)$, and the gluino mass as $m(\tilde{\chi}_2^0 bb)$. For the SPS1a point the statistical uncertainty on the gluino-mass peak position is ~ 4 GeV for 100 fb^{-1}. There is one more complication in the sbottom sector, where the two mass eigenstates $\tilde{b}_{1,2}$ with a mass difference of ~ 30 GeV contribute to the same decay. With high statistics and excellent experimental control of the detector response for bottom jets it might be possible to separate the two sbottom peaks in the $m(\tilde{\chi}_2^0 bb) - m(\tilde{\chi}_2^0 b)$ distribution and measure both sbottom masses.

As a final example for the endpoint method the decay $\tilde{q}_L \to \tilde{\chi}_4^0 q \to \tilde{\ell}_L^\pm \ell^\mp q$ has a branching fraction of a few 10^{-4} in our parameter point. This gives rise to a visible signal, where using the invariant mass of the two leptons we can measure the $\tilde{\chi}_4^0$ mass, provided the masses of the other sparticles in the chain are known.

In addition, we can use the m_{T2} method to measure the masses of $\tilde{\ell}_L$ and $\tilde{q}_R$ masses which decay respectively as $\tilde{\ell}_L \to \ell\tilde{\chi}_1^0$ and $\tilde{q}_R \to q\tilde{\chi}_1^0$. Direct production of $\tilde{\ell}_L$ or $\tilde{q}_R$ pairs we can select by requiring events with two hard leptons or jets, large missing energy and no additional hadronic or leptonic activity in the event. These selection criteria ensure a direct decay of these particles to the invisible LSP. If we compute m_{T2} for these events and input the $\tilde{\chi}_1^0$ mass from the $\tilde{q}_L$ decay we find a clear edge corresponding to the value of the produced sparticle. It should be noted that the $\tilde{\ell}_R$ mass should in principle be accessible by the same method, but the signal for $\tilde{\ell}_R$ production is hidden by the irreducible Standard Model WW background.

Even though they are strictly speaking not supersymmetric particles, for the SPS1a parameter point we will discover the light Higgs boson at the LHC, both through the decay $h \to \gamma\gamma$ and through vector–boson fusion with a subsequent decay to tau leptons [144; 145; 237]. Unfortunately, the moderate value of $\tan \beta = 10$ does not provide enough of an enhancement of the heavy Higgs' Yukawa coupling to bottom quarks, which means that the essentially decoupled light Higgs boson will be the only sign of the extended MSSM Higgs sector.

13.2.2 Rates

In general, we can of course extract model parameters from production cross-sections. This is particularly tempting for strongly interacting new states, because their colour charge is typically independent of the model's details. For example, the production of squark pairs, gluino pairs, stop pairs or even leptoquark pairs only depends on the masses and colour charges of the produced particles. The colour charge is given by either the fundamental or the adjoint colour representations. However, there are two problems in such an analysis.

First, we hardly ever measure the actual production cross-section. While we can of course formulate an exclusion limit for a new particle under the explicit assumption that this particle only decays into one final state, this assumption is unlikely to realistically describe as complex a model as the MSSM. Therefore, we have to analyse cross-sections multiplied by branching ratio assuming we see only a few allowed final states. Measurements of production cross-sections and branching ratios will usually appear as ratios $(\sigma \mathrm{BR})_1/(\sigma \mathrm{BR})_2$. For example, in the squark cascade discussed above the relative branching ratios to $e + \mu$ and τ leptons contain useful information on the stau sector [1516]. One notable exception is the top squark in the MSSM, which with increasing mass dominantly decays through loop-induced amplitudes to charm-neutralino, then on-shell to bottom-chargino mediated by its weak charge and finally to top-gluino via its colour charge.

Second, we know (for example from top-pair production) that higher-order correction of QCD cross-sections are by no means of the order $\alpha_s/(4\pi) \sim 0.01$. Collinear logarithms lead to next-to-leading-order corrections of typically 30–80%, depending on the colour charges of the initial state and the produced particle. Even for weakly interacting Drell–Yan-type processes like neutralino–chargino production, we observe corrections of the order of 30%. For production processes which are (at leading order) proportional to α_s^2 the theoretical uncertainty due to higher-order corrections is roughly covered by the variation of the renormalization and factorization scales. The argument for this error measure is simply that these scales are artifacts of perturbative QCD, which means that to all orders in perturbation theory the scale

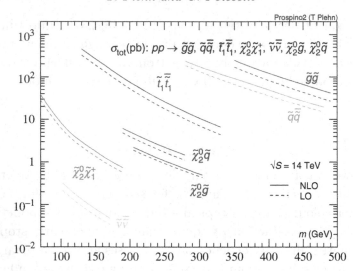

Fig. 13.4. LHC production cross-sections for supersymmetric particles. All rates are shown as a function of the average final-state mass. LO, leading order; NLO, non-leading order.

dependence should vanish. However, for leading-order processes of the order α^2 this error estimate undershoots the known higher-order corrections by large factors. In this situation it is not only crucial to know the theoretical error on a cross-section prediction, but to actually use the higher-order results [245; 247; 248; 1287], as shown for example in Fig. 13.4.

Finally, for squarks and gluinos in the MSSM we face one last challenge: the list of production processes includes $\tilde{q}\tilde{q}$ and $\tilde{q}\tilde{q}^*$ production, gluino pairs $\tilde{g}\tilde{g}$ and last but not least the associated $\tilde{q}\tilde{g}$ process. These processes can only be distinguished by the number of jets in the final state. The associated production has a quark–gluon initial state which is particularly well suited for the production of heavy states at the LHC: while valence quarks dominate the large-x region of the available partonic energy, for smaller $x < 0.01$ the gluon densities practically explode. The mixed qg initial state turns out to be the best compromise for average x values around 0.1.

13.2.3 Quantum numbers

The measurement of discrete quantum numbers, such as the spin of new particles, is always hard at hadron colliders. The classical threshold behaviour is not visible at a hadron collider, in particular if the final state includes missing transverse energy. Instead, we have to rely on angular correlation on the decays side.

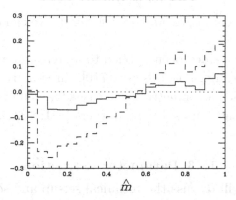

Fig. 13.5. Asymmetry in the normalized invariant mass $m_{j\ell}$ for supersymmetric particles (dashed line) and universal extra dimensions (solid line). The mass spectrum is assumed to be hierarchical, as expected in supersymmetry. Figure taken from ref. [1771].

Recently, a method involving three steps has been developed, for example for the squark decay chain given in Eq. (13.1) [215; 1771].

First, instead of trying to measure spins in a general parameterization we notice that cascade decays radiate particles with known spins. This is most obvious for gluino decays, where we know that the radiated bottom quarks as well as muons are fermions. Hence, the spins inside the decay chain alternate between fermions and bosons. In supersymmetry this fermion/boson nature is switched compared with the corresponding Standard Model particle, so we can make our lives much easier by comparing supersymmetry with a second hypothesis where the spins in the decay chain follow the Standard Model assignments. An example of such a model is Universal Extra Dimensions (UED), where each Standard Model particle has a Kaluza–Klein partner from propagating into the additional dimensions.

Second, both the thresholds and the edges of all invariant masses of the radiated fermions are completely determined by the masses inside the decays chain. The shape of the distribution between the endpoints, however, is nothing but an angular correlation (in some rest frame). This means that, for example, the well-established $m_{j\ell}$ distribution allows us to analyse spin correlations in squark decays in a Lorentz-invariant way.

Last but not least, a proton–proton collider produces considerably more squarks than antisquarks in the squark–gluino associated channel. A decaying squark radiates a quark while an antisquark radiates an antiquark, which means that we can define a non-zero production-side asymmetry between $m_{j\ell^+}$ and $m_{j\ell^-}$. Such an asymmetry is shown in Fig. 13.5, for the SUSY and for the UED hypotheses. We see that in this case where the masses in the decay chain are not too degenerate we can distinguish the two hypotheses.

This basic idea has since been applied to a great many similar situations, such as decays including gauge bosons [1770], three-body decays [595], gluino decays with decay-side asymmetries [74], cascades including charginos [1921], weak-boson-fusion signatures [78] and others [141; 1231].

13.3 Parameter extraction

In this section we will discuss the technical set-up and some physics results concerning the extraction of model parameters from LHC data. Most of the results shown are produced with the SFitter programme [1298], but we emphasize that the alternative approach using Fittino [243; 244] gives equivalent results.

Before we talk about new-physics parameters we need to discuss the proper treatment of experimental and theory errors. Statistical experimental errors are uncorrelated in the measured observables while systematic experimental errors, for example from the jet and lepton energy scales [1923], are fully correlated. Because there is no reason why unknown higher-order corrections should be centred around a given value or even around zero, theory errors should not be assumed Gaussian but flat around a central value. In other words, the probability assigned to a measurement does not depend on its value, as long as it is within the interval covered by the range of validity of the perturbative theory. Confronted with seriously poor agreement in an observable we need to rethink the perturbative description of the underlying theory, for example the Standard Model.

The convolution of such a box-shaped theory error with a Gaussian experimental error leads to the difference of two one-sided error functions which does have a maximum and hence knows about the central value of theoretical prediction. A better solution is a flat distribution within the theoretically acceptable interval which outside this interval drops off like a Gaussian with the width of the experimental error. The log-likelihood in this RFit scheme, given a set of measurements $\mathbf{d}$ and a general correlation matrix C, reads [523; 1110]

$$\log \mathcal{L} = -\frac{1}{2}\chi^2 = -\frac{1}{2}\, \chi_d{}^T\, C^{-1}\, \chi_d$$

$$\chi_{d,i} = \begin{cases} 0 & |d_i - \bar{d}_i| < \sigma_i^{(\text{theo})} \\ \dfrac{|d_i - \bar{d}_i| - \sigma_i^{(\text{theo})}}{\sigma_i^{(\text{exp})}} & |d_i - \bar{d}_i| > \sigma_i^{(\text{theo})}. \end{cases} \tag{13.4}$$

To evaluate the error propagation we rely on toy measurements: for each of a set of smeared toy measurements we determine the best-fit value. The distribution of all best-fit values gives the error on a parameter. This time-consuming option is necessary to obtain correct confidence level intervals.

13.3.1 mSUGRA toy model

Because we have no way of estimating the systematic error due to people making guesses at high-scale theories while sitting in their offices, no model for supersymmetry breaking should be assumed in an LHC analysis. Instead, the breaking mechanism will be inferred from data. Nevertheless, it is instructive to discuss the main features of proper parameter extraction in the more compact mSUGRA toy model.

The relevant frequentist or Bayesian results we present in three steps: first, we compute a completely exclusive log-likelihood map of the parameter space. Second, we rank the local likelihood maxima in this map and identify the global maximum. Last, we compute profile likelihood or Bayesian probability maps of lower dimensionality, down to one-dimensional distributions. Only in this final step do frequentist and Bayesian approaches need to be distinguished.

Looking at the parameter point SPS1a at the LHC, some parameters are heavily correlated, others are poorly constrained, and distinct maxima in the likelihood should appear. Therefore, we would like to study 'likelihoods' over parts of the model-parameter space. Unwanted dimensions of the parameter space we need to eliminate. The problem is that a likelihood map can only be integrated once with a measure in model space. This measure is a prior and defines the Bayesian probability distribution. Alternatively, a profile-likelihood projection most notably guarantees that the best-fit points always survive to the final representation. Note, however, that eliminating a dimension in parameter space always means loss of information. Therefore, producing low-dimensional distributions from the completely exclusive likelihood map might not automatically be the best thing to do.

Our starting point is a Markov chain representing a likelihood map over the entire parameter space given a smeared LHC data set. In mSUGRA it covers the model parameters m_0, $m_{1/2}$, A_0, B, m_t, where B is later traded for the weak-scale $\tan\beta$. Because the resolution of the Markov chain is not sufficient to resolve each local maximum in the log-likelihood map, an additional maximization algorithm starts at the best points of the Markov chains to identify the local maxima. As shown in the table in Fig. 13.6 a general pattern of four distinct mSUGRA maxima emerges: the trilinear coupling

m_0	$m_{1/2}$	$\tan\beta$	A_0	μ	m_t
102	254	11.5	-95	$+$	172.4
105	242	12.9	-174	$-$	172.3
108	266	14.6	742	$+$	173.7
112	261	18.0	633	$-$	173.0
...					

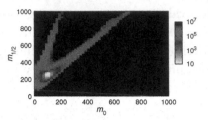

Fig. 13.6. Toy results for mSUGRA in SPS1a. Left: List of the best log-likelihood values over the parameter space. Right: Profile likelihood over the m_0–$m_{1/2}$ plane (brighter areas indicate higher likelihood). All masses in GeV. The χ^2 values of the best points are (from the top) $0.09, 1.5, 73, 139$.

can assume the correct value around -100 GeV, but it can also become large and positive. This degeneracy is correlated with a slight shift in the top mass. Globally, there exists a parabola-shaped correlation between the m_t and A_0 which vanishes if the top quark mass is not part of the parameter set. It has its root in the light Higgs mass and largely depends on the theory error assumed for this mass. Second, the sign of μ is correlated with a slight shift in $\tan\beta$, and these two factors compensate each other in the neutralino–chargino sector. Such a degeneracy is expected because for SPS1a at the LHC we only observe one of the two heavy neutralinos. A third correlation occurs in the m_0–$m_{1/2}$ plane: the maximum starts from the true values for m_0 and $m_{1/2}$ and extends into two branches. They reflect the fact that extracting masses from kinematic endpoints involves quadratic equations.

Since in Fig. 13.6 only one smeared set of measurements is used, the local likelihood maxima do not have to coincide with the correct values. As a matter of fact, changing the shape of the theory errors has an effect on the ranking of the maxima: for Gaussian theory errors the χ^2 values of 4.35, $26.1, 10.5, 22.6$ appear in the order shown in the table in Fig. 13.6.

A likelihood analysis as described above will not answer questions of the following kind: in the light of current data, what sign of μ is preferred in mSUGRA [58; 63]? Note that this is not the same question as asking what is the difference in likelihood for the two best points on each side of μ. Asking for reduced-dimensionality probability densities will be a very typical situation in the LHC era, with questions like: what kind of linear collider should we build? How does dark matter annihilate? How do we detect dark matter?

Shifting from a likelihood to a Bayesian pdf does not affect most of the Markov chain analysis. In complete analogy to the likelihood analysis the

Table 13.2. *Best-fit results with absolute errors in GeV. The columns correspond to mass and endpoint measurements with neglected, (approximate) Gaussian or proper flat theory errors.*

	SPS1a	$\Delta_{zero}^{theo-exp}$	$\Delta_{zero}^{expNoCorr}$	$\Delta_{zero}^{theo-exp}$	$\Delta_{gauss}^{theo-exp}$	$\Delta_{flat}^{theo-exp}$
		Masses		Endpoints		
m_0	100	4.11	1.08	0.50	2.97	2.17
$m_{1/2}$	250	1.81	0.98	0.73	2.99	2.64
$\tan\beta$	10	1.69	0.87	0.65	3.36	2.45
A_0	-100	36.2	23.3	21.2	51.5	49.6
m_t	171.4	0.94	0.79	0.26	0.89	0.97

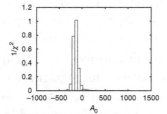 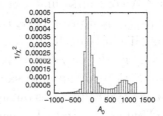

Fig. 13.7. Profile likelihood (left) and Bayesian pdf (right) for A_0. For illustration purposes the parameters m_0 and $m_{1/2}$ are only marginalized around their best-fit values and the sign of μ is fixed.

study of A_0 versus m_t in a Bayesian pdf shows how marginalizing parameters can affect our understanding of the parameter space. Figure 13.7 compares the two approaches for A_0 alone, showing that volume effects significantly enhance the relative weight of the broad secondary maximum.

Once a best-fit point has been determined our attention turns to the precision of the parameter measurement. Two different data sets describe LHC measurements: either kinematic endpoints or particle masses, from a fit to the endpoints without model assumptions [62; 149; 963]. They are only equivalent if the complete correlation matrix is taken into account. For the experimental errors, mass observables introduce non-trivial correlations, whereas the theory is essentially uncorrelated in the masses, but non-trivially correlated in the endpoints. Following Table 13.2, changing from mass to endpoint measurements improves the (experimental-only) errors by a large factor, indicating that to obtain the best precision from the LHC data we need to estimate the correlation between observables correctly.

Theory errors cannot be neglected. For example, the light Higgs mass has to be computed in perturbation theory [642; 879; 1047], with 2 GeV error due to unknown higher-order terms. In addition, mSUGRA theory errors arise from the perturbative renormalization group running [54; 846; 1160]. The impact of the combined theory errors of 1% (3%) for weakly (strongly) interacting particles [891; 1533; 1577] is also shown in Table 13.2. As a consistency check we can compute all observables with SoftSUSY, but determine the parameters with SuSpect [681; 1254]. Indeed, the central values turn out to be statistically compatible, thus giving confidence in the estimated theory errors.

For the LHC Table 13.1 involves only three out of four neutralinos – the third-heaviest neutralino will be missed because of its Higgsino nature. What happens if the fourth-heaviest neutralino is wrongly labelled as third-heaviest? We of course find a best-fitting parameter point, slightly shifted in m_0 and $m_{1/2}$, by up to 1 GeV, and with a reasonable χ^2. While at first sight this looks like a bona fide alternative minimum, it can easily be discarded: the 'wrong' model parameters predict more squark decays to $\tilde{\chi}_4$ than to $\tilde{\chi}_3$, in contradiction to the data sample. Unfortunately, distinguishing discrete alternatives typically relies on signatures which 'should have to be seen'. At the LHC, what can and what cannot be seen is mostly determined by backgrounds and detector effects, which makes automated answering algorithms unrealistic.

13.3.2 Weak-scale MSSM

If the LHC should see physics beyond the Standard Model, the MSSM defined at the weak scale would be (one of) the appropriate model(s) to test. Of the huge MSSM parameter space the LHC will only be sensitive to a limited fraction. For example, CP-violating phases [207], non-minimal flavour violation or light-generation trilinear couplings are unlikely to be understood from LHC data. This leaves us with the effective 19-dimensional parameter space listed in Table 13.3. Obviously, the assumption of parameters being irrelevant for the MSSM likelihood map can and has to be tested.

Because computing the mass spectrum in the weak-scale MSSM does not require any shift in scales, i.e. it does not involve renormalization group running, a smaller theory error for the particle masses can be assumed. As a rough estimate we use a relative error of 1% for the masses of strongly interacting particles and 0.5% for weakly interacting particles [891; 1533; 1577],

Table 13.3. *Best-fitting points in the MSSM likelihood map with two solutions for A_t. All points have approximately the same χ^2 (all masses in GeV). The standard parameter point SPS1a corresponds to column five of the eight solutions.*

	$\mu < 0$				$\mu > 0$			
M_1	96.6	175.1	103.5	365.8	98.3	176.4	105.9	365.3
M_2	181.2	98.4	350.0	130.9	187.5	103.9	348.4	137.8
μ	-354.1	-357.6	-177.7	-159.9	347.8	352.6	178.0	161.5
$\tan \beta$	14.6	14.5	29.1	32.1	15.0	14.8	29.2	32.1
M_3	583.2	583.3	583.3	583.5	583.1	583.1	583.3	583.4
$M_{\tilde{\tau}_{\mathrm{L}}}$	114.9	2704.3	128.3	4794.2	128.0	229.9	3269.3	118.6
$M_{\tilde{\tau}_{\mathrm{R}}}$	348.8	129.9	1292.7	130.1	2266.5	138.5	129.9	255.1
$M_{\tilde{\mu}_{\mathrm{L}}}$	192.7	192.7	192.7	192.9	192.6	192.6	192.7	192.8
$M_{\tilde{\mu}_{\mathrm{R}}}$	131.1	131.1	131.1	131.3	131.0	131.0	131.1	131.2
$M_{\tilde{e}_{\mathrm{L}}}$	186.3	186.4	186.4	186.5	186.2	186.2	186.4	186.4
$M_{\tilde{e}_{\mathrm{R}}}$	131.5	131.5	131.6	131.7	131.4	131.4	131.5	131.6
$M_{\tilde{q}3_{\mathrm{L}}}$	497.1	497.2	494.1	494.0	495.6	495.6	495.8	495.0
$M_{\tilde{t}_{\mathrm{R}}}$	1073.9	920.3	547.9	950.8	547.9	460.5	978.2	520.0
$M_{\tilde{b}_{\mathrm{R}}}$	497.3	497.3	500.4	500.9	498.5	498.5	498.7	499.6
$M_{\tilde{q}_{\mathrm{L}}}$	525.1	525.2	525.3	525.5	525.0	525.0	525.2	525.3
$M_{\tilde{q}_{\mathrm{R}}}$	511.3	511.3	511.4	511.5	511.2	511.2	511.4	511.5
$A_t\,(-)$	-252.3	-348.4	-477.1	-259.0	-470.0	-484.3	-243.4	-465.7
$A_t\,(+)$	384.9	481.8	641.5	432.5	739.2	774.7	440.5	656.9
m_A	350.3	725.8	263.1	1020.0	171.6	156.5	897.6	256.1
m_t	171.4	171.4	171.4	171.4	171.4	171.4	171.4	171.4

plus a 2% non-parametric error on the light MSSM Higgs boson [642; 879; 1047].

As described above, we globally scan the parameter space using Markov chains (with a flat or a Breit–Wigner proposal function) and use a hill climber to identify the local likelihood maxima. This approach can be applied to any problem involving a high-dimensional parameter space, but the details of course have to be adjusted.

Table 13.3 lists the eight best local maxima in the likelihood map. The main feature of these best-fitting points is the structure of the neutralino sector. For each sign of μ four equally good solutions can be classified by an ordering like $M_1 < M_2 < |\mu|$. The only neutralino which cannot be a Higgsino is the lightest neutralino. The physical masses of the three visible neutralinos are the same in all best-fitting points, as is the light Higgs mass.

The shift in $\tan \beta$ for the correct SPS1a parameter point is an effect of the smeared data set combined with the rather poor constraints on this

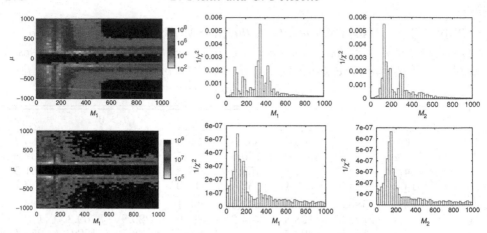

Fig. 13.8. Profile likelihoods (upper) and Bayesian probabilities (lower) for the MSSM. The distributions of the neutralino sector are derived from the log-likelihood map of the neutralino sector alone.

parameter and turns out to be within the error bar. Looking at the neutralino and sfermion mixing matrices, any effect in changing $\tan\beta$ can always be accommodated by a corresponding change in another parameter. This is particularly obvious in the stau sector. There only the lighter of $M_{\tilde{\tau}_{LR}}$ is determined from the kinematic endpoint of $m_{\tau\tau}$. The heavier mass parameter and $\tan\beta$ can compensate each other's effects freely.

A similar effect appears with the two values for A_t. One of them is a solution of the minimization procedure, while the equally good flipped sign is generated by hand. While in general the appearance of secondary maxima is nothing new, compared with the mSUGRA case we observe one significant change: in the MSSM there are enough parameters to compensate for shifts in the alternative solutions that all points shown in Table 13.3 have the same (perfect) likelihood value.

In Fig. 13.8 we first show the profile likelihoods for the neutralino–chargino sector. The M_1–μ plane shows six solutions (lighter regions), three for each sign of μ. They correspond to the three observed neutralinos, each of which can be governed by M_1. The one-dimensional profile likelihood for M_1 shows these three options with peaks around 100, 200 and 350 GeV. The peak above 400 GeV is a statistical fluctuation. For M_2 there is again the 100 GeV peak, where the LSP is a wino. The correct solution around 200 GeV is merged with the first maximum, while the third peak around 300 GeV corresponds to at least one light Higgsino.

We can also study the same parameters using Bayesian probability distributions. While the structures in the two-dimensional M_1–μ planes are very

Table 13.4. *Sample result for the general MSSM parameter determination. The LHC measurements are the kinematic endpoints. The International Linear Collider (ILC) column assumes additional precision mass measurements. All masses in GeV.*

	LHC		ILC		LHC+ILC		SPS1a
$\tan\beta$	$10.0\pm$	4.5	$12.1\pm$	7.0	$12.6\pm$	6.2	10.0
M_1	$102.1\pm$	7.8	$103.3\pm$	1.1	$103.2\pm$	0.95	103.1
M_2	$193.3\pm$	7.8	$194.1\pm$	3.3	$193.3\pm$	2.6	192.9
M_3	$577.2\pm$	14.5	fixed 500		$581.0\pm$	15.1	577.9
$M_{\tilde{\tau}_L}$	$227.8\pm\mathcal{O}(10^3)$		$190.7\pm$	9.1	$190.3\pm$	9.8	193.6
$M_{\tilde{\tau}_R}$	$164.1\pm\mathcal{O}(10^3)$		$136.1\pm$	10.3	$136.5\pm$	11.1	133.4
$M_{\tilde{\mu}_L}$	$193.2\pm$	8.8	$194.5\pm$	1.3	$194.5\pm$	1.2	194.4
$M_{\tilde{\mu}_R}$	$135.0\pm$	8.3	$135.9\pm$	0.87	$136.0\pm$	0.79	135.8
$M_{\tilde{q}_{3L}}$	$481.4\pm$	22.0	$499.4\pm$	$\mathcal{O}(10^2)$	$493.1\pm$	23.2	480.8
$M_{\tilde{t}_R}$	$415.8\pm\mathcal{O}(10^2)$		$434.7\pm\mathcal{O}(4\times10^2)$		$412.7\pm$	63.2	408.3
$M_{\tilde{b}_R}$	$501.7\pm$	17.9	fixed 500		$502.4\pm$	23.8	502.9
$M_{\tilde{q}_L}$	$524.6\pm$	14.5	fixed 500		$526.1\pm$	7.2	526.6
$M_{\tilde{q}_R}$	$507.3\pm$	17.5	fixed 500		$509.0\pm$	19.2	508.1
A_τ	fixed 0		$613.4\pm$	$\mathcal{O}(10^4)$	$764.7\pm\mathcal{O}(10^4)$		-249.4
A_t	$-509.1\pm$	86.7	$-524.1\pm$	$\mathcal{O}(10^3)$	$-493.1\pm$	262.9	-490.9
A_b	fixed 0		fixed 0		$199.6\pm\mathcal{O}(10^4)$		-763.4
m_A	$406.3\pm\mathcal{O}(10^3)$		$393.8\pm$	1.6	$393.7\pm$	1.6	394.9
μ	$350.5\pm$	14.5	$354.8\pm$	3.1	$354.7\pm$	3.0	353.7
m_t	$171.4\pm$	1.0	$171.4\pm$	0.12	$171.4\pm$	0.12	171.4

similar, the one-dimensional histograms show two significant differences: first, the local Bayesian structures in $M_{1,2}$ are much wider, owing to noise and volume effects. On the other hand, the Bayesian pdf correctly answers the question: which neutralino is most likely to be bino-like? This discussion illustrates how the combined analysis of the likelihood and Bayesian approaches can be combined to extract all the available information.

For the MSSM case a proper error analysis is vital, because at the end of the day it will determine if and how well we can extract information on the SUSY breaking mechanism. For the true best-fit parameter point, we show the local error bars in Table 13.4. Indeed, the LHC is not sensitive to several parameters. Some of them, like the light trilinear mixing terms, are fixed from the beginning. Others, like the heavier stau-mass and stop-mass parameters or the pseudoscalar Higgs mass, come out essentially unconstrained. Because the heavy Higgses are for all practical purposes decoupled at the LHC, the parameters in the Higgs sector are $\tan\beta$ and the lightest stop mass. As expected for the slepton sector, the ILC improves the precision by an order of magnitude.

It is instructive to quote some effects of theory errors on the MSSM parameter determination. While the ILC loses a factor of five in precision, going from a per mil determination to half a per cent, the LHC loses less than a factor of two. The naive expectation would have called for only the ILC measurement to be affected, but we see that for example the theory error on $m_{\ell\ell}$ dominates over the experimental LHC error.

While for the LHC and ILC separately not all parameters can be determined, the combination of the two machines allows us to determine all parameters (with the exception of the first and second generation trilinear couplings). The combination of LHC and ILC measurements can be particularly useful to determine the link to dark matter observables [71; 116; 176; 455; 456; 622; 778; 1653; 1867; 1868].

Once the parameters of the weak-scale MSSM Lagrangian have been determined, the next step is to extrapolate the parameters to the Planck or GUT scales. Structures at high scales can give hints for example about supersymmetry breaking [366]. For two reasons, the prime candidates for unification in supersymmetry are the gaugino masses: first, in contrast to the scalar masses, the three gaugino masses can well be argued to belong to the same sector of physics, being the partners of gauge bosons of a possibly unified gauge group. Second, interactions between the hidden SUSY breaking sector and the MSSM particle content can affect the unification pattern, in particular for scalars. In that case, scalar mass unification might be replaced by much less obvious sum rules for scalar masses at some high scale [564].

Technically, upwards renormalization group running is considerably more complicated [1254] than starting from a unification scale and testing the unification hypothesis by comparing to the weak-scale particle spectrum. For example, it is by no means guaranteed that the renormalization group running will converge for weak-scale input values far away from the top-down prediction. A proper bottom-up renormalization group analysis using SuSpect and SFitter and including all errors is currently on the way [871].

13.3.3 Combining measurements

As an example of how adding additional information can improve the kind of parameter extraction described above, we can use the anomalous magnetic moment of the muon. This parameter has been evaluated both in experiment and in theory to unprecedented precision and might well point to physics

Table 13.5. *Combined LHC and $(g-2)_\mu$ analysis. All masses in GeV.*

	No theory errors				Flat theory errors			
	LHC		LHC $\otimes (g-2)$		LHC		LHC $\otimes (g-2)$	
$\tan\beta$	9.8±	2.3	9.7±	2.0	10.0±	4.5	10.3±	2.0
M_1	101.5±	4.6	101.1±	3.6	102.1±	7.8	102.7±	5.9
M_2	191.7±	4.8	191.4±	3.5	193.3±	7.8	193.2±	5.8
M_3	575.7±	7.7	575.4±	7.3	577.2±	14.5	578.2±	12.1
$M_{\tilde{\mu}_L}$	192.6±	5.3	192.3±	4.5	193.2±	8.8	194.0±	6.8
$M_{\tilde{\mu}_R}$	134.0±	4.8	133.6±	3.9	135.0±	8.3	135.6±	6.3
m_A	446.1±$\mathcal{O}(10^3)$		473.9±$\mathcal{O}(10^2)$		406.3±$\mathcal{O}(10^3)$		411.1±$\mathcal{O}(10^2)$	
μ	350.9±	7.3	350.2±	6.5	350.5±	14.5	352.5±	10.8

beyond the Standard Model [1050; 1447]:

$$a_\mu^{(\text{exp})} \equiv (g-2)_\mu/2 = 116592080(63) \times 10^{-11}$$
$$a_\mu^{(\text{SM})} = 116591785(61) \times 10^{-11}, \tag{13.5}$$

where we quote e^+e^- data.

For our example we again use the SPS1a parameter point. It predicts $a_\mu^{(\text{SPS1a})} = a_\mu^{(\text{SM})} + 282 \times 10^{-11}$. This is a deviation from the experimentally observed value of -13×10^{-11}, well below the experimental error bounds, so we can safely use the experimental value to ask what such a measurement would say about the MSSM parameter space.

The anomalous magnetic moment of the muon adds important information to our LHC analysis. First, the deviation from the Standard Model prediction is proportional to the sign of μ [606; 1809]. Including $(g-2)$ data will clearly favour one sign of μ, namely $\mu > 0$, thereby reducing the degeneracy by a factor of two. Moreover, $(g-2)$ at leading order is proportional to $\tan\beta$ which should significantly improve the extraction of the notorious $\tan\beta$.

Table 13.5 shows the result of our SPS1a analysis using Eq. (13.5). For comparison we include the result without $(g-2)$ data from Table 13.4, all with experimental errors only as well as including theory errors. The effect of the additional information on the accuracy of the parameter determination is clearly visible, in particular for $\tan\beta$, which is not well determined by the LHC measurements. The leading linear dependence $(g-2)$ improves the uncertainties on $\tan\beta$ errors by more than a factor of two.

Without $(g-2)$ the best source of information on $\tan\beta$ at the LHC is the light Higgs mass [642; 879; 1047], but this observable strongly relies on the assumed minimal structure of the Higgs sector, on the knowledge of many other MSSM parameters and on the estimate of the theory errors due to higher orders. Additional sources of a $\tan\beta$ measurement are the production rate for heavy Higgs bosons [1239] and rare decays like $B_s \to \mu^+\mu^-$, but both of them work mainly for large values of $\tan\beta$. The improvement in $(g-2)$ impacts all other parameters which are usually re-rotated when $\tan\beta$ is changed. Correlations and loop corrections propagate the improvement over almost the complete parameter space.

13.4 Dark matter and the LHC

As discussed in the previous sections, we cannot constrain all parameters of the 19-dimensional MSSM with LHC measurements. However, it is still of interest to see how well the neutralino relic density $\Omega_\chi h^2$, for example, in the SPS1a parameter point can be constrained. This exercise is performed in detail in ref. [1516], again starting from Table 13.1 and adding the ratio of branching ratios

$$\mathrm{BR}(\tilde{\chi}_2^0 \to \tilde{\ell}_R\ell)/\mathrm{BR}(\tilde{\chi}_2^0 \to \tilde{\tau}_1\tau), \tag{13.6}$$

with a 10% error [683; 1485].

Since $\tan\beta$ is largely unconstrained at the LHC, this study keeps it fixed at the input value of 10 and takes the variation of the relic density prediction with $\tan\beta$ as a systematic uncertainty. We also assume $\mu > 0$ and that the heavier neutralino is indeed the $\tilde{\chi}_4^0$. Under these assumptions we can extract the neutralino mass matrix, and thence determine the LSP's gaugino and Higgsino fractions. Already at this level we can see that the Higgsino component of the $\tilde{\chi}_1^0$ is small, and its relic density will be essentially determined by the annihilation through light sfermions.

The next step is therefore to determine the stau parameters. With $\tan\beta$ fixed the ratio in Eq. (13.6) only depends on the $\tilde{\tau}_1$ and $\tilde{\chi}_2^0$ masses and on the stau mixing angle. We can determine θ_τ to an uncertainty of $\sim$35% by varying $\tan\beta$ between 3 and 30. Finally, we need to make sure that the heavy Higgs bosons are not in the region $m_A \sim 2m_{\tilde{\chi}_1^0}$, which would cause a rapid s-channel annihilation. This can probably be done by measuring or by excluding the decay of A/H into charginos and neutralinos, which provides a lower limit of roughly 300 GeV on the Higgs masses, but detailed experimental studies are needed to substantiate this claim.

Now, we can predict the relic density using micrOMEGAs [261]. For fixed values of $\tan\beta$, m_A and $m_{\tilde{\tau}_2}$, the error on $\Omega_\chi h^2$ is $\sim$20%, dominated by the conservative 5 GeV error bar of the $m_{\tau\tau}$ edge, quoted in Table 13.1. Varying these three parameters within the experimental constraints gives us an additional cumulative uncertainty of $(+2\%, -12\%)$. The study of less favourable model points shows, however, that in general additional information from a linear collider would be needed in order to discriminate different possible solutions giving very different predictions for the dark matter relic density.

13.5 Outlook

In this brief overview we have illustrated how at the LHC we can measure properties of new-physics particles. They include discrete properties such as spins as well as continuous masses. The key signatures for such measurements are cascade decays from strongly interacting states produced in proton–proton collisions to weakly interacting particles which might include a dark matter agent.

Starting from such measurements we have made a case that the LHC should be able to determine major parts of the TeV-scale Lagrangian of new physics including a proper error analysis. This reconstructed Lagrangian we can combine with cosmological observations and/or evolve to a high scale, to convincingly determine the nature of electroweak symmetry breaking and the generation of the weak scale.

We emphasize that in this weak-scale analysis no supersymmetry properties were crucial, so that the results should easily be generalized to any model with strongly as well as weakly interacting new physics that includes a dark matter candidate.

14

Supersymmetric dark matter at colliders

Marco Battaglia and Michael E. Peskin

14.1 Introduction

The study of the interactions of elementary particles at high-energy accelerators over the past 40 years has led us to a paradoxical situation. On one hand, these studies have apparently solved the problem of the nature of the strong and weak nuclear interactions. The precision data from LEP, SLC, the Tevatron and the B-factories has confirmed the leading theory of the strong interactions – QCD – to per cent accuracy and the leading theory of the electroweak interactions $SU(2) \times U(1)$ Yang–Mills theory to the accuracy of parts per mil. On the other hand, there are important phenomena in Nature that are completely outside the scope of this 'Standard Model'. Dark matter, which makes up 80% of the matter in the Universe and cannot be composed of any Standard Model particle, provides the most striking example.

Most of the information that we have now on dark matter relates to the properties we can learn from gravitational measurements. We know the overall cosmic density of dark matter, and the local density of dark matter on the scale of galaxies and clusters. Soon we can hope to have measurements of dark matter at the particle level, of the rates of dark matter annihilation in the Galaxy and of dark matter scattering in underground detectors. To learn what the dark matter particle is and how it fits into a more general theory of Nature, it will be important to interpret these measurements in terms of data obtained from particle physics experiments. At the broadest level, we would like to produce candidate dark matter particles directly in the laboratory and compare the properties of those particles to the properties of dark matter obtained from astrophysical sources.

Particle Dark Matter: Observations, Models and Searches, ed. Gianfranco Bertone. Published by Cambridge University Press. © Cambridge University Press 2010.

This programme would be interesting even if dark matter were an isolated particle produced in an elementary process. However, the most compelling theories of dark matter do not have this character. Dark matter must be stable over the lifetime of the Universe. The simplest way to assure this is to assume that dark matter particles carry a new conserved quantum number that is not shared by the particles of the Standard Model. Then there would probably be other particles that carry this quantum number; these would be unstable and decay to the dark matter particle. These heavier particles should also appear in particle physics experiments. To understand the role of dark matter in the basic model of Nature, it will be necessary to understand the structure of the whole sector of particles carrying the new quantum number.

From here on, we will assume that dark matter is composed of a neutral, stable particle with interactions having, very roughly, the intrinsic strength of the electroweak interactions. We will refer to a dark matter candidate particle of this type as a 'Weakly Interacting Massive Particle' or WIMP. These assumptions typically lead to the conclusion that dark matter was in thermal equilibrium in the early stages of the Universe. From this, as we will review in Section 14.2.1, it follows that the mass scale of dark matter particles is a few hundred GeV. This observation potentially relates the new particles associated with dark matter to the particles needed to explain another important unexplained phenomenon in particle physics, the spontaneous symmetry breaking that gives mass to the W and Z bosons and the quarks and leptons. Conversely, models proposed for weak interaction symmetry breaking often contain new conserved quantum numbers. Such quantum numbers can arise from symmetries of postulated new interactions. They can also isolate the new particles from the Standard ones, so that new particles are protected from giving large corrections to the observables that appear in precision tests of the Standard Model. By one means or the other, most of the models that have been proposed to explain weak interaction symmetry breaking also contain a neutral, stable, weakly interacting particle that could be a perfect candidate for the particle of dark matter.

There are many different classes of models that fit this description. Among these are 'Little Higgs' models that add new gauge groups, and models that add extra dimensions, such as the 'Universal Extra Dimension' and Randall–Sundrum models. The best-known model of this type is the Minimal Supersymmetric Standard Model (MSSM). The MSSM adds a new particle with opposite statistics for every Standard Model particle. A new symmetry, supersymmetry, controls the properties of the new particles according to their $SU(2) \times U(1)$ quantum numbers. The structure of these models and their relation to dark matter are described in the chapters of Part II of this book.

In this chapter, we will review experimental methods for discovering a sector of new particles associated with dark matter, for measuring the properties of new particles in that sector, and, eventually, for determining microscopic properties of the dark matter particle that can be directly compared to astrophysical dark matter measurements. Although we keep an open mind about which general class of models is realized in Nature, in this review we concentrate on the example of supersymmetry. The phenomenology of supersymmetric particles at high-energy colliders has been studied for more than 20 years. In that time, many detailed analyses have been carried out, and many subtle effects have been clarified. Thus, we regard supersymmetry, and, more specifically, the MSSM, as a template that illustrates how we can learn experimentally about the new particle sector associated with dark matter.

In fact, we will make a further restriction. Supersymmetry offers many different possibilities for the dark matter particle. This particle could be a scalar neutrino or a super-weakly interacting particle such as the gravitino. Each choice has its distinctive phenomenology. These are reviewed in Chapter 8 of this book. In this review, we will concentrate on the scenario in which the dark matter particle is the lightest neutralino, that is, the lightest fermionic partner of the gauge and Higgs bosons, and is kept stable by having a non-trivial value of a conserved parity symmetry, R-parity.

In Section 14.2, we will discuss the general issues involved in determining the properties of the WIMP that are most important for the interpretation of astrophysical measurements. We will see that there are many ambiguities and difficulties that must be overcome. These difficulties will appear in our discussion as complications of the MSSM and the determination of its parameters. We caution that analogous difficulties will arise in any other model of WIMP dark matter that is studied in the detail that is now seen in the analyses of supersymmetry. Thus, although our discussion will be cast in the language of supersymmetry and will use specific properties of the MSSM, we believe that the general conclusions apply to any model of WIMP dark matter.

In Sections 14.3 and 14.4, we will discuss how the difficulties presented in Section 14.2 can be overcome. Section 14.3 will discuss determinations of the properties of the WIMP from hadron collider experiments. Section 14.4 will present the further steps toward the measurement of the WIMP properties that can be made at future electron–positron colliders. Section 14.5 will conclude with some general remarks on the feedback from these microscopic measurements into astrophysics.

If the new particles associated with dark matter are in fact at the hundred GeV mass scale, we will discover these particles and the dark matter particle at the CERN Large Hadron Collider (LHC), which is just beginning operation. This adds interest – and urgency – to the story of dark matter experiments at high-energy colliders.

14.2 Questions for the collider experiments

14.2.1 Basic microscopic properties of dark matter particles

In this section, we review the basic objectives of a collider physics programme on dark matter, and some of the subtleties that arise in extracting the properties of the dark matter particle from collider data. These issues are discussed in more detail in Section 2 of ref. [176].

Particle physicists naturally are curious about the microscopic properties of dark matter. They would like to know the mass of the dark matter particle and the strength of its interactions, as means of pursuing the more basic question of how this particle fits together with the well-known particles of the Standard Model into a more complete picture of the laws of physics.

Astrophysicists also have more specific goals. At the moment, we know about dark matter only from its macroscopic gravitational effects. To understand the distribution of dark matter on scales of the size of galaxies and smaller, they would like to be able to detect dark matter particles individually. Experiments that detect dark matter depend crucially on the mass of the dark matter particle and on the particular cross-sections relevant to detection. These include the scattering cross-section for dark matter particles from nucleons, for direct detection, and the annihilation cross-sections of pairs of dark matter particles into specific final states, for indirect detection. This point of view motivates a laboratory programme to determine these cross-sections, just as the study of nucleosynthesis in stars and in the early Universe motivates a laboratory programme to study low-energy nuclear interactions.

In this review, we will concentrate on three specific quantities,

$$m_\chi, \qquad \sigma_{\chi\chi}, \qquad \sigma_{\chi p}, \qquad (14.1)$$

respectively, the mass of the dark matter particle, the total pair annihilation cross-section of dark matter particles, and the cross-section for dark matter-proton scattering at low energy.

The dark matter annihilation cross-section is important not only for detection experiments but also for the physics that determined the current abundance of dark matter. Under the assumption that dark matter was in thermal equilibrium with Standard Model particles in the very early Universe – an assumption valid for most but not all models of neutralino dark matter – we can integrate the Boltzmann equation to compute the density of dark matter that remains after dark matter decouples, the 'thermal relic density'. To a first approximation, this analysis gives for the current fraction of the energy of the Universe in dark matter [1701]

$$\Omega_\chi h^2 = \frac{s_0}{\rho_c/h^2} \left(\frac{45}{\pi g_*}\right)^{1/2} \frac{x_f}{M_{Pl}} \frac{1}{\langle \sigma_{\chi\chi} v \rangle}, \tag{14.2}$$

where s_0 is the current entropy density of the Universe, ρ_c is the critical density, h is the (scaled) Hubble constant, g_* is the number of relativistic degrees of freedom at the time that the dark matter particle goes out of thermal equilibrium, M_{Pl} is the Planck mass and $\langle \sigma_{\chi\chi} v \rangle$ is the thermal average of the dark matter pair annihilation cross-section times the relative velocity. The dominant temperature for the final annihilation is one at which the annihilating neutralinos are non-relativistic: $T/m_\chi \equiv 1/x_f \sim 1/25$. The one unknown on the right-hand side is $\sigma_{\chi\chi}$, so we can use this relation, or a more sophisticated version of the analysis, to predict the current dark matter density.

It is interesting to ask what value of $\sigma_{\chi\chi}$ is required to produce the value of Ω_χ that is observed in cosmic microwave background experiments. Most of the quantities in (14.2) are numbers with large exponents. However, combining them and equating the result to $\Omega_\chi = 0.2$, we obtain

$$\langle \sigma_{\chi\chi} v \rangle = 1 \text{ pb}. \tag{14.3}$$

Interpreting this in terms of a mass, using $\langle \sigma_{\chi\chi} v \rangle = \pi \alpha^2/8M^2$, we find $M \sim 100$ GeV. It is remarkable that this estimate, purely from astrophysical data, places the mass scale of dark matter at the mass scale of electroweak symmetry breaking. This is a strong argument connecting these two phenomena and thus motivating a connection between dark matter and supersymmetry.

In this chapter, we will explain, for the case in which the dark matter particle is the neutralino, how the dark matter particle mass and cross-sections can be determined from data that could actually be collected in particle physics experiments.

This programme must confront many difficulties. First of all, it requires an accelerator with sufficient energy that can reach the mass scale of supersymmetry and produce supersymmetric particles. Further, because the dark matter particle is necessarily weakly interacting, this particle will not be detected by the sorts of detectors used in high-energy collider physics. Thus, we must reconstruct the properties of the dark matter particle from other information we can gather about the events containing supersymmetric particles. For the mass of the dark matter particle, there are well-conceived solutions to this problem that we will present in Sections 14.3.3 and 14.4.2.

For the dark matter cross-sections, however, there are further difficulties. We do not, even conceptually, have any way to produce a beam of dark matter particles that could be used to measure these cross-sections directly. To determine these cross-sections from microscopic information, a very different strategy is needed. We must first understand the basic theory of the interactions of dark matter particles, then learn what specific masses and couplings make the most important contributions to these cross-sections, then devise observables that measure these masses and couplings. In this review, we will assume, as a first step, that the underlying theory is the Minimal Supersymmetric Standard Model. This still leaves many possibilities for the dominant reactions that contribute to neutralino annihilation and scattering. In the remainder of this section, we will review these possibilities and their connection to the spectrum of supersymmetric particles.

14.2.2 Mechanisms of dark matter annihilation

Because supersymmetry introduces a large number of new particles – one for every Standard Model particle – the theory contains a large number of possible mechanisms for pair annihilation of dark matter particles. In principle, any Standard Model partner can be exchanged between neutralinos, converting them to a pair of the corresponding Standard Model states.

Four of these mechanisms are illustrated in Fig. 14.1. The exchange of a slepton, Fig. 14.1(a), allows neutralino annihilation to a lepton pair. A similar diagram would allow neutralino annihilation to a quark–antiquark pair. An alternative mechanism of annihilation to fermion–antifermion states is through an s-channel Higgs boson, Fig. 14.1(b). The exchange of a chargino, Fig. 14.1(c), allows neutralino annihilation to a pair of W or Z bosons, or to a Z boson and a Higgs boson. It is also possible that, at the temperature at which the dark matter density is established, other supersymmetric particles are present in equilibrium with the neutralino, and these contribute to the annihilation of supersymmetric particles. Fig. 14.1(d) illustrates the

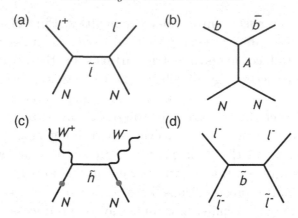

Fig. 14.1. Four mechanisms for the annihilation of neutralino dark matter: (a) annihilation by slepton exchange, (b) annihilation through a Higgs boson resonance, (c) annihilation by chargino exchange, (d) co-annihilation by sleptons.

annihilation of a pair of sleptons to a pair of leptons. Such 'co-annihilation' reactions have been studied also for charginos and for top squarks.

The possible behaviours of supersymmetry models are often studied in the 'constrained MSSM' or 'mSUGRA' parameter space. This is a restricted four-parameter subspace of the MSSM, making the assumption that soft-supersymmetry breaking parameters are unified in the simplest way at the grand unification scale. Within this space, sleptons are lighter than squarks, neutralinos and charginos are much lighter than gluinos, and the lightest neutralinos and charginos are dominantly gauge boson rather than Higgs boson partners. In this context, the mechanisms illustrated in Fig. 14.1 appear in specific parameter regions called, respectively, the 'bulk region', the 'funnel region', the 'focus point region' and the 'co-annihilation region'. The systematics of these regions and the logic of the names are explained in Chapter 8 of this book.

When we extract information about dark matter from experiment, we must take into account that the full parameter space of supersymmetry is much larger than the mSUGRA space and opens more possibilities. Even in the context of simple unification models, it is possible to have spectra in which the mass relations of sfermions and gauginos can be very different from that in mSUGRA. The lightest neutralino can have substantial Higgsino or $SU(2)$ gaugino content. Particular squarks, especially those of the third generation, can be very light. Even with the assumption that the supersymmetry breaking terms conserve flavour and CP, the MSSM has 24 parameters.

The full space includes large regions where each of the mechanisms described above is realized.

Much of the uncertainty about the dominant mechanism of neutralino annihilation will be resolved when the mass spectrum of supersymmetric particles is known. Typically, the dominant contribution to the annihilation cross-section comes from exchange of the lightest particles. But the annihilation cross-sections also depend crucially on the mixing angles associated with the mass eigenstates of supersymmetric particles. We have already pointed out that the presence of a neutralino annihilation to W and Z bosons depends on the admixture of Higgsino and SU(2) gaugino states into the lightest neutralino. Similarly, the cross-sections for tau slepton or top squark annihilation depend strongly on the admixture of the partners of the left- and right-handed fermions that composes the lighter mass eigenstate. In order for collider data to provide a value of the annihilation cross-section useful for astrophysics, the full nature of the exchanged particles should be resolved by experimental measurements.

The neutralino annihilation cross-section also enters the analysis of dark matter indirect detection. Here the goal is simpler. A signal of dark matter annihilation in energetic particles is proportional to $\sigma_{\chi\chi}\rho_\chi^2$, so we need a value of $\sigma_{\chi\chi}$ to interpret annihilation signals in terms of a local density ρ_χ of dark matter. It might seem that we can use (14.3) as a reasonable approximation for this purpose.[1]

However, this again depends on the particular scenario. Neutralino annihilation to fermion pairs proceeds dominantly in the P-wave. For indirect detection, we need the value of this cross-section very close to threshold, and there it is suppressed by two or more orders of magnitude relative to the value at the temperature that determines the relic density. For the same reason, even if the dominant reaction in the early Universe was annihilation to sleptons, the dominant reaction today would be annihilation to $b\bar{b}$. In co-annihilation scenarios, the co-annihilating particle is no longer present, and so the current annihilation cross-section is determined by a subdominant reaction and would be much smaller than (14.3).

On the other hand, annihilation to vector boson pairs and annihilation through Higgs boson resonances have large S-wave components (in the latter case, by annihilation through the CP-odd boson A^0). For these mechanisms, the estimation by (14.3) is correct to better than a factor of two. Even if particle physics experiments could only clarify qualitatively that an S-channel mechanism dominates, they would justify using (14.3) together

---

[1] In this context, Eq. (14.3) is usually written: $\langle \sigma_{\chi\chi} v \rangle = 3 \times 10^{-26} \, \mathrm{cm}^3 \, \mathrm{s}^{-1}$.

with astrophysical observations to make quantitative measurements of the density of dark matter responsible for an annihilation signal.

14.2.3 Mechanisms of dark matter elastic scattering

Many of the same ambiguities that arise in the determination of the annihilation cross-section arise in the determination of the direct detection cross-section $\sigma_{\chi p}$. The direct detection cross-section has two components, a spin-dependent amplitude due to s-channel squark exchange and t-channel Z^0 exchange, and a spin-independent amplitude due to s-channel squark exchange and t-channel Higgs boson exchange. For a large nucleus, the spin-independent cross-section adds coherently over the nucleons, giving a cross-section proportional to A^2. Therefore, it is this contribution that is probed most sensitively by experiments. Because of the strong experimental lower bounds on the masses of squarks, this spin-independent amplitude is likely to be dominated by Higgs exchange. At large values of $\tan \beta$, the heavy Higgs boson H^0 makes the largest contribution.

Thus it is relevant that the coupling of the lightest neutralino to the H^0 boson depends crucially on the neutralino mixing angles. This coupling arises from the supersymmetric gauge coupling of the Higgs field. It vanishes when the lightest neutralino is a pure gaugino, and also when this particle is a pure Higgsino. At the other vertex, the coupling of the H^0 boson to quarks depends on the value of $\tan \beta$. To predict the spin-independent direct detection cross-section, collider experiments must measure these parameters and angles – and, of course, they must measure the mass of the H^0.

There is one more serious problem that blocks the route to using collider data to interpret a direct detection observation in terms of a flux of dark matter particles. The Higgs boson coupling to the nucleon has a large uncertainty from low-energy QCD, owing to our poor knowledge of the nucleon's strange quark content. This gives an uncertainty of about a factor of 4 in any calculation of the direct detection cross-section [396]. This issue is reviewed in [176]. Recently, there have been new results on the strangeness content of the proton from lattice gauge theory [1524]. The preliminary results indicate a small value of the strange quark content; hopefully, further work will determine this parameter accurately.

14.2.4 Decoupling of heavy SUSY particles

Depending on model parameters, the spectrum of supersymmetric particles can extend up to masses of 1 TeV and above. It is unlikely that the masses and couplings of such heavy particles could be measured with the same accuracy as for particles of lower masses, and some of these particles may not be visible at all at the next generation of colliders. So we would like to emphasize that, despite this, accurate predictions of the relic density can be made once the lightest states are observed and accurately measured. In t-channel exchange, the contributions of particles with the heavy mass M are suppressed by the factor $1/M^4$. As long as the light particles contributing to neutralino annihilation can be observed and characterized, the heavy particles decouple and give negligible corrections. The study of dark matter thus gives special importance to precision measurements on the lightest states of the new particle spectrum. Ultimately, it is this precision and specificity, rather than the attainment of the highest possible energies, that determines what influence the next generation of colliders will have on our knowledge of dark matter.

14.3 Dark matter at hadron colliders

14.3.1 Landscape of future hadron colliders

The highest-energy hadron collider now operating is the Fermilab Tevatron. This is a $p\bar{p}$ collider operating at a centre of mass energy of 1.96 TeV with a luminosity that has exceeded $3 \times 10^{32}\,\mathrm{cm}^{-2}\,\mathrm{s}^{-1}$. The Large Hadron Collider, now beginning operation at CERN, will provide pp collisions at energies up to 14 TeV, with a luminosity that is projected to reach $10^{33}\,\mathrm{cm}^{-2}\,\mathrm{s}^{-1}$ in the initial stage.

The Tevatron collider now puts the strongest lower limits on the masses of squarks and gluinos, and puts significant constraints on the masses of the colour-singlet states of supersymmetry. Recent searches from the CDF and DØ experiments exclude squarks and gluinos with masses less than about 300 GeV [3; 10] and charginos with masses less than about 120 GeV [5; 12]. Specific searches for top squarks give limits of about 160 GeV [4; 11]. We caution that the experiments search for specific signatures whose strength depends on many details of the supersymmetry spectrum, and that limits are often quoted only within the mSUGRA parameter space or a subset thereof. Thus, these values should be considered representative rather than strict bounds.

The CDF and DØ experiments also put limits on the presence of the heavy Higgs bosons of supersymmetry, relying on the enhancement of the production cross-sections for these particles at large values of $\tan\beta$. As we have discussed in Section 14.2.3, these Higgs bosons can give an important contribution to the neutralino direct detection cross-section. Significant limits, for example $m_{H,A} > 110$ GeV for $\tan\beta = 60$, have been placed in the analyses [9; 18].

The LHC will offer a major step in energy beyond the Tevatron. Whereas one could reasonably argue that a supersymmetry sector responsible for dark matter could be too heavy to be observed at the Tevatron, the LHC will be sensitive to all but a tiny fraction of the possible parameter space. At the LHC, we thus expect to observe the supersymmetric particles and to study them in some detail. We will survey that programme in the remainder of this section. Another perspective on many of these issues can be found in Chapter 13.

In the further future, plans for hadron colliders aim at increasing both the luminosity and the energy of proton collisions. The design luminosity of the LHC, to be reached in a few years of operation, is $10^{34}\,\mathrm{cm}^{-2}\,\mathrm{s}^{-1}$. Plans are already under way, in the SLHC project, for luminosity increases of an additional order of magnitude. The higher luminosity potentially increases the mass reach by 20% by allowing greater access to the most energetic partons. However, this higher sensitivity comes with a more challenging experimental environment, with about 200 proton–proton collisions per bunch crossing. Plans have been put forward for a very large storage ring (VLHC) that would allow collisions up to 100 TeV. Unfortunately, the luminosity is limited to about $10^{34}\,\mathrm{cm}^{-2}\,\mathrm{s}^{-1}$ because of synchrotron radiation [1697].

14.3.2 Discovery of the WIMP

We have emphasized that, in models of WIMP dark matter associated with new physics at the TeV scale, the dark matter particle is only one element in a complete spectrum of new particles. It is highly motivated that some of these new particles have QCD quantum numbers. In particular, if the new particles also address the problem of symmetry breaking in the electroweak interactions, the new particles should contain one or more partners of the top quark that are colour triplets under QCD.

The presence of these particles changes completely the strategy for discovering dark matter at a hadron collider. The cross-sections for the direct production of pairs of dark matter particles are small, typically of the order of 100 fb at the LHC. The dark matter particles are invisible, so that one must also require observable initial state radiation to detect the events. This is a very difficult signature to detect. On the other hand, the pair-production of particles with QCD colour has a large cross-section, typically of the order of 10 pb, both because this is a strong-interaction process and because the reaction gets a contribution from the gluon–gluon initial state. The heavy

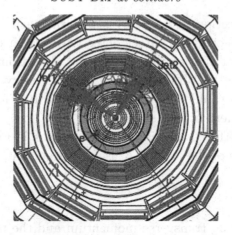

Fig. 14.2. Simulation of an event with supersymmetric particle production in the CMS detector at the LHC. Figure taken from http://cms.web.cern.ch/cms/ Resources/Website/Physics/SUSY.

coloured particles then decay down through the new particle spectrum, emitting quarks and leptons with large transverse momenta. The dark matter particle is the lightest particle in the spectrum that carries the conserved discrete quantum number. The complete event then contains a number of quark jets, possibly also hard leptons, and, finally, unbalanced ('missing') transverse momentum carried away by the unobserved dark matter particles. This is a striking signature that differs in many respects from QCD events at a hadron collider. A typical event of this type, as reconstructed in the CMS detector, is shown in Fig. 14.2.

The visibility of these events at the LHC has been studied in the ATLAS and CMS Technical Design Reports [143; 237] and is summarized in more current reviews [1860; 1952]. The most important backgrounds turn out to be, not pure QCD reactions, but, rather, production of heavy Standard Model particles – W, Z and t – in association with a number of jets from QCD radiation. Techniques for estimating these backgrounds are reviewed in [1383]. The search reach of the LHC depends on the masses of the primary particles produced. In any case, the power of the LHC is quite impressive, with a predicted sensitivity to coloured states of about 3 TeV with 100 fb^{-1} of data and a sensitivity to states up to 1 TeV in the first 1 fb^{-1}.

14.3.3 Estimation of the WIMP mass

Once we observe events with multiple jets or leptons at the LHC, the first task will be to estimate the masses of the new particles, and, in particular, the mass of the WIMP. The WIMP mass can be directly compared to mass estimates from detection of astrophysical dark matter, for example from the recoil energy spectrum in direct detection or from the endpoint of the gamma ray or positron spectrum in indirect detection. Those techniques are expected to give the WIMP mass with an accuracy of about 20%, so if collider experiments could achieve an accuracy of 10% for the WIMP mass, that would be sufficient at this stage.

The first method for determining these masses is to look at the gross kinematical features of the events. A useful variable is

$$m_{\text{eff}} = \not{p}_{\text{T}} + \sum_{1,\ldots,4} p_{\text{T}i}, \tag{14.4}$$

the sum of the missing transverse momentum and the transverse momenta of the four highest-p_{T} jets. In simulation, the distribution of m_{eff} shows a sharp peak whose position is correlated with the mass difference between the mass of the primary coloured particle $M_{\tilde{p}}$ and the WIMP [1097]:

$$m_{\text{eff}} \approx 2(M_{\tilde{p}} - m_{\chi}) \tag{14.5}$$

This leaves a freedom to adjust the overall mass scale. That freedom can be fixed to some extent by fitting the shape of the missing p_{T} spectrum. By combining these two observables, we can constrain the mass of the WIMP with an uncertainty of about 30% [112].

To achieve higher accuracy, we must do more detailed kinematic fitting. The correct strategy depends on the form of the new particle spectrum. In supersymmetric models in which the primary coloured particle is a squark, one often finds the decay chain

$$\tilde{q} \to q + \tilde{\chi}_2^0 \to \ell^+ \ell^- \tilde{\chi}_1^0, \tag{14.6}$$

where the final particle is the WIMP. The lepton pair can be produced by an off-shell Z boson or slepton or from a two-stage decay to an intermediate on-shell slepton

$$\tilde{\chi}_2^0 \to \ell^+ \tilde{\ell}^- \to \ell^- \tilde{\chi}_1^0. \tag{14.7}$$

The two cases can be distinguished by the shape of the endpoint in the $\ell^+ \ell^-$ mass distribution [1568].

In the off-shell case, the spectrum of $\ell^+ \ell^-$ pairs fills out the mass range between the $\tilde{\chi}_2^0$ and the $\tilde{\chi}_1^0$, so that the endpoint of the spectrum directly gives the $\tilde{\chi}_2^0 - \tilde{\chi}_1^0$ mass difference. At the endpoint, the lepton pair is absorbing the maximum amount of energy from the decay, so the χ_1^0 must be at rest in the frame of the lepton pair. Then the momentum vector of the χ_1^0 is determined, and it is possible to add quark jets and search for the squarks as a bump in the invariant mass distribution. There are three unknown masses in the problem, the masses of $\tilde{q}$, $\tilde{\chi}_2^0$ and $\tilde{\chi}_1^0$, constrained by the dilepton endpoint and the squark resonance position. With one additional piece of information, for example the endpoint in $q\ell$ or $q\ell^+\ell^-$, the system is fully constrained and one can solve for the masses. An example of such an analysis is given in [1097].

In the on-shell case, the spectrum of $\ell^+\ell^-$ pairs has a sharp endpoint at the value

$$m(\tilde{\chi}_2)\sqrt{1 - \frac{m^2(\tilde{\ell})}{m^2(\tilde{\chi}_2)}}\sqrt{1 - \frac{m^2(\tilde{\chi}_1)}{m^2(\tilde{\ell})}}. \qquad (14.8)$$

The $q\ell$ and $q\ell^+\ell^-$ distributions have endpoints given by similar formulas. Now there are four unknown masses, but there are a number of visible kinematic endpoints, so again it is possible to solve for the mass values. An example of an analysis of this type is given in [62].

Another very useful variable for mass determinations at hadron colliders is the 'stransverse mass' variable $m_{\mathrm{T}2}$ of Lester and Summers [1327]. The idea of this variable is to partition the missing transverse momentum into two pieces, assigning one piece to each of the unobserved WIMPs. The definition of $m_{\mathrm{T}2}$ is

$$m_{\mathrm{T}2} = \min_{\not{p}_{\mathrm{T}1}+\not{p}_{\mathrm{T}2}=\not{p}_{\mathrm{T}}} \{\max(m_{\mathrm{T}}(p_1, \not{p}_{\mathrm{T}1}), m_{\mathrm{T}}(p_2, \not{p}_{\mathrm{T}2}))\}, \qquad (14.9)$$

where $m_{\mathrm{T}}(p, \not{p}_{\mathrm{T}})$ is the transverse mass of an intermediate particle computed from the observed momentum p and the missing transverse momentum $\not{p}_{\mathrm{T}}$. This quantity also depends on the mass of the missing particle $m(\tilde{\chi}_1)$. It can be shown that $m_{\mathrm{T}2}$ provides a lower bound to the mass of the intermediate particle; thus, the kinematic endpoint in $m_{\mathrm{T}2}$ is an estimator of this mass. The kinematic endpoint in $m_{\mathrm{T}2}$ can be used as an additional ingredient in the fitting procedures just discussed. For an example, see [1243]. Some more sophisticated fitting methods, using more of the information in the event, are given in [534; 1267; 1517].

The general result of these fits, for a WIMP of about 100 GeV and a parent squark of about 500 GeV, is to determine the parent squark mass to a few per cent and the WIMP mass to about 10% with roughly 10 pb^{-1} of LHC data. In specific examples, with all spectral features working in one's favour, it is possible to reach to part-per-mil level [218] in 'asymptotic' LHC analyses with very large data samples.

14.3.4 Identification of the new physics model

Once the masses of the WIMP and other particles of the new spectrum have been measured, the next question is that of determining the model into which these particles fit. A major part of this issue is determining the spin and statistics of the WIMP. In supersymmetry, the WIMP is a spin-$\frac{1}{2}$ fermion, and the partners of the quarks that might provide the primary production of new particles are scalars. In most of the alternative models mentioned in Section 14.1, the partners of the quarks have spin-$\frac{1}{2}$ and the WIMP is a boson, either spin 0 or spin 1. Can we tell the difference from measurements at a hadron collider?

The most obvious difference between the properties of particles with different spin is that the pair-production cross-section is higher for higher spin, about a factor 5 higher, for example, for spin-$\frac{1}{2}$ as opposed to spin-0 particles of the same mass. However, this cross-section depends strongly on the mass, roughly as m^{-6}. Thus, the primary new particle mass must be well determined from kinematics to make this discrimination. Other factors might also make the discrimination more ambiguous. In supersymmetry, it often happens that all squarks have the same mass to within a few GeV. Then the multiplicity of squarks being produced with a given threshold affects the cross-section measurements. At hadron colliders, we observe specific signatures of new particle production, not total cross-sections. Then the branching ratio to the observable decay modes must be included. These depend on the details of the new particle spectrum.

The possible confusion between a spin-0 primary with a lower mass and a spin-$\frac{1}{2}$ primary with a higher mass can be resolved by looking at other features of the kinematic distributions. In a very illuminating paper, Meade and Reece contrasted simple models with spin-0 and spin-$\frac{1}{2}$ partners of the top quark, choosing values of the new particle masses that produced the same cross-section [1415]. The scalar particle production is typically more highly excited away from threshold, resulting in a broader distribution of the decay products in rapidity than in the spin-$\frac{1}{2}$ case. Similarly, a recent study by Hubisz, Lykken, Pierini and Spiropulu found that the scalar case can be distinguished by a broader tail in the p_T distribution of the decay products [1140]. A recent discussion of kinematic spin determination that brings in additional observables can be found in [1204].

In principle, the decays of particles with different spin should have characteristic angular distributions that identify the spin in a specific way. However, because it is difficult to completely reconstruct events with two missing particles, it is difficult to find the rest frame of a decaying particle to identify these characteristic distributions. The problem is not only in finding enough constraints to solve for the momentum of the decaying particle but also in overcoming the combinatoric ambiguity in assigning jets to different stages of a sequential decay. Some specific examples of spin discrimination on the basis of decay angular distributions have been presented in [74; 215; 991; 1771].

14.3.5 Estimation of the WIMP relic density

Once we have measured the most important masses in the new particle spectrum and understood the scenario, we could attempt to estimate properties of the WIMP relevant to astrophysical observations. The quantity of most interest here is the WIMP relic density $\Omega_\chi h^2$. If we could predict this density from microscopic information and obtain a result that agrees with observations from the cosmic microwave background, that would be a remarkable achievement toward an understanding of dark matter.

As we have discussed in Section 14.2.2, the task of predicting the WIMP relic density from collider measurements is much more difficult than that of determining the WIMP mass. The relic density is connected to the WIMP pair-annihilation cross-section. This depends on the mass of the WIMP and the lighter new particles that decay to the WIMP, and on the spins and quantum numbers of the new particles. It also depends on the details of the WIMP coupling to these particles. In addition, many different mechanisms can dominate the annihilation process in different regions, so it is necessary to determine the dominant process in the region actually chosen in Nature.

We do not know a general method for determining $\Omega_\chi h^2$ from hadron collider data. However, there are two cases in which a plausible argument has been made that this can be done. We will now review them.

In [1516], Nojiri, Polesello and Tovey (NPT) studied the example of the MSSM benchmark point SPS1a'. This is a point close to the bulk region in which the WIMP annihilation is dominated by annihilation to lepton pairs through slepton exchange. An earlier point SPS1a, whose collider physics was studied in detail in [1923], predicted $\Omega_\chi h^2 \sim 0.2$. The modified point is obtained by pushing the lighter tau slepton to slightly lower mass, allowing some co-annihilations that enhance the annihilation rate for supersymmetry. At SPS1a', the mass of the $\widetilde{\chi}_1^0$ is 97 GeV and $\tan\beta = 10$.

To predict the WIMP relic density in this model, it is necessary to have precise values for the neutralino and slepton masses. This model has on-shell decays to sleptons, and so masses can be extracted from the locations of kinematic endpoints, as described in Section 14.3.3. Three of the four neutralinos can be observed in decay chains involving the sleptons. Mass differences can be determined somewhat more accurately than absolute masses. Because co-annihilation physics is involved, the relic density is especially sensitive to the $\tilde{\tau}_1 - \tilde{\chi}_1^0$ mass difference. NPT estimate the error on this difference to be 2.5 GeV.

The WIMP relic density also depends on the exact quantum numbers of the lightest neutralino and the lightest stau. The neutralino mixing angles are determined from an analysis of the mass matrix. Even in the general MSSM, this matrix depends only on the four parameters m_1, m_2, μ and $\tan \beta$. The Higgs sector of this model is not expected to be distinguishable from that of the Standard Model at the LHC. Within the MSSM, that information implies the limits $3 < \tan \beta < 15$. For fixed $\tan \beta$, one can solve for the remaining parameters from the three neutralino masses. Then the mixing angles in the neutralino system are constrained, subject to the uncertainty in $\tan \beta$. The stau mixing angles are determined from the measurement of the ratio of branching ratios $BR(\tilde{\chi}_2^0 \to e\tilde{e})/BR(\tilde{\chi}_2^0 \to \tau\tilde{\tau})$.

Combining these sources of error, NPT find an expected uncertainty in the prediction of $\Omega_\chi h^2$ of 19%. This uncertainty improves to 10% with an improvement in the measurement of the kinematic endpoint in $\tau^+\tau^-$. A technique to improve this measurement has been put forward by Arnowitt, Dutta, Kamon, Toback and collaborators [124; 125]. These authors note that not only is the endpoint of the $\tau^+\tau^-$ mass distribution given by (14.8), but also the shape of this distribution is a linear function of $m(\tau^+\tau^-)$. Some energy from each τ is carried away by neutrinos, but, if we observe the τs in their hadronic decays, the distribution of visible energy fraction has a well-understood shape. Folding this shape with the linear distribution of masses, we see that the endpoint position is directly correlated with the position of the peak in the visible energy distribution from $\tau^+\tau^-$, a quantity that is much easier to measure.

Arnowitt *et al.* developed this method to study a point in the MSSM with heavier neutralinos in which the dominant annihilation process is co-annihilation [125; 126]. At the point they have chosen, the mass of the $\tilde{\chi}_1^0$ is 141 GeV and $\tan \beta = 40$. It is even more important in this case to measure the stau mass accurately, since the annihilation cross-section depends strongly on the stau–neutralino mass difference. Their full analysis, though, requires the use of special features of the supersymmetry spectrum in mSUGRA. As a first step, they make use of the assumption that the lightest neutralinos are mainly gaugino and that the three gaugino masses obey the grand unification relation. This fixes the masses of the $\tilde{\chi}_1^0$ and $\tilde{\chi}_2^0$ in terms of the observed mass of the gluino. Using this information, the stau–neutralino mass difference of 10 GeV can be determined to an accuracy of 1 GeV.

The masses of the stau and b squark are used to determine the remaining mSUGRA parameters $\tan\beta$ and A_0. That analysis requires the full set of assumptions of mSUGRA, including the assumption that the Higgs field and sfermion masses are unified. Within the mSUGRA subspace, though, their analysis gives a prediction of $\Omega_\chi h^2$ to 5% accuracy with 50 fb^{-1} of LHC data.

14.3.6 Problem of A^0

Any attempt to predict the WIMP relic density in supersymmetry must take account of the Higgs sector. As we have explained in Section 14.2.2, one possible mechanism for neutralino pair annihilation is resonant annihilation through one of the Higgs bosons to the $b\bar{b}$ final state. The dominant contribution comes from the CP-odd boson A^0, which allows neutralino annihilation in the S-wave. A simple estimate, assuming that the neutralino is exactly on resonance at rest, $m(A) = 2m(\widetilde{\chi}_1^0)$, gives an annihilation cross-section of about 100 pb. The desired cross-section of 1 pb is obtained when the mass of the A^0 is 10–20 GeV away from the neutralino threshold.

This introduces an element of randomness into the predictions for the neutralino relic density from the LHC. The A^0 can potentially be observed at the LHC in its decay to $\tau^+\tau^-$, but only if $\tan\beta$ is sufficiently large. If $\tan\beta$ is small, it is possible that the A^0 boson could be unseen but still have the correct mass to dominate the annihilation rate. If $\tan\beta$ is large, the A^0 could be seen, but the contribution from the A^0 could still be uncertain if, as is usually the case, $\tan\beta$ and the neutralino mixing angles are not well determined by the data.

It is thus an essential element of the study of dark matter at colliders to discover and measure the properties of the heavy Higgs bosons. For the relic density, the measurement of the properties of these bosons removes a major source of uncertainty in the prediction. For the direct detection cross-section, these measurements will be even more important, since the dominant contribution to the direct detection cross-section often comes from Higgs boson exchange. We will see both of these influences in the analyses presented in the next section.

14.4 Dark matter at lepton colliders

14.4.1 Landscape of future lepton colliders

The proton is not an elementary particle. Many of the complicating features of events in hadron–hadron collisions come from the fact that they use the composite proton to produce processes initiated by quarks and gluons. In contrast, electrons and positrons are elementary, and their interactions can be studied directly at lepton–lepton colliders. As a consequence of this, lepton colliders offer many advantages in studying the properties of new particles. These advantages could well be crucial to gather the information needed to determine unambiguously the properties of the dark matter particle.

There are three special features of lepton collisions that will play an important role in the discussion to follow. First, in lepton–lepton annihilation, all of the energy of the initial state is transferred to the final state particles. This simplifies the kinematics, and that is particularly important when we study invisible particles. Second, all annihilation cross-sections, standard and non-standard, have roughly the same strength. Thus, new particle production is not obscured by strong-interaction backgrounds, and all major decay channels of new particles become visible. Finally, it is possible to tune the centre of mass energy of the annihilation and the polarizations of the annihilating particles. This can be used powerfully to clarify which particles of the new spectrum are being produced and what the quantum numbers of those particles are.

At the moment, the strongest lower limits on supersymmetric particles from lepton colliders come from the LEP collider at CERN, which ran at a maximum energy of 208 GeV. Experiments at this collider restrict electrically charged supersymmetric particles to have masses above 100 GeV [1031]. and also restrict the Higgs boson to have a mass above 114 GeV [1187]. There are particular supersymmetry spectra to which these limits do not apply, but, unlike the situation for hadron colliders, these are special situations tailored to evade robust searches.

The energy attained at LEP was much lower than that of the Tevatron, and this highlights a difficulty for lepton colliders. Because the mass of the electron is much lower than the mass of the proton, energy loss by synchrotron radiation is a problem even for low-energy colliders. The energy loss from this source scales as γ^4 and eventually becomes prohibitive. In the tunnel of radius 4.3 km used for both LEP and LHC, a 250 GeV electron beam would lose 80 GeV per turn, while the LHC proton beam of 7 TeV loses only 4.4 keV per turn.

One solution proposed for this problem is the e^+e^- linear collider, a facility with two linear accelerators arranged to collide small, intense bunches of electrons and positrons. The concept was advanced in a seminal 1965 paper by Tigner [1853] and proved in the operation of the Stanford Linear Collider at SLAC [1574]. In the past decade, advances in the technology of linear accelerators have led to the design of the International Linear Collider [438]. The ILC, based on the use of superconducting RF cavities, is designed for operation at centre of mass energies from 250 GeV to 500 GeV, and to be upgradable to about 1 TeV. The studies that we will discuss in this section were carried out as prospective studies of the ILC capabilities.

For the further future, a number of different strategies have been proposed to carry out lepton–lepton collisions at higher energies. The concept of a circular storage ring can be used at TeV energies if, instead of electrons, we accelerate muons. The current status of the $\mu^+\mu^-$ collider is discussed in [98]. Current research focuses on the problem of collecting the muons needed to achieve high luminosity and cooling them to the small phase space required for injection into a synchrotron. The fact that muons decay in the collider also introduces severe and troublesome backgrounds. To pursue the path of linear e^+e^- colliders, new acceleration methods are required that can achieve higher accelerating gradients than the $30\,\mathrm{MeV\,m^{-1}}$ expected for the ILC. The CLIC technology, which is based on the transfer of energy from an intense, low-energy drive beam to the high-energy, low-current main beam, has demonstrated gradients up to $150\,\mathrm{MeV\,m^{-1}}$ [232]. A linear collider based on the CLIC technology aims at collisions from 1 TeV up to 3 TeV with high luminosity. In a more distant future, laser wakefield accelerators, which have been demonstrated to produce accelerating fields of $50\,\mathrm{GeV\,m^{-1}}$ [921; 1849], may open a path towards ultra-high-energy e^+e^- collisions.

Although in this review we concentrate on dark matter candidates of mass about 100 GeV, we should note that data from lower-energy e^+e^- colliders place significant constraints on lighter candidate dark matter particles. The asymmetric B-factories at 10 GeV in the centre of mass – KEKB at KEK and PEP-II at SLAC – have generated huge data sets that allow very sensitive searches. In particular, one can search for light dark matter candidates in the process

$$e^+e^- \to \Upsilon(n\mathrm{S}) \to \pi^+\pi^-\Upsilon(1\mathrm{S}), \quad \Upsilon(1\mathrm{S}) \to \text{invisible}. \qquad (14.10)$$

The first dedicated analysis has been published by the Belle collaboration, based on 2.9 fb^{-1} of luminosity, corresponding to $1.1 \times 10^7\ \Upsilon(3\mathrm{S})$ decays. There is no observed excess of events over the background expected from $\Upsilon(1\mathrm{S}) \to \ell^+\ell^-$, $\ell = e$ or μ, with the two leptons escaping detection. This sets an upper limit $\mathrm{BR}(\Upsilon(1\mathrm{S}) \to \text{invisible}) < 2.5 \times 10^{-3}$ [1831]. At the end of PEP-II operation in 2008, the Babar experiment collected 30 fb^{-1} at the $\Upsilon(3\mathrm{S})$ and 14 fb^{-1} at the $\Upsilon(2\mathrm{S})$ resonances. The analysis of these data should soon provide us with significantly stronger bounds on scenarios with dark matter candidates with mass of order 1 GeV.

14.4.2 Estimation of the WIMP mass

As at hadron colliders, the WIMP is not observable in lepton collider experiments, so its mass must be inferred from measurements of the other particles produced in supersymmetry events. At an e^+e^- collider, however, the kinematics that determine the WIMP mass can be very simple.

At an e^+e^- collider, we can tune to a centre of mass (CM) energy at which the dominant supersymmetry process is the pair production of the lightest charged supersymmetric particle. Often, this particle has a simple two-body decay. The process of slepton pair production and decay, with the two-body decay $\tilde{\ell} \to \tilde{\chi}_1^0 \ell$, involves only two unknown masses, m_χ and $m_{\tilde{\ell}}$. The energy distribution of the final leptons has sharp kinematic endpoints at

$$E_\ell = \gamma(1 \pm \beta)\frac{m_{\tilde{\ell}}^2 - m_\chi^2}{2m_{\tilde{\ell}}}, \qquad (14.11)$$

where $\gamma = E_{\rm CM}/m_{\tilde{\ell}}$ and β is the corresponding velocity. From the measured endpoint positions, one can solve for the unknown masses m_χ and $m_{\tilde{\ell}}$. A similar method can be used with the three-body chargino and neutralino decays $\tilde{\chi}_1^{+,0} \to \tilde{\chi}_1^0 q\bar{q},\ \tilde{\chi}_1^0 \ell^+\ell^-$, using the measured mass of the $q\bar{q}$ or $\ell^+\ell^-$ system to solve for two unknown masses.

Still higher accuracy is possible using a method proposed by Feng and Finnell [828]. Because we know the total energy in the final state and the energy of the observable decay products, we know the total missing 4-momentum. This can be divided into the momentum vectors of the two WIMPs in a two-parameter family of ways. This division is similar to that in (14.9) and, indeed, was a precursor of that method. Among the possible ways to divide the missing energy vector, one gives the minimum possible value of the primary particle mass. Feng and Finnell observed that the values of this minimum mass, which must be less than the true particle mass, accumulate at this maximum value.

In both methods, the measurement requires a precise determination of the energies of the detected particles and of the initial beam energy. The endpoint energies are measured to a particularly high accuracy in two-body slepton decays; for the ILC, the anticipated detector momentum resolution, $\delta p/p^2 \simeq 10^{-5}$, is an order of magnitude better than that of the LEP and LHC detectors. The main resolution effect comes from beamstrahlung, the radiation of incoming electrons and positrons in the intense electric field of the opposite beam. It is planned that this effect will be measured directly using specific beam diagnostics.

An e^+e^- collider has another, even more precise, technique for measuring the masses of new particles. Since we can tune the centre of mass energy, we can scan for the threshold of the $e^+e^- \to X\bar{X}$ pair production process. The particle mass, and also its width, can be extracted from a fit to the signal event yield as a function of $\sqrt{s}$ [835]. It is expected that the experiments will determine the mass of the supersymmetric particle X using a threshold scan in a short dedicated run at $\sqrt{s} \sim 2M_X$ and then determine the mass relation between the lightest neutralino and X by running at a higher centre of mass energy, where the production cross-section is larger and other aspects of the physics program can be done in parallel. In most scenarios, the mass of the WIMP can be determined to an accuracy of about 1 part per mil.

14.4.3 Identification of the new physics model

Even with a precise knowledge of the supersymmetry spectrum, we cannot determine the important cross-sections of the dark matter particle without a qualitative identification of the new physics model. In Section 14.3.4, we explained how this could be done in hadron collider experiments. Although much information will be available, the methods presented there did not give unambiguous information. In contrast, lepton colliders provide precise identification of the quantum numbers of new particles. The pair production of new particles in lepton annihilation has a specific dependence on the polar angle of production, with

$$\frac{d\sigma}{d\cos\theta} \sim (1 - \cos^2\theta) \tag{14.12}$$

for production of scalars and (simplifying for centre-of-mass energies far above threshold)

$$\frac{d\sigma}{d\cos\theta} \sim A(1 + \cos\theta)^2 + B(1 - \cos\theta)^2 \tag{14.13}$$

for production of spin-$\frac{1}{2}$ fermions. The total cross-sections and the shape coefficients A and B are precisely predicted in terms of the Standard Model quantum numbers of the pair-produced particles, with a characteristic dependence on beam polarization. Because the kinematics of pair-production at lepton colliders is very simple, these observables can be measured to parts-per-mil accuracy.

For particles with mixing of weak-interaction eigenstates, these measurements can precisely determine the mixing angles. We have emphasized in Section 14.2 that the values of the WIMP annihilation and scattering cross-sections can depend strongly on the mixing angles. Figure 14.3 shows an

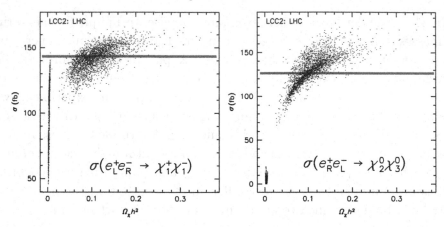

Fig. 14.3. Supersymmetry models with the same neutralino mass spectrum as that of the model LCC2, presented as a scatter plot in the plane of a particular ILC cross-section σ versus the neutralino relic density $\Omega_\chi h^2$, from the analysis of [176]. The cross-sections used are polarized-beam cross-sections for gaugino pair production: on the left, $\sigma(e_L^+ e_R^- \to \tilde{\chi}_1^+ \tilde{\chi}_1^-)$, on the right, $\sigma(e_R^+ e_L^- \to \tilde{\chi}_2^0 \tilde{\chi}_3^0)$. The horizontal bands show the ILC cross-section measurements, with errors, expected for LCC2.

example in which pair-production cross-sections in $e^+ e^-$ can play an important role in solving the mixing of neutralinos [176]. In principle, the lightest neutralino in a supersymmetry model can be mainly composed of the U(1) gaugino, or the SU(2) gaugino, or the Higgsinos. These possibilities lead to dramatically different annihilation cross-sections and, thus, different predictions for the relic density. The points in the figure were generated for the supersymmetry parameter point LCC2, to be defined in the next section, and for other supersymmetry models with the same neutralino mass spectrum. The alternative possibilities for the content of the lightest neutralino are shown in the figure as clusters of points at different values of the horizontal coordinate, which is the predicted relic density. The vertical coordinate in the two plots gives the value of a polarized chargino or neutralino pair-production cross-section at the ILC. The horizontal band in each plot indicates the expected ILC cross-section measurement. These measurements resolve the qualitative ambiguity and go a long way towards generating a precise prediction of the WIMP relic density.

14.4.4 Estimation of the WIMP relic density

We have now seen that $e^+ e^-$ colliders, with precision mass and cross-section measurements, offer many pieces of information that can resolve the many ambiguities in the determination of the WIMP annihilation and scattering cross-sections from collider data. We will now present some examples in which this information is assembled globally to estimate these key properties of the WIMP. These analyses are described in more detail in [176].

As illustrative examples, we chose for this analysis four parameter points from the mSUGRA parameter space. These points were chosen to represent the variety of neutralino annihilation mechanisms discussed in Section 14.2.2. The four points are labelled LCC1 (in the bulk region), LCC2 (in the focus point region), LCC3 (in the stau co-annihilation region) and LCC4 (in the A^0 funnel region). For each of these points, we compiled the set of measurements to be expected at the LHC and at the ILC. The results used for the ILC were based on studies [231; 1005; 1228; 1923] based on realistic simulation of the expected detectors and experimental conditions.

The wealth of information available from the e^+e^- collider makes it reasonable to extract a probability density function for the WIMP properties as a function of coordinates in the full 24-dimensional parameter space of the MSSM. This space contains all of the alternative interpretations of the collider data and their ambiguities. If the distribution function becomes strongly peaked about a particular value of a WIMP cross-section, this can be taken as a model-independent determination of the cross-section from collider data.

The four benchmark points for this study were chosen such that the lightest charged supersymmetric particles could be observed in e^+e^- collisions at 500 GeV. The full supersymmetry spectra extend to higher masses. In all four examples, collider operation at 1 TeV was necessary to fully constrain the supersymmetry scenario.

At the point LCC1, the dominant WIMP annihilation mechanism is through t-channel slepton exchange. The precision measurement of the masses of the χ_1^0 and the sleptons – in particular the lightest of them, the $\tilde{\tau}_1$ – is important for obtaining a very accurate determination of $\sigma_{\chi\chi}$. In the simulations [1923], the mass of the WIMP can be determined with an accuracy of 0.05% and the mass difference between this state and the $\tilde{\tau}_1$ is measured to ± 0.3 GeV. These results improve by factors of 100 and 20, respectively, on what would be obtainable from the LHC data. As a result, $\Omega_\chi h^2$ can be inferred to an accuracy of 1.8%. Most of the residual error comes from the uncertainty due to the possible influence of the A^0 boson. Running the e^+e^- collider at 1 TeV allows the discovery of this boson and the measurement of its mass. The error on the $\sigma_{\chi\chi}$ is then reduced to 0.2%, so that the prediction for the neutralino relic density would be comparable in accuracy to the best measurements from the cosmic microwave background in that era.

At the point LCC2, the dominant physics of dark matter annihilation involves the neutralinos and charginos. The mass of the gluino is 850 GeV, so the gaugino part of the supersymmetry spectrum can be studied at the LHC. The mass spectrum of neutralinos can be measured in decay cascades from the gluino. Measurements at a linear collider at 500 GeV and 1 TeV will improve these mass measurements to the accuracy of a few parts per mil. More importantly, these measurements will resolve the quantum numbers and mixing of the neutralinos, in the manner that we discussed at the end of the previous section. The squarks and sleptons are so heavy, at this parameter point, that they most probably would not be observed at the LHC. But we have noted that such heavy particles decouple and do not affect the important dark matter cross-sections. Taking all of these points into account, the ILC operating at centre-of-mass energies up to 1 TeV can determine $\sigma_{\chi\chi}$ and predict the neutralino relic density to an accuracy of 8%.

At the point LCC3, the relic density is determined by co-annihilation reactions. The predicted value depends on the number density of tau sleptons present at the temperature at which this density is established. This depends sensitively on the small mass difference between the lightest neutralino and the lightest stau. At LCC3, $M_{\tilde{\tau}} - M_\chi = 10.8$ GeV. This value can be measured with a 10% accuracy at a linear collider at 0.5 TeV. This accuracy is similar to that claimed in the LHC analyses in [126] but does not require the same strong assumptions of unification of the gaugino masses. The WIMP mass can be measured to 0.07% relative accuracy, as opposed to 10% at the LHC. Still, to fix the relic density more information is needed. Most importantly, we need to fix the neutralino mixing angles and the value of $\tan\beta$. A similar situation applies at the point LCC4. Here the annihilation cross-section is dominated by annihilation through the A^0 resonance. Although the A^0 can be discovered at the LHC for this set of supersymmetry parameters, the determination of the annihilation cross-section also requires knowledge of the neutralino mixing angles and the value of $\tan\beta$. In both cases, these parameters can be obtained at the ILC in running at 1 TeV, from studies of neutralino and chargino pair production and associated $H^0 A^0$ Higgs pair production. In particular, the heavy Higgs bosons in these models have masses near 420 GeV. The masses of these bosons can be measured to a relative accuracy of 0.24%, and the widths can also be measured to 19% accuracy [231]. The decay branching fractions to b and τ pairs are also measured, constraining the contribution of additional final states to the total width; then this study alone gives $\tan\beta$ in a model-independent way to about 10%. The final result is that an e^+e^- collider operating up to 1 TeV can predict $\Omega_\chi h^2$ with an accuracy of 20% in both cases.

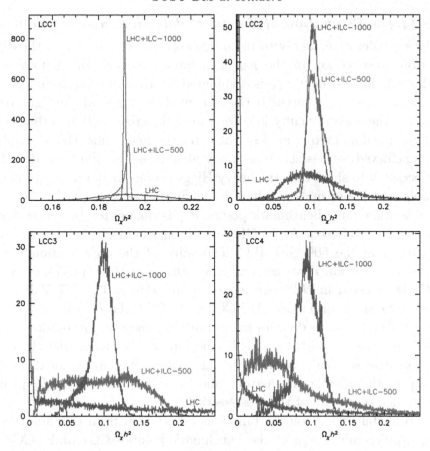

Fig. 14.4. Probability distributions of the microscopic prediction for $\Omega_\chi h^2$ in the analyses of [176] for the four benchmark models LCC1–LCC4. In each plot, the three curves show the results for expected LHC measurements, from these plus expected ILC measurements up to 500 GeV and from these plus expected ILC measurements up to 1 TeV.

The probability distribution functions for the neutralino relic density at the four benchmark points are compared in Fig. 14.4. In each case, we show our estimate of the distribution that would be obtained from LHC data, from the ILC at 500 GeV and from the ILC at 1 TeV, in a model-independent analysis within the 24-parameter MSSM. Despite the very different nature of the dominant annihilation processes in the different regions, necessitating the measurement of different sets of masses and cross-sections, we find that in all cases an e^+e^- collider of sufficient energy can predict $\Omega_\chi h^2$ with an accuracy comparable to that currently available from astrophysical data.

14.4.5 Estimation of the direct detection cross-section

Similar considerations apply to the microscopic determination of the direct detection cross-section in the four benchmark models. In all four models, the spin-independent $\sigma_{\chi p}$ is dominated by Higgs boson exchanges. At LCC2, the dominant process is the t-channel exchange of the light Higgs boson h^0. The main difficulty in determining the cross-section is that of fixing the neutralino mixing angles that enter the neutralino–Higgs coupling. This is achieved to reasonable accuracy already at the 500 GeV ILC. It is also necessary to show that the heavy Higgs bosons are so heavy that they decouple and do not make significant contributions.

At the other three benchmark points, $\sigma_{\chi p}$ is dominated by the exchange of the heavy H^0 boson. At LCC1, this boson would not be discovered at the LHC or at the 500 GeV ILC. The value of the cross-section is then completely uncertain if we use only the data from these machines. When the H^0 is observed in $H^0 A^0$ associated production at the 1 TeV ILC, the correct value snaps into place. At LCC3 and LCC4, the H^0 can be observed at the LHC and its mass determined accurately. However, without knowledge of the neutralino mixing angles, it is again impossible to predict the direct detection cross-section accurately. In these cases, the data from the 1 TeV ILC supply the missing information, and the e^+e^- data eventually give an accurate determination of the cross-section.

The probability distribution functions determined from this analysis for the spin-independent $\sigma_{\chi p}$ at the benchmark points LCC1 and LCC3 are shown in Fig. 14.5. As we saw for $\Omega_\chi h^2$, the data from an e^+e^- collider will eventually be sufficiently incisive to overcome all of the difficulties that we presented in Section 14.2 of this review. This conclusion follows from the general ability of experiments at an e^+e^- collider to perform kinematic mass measurements at the accuracy of parts per mil, to measure complete patterns of branching ratios for new particles, and to measure elementary cross-sections from polarized initial-state leptons.

Subsequent studies have shown that the estimates of accuracies assumed in the ILC analyses presented above remain virtually unchanged when the analyses are performed using full simulation of the detector and detailed reconstruction, as compared to the use of a simpler parameterized detector response. This is due to the relatively simple topologies of e^+e^- events and the high granularity and resolution currently envisaged for the ILC detectors. Both features ensure a large set of measurements performed with large signal-to-background ratios and high analytical power.

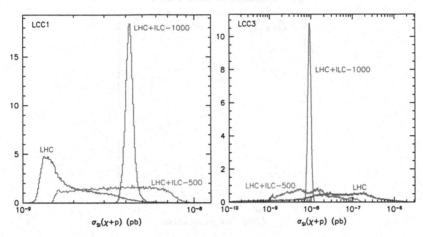

Fig. 14.5. Probability distributions of direct detection cross-section $\sigma_{\chi p}$ in the analyses of [176] for the benchmark models LCC1 and LCC3. In each plot, the three curves show the results for expected LHC measurements, from these plus expected ILC measurements up to 500 GeV and from these plus expected ILC measurements up to 1 TeV.

14.5 Collider measurements and astrophysical questions

In this chapter, we have explained how the most important properties of a WIMP dark matter candidate – the mass, the thermal relic density, the annihilation cross-section and the direct detection cross-section – can be determined by particle physics measurements at high-energy colliders. The physics of dark matter particles is often discussed as if the major questions about astrophysical dark matter came from questions about the particle physics. That is certainly justified today, since we are completely ignorant about the particle physics of dark matter, while the astrophysics is merely highly uncertain.

But we have argued in this chapter that the era is coming in which we will have observed credible dark matter candidate particles in the laboratory and determined their masses and couplings. In that era, it will be interesting to work in the other direction, taking the particle physics of dark matter as given and asking what implications this has for the history of dark matter and its distribution in the Universe.

There are many ways in which the thermal relic density predicted from particle physics could differ from the density observed in cosmology. In WIMP models, the relic density is produced at a time about 10^{-9} seconds after the Big Bang, so the comparison between these quantities tests standard cosmology back to this very early time. If the predicted density is lower,

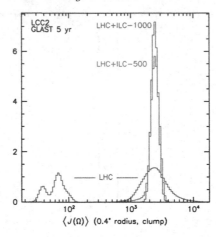

Fig. 14.6. Probability distributions for $\int \mathrm{d}z\rho_\chi^2$, the integral of the WIMP density along the line of sight through a $10^6 M_\odot$ mass clump of dark matter in the Galaxy, based on a 5-year observation by the Fermi Large Area Telescope. The absolute scale of ρ_χ in this analysis [176] is obtained from the annihilation cross-section extracted from collider data from the LHC and two stages of the ILC.

the observed WIMP could be only one component of the total dark matter. It is also possible that WIMPs were produced by a non-thermal mechanism, for example the decay of coherent scalar fields set up in a phase transition early in the history of the Universe. If the predicted density is higher, this might signal the presence of phase transitions or late decays of heavy particles that increased the entropy of the Universe since the WIMP relic density was formed, or, alternatively, the decay of WIMPs to a super-weakly interacting particle.

Even if the dark matter is made entirely out of the observed WIMPs, the local density of dark matter is highly uncertain. The density of dark matter in the disk of our Galaxy is uncertain by a factor of 3, even in models in which the dark matter halo is smooth. This uncertainty affects predictions of the direct detection rate. Indirect detection is expected to be dominated by the annihilation of dark matter in clumps of high density. This clumping is estimated from numerical simulations, as described in Chapter 2 of this book. The central densities of dark matter clumps are uncertain by many orders of magnitude. With definite values for the direct detection and annihilation cross-sections, we can use data from indirect detection to measure these densities. Figure 14.6 shows an example of this. The analysis shown here combines LHC and ILC data on the point LCC2 with data that would be obtained from gamma-ray observations with the Fermi Large Area

Telescope to fix the central density of a dark matter clump in the Galaxy to 20% accuracy.

Thus, just as the study of dark matter particles at colliders will answer important questions about particle physics, it will give us new tools to attack important astrophysical questions, including central questions about the formation of our Galaxy.

With the start of the LHC experiments, we expect that the era of direct experimental observation of dark matter is about to begin. The results should give us not only a new view of particle physics, but also a new tool for viewing the Universe.

15

Extra dimensions at the LHC

Kyoungchul Kong, Konstantin Matchev and Géraldine Servant

In models with extra dimensions, the usual $(3 + 1)$-dimensional space-time $x^\mu \equiv (x^0, x^1, x^2, x^3)$ is extended to include additional spatial dimensions parameterized by coordinates $x^4, x^5, \ldots, x^{3+N}$. Here N is the number of extra dimensions. String theory arguments would suggest that in principle N can be as large as 6 or 7. In this chapter, we are interested in extra-dimensional (ED) models where all particles of the Standard Model (SM) are allowed to propagate in the bulk, i.e. along any of the x^{3+i} $(i = 1, \ldots, N)$ directions [100]. In order to avoid a blatant contradiction with the observed reality, the extra dimensions in such models must be extremely small: smaller than the smallest scale which has been currently resolved by experiment. Therefore, the extra dimensions are assumed to be suitably compactified on some manifold of sufficiently small size (see Fig. 15.1).

Depending on the type of metric in the bulk, the ED models fall into one of the following two categories: flat, also known as 'universal' extra dimensions (UED) models, discussed in Section 15.1, or warped ED models, discussed in Section 15.2. As it turns out, the collider signals of the ED models are strikingly similar to the signatures of supersymmetry (SUSY) discussed in Chapter 13 [537]. Section 15.3 outlines some general methods for distinguishing an ED model from SUSY at high-energy colliders.

15.1 Flat extra dimensions (UED)

15.1.1 Definition

In this section, we choose the metric on the extra dimensions to be flat. For simplicity, we shall limit our discussion to models with $N = 1$ or $N = 2$ UEDs. In the simplest case of $N = 1$, a compact extra dimension x^4 would

Particle Dark Matter: Observations, Models and Searches, ed. Gianfranco Bertone. Published by Cambridge University Press. © Cambridge University Press 2010.

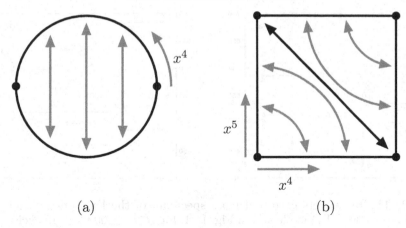

(a)   (b)

Fig. 15.1. (a) Compactification of $N = 1$ extra dimension on a circle with opposite points identified. (b) Compactification of $N = 2$ extra dimensions on the chiral square with adjacent sides identified. In each case, the arrows indicate the corresponding identification. The dots represent fixed (boundary) points.

have the topology of a circle S^1. However, in order to implement the chiral fermions of the SM in a UED framework, one must use a manifold with endpoints, e.g. the orbifold S^1/Z_2, pictorially represented in Fig. 15.1(a). The opposite sides of the circle are identified, as indicated by the vertical arrows. The net result is a single line segment with two endpoints, denoted by the dots. The size of the extra dimension in this case is simply parameterized by the radius of the circle R.

In the case of two extra dimensions ($N = 2$) there are several possibilities for compactification. One of them is the so-called 'chiral square' and corresponds to a T^2/Z_4 orbifold [686]. It can be visualized as shown in Fig. 15.1(b). The two extra dimensions have equal size, and the boundary conditions are such that the adjacent sides of the 'chiral square' are identified, as indicated by the arrows. The resulting orbifold endpoints are again denoted by dots.

An important concept in any UED model is the notion of Kaluza–Klein (KK) parity, whose origin can be traced back to the geometrical symmetries of the compactification. For example, in the $N = 1$ case of Fig. 15.1(a), KK parity corresponds to the reflection symmetry with respect to the centre of the line segment. Similarly, in the $N = 2$ case of Fig. 15.1(b), KK parity is due to the symmetry with respect to the centre of the chiral square.[1] In general, UED models respect KK parity, and this fact has important consequences for their collider and astroparticle phenomenology.

---

[1] Notice that there is only one KK parity since the two directions x^4 and x^5 of the chiral square are related to each other through the boundary condition.

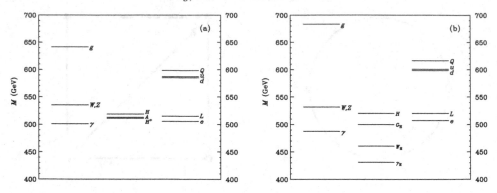

Fig. 15.2. The radiatively corrected mass spectrum of the level one ($n = 1$) KK particles in the two UED models from Fig. 15.1, for $R^{-1} = 500$ GeV. In each panel, from left to right we list the KK particles of spin-1, spin-0 and spin-$\frac{1}{2}$. Figures taken from [470] and [538].

15.1.2 Mass spectrum

Since the extra dimensions are compact, the extra-dimensional components of the momentum of any SM particle are quantized in units of $\frac{1}{R}$. From the usual four-dimensional point of view, those momentum components are interpreted as masses. Therefore in UED models each SM particle is accompanied by an infinite tower of heavy KK particles with masses $\frac{n}{R}$, where the integer n counts the number of quantum units of extra-dimensional momentum. All KK particles at a given n are said to belong to the n-th KK level, and at leading order appear to be exactly degenerate.

However, the masses of the KK particles receive corrections from several sources, which will lift this degeneracy. First, there are tree-level corrections arising from electroweak symmetry breaking through the usual Higgs mechanism. More importantly, there are one-loop mass renormalization effects due to the usual SM interactions in the bulk [538]. Finally, there may be contributions from boundary terms which live on the orbifold fixed points (the dots in Fig. 15.1) [538; 858]. In the so-called 'minimal UED' models, the last effect is ignored and the resulting one-loop radiatively corrected spectrum of the first level KK modes is as shown in Fig. 15.2.

In the case of $N = 1$ minimal UED shown in Fig. 15.2(a), we find that the SM particle content is simply duplicated at the $n = 1$ level. The mass splittings among the KK particles arise mainly because of radiative corrections, which are largest for the coloured particles (KK quarks and KK gluon). The lightest KK particle (LKP) at level one in this case is denoted by γ_1 and represents a linear superposition of the KK modes of the hypercharge gauge boson B_1 and the neutral component of the SU(2) gauge boson W_1^0.

Figure 15.2(b) reveals that the $n = 1$ KK mass spectrum is somewhat more complex in the case of two extra dimensions ($N = 2$). This is because gauge bosons propagating in 5+1 dimensions may be polarized along either of the two extra dimensions. As a result, for each spin-1 KK particle associated with a gauge boson, there are two spin-0 fields transforming in the adjoint representation of the gauge group. One linear combination of those becomes the longitudinal degree of freedom of the spin-1 KK particle, while the other linear combination remains as a physical spin-0 particle, called the spinless adjoint. In Fig. 15.2(b) the spinless adjoints are designated by an index H. Figure 15.2(b) also reveals that in the minimal $N = 2$ UED model, the LKP is the spinless photon γ_H [470; 1586].

15.1.3 Collider signals

In terms of the KK level number n,[2] KK parity can be simply defined as $(-1)^n$. The usual Standard Model particles do not have any extra-dimensional momentum, and therefore have $n = 0$ and positive KK parity. On the other hand, the KK particles can have either positive or negative KK parity, depending on the value of n. Notice that the lightest KK particle at $n = 1$ (i.e. the LKP) has negative KK parity and is absolutely stable, since KK parity conservation prevents it from decaying into SM particles. The collider phenomenology of the UED models is therefore largely determined by the nature of the LKP. In both of the minimal UED models shown in Fig. 15.2, the LKP is a neutral weakly interacting particle, whose signature will be missing energy in the detector. At hadron colliders the total parton level energy in the collision is a priori unknown, hence the presence of LKP particles in the event must be inferred from an imbalance in the total transverse momentum.

The collider phenomenology of the minimal UED models from Fig. 15.2 has been extensively investigated at both linear colliders [229; 230; 350; 886; 941; 943; 1644] and hadron colliders [470; 537; 614; 685; 941; 942; 1367; 1649]. Owing to KK parity conservation, the KK particles are always pair-produced, and then each one undergoes a cascade decay to the LKP, as illustrated in Fig. 15.3 [537; 685]. It is interesting to notice that the decay patterns in UED look very similar to those arising in R-parity conserving supersymmetry (see Chapter 13). The typical UED signatures include a

---

[2] In the case of $N = 2$, each KK particle is characterized by two indices, n_1 and n_2, counting the quantum units of momentum along each extra dimension. The KK parity is then defined in terms of the total KK level number $n_1 + n_2$ as $(-1)^{n_1+n_2}$.

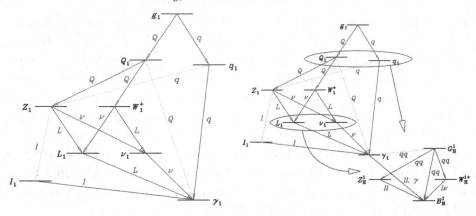

Fig. 15.3. Qualitative sketch of the level 1 KK spectroscopy depicting the dominant (solid) and rare (dotted) transitions and the corresponding decay product for the $N = 1$ (left) and $N = 2$ (right) UED models from Fig. 15.2. Figures taken from refs. [537] and [886].

certain number of jets, a certain number of leptons and photons, plus missing energy $\not{E}_T$.

Which particular signature among all these offers the best prospects for discovery? The answer to this question depends on the interplay between the predicted signal rates in UED and the expected SM backgrounds. For example, at lepton colliders the SM backgrounds are firmly under control, and the best channel is typically the one with the largest signal rate. Since at lepton colliders the new particles are produced through the same (electroweak) interactions, the largest rates are associated with the lightest particles in the spectrum – the KK leptons and the electroweak KK gauge bosons. In contrast, at hadron colliders the dominant production is through strong interactions, and the largest cross-sections belong to the coloured KK particles, which typically decay through jets. Unfortunately, the SM QCD backgrounds to the jetty UED signatures are significant, and in that case there is a substantial benefit in looking for leptons instead. As seen in Fig. 15.3, the decays of the weakly interacting KK quarks Q_1 proceed through SU(2) KK gauge bosons $W_1^\pm$ and Z_1, whose decays are often accompanied by leptons (electrons or muons). The inclusive pair production of a $Q_1\bar{Q}_1$ pair therefore may yield up to four leptons, plus missing energy. Figure 15.4 shows the corresponding discovery reach for the minimal $N = 1$ UED model at the Tevatron and the LHC in the $4\ell + \not{E}_T$ channel [537]. Recently the CDF collaboration performed a search for the minimal $N = 1$ UED model in a multilepton channel, based on 100 pb^{-1} of data at $\sqrt{s} = 1.8$ TeV. That analysis placed a lower limit on the UED scale R^{-1} of 280 GeV (at 95% C.L.) [1336].

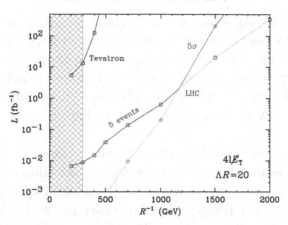

Fig. 15.4. Discovery reach for the minimal $N = 1$ UED model at the Tevatron and the LHC in the $4\ell + \not{E}_T$ channel. The plot shows the total integrated luminosity per experiment which is required for a 5σ discovery, given the observation of at least five events. Figure taken from ref. [537].

From Fig. 15.2(b) one can notice that owing to the presence of additional KK particles (spinless adjoints) at each level, the cascade decays become longer and may yield events with even more leptons. Such events with very high lepton multiplicity would be a smoking gun for the $N = 2$ UED model [685].

15.2 Warped extra dimensions

The UED models discussed in the previous section are very peculiar: the extra dimension is an interval with a flat background geometry, and KK parity is realized as a geometric reflection about the midpoint of the extra dimension. It is important to note that KK parity has a larger parent symmetry, KK number conservation, which is broken only by the interactions living on the orbifold boundary points (the dots in Fig. 15.1). In the literature on UED models, it is usually assumed that the boundary interactions are symmetric with respect to the Z_2 reflection about the midpoint, so that KK parity is an exact symmetry. It is also assumed that they are suppressed (loop-induced), implying that KK number is still an approximate symmetry. These assumptions have very important phenomenological implications, as both KK parity and the approximate KK number conservation are needed to evade precision electroweak constraints for UED models. KK parity eliminates couplings of a single odd KK mode with the SM field, whereas the approximate KK number conservation suppresses certain interactions among the even-level KK modes, such as single coupling of the second

KK mode with the SM, which are not forbidden by KK parity. In the end, both the odd and even KK modes are allowed to have masses well below 1 TeV. If there were only KK parity and not the approximate KK number conservation, experimental constraints would have required the second KK mass to be higher than 2–3 TeV and, therefore, the compactification scale to be around 1 TeV or higher (recall that in flat geometry KK modes are evenly spaced).

The flatness of wavefunction profiles in UED is not natural and reflects the fact that electroweak symmetry breaking is not addressed but just postulated. A model of dynamical symmetry breaking in UED would typically spoil the flatness of the Higgs profile and constraints on the KK scale would have to be re-examined accordingly. The virtue of UED is that mass scales of new particles are allowed to be very close to the electroweak scale at a few hundred GeV, allowing for easy access at the LHC, and offering an interesting benchmark model for LHC searches. However, the UED model does not address the hierarchy between the Planck and the weak scale, nor does it address the fermion mass hierarchy. In contrast, as shown by Randall and Sundrum [1625], warped extra dimensions have provided a new framework for addressing the hierarchy problem in extensions of the Standard Model.

15.2.1 *Generic features of warped space-time*

The Randall–Sundrum (RS) solution [1625] is based on a slice of five-dimensional anti de Sitter space AdS$_5$ bounded by two three-branes, the UV and IR branes. The background space-time metric of AdS$_5$ is

$$\mathrm{d}s^2 = \mathrm{e}^{-2ky}\eta_{\mu\nu}\mathrm{d}x^\mu\mathrm{d}x^\nu - \mathrm{d}y^2, \tag{15.1}$$

where k is the AdS curvature scale of order the Planck scale (fixed by the bulk cosmological constant). The y dependence in the metric is known as the 'warp' factor. The UV (IR) brane is located at $y = 0$ ($y = \pi R$) (see Fig. 15.5). The key point is that distance scales are measured with the non-factorizable metric of AdS space. Hence, energy scales are location-dependent along the fifth dimension and the hierarchy problem can be redshifted away. The RS model supposes that all fundamental mass parameters are of order the Planck scale. Owing to the warped geometry, the 4D (or zero-mode) graviton is localized near the UV/Planck brane, whereas the Higgs sector can be localized near the IR brane where the cutoff scale is scaled down to $M_{\mathrm{Pl}}\mathrm{e}^{-\pi kR}$. According to the AdS/CFT correspondence [1377; 1941], AdS$_5$ is dual to a 4D strongly coupled conformal field theory (CFT). Thus, the RS solution is conjectured to be dual to composite Higgs

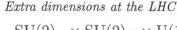

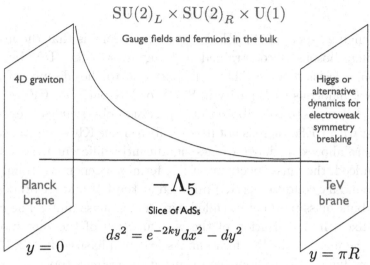

Fig. 15.5. Randall–Sundrum set-up.

models [117; 574; 1629] where the TeV scale is hierarchically smaller and stable compared with the UV scale.

In the original RS model, the entire SM was assumed to be localized on the TeV brane. It was subsequently realized that when the SM fermions [937; 1027] and gauge fields [517; 620; 1584] are allowed to propagate in the bulk, such a framework not only solves the Planck–weak hierarchy, but can also address the flavour hierarchy. The idea is that light SM fermions (which are zero modes of 5D fermions) can be localized near the UV brane, whereas the top quark is localized near the IR brane, resulting in small and large couplings respectively to the SM Higgs localized near the IR brane. In the CFT language, the Standard Model fermions and gauge bosons are partly composite to varying degrees, ranging from an elementary electron to a composite top quark. Finally, a central requirement in these constructions is having an approximate 'custodial isospin' symmetry of the strong sector to protect the electroweak (EW) ρ parameter. This is ensured by extending the EW gauge group to $SU(2)_L \times SU(2)_R \times U(1)$ [28].

In the RS set-up, the KK modes are generically localized towards the IR brane. On the other hand, the zero-mode gauge bosons have a flat profile along the extra dimension while the zero-mode fermions can be arbitrarily localized in the bulk. An important consequence for collider searches is that light fermions have small couplings to KK modes (including the KK graviton) while the top quark and the Higgs have a large coupling to KK modes.

15.2.2 Mass spectrum

The challenging aspect of RS collider phenomenology is that the mass scale of KK gauge bosons is constrained to be at least a few TeV by the EW and flavour precision tests. This is in part due to the absence of a parity symmetry (analogous to R-parity in SUSY or KK parity in UED), allowing tree-level exchanges to contribute to the precision observables. Nevertheless, in contrast with UED, there is not necessarily a single KK scale and some KK fermions are allowed to have a mass significantly different from KK gauge bosons. Indeed, the mass spectrum of KK fermions depends strongly on the type of boundary conditions (BC) imposed at the UV and IR branes as well as on the bulk mass parameter, called c in Planck mass units. The c parameter also fixes the localization of the wavefunction of the zero modes and therefore the mass of the SM fermion. As first emphasized in [32; 33], there can be very light KK fermions as a consequence of the top compositeness. BC are commonly modelled by either Neumann $(+)$ or Dirichlet $(-)$ BC[3] in orbifold compactifications. Five-dimensional fermions lead to two chiral fermions in 4D, only one of which has a zero mode that can reproduce the chiral SM fermion. Standard Model fermions are associated with $(++)$ BC (first sign is for Planck brane, second for TeV brane). The other chirality is $(--)$ and does not have a zero mode. In the particular case of the breaking of the grand unified gauge group (GUT) to the SM, fermionic GUT partners which do not have zero modes are assigned Dirichlet boundary conditions on the Planck brane, i.e. they have $(-+)$ boundary conditions.[4] When computing the KK spectrum of $(-+)$ fermions one finds that for $c < 1/2$ the lightest KK fermion is lighter than the lightest KK gauge boson. For the particular case $c < -1/2$, the mass of this KK fermion is exponentially smaller than that of the gauge KK mode. Figure 15.6 shows the mass of the lightest $(-+)$ KK fermion as a function of c and for different values of the KK gauge boson mass M_{KK}. There is an intuitive argument for the lightness of the KK fermion: for $c \ll 1/2$, the zero mode of the fermion with $(++)$ boundary condition is localized near the TeV brane. Changing the boundary condition to $(-+)$ makes this 'would-be' zero mode massive, but since it is localized near the TeV brane, the effect of changing the boundary condition on the Planck brane is suppressed, resulting in a small mass for the would-be zero mode.

Therefore, the scale for KK fermions can be different from the scale of KK gauge bosons. The lightest KK fermion is the one with the smallest

---

[3] For a comprehensive description of boundary conditions of fermions on an interval, see [594].

[4] Consistent extra-dimensional GUT models require a replication of GUT multiplets as the zero-mode SM particles are obtained from different multiplets.

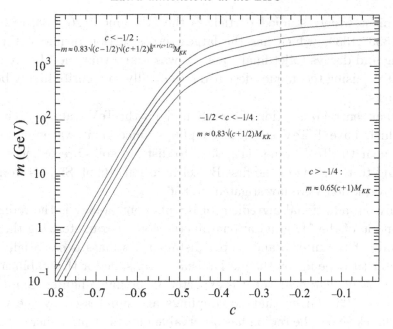

Fig. 15.6. Mass of the $(-+)$ KK fermion as a function of its c parameter for different values of the KK gauge boson mass $M_{KK} = 3, 5, 7, 10$ TeV (from bottom to top). For large and negative c, the KK fermion can be infinitely light.

c parameter. For example, in the warped SO(10) models of refs. [32; 33], the lightest KK fermion will come from the GUT multiplet which contains the top quark. Indeed, the top quark, being the heaviest SM fermion, is the closest to the TeV brane. This is achieved by requiring a negative c. Thus, all $(-+)$ KK fermions in the GUT multiplet containing the SM top quark are potentially light. Mass splittings between KK GUT partners of the top quark can have various origins, in particular due to GUT breaking in the bulk [32; 33]. Direct production at the LHC of light KK quarks leading to multi W final states was studied in [575; 647].

15.2.3 Collider signals

As a result of wavefunction localization, SM gauge bosons and light fermions couple weakly to the KK states, whereas the KK states mostly decay to top quarks and longitudinal W/Z/Higgs. Hence, the golden decay channels such as resonant signals of dileptons or diphotons are suppressed. Besides, given that the KK mass scale is typically high (a few TeV) the top quarks or W or Z bosons resulting from the decays of these KK states are highly boosted.

The most widely studied particle is the KK gluon [26; 488; 682; 1033; 1034; 1333; 1334], which has the largest cross-section owing to the QCD coupling and decays to jet final states. It was found that the LHC reach can be ~4 TeV, using techniques designed specifically to identify highly boosted top quarks.

Another central prediction of the RS model is the TeV Kaluza–Klein gravitons which have 1/TeV strength couplings since their wavefunctions are peaked near the TeV brane. They lead to distinct spin-2 resonances spaced according to the roots of the first Bessel function [1626]. Signals from their direct production were investigated in [856].

Finally, an additional ingredient of RS phenomenology is the radion, the scalar mode of the 5D gravitational fluctuations, parameterizing the vibration mode of the inter-brane proper distance. Its mass is essentially a free parameter (it depends on the mechanism responsible for the stabilization of the inter-brane distance) but it is expected to be much lighter than the KK excitations [593]. Given that its couplings are suppressed by the warped-down Planck scale, the radion has observable effects at high-energy colliders [593; 596; 695; 696; 1038; 1091]. Like the Higgs, the radion is located near the IR brane and its interactions are proportional to the mass of the field it couples to (through the trace of the energy–momentum tensor). The Higgs and radion can mix through a gravitational kinetic mixing term [959] with interesting consequences [593; 695; 696; 1038; 1091]. Signals from direct production of the radion have been studied in [596; 1650; 1857].

In addition to the searches for the radion, the KK graviton and the KK gluons, other studies of warped phenomenology at the LHC have dealt with KK neutral electroweak gauge bosons [30] and KK (heavy) fermions [621]. All these signatures are common to RS models. Besides, new characteristic predictions appear in more specific models. We mentioned the light KK fermions which appear, for instance, in warped GUT models with Z_3 symmetry and LZP dark matter [32; 33] but more generically in models where the Higgs is a pseudo-Goldstone boson [574; 575]. They are for instance predicted in the gauge–Higgs unification models of ref. [1544] which contain a Z_2 mirror symmetry. The associated signatures are jets and missing energy and benefit from a large cross-section (vector-like quarks are pair-produced via the standard QCD interactions). Other interesting warped models with distinctive phenomenologies have been proposed, such as warped supersymmetric models [1520]. A recent finding and a generic prediction of 5D models is the existence of stable skyrmion configurations [1585] with phenomenological consequences that remain to be investigated.

15.3 SUSY–UED discrimination at the LHC

As discussed in Section 15.1.3, the generic collider signatures of the minimal UED models involve jets, leptons and missing energy, just like the supersymmetry signatures of Chapter 13. In addition, the couplings of the KK partners are equal to those of their SM counterparts. The same property is shared by the superpartners in SUSY models. A natural question, therefore, is whether a given UED model can be experimentally differentiated from supersymmetry and vice versa. This issue is the subject of this section.

In general, there are two fundamental differences between UED and SUSY.

(i) The spins of the SM particles and their KK partners are the same, while in SUSY they differ by $\frac{1}{2}$.

(ii) For each particle of the Standard Model, the UED models predict an infinite tower of new particles (Kaluza–Klein partners). In contrast, the simplest SUSY models contain only one partner per SM particle.

Thus the best way to discriminate between UED and SUSY is either to measure the spins of the new particles or to explore the higher-level states of the KK tower.

Spin determinations in missing energy events are rather challenging (especially at hadron colliders), owing to the presence of at least two invisible particles in each event, whose energies and momenta are not measured. Ideally, one would like to be able to reconstruct the energies and momenta of the escaping particles, in which case the spins can be determined in one of several ways (see Sections 15.3.3, 15.3.4 and 15.3.5). However, when the momenta cannot be reliably determined, we are limited to studying only the properties of the particles which are visible in the detector, e.g. their invariant mass distributions (see Section 15.3.1).

On the other hand, it is easier to find the higher KK modes, as long as they are produced abundantly. In particular, the $n = 2$ excited KK states have positive KK parity and can directly decay into a pair of SM particles. Such higher-level KK particles can then be looked for via traditional resonance searches (see Section 15.3.2). The observation of a rich resonance structure would be quite indicative of UED.

15.3.1 Spin measurements from invariant mass distributions

Consider the three-step decay chain exhibited in Fig. 15.7, which is rather typical in both UED and SUSY models. For example, in UED this chain may arise from the $Q_1 \to Z_1 \to L_1 \to \gamma_1$ transitions shown in Fig. 15.3. The measured visible decay products are a quark jet j and two opposite

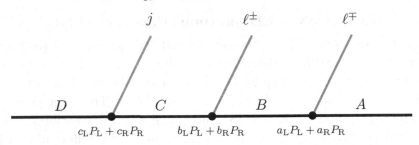

Fig. 15.7. The typical UED or SUSY cascade decay chain under consideration. At each vertex we assume the most general fermion couplings (see ref. [479] for the exact definitions).

sign leptons ℓ^+ and ℓ^-, while the end product A is invisible in the detector. Given this limited amount of information, in principle there are six possible spin configurations for the heavy partners D, C, B and A: $SFSF$, $FSFS$, $FSFV$, $FVFS$, $FVFV$ and $SFVF$, where S stands for a spin-0 scalar, F stands for a spin-$\frac{1}{2}$ fermion and V stands for a spin-1 vector particle. The main goal of the invariant mass analysis here will be to discriminate among these six possibilities, and in particular between $SFSF$ (SUSY) and $FVFV$ (minimal UED).

It is well known that the invariant mass distributions of the visible particles (the jet and two leptons in our case) already contain some information about the spins of the intermediate heavy particles A, B, C and D [141; 215; 1771; 1921]. Unfortunately, the invariant mass distributions are also affected by a number of additional factors, which have nothing to do with spins, such as: the chirality of the couplings at each vertex [479; 1231]; the fraction of events f in which the cascade is initiated by a particle D rather than its antiparticle $\bar{D}$ [215]; and finally, the mass splittings among the heavy partners [1771]. Therefore, in order to do a pure and model-independent spin measurement, one has to somehow eliminate the effect of those three extraneous factors.

Fortunately, the masses of A, B, C and D can be completely determined ahead of time, for example by measuring the kinematic endpoints of various invariant mass distributions made out of the visible decay products in the decay chain of Fig. 15.7 [62; 962; 963; 1097], or through a sufficient number of transverse mass measurements [480; 1327]. Once the mass spectrum is thus determined, we are still left with a complete lack of knowledge regarding the coupling chiralities and particle fraction f. In spite of this residual ambiguity, the spins can nevertheless be determined, at least as a matter of principle [479]. To this end, one should not make any a-priori assumptions

Table 15.1. *Two study points from SUSY (SPS1a) and UED (UED500), characterized by particle masses, chirality coefficients and particle–antiparticle fractions f and $\bar{f}$.*

	SPS1a	UED500
(m_A, m_B, m_C, m_D) in GeV	(96, 143, 177, 537)	(501, 515, 536, 598)
$(f, \bar{f})$	(0.7, 0.3)	(0.66, 0.34)
$(a_{\mathrm{L}}, a_{\mathrm{R}}, b_{\mathrm{L}}, b_{\mathrm{R}}, c_{\mathrm{L}}, c_{\mathrm{R}})$	(0, 1, 0, 1, 1, 0)	(1, 0, 1, 0, 1, 0)

and instead consider the most general fermion couplings at each vertex in Fig. 15.7 and any allowed value for the parameter f. Then, the invariant mass distributions should be used to make separate independent measurements of the spins, on one hand, and of the couplings and f fraction, on the other. Following the analysis of [479] here we shall illustrate this procedure with two examples – one from supersymmetry and one from UED. The corresponding mass spectra, chirality parameters and particle–antiparticle fraction f and $\bar{f}$ for each case are listed in Table 15.1.

Given the three visible particles from the decay chain of Fig. 15.7, one can form three well-defined two-particle invariant mass distributions: one dilepton ($\ell^+\ell^-$), and two jet–lepton ($j\ell^+$ and $j\ell^-$) distributions. For the purposes of the spin analysis, it is actually more convenient to consider the sum and the difference of the two jet–lepton distributions [479]. The shapes of the resulting invariant mass distributions are given schematically by the following formulas [479]:

$$\left(\frac{\mathrm{d}N}{\mathrm{d}m_{\ell\ell}^2}\right)_S = F_{S;\delta}^{(\ell\ell)}(m_{\ell\ell}^2) + \alpha\, F_{S;\alpha}^{(\ell\ell)}(m_{\ell\ell}^2) \tag{15.2}$$

$$\left(\frac{\mathrm{d}N}{\mathrm{d}m_{j\ell^+}^2}\right)_S + \left(\frac{\mathrm{d}N}{\mathrm{d}m_{j\ell^-}^2}\right)_S = F_{S;\delta}^{(j\ell)}(m_{j\ell}^2) + \alpha\, F_{S;\alpha}^{(j\ell)}(m_{j\ell}^2) \tag{15.3}$$

$$\left(\frac{\mathrm{d}N}{\mathrm{d}m_{j\ell^+}^2}\right)_S - \left(\frac{\mathrm{d}N}{\mathrm{d}m_{j\ell^-}^2}\right)_S = \beta\, F_{S;\beta}^{(j\ell)}(m_{j\ell}^2) + \gamma\, F_{S;\gamma}^{(j\ell)}(m_{j\ell}^2), \tag{15.4}$$

where the functions F, given explicitly in [479], are known functions of the masses of particles A, B, C and D. As indicated by the index S, there is a separate set of F functions for each spin configuration: $S = \{SFSF, FSFS, FSFV, FVFS, FVFV, SFVF\}$. Thus the functions F contain the pure spin information. On the other hand, the coefficients α, β

and γ encode all of the residual model dependence, namely the effect of the coupling chiralities and particle–antiparticle fraction f. Since the coefficients α, β and γ are a priori unknown, they will need to be determined from experiment, by fitting the predicted shapes (15.2–15.4) to the data. The results from this exercise for the two study points from Table 15.1 are shown in Fig. 15.8. The solid lines in each panel represent the input invariant mass distributions which will be presumably measured by experiment. The other (dotted or dashed) lines are the best fits to these data, for each of the remaining five spin configurations S. The line code is the following. A dashed line indicates that the trial model fits the input data perfectly, while a dotted line implies that the fit fails to match the input data. The best fit values for the relevant coefficients (α, β and γ) for each case are also shown, except for those cases (labelled by 'NA') where they are left undetermined by the fit.

The results from Fig. 15.8 show that the success of the spin measurement method depends on the type of new physics which happens to be discovered. In the case of supersymmetry (panels (a), (c), (e)), the spin chain can be unambiguously determined to be $SFSF$. Furthermore, this can be done solely on the basis of the distribution (15.4), which is sufficiently powerful to rule out all of the remaining five spin chain candidates.[5] On the other hand, the UED case (panels (b), (d), (f)) is more challenging, and the end result is inconclusive – both models $FVFS$ and $FVFV$ are able to fit all three distributions (15.2–15.4) perfectly. This confusion is not related to the specific choice of our UED study point, but is a general feature of any $FVFV$ and $FVFS$ (as well as $FSFV$ and $FSFS$) pair of models [479]. In summary, it appears that through studies of the shapes of the invariant mass distributions of the visible decay products in a chain such as the one in Fig. 15.7, one should be able to discriminate between SUSY and UED, while the specific type of UED model may remain uncertain.

15.3.2 Higher-level KK resonance searches

As mentioned earlier, the discovery of the higher-level KK resonances would be another strong indication of the UED scenario. At hadron colliders like the LHC, resonance searches are easiest in the dilepton (dimuon or dielectron) channels. The corresponding 5σ discovery reach for (a) γ_2 and (b) Z_2 is shown in Fig. 15.9 [614]. In each plot, the upper set of lines labelled 'DY' makes use of the single γ_2, or Z_2 production only, while the lower set of lines (labelled 'all processes') includes in addition indirect γ_2 and Z_2 production

---

[5] The distribution (15.4) is closely related to the lepton charge asymmetry proposed in [215].

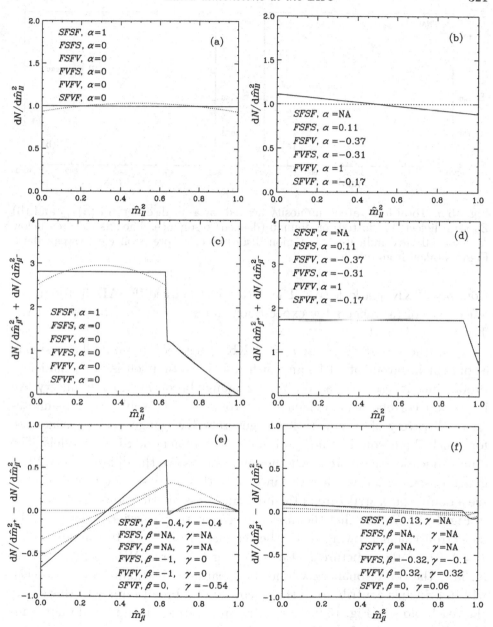

Fig. 15.8. Best fits to the three invariant mass distributions predicted for the SUSY (a,c,e) and UED (b,d,f) study points from Table 15.1. The solid line in each plot represents the input 'data' while the other (dotted or dashed) lines are the best fits to the data for each of the remaining five spin configurations. Dashed lines indicate a perfect match to the data, while dotted lines imply that the best fit fails to perfectly reproduce the data. The best fit values for the relevant coefficients (α, β and γ) are also shown.

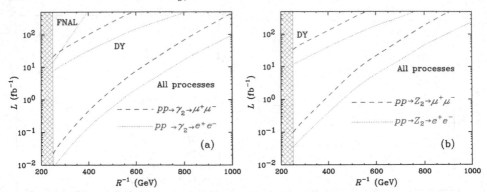

Fig. 15.9. Total integrated luminosity needed for a 5σ discovery of (a) γ_2 and (b) Z_2 as a dielectron (dotted) or dimuon (dashed) resonance. The shaded area below $R^{-1} = 250$ GeV indicates the region disfavoured by precision electroweak data. Figures taken from ref. [614].

from $n = 2$ KK quark decays. The dotted line marked 'FNAL' in the upper left corner of (a) reflects the expectations for a $\gamma_2 \to e^+e^-$ discovery at the Tevatron in Run II.

While the discovery of an $n = 2$ KK gauge boson would be a strong argument in favour of UED, any such resonance by itself is not a sufficient proof, since it resembles an ordinary Z' gauge boson in supersymmetry. An important corroborating evidence in favour of UED would be the simultaneous discovery of several rather degenerate KK gauge boson resonances, for which there would be no good motivation in generic SUSY models. The crucial question therefore is whether one can resolve the different $n = 2$ KK gauge bosons as individual resonances. For this purpose, one would need to see a double peak structure in the invariant mass distributions, as illustrated in Fig. 15.10. We see that the diresonance structure is easier to detect in the dielectron channel, owing to the better mass resolution. In dimuons, with $L = 100\,\text{fb}^{-1}$ the structure is also beginning to emerge. We should note that initially the two resonances will not be separately distinguishable, and each will in principle contribute to the discovery of a broad bump. In this sense, the reach plots in Fig. 15.9 are rather conservative, since they do not combine the two signals from Z_2 and γ_2, but show the reach for each resonance separately.

15.3.3 *Spin measurements from production cross-sections*

The spin of the new particles can also be inferred from the threshold behaviour of their production cross-section [229; 230]. For an s-channel diagram mediated by a gauge boson, the pair-production cross-section for a

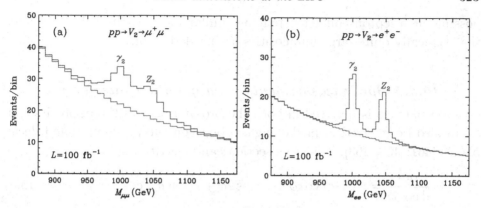

Fig. 15.10. The $V_2 \equiv \gamma_2, Z_2$ diresonance structure in UED with $R^{-1} = 500$ GeV, for the (a) dimuon and (b) dielectron channel at the LHC with $L = 100\,\text{fb}^{-1}$. The SM background is shown with the continuous underlying histogram. Figures taken from ref. [614].

spin-0 particle behaves like $\sigma \sim \beta^3$ while the cross-section for a spin-$\frac{1}{2}$ particle behaves as $\sigma \sim \beta$, where $\beta = \sqrt{1 - \frac{4m^2}{s}}$, and $\sqrt{s}$ is the total centre-of-mass energy, while m is the mass of the new particle. At lepton colliders the threshold behaviour can be easily studied by varying the beam energy and measuring the corresponding total cross-section, without any need for reconstructing the kinematics of the missing particles. In contrast, at hadron colliders the initial state partons cannot be controlled, so in order to apply this method, one must fully reconstruct the final state, which is difficult when there are two or more missing particles.

The total production cross-section may also be used as an indicator of spin [613; 1204]. For example, the total cross-sections of the fermion KK modes in UED are 5–10 times as large as the corresponding cross-sections for scalar superpartners of the same mass. However, the measurement of the total cross-section necessarily involves additional model-dependent assumptions regarding the branching fractions, the production mechanism, etc.

15.3.4 Spin measurements from angular distributions

Perhaps the most direct indication of the spin of the new particles is provided by the azimuthal angular distribution at production [229; 230]. Assuming production through an s-channel gauge boson, the angular distribution for a spin-0 particle is $\sim (1 - \cos^2 \theta)$, where θ is the azimuthal production angle in the centre-of-mass frame. In contrast, the distribution for a spin-$\frac{1}{2}$ particle is $\sim (1 + \cos^2 \theta)$. Unfortunately, reconstructing the angle θ generally requires a good knowledge of the momentum of the missing particles, which is only possible at a lepton collider. Applying similar ideas at the LHC, one finds that typically quite large luminosities are needed [216].

15.3.5 *Spin measurements from quantum interference*

When a particle is involved in both the production and the decay, its spin s can also be inferred from the angle ϕ between the production and decay planes [461; 462; 463]. The cross-section can be written as

$$\frac{\mathrm{d}\sigma}{\mathrm{d}\cos\phi} = a_0 + a_1 \cos\phi + a_2 \cos 2\phi + \cdots + a_{2s} \cos 2s\phi \,. \qquad (15.5)$$

By measuring the coefficient a_{2s} of the highest cos mode, one can in principle extract the spin s of the particle. This method is especially useful since it does not rely on the particular production mechanism, and is equally applicable to s-channel and t-channel processes. However, its drawback is that the ϕ dependence results from integrating out all other degrees of freedom, which often leads to a vanishing coefficient as a result of cancellations, for instance in the case of a purely vector-like coupling, or in the case of a pp collider like the LHC. As a result, the practical applicability of the method is rather model-dependent.

16

SUSY tools

Fawzi Boudjema, Joakim Edsjö and Paolo Gondolo

The long-awaited Large Hadron Collider (LHC) is expected to start taking data in 2009. The LHC research programme has traditionally been centred around the discovery of the Higgs boson. However, the Standard Model description of this particle calls for New Physics. Until a few years ago, the epitome of this New Physics has been supersymmetry, which when endowed with a discrete symmetry called R-parity furnishes a good dark matter candidate. Recently a few alternatives have been put forward. Originally, they were confined to solving the Higgs problem, but it has been discovered that, generically, their most viable implementation (in accord with electroweak precision data, proton decay, etc.) fares far better if a discrete symmetry is embedded in the model. The discrete symmetry is behind the existence of a possible dark matter candidate.

From another viewpoint, stressed in many parts of this book, the past few years have witnessed spectacular advances in cosmology and astrophysics confirming that ordinary matter is a minute part of what constitutes the Universe at large. At the same time in which the LHC will be gathering data, a host of non-collider astrophysical and cosmological observations with ever-increasing accuracy will be carried out in search of dark matter. For example, the upcoming PLANCK experiment will make cosmology enter the era of precision measurements, akin to what we witnessed with the LEP experiments.

The emergence of this new paradigm means it is of utmost importance to analyse and combine data from these upcoming observations with those at the LHC. This will also pave the way to search strategies for the next Linear Collider, LC. This important programme is only possible if a cross-border particle–astroparticle collaboration is set up, having at its disposal common

Particle Dark Matter: Observations, Models and Searches, ed. Gianfranco Bertone. Published by Cambridge University Press. © Cambridge University Press 2010.

or complementary tools to conduct global searches and analyses. Many groups, from erstwhile diverse communities, are now developing, improving, generalizing, interfacing and exploiting such tools for the prediction and analysis of dark matter signals from a combination of terrestrial and non-terrestrial observations, paying particular attention to the astrophysical uncertainties. Most of this work has been conducted in the context of supersymmetry, but the latest numerical tools are not limited to it.

In this chapter, we will go through some of the tools available both from a dark matter point of view and at accelerators. For natural reasons, we will focus on public tools, even though there are some rather sophisticated private tools as well. For supersymmetric dark matter calculations, one of the first public tools available was Neutdriver [1181; 1184]. It was a precursor in the field, but has by now been superseded by other more sophisticated tools. There are currently three publicly available codes for calculations of dark matter densities and dark matter signals: Dark-SUSY [978; 979; 980], micrOMEGAs [260; 261; 262; 263] and IsaRED (part of ISASUSY/Isajet [1539]).

16.1 Annihilation cross-section and the relic density

The general theory behind relic density calculations of dark matter particles is given in Chapter 7. Here we will focus on supersymmetric dark matter (neutralinos) and various tools available for calculating the relic density. All three of the publicly available codes for calculating the relic density of neutralinos, DarkSUSY, micrOMEGAs and IsaRED, are capable of reading (and sometimes writing) SUSY Les Houches Accord (SLHA) files [53; 1767], which allows for an easy interface between these codes and other tools to be described in Section 16.5.

We will in the following refer to these codes and how they calculate the relic density. We will use the notation of Chapter 7 and only write down the equations needed to facilitate our discussions here.

16.1.1 The Boltzmann equation

In most supersymmetric models of interest for dark matter phenomenology, the lightest neutralino, χ, is the lightest supersymmetric particle and our dark matter candidate. As such, we want to calculate the relic density of neutralinos in the Universe today as accurately as possible, which means that we need to solve the Boltzmann equation

$$\frac{dn_\chi}{dt} = -3Hn_\chi - \langle \sigma_{\text{ann}} v \rangle \left(n_\chi^2 - n_{\chi,\text{eq}}^2 \right), \tag{16.1}$$

where n_χ is the number density of neutralinos, H is the Hubble parameter, $\langle \sigma_{\mathrm{ann}} v \rangle$ is the thermally averaged annihilation cross-section and $n_{\chi,\mathrm{eq}}$ is the equilibrium number density of neutralinos. This equation needs to be solved over time (or temperature) properly calculating the thermal average at each time step. When the neutralinos no longer can follow the chemical equilibrium density $n_{\chi,\mathrm{eq}}$, they are said to freeze out. There are several complications in solving Eq. (16.1); for example, we may have resonances and thresholds in our annihilation cross-section. The solution to this is to calculate the annihilation cross-section in general relativistic form, for arbitrary relative velocities, v. Another complication is that other supersymmetric particles of similar mass will be present during freeze-out of the neutralinos. To solve this we need to take into account the so-called co-annihilations between all the SUSY particles that are almost degenerate in mass with the neutralino (in practice, it is often enough to consider co-annihilations between all SUSY particles up to about 50% heavier than the neutralino). Following the discussion in Chapter 7, we can solve for the total number density of SUSY particles,

$$n = \sum_{i=1}^{N} n_i \tag{16.2}$$

instead of only the neutralino number density. It is also advantageous to rephrase the Boltzmann equation in terms of the abundance $Y = n/s$ and use $x = m_\chi/T$ as independent variable instead of time or temperature. When co-annihilations are included, the Boltzmann equation (7.10) can then be written as

$$\frac{\mathrm{d}Y}{\mathrm{d}x} = -\left(\frac{45}{\pi M_{\mathrm{P}}^2}\right)^{-1/2} \frac{g_*^{1/2} m_\chi}{x^2} \langle \sigma_{\mathrm{eff}} v \rangle \, (Y^2 - Y_{\mathrm{eq}}^2), \tag{16.3}$$

where $\langle \sigma_{\mathrm{eff}} v \rangle$ is given by Eq. (7.18).

16.1.2 Solving the Boltzmann equation

To solve the Boltzmann equation (16.3) we need to calculate the thermally averaged annihilation cross-section $\langle \sigma_{\mathrm{ann}} v \rangle$ for each given time (temperature). This is typically quite CPU-intensive, and we therefore need to use some tricks. In DarkSUSY [980], the solution is speeded up by tabulating W_{eff} in Eq. (7.19), but using the momentum of the χ, p_{eff} as independent variable instead of s. This tabulation takes extra care of thresholds and resonances making sure that they are tabulated properly. This tabulated W_{eff}

is then used to calculate the thermal average $\langle \sigma_{\mathrm{eff}} v \rangle$ for each time (temperature), using Eq. (7.18). The advantage with this method is that W_{eff} does not depend on temperature and instead the temperature dependence of $\langle \sigma_{\mathrm{eff}} v \rangle$ is completely taken care of by the other factors in Eq. (7.18). Numerically, one needs to take special care of the modified Bessel functions K_1 and K_2 which both contain exponentials that need to be handled separately to avoid numerical underflows. The Boltzmann equation (16.3) is then solved with a special implicit method with adaptive stepsize control, which is needed because the equation is stiff and develops numerical instabilities unless an implicit method is used.

The details of the DarkSUSY method are as follows (see DarkSUSY manual [980]). The derivative dY/dx in Eq. (16.3) is replaced with a finite difference $(Y_{n+1} - Y_n)/h$, where $Y_n = Y(x_n)$ and $x_{n+1} = x_n + h$. Then Y_{n+1} is computed in two ways: first, the right-hand side of Eq. (16.3) is approximated with $-\frac{1}{2}[\lambda_n(Y_n^2 - Y_{\mathrm{eq},n}^2) + \lambda_{n+1}(Y_{n+1}^2 - Y_{\mathrm{eq},n+1}^2)]$, where $\lambda(x) = (45/\pi M_{\mathrm{P}}^2)^{-1/2} (g_*^{1/2} m/x^2) \langle \sigma_{\mathrm{eff}} v \rangle$, and an analytic solution is used for the resulting second-degree algebraic equation in Y_{n+1}; second, the right-hand side of Eq. (16.3) is approximated with $-\lambda_{n+1}(Y_{n+1}^2 - Y_{\mathrm{eq},n+1}^2)$ and an analytic solution of the algebraic equation for Y_{n+1} is used. The stepsize h is reduced or increased to maintain the difference between the two approximate values of Y_{n+1} within a specified error. Overall, the solution of the Boltzmann equation and the tabulation of W_{eff} solves for the relic density to within about 1%. If needed, higher accuracy can also be chosen as an option.

In micrOMEGAs [262], $\langle \sigma_{\mathrm{ann}} v \rangle$ at a given temperature T is arrived at by performing a direct integration and does not therefore rely on a tabulation of the matrix elements squared. Two modes are provided to perform the integration. In the accurate mode the programme evaluates all integrals by means of an adaptive Simpson integration routine. It automatically detects all singularities of the integrands and checks the precision of the calculation increasing the number of points until an accuracy of 10^{-3} is reached. In the default mode (fast mode) the accuracy is not checked but a set of points is sampled according to the behaviour of the integrand: poles, thresholds and Boltzmann suppression at high momentum. The first integral over scattering angles is performed by means of a five-point Gauss formula. The accuracy of this mode is generally about 1%. The user can also test the precision of the approximation based on expanding the cross-section in terms of its s- and p-wave components.

In the Boltzmann equation, we need to know $g_{\mathrm{eff}}(T)$ and $h_{\mathrm{eff}}(T)$ that enter $g_*^{1/2}$ through Eq. (7.9). In DarkSUSY, the default is to use the estimates in

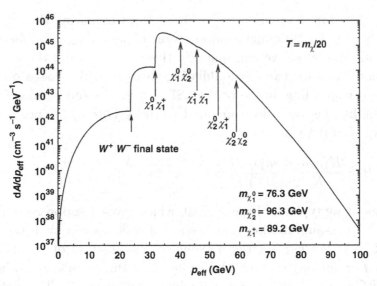

Fig. 16.1. Total differential annihilation rate per unit volume dA/dp_{eff} for the same model as in Fig. 7.3, evaluated at a temperature $T = m_\chi/20$, typical of freeze-out. Notice the Boltzmann suppression at high p_{eff}.

ref. [1099], but other options are also available. Typically, different estimates of $g_{\text{eff}}(T)$ and $h_{\text{eff}}(T)$ translate into relic densities that differ by a few per cent.

IsaRED on the other hand does not solve the Boltzmann equation numerically; instead it finds the freeze-out temperature (the temperature when the annihilation rate equals the expansion rate of the Universe) and calculates the relic density from that (including remnant annihilations at later times). For the thermal averaged annihilation cross-section, it uses the same relativistic treatment as DarkSUSY and micrOMEGAs.

16.1.3 Co-annihilation criteria

In principle, one should include all SUSY particle co-annihilations when calculating the relic density. However, the heavier they are, the less abundant they will be and they can thus be neglected. This is important to speed up the calculation, as we would otherwise spend most of our CPU cycles on calculating non-important co-annihilation cross-sections. One can estimate which particles to include, by investigating Eq. (7.18). The modified Bessel function K_1 contains an exponential, the so-called Boltzmann suppression, that will suppress all heavier particles. In Fig. 16.1, we show dA/dp_{eff} for the same model as in Fig. 7.3. Essentially (apart from normalization) dA/dp_{eff}

is the integrand in the numerator in Eq. (7.18). Comparing the two figures, we clearly see the Boltzmann suppression of larger p_{eff}, i.e. heavier co-annihilating particles. We can quantify this by comparing the Boltzmann suppression factor for two co-annihilating particles with masses m_i and m_j with the corresponding factor for the LSP, the χ. The suppression factor for the co-annihilating particles compared to the χ is roughly (neglecting the p_{eff}^2 in Eq. (7.18)) given by

$$B = \frac{K_1((m_i + m_j)/T)}{K_1(2m_\chi/T)} \simeq e^{-x\frac{m_i + m_j - 2m_\chi}{m_\chi}} ; \quad x = \frac{m_\chi}{T}. \tag{16.4}$$

At freeze-out we typically have $x \simeq 20$, which gives a suppression factor of $B \simeq 10^{-6}$ for co-annihilation particles about 50% heavier than the LSP, χ. In DarkSUSY, one can set the maximum mass fraction of co-annihilation particles $f = m_i/m_\chi$ that will be included in the calculation, whereas in micrOMEGAs one sets instead a minimum value for B. The defaults in DarkSUSY ($f = 1.5$) and micrOMEGAs ($B_{\text{min}} = 10^{-6}$) are roughly equivalent. One should remember, though, that the value to choose depends on the particle physics model. For example, for chargino co-annihilations, the $\chi^+\chi^-$ co-annihilation cross-section can be orders of magnitudes larger than the $\chi\chi$ annihilation cross-sections, and one should choose f or B_{min} so that one does not accidentally neglect co-annihilations that are important. For the MSSM, the default values of DarkSUSY and micrOMEGAs are typically sufficient for all interesting cases. IsaRED instead includes a preset collection of particles that are of relevance for the mSUGRA set-up.

16.1.4 Annihilation cross-section

At the heart of the relic density calculation are the annihilation and co-annihilation cross-sections. In the MSSM there are over 2800 sub-processes (not counting charged-conjugate final states) that can in principle contribute in the relic density calculation. It appears at first sight to be a daunting task to provide such a general code.

In DarkSUSY, all annihilation and co-annihilation cross-sections for the MSSM[1] are calculated at tree level by hand with the help of symbolic programmes like Reduce, Form or Mathematica. The calculations are performed with general expressions for the vertices for the Feynman rules and the results are converted to Fortran code. The vertices are then calculated numerically for any given MSSM model. The analytically calculated cross-sections are

---

[1] Gluino co-annihilations are currently not included.

differential in the angle of the outgoing particles, and the integration over the outgoing angle is performed numerically.

In micrOMEGAs, on the other hand, any annihilation and co-annihilation cross-sections are calculated automatically and generated on the fly. This is possible thanks to an interface to CalcHEP [1614], which is an automatic matrix element/cross-sections generator. This automation is carried one step further in that CalcHEP itself reads its MSSM model file (Feynman rules) from LanHEP [1725], which outputs the complete set of Feynman rules from a simple coding of the Lagrangian (see Section 16.5). In the first call to micrOMEGAs only those subprocesses needed for the given set of the MSSM parameters are generated. The corresponding 'shared' library is stored on the user disk space and is accessible for all subsequent calls, thus each process is generated and compiled only once. This library is then filled with more and more processes whenever the user needs new processes for different MSSM scenarios. This avoids having to distribute a huge code with all the possible 2800 processes.

Both methods have advantages and disadvantages. In the DarkSUSY set-up, no recalculation of the (analytical) annihilation cross-sections is needed, which can speed things up. Also, the analytically calculated annihilation cross-sections can be optimized to be faster. On the other hand, the micrOMEGAs set-up makes it easier to adapt the code to non-MSSM cases. In both codes, though, the actual Boltzmann equation solver is very general and works for any kind of WIMP dark matter, not only SUSY dark matter. In IsaRED, CompHEP [383] is used to calculate the annihilation and co-annihilation cross-sections for a subset of SUSY particles of relevance mostly for mSUGRA (the two lightest neutralinos, the lightest chargino, the left-handed eigenstates of sleptons and squarks, and gluinos). The expressions for the annihilation cross-sections in IsaRED are not calculated on the fly, but are instead precalculated and included with the code.

16.2 Direct detection

Detailed expressions for detection rates in direct detection experiments are presented in Chapter 17. Here we focus on characteristics of the elastic scattering cross-sections and event rates as they are implemented in numerical tools. Direct detection rates depend on the differential elastic WIMP-nucleus cross-section $d\sigma_{WN}/dE_R$, where E_R is the energy of the recoiling nucleus:

$$\frac{d\sigma_{WN}}{dE_R} = \frac{m_N}{2\mu_N^2 v^2} \left(\sigma_0^{SI} F_{SI}^2(E_R) + \sigma_0^{SD} F_{SD}^2(E_R) \right). \tag{16.5}$$

Here m_N is the nucleus mass, $\mu_N = m_\chi m_N/(m_\chi + m_N)$ is the WIMP–nucleus reduced mass (m_χ being the WIMP mass), $\mathbf{v}$ is the WIMP–nucleus relative velocity before the collision, $\sigma_0^{\mathrm{SI,SD}}$ are the spin-independent and spin-dependent cross-sections at zero momentum transfer and $F_{\mathrm{SI,SD}}^2(E_R)$ are the squares of the corresponding form factors (also called structure functions). In terms of these quantities, the directional and non-directional direct detection rates read

$$\frac{\mathrm{d}R}{\mathrm{d}E_R \mathrm{d}\Omega_q} = \frac{\rho_0}{8\pi\mu_N^2 m_\chi} \sigma_{WN}(E_R)\, \hat{f}(v_{\min}, \Omega_q), \qquad (16.6)$$

$$\frac{\mathrm{d}R}{\mathrm{d}E_R} = \frac{\rho_0}{2\mu_N^2 m_\chi} \sigma_{WN}(E_R)\, \eta(v_{\min}), \qquad (16.7)$$

where

$$\sigma_{WN}(E_R) = \sigma_0^{\mathrm{SI}} F_{\mathrm{SI}}^2(E_R) + \sigma_0^{\mathrm{SD}} F_{\mathrm{SD}}^2(E_R), \qquad (16.8)$$

$$\eta(v_{\min}) = \int_{v > v_{\min}} \frac{f(\mathbf{v})}{v}\, \mathrm{d}^3 v, \qquad (16.9)$$

$v_{\min} = [(m_N E_R)/(2\mu_N^2)]^{1/2}$ is the minimum WIMP speed which can cause a recoil of energy E_R, Ω_q is the direction of the nucleus recoil momentum of magnitude $q = \sqrt{2m_N E_R}$ and $\hat{f}(v_{\min}, \Omega_q)$ is the Radon transform of the velocity distribution function $f(\mathbf{v})$. Equation (16.7) generalizes Eq. (17.1) of Chapter 17 to anisotropic WIMP velocity distributions $f(\mathbf{v})$.

An important property of Eqs. (16.6) and (16.7) is the factorization of the particle physics properties, $\sigma_{WN}(E_R)$, and the astrophysics properties, $\rho_0 \hat{f}(v_{\min}, \Omega_q)$ and $\rho_0 \eta(v_{\min})$.

Dark matter codes such as DarkSUSY, IsaRED/RES and micrOMEGAs compute the particle physics and the astrophysics factors to various levels of precision and offer a number of choices for the form factors and the velocity distribution (more and more as they are upgraded). All codes provide the zero-momentum transfer cross-sections (although some are still limited to the axial and scalar couplings of supersymmetric neutralinos), and most of the codes provide routines for direct detection rates off composite targets besides single nuclei.

For the spin-independent part, dark matter codes use the factorized form $\sigma_{WN}^{\mathrm{SI}}(E_R) = \sigma_0^{\mathrm{SI}} F_{\mathrm{SI}}^2(E_R)$ with, in the notation of Chapter 17,

$$\sigma_0^{\mathrm{SI}} = \frac{4\mu_n^2}{\pi} \left[\left(Z f^p + (A - Z) f^n \right)^2 + \frac{B_N^2}{256} \right]. \qquad (16.10)$$

Various expressions for the spin-independent form factor $F_{\mathrm{SI}}^2(E_R)$ are typically available. For example, DarkSUSY 5 automatically selects the best available form factor among Sums-of-Gaussians, Fourier-Bessel and Helm parameterizations (see [722] for a comparison of these approximations).

The spin-dependent part is often not factorized, so as to use the same functions provided by detailed simulations at zero and non-zero momentum transfer. With the by-now-standard normalization of the spin structure functions in [793], one has

$$
\sigma_0^{\text{SD}} F_{\text{SD}}^2(E_{\text{R}}) = \frac{32\mu_N^2 G_{\text{F}}^2}{2J+1} \left[a_0^2 S_{00}(E_{\text{R}}) + a_1^2 S_{11}(E_{\text{R}}) + a_0 a_1 S_{01}(E_{\text{R}}) \right]
$$

(16.11)

$$
= \frac{32\mu_N^2 G_{\text{F}}^2}{2J+1} \left[a_p^2 S_{pp}(E_{\text{R}}) + a_n^2 S_{nn}(E_{\text{R}}) + a_p a_n S_{pn}(E_{\text{R}}) \right],
$$

(16.12)

where $a_0 = a_p + a_n$, $a_1 = a_p - a_n$,

$$
S_{pp}(E_{\text{R}}) = S_{00}(E_{\text{R}}) + S_{11}(E_{\text{R}}) + S_{01}(E_{\text{R}}), \tag{16.13}
$$
$$
S_{nn}(E_{\text{R}}) = S_{00}(E_{\text{R}}) + S_{11}(F_{\text{R}}) - S_{01}(E_{\text{R}}), \tag{16.14}
$$
$$
S_{pn}(E_{\text{R}}) = 2\left[S_{00}(E_{\text{R}}) - S_{11}(E_{\text{R}}) \right]. \tag{16.15}
$$

When the nuclear spin is approximated by the spin of the odd nucleon only, one finds [50]

$$
S_{pp}(E_{\text{R}}) = \frac{\lambda_N^2 J(J+1)(2J+1)}{\pi}, \quad S_{nn}(E_{\text{R}}) = 0, \quad S_{pn}(E_{\text{R}}) = 0, \quad (16.16)
$$

for a proton-odd nucleus, and

$$
S_{pp}(E_{\text{R}}) = 0, \quad S_{nn}(E_{\text{R}}) = \frac{\lambda_N^2 J(J+1)(2J+1)}{\pi}, \quad S_{pn}(E_{\text{R}}) = 0, \quad (16.17)
$$

for a neutron-odd nucleus. Here λ_N is conventionally defined through the relation $\langle N|\mathbf{S}|N\rangle = \lambda_N \langle N|\mathbf{J}|N\rangle$, where $|N\rangle$ is the nuclear state, $\mathbf{S}$ is the nuclear spin and $\mathbf{J}$ is the nuclear total angular momentum. Tables of $\lambda_N^2 J$ $(J+1)$ values for several nuclei can be found in [756] and [1774].

The quantities f^p, f^n, B_N, a_0, a_1 are sums of products of the WIMP–quark and WIMP–gluon coupling constants $\alpha_{q,G}^{\text{S,V,A,P,T}}$ (for scalar, vector, axial, pseudoscalar and tensor currents) and of the contributions f_{TG}, f_{Tq} and Δ_q of the gluons and each quark flavour to the mass and spin of protons and neutrons. Values for the nucleonic matrix elements of gluons and quarks, in practice values for f_{TG}, f_{Tq} and Δ_q, are either hardcoded or can be set by the user. Values for the effective coupling constants $\alpha_q^{\text{S,V,A,P,T}}$ are either precomputed analytically (DarkSUSY) or computed numerically on the fly (micrOMEGAs). For example, the effective Lagrangian at the zero momentum transfer for the interaction of a fermionic WIMP χ with quarks q reads

$$\mathcal{L}_F = \alpha_q^S \, \bar{\chi}\chi \, \bar{q}q + \alpha_q^V \, \bar{\chi}\gamma_\mu\chi \, \bar{q}\gamma^\mu q,$$
$$+ \alpha_q^P \, \bar{\chi}\gamma_5\gamma_\mu\chi \, \bar{q}\gamma_5\gamma^\mu q + \tfrac{1}{2}\alpha_q^T \, \bar{\chi}\sigma_{\mu\nu}\chi \, \bar{q}\sigma^{\mu\nu}q, \tag{16.18}$$

In the case of a Majorana WIMP, like the neutralino in the MSSM, only operators even under $\chi \leftrightarrow \bar{\chi}$ are possible (i.e. $\alpha_q^V = \alpha_q^T = 0$). In micrOMEGAs, the numerical values of the coefficients α_q are obtained combining appropriate matrix elements for $\chi q \to \chi q$ and $\bar{\chi}q \to \bar{\chi}q$ scattering at zero momentum transfer. For example,

$$\alpha_q^S + \alpha_q^V = -\mathrm{i}\,\frac{\langle q(p_1), \chi(p_2)|\hat{S}\mathcal{O}_S|q(p_1), \chi(p_2)\rangle}{\langle q(p_1), \chi(p_2)|\mathcal{O}_S\mathcal{O}_S|q(p_1), \chi(p_2)\rangle},$$
$$\alpha_q^S - \alpha_q^V = -\mathrm{i}\,\frac{\langle \bar{q}(p_1), \chi(p_2)|\hat{S}\mathcal{O}_S|\bar{q}(p_1), \chi(p_2)\rangle}{\langle \bar{q}(p_1), \chi(p_2)|\mathcal{O}_S\mathcal{O}_S|\bar{q}(p_1), \chi(p_2)\rangle}, \tag{16.19}$$

where $\mathcal{O}_S = \bar{\chi}\chi \, \bar{q}q$, the S-matrix $\hat{S} = 1 - \mathrm{i}\mathcal{L}$ is obtained from the complete Lagrangian at the quark level, and the scattering matrix elements on the right-hand sides are computed with CalcHEP. More general cases, including a generic local WIMP–quark operator and WIMPs with spin-0 and spin-1, are presented in [263].

Loop contributions are essential in the treatment of the WIMP–quark and especially WIMP–gluon coupling constants. For example, for neutral WIMPs like the supersymmetric neutralino, there is no neutralino–gluon coupling at the tree-level and the gluon contribution to α^S arises at the one-loop level. Complete analytic one-loop calculations for neutralino–quark and neutralino–gluon couplings were performed in [713; 714]; these formulas are incorporated in DarkSUSY. Automatic numerical calculations of all $\alpha_q^{S,V,A,P,T}$ at one-loop from user-specified generic Lagrangians (with approximate treatment of some of the loop corrections, see [263]) are currently available in micrOMEGAs.

16.3 Indirect detection

Indirect detection methods are many and varied. Here we focus on the following traditional methods: neutrinos from the Sun and the Earth, and gamma-rays, neutrinos and charged cosmic rays (positrons, antiprotons and antideuterons) from annihilations in the Galactic halo. There are also other indirect signals, like synchrotron emission, signals from cosmological haloes (giving a diffuse flux) and indirect consequences of the presence of dark matter in stars (Chapter 29), but we will not focus on them here. Most of the theory needed for this discussion is found in Chapters 24, 25 and 26; we use the notation in those chapters and elaborate on the formulas given there when needed.

The main public tools available to calculate indirect rates are Dark-SUSY [978; 979; 980] and micrOMEGAs [260; 261; 262; 263]. In addition, there are also approximate simple formulas and parameterizations available that can be used, but we will focus here on the numerical codes.

16.3.1 Neutrinos from the Sun/Earth

To calculate the neutrinos from the Sun/Earth, we need to calculate the capture rate of neutralinos in the Sun/Earth, then solve the evolution equation for capture and annihilation in the Sun/Earth, let the neutralinos annihilate in the centre of the Sun/Earth to produce neutrinos and finally let the neutrinos propagate to the neutrino detector at the Earth (taking interactions and oscillations into account).

In Chapter 25, approximate formulas are given for the capture rate in the Sun. These formulas are good for quick calculations, but they include several approximations; with numerical codes, we can actually do better. DarkSUSY is currently the only public code that includes neutrino fluxes from the Sun/Earth and uses the full expressions in ref. [994], where the capture is integrated over the full Sun/Earth including capture on the 16 main elements for the Sun (and 11 for the Earth). In DarkSUSY, an arbitrary velocity distribution can be used if desired in place of the commonly assumed Maxwell–Boltzmann distribution. For example, the Earth does not capture WIMPs directly from the Galactic halo; instead it captures from a distribution that has diffused into the Solar System by gravitation interactions [996]. DarkSUSY uses a velocity distribution at the Earth based on numerical simulations that take this diffusion into account [1354]. There are also indications from more recent numerical simulations of WIMP diffusion in the Solar System [1570] that heavier WIMPs will have a reduced capture rate in the Sun owing to gravitational effects due to Jupiter. The DarkSUSY user can optionally include these effects.

After capture, the evolution equation for the number density accumulated in the Sun/Earth is solved to give the annihilation rate today. Once the WIMPs have accumulated in the centre of the Sun/Earth, they annihilate and eventually produce neutrinos. In DarkSUSY, the annihilation and propagation of neutrinos is handled by a separate code, WimpSim [369; 734]. WimpSim takes care of annihilations to Standard Model particles in the central regions of the Sun/Earth with the help of Pythia [1765]. Energy losses and stopping of particles in the dense environments at the centre of the Sun/Earth are also included. All flavours of neutrinos (and antineutrinos) are then propagated out of the Sun/Earth, taking oscillations and interactions (the latter only relevant for the Sun) into account. This is done in a full three-neutrino-flavour set-up [369]. Once at the detector, the neutrinos are allowed to interact and produce charged leptons and hadronic showers. WimpSim has been run for a range of annihilation channels and masses from 10 GeV to 10 TeV, and the results have been summarized as yield tables that are read and interpolated by DarkSUSY. These results agree very well with a similar analysis in ref. [556], where parameterizations and downloadable data files are also given. For annihilation channels that are dependent on particle physics models (like annihilation to Higgs bosons), the Higgs bosons are allowed to decay in flight in DarkSUSY and the resulting fluxes are calculated from their decay products (properly Lorentz boosted).

The routines in DarkSUSY are also general enough to be easily adapted to other particle dark matter candidates, such as Kaluza–Klein dark matter.

16.3.2 Charged cosmic rays

The theory behind propagation of charged cosmic rays in the Galaxy is presented in Chapter 26. We will use the notation in that chapter. In principle, what we have to do is to solve the master equation (26.4) with appropriate diffusion coefficients, energy loss terms, source terms and boundary conditions. Currently, micrOMEGAs includes the source spectra for arbitrary SUSY models, but does not include the spectra after propagation. This will, however, be included in future versions, using the results in ref. [453]. In DarkSUSY, both the source spectra and the spectra after propagation in various propagation models are included. DarkSUSY implements axisymmetric propagation models and spherically symmetric (or at least axisymmetric) halo models. In DarkSUSY, diffusion is assumed to take place only in space (i.e. the term K_{EE} in Eq. (26.4) is assumed to be negligible). However, it also offers a full interface and integration with the leading cosmic ray propagation code GALPROP [1820] where more sophisticated propagation models can be used. For the halo density, several preset profiles are available, the default being an NFW profile [1501]. However, the user can supply her/his own halo profile; if it is given in the form of Eq. (26.15), it is particularly simple to do so. For solar modulation, DarkSUSY offers a standard spherical force-field approximation as explained in Chapter 26.

For the source spectra (i.e. the spectra before propagation), DarkSUSY uses a similar set-up as for the neutrino fluxes from the Sun/Earth given above. A large set of annihilation channels are simulated with Pythia [1765] in vacuum, for a range of masses, and the yields of antiprotons, positrons, gamma-rays and neutrinos are stored as data tables. These tables are then read and interpolated by DarkSUSY at run-time. Higgs boson decays are included stepping down the decay chain. micrOMEGAs currently uses the same data files as DarkSUSY, but both codes are planning on using new updated simulations in future releases.

For more details about the DarkSUSY implementation of antiproton fluxes, see ref. [297]; for the positron fluxes, see ref. [179]. The antideuteron fluxes are calculated from the antiproton fluxes with the method given in ref. [701].

16.3.3 Gamma-rays and neutrinos

Gamma-ray and neutrino spectra from annihilation in the Galactic halo can be calculated with both DarkSUSY and micrOMEGAs. As mentioned above, the DarkSUSY spectra are based upon Pythia simulations, which are then read in and interpolated by DarkSUSY. Currently, micrOMEGAs uses the same tables as DarkSUSY. DarkSUSY also includes internal bremsstrahlung photons [444] that can be very important in some parts of the parameter space (e.g. in the stau co-annihilation region). DarkSUSY also includes the monochromatic gamma-ray lines from annihilation to $\gamma\gamma$ [302] and $Z\gamma$ [1879] that occur at loop level. Currently, the DarkSUSY and the micrOMEGAs neutrino spectra from halo annihilations do not include neutrino oscillations, but this will be addressed in future versions.

Gamma-rays and neutrinos are not affected by propagation, and hence the flux can be written as

$$\frac{\mathrm{d}^2\Phi_{\gamma/\nu}(\psi)}{\mathrm{d}E\mathrm{d}\Omega} \simeq 9.395 \times 10^{-12} \left(\frac{\mathrm{d}N_{\gamma/\nu}}{\mathrm{d}E}\right) \left(\frac{\sigma v}{10^{-29}\,\mathrm{cm}^3\,\mathrm{s}^{-1}}\right) \left(\frac{10\,\mathrm{GeV}}{m_\chi}\right)^2$$
$$\times J(\psi)\ \mathrm{cm}^{-2}\,\mathrm{s}^{-1}\,\mathrm{sr}^{-1}, \tag{16.20}$$

where we have defined the dimensionless function

$$J(\psi) = \frac{1}{8.5\,\mathrm{kpc}} \cdot \left(\frac{1}{0.3\,\mathrm{GeV}\,\mathrm{cm}^{-3}}\right)^2 \int_{\mathrm{line\ of\ sight}} \rho_\chi^2(l)\,\mathrm{d}\,l(\psi), \tag{16.21}$$

with ψ being the angle from the Galactic centre direction to the direction of observation, and $\mathrm{d}N_{\gamma/\nu}/\mathrm{d}E$ being the spectrum of gamma-rays or neutrinos. The line-of-sight integral, Eq. (16.21), can be calculated for any spherically symmetric profile in DarkSUSY (whereas micrOMEGAs currently implements an isothermal profile only).

16.4 Exploring the parameter space

One problem that arises when exploring a specific supersymmetric model set-up (e.g. mSUGRA or a low-energy MSSM model) is how to scan the parameter space. The dimensions of this parameter space, i.e. the number of free independent parameters, can be large. For example, the general MSSM model has 124 parameters (MSSM-124), 18 of which define the Standard Model (SM). If one assumes *CP*-conservation, the number of parameters reduces to 63 (MSSM-63), which is still a large number. Typically, one reduces the number of parameters still further with inspired theoretical insights (see Chapter 8). In minimal supergravity, for example, unification of coupling constants, of gaugino masses and of scalar masses leaves only 23 parameters, i.e. SM+5, one of which is just a positive or negative sign. An intermediate model often used in neutralino dark matter studies (MSSM-25) has 25 parameters, i.e. SM+7.

One typically wants to find parameter values that are theoretically consistent, have a preferred relic density and are not already excluded by other searches (e.g. rare decays or other accelerator searches). The brute force method would be to scan over the parameter space with some kind of grid scan. One soon realizes that one can typically get a better efficiency in the scans (i.e. more points that pass the cuts, or a better sampling of different interesting regions in the parameter space) by scanning in the logarithm of the mass parameters instead of the mass parameters directly. For higher-dimensional parameter spaces, it is often also more advantageous to scan randomly instead of on a fixed grid.

However, none of these methods are very effective in finding regions of parameter space that pass all the cuts. The relic density cut alone discards many models because of the high precision with which we know the relic density of dark matter today. Hence, more sophisticated methods have evolved that are efficient in generating points inside the interesting regions. Most of these use a Markov Chain Monte Carlo (MCMC) [61; 176; 182; 622; 728; 945] to generate points according to a goal distribution specified by the researcher. The goal distribution can be a function without direct physical meaning (e.g. a Gaussian distribution that peaks in the desired region) or could have a statistical meaning (e.g. a likelihood function or a prior distribution in a Bayesian analysis). A recent public code to perform these tasks, which is linked to DarkSUSY, micrOMEGAs and other codes, is

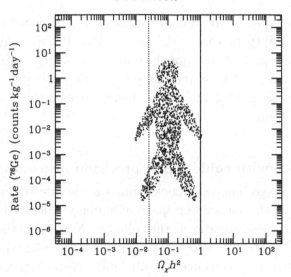

Fig. 16.2. A humorous scatter plot showing correlations that may arise with appropriate choices of priors while scanning parameter space: in this case, the priors reflect the anthropic principle.

SuperBayeS [622]. These advanced methods can be very effective in finding the interesting regions of the parameter space. However, when interpreting the distribution of points that these methods produce, one should be very careful. The definition of 'interesting' is different for different investigators, and the way points are generated always involves a prior in parameter space (even grid methods can be said to have a prior, namely a series of Dirac delta functions at each grid point). One could go to the extreme of producing any kind of results by choosing appropriate priors. Figure 16.2, for example, shows the correlation between the direct detection rate and the relic density jokingly obtained with priors that reflect the anthropic principle. When parameter space scans with priors are used to compute statistical inferences on data, e.g. likelihood contours [1869], one should keep in mind their rather severe dependence on the priors, especially the priors on very poorly known parameters for which there are little or no experimental data (supersymmetric masses, for example). This dependence arises from the use of Bayes' theorem, which gives the probability of the model parameter given the data (the likelihood) in terms of the probability of the data given the model parameters (the assumed distribution of experimental and theoretical errors) as

$$\mathrm{Prob(model|data)} = \mathrm{Prob(data|model)} \frac{\mathrm{Prob(model)}}{\mathrm{Prob(data)}}. \tag{16.22}$$

While the probability of the data Prob(data) is just a normalization constant, the probability of the model parameters Prob(model) is the prior representing the degree of belief or the relative preference the researcher has in specific values of the model parameters. When real experimental data on particle dark matter models are in, the dependence on the priors is expected to become less severe.

16.5 Interface with collider and precision measurements codes

Until a few years ago, one used constraints on the inferred amount of dark matter to delimit the parameter space of supersymmetry in order to narrow searches of supersymmetry at the colliders. Now, with the improvement of the precision on the cosmological parameters, one asks whether the LHC and LC can match the precision of the upcoming cosmology and astrophysics experiments in, for example, reconstructing the relic density once supersymmetry is identified [57; 176]. This may even bring a bonus in that one can test some cosmological and astrophysical assumptions, like indirectly 'probing' the history of the early Universe. For this programme to be feasible one needs to control the particle physics component with as much accuracy as possible. To be able to conduct a cohesive and self-consistent precision test of the origin of DM from the particle physics point of view, one needs to calculate not only those dark matter cross-sections but also the observables at the colliders that are predicted for the same dark matter model. Ideally, therefore, one would like a common tool that performs this task. In many instances this also requires that one goes beyond calculations at tree level. This is especially true in the case of supersymmetry where it is known that radiative corrections can be large. Some progress in this direction has also been made recently within the SloopS Collaboration [212; 213; 411; 412; 887].

The dark matter codes for supersymmetry such as DarkSUSY, IsaRED/RES and micrOMEGAs include some higher-order effects. Moreover, because of the complexity of the MSSM which has a large number of parameters and a large array of predictions, these codes also rely on other more specific codes that predict various other observables in supersymmetry. This concerns for example codes for the calculation of the spectrum based on the renormalization group equations (RGE) that predict the low-energy physical masses from an input at the unification scale in some constrained model of supersymmetry breaking. Spectrum calculators include Spheno [1587], Softsusy [55], suspect [681] and ISASUGRA (part of ISAJET) [1539]. These codes themselves may borrow from more specialized codes like those for the calculation of the Higgs masses, such as Feynhiggs [1082]. The codes for the mass spectra may also feed in stand-alone codes for the calculation of precision measurements, like the calculation of $(g-2)_\mu$ and B observables. Examples of such codes or 'flavour calculators' are SUSYbsg [641] and superiso [1373].

To make contact with LHC and LC observables one of course needs matrix elements for cross-sections and decays of the supersymmetric particles. Many multipurpose matrix element generators exist for supersymmetry, among them Amegic [1276], CalcHEP [1614], CompHEP [383], Grace-SUSY [895], Omega [1230] and Madgraph [76]. Multipurpose matrix element generators can return results for any cross-section or decay of a supersymmetric particle at the tree level. More dedicated and specialized codes in this category (cross-sections and decays) usually improve by going beyond tree level, among them PROSPINO [246] for the production of superparticles at a hadronic collider, HDECAY [680] for the decay of the Higgs and SDECAY [1485] for the decay of other sparticles. Automatic codes for generic one-loop cross-sections and decays with supersymmetry have also been completed recently as concerns the electroweak corrections and some classes of QCD corrections: Grace-susy-1loop [896] and SloopS [212; 213].

For simulations at the colliders one still needs to incorporate the matrix elements for the production (the hard process) and the decays (of the unstable superparticle resonances) into fully fledged Monte Carlo generators. The latter include (i) parton shower (radiation), (ii) multiple interaction and beam remnant in hadronic machines and (iii) hadronization. The main Monte Carlo event generators are currently Herwig++ [168], ISAJET [1539], Pythia [1765] and SHERPA [965]. Figure 16.3 shows the ingredients that go into a Monte Carlo event generator. The mass spectrum module is what defines the model here, so its content is also encoded in the dark matter codes. One can use these generators to simulate the signatures of a particular model at the colliders and combine these findings with the manifestation of the same model in dark matter searches (direct and/or indirect), the prediction of the relic density in a particular cosmological model. One can constrain or reconstruct the model even further by taking into account observables from indirect precision measurements encoded in the flavour calculators. Codes ('Fitters') that perform these fits or constraints have been written specifically with supersymmetry in mind; we can mention Fittino [244], SFitter [1299] and SuperBayeS [622].

As we have seen, there is a very large variety of codes that cover different aspects of the phenomenology of supersymmetry. A recent compendium of these codes can be found in [56]. Because of the large number of parameters in a general supersymmetric model and because many modules are fed into

F. Boudjema, J. Edsjö and P. Gondolo

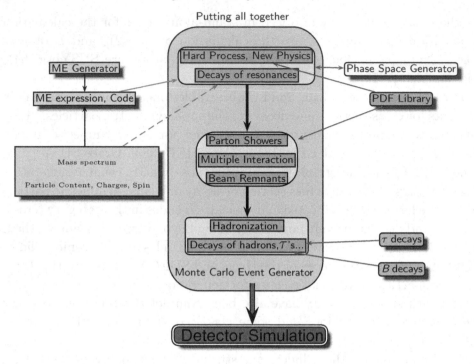

Fig. 16.3. Modules that are involved in the construction of a Monte Carlo event generator for particle collider physics.

other modules, it is best to avoid errors as much as possible when passing parameters from one code to another. Some of these errors can be as trivial as a problem of sign convention. The SUSY Les Houches Accord (SLHA) [1767] allows an easy parsing for the MSSM parameters. This accord has been extended [53] to deal with more general supersymmetric models, like the NMSSM for which a version of micrOMEGAs exists that uses or can be used with the NMHDECAY code [789], or with the inclusion of CP violation via CPSuperH [1320].

Figure 16.4 shows how the different codes for the calculation of super-symmetric observables both at the colliders and in dark matter searches or the evaluation of the relic density are interrelated. The calculation of the matrix elements needed for these codes requires first of all reading the Feynman rules. This in itself is a titanic endeavour because of the complexity of the MSSM and its extensions. For the dark matter codes, where a very large number of processes are involved, especially in the calculation of the relic density, practically the whole set of rules is called for. This is even more so for one-loop calculations. Special tools now exist to achieve this task

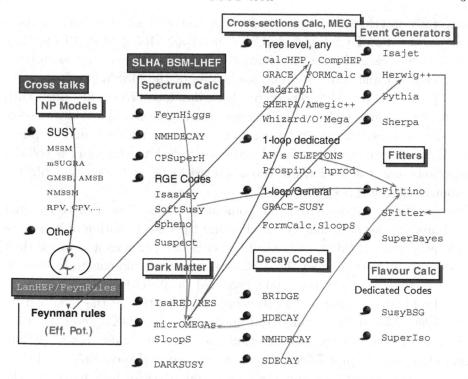

Fig. 16.4. A collection of codes for the calculation of observables for supersymmetric particles. The event generators and the tree-level matrix elements codes cover also the Standard Model and in some instances other models of New Physics. The arrows show how the output of one code is fed into another code, which can itself be a source for other codes.

whereby the model file (containing the Feynman rules) is generated automatically from just coding the Lagrangian in a manner as close as possible to the calculation by hand. This was first done more than 10 years ago by LanHEP [1725] based on C for easy interface with CompHEP. The implementation here is very similar to the canonical coordinate representation. Use of multiplets and the superpotential is built in to minimize human error. The ghost Lagrangian is derived directly from the BRST transformations. Very recently FeynRules [547] based on Mathematica enables the same task to be performed and can output to different matrix element generators. The rapid development of MicrOMEGAs was in large part made possible because of the extensive automation based on LanHEP and CalcHEP/CompHEP. LanHEP has also been greatly extended to automatically implement a model file for calculations at one loop. It is now possible to shift fields and parameters and thus generate counterterms most efficiently. LanHEP has been successfully interfaced with the highly efficient and automatized one-loop packages based on FeynArts/FeynCalc/LoopTools [1053; 1054; 1055; 1056]. The SloopS package is the combination of LanHEP and FeynArts, after a fully consistent and complete renormalization of the MSSM has been completed [212; 213; 411; 412]. SloopS has been developed from the outset such that it is applicable to both high-energy collider observables and processes occurring at very low velocity, such as is the case with dark matter particles. This will bring the cross breeding between dark matter and the collider predictions to a new level of precision, at least as far as the particle physics component is concerned.

We are seeing intense activity in combining codes to be able to conduct global analyses for dark matter searches and the determination of its microscopic properties. If future colliders discover SUSY particles and probe their properties, one could predict the dark matter density and constrain cosmology with the help of precision data provided by WMAP and PLANCK. It would be highly exciting if the precision reconstruction of the relic density from observables at the colliders did not match PLANCK's determination. This would mean that the post-inflation era is most probably not radiation-dominated (see Chapter 7 for a discussion of alternative cosmologies before Big Bang nucleosynthesis and their effect on particle dark matter). The same collider data on the microscopic properties of DM, when put against a combination of data from direct/indirect detection, can also give strong constraints on the astrophysical properties of DM such as its distribution and clustering. These properties reveal much about galaxy formation [57; 176]. For this programme to be carried through successfully, tools developed in the cosmology/astrophysics community and tools developed within the particle physics collider community (and their interfaces) are essential.

Acknowledgements

This work was supported in part by the GDRI-ACPP of CNRS and by the French ANR project ToolsDMColl, BLAN07-2-194882.

Part IV

Direct detection

17

Direct detection of WIMPs

David G. Cerdeño and Anne M. Green

17.1 Introduction

If the Milky Way's DM halo is composed of WIMPs, then the WIMP flux on the Earth is of the order of $10^5(100\,\text{GeV}/m_\chi)\,\text{cm}^{-2}\,\text{s}^{-1}$. This flux is sufficiently large that, even though the WIMPs are weakly interacting, a small but potentially measurable fraction will elastically scatter off nuclei. Direct detection experiments aim to detect WIMPs via the nuclear recoils, caused by WIMP elastic scattering, in dedicated low background detectors [988]. More specifically they aim to measure the rate, R, and energies, E_R, of the nuclear recoils (and in directional experiments the directions as well).

In this chapter we overview the theoretical calculation of the direct detection event rate and the potential direct detection signals. Section 17.2 outlines the calculation of the event rate, including the spin-independent and spin-dependent contributions and the hadronic matrix elements. Section 17.3 discusses the astrophysical input into the event rate calculation, including the local WIMP velocity distribution and density. In Section 17.4 we describe the direction detection signals, specifically the energy, time and direction dependence of the event rate. Finally in Section 17.5 we discuss the predicted ranges for the WIMP mass and cross-sections in various particle physics models.

17.2 Event rate

The differential event rate, usually expressed in terms of counts $\text{kg}^{-1}\,\text{day}^{-1}$ keV^{-1} (a quantity referred to as a differential rate unit or dru), for a WIMP

Particle Dark Matter: Observations, Models and Searches, ed. Gianfranco Bertone. Published by Cambridge University Press. © Cambridge University Press 2010.

with mass m_χ and a nucleus with mass m_N is given by

$$\frac{\mathrm{d}R}{\mathrm{d}E_\mathrm{R}} = \frac{\rho_0}{m_N\,m_\chi} \int_{v_{\min}}^{\infty} v f(v) \frac{\mathrm{d}\sigma_{WN}}{\mathrm{d}E_\mathrm{R}}(v, E_\mathrm{R})\,\mathrm{d}v\,, \qquad (17.1)$$

where ρ_0 is the local WIMP density, $\frac{\mathrm{d}\sigma_{WN}}{\mathrm{d}E_\mathrm{R}}(v, E_\mathrm{R})$ is the differential cross-section for the WIMP–nucleus elastic scattering and $f(v)$ is the WIMP speed distribution in the detector frame normalized to unity.

Since the WIMP–nucleon relative speed is of order $100\,\mathrm{km^{-1}\,s^{-1}}$, the elastic scattering occurs in the extreme non-relativistic limit, and the recoil energy of the nucleon is easily calculated in terms of the scattering angle in the center of mass frame, θ^*

$$E_\mathrm{R} = \frac{\mu_N^2 v^2 (1 - \cos\theta^*)}{m_N}\,, \qquad (17.2)$$

where $\mu_N = m_\chi m_N/(m_\chi + m_N)$ is the WIMP–nucleus reduced mass.

The lower limit of the integration over WIMP speeds is given by the minimum WIMP speed which can cause a recoil of energy E_R: $v_{\min} = \sqrt{(m_N E_\mathrm{R})/(2\mu_N^2)}$. The upper limit is formally infinite, but the local escape speed v_{esc} (see Section 17.3.2) is the maximum speed *in the Galactic rest frame* for WIMPs which are gravitationally bound to the Milky Way.

The total event rate (per kilogram per day) is found by integrating the differential event rate over all the possible recoil energies:

$$R = \int_{E_\mathrm{T}}^{\infty} \mathrm{d}E_\mathrm{R} \frac{\rho_0}{m_N\,m_\chi} \int_{v_{\min}}^{\infty} v f(v) \frac{\mathrm{d}\sigma_{WN}}{\mathrm{d}E_\mathrm{R}}(v, E_\mathrm{R})\,\mathrm{d}v\,, \qquad (17.3)$$

where E_T is the threshold energy, the smallest recoil energy which the detector is capable of measuring.

The WIMP–nucleus differential cross-section encodes the particle physics inputs (and associated uncertainties) including the WIMP interaction properties. It depends fundamentally on the WIMP–quark interaction strength, which is calculated from the microscopic description of the model, in terms of an effective Lagrangian describing the interaction of the particular WIMP candidate with quarks and gluons. The resulting cross-section is then promoted to a WIMP–nucleon cross-section. This entails the use of hadronic matrix elements, which describe the nucleon content in quarks and gluons, and are subject to large uncertainties. In general, the WIMP–nucleus cross-section can be separated into a spin-independent (scalar) and a spin-dependent contribution,

$$\frac{\mathrm{d}\sigma_{WN}}{\mathrm{d}E_\mathrm{R}} = \left(\frac{\mathrm{d}\sigma_{WN}}{\mathrm{d}E_\mathrm{R}}\right)_{\mathrm{SI}} + \left(\frac{\mathrm{d}\sigma_{WN}}{\mathrm{d}E_\mathrm{R}}\right)_{\mathrm{SD}}. \qquad (17.4)$$

Finally, the total WIMP–nucleus cross-section is calculated by adding coherently the above spin and scalar components, using nuclear wavefunctions. The form factor, $F(E_{\mathrm{R}})$, encodes the dependence on the momentum transfer, $q = \sqrt{2m_N E_{\mathrm{R}}}$, and accounts for the coherence loss which leads to a suppression in the event rate for heavy WIMPs or nucleons. In general, we can express the differential cross-section as

$$\frac{\mathrm{d}\sigma_{WN}}{\mathrm{d}E_{\mathrm{R}}} = \frac{m_N}{2\mu_N^2 v^2}\left(\sigma_0^{\mathrm{SI}} F_{\mathrm{SI}}^2(E_{\mathrm{R}}) + \sigma_0^{\mathrm{SD}} F_{\mathrm{SD}}^2(E_{\mathrm{R}})\right), \qquad (17.5)$$

where $\sigma_0^{\mathrm{SI,\,SD}}$ are the spin-independent and -dependent cross-sections at zero momentum transfer.

The origin of the different contributions is best understood at the microscopic level, by analysing the Lagrangian which describes the WIMP interactions with quarks. The contributions to the spin-independent cross-section arise from scalar and vector couplings to quarks, whereas the spin-dependent part of the cross-section originates from axial-vector couplings. These contributions are characteristic of the particular WIMP candidate (see for example [205]) and can be potentially useful for their discrimination in direct detection experiments.

17.2.1 Spin-dependent contribution

The contributions to the spin-dependent (SD) part of the WIMP–nucleus scattering cross-section arise from couplings of the WIMP field to the quark axial current, $\bar{q}\gamma_\mu\gamma_5 q$. For example, if the WIMP is a (Dirac or Majorana) fermion, such as the lightest neutralino in supersymmetric models, the Lagrangian can contain the term

$$\mathcal{L} \supset \alpha_q^A (\bar{\chi}\gamma^\mu\gamma_5\chi)(\bar{q}\gamma_\mu\gamma_5 q). \qquad (17.6)$$

If the WIMP is a spin-1 field, such as in the case of LKP and LTP, the interaction term is slightly different,

$$\mathcal{L} \supset \alpha_q^A \epsilon^{\mu\nu\rho\sigma}(B_\rho \overleftrightarrow{\partial_\mu} B_\nu)(\bar{q}\gamma^\sigma\gamma_5 q). \qquad (17.7)$$

In both cases, the nucleus, N, matrix element reads

$$\langle N|\bar{q}\gamma_\mu\gamma_5 q|N\rangle = 2\lambda_q^N\langle N|J_N|N\rangle, \qquad (17.8)$$

where the coefficients λ_q^N relate the quark spin matrix elements to the angular momentum of the nucleons. They can be parameterized as

$$\lambda_q^N \simeq \frac{\Delta_q^{(p)}\langle S_p\rangle + \Delta_q^{(n)}\langle S_n\rangle}{J}, \qquad (17.9)$$

where J is the total angular momentum of the nucleus, the quantities Δq^n are related to the matrix element of the axial-vector current in a nucleon, $\langle n|\bar{q}\gamma_\mu\gamma_5 q|n\rangle = 2s_\mu^{(n)}\Delta_q^{(n)}$, and $\langle S_{p,n}\rangle = \langle N|S_{p,n}|N\rangle$ is the expectation value of the spin content of the proton or neutron group in the nucleus.[1] Adding the contributions from the different quarks, it is customary to define

$$a_p = \sum_{q=u,d,s} \frac{\alpha_q^A}{\sqrt{2}G_F}\Delta_q^p ; \quad a_n = \sum_{q=u,d,s} \frac{\alpha_q^A}{\sqrt{2}G_F}\Delta_q^n , \tag{17.10}$$

and

$$\Lambda = \frac{1}{J}\left[a_p\langle S_p\rangle + a_n\langle S_n\rangle\right] . \tag{17.11}$$

The resulting differential cross-section can then be expressed (in the case of a fermionic WIMP) as

$$\left(\frac{d\sigma_{WN}}{dE_R}\right)_{SD} = \frac{16m_N}{\pi v^2}\Lambda^2 G_F^2 J(J+1)\frac{S(E_R)}{S(0)} , \tag{17.12}$$

(using $d|\mathbf{q}|^2 = 2m_N dE_R$). The expression for a spin-1 WIMP can be found, for example, in ref. [205].

In the parameterization of the form factor it is common to use a decomposition into isoscalar, $a_0 = a_p + a_n$, and isovector, $a_1 = a_p - a_n$, couplings

$$S(q) = a_0^2 S_{00}(q) + a_0 a_1 S_{01}(q) + a_1^2 S_{11}(q) , \tag{17.13}$$

where the parameters S_{ij} are determined experimentally.

17.2.2 Spin-independent contribution

Spin-independent (SI) contributions to the total cross-section may arise from scalar–scalar and vector–vector couplings in the Lagrangian:

$$\mathcal{L} \supset \alpha_q^S \bar{\chi}\chi\bar{q}q + \alpha_q^V \bar{\chi}\gamma_\mu\chi\bar{q}\gamma^\mu q . \tag{17.14}$$

The presence of these couplings depends on the particle physics model underlying the WIMP candidate. In general one can write

$$\left(\frac{d\sigma_{WN}}{dE_R}\right)_{SI} = \frac{m_N \sigma_0 F^2(E_R)}{2\mu_N^2 v^2} , \tag{17.15}$$

---

[1] These quantities can be determined from simple nuclear models. For example, the single-particle shell model assumes the nuclear spin is solely due to the spin of the single unpaired proton or neutron, and therefore vanishes for even nuclei. More accurate results can be obtained by using detailed nuclear calculations.

where the nuclear form factor for coherent interactions $F^2(E_R)$ can be qualitatively understood as a Fourier transform of the nucleon density and is usually parameterized in terms of the momentum transfer as [793; 1084]

$$F^2(q) = \left(\frac{3 j_1(q R_1)}{q R_1} \right)^2 \exp\left[-q^2 s^2 \right], \qquad (17.16)$$

where j_1 is a spherical Bessel function, $s \simeq 1$ fm is a measure of the nuclear skin thickness and $R_1 = \sqrt{R^2 - 5 s^2}$ with $R \simeq 1.2\, A^{1/2}$ fm. The form factor is normalized to unity at zero momentum transfer, $F(0) = 1$.

The contribution from the scalar coupling leads to the following expression for the WIMP–nucleon cross-section,

$$\sigma_0 = \frac{4 \mu_N^2}{\pi} \left[Z f^p + (A - Z) f^n \right]^2, \qquad (17.17)$$

with

$$\frac{f^p}{m_p} = \sum_{q=u,d,s} \frac{\alpha_q^S}{m_q} f_{Tq}^p + \frac{2}{27} f_{TG}^p \sum_{q=c,b,t} \frac{\alpha_q^S}{m_q}, \qquad (17.18)$$

where the quantities f_{Tq}^p represent the contributions of the light quarks to the mass of the proton, and are defined as $m_p f_{Tq}^p \equiv \langle p | m_q \bar{q} q | p \rangle$. Similarly the second term is due to the interaction of the WIMP and the gluon scalar density in the nucleon, with $f_{TG}^p = 1 - \sum_{q=u,d,s} f_{Tq}^p$. They are determined experimentally,

$$f_{Tu}^p = 0.020 \pm 0.004, \quad f_{Td}^p = 0.026 \pm 0.005, \quad f_{Ts}^p = 0.118 \pm 0.062, \quad (17.19)$$

with $f_{Tu}^n = f_{Td}^p$, $f_{Td}^n = f_{Tu}^p$ and $f_{Ts}^n = f_{Ts}^p$. The uncertainties in these quantities, among which the most important is that in f_{Ts}, mainly stem from the determination of the π-nucleon sigma term.

The vector coupling (which is present, for example, in the case of a Dirac fermion but vanishes for Majorana particles) gives rise to an extra contribution. Interestingly, the sea quarks and gluons do not contribute to the vector current. Only valence quarks contribute, leading to the following expression

$$\sigma_0 = \frac{\mu_N^2 B_N^2}{64 \pi}, \qquad (17.20)$$

with

$$B_N \equiv \alpha_u^V (A + Z) + \alpha_d^V (2A - Z). \qquad (17.21)$$

Thus, for a general WIMP with both scalar and vector interactions, the spin-independent contribution to the scattering cross-section would read

$$\left(\frac{d\sigma_{WN}}{dE_R}\right)_{SI} = \frac{2\,m_N}{\pi v^2}\left[[Zf^p + (A-Z)f^n]^2 + \frac{B_N^2}{256}\right]F^2(E_R). \qquad (17.22)$$

In most cases the WIMP coupling to neutrons and protons is very similar, $f^p \approx f^n$, and therefore the scalar contribution can be approximated by

$$\left(\frac{d\sigma_{WN}}{dE_R}\right)_{SI} = \frac{2\,m_N\,A^2(f^p)^2}{\pi v^2}F^2(E_R). \qquad (17.23)$$

The spin-independent contribution basically scales as the square of the number of nucleons (A^2), whereas the spin-dependent one is proportional to a function of the nuclear angular momentum, $(J+1)/J$. Although in general both have to be taken into account, the scalar component dominates for heavy targets ($A > 20$), which is the case for most experiments (usually based on targets with heavy nuclei such as silicon, germanium, iodine or xenon). Nevertheless, dedicated experiments exist that are also sensitive to the SD WIMP coupling through the choice of targets with a large nuclear angular momentum.

As we have seen, the WIMP direct detection rate depends on both astrophysical input (the local DM density and velocity distribution, in the lab frame) and particle physics input (nuclear form factors and interaction cross-sections, which depend on the theoretical framework in which the WIMP candidate arises). We will discuss these inputs in more detail in Sections 17.3 and 17.5 respectively.

17.2.3 Hadronic matrix elements

The effect of uncertainties in the hadronic matrix elements has been studied in detail for the specific case of neutralino dark matter [394; 396; 754; 781; 783]. Concerning the SI cross-section, the quantities f_{Tq}^p in Eq. (17.19) can be parameterized in terms of the π-nucleon sigma term, $\Sigma_{\pi N}$ (see in this respect, for example, refs. [396; 783]), which, in terms of the u and d quark masses, reads

$$\Sigma_{\pi N} = \frac{1}{2}\left(m_u + m_d\right)\langle N|\bar{u}u + \bar{d}d|N\rangle, \qquad (17.24)$$

and is related to the strange quark scalar density in the nucleon. The largest source of uncertainty in f_{Tq}^p stems from the determination of this quantity, for which the current data imply $\Sigma_{\pi N} = (64 \pm 8)\,\text{MeV}$ [1549], which translates into a variation of a factor of 4 in f_{Ts}. Notice that in general the WIMP interaction with strange quarks would be the leading contribution to the SI cross-section, owing to its larger Yukawa coupling. In this case, σ_0 is roughly proportional to f_{Ts}^2 and the above uncertainty in the strange quark content leads to a variation of more than one order of magnitude in the resulting SI cross-section [396; 783]).

Similarly, for the SD cross-section the uncertainties in the strange spin contribution Δ_s are the dominant contribution to the error in σ_0. However, in the case of the neutralino, this can imply a correction of as much as a factor 2 in the resulting cross-section [783], being therefore much smaller than the above uncertainty for the SI cross-section. It should be emphasized, however, that uncertainties in the determination of the spin form factors $S(q)$ would also affect the theoretical predictions for the dark matter detection rate.

17.3 Astrophysics input

17.3.1 Local DM density

The differential event rate is directly proportional to the local WIMP density, $\rho_0 \equiv \rho(r = R_0)$, where $R_0 = (8.0 \pm 0.5)\,\mathrm{kpc}$ [1636] is the solar radius. Any observational uncertainty in ρ_0 therefore translates directly into an uncertainty in the event rate and the inferred constraints on, or measurements of, the scattering cross-sections.

Exclusion limits are traditionally calculated assuming a canonical local WIMP density, $\rho_0 = 0.3\,\mathrm{GeV\,cm^{-3}}$. The local WIMP density is calculated by applying observational constraints (including measurements of the rotation curve) to models of the Milky Way and the values obtained can vary by a factor of order 2 depending on the models used [164; 304; 484; 918]. A recent study [1933] using spherical halo models with a cusp ($\rho \propto r(r)^{-\alpha}$ as $r \to 0$) finds $\rho_0 = (0.30 \pm 0.05)\,\mathrm{GeV cm^{-3}}$.

17.3.2 Speed distribution

The standard halo model, conventionally used in calculations of exclusion limits and signals, has an isotropic, Gaussian velocity distribution (often referred to as Maxwellian)

$$f(\mathbf{v}) = \frac{1}{\sqrt{2\pi}\sigma} \exp\left(-\frac{|\mathbf{v}|^2}{2\sigma^2}\right). \tag{17.25}$$

The speed dispersion is related to the local circular speed by $\sigma = \sqrt{3/2}v_c$ and $v_c = (220 \pm 20)\,\mathrm{km\,s^{-1}}$ [1223] (see Section 17.3.3) so that $\sigma \approx 270\,\mathrm{km\,s^{-1}}$. This velocity distribution corresponds to an isotropic singular isothermal sphere with density profile $\rho(r) \propto r^{-2}$. The isothermal sphere is simple, and not unreasonable as a first approximation, but it is unlikely to be an accurate model of the actual density and velocity distribution of the Milky Way. Observations and numerical simulations (see Chapter 2) indicate that dark matter haloes do not have a $1/r^2$ density profile and are (to some extent at least) triaxial and anisotropic.

If the velocity distribution is isotropic there is a one-to-one relation between $f(\mathbf{v})$ and the spherically symmetric density profile given by Eddington's formula [733], see refs. [1881; 1900]. In general the steady state phase-space distribution of a collection of collisionless particles is given by the collisionless Boltzmann equation and the velocity dispersions of the system are calculated via the Jeans equations (e.g. ref. [358]). Solving the Jeans equations requires assumptions to be made, and therefore even for a specific density distribution the solution is not unique. Several specific models have been used in the context of WIMP direct detection signals. The logarithmic ellipsoidal model [802] is the simplest triaxial generalization of the isothermal sphere and has a velocity distribution which is a multivariate Gaussian. Osipkov–Merritt models [1420; 1536] are spherically symmetric with radially dependent anisotropic velocity distributions. Fitting functions for the speed distributions in these models are available, for a selection of density profiles, in ref. [1932]. Velocity distributions have also been extracted from cosmological simulations, with both multivariate Gaussian [1085] and Tsallis [1870] distributions [1072] being advocated as fitting functions. While it is not known whether any of these models provide a good approximation to the real local velocity distribution function, the models are nonetheless useful for assessing the uncertainties in the direct detection signals.

Particles with speed, in the Galactic rest frame, greater than the local escape speed, $v_{\rm esc} = \sqrt{2|\Phi(R_0)|}$ where $\Phi(r)$ is the potential, are not gravitationally bound. Many of the models used, in particular the standard halo model, formally extend to infinite radii and therefore their speed distribution has to be truncated at $v_{\rm esc}$ 'by hand' (see for example ref. [719]). The standard value for the escape speed is $v_{\rm esc} = 650\,{\rm km\,s^{-1}}$. A recent analysis, using high-velocity stars from the RAVE survey, finds $498\,{\rm km\,s^{-1}} < v_{\rm esc} < 608\,{\rm km\,s^{-1}}$ with a median likelihood of $544\,{\rm km\,s^{-1}}$ [1772].

In Section 17.4 we discuss the impact of uncertainty in the speed distribution on the direct detection signals.

17.3.3 Earth's motion

The WIMP speed distribution in the detector rest frame is calculated by carrying out a time-dependent Galilean transformation: $\mathbf{v} \to \tilde{\mathbf{v}} = \mathbf{v} + \mathbf{v}_{\rm E}(t)$.

The Earth's motion relative to the Galactic rest frame, $\mathbf{v}_E(t)$, is made up of three components: the motion of the local standard of rest (LSR), the Sun's peculiar motion with respect to the LSR, $\mathbf{v}_\odot^p$, and the Earth's orbit about the Sun, $\mathbf{v}_E^{orb}$.

If the Milky Way is axisymmetric then the motion of the LSR is given by the local circular velocity $(0, v_c; 0)$, where $v_c = 220 \, \mathrm{km \, s^{-1}}$ is the standard value. Kerr and Lynden-Bell found, by combining a large number of independent measurements, $v_c = (222 \pm 20) \, \mathrm{km \, s^{-1}}$ [1223]. A more recent determination, using the proper motions of Cepheids measured by Hipparcos [826], is broadly consistent: $v_c = (218 \pm 7) \, \mathrm{km \, s^{-1}} (R_0/8 \, \mathrm{kpc})$.

The Sun's peculiar motion, determined using the parallaxes and proper motions of stars in the solar neighbourhood from the Hipparcos catalogue, is $\mathbf{v}_\odot^p = (10.0 \pm 0.4, 5.2 \pm 0.6, 7.2 \pm 0.4) \, \mathrm{km \, s^{-1}}$ [643] in Galactic coordinates (where x points towards the Galactic centre, y is the direction of Galactic rotation and z points towards the North Galactic Pole).

A relatively simple, and reasonably accurate, expression for the Earth's motion about the Sun can be found by ignoring the ellipticity of the Earth's orbit and the non-uniform motion of the Sun in right ascension [925]: $\mathbf{v}_E^{orb} = v_E[\mathbf{e}_1 \sin \lambda(t) - \mathbf{e}_2 \cos \lambda(t)]$ where $v_E = 29.8 \, \mathrm{km \, s^{-1}}$ is the orbital speed of the Earth, $\lambda(t) = 2\pi(t - 0.218)$ is the Sun's ecliptic longitude (with t in years) and $\mathbf{e}_{1(2)}$ are unit vectors in the direction of the Sun at the Spring equinox (Summer solstice). In Galactic coordinates $\mathbf{e}_1 = (-0.0670, 0.4927, -0.8676)$ and $\mathbf{e}_2 = (-0.9931, -0.1170, 0.01032)$. More accurate expressions can be found in ref. [1015].

The main characteristics of the WIMP signals can be found using only the motion of the LSR, and for the time dependence the component of the Earth's orbital velocity in that direction. However, accurate calculations, for instance for comparison with data, require all the components described above to be taken into account.

17.3.4 Ultra-fine structure

Most of the WIMP velocity distributions discussed in Section 17.3.2 are derived by solving the collisionless Boltzmann equation, which assumes that the phase-space distribution has reached a steady state. However, this may not be a good assumption for the Milky Way; structure formation in CDM cosmologies occurs hierarchically and the relevant dynamical timescales for the Milky Way are not many orders of magnitude smaller than the age of the Universe.

Both astronomical observations and numerical simulations (owing to their finite resolution) typically probe the dark matter distribution on roughly kpc scales. Direct detection experiments probe the DM distribution on sub-milliparsec scales (the Earth's speed with respect to the Galactic rest frame is $\approx 0.2\,\mathrm{mpc\,yr^{-1}}$). It has been argued (see for example refs. [813; 1470; 1808]) that on these scales the DM may not have yet reached a steady state and could have a non-smooth phase-space distribution. On the other hand it has been argued that the rapid decrease in density of streams evolving in a realistic, ellipsoidal, Galactic potential means that there will be a large number of overlapping streams in the solar neighbourhood producing an effectively smooth DM distribution [1909].

If the local DM distribution consists of a small number of streams, rather than a smooth distribution, then there would be significant changes in the signals which we will discuss in Section 17.4. This is currently an open issue; directly calculating the DM distribution on the scales probed by direct detection experiments is a difficult and unresolved problem.

It has been suggested that a tidal stream from the Sagittarius (Sgr) dwarf galaxy, which is in the process of being disrupted, passes through the solar neighbourhood with the associated DM potentially producing distinctive signals in direct detection experiments [882; 883]. Subsequent numerical simulations of the disruption of Sgr along with observational searches for local streams of stars suggest that the Sgr stream does not in fact pass through the solar neighbourhood (see for example ref. [1715]). Nonetheless, the calculations of the resulting WIMP signals in refs. [882; 883] provide a useful illustration of the qualitative effects of streams.

17.4 Signals

We have already seen in Section 17.2 that the recoil rate is energy dependent owing to the kinematics of elastic scattering, combined with the WIMP speed distribution. Owing to the motion of the Earth with respect to the Galactic rest frame the recoil rate is also both time- and direction-dependent. In this section we examine the energy, time and direction dependence of the recoil rate and the resulting WIMP signals. In each case we first focus on the signal expected for the standard halo model, with a Maxwellian velocity distribution, before discussing the effect on the signal of changes in the WIMP velocity distribution.

17.4.1 Energy dependence

The shape of the differential event rate depends on the WIMP and target masses, the WIMP velocity distribution and the form factor. For the standard halo model the expression for the differential event rate, Eq. (17.1), can be rewritten approximately (cf. ref. [1330]) as

$$\frac{\mathrm{d}R}{\mathrm{d}E_{\mathrm{R}}} \approx \left(\frac{\mathrm{d}R}{\mathrm{d}E_{\mathrm{R}}}\right)_0 F^2(E_{\mathrm{R}}) \exp\left(-\frac{E_{\mathrm{R}}}{E_{\mathrm{c}}}\right), \qquad (17.26)$$

where $(\mathrm{d}R/\mathrm{d}E_{\mathrm{R}})_0$ is the event rate in the $E \to 0$ keV limit. The characteristic energy scale is given by $E_{\mathrm{c}} = (c_1 2\mu_N^2 v_{\mathrm{c}}^2)/m_N$, where c_1 is a parameter of order unity which depends on the target nuclei. If the WIMP is much lighter than the target nuclei, $m_\chi \ll m_N$, then $E_{\mathrm{c}} \propto m_\chi^2/m_N$ while if the WIMP is much heavier than the target nuclei $E_{\mathrm{c}} \propto m_N$. The total recoil rate is directly proportional to the WIMP number density, which varies as $1/m_\chi$.

In Fig. 17.1 we plot the differential event rate for Ge and Xe targets and a range of WIMP masses. As expected, for a fixed target the differential event rate decreases more rapidly with increasing recoil energy for light WIMPs. For a fixed WIMP mass the decline of the differential event rate is steepest for heavy target nuclei. The dependence of the energy spectrum on the WIMP mass allows the WIMP mass to be estimated from the energies of detected events (e.g. ref. [1011]). Furthermore, the consistency of energy spectra measured by experiments using different target nuclei would confirm that the events were due to WIMP scattering (rather than, for instance, neutron backgrounds) [1330]. In particular, for spin-independent interactions, the total event rate scales as A^2. This is sometimes referred to as the 'materials signal'.

The WIMP and target mass dependence of the differential event rate also have some general consequences for experiments. The dependence of the total event rate on m_χ means that, for fixed cross-section, a larger target mass will be required to detect heavy WIMPs than lighter WIMPs. For very light WIMPs the rapid decrease of the energy spectrum with increasing recoil energy means that the event rate above the detector threshold energy, E_{T}, may be small. If the WIMP is light, $<\mathcal{O}(10\,\mathrm{GeV})$, a detector with a low threshold energy, $<\mathcal{O}(\,\mathrm{keV})$, will be required.

The most significant astrophysical uncertainties in the differential event rate come from the uncertainties in the local WIMP density and circular velocity. As discussed in Section 17.3.1, the uncertainty in the local DM density translates directly into an uncertainty in constraints on (or in the future measurements of) the scattering cross-section. The *time-averaged* differential event rate is found by integrating the WIMP velocity distribution, therefore it is only weakly sensitive to changes in the shape of the WIMP velocity distribution. For the smooth halo models discussed in Section 17.3.2

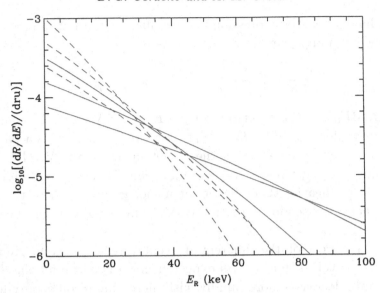

Fig. 17.1. The dependence of the spin-independent differential event rate on the WIMP mass and target. The solid and dashed lines are for Ge and Xe respectively and WIMP masses of (from top to bottom at $E_R = 0\,\text{keV}$) 50, 100 and 200 keV. The scattering cross-section on the proton is taken to be $\sigma_p^{SI} = 10^{-8}$ pb.

the time-averaged differential event rates are fairly similar to that produced by the standard halo model [702; 1200]. Consequently exclusion limits vary only weakly [702; 1009] and there would be a small (of order a few per cent) systematic uncertainty in the WIMP mass deduced from a measured energy spectrum [1012]. With multiple detectors it would in principle be possible to measure the WIMP mass without any assumptions about the WIMP velocity distribution [717].

In the extreme case of the WIMP distribution being composed of a small number of streams the differential event rate would consist of a series of steps (sloping because of the form factor). The positions of the steps would depend on the stream velocities and the target and WIMP masses, while the relative heights of the steps would depend on the stream densities.

17.4.2 Time dependence

The Earth's orbit about the Sun leads to a time dependence, specifically an annual modulation, in the differential event rate [719; 881]. The Earth's speed with respect to the Galactic rest frame is largest in summer when the component of the Earth's orbital velocity in the direction of solar motion is largest. Therefore the number of WIMPs with high (low) speeds in the

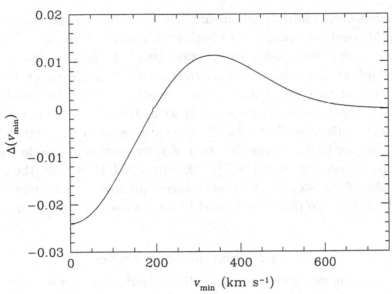

Fig. 17.2 Dependence of the amplitude of the annual modulation, $\Delta(v_{\min})$, on $v_{\min}$.

detector rest frame is largest (smallest) in summer. Consequently the differential event rate has an annual modulation, with a peak in winter for small recoil energies and in summer for larger recoil energies [1604]. The energy at which the annual modulation changes phase is often referred to as the 'crossing energy'.

Since the Earth's orbital speed is significantly smaller than the Sun's circular speed the amplitude of the modulation is small and, to a first approximation, the differential event rate can, for the standard halo model, be written as a Taylor series:

$$\frac{\mathrm{d}R}{\mathrm{d}E_{\mathrm{R}}} \approx \left(\frac{\bar{\mathrm{d}R}}{\mathrm{d}E_{\mathrm{R}}}\right)[1 + \Delta(E_{\mathrm{R}})\cos\alpha(t)], \qquad (17.27)$$

where $\alpha(t) = 2\pi(t - t_0)/T$, $T = 1$ year and $t_0 \sim 150$ days. In Fig. 17.2 we plot the energy dependence of the amplitude in terms of $v_{\min}$ (recall that $v_{\min} \propto E_{\mathrm{R}}^{1/2}$ with the constant of proportionality depending on the WIMP and target nuclei masses). The amplitude of the modulation is of order 1–10%.

The Earth's rotation provides another potential time dependence in the form of a diurnal modulation as the Earth acts as a shield in front of the detector [571; 1079], but the amplitude of this effect is expected to be small (<1% [1079]).

There has been a substantial amount of work on the annual modulation for the non-Maxwellian velocity distributions described in Section 17.3.2 [270; 440; 805; 1008; 1010; 1881; 1898; 1899; 1900]. In contrast to the time-averaged differential event rate, both the phase and amplitude of the annual modulation can vary substantially. Consequently the regions of WIMP mass-cross-section parameter space consistent with an observed signal can change significantly [269; 270]. Note that if the components of the Earth's orbital velocity perpendicular to the direction of solar motion are neglected, then the phase change will be missed [1010]. For a WIMP stream the position and height of the step in the energy spectrum would vary annually (e.g. ref. [1694]). See also Chapters 18 and 19 for a more detailed discussion.

17.4.3 Direction dependence

The detector motion with respect to the Galactic rest frame also produces a directional signal. The WIMP flux in the lab frame is sharply peaked in the direction of motion of the Earth, and hence the recoil spectrum is also peaked in this direction (albeit less sharply because of the elastic scattering).

The directional recoil rate is most compactly written as [976]

$$\frac{\mathrm{d}R}{\mathrm{d}E_{\mathrm{R}}\mathrm{d}\Omega} = \frac{\rho_0 \sigma_0 A^2}{4\pi \mu_p^2 m_\chi} F^2(E_{\mathrm{R}}) \hat{f}(v_{\min}, \hat{\mathbf{q}}), \qquad (17.28)$$

where $\mathrm{d}\Omega = \mathrm{d}\phi\,\mathrm{d}\cos\theta$, $\hat{\mathbf{q}}$ is the recoil direction and $\hat{f}(v_{\min}, \hat{\mathbf{q}})$ is the three-dimensional Radon transform of the WIMP velocity distribution $f(\mathbf{v})$

$$\hat{f}(v_{\min}, \hat{\mathbf{q}}) = \int \delta(\mathbf{v} \cdot \hat{\mathbf{q}} - v_{\min}) f(\mathbf{v}) \mathrm{d}^3 v. \qquad (17.29)$$

Geometrically the Radon transform is the integral of the function $f(\mathbf{v})$ on a plane orthogonal to the direction $\hat{\mathbf{q}}$ at a distance $v_{\min}$ from the origin. See ref. [577] for an alternative, but equivalent, expression.

For the standard halo model the direction dependence is approximately given by [1783]

$$\frac{\mathrm{d}R}{\mathrm{d}E_{\mathrm{R}}\,\mathrm{d}\cos\gamma} \propto \exp\left[-\frac{(v_\odot \cos\gamma - v_{\min})^2}{v_c^2}\right], \qquad (17.30)$$

where γ is the angle between the recoil and the mean direction of solar motion. The distribution of recoil directions peaks in the mean direction of motion of the Sun (towards the constellation CYGNUS [1775]) with the event rate in the forward direction being roughly an order of magnitude larger than that in the backward direction [1783], since the Sun's speed is

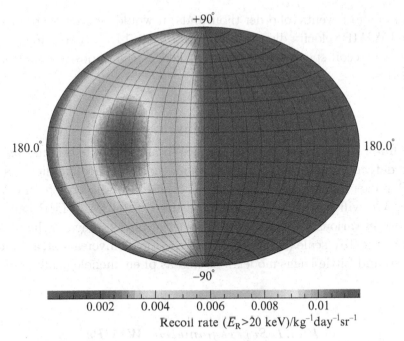

+90°

180.0°   180.0°

−90°

| 0.002 | 0.004 | 0.006 | 0.008 | 0.01 |

Recoil rate $(E_R > 20\text{ keV})/\text{kg}^{-1}\text{day}^{-1}\text{sr}^{-1}$

Fig. 17.3. Hammer–Aitoff projection of the directional recoil rate for S, above $E_T = 20\text{ keV}$, due to a standard halo of WIMPs with $m_\chi = 100\text{ GeV}$. This figure was generated using the HADES directional dark matter simulation code written by Ben Morgan.

comparable to the mean WIMP speed. The directional recoil rate of $m_\chi = 100$ GeV WIMPs on S is shown in Fig. 17.3.

With an ideal directional detector, with 3D readout and capable of measuring the senses of the recoils (i.e. distinguishing between the head and tail of each recoil), it would be possible to distinguish a WIMP signal from isotropic backgrounds with only of order 10 events [576; 577; 1477]. This number increases significantly (by roughly an order of magnitude) if either the senses cannot be measured or the readout is only 2D [579; 1014; 1476]. Another potential directional signal is the rotation of the mean recoil direction in the lab over a sidereal day owing to the motion of the Earth [1775]. See Chapter 22 for the principles and practice of directional detection experiments.

For plausible smooth halo models changes in the WIMP velocity distribution affect the detailed angular recoil rate. However, the rear–front asymmetry is robust and the number of events required by an ideal detector only varies by order 10% [576; 577; 1477]. For non-ideal detectors the variation in the number of events required can be larger [1476; 1477]. With a large number of events (of order thousands) it would be possible to probe the detailed WIMP velocity distribution [1134; 1476; 1477]. A stream of WIMPs produces a recoil spectrum which is peaked in the opposite direction (e.g. ref. [50]).

17.5 Particle physics input

Let us finally address the particle physics input to the determination of the WIMP detection rate, which enters through the theoretical predictions to the WIMP–nucleus scattering cross-section. These are sensitive to the WIMP nature. We will here briefly summarize the results for WIMP candidates that arise in various well-motivated theories of physics beyond the Standard Model at the TeV scale (supersymmetric theories, universal extra dimension scenarios and Little Higgs models), as well as phenomenologically motivated scenarios.

17.5.1 Supersymmetric WIMPs

The canonical and best-studied supersymmetric WIMP is the lightest neutralino, χ_1^0 (see Chapter 8). Its detection properties are very dependent on its composition. More specifically, within the MSSM framework, in the expressions for the scattering amplitudes [713; 714; 754; 932; 1792], the SI part of the neutralino–nucleon cross-section receives contributions from Higgs exchange in a t-channel and squark exchange in an s-channel. The latter also contributes to the SD part of the cross-section, together with a Z boson exchange in a t-channel.

The dependence of the cross-sections and detection prospects on the neutralino composition are well known. For example, a large Higgsino component induces an enhancement of both the Higgs and Z boson exchange diagrams, thereby leading to an increase in both the SD and SI cross-sections. On the other hand, the presence of light squarks (if they become almost degenerate with the neutralino) can lead to an enhancement of (mainly) the SD cross-section.

Analyses of general supersymmetric scenarios with parameters defined at low energy reveal that the neutralino SI cross-section can be as large as 10^{-5} pb for a wide range of neutralino masses up to 1 TeV [781; 1237]. Interestingly, when gaugino masses not fulfilling the GUT relation are allowed, very light neutralinos with masses $m_{\chi_1^0} \gtrsim 7$ GeV, and potentially large cross-sections, can be obtained [399; 409]. It has been argued that these neutralinos could account for the DAMA/LIBRA annual modulation signal without

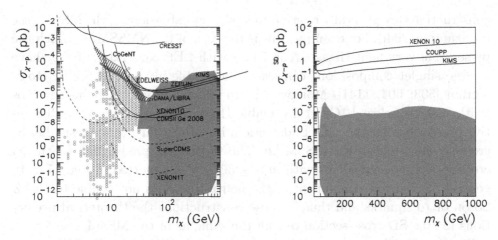

Fig. 17.4. Left: Theoretical predictions for neutralino–nucleon SI cross-section as a function of the neutralino mass obtained by combining the scans of the MSSM parameters from refs. [399; 401; 781; 1237]. The theoretical predictions for the SI cross-section of very light neutralinos in the NMSSM from ref. [2] are shown by means of empty grey circles. Present and projected experimental sensitivities are displayed using solid and dashed lines respectively. Right: Theoretical predictions for neutralino–nucleon SD cross-section as a function of the neutralino mass, using the data from the supergravity scan of [334].

contradicting the null results from CDMS and XENON10 [401], but this interpretation is now more constrained [2]. All these features are illustrated in Fig. 17.4.

Analyses have also been done from the point of view of supergravity theories, where the SUSY parameters are defined at the GUT scale. In the simplest case, the CMSSM, the cross-sections are generally small, since χ_1^0 is mostly bino. The largest cross-sections, $\sigma^{\mathrm{SI}} \approx \mathcal{O}(10^{-8}\,\mathrm{pb})$, are obtained in the focus point region, where the neutralino becomes a mixed bino–Higgsino state [160; 736]. Interestingly, this region seems to be favoured by recent Bayesian analyses of the CMSSM parameter space [60; 622; 1656; 1869]. Moreover, the predicted SD cross-section is also sizable in the focus point region, approximately reaching $10^{-4} - 10^{-3}$ pb. In more general supergravity scenarios the predicted cross-sections can be significantly larger through the inclusion of non-universal values for the scalar masses [19; 121; 127; 155; 161; 287; 391; 510; 650; 712; 716; 751; 755; 773; 782; 1500] (non-universalities in the Higgs mass parameters being the most effective), the gaugino masses [123; 330; 362; 510; 528; 581], or both [512; 1542].

The detection prospects of the lightest neutralino in extended supersymmetric models may be significantly different, mostly because of the changes in the Higgs sector and the presence of new neutralino states. These constructions are generally referred to as singlet extensions of the MSSM (see for example [204]). For example, this is the case of the NMSSM, in which the presence of very light Higgses (consistent with LEP constraints if they have a large singlet composition) can lead to a sizable increase of the SI cross-section [503; 504; 1141]. Moreover, in the NMSSM very light neutralinos (with masses below 10 GeV) are viable [1037] and can have very distinctive predictions for their direct detection, including, for example, smaller SI cross-section than in the MSSM [2]. The theoretical predictions for the SI cross-section of neutralinos with $m_{\chi_1^0} \leq 30$ GeV are plotted in Fig. 17.4. In general, the singlet component of the neutralino does not couple to the Z boson or to squarks and thus in these constructions the theoretical predictions for the SD cross-section remain the same as in the MSSM.

Finally, there is another viable supersymmetric WIMP candidate for dark matter, the lightest sneutrino. The left-handed sneutrino in the MSSM is excluded given its sizable coupling to the Z boson. They therefore either annihilate too rapidly, resulting in a very small relic abundance, or have large scattering cross-sections which have already been excluded by direct detection experiments [807]. Several models have been proposed to revive sneutrino DM by reducing its coupling with the Z boson. This can be achieved by introducing a mixture of left- and right-handed sneutrinos [108; 109; 115; 1124], or by considering a purely right-handed sneutrino in models with an extended gauge sector [1319] or Higgs sector [648; 917] such as the NMSSM [505]. In the first class of models, the elastic scattering of sneutrinos with quarks would take place through the t-exchange of Z bosons, whereas in the second class it would mostly be due to the exchange of Higgs bosons. The resulting SI cross-section in these cases can be within the reach of future detectors for a wide range of sneutrino masses [505; 648]. Being a scalar particle, the SD cross-section vanishes for the sneutrino.

17.5.2 Kaluza–Klein dark matter in UED

Models of universal extra dimensions, in which all fields are allowed to propagate in the bulk [100], also provide well-motivated candidates for WIMP dark matter in the form of the lightest Kaluza–Klein particle (LKP), which is likely to be associated with the first KK excitation of the hypercharge gauge boson, $B_0^{(1)}$ [538; 1730] (see also the discussion in Chapter 9).

The elastic scattering of $B_0^{(1)}$ with quarks takes place through the exchange of KK quarks along t- and s-channels, which contribute to both the SD and SI cross-section, and a Higgs exchange along a t-channel which only gives an SI contribution [533; 1526; 1729]. The theoretical predictions for the elastic

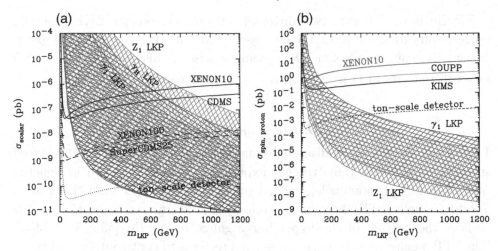

Fig. 17.5. (a) Theoretical predictions for the SI LKP–nucleon scattering cross-section as a function of the LKP mass for the LKP candidates discussed in the text. (b) Predictions for the SD LKP–proton cross-section. Figures reproduced with permission from [128].

scattering cross-sections of $B_0^{(1)}$ are very dependent on the mass splitting between it and the KK quark, Δ_q. In particular, both the SI and SD contributions increase when Δ_q becomes small, as a consequence of the enhancement of the contribution from KK quarks. This is relevant, since in the UED scenario one expects a quasi-degenerate spectrum, in which the splittings between the masses are only induced by radiative corrections [538]. The SI cross-section can also be larger in the presence of light Higgses and for small LKP masses. However, the Higgs mass is generally larger than in the MSSM and its contribution is suppressed with respect to that of KK quarks.

It has been shown that the theoretical predictions for the SI cross-section of $B_0^{(1)}$ can be as large as $\sigma^{\rm SI} \approx 10^{-6}$ pb for masses ranging from 500 GeV to 1 TeV, when $\Delta_q \approx 0.01$, for which the correct relic density can be obtained [478; 1268]. Under the same conditions, the predicted SD cross-sections are smaller, and, for masses up to 1 TeV, ton-scale detectors would be required to detect them. These predictions are illustrated in Fig. 17.5, from [128].

Other LKP candidates are possible within the UED model if non-vanishing boundary terms are allowed. More specifically, one may consider the first excited states associated with either the Z boson, $Z^{(1)}$, or the Higgs, $H^{(1)}$ [128]. The detection properties of the $Z^{(1)}$ are very similar to those of the $B_0^{(1)}$ [128] (although for the $Z^{(1)}$ the neutron and proton spin-dependent cross-sections are exactly the same, contrary to the case of the $B_0^{(1)}$).

Finally, in models with two universal extra dimensions the LKP generally corresponds to the KK excitation associated with the hypercharge gauge boson, the spinless photon [684]. Being a scalar, this particle has no SD cross-section. Its SI cross-section can be similar to that of the $B_0^{(1)}$.

17.5.3 Little Higgs dark matter

In these constructions a discrete symmetry, called T-parity, is introduced in order to alleviate the stringent experimental constraints on low-energy observables. A phenomenological consequence of T-parity is that the lightest T-odd particle (LTP) becomes absolutely stable. Interestingly, the LTP is usually the partner of the hypercharge gauge boson B_H [535; 536; 1353]. The LTP–nucleon scattering cross-section receives SI contributions via Higgs and heavy quark exchange, the latter being the only contribution to the SD part [361]. Thus, the resulting expressions are very similar to the case of KK dark matter in UED. However, unlike the UED scenario, there is no reason for the heavy quarks to be degenerate in mass with A_H. This, together with the smallness of the heavy quark Yukawa couplings, implies that their scattering cross-sections are in fact very suppressed. The SI cross-section, being dominated by the Higgs exchange t-channel, increases slightly when the Higgs mass is small but is nevertheless generally below 10^{-10} pb. The theoretical predictions for the SD cross-section are also very small. In summary, the direct detection of Little Higgs dark matter is much more difficult than the SUSY and UED cases.

17.5.4 Minimal models and other approaches for dark matter

Instead of considering DM candidates arising in existing theories beyond the SM, a bottom-up approach can be adopted in which minimal additions to the SM are considered, involving the inclusion of a WIMP field (usually a new singlet) and new symmetries that protect their decay (in some cases, also a new 'mediator' sector that couples the WIMP to the SM). Examples in this direction include WIMPs with singlet mediation [471; 1347; 1408], models with an extended electroweak sector [196; 557; 1352], models with additional gauge groups, and the 'secluded dark matter' scenario [1595] in which WIMPs could escape direct detection.

The theoretical predictions for the direct detection of WIMPs in this class of models are very dependent on the mediator sector, since it determines the couplings of WIMPs to ordinary matter (quarks). For example, scalar WIMPs interact with ordinary quarks through the exchange of Higgs bosons in a t-channel [471; 1408]. The singlet coupling to the Higgs is constrained in order to reproduce the correct relic abundance, thus leaving only the WIMP and Higgs masses as free parameters. The resulting cross-sections increase as both masses decrease and can be as large as 10^{-6} pb for very light WIMPs of order 10 GeV but are reduced to be below 10^{-8} pb when the DM candidate has a mass above 100 GeV.

Other scenarios can have more general couplings. In the 'inert doublet model', where elastic scattering proceeds only through the exchange of a Higgs or a Z boson along a t-channel, the resulting SI cross-section is rather small. Only for light WIMPs, with masses below 100 GeV, is the predicted SI cross-section large enough to be experimentally tested (from about 10^{-10} pb to as much as 10^{-7} pb) whereas the predictions for heavy WIMPs are well below the sensitivity of ton-size experiments, usually of order 10^{-13} pb [1352].

In the minimal DM approach of [557], WIMP candidates with direct couplings to the Z boson are already excluded by direct DM searches. However, some fermionic candidates are still viable which interact with quarks through the exchange of W (and Higgs) bosons. These particles (with masses of several TeV) can have an SI cross-section of order 10^{-8} pb.

17.5.5 Inelastic cross-section

It is finally worth mentioning that the WIMP–nucleon cross-section can also receive a contribution from inelastic scattering by creating either an excited nuclear [757] or electronic state [1795] or even by creating an excited WIMP state [518; 1871; 1872]. The last possibility is particularly interesting if the mass difference, δ, between the excited dark matter candidate χ_2 and WIMP χ_1 is of order of the WIMP kinetic energy (i.e. about 100 keV). In that case, the inelastic scattering off nuclei $\chi_1 N \to \chi_2 N$ can occur and the only kinematic change is in the minimal WIMP velocity that can trigger a specific recoil energy, which is increased by $\Delta v_{\min} = \delta/\sqrt{(2m_N E_{\mathrm{R}})}$. This clearly favours detection in heavy targets such as iodine (since $\Delta v_{\min}$ is smaller) and might provide a possible explanation for the DAMA/LIBRA signal compatible with the null results in other experiments [518].

17.5.6 Discrimination of dark matter candidates

As illustrated by Figs. 17.4 and 17.5, current experiments are already probing the masses and cross-sections predicted for various WIMP candidates. Furthermore, future experiments will be sensitive to a substantial fraction of

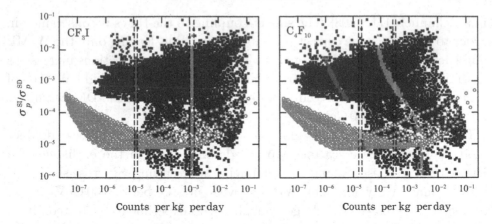

Fig. 17.6. Left: The detection with a given target such as CF_3I can only loosely constrain models for the dark matter (squares for neutralinos, circles for the LKP) in the $\sigma_p^{SI}/\sigma_p^{SD}$ versus count-rate plane. Dots show the many models consistent with a measurement of $\sim 10^{-5}$ (10^{-3}) counts per kg per day. Right: Measurement of the event rate in a second target such as C_4F_{10}, with lower sensitivity to spin-independent couplings, effectively reduces the remaining number of allowed models (dots) and generally allows discrimination between the neutralino and the LKP. Figures reproduced with permission from [334].

the parameter space. If any of these experiments succeed in detecting dark matter particles, the next objective will be to identify its particle nature.

In this sense, the simultaneous measurement of both the SI and SD dark matter couplings, through experiments which are sensitive to both signals, can provide very valuable information [334]. This is illustrated in Fig. 17.6, where the ratio of SI to SD cross-section is plotted for the neutralino and LKP cases versus the event rate for two complementary choices of target materials (that could be used in the COUPP experiment). As shown in the left panel, the measurement of an event rate in a single detector does reduce the number of allowed models, but does not generally place significant constraints on coupling parameters or on the nature of the dark matter detected. However, as shown in the right panel, a subsequent detection using a second complementary target does substantially reduce the allowed range of coupling parameters, and allows, in most cases, an effective discrimination between neutralino and LKP candidates (see Chapter 15 for a discussion on the strategies of discrimination with accelerator searches).

This analysis can be extended to other dark matter candidates. Moreover, in order to eliminate the large astrophysical and theoretical uncertainties which affect the dark matter rates, the ratio of WIMP–proton and WIMP–neutron amplitudes can be used [264] and compared for different target materials. For example, the comparative study of the ratios of SD neutron to proton amplitudes can provide good discrimination of dark matter models by distinguishing candidates for which the SD cross-section is dominated by Z boson exchange (such as the neutralino in some regions of the parameter space) from those where the dominant channel is squark or KK quark exchange (such as the LKP or LTP). A similar analysis for the SI neutron to proton ratio can be used to disentangle models with dominant Higgs or Z boson exchange, but in practice the different target materials are less sensitive to these differences. Finally, the mass determination techniques described in Section 17.4.1 can provide complementary information that could lead to more effective discrimination between the various dark matter models.

18

Annual modulation signature with large mass highly radiopure NaI(Tl)

Rita Bernabei and Pierluigi Belli

In this chapter we discuss the prospects for detecting dark matter (DM) by means of the model-independent annual modulation signature, using large-mass highly radiopure NaI(Tl) detectors at the Gran Sasso National Laboratory of the INFN.

18.1 The annual modulation signature and the target material

A model-independent approach is necessary in order to find the presence of DM particles in the Galactic halo. In principle, two main possibilities exist; they are based on the correlation between the distribution of the events, detected in a suitable underground set-up, and the Galactic motion of the Earth.

The first one (which mainly applies just to WIMP or WIMP-like DM candidates) correlates the direction of WIMP-induced nuclear recoils with that of the Earth's velocity. This directionality signature is, however, difficult to exploit in practice, mainly because of technical difficulties in reliably and efficiently detecting the short recoil track and in realizing suitably large mass detectors; this will be discussed elsewhere in this volume.

Another possibility is the DM annual modulation signature, which is the only feasible one at present; it is sensitive to wide ranges both of DM candidates and of interactions, and it is also able to test a large interval of cross-sections and of halo densities. This was originally suggested in the 1980s in refs. [719; 881]. Such a signature exploits the effect of the Earth's revolution around the Sun on the number of events induced by the DM particles in a suitable low-background set-up placed deep underground. In fact, as a consequence of its annual revolution, the Earth should be crossed by a

Particle Dark Matter: Observations, Models and Searches, ed. Gianfranco Bertone. Published by Cambridge University Press. © Cambridge University Press 2010.

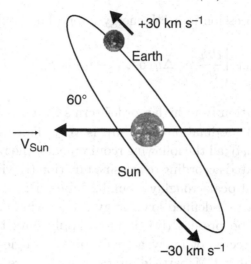

+30 km s^{-1}

Earth

60°

V_{Sun}

Sun

−30 km s^{-1}

Fig. 18.1. Schematic view of the Earth's motion around the Sun, which is moving in the Galaxy.

larger flux of dark matter particles around 2 June (when its rotational velocity is added to that of the Solar System with respect to the Galaxy) and by a smaller flux around 2 December (when the two velocities are subtracted).

The expected differential rate as a function of the detected energy, dR/dE (see refs. [270; 319; 320; 322; 323; 325; 326; 327] for detailed discussions), depends on the velocity distribution of the DM particles and on the Earth's velocity in the Galactic frame, $\mathbf{v_E}(t)$ (see Fig. 18.1). Projecting $\mathbf{v_E}(t)$ on the Galactic plane, one can write: $v_E(t) = v_\odot + v_\oplus \cos\gamma\cos\omega(t - t_0)$. Here $v_\odot$ is the Sun's velocity with respect to the Galactic halo ($v_\odot \simeq v_0 + 12\,\mathrm{km\,s^{-1}}$ and v_0 is the local velocity, whose value is in the range 170–270 km s^{-1} [270; 592; 597; 1324]; $v_\oplus = 30\,\mathrm{km\,s^{-1}}$ is the Earth's orbital velocity around the Sun on a plane with inclination $\gamma = 60°$ with respect to the galactic plane. Furthermore, $\omega = 2\pi/T$ with $T = 1$ year and t_0 is roughly 2 June (when the Earth's speed is a maximum). The Earth's velocity can be conveniently expressed in units of v_0: $\eta(t) = v_E(t)/v_0 = \eta_0 + \Delta\eta\cos\omega(t - t_0)$, where – depending on the assumed value of the local velocity – $\eta_0 = 1.04$–1.07 is the yearly average of η and $\Delta\eta = 0.05$–0.09. Since $\Delta\eta \ll \eta_0$, the expected counting rate can be expressed by the first-order Taylor approximation:

$$\frac{dR}{dE}[\eta(t)] = \frac{dR}{dE}[\eta_0] + \frac{\partial}{\partial\eta}\left(\frac{dR}{dE}\right)_{\eta=\eta_0} \Delta\eta\cos\omega(t - t_0). \qquad (18.1)$$

Averaging this expression in a kth energy interval one obtains:

$$S_k[\eta(t)] = S_k[\eta_0] + \left[\frac{\partial S_k}{\partial \eta}\right]_{\eta_0} \Delta\eta \cos\omega(t - t_0) = S_{0,k} + S_{m,k}\cos\omega(t - t_0),$$

$$(18.2)$$

where the contribution from higher-order terms is less than 0.1%.

The DM annual modulation signature is very distinctive since it must simultaneously satisfy all the following requirements: the rate must contain a component modulated according to a cosine function (i) with one-year period (ii) and a phase that peaks roughly around 2 June (iii); this modulation must only be found in a well-defined low-energy range, where events induced by DM particles can be present (iv); it must apply only to those events in which just one detector of many actually 'fires' (single-hit events), since the probability that a DM particle interacts with more than one detector is negligible (v); for the halo distributions usually adopted, the modulation amplitude in the region of maximal sensitivity must be $\lesssim 7\%$ (vi), but it can be larger in the case of some possible scenarios such as those in refs. [883; 1871; 1872]. Only systematic effects able to fulfil all these requirements and to account for the whole observed modulation amplitude could mimic this signature; thus, no other effect investigated so far in the field of rare processes offers so stringent and unambiguous a signature.

With current technology, the annual modulation remains at present the main signature of the DM particles' signal. It is worth noting that highly radiopure set-ups are necessary since no approach for rejection of electromagnetic component of the counting rate can be applied contemporaneously to the data. This is because: (i) all similar approaches have a statistical nature (which will affect the investigation of the DM annual modulation signature), and well-known side processes inducing recoils exist as well; (ii) the experiment will be insensitive to the many possible DM candidates which provide part or all of the signal in electromagnetic form (also including WIMPs in some scenarios). On the other hand the annual modulation analysis also acts as a very efficient background rejection.

Highly radiopure NaI(Tl) scintillators offer many competitive features for effectively investigating this signature, such as high duty cycle; well-known technology; feasibility of large detector masses; no safety problems; lower cost than every other considered technique; need for a relatively small underground space; high radiopurity reachable by suitable material selections and protocols, by chemical/physical purifications, etc.; feasibility of well-controlled operational conditions and monitoring; feasibility of routine calibrations down to few keV in the same conditions as the production runs;

high light response (that is, keV threshold really reachable); no need for re-purification or cooling down/warming up procedures (implying high reproducibility, high stability, etc.); absence of microphonic noise and an effective noise rejection at threshold (time decay of NaI(Tl) pulses is hundreds of nanoseconds, while that of noise pulses is tens of nanoseconds); wide sensitivity to both high- and low-mass DM candidates and to many different interaction types and astrophysical, nuclear and particle physics scenarios; possibility of effectively investigating all the aspects of interest for the DM annual modulation signature; and possibility of achieving significant results on several other rare processes.

However, neither commercial low-background NaI(Tl) detectors nor NaI(Tl) detectors grown with old technology (even after 'revision') can reliably reach the needed sensitivity. Thus, suitable R&D must be devoted to selecting all the involved materials and procedures, as was done by DAMA during about 20 years of developments.

18.2 The DAMA/NaI and DAMA/LIBRA experiments

The aforementioned arguments motivated the development and the use of highly radiopure NaI(Tl) scintillators as target detectors for the DAMA/NaI and DAMA/LIBRA set-ups. Their competitivity is based on factors such as the intrinsic radiopurity reached; the calibration near energy threshold; the routine calibrations in the same condition as the production runs; the high sensitivity to many of the DM candidates, interaction types and astrophysical, nuclear and particle physics scenarios; the granularity of the set-ups; data taking up to the MeV scale (even though the optimization is made for the lowest energy region); and the full control of the running conditions.

The first-generation DAMA/NaI [313; 314; 319; 320] and the second-generation DAMA/LIBRA [324; 328] set-ups are part of the DAMA project,[1] and they were and are located deep underground in the Gran Sasso National Laboratory of the INFN. Their main aim is direct detection of dark matter particles in the Galactic halo through the model-independent annual modulation signature. The DAMA project, which is mainly based on the development and use of low-background scintillators, also includes other low-background set-ups,[2] such as DAMA/LXe ($\sim$6.5 kg pure liquid xenon scintillator); DAMA/R&D, devoted to tests on prototypes and small-scale

---

[1] The DAMA proposal to the INFN Scientific Committee II on 24 April 1990 was made by P. Belli, R. Bernabei, C. Bacci, A. Incicchitti, R. Marcovaldi and D. Prosperi.

[2] For some references see the DAMA web page: http://people.roma2.infn.it/dama/

experiments; and the DAMA/Ge detector for sample measurements and small-scale experiments.

The DAMA/NaI set-up and its performance are described in refs. [313; 314; 319; 320], while the DAMA/LIBRA set-up and its performance are described in detail in ref. [328]. Here we just remind the reader that: (i) the detectors' responses range from 5.5 to 7.5 photoelectrons per keV; (ii) the hardware threshold of each photomultiplier is at the single photoelectron level (each detector is equipped with two low-background photomultipliers working in coincidence); (iii) energy calibrations with X-rays or gamma-ray sources are regularly carried out down to a few keV; (iv) the energy threshold of the experiment is 2 keV. The DAMA/NaI experiment collected an exposure of 0.29 ton×yr over seven annual cycles [314; 319; 320], while DAMA/LIBRA has released results so far on an exposure of 0.53 ton×yr collected over four annual cycles [324]; thus, the total exposure of the two experiments is 0.82 ton×yr, which is orders of magnitude larger than the exposure typically collected in the field.

18.3 The model-independent results

Several analyses on the model-independent investigation of the DM annual modulation signature have been performed (see ref. [324] and references therein); here we mention just a few arguments. In particular, Fig. 18.2 shows the time behaviour of the experimental residual rates for single-hit events collected by DAMA/NaI and by DAMA/LIBRA in the (2–6) keV energy interval. The superimposed curve represents the cosinusoidal function behaviour $A \cos \omega(t - t_0)$ with a period $T = \frac{2\pi}{\omega} = 1$ yr and with a phase $t_0 = 152.5$ day (2 June), while the modulation amplitude, A, has been obtained by best fit over the DAMA/NaI and DAMA/LIBRA data. When the period and the phase parameters are released in the fit, values well compatible with those expected for a DM-particle-induced effect are also obtained [324]. Similar arguments hold, for example, for the (2–4) and (2–5) keV energy intervals.

The data of Fig. 18.2 have also been investigated by Fourier analysis, and a clear peak corresponding to a period of 1 year has been found in the (2–6) keV energy interval [324]. For comparison, the power spectrum of the (6–14) keV energy interval has also been investigated; it shows only aliasing peaks. Similar results are also obtained when comparing the single-hit residuals in the (2–6) keV energy interval with those in the (6–14) keV energy interval; in fact, a clear modulation is present in the lowest energy interval, while it is absent just above. In particular, in order to verify absence of annual

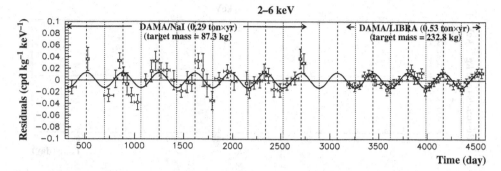

Fig. 18.2. Experimental model-independent residual rate of the single-hit scintillation events, measured by DAMA/NaI and DAMA/LIBRA in the (2–6) keV energy
interval as a function of time. The zero of the timescale is 1 January of the first
year of data taking by DAMA/NaI. The experimental points present the errors as
vertical bars and the associated time bin width as horizontal bars. The superimposed curve is the cosinusoidal function behaviour $A \cos \omega(t - t_0)$ with a period
$T = \frac{2\pi}{\omega} = 1$ yr, with a phase $t_0 = 152.5$ day (2 June) and with a modulation
amplitude, A, equal to the central values obtained by best fit over the whole data,
that is: (0.0129 ± 0.0016) cpd kg^{-1} keV^{-1}. The dashed vertical lines correspond to
the maximum of the signal (2 June), while the dotted vertical lines correspond to
the minimum. The total exposure is 0.82 ton × yr. For details, more information
and data analyses see [324].

modulation in other energy regions and thus also verify the absence of any
significant background modulation, the energy distribution measured during the data-taking periods in energy regions not of interest for DM detection has also been investigated. The background in the lowest energy region
is essentially due to 'Compton' electrons, X-rays and/or Auger electrons,
muon-induced events, etc., which are strictly correlated with the events in
the higher-energy part of the spectrum. Thus, if a modulation detected in
the lowest-energy region were due to a modulation of the background (rather
than to a signal), an equal or larger modulation in the higher energy regions
should be present. The analyses have allowed the exclusion of any possible
background modulation in the whole energy spectrum down to levels much
lower than the effect found in the lowest-energy region for the single-hit
events [324].

A further relevant investigation has been done by applying the same hardware and software procedures used to acquire and to analyse the single-hit
residual rate to the multiple-hits in the (2–6) keV energy interval. Since the
probability that a DM particle interacts in more than one detector is negligible, a DM signal can only be present in the single-hit residual rate. Thus,
this allows a test of the background behaviour in the same energy interval
as the observed positive effect. In particular, Fig. 18.3 shows the residual

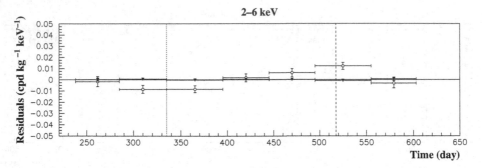

Fig. 18.3. Experimental residual rates over the four DAMA/LIBRA annual cycles for single-hit events (open circles) (class of events to which DM events belong) and for multiple-hits events (filled triangles) (class of events to which DM events do not belong), in the energy region (2–6) keV. They have been obtained by considering for each class of events the data as collected in a single annual cycle and by using in both cases identical hardware and identical software procedures. The initial time of the scale is taken on 7 August. The experimental points present the errors as vertical bars and the associated time bin width as horizontal bars. For details and more information see ref. [324]. Analogous results were obtained for the DAMA/NaI data [320].

rates of the single-hit events measured over the four DAMA/LIBRA annual cycles, as collected in a single annual cycle, together with the residual rates of the multiple-hits events. A clear modulation is present in the single-hit events, whereas the fitted modulation amplitudes for the multiple-hits residual rate are well compatible with zero: $-(0.0004 \pm 0.0006)$ cpd kg^{-1} keV^{-1}. For more information see [324]. Similar results were previously obtained for the DAMA/NaI case [320]. Thus, again, evidence of annual modulation with the features required of the DM annual modulation signature is present in the single-hit residuals (event class to which the DM particle-induced events belong), while it is absent in the multiple-hits residual rate (event class to which only background events belong). Since the same hardware and software procedures were used to analyse the two classes of events, the result offers additional strong support for the presence of a DM particle component in the Galactic halo, further excluding any side effects from hardware or software procedures or from background.

The annual modulation present at low energy can also be shown by depicting the differential modulation amplitudes, $S_{m,k}$, as a function of the energy; the $S_{m,k}$ is the modulation amplitude of the modulated part of the signal (see above) obtained by maximum likelihood method over the data, considering $T = 1$ yr and $t_0 = 152.5$ day. In Fig. 18.4 the measured amplitudes $S_{m,k}$ for the total exposure (0.82 ton×yr, DAMA/NaI and DAMA/LIBRA)

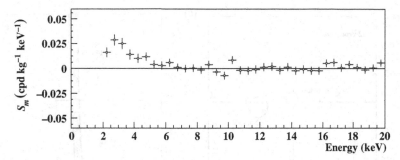

Fig. 18.4. Energy distribution of the $S_{m,k}$ for the total exposure of DAMA/NaI and DAMA/LIBRA: 0.82 ton×yr. A clear modulation is present in the lowest-energy region, while $S_{m,k}$ values compatible with zero are present just above. In fact, the $S_{m,k}$ values in the (6–20) keV energy interval have random fluctuations around zero with χ^2/d.o.f. equal to 24.4/28 (where d.o.f. is degrees of freedom). For more information see ref. [324].

are reported as a function of the energy. It can be inferred that a positive signal is present in the (2–6) keV energy interval, while $S_{m,k}$ values compatible with zero are present just above. In fact, the $S_{m,k}$ values in the (6–20) keV energy interval have random fluctuations around zero with χ^2 equal to 24.4 for 28 degrees of freedom. All this confirms the other analyses.

It has also been verified that the measured modulation amplitudes are statistically well distributed in all the crystals, in all the annual cycles and in the energy bins; these and other discussions can be found in ref. [324].

It is also interesting to look at the results of the analysis performed by releasing the phase from the value $t_0 = 152.5$ day in the maximum likelihood procedure to evaluate the modulation amplitudes from the data of the seven annual cycles of DAMA/NaI and the four annual cycles of DAMA/LIBRA. In this case the signal has to be written as: $S_{0,k} + S_{m,k} \cos \omega(t - t_0) + Z_{m,k} \sin \omega(t - t_0) = S_{0,k} + Y_{m,k} \cos \omega(t - t^*)$. Obviously, for signals induced by DM particles one would expect: (i) $Z_{m,k} \sim 0$ (because of the orthogonality between the cosine and the sine functions); (ii) $S_{m,k} \simeq Y_{m,k}$; (iii) $t^* \simeq t_0 = 152.5$ day. In fact, these conditions hold for most of the dark halo models; however, it is worth noting that slight differences can be expected in the case of possible contributions from non-thermalized DM components, such as the SagDEG stream [321] and the caustics [1345].

Figure 18.5(a) shows the 2σ contours in the plane (S_m, Z_m) for the (2–6) keV and (6–14) keV energy intervals and Fig. 18.5(b) shows, instead, those in the plane (Y_m, t^*). The best fit values for the (2–6) keV energy interval are (1σ errors): $S_m = (0.0122 \pm 0.0016)$ cpd kg^{-1} keV^{-1}; $Z_m = -(0.0019 \pm 0.0017)$ cpd kg^{-1} keV^{-1}; $Y_m = (0.0123 \pm 0.0016)$ cpd kg^{-1} keV^{-1};

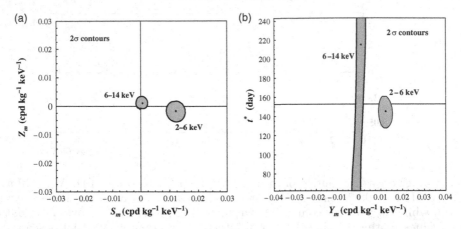

Fig. 18.5. 2σ contours in the plane (S_m, Z_m) (a) and in the plane (Y_m, t^*) (b) for the (2–6) keV and (6–14) keV energy intervals. The contours have been obtained by the maximum likelihood method, considering the seven annual cycles of DAMA/NaI and the four annual cycles of DAMA/LIBRA all together. A modulation is present in the lower energy intervals and the period and the phase agree with those expected for DM induced signals. See ref. [324].

$t^* = (144.0 \pm 7.5)$ day; while those for the (6–14) keV energy interval are: $S_m = (0.0005 \pm 0.0010)$ cpd kg^{-1} keV^{-1}; $Z_m = (0.0011 \pm 0.0012)$ cpd kg^{-1} keV^{-1}; $Y_m = (0.0012 \pm 0.0011)$ cpd kg^{-1} keV^{-1} and t^* obviously not determined (see Fig. 18.5). These results confirm those achieved by other kinds of analyses. In particular, a modulation amplitude is present in the lower energy intervals for single-hit events, and the period and the phase agree with those expected for DM-induced signals. Several other analyses have been performed; for more information see ref. [324].

In conclusion, the data of the first four annual cycles of DAMA/LIBRA, like those of DAMA/NaI previously, fulfil all the requirements of the DM annual modulation signature. As previously done for DAMA/NaI [319; 320], careful investigations on absence of any significant systematics or side reaction effect have been quantitatively carried out and reported in detail in ref. [324]. No modulation has been found in any possible source of systematics or side reactions for DAMA/LIBRA as well; thus, cautious upper limits (90% C.L.) on the possible contributions to the DAMA/LIBRA measured modulation amplitude have been estimated [324]; these cautious upper limits are – at maximum – below 1% of the observed modulation amplitude. In addition, it is worth stressing that neither systematics nor side effects – able to satisfy all the requirements of the signature simultaneously and to account for the measured modulation amplitude – have been found or suggested by anyone over more than a decade.[3] For detailed quantitative discussions on all the related topics and for results see ref. [324] and references therein.

Summarizing, the cumulative analysis of the single-hit residual rate favours the presence of a modulated cosine-like behaviour with the features expected for a DM-induced effect, at 8.2σ C.L. [324].

18.4 The corollary quests for the candidate particle(s)

The corollary quests on the nature of the DM particle(s) (detected by means of the DM annual modulation signature) and on the astrophysical, nuclear and particle physics scenarios require subsequent model-dependent corollary analyses, as performed for example in refs. [319; 320; 321; 322; 323; 325; 326; 327]. Let us point out that no approach in direct and indirect DM searches exists that can offer information about the nature of the candidate independently on assumed astrophysical, nuclear and particle physics scenarios. It has been shown – on the basis of the DAMA/NaI result – that the result can be compatible with a wide set of possibilities (see e.g. refs. [319; 320; 322; 323; 325; 326; 327]). Obviously, this is also the case when the DAMA/NaI and DAMA/LIBRA data are considered all together (see e.g. [324]); previous corollary investigations are being updated and some new ones are in progress.

In particular, many DM candidates have been considered; most of them, relics from the Big Bang, were non-relativistic at decoupling time. They have to be neutral or – if charged – neutralized with charged particles (such as electrons and protons), stable or quasi-stable (e.g. with a decay time of order of the age of the Universe) and have to interact weakly with ordinary matter. These features are respected by the axions, by a class of candidates named WIMPs (which have similar but not identical phenomenology to each other), by light dark matter particles or particles on the MeV scale, by the light ($\sim$keV mass) bosonic candidate (either with pseudoscalar or with scalar axion-like coupling) and by others. Therefore possible DM candidates range from heavy particles, such as the neutralino or sneutrino from SUSY models, Kaluza–Klein particles from theories with extra dimensions, mirror matter, heavy neutrinos, etc., to lighter particles such as the keV-scale sterile neutrino, axino, gravitino, moduli fields, majorons, familons, etc. Just as an

---

[3] We take this opportunity to point out that the DM annual modulation is not a 'seasonal' variation, nor is it a 'winter–summer' effect, as sometimes naively mentioned. In fact, the DM annual modulation is not related to the relative position of the Sun, but it is related to the Earth's velocity in the Galactic frame. Thus, for example, the phase of the DM annual modulation (roughly 2 June) is different from those of physical quantities (such as temperature of atmosphere, pressure, other meteorological parameters or cosmic-ray flux) that are correlated with the seasons instead.

example, we mention the case of the WIMP class of DM candidate particles, which can span from low- to high-mass candidates and with distinctly different phenomenology from each other. In particular, many possible scenarios exist for WIMP interaction with ordinary matter: (i) mixed SI and SD coupling; (ii) dominant SI coupling; (iii) dominant SD coupling; (iv) preferred SI inelastic scattering; (v) dominant electron interaction, etc. In addition, an electromagnetic contribution is present for WIMPs even in the case that the interaction with nuclei is assumed to be dominant since excitation of bound electrons in scatterings on nuclei can occur (the well-known Migdal effect). There also exist candidates, such as light bosons, whose direct detection process in the target material is based on the total conversion into electromagnetic radiation of the mass of the absorbed bosonic particle; thus, in these processes the recoil of the target nucleus is negligible and it is not involved in the detection process at all. It is worth noting that signals giving an electromagnetic contribution to the counting rate are lost in activities applying rejection procedures of the electromagnetic component of their counting rate. Phenomenological properties of some basic interaction mechanisms induced by DM particles are discussed, for instance, in refs. [194; 319; 320; 322; 323; 325; 326; 327; 388; 932; 948; 988; 1016; 1017; 1602; 1871; 1872] and in the literature. In particular, the results of the DAMA experiments also fit quite well with the model discussed in refs. [399; 401; 409].

18.5 Comparison with other activities

No other experiment exists whose results can be directly compared in a model-independent way with those of DAMA/NaI and DAMA/LIBRA. In particular, let us point out that results obtained with different target materials and/or different approaches cannot intrinsically be directly compared among themselves even when considering the same kind of candidate and coupling.

In particular, we remark that some of the other activities presented elsewhere in this volume are insensitive to the annual modulation signature; use different target materials; often provide model-dependent exclusion limits without accounting for the existing experimental, theoretical and phenomenological uncertainties and for the existing alternative choices; often use crude approximation in the calculations; have lower sensitivity than DAMA/NaI and DAMA/LIBRA in several scenarios; moreover, scenarios exist (see literature) to which the others are not only disfavoured with respect to the DAMA experiments, but even blind. Furthermore, considering the difficulties of exploiting relatively new techniques in this very low-energy and low-rate field, additional limitations in the model-dependent sensitivities (which are specific to 'nuclear recoils' and to a single assumed scenario and parameters set) can arise if, for example, (i) the physical energy threshold is unproven with suitable keV source calibrations; (ii) the energy scale is extrapolated from higher energy; (iii) the stability of the running parameters and of all the used 'rejection' windows over long term at the needed level of precision is unproven; (iv) the DM candidates provide part (even for WIMPs in some scenarios) or all of the signals in electromagnetic form; (v) marginal exposures are considered; (vi) the efficiencies in each one of the many procedures of data handling are not proven at the level needed for the claimed precision (a control of systematics at levels of 10^{-4} to 10^{-8} is required); (vii) the detector response has a lack of uniformity in the light/charge collection e.g. in two-phase liquid xenon detectors. Moreover, the implications of the DAMA model-independent results are generally quoted in an incorrect and partial way, without updating. Some arguments have been addressed, in, for example, refs. [307; 308; 319; 320; 321; 322; 323; 325; 326; 327].

In conclusion, claims for contradiction have no robust scientific basis.

Regarding the indirect detection searches, let us stress that, again, no direct model-independent comparison can be performed between the results obtained in direct and indirect activities. Anyhow, if the positron flux and/or gamma-ray flux from the centre of the Galaxy were higher than expected on the basis of some simulations of hypothesized background contribution, and these excesses were interpreted in terms of dark matter, this still may not be in conflict with the effect observed by DAMA experiments.

18.6 Future perspectives

The merits of suitable highly radiopure set-ups using NaI(Tl) detectors have been demonstrated by the DAMA set-ups which have been/are the highest radiopure set-ups available in the field so far. They have effectively pursued a model-independent approach to investigate the presence of DM particles in the Galactic halo by collecting exposures several orders of magnitude larger than those usually available in the field and have obtained or are in the process of obtaining many other complementary or by-product results.

In 1996, while running DAMA/NaI, DAMA proposed to INFN that they should realize a ton set-up [306; 309]. A new R&D project for highly radiopure NaI(Tl) detectors was funded at that time and carried out for several years in order to realize the second-generation DAMA/LIBRA set-up, as an intermediate step. During September 2008, this set-up has been upgraded, in order to restore a photomultiplier, replace the transient digitizers with those of higher performance and install a new higher-performing DAQ system with optical fibres. Moreover, the replacement of all the photomultipliers with new ones having larger quantum efficiency is foreseen around the end of 2009; this would allow us to lower the energy threshold of the experiment to below 2 keV. An increase of the exposure and a lowering of the energy threshold will allow us to investigate the model-independent evidence with increased sensitivity; moreover, it will improve the discrimination capability among different astrophysical, nuclear and particle physics scenarios.

At present a third-generation R&D effort towards a possible highly radiopure NaI(Tl) ton set-up has been funded and the DAMA collaboration has already performed various related works.

The collection of larger exposures (with DAMA/LIBRA or with the possible DAMA/1 ton) will allow the investigation of several open aspects of the nature of the candidate particle(s) and the various related astrophysical, nuclear and particle physics as well as high-sensitivity investigation of other DM features and second-order effects. In particular, some of the many topics (not yet well known at present and affecting any model-dependent result and comparison) to be investigated are: (i) velocity and spatial distribution of the dark matter particles in the Galactic halo; (ii) effects induced on the dark matter particles distribution in the Galactic halo by contributions from satellite galaxies tidal streams; (iii) effects induced on the dark matter particles distribution in the galactic halo by the existence of caustics; (iv) detection of possible 'solar wakes'; (v) investigation of possible diurnal effects (due for example to the contribution of the Earth's rotation velocity); (vi) study of possible structures such as clumpiness with small scale size; (vii) nature of any coupling(s) of the dark matter particle with the ^{23}Na and ^{127}I; and (viii) scaling laws and cross-sections.

Finally, let us note that ultra-low-background NaI(Tl) scintillators can also offer the possibility of achieving significant results on several other rare processes, as the former DAMA/NaI set-up has already shown.

19

Particle dark matter and the DAMA/NaI and DAMA/LIBRA annual modulation effect

Nicolao Fornengo

19.1 The DAMA annual modulation effect

The DAMA experiment [311; 314; 319; 320; 328] consists of low-background, highly radiopure NaI(Tl) scintillators with very stable response over long time periods. These characteristics allow the DAMA experiment to look for the annual modulation [719; 881] of the counting rate: this peculiar effect is expected if the Galactic halo is composed of DM particles interacting with the target detector and represents a specific signature for the direct search of DM.

The experiment operated in the first phase as DAMA/NaI for 7 years with a mass of about 100 kg and collected a total exposure of 0.29 ton×yr [310; 312; 315; 319; 320]. The second phase is now under operation as DAMA/LIBRA, which recently released data on an additional four annual cycles with a mass of about 250 kg, adding 0.53 ton×yr to the total exposure [324]. The cumulative analysis of the single-hit residual rate of the DAMA/NaI and DAMA/LIBRA detectors favours at 8.2σ C.L. the presence of annual modulation with the features which are expected from a DM candidate [Chapter 18; 319; 324].

In the case of WIMP DM which scatters off the nuclei of the detector, the DAMA annual modulation effect provides direct information on two basic properties of the DM particle: the particle mass m_χ and its elastic scattering cross-section off nuclei. The latter is then usually and more conveniently translated into the cross-section on a single nucleon $\sigma_{\text{scalar}}^{(\text{nucleon})}$ [405], in the case of both coherent and spin-dependent interactions.

In addition to the particle properties, the direct detection rate critically depends on other relevant quantities. Nuclear physics inputs are

Particle Dark Matter: Observations, Models and Searches, ed. Gianfranco Bertone. Published by Cambridge University Press. © Cambridge University Press 2010.

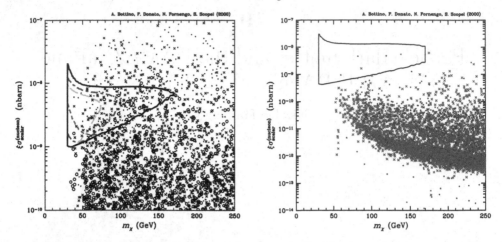

Fig. 19.1. Left: DAMA/NaI annual modulation region from the data of ref. [315], for an isothermal DM distribution with $\rho_0 = 0.3$ GeV cm^{-3} and isotropic Maxwellian $f_0(v)$ with: $v_0 = 170$ (short-dashed line), 220 (long- and short-dashed) and 270 (long-dashed line) km s^{-1}. The scatter plot shows theoretical predictions for a low-energy MSSM. From [393]. Right: DAMA/NaI region from the data of [315], for an isothermal DM distribution with $(170 < v_0 < 220)$ km s^{-1} and $(0.2 < \rho_0 < 0.7)$ GeV cm^{-3}. The scatter plot is calculated in a SUGRA scheme with non-universality in the Higgs sector. From [395]. In both panels, the analysis of the data has been performed for WIMPs with a mass larger than 30 GeV (as seen from the vertical cutoff line at $m_\chi = 30$ GeV).

required: form factors for both coherent [1084] and spin-dependent [794] interactions have to be modelled in order to extract information on the particle physics properties from the experimental data [405]; the recoil response may additionally be affected by the channelling effect, which occurs when an ion traverses a detector with a crystalline structure [327]. Astrophysics affects the detection rate through the dark matter distribution in the local neighbourhood: both the local dark matter density ρ_0 and the local velocity distribution function (DF) $f_0(\mathbf{v})$ [269; 270] influence the response of the detector to a DM particle scattering. The actual dependence of the direct detection rate on all the above quantities may be found in refs. [269; 506].

From the time and energy behaviour of the counting rate, an allowed region in the m_χ–$\sigma_{\text{scalar}}^{\text{(nucleon)}}$ plane is obtained: a DM particle with a mass and cross-section inside that region produces an effect in the DAMA detector which explains the observed annual modulation of the rate. This region is then dependent on the assumptions under which it has been obtained, namely the nuclear physics inputs and the Galactic DM halo model.

Figure 19.1(left) [393] shows the allowed region in the m_χ–$\sigma_{\text{scalar}}^{\text{(nucleon)}}$ plane for coherent scattering of WIMPs and for three different velocity DFs (and

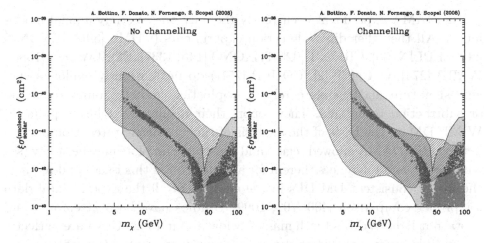

Fig. 19.2. The hatched region shows the combined DAMA/NaI and DAMA/LIBRA annual modulation region for WIMP DM [324; 401]. In (a) the channelling effect is not included, while it is considered in (b). A large set of astrophysical uncertainties is included. The scatter plot shows theoretical predictions for a gaugino non-universal MSSM with light neutralinos. The band around the scatter plot shows the effect of the nuclear uncertainties in the hadronic matrix elements. From [401].

for DM particles with $m_\chi > 30$ GeV), as obtained from the DAMA/NaI data with a total exposure of 57 986 kg×day [315]. In Fig. 19.1(right) [395] the allowed region from the same set of data is extended to include a possible (although not full) uncertainty in the local DM density. Figure 19.2 [401] shows the combined analysis of the recent DAMA/LIBRA and the full set of DAMA/NaI data, and extends the study to lower WIMP masses. In Fig. 19.2 [401] a large variety of DM DFs have been included [269]. Moreover, the influence on the extracted allowed region from the channelling effect [327] is explicitly shown.

The impacts of astrophysical modelling and uncertainties on the DAMA effect have been discussed in refs. [390; 391; 392; 393; 395; 401; 403; 440; 578; 1008; 1010] and specifically studied in refs. [269; 270]. The effect of DM streams has been discussed in refs. [519; 831; 882; 883; 983; 1694]; while caustic rings have been considered in ref. [1007]. A specific analysis of the DAMA data in the case of substructures has been performed in ref. [321]. Allowed regions for selected or combined DM DFs can be found in refs. [269; 310; 312; 315; 319; 320] for DAMA/NaI and in ref. [401] for DAMA/NaI + DAMA/LIBRA. The variation of the allowed region may be sizable: the various analyses show that the DAMA effect is compatible with WIMPs with a mass ranging from a few GeV [401] up to 1 TeV (for corotating halo modes [269]) and cross-section from 10^{-43} cm^2 to 10^{-38} cm^2, depending on the mass.

The DAMA experiment is currently the only one to study annual modulation. All the other direct detection experiments (CDMS [39], XENON10 [90], ZEPLIN [66], CRESST [92], TEXONO [146; 1337], EDELWEISS [1683], WARP [274], CoGeNT [2], COUPP [251]) do not currently exploit a specific signature and therefore rely on sophisticated background reduction or subtraction techniques. They report their results as exclusion plots for WIMP DM on the basis of the analysis of their counting rates. Comparison between the DAMA allowed region and the exclusion plots may be tricky (see Chapter 18 and references therein for a discussion of this issue) and requires the use of consistent DM DFs in the analysis of all the experimental data for proper comparison [400; 702; 1009]. Studies have been attempted and show that light WIMPs (with masses below about 10–20 GeV) are perfectly compatible among all direct detection experiments: attention to the low-mass sector was raised in refs. [397; 399; 400; 409] and also pursued in refs. [519; 805; 1289; 1572; 1695]. Spin-dependent interactions have also been discussed in [1696]. For heavier DM, recent experiments report upper bounds which may constrain the DAMA region for the specific case of a spherical isothermal galactic halo [39; 90] (although a thorough analysis which takes into account a wide variation of Galactic models is not currently available). A mechanism which could be at work for the intermediate mass range is offered by inelastic DM, where WIMPs can scatter off nuclei only by making a transition to a nearly degenerate state that is roughly 100 keV heavier [1871; 1872]. In this case different detectors, which employ different nuclei and possess different energy thresholds, may respond differently: it has been shown [518; 1871; 1872] that while DAMA could see the signal, other experiments may have strongly suppressed responses.

Finally we mention that DM particles which induce partly or only electromagnetic contributions to the counting rate would currently be seen only in the DAMA detectors and not in detectors that apply rejection techniques to the electromagnetic component of their counting rate [323; 325].

DAMA data have been specifically analysed also for spin-dependent or mixed scenarios [316], inelastic scattering [318], light scalar and pseudoscalar particles [322; 326], electromagnetic [323] or electron [325] interactions.

19.2 Supersymmetric candidates

The DAMA region in the mass–cross-section plane can then be compared with theoretical predictions for DM candidates. For any candidate, the scattering cross-section is calculated, together with the candidate relic abundance. In the case of relic abundance sizably lower than the one that corresponds to the amount of DM in the Galaxy, we need to rescale the local DM density proportionally [387; 388; 402; 909]: the rescaling procedure introduces into the theoretical calculation the scaling factor $\xi \leq 1$, and the comparison of the experimental results is done on the quantity $\xi \sigma_{\text{scalar}}^{(\text{nucleon})}$ [387; 388; 401].

The most studied WIMP DM candidate is the neutralino, in various schemes of supersymmetric theories. Neutralinos typically offer a sizable coherent scattering off nuclei (although they can also produce spin-dependent scatterings). The first analysis of the DAMA effect in terms of WIMPs scattering was performed in ref. [389] on the first data release of DAMA/NaI with a 4549 kg×day exposure [310]. In ref. [389] it was shown that neutralinos in a low-energy MSSM were compatible with the observed effect. The analysis was then extended in ref. [392], on the second DAMA/NaI release with an exposure of 14 962 kg × day [312]. In ref. [392], attention was drawn both to the impact of (some) astrophysical uncertainties and to the nuclear physics uncertainties in the theoretical calculation of the scattering cross-section [394; 396; 404; 783]. Properties of the neutralinos which are compatible with the DAMA effect were derived. The analysis was then extended in ref. [391] to SUGRA models (both universal and non-universal in the Higgs sector [287], a model which contained the mechanism that was later dubbed 'focus-point') and compatibility was shown to be achieved for large values of $\tan \beta$ and/or for specific patterns of Higgs non-universality. The SUSY configurations compatible with the DAMA effect were then analysed in terms of indirect detection signals in ref. [390], namely antiproton searches and neutrino fluxes from the Earth and the Sun. The effect of astrophysical uncertainties on the reconstruction of the DAMA signal (DAMA/NaI with 19 511 kg×day exposure [312]) and their impact on the derived properties of neutralinos were discussed in ref. [270]. A detailed analysis of the interplay with the indirect detection signals (antiprotons, neutrino fluxes) in connection with the astrophysical uncertainties, which affect both direct and indirect detection rate although in different ways, has been performed in ref. [393] on the DAMA/NaI 57 986 kg×day exposure data [315]. One example for the neutrino flux coming from the Sun is reported in Fig. 19.3(a). A detailed study of the impact of the uncertainties arising from the coupling to the nucleon was made in refs. [394; 395; 396], both for the MSSM and for (universal and non-universal) SUGRA models. An interesting implication for physics at accelerators, which will have the chance to be checked at the LHC, was derived in ref. [408]. Studies of MSSM or SUGRA schemes relevant for the DAMA effect, including non-universal soft scalar and gaugino mass models or models from string theory, have also been performed in [19; 127; 153; 181; 259;

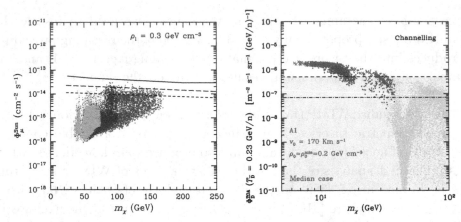

Fig. 19.3. Left: Predictions for the upgoing muon flux from neutralino annihilation in the Sun, for MSSM configurations compatible with the DAMA effect from the data of ref. [315]. The three bands refer to three different Galactic halo models. The median and lower horizontal lines show the upper limits from MACRO [83] and SuperK [1049]. Adapted from [393]. Right: Predictions for the antideuteron flux calculated for gaugino non-universal MSSM configurations compatible with the DAMA effect from the data of ref. [324]. The upper and lower horizontal lines show the estimated sensitivities of AMS and GAPS. From [401].

356; 503; 508; 509; 511; 512; 525; 526; 527; 528; 833; 908; 974; 1129; 1237]. Neutrino signals have also been discussed in refs. [651; 830; 1125; 1695; 1882]. An example of the comparison between the DAMA regions (under various astrophysical assumptions) and neutralino scattering is shown in Fig. 19.1 for a low-energy MSSM [393] and for a non-universal SUGRA scheme [395].

   More recently there has been a lot of attention given to light neutralinos, which can arise, for example, in SUSY models where gaugino universality is violated [397; 409] or in models with additional gauge singlets [504; 1037]. The interest in light neutralinos was first specifically raised in refs. [397; 409]. In ref. [397] it was noticed that these light neutralinos may have relevance for the DAMA modulation effect. Detailed studies were performed in ref. [399] for the last DAMA/NaI data release (108 000 kg × day exposure [319]), where the data were analysed for the whole mass range, including light DM particles. The effect of channelling was introduced into the study of light neutralinos in ref. [403] and the impact on the predictions of antiproton and antideuteron [698; 701] fluxes for indirect searches was detailed. Predictions for LHC studies for the light neutralino configurations compatible with DAMA have been derived in ref. [407]. Models where light neutralinos are naturally realized and may explain the DAMA effect have been also discussed in [504; 723; 1037; 1608].

The most recent study of light neutralinos in relation to the DAMA effect has been performed in ref. [401] for the recent combined analysis of the DAMA/NaI and the first DAMA/LIBRA data [324]. The comparison between the DAMA region and the predictions for light neutralinos is shown in Fig. 19.2. In ref. [401] the impact of the channelling effect and of astrophysical and nuclear uncertainties has been detailed for the reconstruction of light neutralino properties from the DAMA effect. Predictions for indirect searches (antiprotons and antideuterons) have been discussed [401]. The case of antideuteron searches is reported in Fig. 19.3(b).

Recently sneutrino DM has also been reconsidered [109]. Left–right mixed sneutrinos possess suppressed coupling which allow acceptable relic abundance and direct detection rates compatible with DAMA [109; 114; 115; 386; 1872]. Lepton-number violating terms may also allow sneutrinos to be viable and compatible with the DAMA effect [109], representing a natural realization of inelastic DM [109]. In ref. [109], antiproton, antideuteron, gamma-ray and neutrino indirect signals from sneutrinos have been analysed. Other recent models for sneutrinos which can produce scattering cross-sections compatible with DAMA and connect the solution of the DM problem to the solution of the neutrino mass have been discussed in refs. [108; 505].

19.3 Additional candidates

DM candidates other than neutralinos and sneutrinos have been proposed to explain the DAMA effect. These candidates arise in various extensions of the Standard Model (either supersymmetric or not) ranging from minimal addition of scalar or fermionic fields to particle candidates obtained from more general theories, involving extra dimensions or string theory.

An interesting possibility is offered by mirror dark matter [284; 553; 554; 860; 861; 862; 863; 864; 865; 866; 867; 869]. Light (keV) scalar and pseudoscalar particles [322] have been discussed. Light (10–100 keV) super-weak bosons were studied in [1594], together with the limits coming from gamma-background and stellar cooling. A keV axion-like particle has been constrained by solar physics [984], although uncertainties or non-minimal scenarios may still allow it [914]. Scalar DM in the GeV range has been shown to be compatible with the DAMA result [89]. WIMPless dark matter from gauge-mediated SUSY breaking [831], gauge-singlet scalar fields [1041], fourth-generation fermions [815; 1845] or other realizations of fermionic DM

[1077; 1236], singlet extensions of the universal extra dimension model [170], exotic candidates from conformal strong coupling theories [1273] and models with suppressed elastic scattering [854] have been proposed. Inelastic DM may be realized in string theory [1288], models with hidden gauge symmetries [113; 118] or non-chiral DM [801]. Self-interacting DM [1455] or composite DM states [1226] have also been proposed.

20

Cryogenic detectors

Gilles Gerbier and Jules Gascon

20.1 Introduction

We consider here as cryogenic detectors the solid state or superfluid ^{3}He detectors operated at temperature lower than 77 K. The principle of operation of these detectors is presented in Section 20.2. Liquid xenon, argon and neon detectors operated respectively at 165 K, 87 K and 20 K are described in the following chapter.

The first searches for dark matter particles were performed with ultra-pure semiconductors, operated at liquid nitrogen temperature and installed in pure lead and copper shields in underground environments. Combining a-priori excellent energy resolutions (low energy thresholds) and very pure detector material, they produced the first limits on WIMP searches in the 1980s. It turns out that about 20 years later, such germanium detectors, reaching sub-keV thresholds, have produced the best limits at very low WIMP masses, i.e. lower than 10 GeV. Performances and results of these detectors are described in Section 20.3.1.

In the 1980s, the idea was also put forward in the United States, Europe and Japan of using very low temperature detectors to achieve the required excellent energy resolution and low threshold characteristic of semiconductors, but with different materials and then different nuclei. The idea was to detect particles in a crystal by measuring the increase of temperature induced by the energy deposition. As the heat capacity approximately follows a Debye law with a T^3 dependence, it is possible to consider real calorimetric measurements down to very small energy deposition, by using the proper absorber and low enough temperature. With single parameter (heat or phonon) measurement, such detectors produced the best limits for very

Particle Dark Matter: Observations, Models and Searches, ed. Gianfranco Bertone. Published by Cambridge University Press. © Cambridge University Press 2010.

low WIMP masses, owing to their sub-keV threshold. This is described in Section 20.3.2.

However, to make further progress relative to these first results, and also to be in a position to identify a signal rather than setting limits, it quite rapidly became clear that dark matter detectors should also have the capability to reject the overwhelming electromagnetic background induced by gamma-rays or X-rays from radioactive contaminants, in order to select the interactions of interest, namely the nuclear recoils. This is possible by taking advantage of the very different energy deposition density in the case of nuclear recoils, or electron recoils, and by measuring two different quantities reflecting this difference. This has been achieved in double parameter measurement bolometers, by measuring two signals simultaneously for each interaction: charge and phonon in ionization detectors, and light and phonon in scintillating bolometers. These two techniques are described respectively in Sections 20.4 and 20.5. This approach produces the best actual limits and shows great potential for improvement.

To complete this overview of very-low-temperature detectors, it is worth mentioning the novel way of detecting particles using superfluid ^{3}He that has been put forward by the group MACHe3. The extreme purity of this material led them to suggest a dark matter detector where the interaction of a particle creates quasiparticles which are detected by the attenuation of the vibration of a wire inside the liquid. The R&D related to this project is summarized in Section 20.6.

Finally, the prospects for very large arrays with more than 1000 kg of detectors are described in Section 20.7.

The main focus on this review is on results from the past five years. Description of earlier work can be found in refs. [85; 521; 911].

20.2 Principles of operation of solid state cryogenic detectors

20.2.1 Physics processes

Cryogenic detectors are designed to measure accurately the kinetic energy of an ion recoiling after an elastic collision with a WIMP from our Galactic halo. This kinetic energy being in the range from 1 to 100 keV, an ion slowing down in a crystal lattice is stopped in a range of less than 100 nm. The ion dissipates its energy in a series of collisions with electrons and ions of the lattice. Three signals proportional to the recoil kinetic energy can be extracted from this process: phonons, ionization and scintillation.

Phonon excitations of the crystal matrix arise as the multiple collisions convert the kinetic energy of the initial recoil into collective excitation of the crystal. Athermal phonon sensors detect them at a very early stage in this process; thermal sensors wait for the full thermalization of the phonon distribution and measure the resulting elevation of temperature of the crystal. The ionization signal depends on the fraction of the initial kinetic energy that is dissipated in collisions with electrons. For a 20 keV germanium recoil in a germanium lattice, this fraction is approximately a third. This energy is ultimately converted back into phonon excitations, unless the ionized electrons do not immediately recombine but are instead made to drift out of the lattice by means of an electric field. In scintillating material, a fraction of the ionization energy is lost as photons emitted by electron de-excitation can escape the crystal. In scintillating material, this fraction is typically a small percentage of the total recoil energy, but it can reach 12% at low temperature for some scintillators.

The relative importance of the phonon, ionization and scintillation signals is substantially different when the initially recoiling particle is not an ion, but instead is an electron as is the case for gamma-ray and beta-ray interactions. The relative yields of the three signals can thus be used to distinguish these interactions from those where the initial recoil is an ion, as is the case for WIMP or fast-neutron interactions in the crystal. The relative yields of these signals normalized to the response to a gamma-ray interaction is called the quenching factor. By simultaneous measurement of two of the three possible signals, one can extract both the energy and the quenching of individual events, and thus identify the nature of the initial recoil.

In the context of rare events searches, an important issue is the reduction of the rate due to the natural radioactivity of the detector and its immediate surroundings, and in particular its mechanical support and its front-end electronics. This puts stringent constraints on the technological choices for the extraction of the photon, ionization or scintillation signals.

20.2.2 Charge collection

In semiconductor material such as silicon and germanium at cryogenic temperatures, the number of electron–hole pairs created following a gamma-ray interaction is strictly proportional to the incident energy. For example, in germanium, it takes on average 3 eV to create one pair, with very little dispersion from statistical fluctuations, for gamma-ray energies from a fraction

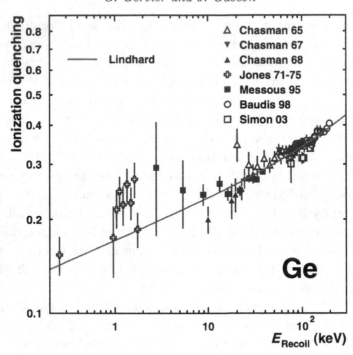

Fig. 20.1. Quenching of the ionization signal for germanium ions recoiling in germanium, as a function of the recoil energy, taken from the review in ref. [280]. The line corresponds to the prediction from the Lindhard theory [1344].

of a keV up to the MeV range. In these detectors, the surface of the detector is covered with a pair of oppositely polarized electrodes to collect both the electrons and holes. The electrodes are either implanted or vaporized on the surface. The charge collected on the electrodes is read out using charge amplifiers. Charge collection times are of the order of 10 to 100 ns, depending on the detector temperature and the applied field.

Figure 20.1 shows measurements of the quenching factor of the ionization signal for germanium recoils in a germanium detector. The energy dependence of this quantity is very well described with the Lindhard theory for the stopping power of ions in matter [1344], which also correctly estimates its overall normalization. The smooth energy dependence extends down to the keV region, and thus is relevant for the entire energy range expected from germanium recoils arising from WIMP collisions. Thus, the ionization yield of a 50 keV recoil will be lower than that of a 50 keV electron by a factor of 3. This ratio becomes 4 at approximately 10 keV recoil.

In order to use the ratio of the collected charge to the thermal signal to reject the overwhelming background from electron-induced events, it is important that the charge collection be extremely reliable. An electron recoil where two-thirds of the charge is not collected cannot be distinguished from a nuclear recoil on the basis of its quenching alone. Two processes may alter the charge collection: recombination and trapping. The charges are created in a dense region and may shield each other from the external field. The applied field must be large enough to prevent electron–hole recombination at this early stage. Then, as the charges drift toward the electrodes, they may be trapped in crystal defects. This second effect can be reduced by using semiconductor material with a very low density of impurities and defects.

Detectors at 77 K are polarized at voltages of the order of 1 kV. Such high voltages are not suited for the simultaneous measurement of phonon signal. The charge current in the detector induces a Joule heating of the crystal which is proportional to the applied voltage. This effect can be used to amplify the thermal signal. It can be a problem in detectors designed to use the difference in quenching of ion and electron recoils, as this information is diluted by the contribution of the phonons generated by the Joule heating process. As a consequence, polarization fields in millikelvin detectors are kept to a few volts, and more care must be devoted to the question of detector purity and defects.

As a large number of defects occur near the surface of the detector, the transition region between the bulk and the electrodes is a key issue in the design of these detectors. This strongly affects the response to energy deposits occurring close to the detector surface. For these events, dead-layer problems may be important. In addition, the early stage of the drift occurs in a region where the density of traps is high, and there is also a non-negligible probability that some of the charge can diffuse on the wrong-sign electrode. Particularly sensitive to these effects are interactions due to β-emitting contaminants on the detector surface or on adjacent surfaces. As an example of corrective design, the presence of an amorphous layer under the electrodes has been shown to improve charge collection [278; 1748]. A layer with an appropriate thickness can constitute a small energy barrier that can prevent the random diffusion of charge toward the wrong-sign electrode, while those drifting along the electric field lines ultimately reach the right-sign electrode. These benefits may outweigh the drawback of increasing the dead-layer thickness.

Recently, a new technique using interleaved concentric electrodes [450; 638] has been successfully demonstrated as an efficient means to tag near-surface events.

20.2.3 Phonon and heat signals

There are two families of sensors for the phonon signal: thermal and athermal. In thermal sensors, the aim is to wait for the full thermalization of the phonons within the closed system constituted by the bulk of the detector and the sensor itself. Because of this, these detectors have intrinsically slow time response (typically 0.1 to 10 ms). The temperature increase ΔT is equal to E/C, where E is the deposited energy, and C is the heat capacity of the closed system. For germanium detectors cooled to 20 mK, the elevation of temperature is typically of the order of $1\,\mu$K. The two most widely used techniques to measure these small temperature variations are neutron transmutation doped germanium sensors (NTD) and transition edge sensors (TES).

An NTD sensor is a small germanium semiconductor crystal ($\sim$mm^3) that has been exposed to a thermal neutron flux in order to make a large but controlled density of impurity. The aim is to obtain that the transition between the semiconducting and resistance regime be around the desired temperature T_0, with a dependence of the resistance with temperature T as $\exp(-\sqrt{(T/T_0)})$. The NTD can be used as a thermistance to measure small temperature variations relative to T_0. Its resistance is continuously measured by flowing a current through it and measuring the resulting voltage V. The sensor is attached to the bulk of the detector using a thermally conducting glue. The signal rise time is a function of the thermalization time as well as the coupling between the detector and the sensor. The return of the detector and its sensors to the initial temperature is made slower by reducing as much as possible the thermal link with the refrigerator. The required precision for the measurement of the voltage variation across the NTD is a small fraction of a microvolt. This must be achieved without bringing warm and/or radioactive electronics components close to the detector. A common solution is to amplify the signal in field effect transistors chosen for their ability to operate in a relatively cold environment ($\sim$150 K), located at a few tens of centimetres from the detector itself.

A TES is made of a thin superconducting tungsten film operated near the temperature for its transition to the conductor state. A heater with an electrothermal feedback system is used to maintain the film at the temperature on the transition edge, where the slope of the resistance as a function of temperature is large. A thermal pulse is detected by a change in the feedback current, collected by a SQUID. This low-impedance signal processing device is not as strongly affected by stray capacitance as the FET readout system, and can be installed at a safe distance from the detector itself.

To increase the phonon collection efficiency, the TES can be coupled to larger-area superconducting aluminium and tungsten films (as in the

CRESST experiment [93]). These large-area films do not undergo a transition when absorbing phonons. Instead, quasi-particles are excited, and these in turn diffuse inside the TES proper. The CDMS collaboration [41] have refined this technique by equipping a single Ge detector with an array of 4144 TES, each connected to eight aluminium quasi-particle traps.

This array entirely covers one side of the detector and collects phonons very efficiently, before their full thermalization. This makes it possible to observe the build-up in time of the phonon signal at the microsecond scale. Significant differences are observed depending on whether the event occurs in the bulk of the detector or close to its surface. This additional information is extremely useful to reject the latter class of events, which are subject to charge collection problems. In addition, the array is read out in four independent quadrants, and the relative signal amplitudes provide a measurement of the position of the interaction in the plane parallel to the array.

Detectors based on thin NbSi films in the transition from metal to insulator have also been studied, as an alternative method sensitive to athermal phonons [1179]. The technique has not demonstrated yet that it can attain the energy resolution necessary for obtaining a recoil energy threshold of the order of 10 keV.

There is some uncertainty over whether the thermal signal is subject to the quenching effect, as is the case for the ionization yield [280]. In principle, in a truly closed-system detector, and after full thermalization of the initial energy, all energy should eventually degrade to thermal energy, no matter what its initial source. Notable exceptions are when the material is scintillating, and when some energy is lost in crystal defects. However, the effect on the quenching is bound to be modest, as it depends on the difference in the behaviour of electron and ion recoils. All available data (cf. review in ref. [280]) suggest that any effect on the phonon signal is small and has a negligible effect on the sensitivity to WIMP-induced recoils.

20.2.4 Light collection

The collection of the scintillation photons coming out of a cryogenic crystal using conventional techniques such as photomultipliers or avalanche photodiodes is not readily possible because these devices cannot operate efficiently at millikelvin temperatures. In addition, their natural radioactivity is large and their presence cannot be tolerated close to the WIMP detectors chosen for their very high radiopurity.

An elegant solution, pioneered by CRESST [439; 1718] and ROSEBUD [82; 371], is to use another phonon detector as a light sensor. This device is a thin Si or Ge wafer, equipped with either a TES or NTD. In this case the signal amplitude depends on the light collection efficiency of the sensor at the wavelength of the emitted scintillation. The bulk detector must be encased in a reflective coating in order to redirect as much light as possible to the light sensor, which should cover as much solid angle as possible.

The fraction of the initial energy converted into scintillating photons may be as high as 12% in the best scintillators studied at cryogenic temperature, but effective measured light yields are of the order of 1% (see Section 20.5).

The energy resolution is strongly affected by the statistical fluctuations in the number of emitted – and detected – photons. It is thus important that the phonon measurement of the light sensor be significantly more precise than that of the bulk detector. This is achieved by using a light detector as thin as possible in order to reduce its heat capacity. Amplification schemes using polarized detectors have also been studied [1157].

20.3 Single parameter detectors

20.3.1 Germanium detectors at 77 K

The best sensitivities obtained with germanium detectors operated at liquid nitrogen temperature have been achieved by the IGEX collaboration [1155; 1475], after use of improved neutron shielding (Fig. 20.2). The effective

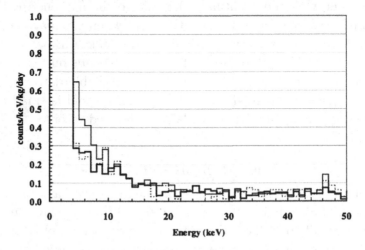

Fig. 20.2. Normalized low-energy spectrum of the IGEX RG-II detector corresponding to the three different neutron shielding conditions: 236 kg d with 20 cm of polyethylene (thin solid line), 194 kg d with 40 cm of polyethylene and borated water (thick solid line) and 82 kg d with 80 cm of polyethylene and borated water (dashed line)[1155].

threshold is 4 keV, equivalent to a threshold in recoil energy four times larger, when taking into account the quenching of the ionization signal for recoiling ions. The rate close to threshold is approximately 0.3 cpd keV^{-1}kg^{-1}, corresponding to an upper limit on the scattering cross-section of a 60 GeV/c^2 WIMP on a nucleon of $\sim 10^{-5}$ pb. The residual background rate was tentatively attributed in part to tritium contamination induced by cosmogenic activation.

Recent results from this type of detector have been obtained in 2007 and 2008 by the TEXONO and CoGeNT groups, achieving sub-keV thresholds.

The TEXONO collaboration uses four ULGe (ultra-low-energy Ge) detectors of mass 5 g, surrounded by an anti-Compton system and a muon veto. The experiment is operated at the Kuo-Sheng (KS) Reactor Laboratory in Taiwan at an equivalent depth of 30 mwe (metres water equivalent). The measured count rate at a threshold energy of 0.2 keV is approximately 200 cpd keV^{-1}kg^{-1} [1337]. This is much larger than the IGEX rate, but is rather low for the shallowness of the experimental site. Once efficiency and quenching effects are taken into account, it turns out that this low-energy measurement is the most sensitive one for WIMP masses between 3 and 6 GeV/c^2 [1338].

It should be noted that quenching effects are rather large at such low energy. The effective threshold for a 50% efficiency at 0.2 keV corresponds to a recoil energy of approximately 1 keV [188] according to a dedicated quenching factor measurement performed in this very low-energy domain [189].

A new result on low-mass WIMPs also came out in 2008 from a single massive detector of 475 g of germanium, specially equipped with a point-like electrode and dedicated low-noise FET [188]. Installed in a pumping station in the city of Chicago, at 330 mwe, equipped with anti-Compton and muon vetos together with multiple shields against gammas and neutrons, this detector had been operated at that time for only a couple of weeks. Although a threshold of a few hundred eV is in principle reachable, the experimental threshold appears to be at 0.5 keV, at which energy the rate is roughly 50 cpd keV^{-1} kg^{-1}. From these data were extracted limits in the usual diagram of cross-section against WIMP mass (see Figure 20.3, left) that appear to be better than previous limits for masses between 4 and 8 GeV [2].

Furthermore, these data allow one to test and reject the hypothesis according to which the annual modulation observed by the DAMA experiment might be induced by axion-like pseudoscalars (see Fig. 20.3 right). They would appear as a visible peak at mass m_a superimposed on the background in the 0.5–8 keV region.

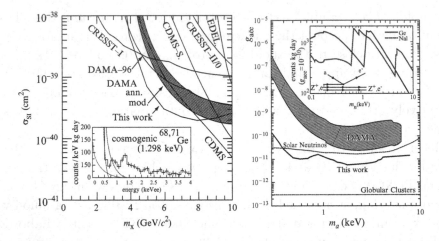

Fig. 20.3. Left: Lines delimit the coupling (σ_n^{SI}) vs. WIMP mass (m_χ) regions excluded by experiments. Inset: spectrum used for the extraction of present limits. Lines display the signals expected from some reference WIMP candidates (dotted: $m_\chi = 8\,\mathrm{GeV}$, $m_\chi = 10\,\mathrm{GeV}$, dashed: $m_\chi = 6\,\mathrm{GeV}$, dash-dotted: $m_\chi = 4\,\mathrm{GeV}$); Right: Hatched region: viable parameter space in an interpretation of the DAMA modulation involving an axio-electric coupling from pseudoscalars composing a dark isothermal halo. The solid line indicates present limits, dashed lines recent astrophysical bounds. Inset: expected pseudoscalar interaction rates in Ge and NaI, for a fixed value of g, as a function of pseudoscalar mass m_a.

20.3.2 Sapphire bolometers

The first bolometers studied for dark matter searches were sapphire-based detectors, as the Debye temperature of these crystals is very high, which means a relative high increase of temperature for a given energy deposition. Because of the possibility of obtaining a sub-keV threshold, such detectors, measuring only the total energy deposited in the detector, produced the best limits for low-mass WIMPs [92], and were only superseded recently by germanium detectors operated at liquid nitrogen temperature (see Section 20.3.1).

CRESST, a German–UK collaboration, has obtained a low threshold of 580 eV, using transition edge sensors (TES) coupled to a 240 g Al_2O_3 crystal. The spectrum is shown in Fig. 20.4. It exhibits a rate of a few 10s of cpd keV^{-1}kg^{-1} at around 1 keV. As the phonon signal is not strongly affected by quenching effects, the energy scale here directly refers to recoil energy, in contrast with the results from ionization-only devices discussed in Section 20.3.1. Even after correcting for the ionization quenching, the count rate obtained with germanium detectors is an order of magnitude lower than the rate observed in this experiment. This is presumably due to the higher

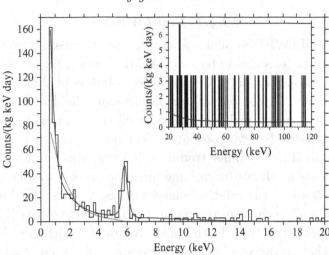

Fig. 20.4. Energy spectrum of a 240 g Al_2O_3 detector during the dark matter run (1.51 kg days) in 200 eV bins. The insert shows the spectrum at higher energies. The fully drawn curve is an empirical fit to the experimental spectrum which serves for extracting dark matter limits. For illustration a 5 GeV WIMP excluded at 90% C.L. is shown as a dashed curve [92].

intrinsic purity of germanium and also its relatively smaller gamma absorption length, which tends to harden the recorded energy spectrum and reduce the population at low energy.

No significant effort has been made to increase the performances of these detectors. As underlined in the introduction of this chapter, the focus has recently been shifted to double-parameter measurement detectors with active discrimination of the electron recoil background (Section 20.5). The thresholds achieved so far with double-parameter detectors are not yet competitive with those of the best single-parameter devices. However, this advantage is only relevant for searches for WIMPs with masses below 10 GeV, far below the values favoured by conventional SUSY models.

20.4 Ionization phonon bolometers

Two groups, CDMS [39] and EDELWEISS [1683], have actively developed detectors with simultaneous measurement of ionization and phonon signals. EDELWEISS uses the thermal signal from an NTD sensor, while CDMS exploits the athermal phonon signal from a quasi-electron trap (QET) array to reject surface events.

20.4.1 EDELWEISS

In 2005, the EDELWEISS collaboration reported the results from three cryo-genic germanium detectors in terms of WIMP limits [279; 1683] and residual backgrounds [855]. The detectors were Ge cylinders of 2 cm in height and 7 cm in diameter. To homogenize the electrostatic field in the periphery of the detector, the cylindrical edges were bevelled at an angle of 45°. The top surface was covered by two concentric electrodes, collecting separately the ionization created in the inner (radius < 2.5 cm) and outer volumes of the detector. Events in the outer volume are more prone to charge collection problems, because of the relatively large surface that cannot be covered by the electrode, and because of the natural radioactivity of the NTD, connec-tors and support structure. In the inner volume (180 g), the charge collec-tion is such that no defect of charge collection is observed in samples of 10^4 gamma-ray interactions.

Figure 20.5 shows the nuclear recoil discrimination in the EDELWEISS detectors. For each event (recorded here during a calibration with a ^{252}Cf neutron source) the yield is displayed as a function of recoil energy. The pho-ton and neutron interactions are separated into two bands: the regions where 90% of them are expected are shown. Most of the nuclear recoils arising from WIMP collisions should appear inside the 90% efficient nuclear recoil band determined using the neutron calibration data. Most of the events appearing outside the electron and nuclear recoil band in Fig. 20.5 are caused by pile-ups of these two types of recoils due to inelastic collisions. The efficiency of the rejection of gamma interactions depends on the energy resolution. Rejection of 99.9% of photons down to energies of 15 keV can be routinely obtained [1683].

The experiment was installed underground, in the Laboratoire Souterrain de Modane, where the 4800 mwe rock overburden provides a reduction of six orders of magnitude of the cosmic muon background. A 10 cm Cu and 15 cm Pb shield reduced the gamma-ray background rate to a level of 2 events per kilogram per day and per keV in the range relevant for WIMP searches. After rejecting coincidences between detectors and interaction in the outer volume, the gamma-ray background rate is 1 event kg^{-1} day^{-1} keV^{-1}. An experi-mental threshold of 10 keV could be achieved, with the efficiency reaching 50% of its plateau below 15 keV. In 2001, the experiment was able to set the most stringent limits on WIMP interactions at that time, but no fur-ther improvements were possible because of the appearance of two sources of background: neutrons and surface events. Neutron interactions were iden-tified as coincidence events between two detectors, with both interactions

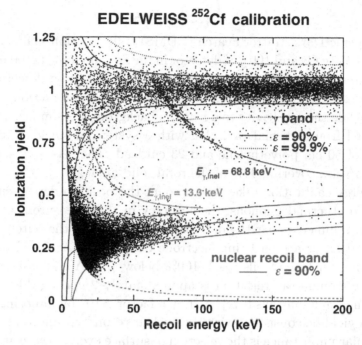

Fig. 20.5. Distribution of the ionization yield (normalized to one for gamma interactions) as a function of recoil energy recorded in an EDELWEISS detector exposed to a ^{252}Cf neutron source. The thick lines represent the 90% nuclear and electronic recoil zones. The dotted line corresponds to the ionization threshold curve. The dashed lines show where events associated with the inelastic scattering of neutrons on ^{73}Ge (13.26 and 68.75 keV excited levels) are expected.

being tagged as nuclear recoils. Surface events were found to be due to ^{210}Pb contamination of the surface, arising from the exposition of the detector to ambient radon. The detector set-up has since been upgraded, notably by increasing the polyethylene neutron shield from 30 to 50 cm and by improving the monitoring and control of low-level radioactivity during detector handling and storage (clean room operations, de-radonized air). To reduce the neutron background to the levels required for a sensitivity to count rates below 0.001 event kg^{-1} day^{-1}, a muon veto covers more than 98% of the solid angle of the experimental set-up. With the improvement in material selections and shielding design, the gamma-ray background has been reduced by a factor of two. More importantly, detectors with an interleaved electrode design are now being deployed, for rejecting the surface events that limited the previous performance of the experiment. The new set-up that has just been commissioned can house up to 40 kg of germanium detectors.

20.4.2 CDMS

The Cryogenic Dark Matter Search (CDMS) collaboration [39] is operating an array of 250 g Ge and 100 g Si detector in the Soudan mine (2100 mwe) [41]. The Si detectors are mostly used as comparison, as these detectors would have a negligible sensitivity to spin-independent WIMP interactions compared with Ge, while the sensitivity to a neutron background is similar. The set-up is protected from external radiation with a system of shields with a total thickness of 50 cm polyethylene and 23 cm lead. The set-up is surrounded by a muon veto, covering 99% of the total solid angle.

The energy calibration takes into account the position dependence of the response to athermal phonons; the accuracy in that measurement is of the order of a millimetre near the detector axis. Events in the outer perimeter are rejected using a guard ring electrode design. The achieved efficiencies for detecting nuclear recoils reach 100% below 10 keV. The average event rate before gamma-ray rejection is approximately 2 cpd $kg^{-1}\,keV^{-1}$. Bulk interaction of gamma-rays is rejected by a factor $>10^5$ by imposing that the ionization yield be consistent with the nuclear recoil hypothesis at 90% C.L. Of particular importance is the rejection of surface events: these are rejected using a cut (the vertical line) on phonon timing parameter, derived from the rise time of the phonon signal and its delay relative to the ionization signal (Fig. 20.6). By adjusting this cut, the expected background from surface events in a given total exposure can be reduced to a fraction of count.

A total of 19 Ge detectors are currently in use [39], and 15 of these were used to collect an exposure of 398 kg days in 2006 and 2007. After the cut on the data quality, the fiducial volume and the phonon timing parameter (Fig. 20.6), the remaining effective exposure is 121 kg days. With this data set, combined with earlier data sets, the experiment was able to set an upper limit on the spin-independent cross-section for the scattering of a 60 GeV/c^2 WIMP on a nucleon of 4.6×10^{-8} picobarn, at 90% C.L. (see Fig. 20.7). Up to now, this experiment has achieved the best sensitivity for WIMP masses above 44 GeV/c^2.

20.5 Scintillation phonon bolometers

20.5.1 CRESST

Two groups, CRESST [439; 1718] and ROSEBUD [82; 371], have been testing the phonon-scintillation discrimination scheme, and obtained results in low-background environments at underground sites.

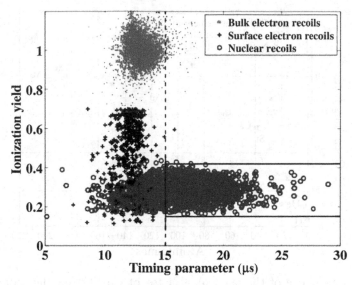

Fig. 20.6. Ionization yield versus phonon timing parameter for CDMS cryogenic Ge detector [39]. The open circles correspond to low-yield nuclear recoils from a calibration with a ^{252}Cf source. The high-yield events from electron recoils recorded in a calibration with a ^{133}Ba gamma-ray source are shown as dots, and the low-yield events, associated to surface interactions, by crosses.

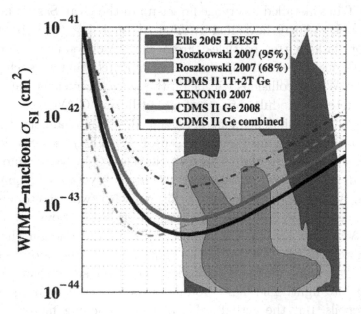

Fig. 20.7. Spin-independent WIMP–nucleon cross-section upper limits (90% C.L.) versus WIMP mass from the CDMS experiment compared with theoretical expectations and the limit obtained by the XENON10 experiment (taken from ref. [39]).

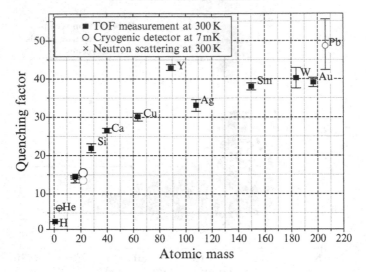

Fig. 20.8. Dependence of the quenching factor of $CaWO_4$ on the atomic mass of selected elements. The full squares are the measurements of ref. [1510] (TOF, time of flight). The neutron scattering result at room temperature from [580] is shown as a cross. The open circles are measurements done with alpha-decay events in cryogenic detectors at a temperature of 7 mK, taken from ref. [93].

CRESST has installed its cryogenic set-up in the Gran Sasso underground laboratory. Its efforts have mainly focused on using $CaWO_4$ crystals as targets for dark matter particles. From the A^4 dependence of the coherent WIMP–nucleon scattering rate, it is expected that in such a crystal, most of the WIMP collisions will be on the heaviest nucleus, tungsten. One CRESST detector module consists of a scintillating $CaWO_4$ crystal of 300 g equipped with two independent cryogenic detector channels: the phonon channel, a tungsten transition edge sensor (W-TES) (see Section 20.2.3) deposited directly onto the surface of the $CaWO_4$ crystal, and the light channel, an SOS (silicon on sapphire) or silicon crystal also equipped with a W-TES. A dark matter run was performed with two detectors of 300 g in 2003, lasting 53 days [93]. The following energy resolutions were obtained: 1 keV FWHM for the 46 keV line of ^{210}Pb in the phonon channel, and approximately 12 keV FWHM in the light channel. The poorer energy resolution for the light channel is a consequence of the small fraction of energy converted into light. The light production is expected to be quenched for nuclear recoils, but the actual value of the quenching factor depends on the nature of the recoiling nucleus. The results of the extensive calibrations performed by the CRESST collaboration [1510] are illustrated in Fig. 20.8. These measurements include neutron beam calibration results, as well as

those from an original method using low-energy ions obtained by laser abla-
tion followed by acceleration up to 18 keV. These ions are then focused on
a small $CaWO_4$ crystal viewed by a photomultiplier tube. The response to
electron recoils was calibrated with a ^{55}Fe source. In that experiment, the
number of detected photoelectrons per incident energy was measured to be
approximately 4 per keV. The measured quenching factor for the heaviest
nucleus, tungsten, is $(40)^{-1}$, that is 2.5%. From this large quenching effect,
one would anticipate an excellent discrimination between electron and W
recoils. This advantage is mitigated by the low light production that results
in the inability to distinguish the expected light signal from W recoils from
a non-scintillating event, for recoil energies below 40 keV. Such events can
occur, for example, as a consequence of the crystal relaxation from internal
stresses [137].

Knowledge of the values of the quenching factor for nuclear recoils then
allows one to define where to expect the WIMP signal in the discrimination
parameter scatter plot of Fig. 20.9. From the absence of events below the
full line, where tungsten recoils are expected, a limit has been derived
in the diagram of mass versus cross-section. This limit was comparable with
the one obtained by EDELWEISS in 2002 [278].

More recent results have been obtained in an updated set-up that now
includes a neutron shield, during a commissioning run reported in [1510],
with similar detectors. A factor-of-four increase in sensitivity has been
obtained, still giving limits higher than best actual reported limits [39].

Further developments are under way. The light yield can be increased
by using new compounds, for instance $ZnWO_4$, which seems a promising
though delicate crystal [236]. Other developments are being conducted to
build cryogenic composite detectors, more adapted to the production of
large numbers of detectors. The main issue here is to avoid the deposition
of tungsten TES directly on the crystal, a process that needs heating cycles
which are likely to damage the absorber [1657].

20.5.2 ROSEBUD

ROSEBUD is a collaboration between groups from the IAS in France and
Saragoza University, dedicated to the development of scintillating bolometers
for dark matter detection [80; 81]. Some of the crystals studied are BGO,
Al_2O_3 and LiF. ROSEBUD has characterized these scintillators at very low
temperature. An important issue is the light yield, essential for obtaining
baseline resolutions compatible with the goal of being able to discriminate

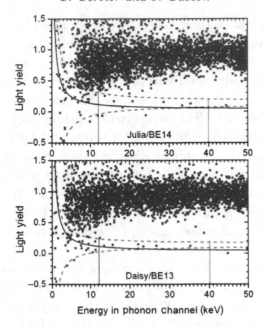

Fig. 20.9. Low-energy event distributions in the dark matter run for the two CRESST modules Julia/BE14 (top) and Daisy/BE13 (bottom). The vertical axis represents the light yield expressed as the ratio of energy in the light channel to energy in the phonon channel, and the horizontal axis the energy in the phonon channel. The region between the dashed curves contains 90% of events with a quenching value of $(7.4)^{-1}$, the expected value for oxygen and calcium nuclear recoils. The region below the solid curves contains 90% of tungsten recoils, assuming a quenching factor of $(40)^{-1}$. The vertical lines indicate the energy range used for the WIMP analysis.

between nuclear recoils and electron recoils down to a recoil energy of 10 keV. The BGO light yield has been shown to increase with decreasing temperature and is found to be 23 700 photons per MeV [955] at 6 K. However, the measured fraction of deposited energy in the form of scintillation, taking into account light collection and possible trapping effects, is 0.75%. A higher measured light yield has been measured for Al_2O_3 (1.35%). Observed anticorrelation between light and heat signals in Al_2O_3, possibly caused by inhomogeneities in the crystal, accounts for the poor light energy resolution in Al_2O_3.

Quenching factors of nuclear recoils relative to electron recoils have also been measured to be approximately 1/17, 1/14 and 1/6.5 respectively for Al_2O_3, BGO and LiF. These numbers concern the response of lighter nucleus types present in the crystal.

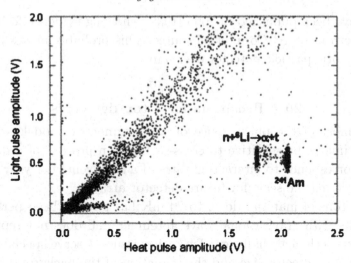

Fig. 20.10. Light versus heat discrimination plot obtained with the 33 g LiF bolometer at IAS Orsay in 12.1 h of live time. The $^6\text{Li}(n,\alpha)^3\text{H}$ events ($Q = 4.78$ MeV) are easily identified and were used to measure the thermal neutron flux at Orsay. Events from an ^{241}Am inner source (α of 5.44 and 5.48 MeV) are also indicated.

The isotope ^{6}Li has a very high cross-section for thermal neutron capture. With a fraction ^{6}Li in natural Li of 7.6%, LiF is a very sensitive thermal neutron detector. Thanks to the discrimination provided by the double measurement of heat and scintillation, a LiF detector can provide monitoring of the thermal neutron environment with an excellent signal-to-noise ratio. This is shown in Fig. 20.10. Lithium-6 can also be used to measure fast neutrons but with much lower sensitivity.

20.6 MACHe3

In terms of extreme purity, superfluid helium-3 and helium-4 surpass even semiconductor materials. Nothing can dissolve in ^{3}He at 100 μK. This has been a motivation for proposing MACHe3 [1686], a detector where the interaction of a particle creates quasiparticles that are detected by the attenuation of the oscillation of a wire inside liquid ^{3}He. Additional motivations come from the very low Compton cross-section in ^{3}He, making this medium transparent to photons, and from the half-integer spin of the nucleus, making this target complementary to detectors usually sensitive mostly to scalar interaction. However, the drawbacks are the very high cost of ^{3}He, and that no rejection against the electron background seems possible. First energy spectra have been published in 2005 [1939], showing evidence for the detection of electrons of a ^{57}Co internal source. This probably makes this device the most exotic particle detector ever built.

20.7 Prospects for 1 ton detectors

Covering most of the phase space of supersymmetric models will require that experiments be sensitive to cross-sections on protons of around 10^{-10} pb. This corresponds to interaction rates of a few events per year per ton of detector, setting the scale for future detector arrays.

Cooling tons of material down to 20 mK has already been performed by experiments with gravitational wave antennae, and does not represent the most serious technical challenge. Two other issues deserve special attention: the increase of detector size and the reduction of the backgrounds.

The mass of individual detector units must be increased in order to keep the number of readout channels to a reasonable level. The present unit size (250 to 400 g) should be increased to at least 1 kg. This constraint on the number of channels does not arise primarily from the cost of electronics, but from thermal constraints. Each readout channel corresponds to a few wire connections going from 20 mK to 300 K, each contributing to the heat leak between the cold and warm components of the experiment. A larger mass for the detector unit would alleviate this problem. However, the size of the signal in a calorimetric detector is, in principle, inversely proportional to its heat capacity, and therefore its mass. If the baseline electronic noise is constant, an increased mass may degrade the resolution of the heat signal to unacceptable values. This effect has not been a problem with present-day detectors, partly because the total heat capacity of the detector depends also on all the elements coupled to it, such as electrodes, TES or NTD sensors, and amorphous layers for Ge. Although small in mass, these elements have a non-negligible contribution to the total heat capacity. Increasing the size to the kilogram range will require a careful optimization of the detector design. Another issue related to the increase of detector size is the ability to identify some of the neutron interactions by means of coincidences between detector units. Reducing the segmentation of the detector array can degrade performances in this respect, and a compromise must be found.

The other issue is the reduction of backgrounds. The gamma-ray background must be decreased by improving the discrimination capabilities and the selection of materials for the detectors and their surroundings. Here

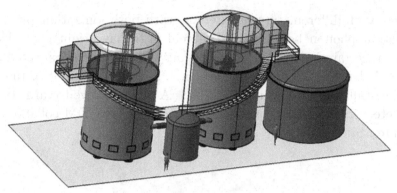

Fig. 20.11. Preliminary layout of the EURECA experiment. The two large cylinders on the left are the water tanks serving as active muon veto. The cryostats hosting the detectors are lowered into these tanks. The cylinder on the right is an auxiliary water reservoir buffer. The water installation is fitted with a radon purifier plant.

again, the neutron background deserves special attention. The ambitious goal is to reduce the neutron flux in the entire detector volume to a few neutrons per year. The neutron flux originating from fission of uranium or α–n reactions in the surrounding rock can be attenuated at the appropriate level with the use of shields made of passive light material such as polyethylene and water. Neutrons from the deep-inelastic scattering of muons in the heavy materials surrounding the detectors can be vetoed by enveloping the experiment with an efficient active shield, such as a large water tank equipped with photomultipliers. With these improved passive and active shields, the remaining significant source of background will probably be neutrons from the fission decay of the small uranium content of the materials in the vicinity of the detectors. To prevent this, the design should emphasize a packed arrangement of the detectors with a very light holding structure.

Of course, all technical issues of cold production, shielding strategy, material selection, radon free air production, wiring and electronics strategy, cryogenics choice and interface with hosting laboratory have to be addressed well in advance, and are partly independent of the choice of detectors. These studies are actually performed within a European collaboration, EURECA [1275], which federates the teams from the two largest cryogenic experiments in Europe, CRESST and EDELWEISS, as well as those from ROSEBUD, together with groups from CERN and other European laboratories. EURECA has started the simulation and technical studies needed to answer the above questions. At the core of the EURECA design is the ability to host detectors with different target materials, with both ionization–phonon and scintillation–phonon bolometers. The goal is to have tight control of systematic uncertainties and, more importantly, to verify the expected atomic number A dependence of a possible signal. A possible layout of the underground installation is shown in Fig. 20.11. A similar move towards ton cryogenic detectors is under consideration within the GEODM collaboration in the United States.

21

Liquid noble gases

Elena Aprile and Laura Baudis

21.1 Noble liquids for dark matter detection

Noble liquids, specifically liquid xenon (LXe), liquid argon (LAr) and liquid neon (LNe), are excellent scintillators and, with the exception of LNe, also very good ionizers in response to the passage of radiation, thus providing an excellent alternative to cryogenic detectors (see Chapter 20). The possibility of simultaneously detecting ionization and scintillation signals in LXe and LAr is a unique feature of these liquids compared with other detection media. This capability, together with the promise of scale-up to large mass at a modest cost compared with semiconductors, has contributed to make LXe and LAr popular targets and detectors for rare physics events such as those associated with dark matter, solar neutrinos and neutrinoless double beta decay interactions. In this section, we first describe the ionization and scintillation mechanism in noble liquids, including recent measurements of the relative scintillation efficiency and ionization yield for nuclear recoils, relevant for dark matter searches. We then describe the background rejection capability of noble liquids, based on pulse shape discrimination of the scintillation light and the ratio between ionization and scintillation signals (S2/S1). At the end of the section, we briefly discuss the key requirement for ultra-high purity of noble liquids for dark matter detection.

21.1.1 Physical properties of noble liquids

Table 21.1 summarizes the physical properties of the three noble liquids being used or planned to be used for dark matter direct detection. The atomic number, density, boiling point temperature, abundance in the atmosphere and intrinsic radioactive isotopes (not shown in table) determine

Particle Dark Matter: Observations, Models and Searches, ed. Gianfranco Bertone. Published by Cambridge University Press. © Cambridge University Press 2010.

Table 21.1. *Physical properties of xenon, argon and neon.*

Properties [unit]	Xe	Ar	Ne
Atomic number	54	18	10
Mean relative atomic mass	131.3	40.0	20.2
Boiling point T_b at 1 atm [K]	165.0	87.3	27.1
Melting point T_m at 1 atm [K]	161.4	83.8	24.6
Gas density at 1 atm & 298 K [$g\,l^{-1}$]	5.40	1.63	0.82
Gas density at 1 atm & T_b [$g\,l^{-1}$]	9.99	5.77	9.56
Liquid density at T_b [$g\,cm^{-3}$]	2.94	1.40	1.21
Dielectric constant of liquid	1.95	1.51	1.53
Volume fraction in Earth's atmosphere [ppm]	0.09	9340	18.2

several practical aspects of a dark matter detector based on the specific noble liquid. The high atomic number and high density make LXe an excellent detector medium for penetrating radiation. Its relatively high temperature, compared with that of LAr and LNe, also facilitates detector handling. In terms of cost, LXe is the most expensive of the three noble liquids, owing to its low fraction in the atmosphere. However, the problem of radioactive [39]Ar present at the level of 1 Bq kg^{-1} in atmospheric Ar will increase the cost of LAr for large dark matter detectors, which will require Ar depleted in [39]Ar by centrifugation or by extracting it from other sources than the atmosphere.

21.1.2 Ionization and scintillation production

The ionization process. The energy loss of an incident particle in noble liquids is shared between the following processes: ionization, excitation and sub-excitation electrons liberated in the ionization process. The average energy loss in ionization is slightly larger than the ionization potential or the gap energy because it includes multiple ionization processes. As a result, the ratio of the W-value, the average energy required to produce an electron-ion pair, to the ionization potential or the gap energy is 1.6–1.7 [102]. Table 21.2 shows the W-values in noble gases (liquid and gaseous states) [102; 691; 1459; 1833]. In general, the W-value in the liquid phase is smaller than in the gaseous phase, and the W-value in liquid xenon is smaller than that in liquid argon and liquid neon. As a consequence, the ionization yield in liquid xenon is the highest of all noble liquids.

The scintillation process. Luminescence emitted from liquids or solids is called scintillation. Scintillation from noble liquids arises in two distinct processes: excited atoms R^* and ions R^+ (both produced by ionizing radiation).

$$R^* + R + R \rightarrow R_2^* + R \qquad (21.1)$$

$$R_2^* \rightarrow 2R + h\nu$$

$$R^+ + R \rightarrow R_2^+ \qquad (21.2)$$

$$R_2^+ + e^- \rightarrow R^{**} + R$$

$$R^{**} \rightarrow R^* + \text{heat}$$

$$R^* + R + R \rightarrow R_2^* + R$$

$$R_2^* \rightarrow 2R + h\nu,$$

where $h\nu$ denotes the vacuum-ultraviolet (VUV) photons emitted in the process, with wavelength of 178 nm, 128 nm and 78 nm for LXe, LAr and LNe, respectively; $R^{**} \rightarrow R^* +$ heat corresponds to a non-radiative transition. In both processes, the excited dimer R_2^*, at its lowest excited level, is de-excited to the dissociative ground state by the emission of a single UV photon. This comes from the large energy gap between the lowest excitation and the ground level, forbidding other decay channels such as non-radiative transitions. The average energy required for the production of a single photon, W_{ph}, for alpha- and beta-particles, is listed in Table 21.2 [692].

The scintillation pulse shape. The scintillation light from pure liquid neon, argon and xenon has two decay components due to de-excitation of singlet and triplet states of the excited dimer $R_2^* \rightarrow 2R + h\nu$. Figure 21.1 [1109; 1283] shows for instance the measured decay shapes of the scintillation light for electrons, alpha-particles and fission fragments in liquid xenon. As expected, the decay shapes for alpha-particles and fission fragments have two components. The shorter decay shape is produced by the de-excitation of singlet states and the longer one by the de-excitation of triplet states. However, scintillation for relativistic electrons has only one decay component. The differences of pulse shape between different types of particle interactions in noble liquids can be used to discriminate these particles effectively. This 'pulse shape discrimination' (PSD) is particularly effective for liquid argon, given the large separation of the two decay components [1109; 1346] (see Section 21.3 for details).

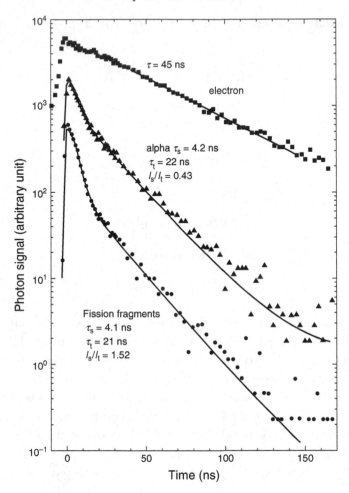

Fig. 21.1. Decay curves of luminescence from liquid xenon excited by electrons, α-particles and fission fragments, without an external electric field [1109; 1283].

The most relevant scintillation and ionization properties of LXe and LAr are listed in Table 21.2.

21.1.3 Relative scintillation efficiency of nuclear recoils

The scintillation light yield of nuclear recoils in noble liquids is quite different from the one produced by electron recoils of the same energy. Knowledge of this ratio, termed relative scintillation efficiency (L_{eff}), is important for the determination of the sensitivity of noble liquids for dark matter detection. To our knowledge, the relative scintillation efficiency of nuclear recoils has so far been measured only for LXe [43; 103; 119; 317] down to 10 keV nuclear

Table 21.2. *Ionization potential and W-value in gaseous argon
and xenon; gap energy and W-value in liquid argon and xenon.*

Gas	Ar	Xe
Ionization potential I (eV)	15.75	12.13
W-value (eV)	26.4^a	22.0^a
Liquid	**LAr**	**LXe**
Gap energy (eV)	14.3	9.28
W-value (eV)	23.6 ± 0.3^a	15.6 ± 0.3^b
$W_{ph}(\alpha)$ (eV)	27.1	17.9
$W_{ph}(\beta)$ (eV)	24.4	21.6

^{a} Ref. [691]
^{b} Ref. [102]

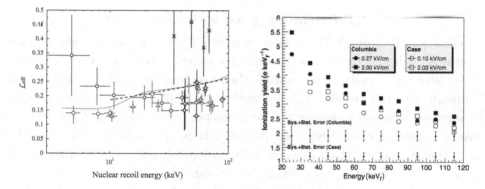

Fig. 21.2. Left: Relative scintillation efficiency of nuclear recoils measured from
different experiments [105]; Right: Ionization yield from nuclear recoils measured
with small-scale two-phase xenon detectors [104].

recoil energy (keV$_r$). A more recent measurement [105] extends the results
down to 5 keV$_r$. The relative scintillation efficiency for nuclear recoils of
5 keV$_r$ is 14% and is constant around this value up to 10 keV$_r$. For higher-
energy recoils, the value is on average about 19% (see Fig. 21.2, left). The
best model to describe the scintillation quenching of nuclear recoils in liquid
xenon is based on a biexcitonic diffusion–reaction mechanism [1107]. Such
a model can also be used to predict the relative scintillation efficiency of
nuclear recoils in LAr, giving a value of about 20% at 10 keV$_r$, gradually
increasing to 25% at 100 keV$_r$ [1106]. A measurement of L_{eff} in LAr and LNe

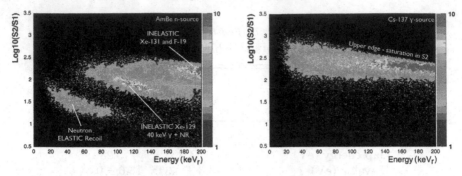

Fig. 21.3. A two-phase xenon detector's response to an AmBe neutron (left) and a
^{137}Cs gamma source (right), at $2\,\mathrm{kV\,cm^{-1}}$ drift field [104].

is currently in progress [F. Cavanna, private communication; D. McKinsey, private communication].

21.1.4 Ionization yield of nuclear recoils

Nuclear recoils from WIMPs (and neutrons) have denser tracks, and are assumed to have greater electron–ion recombination than electron recoils. The collection of ionization electrons thus becomes more difficult for nuclear than for electron recoils. The ionization yield of nuclear recoils in LXe, defined as the number of observed electrons per unit recoil energy ($\mathrm{e^-\,keV_r^{-1}}$), was measured for the first time in [104]. It is shown in Fig. 21.2 (right) as a function of external electric field and recoil energy. The increase in charge yield for low-energy recoils is explained by the lower recombination rate expected from the drop in electronic stopping power at low energies. This behaviour, although theoretically not fully understood, allows the observation of Xe nuclear recoils down to a few $\mathrm{keV_r}$, improving the event rate and the sensitivity for WIMP detection.

The ionization, or S2, of nuclear recoils in LAr has been measured in the WARP experiment [274; 1004]. It also shows significantly lower yield than the one from electron recoils. The ratio of ionization and scintillation for nuclear recoils in liquid xenon was reported in [104]. Figure 21.3 shows the response to neutron and low-energy Compton scattering events from gamma-rays. The logarithm of the ratio S2/S1 is plotted as a function of nuclear recoil energy. The elastic nuclear recoil band is clearly separated from the electron recoil band, providing a basis for discrimination against background gammas and betas in the two-phase operation of noble liquid detectors. In the following we use the convention of $\mathrm{keV_{ee}}$ to refer to the energy of electron recoils.

21.1.5 Electron attachment and light absorption by impurities

A large number of dark matter experiments based on noble liquids rely on the simultaneous detection of the scintillation and ionization signals from an event interaction. In order to achieve a high collection for both signals, the concentration of impurities in the liquid has to be reduced and maintained to a level well below 1 part per 10^9 (part per billion, ppb) oxygen equivalent. The scintillation light signal from LXe and LAr is strongly reduced by the presence of water vapour. The mean length (λ) of a scintillation photon travelling in the liquid is called the 'absorption length'. An absorption length of the order of 1 m has been achieved in LXe [1537]. Future ton-scale or multi-ton-scale noble liquid detectors will require an absorption length longer than 10 m, corresponding to a water vapour contamination well below 10 parts per trillion (ppt).

The detection of the ionization signal is more challenging, as it requires both high purity and a high electric field. Figure 21.4 shows the variation

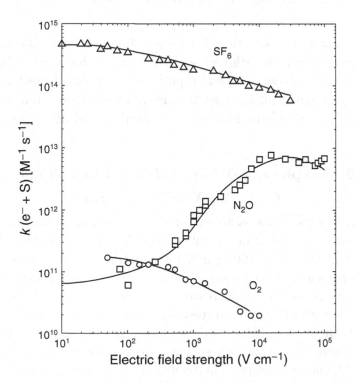

Fig. 21.4. Rate constant for the attachment of electrons in liquid xenon ($T = 167\,^\circ$K) to several solutes: ($\triangle$) SF$_6$, ($\square$) N$_2$O, ($\circ$) O$_2$ [174].

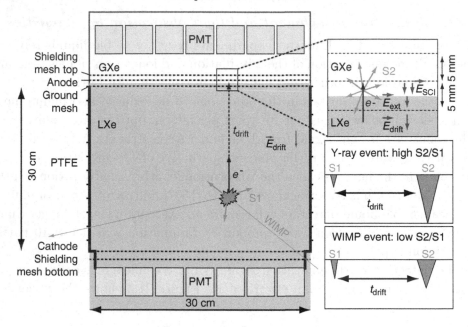

Fig. 21.5. Measurement principle of a two-phase xenon TPC.

of the attachment cross-section in LXe as a function of the applied field for some typical substances with high electronegative affinity, SF_6, N_2O and O_2 [174]. The concentration of these impurities in the liquid must be less than 1 ppb oxygen equivalent, in order to achieve an electron attenuation length longer than 1 m, as required for ton-scale noble liquid detectors.

21.2 Two-phase XeTPCs: XENON, ZEPLIN and LUX

21.2.1 The XENON experiment

The XENON experiment aims at direct detection of WIMPs with a two-phase xenon time projection chamber (XeTPC). It uses a phased approach with TPCs at the 10 kg, 100 kg and 1000 kg fiducial target mass scale. The detector's principle of operation is shown in Fig. 21.5. An event within the active volume of the TPC will create ionization electrons and prompt scintillation photons which are simultaneously detected. The ionization electrons, which drift in the pure LXe under an external electric field, are extracted from the liquid to the gas where they are accelerated by an appropriate electric field, emitting proportional scintillation photons. Two arrays of photomultiplier tubes (PMTs), one in the liquid and one in the gas, are used to detect the prompt scintillation signal (S1) and the delayed (by the drift time in the liquid) proportional scintillation signal (S2). The array of PMTs immersed in the LXe collect the majority of the prompt S1 signal, which is totally reflected downwards at the liquid–gas interface. The ratio of the two signals is different for nuclear recoil events, such as produced by WIMP or neutron interactions, and electron recoil events, such as produced by background beta- and gamma-rays, providing the basis for background discrimination. Since electron diffusion in LXe is small, the proportional scintillation photons carry the x–y information of the interaction site. With the z-information from the drift time measurement, the TPC provides complete 3D-event localization which allows the background to be rejected via fiducial volume cuts. To test the XENON concept and verify the achievable energy threshold, background rejection power and sensitivity, a detector with a fiducial mass of the order of 10 kg (XENON10) was developed and operated at the Gran Sasso National Laboratory (LNGS) [106]. The laboratory is located at a depth of ∼3800 mwe, which reduces the cosmic ray muon flux to a level of 1.1 muons/(m² h).

The PMTs used in XENON10 were Hamamatsu R8520-06-Al, 1-inch square, optimized for Xe 178 nm wavelength and selected for low radioactivity. The millimetre position resolution of the XENON10 TPC, along with the LXe self-shielding, provided additional and important background rejection capability. The total LXe mass in XENON10 was 25 kg and the maximum drift gap in the liquid was 15 cm with a field of 0.7 kV cm^{-1}. The gamma background of XENON10 was at the level of 0.6 events/(kg d keV) in the 5.4 kg central mass region. The energy threshold was 4.5 keV nuclear recoil energy (keV$_r$) with a light yield of 0.7 photoelectrons per keV$_r$. XENON10 was calibrated with both gamma and AmBe neutron source, and stable operation had been demonstrated over many months. Based on calibration, and on a period of non-blind WIMP search data, the WIMP search region was defined between 4.5 and 29.6 keV nuclear recoil energy, 3σ below the mean of the nuclear recoil band (at 50% nuclear recoil acceptance). From a total of 1800 events in the 58.6 live-days of blind WIMP search data, 10 events were observed in the WIMP search region, with $7.0^{+1.4}_{-1.0}$ events expected based on statistical (Gaussian) leakage alone. Given the uncertainty in the number of estimated leakage events from electron recoils (no neutron-induced recoil events were expected for above exposure), conservative limits with no background subtraction were calculated for WIMP–nucleon cross-sections. XENON10 could exclude previously unexplored parameter space, setting a new 90% C.L. upper limit of 4.5×10^{-44} cm² and 5×10^{-39} cm² for the WIMP–nucleon spin-independent and spin-dependent cross-sections, respectively, at a WIMP mass of 30 GeV/c^2 [90; 91]. The results from XENON10

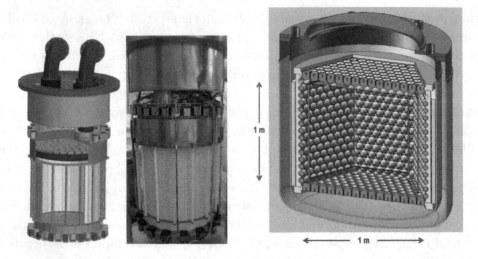

Fig. 21.6. Left: A schematic drawing showing the inner structure and a photo of the XENON100 TPC. Right: A schematic drawing of the XENON1T detector with full coverage of QUPID photodetectors and an OFHC copper cryostat.

have for the first time proved the concept of a XeTPC for dark matter searches.

Currently, a new 100 kg scale detector (XENON100) with a factor of 100 less background than the XENON10 prototype has been built and installed at LNGS. The background reduction goal was achieved thanks to: (a) new PMTs with very low radioactivity level; (b) great reduction in mass and radioactivity of the cryostat with an optimized design and selected stainless steel; (c) careful selection and screening of construction materials and (d) a new cryogenic system design to remove the cryocooler and other known sources of background from the proximity of the target. XENON100 uses many design features successfully tested in XENON10, with the addition of an active LXe veto surrounding the target. A photo of the detector is shown in Fig. 21.6. The target with 65 kg of LXe is instrumented with 178 PMTs in two arrays, 98 in the gas and 80 in the liquid. The PMTs (Hamamatsu R8520-06-Al 1-inch square) are selected for ultra-low radioactivity and high quantum efficiency, compared with those used in XENON10. A PTFE cylinder defines the TPC sensitive volume, with a drift gap of 30 cm (doubling that of XENON10). PTFE is used for its properties as a good VUV reflector and insulator. The PTFE cylinder optically separates the TPC volume from the active LXe veto volume (105 kg), which is viewed by 64 PMTs. A double-walled stainless steel vacuum vessel encloses the target and veto. Cooling is provided by a 170 W pulse tube refrigerator (PTR), originally developed for the MEG experiment [1078]. The PTR is used to liquefy Xe gas and to maintain the liquid temperature during operation.

The estimated background contributions from different sources in XENON100 are based on detailed Monte Carlo simulations and the measured activity of all detector and shield construction materials. These materials were selected for their low intrinsic radioactivity with a dedicated screening facility, consisting of a 2.2 kg high-purity Ge detector in an ultra-low background Cu cryostat and Cu/Pb shield, operated at LNGS. The electron recoil (ER) background is dominated by gamma-rays from decays of ^{238}U, ^{232}Th, ^{40}K and ^{60}Co in detector materials, mostly in the PMTs. The self-shielding of LXe reduces the ER rate significantly in the central part of the target. The intrinsic ER rate due to β decays of ^{85}Kr is reduced to 1×10^{-4} events/(kg d keV). It requires a Kr/Xe level of 5 parts per trillion which will be achieved by using a dedicated cryogenic distillation column installed at LNGS. XENON100 aims to reach a gamma background rate below 10 (3) $\times 10^{-3}$ events/(kg d keV) in 50 (30) kg fiducial mass. The nuclear recoil background due to (α, n) and spontaneous fission reactions in the detector, shielding materials and surrounding rock/concrete is predicted to be less than 1 event/(100 kg yr), while high-energy neutrons induced by cosmic muon spallation in the rocks around the cavern are estimated to contribute <0.3 events/(100 kg yr).

A first result from a WIMP search with XENON100 is expected in 2010, aiming to reach a spin-independent WIMP sensitivity of $\sigma \sim 2 \times 10^{-45}$ cm^2 for 100 GeV/c^2 WIMPs. An upgraded XENON100, with a lower background and larger target mass, aims to accumulate WIMP search data for two years in a 100 kg fiducial target, and to reach a sensitivity of $\sigma \sim 2 \times 10^{-46}$ cm^2 (see Fig. 21.7).

The XENON100 phase is also testing key technologies and performance characteristics relevant for the realization of a ton-scale XENON experiment. Such a detector, called XENON1T, would contain a total of 3 ton of LXe with a volume of 1 m^3. To further reduce the background from detector components, an oxygen-free high-purity copper (OFHC) cryostat and QUPID photodetectors [110] will be used. A full coverage with QUPID sensors is planned (see Fig. 21.6), in order to improve the light collection and thus the detector's energy threshold. To suppress the muon-induced neutron background to insignificant levels, XENON1T will need a deeper underground laboratory or a very large shield. One possibility is to use the Large Volume Detector (LVD) liquid scintillator detector [1724] as an active shield. The design of such an integration between XENON1T and LVD is currently being studied. Other candidate sites for XENON1T include

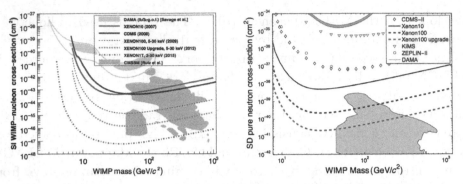

Fig. 21.7. Spin-independent (left) and spin-dependent pure neutron (right) WIMP–nucleon cross-sections as a function of WIMP mass, probed by different stages of the XENON programme. On the left, the low mass WIMP region allowed by DAMA [324; 1695], and the current most stringent limits from XENON10 [90] and CDMS [39] are shown as solid curves. The dashed curves show the predicted sensitivities of XENON100 (upper) and XENON100 upgrade (lower). The dot-dashed curve (left) shows the predicted sensitivity of XENON1T. The filled regions are the allowed parameter space for the lightest neutralino in the constrained minimal supersymmetric Standard Model [1656].

the Modane Underground Laboratory (LSM) in France and the Homestake mine in the United States. The aim of XENON1T is to reach a sensitivity to spin-independent WIMP–nucleon cross-sections of $10^{-47}\,\mathrm{cm}^2$. A multi-ton, multi-target detection (liquid argon and liquid xenon) system for dark matter with similar technology to XENON1T is also being planned [110].

21.2.2 The ZEPLIN experiments

The ZEPLIN II and ZEPLIN III experiments, conceived and built much earlier than the XENON10 prototype, are based on the same principle of a two-phase xenon time projection chamber. Background discrimination is provided by the simultaneous measurement of the scintillation photons and the ionization electrons produced by incident particle interactions. They differ, however, from each other and from XENON10 in the details of the detector's design: cryogenics and Xe purification, choice of PMTs, energy threshold, operating field, total sensitive mass, passive gamma and neutron shielding and thus overall background rate and sensitivity. Both ZEPLIN II and ZEPLIN III experiments are located in the Boulby Underground Laboratory in the United Kingdom, at a depth of 2805 mwe, reducing the cosmic-ray muon flux to a level of about $1.4\,\mathrm{muons}/(\mathrm{m}^2\,\mathrm{h})$.

ZEPLIN II. ZEPLIN II had a target mass of 31 kg of LXe (7.2 kg fiducial mass), viewed from above by an hexagonal array of seven ETL, 13 cm diameter, low-background PMTs (D742QKFLB). The target vessel was made of rolled copper and was surrounded by a stainless steel vacuum vessel. The active volume was defined by a thick PTFE tapered annulus, acting both as a VUV light reflector and as a support for the field shaping rings. A cathode mesh at the bottom and another grid below the liquid surface along with the field shaping rings served to maintain a $1 \, \text{kV cm}^{-1}$ electric field across the drift gap of 14 cm. To extract the drifted electrons from the liquid and to provide the electroluminescence region in the gas phase, a third grid was placed above the liquid surface. The strong electric field of $4.2 \, \text{kV cm}^{-1}$ in the liquid and $8.4 \, \text{kV cm}^{-1}$ in the gas extracted electrons from the liquid with an efficiency of 90%. The secondary yield was $\sim$230 electroluminescence photons per extracted electron, at a mean pressure of 1.5 bar. The energy threshold was $5 \, \text{keV}_{ee}$ [68].

Results from a dark matter search with the ZEPLIN II detector are reported in [68]. With a live time of 31.2 days and a total exposure of 225 kg d after fiducial and stability cuts, a total of 29 events were observed in a 5–$20 \, \text{keV}_{ee}$ window with 50% nuclear recoil acceptance; 28.6 ± 4.3 gamma-ray and radon progeny background events were expected. Using the Feldman–Cousins approach, this gives a 90% C.L. upper limit of 10.4 nuclear recoils, corresponding to a WIMP–nucleon spin-independent cross-section of $6.6 \times 10^{-43} \text{cm}^2$ at a WIMP mass of $65 \, \text{GeV}/c^2$.

ZEPLIN III. The ZEPLIN III detector is based on a substantially different design strategy. It has a shallow drift gap of 3.5 cm for operation at a field of several kV cm^{-1}, compared with the $1 \, \text{kV cm}^{-1}$ used across the $\sim$15 cm drift gap of ZEPLIN II and XENON10. This allows a better separation between the electron and nuclear recoil bands [44]. The cylindrical target volume of 19.3 cm radius and 3.5 cm depth, containing 12 kg of LXe, is viewed from below by an array of 31 2-inch ETL D730/9829Q photomultipliers, yielding a lower energy threshold than in ZEPLIN II. The PMTs, operated in the liquid, are used for both prompt and proportional scintillation signal detection, taking advantage of the flat planar geometry (Fig. 21.8). The drift field and the proportional scintillation region are defined by a cathode, placed above the PMTs, and a solid copper plate ('anode mirror'), 5 mm above the liquid level. The pattern of the proportional signals on the PMTs provides 2D spatial resolution <1 cm. The bulk of the detector parts are made of OFHC copper. Cooling is achieved using liquid nitrogen in a 36 l reservoir located below the target vessel and in contact via two thermal links (Fig. 21.8).

The results of the first dark matter search with ZEPLIN III have recently been reported [1316]. The gamma background rate is at the level of

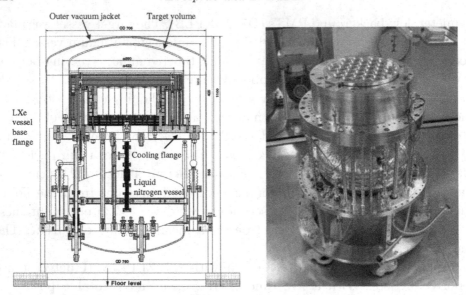

Fig. 21.8. Left: A schematic view of the ZEPLIN III detector. Right: A picture of the assembled ZEPLIN III PMT array above the cryogenic reservoir.

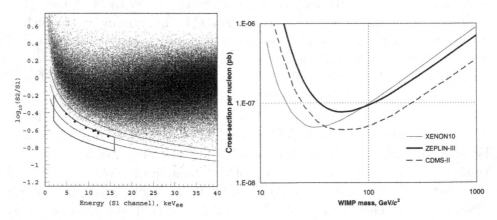

Fig. 21.9. Left: Scatter plot of log(S2/S1) as a function of energy for the first ZEPLIN III science run. Seven events are observed in the WIMP-search region. Right: Upper limit on the WIMP–nucleon elastic scattering cross-section based on zero events being consistent with a WIMP signal [1316].

$\sim$10.5 events/(kg d keV) in the energy region below 50 keV$_{ee}$ and in a fiducial region of 6.7 kg of LXe. An analysis of the 847 kg d of raw data (127 kg d after cuts) has resulted in seven observed events in the WIMP search region between 2 and 16 keV$_{ee}$ (Fig. 21.9, left). These events are compatible with leakage from the electron recoil distribution, the signal lower limit being consistent with zero. The upper limit on the spin-independent WIMP–nucleon

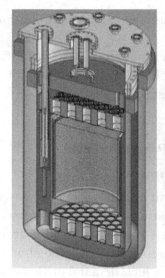 

Fig. 21.10. A schematic view of the LUX detector (left). R8778 Hamamatsu PMT arrays for the 50 kg LUX prototype (right).

cross-section is 7.7×10^{-8} pb at a WIMP mass of $55\,\text{GeV}/c^2$ (Fig. 21.9, right) [1316]. An upgrade of the detector with low-background PMTs and a neutron veto is planned before the next science run.

21.2.3 The LUX experiment

The Large Underground Xenon (LUX) experiment aims at deploying a 300 kg, two-phase XeTPC in a water shield to be constructed at the 4850 feet deep Sanford Underground Laboratory of the Homestake Mine, in South Dakota. Homestake has been selected as the site for a Deep Underground Science and Engineering Laboratory (DUSEL), to be supported jointly by the US National Science Foundation and Department of Energy.

The LUX detector adopts the same principle as the XENON10 detector, and specifically exploits a number of features of LXe to drive backgrounds down. Three hundred kilograms of liquid xenon (with 100 kg in the fiducial region) will be viewed by 122 Hamamatsu R8778 PMTs. Figure 21.10 shows a schematic view of the detector. The event rate due to backgrounds in the 100 kg fiducial target is expected to be suppressed to 0.8×10^{-3} events/ (kg d keV) (dominated by the PMTs) owing to the remaining 200 kg of LXe acting as self-shielding. The water shield surrounding the detector will efficiently remove external and muon-generated neutrons. The goal of the

LUX experiment is to reach a WIMP–nucleon spin-independent cross-section at 7×10^{-46} cm^2 in 10 months of operation.[1]

A 50 kg prototype for the LUX detector is being built and is currently being tested. Figure 21.10 (right) shows the top PMT array for the prototype. The full detector will be installed at Homestake in 2010.

21.3 Two-phase ArTPCs: WArP and ArDM

21.3.1 The WArP experiment

The WArP experiment, located at LNGS, is the first two-phase TPC based on LAr as sensitive target. There are two main features related to the detection of the scintillation signal, which distinguish a two-phase ArTPC from two-phase XeTPC. The first is a disadvantage associated with the shorter wavelength of the Ar scintillation light (128 nm) which demands the use of wavelength shifters, to shift the wavelength of the light to about 442 nm for efficient detection by commercial PMTs [84]. Typically, tetra phenyl butadiene (TPB) is used as wavelength shifter, coated on the inner TPC surfaces and the PMT windows, with some optimization in the procedure to allow for long-term stability and reliability. The second is an advantage associated with the larger time separation between the singlet and triplet state transitions leading to the Ar scintillation. This allows for pulse shape discrimination (PSD) of background events, in addition to the S2/S1 ratio and the 3D event position reconstruction. The decay times differ by two orders of magnitude, being $\tau_s = 7$ ns and $\tau_t = 1.6$ μs for the singlet and triplet state, respectively. The ratio between the emission intensities of the singlet/triplet states depends on the deposited energy per unit path length, dE/dx. It has been measured to be about 0.3 for gammas and in the range of 1.3–3 for alphas and 3 for argon ion recoils [1109; 1346; 1560]. The difference in the pulse shapes is used to distinguish between electronic and nuclear recoils. A discrimination power of $\sim 10^6$ has been achieved for nuclear recoil energies above ~ 50 keV [1346].

The advantage of the PSD has been successfully implemented in the WArP phased programme. In Phase 1, a small (2.31) two-phase ArTPC was built and operated underground, first without and later with PSD and a passive neutron shield. In this phase, ordinary argon with a high background rate was used, which has made it possible to explore the very high rejection rate in a relatively short time. The results have been published in [274].

---

[1] http://luxdarkmatter.org/

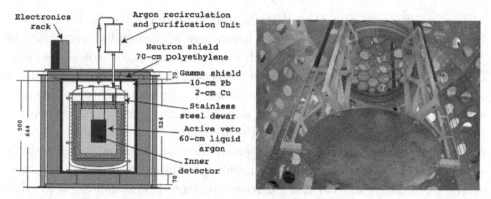

Fig. 21.11. Left: Schematic view of the WArP detector. Right: Copper structure for the active LAr shield, constructed at LNGS.

In Phase 2, the focus has been on reducing ($\sim$1/100) the main background due to the ^{39}Ar natural contamination, by using ^{39}Ar depleted argon. A new 2.3 l ArTPC made with selected low-radioactive materials was filled with a depleted Ar sample produced by isotopic separation. With this new detector about an order of magnitude improvement in sensitivity is expected, compared with the previous result. The main effort of the WArP collaboration is a new 100 l (140 kg) two-phase ArTPC (Phase 3). The detector is under assembly at LNGS and operation is expected by 2009.

The WArP 100 l detector layout is schematically shown in Fig. 21.11 (left). It consists of an external passive shield, a double-walled stainless steel cryostat, selected for low intrinsic radioactivity, containing the inner 100 l TPC, the active veto made of 8 t of LAr and a 10 cm thick polyethylene shield. The inner detector, equipped with field-shaping electrodes for the drift field, grids for extraction of ionization electrons from liquid to the gas phase and proportional light production, and PMTs (about 40 2-inch phototubes, ETL D757UKFLA) for the readout of the primary and secondary light signals, is suspended at the centre of an LAr volume. This volume, with a minimum thickness of 60 cm, works as active shield against gamma and neutron backgrounds.

A total of about 300 3-inch phototubes (ETL D750UKFLA) are used in the active shield for 10% photocathode coverage of the inner surface. They are held in place by PEEK supports connected to a thin copper structure delimiting the external surface of the active shield volume. The basic elements of the waveshifting/reflecting layer (TPB deposited on a highly reflective plastic substrate) are mounted on the Cu structure as well. The inner detector and the active shield are optically separated in order to avoid vetoing of events occurring in the central volume.

The sensitivity of WArP to WIMPs is ultimately limited by the intrinsic background from ^{39}Ar beta decays, which give an integrated event rate of 1 kHz in a 1 ton detector. Argon with low ^{39}Ar content, or isotopic separation, as well as an excellent background rejection ($>10^8$) is therefore required for operating large argon TPCs for low-energy, rare event searches. Isotopic separation would make a multi-ton scale experiment extremely expensive and would require a long production time. Argon from natural gas wells is much more attractive because ^{39}Ar production induced by cosmic rays is strongly suppressed underground. In addition, the large quantity of argon stored in underground natural gas reservoirs would be sufficient to provide material for the construction of a multi-ton WIMP detector. A first measurement of ^{39}Ar in argon from underground natural gas reservoirs has been performed during 2007, within the WArP R&D programme [913].

For WArP Phase 4, the sensitive volume of the detector may be extended to 1 t with minor modifications which will keep the main set-up unchanged, including the outer active veto, depending on the background level attained during Phase 3.

21.3.2 The ArDM experiment

The ArDM (argon dark matter) experiment is also based on a two-phase ArTPC, but with a different readout scheme for the ionization electrons signal rather than the PMTs-based readout used in the experiments discussed so far [1658]. The ArDM charge readout is provided by two LEM (large electron multiplier) plates for charge amplification in the gas by means of a high field generated in small, cylindrical shaped holes. The ionization electrons are drifted to the liquid–gas interface and are extracted into the gas phase and multiplied by the LEMs. Each LEM is composed by a 1.5 mm thick plate of isolating material, usually vetronite, which is covered by a copper layer on top and bottom. Holes of about 500 μm in diameter are homogeneously distributed on the LEM, at a separation of 800 μm. By placing an electric field of ~2 kV mm^{-1} between both sides of the plate, an avalanche of electrons is generated, and charge amplification factors ~10^3–10^4 can be obtained. The conceptual design of the detector is shown in Fig. 21.12. A total 850 kg of liquid argon are contained in a vacuum insulated vessel. The high voltage needed for the electric field over the drift length of 1.2 m is provided by a Greinacher high-voltage circuit, designed to reach up to 4 kV cm^{-1}. A series of electrodes are installed and biased along the full drift path to keep the field uniform at a level of a few per cent. To detect the primary scintillation light an array of 14 PMTs (Hamamatsu 5912 and ETL 9357), sensitive to

Fig. 21.12. Left: Schematic view of the ArDM detector and a picture of the inner TPC. The installed PMTs, field shaping rings and the external part of the wavelength shifter sheets can be seen. Right: The ArDM vessel, installed at CERN. The empty flange on top connects the vessel with the argon purification and recirculation systems.

visible light, are used at the bottom of the drift volume, immersed in LAr. The inside of the field shaping rings is covered with reflecting foils coated with $1 \, \text{mg cm}^{-2}$ TPB [84] to shift to the visible range and to reflect the scintillation light of liquid argon. Figure 21.12 shows pictures of the vessel and the actual detector with the light collection system already mounted.

Like WArP, ArDM expects to be able to reject the dominant background from the internal ^{39}Ar signal using the simultaneous charge and light signals, as well as pulse shape discrimination. The group is also studying the possibility of using ^{39}Ar-depleted argon extracted from underground natural gas wells [913]. The 1 ton prototype is currently under construction at CERN [1211]. After initial tests on site, it is expected that the experiment will be operated at the Canfranc Underground Laboratory, at a depth of 2500 mwe. With a nuclear recoil energy threshold of 30 keV, the sensitivity of the ArDM 1 ton prototype would access the WIMP–nucleon cross-section region of 10^{-42} cm^2. By improving the background rejection power and reducing the background sources, a sensitivity of 10^{-44} cm^2 is projected [1300].

21.4 Single-phase detectors: XMASS, DEAP/CLEAN

21.4.1 The XMASS experiment

The XMASS experiment is based on a single-phase XeTPC aiming at neutrinoless double beta decay in ^{136}Xe and low-energy solar neutrino detection, in addition to dark matter direct detection. The strategy is to use the self-shielding of the liquid in order to achieve an effectively background-free inner volume. The self-shielding works better for the gamma than for the neutron background, given the differences in their mean free paths. The position of an interaction can be reconstructed based on the pulse-height pattern in the photodetectors, and will depend on the size of the photodetectors, as well as on the light yield. In the first phase of the XMASS programme, a 100 kg prototype was deployed at the Kamioka underground laboratory in Japan. The goal was to demonstrate the basic detector technology and to study the feasibility of a larger-scale detector. The second phase, currently under development, consists of an 800 kg LXe detector with the main physics goal of detecting WIMPs. The longer-term goal is to build a 20 ton liquid xenon detector [15].

The 100 kg prototype, with 3 kg of LXe in the fiducial volume, was a 30 cm^3 detector made of OFHC copper, viewed by 54 2-inch low-background Hamamatsu R8778 PMTs, coupled to the sensitive volume with MgF$_2$ windows. The achieved photo coverage was 16%. The prototype was operated in a low-background shield to reduce the environmental gamma and neutron background [1235]. The background of the innermost 10 cm^3 volume amounted to 10^{-2} events/(kg d keV) below 100 keV, dominated by the radioactivity of the PMTs [15]. To remove the internal ^{85}Kr contamination of xenon, a distillation system with the capability of reducing the krypton to 1/1000 of its original concentration at a throughput of 0.6 kg Xe per hour was developed. The technique is based on the different boiling temperatures of xenon (165 K) and krypton (120 K) at 1 atmosphere. The achieved purity level is about 3.3×10^{-12} Kr/Xe [mol/mol] (or 3.3 ppt), with a 99% collection efficiency for the xenon [16].

A schematic view of the 800 kg detector is shown in Fig. 21.13 (left panel). About 1 t of LXe is contained in the volume inside the PMT photocathode plane, which is 45 cm in radius. The inner fiducial region, with a radius of 20 cm, will contain 100 kg of liquid xenon. The advantage of single-phase detectors such as XMASS over the two-phase TPCs discussed earlier is the absence of a drift region and hence biased electrodes, making the design simpler. However, this carries the cost of a large amount of expensive (in the case of xenon, and ^{39}Ar depleted argon) cryogenic liquid which is used

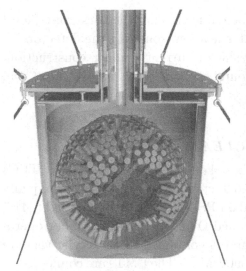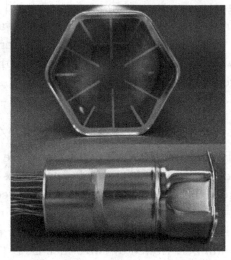

Fig. 21.13. Left: A schematic view of the XMASS800 detector. Right: a Hamamatsu R8778MOD low-background PMT with a hexagonal photocathode, as will be used in XMASS800.

as the shield. A disadvantage is the loss of information from the ionization signal, which, at least for the case of LXe detectors, cannot be compensated by PSD given that the singlet and triplet states are not well separated in time (3 ns versus 27 ns). These detectors are thus relying on absolute suppression of both external and internal electronic recoil background.

As in all detectors relying on PMTs for the detection of the scintillation signal, XMASS's gamma background is dominated by the intrinsic radioactivity of the PMTs. Currently a new version of the Hamamatsu R8778MOD low-background PMT, with a hexagonal photocathode shape, is being developed for XMASS (see Fig. 21.13, right). With a total of 812 hexagonal PMTs, a photo coverage of 67% can be achieved. The scintillation light yield inferred from simulations will amount to 4 photoelectrons per keV.

The radioactivity goal for the new PMTs is a factor of 10 lower than for those used in the 100 kg prototype. The estimated background rate from gamma-rays emitted by the PMTs is below 1×10^{-4} events/(kg d keV) in a fiducial volume of 25 cm radius. This background rate will allow a WIMP–nucleon cross-section sensitivity of 10^{-45} cm^2 for spin-independent couplings.

The requirements for the purity of the liquid xenon itself with respect to ^{238}U, ^{228}Th and ^{85}Kr levels are $<1 \times 10^{-14}$ g/g, $<2 \times 10^{-14}$ g/g and <1 ppt, respectively. These will be achieved by adding a radon filter, and designing a new distillation column with a longer tower length. The environmental gamma and neutron background will be reduced below the background level from the PMTs by a 200 cm thick water layer surrounding the detector.

The 800 kg detector and its water shield, currently under construction, will be installed at the new underground area in the Kamioka mine [15], recently completed.

21.4.2 The DEAP/CLEAN experiments

The DEAP/CLEAN experiments are based on single-phase Ar and Ne TPCs for dark matter and solar neutrinos detection. Like Ar, Ne offers equally good background discrimination based on PSD of the scintillation light [414; 1414]. A small, 7 kg LAr prototype, called DEAP-1, has been under operation at SNOLAB since 2007. The liquid is contained in an acrylic vacuum chamber, viewed by two 5-inch ETL 9390 PMTs. The PMTs are connected to the quartz windows of the chamber via 8-inch long acrylic light guides. The inner chamber surface is covered with TBP wavelength shifter. The demonstrated discrimination between electron and nuclear recoils is at the level of 6×10^{-8} for nuclear recoil energies above 60 keV (assuming a light yield of about 5 photoelectrons per keV). The sensitivity to WIMPs is currently limited by background coming from surface alpha-contamination, mainly from ^{210}Pb nuclei which were implanted on the surfaces during the exposure of detector components to airborne radon. A number of efforts to reduce the surface backgrounds are under way, after which 18 months of physics data, along with calibrations and background runs, are being planned. The aimed sensitivity is a cross-section of 10^{-44}cm^2 for spin-independent WIMP–nucleon couplings [413].

The Mini-CLEAN detector, a $\sim$360 kg prototype, is being built such that it can operate with both liquid neon and argon [1413]. While the liquid neon detector will provide a better sensitivity to *pp* solar neutrinos, the liquid argon version is mainly targeted at WIMP detection. Mini-CLEAN will consist of a spherical vessel with 80 cm diameter, filled with purified liquid neon or argon, at a temperature of 27 K or 87 K, respectively. The centre of the detector will be viewed by 92 PMTs immersed in the liquid. A soccer-ball shaped array of fused silica plates, with the inward-pointing surfaces coated with TPB, will be mounted in the centre of the spherical vessel. The light, after having been shifted from 80 nm in LNe or 128 nm in LAr to 440 nm, will be transported from the wavelength shifter plates to the nearest PMT via aluminized mylar light guides. The enclosed, central liquid mass would amount to about 100 kg. The cooling power will be provided by the second stage of a pulse tube refrigerator. It will continuously liquefy neon

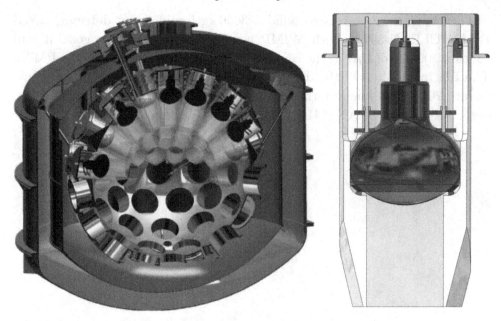

Fig. 21.14. A schematic view of the MiniCLEAN detector (left) and of an optical module (right).

or argon, after it has passed through a purification system. A schematic view of the Mini-CLEAN detector and its optical module is shown in Fig. 21.14.

The operation of PMTs at LNe temperatures, and the pulse shape discrimination capabilities of both neon and argon, have been studied with a 200 g prototype at Yale, Pico-CLEAN [1413].

The immediate goal of Mini-CLEAN is to test the long-term performance of PMTs at low temperatures, the efficiency of light collection and the accuracy of position reconstruction using the PMT hit pattern as well as the liquid argon and neon purification methods for radioactive contaminants and for scintillation light absorbers. After about one year of study at the surface laboratory at Yale University, deep underground operation of the detector is planned. It will allow researchers to characterize the backgrounds and to perform a WIMP search. Based on Monte Carlo simulations, the backgrounds will be dominated by fast neutrons produced in (α, n) interactions due to ^{238}U and ^{222}Th in the PMT glass. The expected rate is ~ 2 events per year in the 100 kg target, after exploiting the tagging capabilities due to subsequent neutron capture on argon with emission of a delayed gamma-ray [1413].

The near-future plan is to build a 3600 kg liquid argon detector, called DEAP/CLEAN-3600. With WIMP searches as its primary goal, it will contain 1000 kg of LAr in the central fiducial volume, viewed by 266 PMTs. The full installation of the detector at SNOLAB is planned to be finished in late 2010. The longer-term goal is to build a 100-ton-scale, liquid neon detector, called CLEAN [1412].

22

Directional detectors

Neil Spooner

22.1 Direct dark matter detection technologies and directionality

The direct detection technologies described in Chapters 18, 20 and 21 are motivated mainly by a desire actively to distinguish hypothesized WIMP-induced nuclear recoil events from backgrounds such as electron recoils. In each case this is achieved by attempting to measure a quantity, or ratio of quantities, such as heat to light, that depends on the dE/dx of the particle. This raises the question, why not measure the dE/dx directly itself, that is, measure the energy loss distribution along the recoil tracks? Achieving this in a suitable medium, for instance by imaging the tracks in some way, could provide the maximum possible information on events, not just the dE/dx and Bragg curve but also the range and perhaps the absolute direction of the recoiling nucleus.

This is the objective of directional WIMP detectors discussed here. However, the advantage of directional sensitivity is not just the prospect of reaching the best feasible discrimination and particle identification, possible because there is no doubt that different species with the same recoil energy will have different ranges for instance, but more importantly the prospect of linking the direction of those recoils to our motion through the Galaxy. This would provide a clear route towards a signature based on the non-terrestrial nature of WIMPs, a potentially definitive signature for WIMPs as dark matter, with maximum model-independency from particle physics and cosmology assumptions. Indeed it can be argued that such a direct link with our motion in the Galaxy is a necessary prerequisite if WIMPs are ever reliably to be claimed as responsible for the Galactic dark matter. The seeking of a Galactic signature is also the motivation behind searches for

Particle Dark Matter: Observations, Models and Searches, ed. Gianfranco Bertone. Published by Cambridge University Press. © Cambridge University Press 2010.

the predicted annual modulation in the total event rate and energy spectrum due to the Earth's orbit around the Sun [328]. However, the directional signature is potentially a far larger effect and, most significantly, one that cannot be mimicked by terrestrial background modulations.

22.2 The directional signature and statistics

The basis of directional detection lies in the standard assumption that there is a finite Galactic orbital velocity of our Solar System, $\sim$220 km s^{-1} relative to the velocity distribution of WIMPs in the Galactic halo, usually assumed to be a Maxwell–Boltzmann distribution with dispersion $\sigma_v = 155$ km s^{-1}. As first pointed out by Spergel in 1988 [1783], for a detector in the Solar System there is thus in effect a WIMP 'wind' experienced such that there is a preference for any resulting induced nuclear recoils to have a velocity vector opposite to this orbital motion through the Galaxy, essentially from the direction Cygnus (declination $\sim$45°). This preference, quantified for instance by the ratio of recoils scattered into the backwards hemisphere relative to the forwards hemisphere, is at least a factor of 10 and typically greater than a factor of 100 for WIMPs above 50 GeV [1476]. So this is potentially a very powerful signal in principle. In practice, since the Earth is itself orbiting the Sun, the expectation is that the mean direction of recoils in a detector fixed on the Earth will oscillate over a sidereal day. For instance, a directional detector on Earth at a latitude $\sim$45° (the Boulby site of the DRIFT experiments is at 54.6°) would see the WIMP wind vector oscillate from pointing south to pointing to the Earth's centre, repeating each sidereal day and going rapidly out of phase with the terrestrial day (see Fig. 22.1), a behaviour that cannot be mimicked by any terrestrial background.

Details of the directional dependence and likely sensitivities to this signal in the case of an idealized detector are described in Chapter 17. There, Eq. 17.30 is given as an approximate expression for directional dependence in terms of the angle between the recoil direction and mean direction of solar motion, assuming the standard halo model. Several authors (see Chapter 17) have explored the expected recoil velocity distributions for this standard halo, but also the possibility of non-standard WIMP halo models such as triaxial and oblate haloes, and the effect of WIMP stream components. A typical simulation result for an example velocity distribution is shown in Fig. 22.2. The obvious deviation this shows from a purely isotropic distribution, as might be expected from a terrestrial source of background neutrons for instance, illustrates the potential power of the directional technique. Indeed, in principle, following detection of WIMPs with a sufficiently

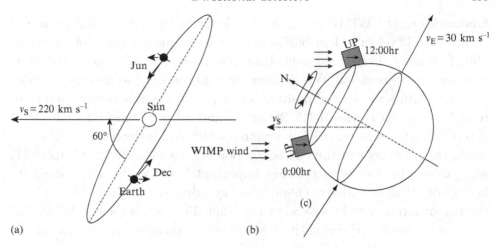

(a)   (b)

Fig. 22.1. (a) The orbit of the Earth with respect to the Sun and the motion of the Solar System. (b) The spin axis of the Earth with respect to the Solar System velocity. A detector at latitude 45° sees the wind vector oscillate from horizontal (south) to vertical over a sidereal day.

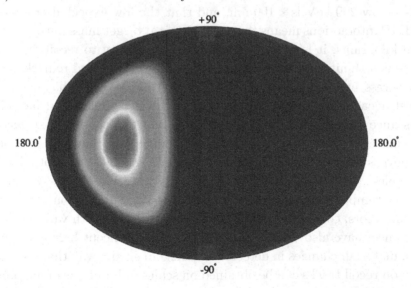

Fig. 22.2. Example simulation of WIMP flux plotted in Galactic coordinates for a Maxwell–Boltzmann halo with $v_0 = 220\,\mathrm{km\ s^{-1}}$. Courtesy of B. Morgan.

sensitive directional detector, it would be possible to go further and determine which of the postulated WIMP halo velocity distributions nature has actually chosen, a type of WIMP astronomy.

Unfortunately, the physics and practicalities of what can be achieved with real detector technology will inevitably detract from the theoretical possibilities described above and in Chapter 17. Important here is, first, the kinematics of the WIMP-recoil elastic scattering process in relation to the actual WIMP velocity distribution and the subsequent physics of ion straggling in matter. These are fundamental processes that will tend to wash out attempts to reconstruct WIMP directions. Second is the detector physics that determines the spatial position sensitivity of any readout, its ability to record ionization tracks in 1D, 2D or 3D and, of particular importance, of determining the sense of recoil directions (which end is the head or tail of the track, the place nearest the interaction point?). As indicated in Chapter 17, sense determination is particularly important because this can contribute factors of 10 to sensitivity. One of the key advances in the field has been the confirmation now by several groups that direction sense can be seen at all. The remainder of this chapter will focus on these issues, describing the practical state of the art for directional detectors.

22.3 Directional detector concepts

Given a starting point that the range of WIMP-induced recoils in solids and liquids below 200 keV is <100 nm, and that the low expected cross-section for WIMP interactions means that a viable total target mass must be at least several kilograms, it is clear that achieving sensitivity to recoil direction in practice is a significant technological challenge. It is indeed remarkable that any progress has been made at all. In fact rapid progress has occurred in the past decade. For solids, schemes can be devised that use surface effects, for instance to collect atoms ejected from one or other thin surface of a target material by a passing WIMP. The huge total surface area needed to achieve sufficient target mass makes these schemes conceptually hard. Nevertheless, such ideas have been explored with some success, for instance using low-temperature silicon wafers [1392; 1919]. Fabrication of multilayer semiconductors, composite scintillators or scintillators with variable doping, for instance, have also been explored [1789]. The concept here is to encode, in bulk material, changes in detector response in such a way that some information on recoil tracks can be obtained on scales of hundreds of nanometres. In practice, however, to construct an experiment of realistic mass with such novel solids would probably require bulk nanofabrication techniques not so far easily available.

An alternative is to use solids that already naturally possess some characteristic response to particle interactions that depend on ionization track orientation. Examples include roton anisotropy in liquid He [185], phonon anisotropy in Si [1139] or BaF_2 crystals [912] and scintillation light anisotropy in certain organic scintillators. However, all these schemes have also encountered significant problems, with only organic scintillators, notably stilbene and anthracene crystals, making some progress. In these crystals it has been known for decades that the scintillation response to alpha-particle tracks, for instance the absolute light output, depends on the orientation of the track relative to the crystal axis. Tests using neutron beams have revealed that this is also true for low-energy nuclear recoils [1551]. Unfortunately, the effect is very small (a few per cent anisotropy) and also degenerate with respect to the recoil direction sense [1788]. Furthermore, the heaviest nucleus involved in all these crystals investigated so far is carbon. The scintillation quench factor for carbon in organic scintillators is known to be poor, typically ~2% [1118] and anyway carbon is a poor kinematic match to expected WIMP masses. The expectation is therefore that any WIMP directional signature would be too small and essentially washed out even if very large crystal masses could be instrumented to compensate for the poor efficiency.

A further concept is to use particular solid materials in which the presence of a recoil track can be extracted by etching and subsequent scanning with a microscope. An early attempt at this involved using natural ancient mica [1777]. Here the great integration time involved (millions of years) compensated for the small mass of samples readily available and the slow rate of scanning, though not sufficiently to produce a competitive technique. Recently, there has been a proposal from Nagoya University to use nuclear emulsions, invigorated by development of new automated scanning techniques developed for particle physics applications [1491]. The submicrometre 3D spatial resolution on offer can in principle allow the direction of very short recoils to be determined. Both these techniques of course still do not provide real-time information and so it would be hard to correlate results with our motion through the Galaxy. Achieving sufficiently high rejection power of low-energy alpha and electron backgrounds in these 'passive' techniques is also difficult.

Given the difficulties with solids, an alternative, much more promising route is to use traditional gas time-projection chamber (TPC) technology but operated at low pressure (<100 Torr) such that the induced nuclear recoil tracks are long enough to be imaged using conventional charge or light readout schemes. At such pressure a typical 100 GeV WIMP of $220 \, \text{km s}^{-1}$ can cause a gas nucleus to recoil 1–2 mm. The obvious disadvantage here is the low density of the target. However, the possibility of room-temperature operation, the relatively conventional and well-understood nature of TPC technology, the reduced requirement for gamma shielding because of the good discrimination power and the low cost of excavation underground mean that the implied large volume of such an experiment is not necessarily costlier even than the conventional non-directional schemes using liquid noble gases or cryogenic

Table 22.1. *Directional detector projects*

Experiment	Readout	Gas	Status
DRIFT	Negative ion MWPC	CS_2/CF_4	R&D, underground (Boulby)
NEWAGE	μ-PIC	CF_4	R&D, underground (Kamioka)
MIMAC	micromegas, pixel anode	$^3He/CF_4$	R&D
DMTPC	CCD, optical	CF_4	R&D

bolometers (see Chapters 18, 20 and 21). Most progress has been made with these TPC detectors, in which the gas acts as both target and detector material. It is these that provide the focus of the remainder of this chapter.

22.4 Gas detector physics – diffusion and straggling

Table 22.1 lists the basic characteristics of the world's current main TPC dark matter detectors. In all cases, the principle is that a charge ionization track, produced by a WIMP-induced nuclear recoil (or other event) as it loses energy in the gas by multiple nuclear scatters, is drifted in a uniform electric field to a readout plane (see Fig. 22.3).

Here a higher electric field is arranged to produce avalanche amplification of the track. Two-dimensional projection reconstruction of the track is then possible by reading out the charge directly, using 2D multi-wire proportional counters (MWPCs) or other charge-sensitive strip or pixel devices, or through production of electroluminescent photons that can be read out by a CCD camera. Pulse arrival time information, more easily available in the former technique, can allow full 3D reconstruction. Figures 22.4 and 22.5 show example results of the application of these techniques used to image alpha tracks: the first shows an early 2D CCD image of an alpha-particle track obtained using the electroluminescence produced in a Gas Electron Multiplier (GEM) readout with CS_2 gas [1312] and the second an alpha-particle track reconstructed in 3D using MWPC charge readout in DRIFT-II [475].

The physics principles and practicalities relevant to such directional concepts are well illustrated by the design and operation of the longest standing effort, the DRIFT experiments (Directional Recoil Identification From Tracks) based at the Boulby underground laboratory, UK (see Fig. 22.6). We therefore mention this detector in some detail in the following sections to illustrate the principles, leaving more specific details of DRIFT results and the other gaseous ideas to Section 22.8. Detailed technical descriptions of DRIFT can be found in [65; 67].

In DRIFT the readout consists of relatively conventional multi-wire proportional counter planes covering $1\,m^2$ and comprising two planes of 512

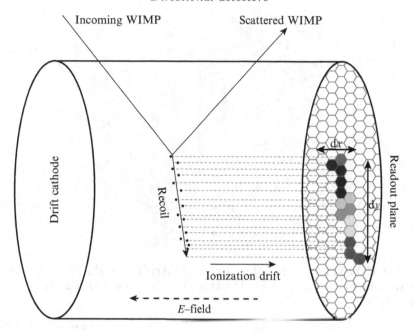

Fig. 22.3. Schematic diagram of nuclear recoil detection in a gas TPC. Ionization charge produced along the track is drifted in a strong electric field to a readout plane where the x and y projections of the track can be measured. Differential charge deposition along the track potentially allows the sense of the track to be determined.

wires with 2 mm pitch. In DRIFT there are two back-to-back detector regions sharing a central cathode plane, each read out by an identical MWPC. The detector volume is ~ 1 m^3 yielding a target mass of 167 g of CS$_2$ at 40 Torr. A high field region in the MWPCs results in avalanche charge amplification between grid and anode planes, these being read out via charge amplifiers to provide x dimension information and, via induced signals, y dimensions. The z dimension, perpendicular to the readout plane, can then be extracted from the charge arrival times. This technique allows full 3D reconstruction, in principle, though note that as there is no event trigger here there is also no measure of the absolute z position (see Section 22.5). The output is in effect x, Δx, y, Δy and Δz. In DRIFT the spatial position resolution reaches in principle $<200\,\mu$m in Δy and Δz, and 2 mm in Δx. Details of DRIFT analysis procedures are given in [474]. The improved resolution in Δy with respect to Δx arises from the possibility of using interpolation of induced pulses across multiple wires in that plane.

In addition to the charge position sensitivity of the readout itself (the spatial resolution) two more important factors in a directional TPC critically affect the ability to determine a directional signature. First, straggling of the recoiling ion will occur in the low-pressure gas such that over successive

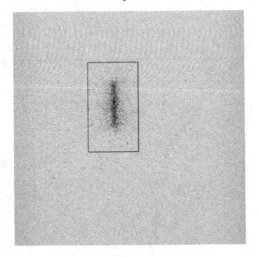

Fig. 22.4. CCD image of 5 MeV alpha in 20 Torr Ar/(10%)CH_4 + CS_2 using imaging of electroluminescence in a GEM. The track is 8 cm long and contains about 3200 photons [1312].

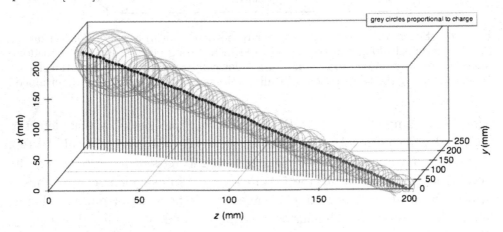

Fig. 22.5. Example 3D reconstruction of an alpha-particle track in DRIFT-II. Grey circles are proportional to the amount of charge deposited at each point. Courtesy of D. Muna.

nuclear scatters the ion will tend to deviate away from its initial direction. Second, diffusion of the charge will occur during drift of the ionization track to the readout plane. For the diffusion, Eq. (22.1) below gives an expression for the rms spatial spread lateral diffusion, σ, for ions or electrons of characteristic energy ϵ_k, drifted distance L in electric field E [1651]:

$$\sigma = \sqrt{\frac{4\epsilon_k L}{3eE}}. \tag{22.1}$$

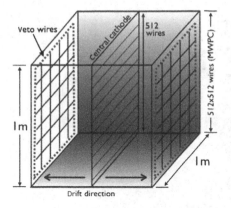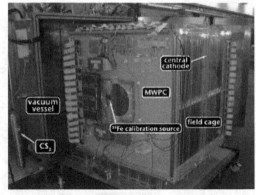

Fig. 22.6. DRIFT-II detector at Boulby mine.

Both straggling and diffusion, given by Eq. (22.1), provide a fundamental physics limitation to the directional sensitivity that must be understood for any chosen target gas, pressure and drift length L in the detector, in relation to the readout resolution. Most notably, the rms diffusion should be kept less than the desired track length to be reconstructed at the required energy threshold. There is little point also in supplying a readout with position resolution smaller than this diffusion. Conversely, the resolution of the readout can be thought to set the scale for the allowed diffusion and hence, through Eq. (22.1), the maximum reasonable drift length. The trade-offs between drift length, diffusion, readout resolution and gas pressure (and hence target mass per detector) are a vital cost and design driver here when considering scale-up to a realistic directional experiment (see Section 22.11). However, in general it is most important to minimize the diffusion so that the largest possible drift distance can be used and hence the minimum number of expensive readout planes per unit target mass, with these built to have position resolution no better than can be justified by the diffusion limit. This is the philosophy behind DRIFT and the use of conventional MWPC technology.

A conventional magnet system could be used in principle to suppress diffusion but this would probably be too cumbersome and expensive for an eventual large-scale experiment underground. An alternative, as introduced in DRIFT, is to operate with the rarely used CS_2 negative ion gas. Here CS_2 acts as both the target and, as CS_2^- ions, the charge carrier, instead of electrons. This special negative ion TPC (NI-TPC) configuration is an important trick, which works by effectively reducing the ϵ_k term in Eq. (22.1), as follows. For both ions and electrons in a gas there will first be diffusion outwards such that, via multiple collisions with the gas atoms, they acquire the thermal energy distribution of the gas. However, in an electric field, owing

N. Spooner

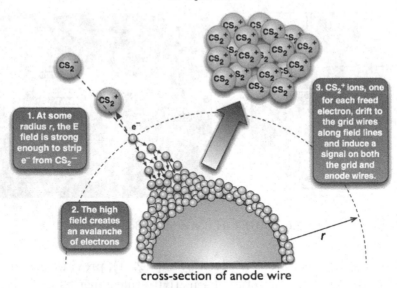

1. At some radius r, the E field is strong enough to strip e^- from CS_2^-

2. The high field creates an avalanche of electrons

3. CS_2^+ ions, one for each freed electron, drift to the grid wires along field lines and induce a signal on both the grid and anode wires.

cross-section of anode wire

Fig. 22.7. The negative ion drift principle. Courtesy of D. Muna.

to their small mass relative to the gas atoms and large mean free path, electrons can subsequently gain substantial energies between collisions such that subsequent collisions cannot efficiently re-thermalize this energy. This yields values for ϵ_k in Eq. (22.1) up to $\sim(2/3)e$. For ions, however, where the masses are similar to the gas atoms, any energy gained between collisions is readily thermalized. It can then be shown, independent of the type of ion, that the average energy tends to remain constant and close to thermal energies, so that $\epsilon_k \sim (3/2)\,k_BT$, even for high values of reduced field (E/P), where P is the pressure. For CS_2, values up to ~28 V cm^{-1}torr^{-1}are found [67].

Thus, from Eq. (22.1), it can be seen that a cloud of drifting ions is subject to a much smaller spatial spread than a cloud of drifting electrons under comparable drift fields and drift distances, typically σ_e being $\sim10^3\sigma_{\text{ions}}$. A further essential point here is that CS_2 is only mildly electronegative, such that although electrons attach to the molecules, they are also easily released by application of a moderate electric field. The full process is illustrated in Fig. 22.7. The operation of CS_2 as a TPC gas was first reported using a CS_2–CO_2 mixture [587] and introduced to DRIFT by Martoff [1460]. Measurements for DRIFT indicate a total diffusion for pure CS_2 at 40 torr of <1.4 mm per m drift length [1525].

Whilst the advantages of achieving this relatively low diffusion are clear, this has not deterred several groups from studying more conventional gases such as CF_4 and He or indeed mixtures [1460], particularly with reference to spin-dependent interactions for which F and ^{3}He provide excellent targets.

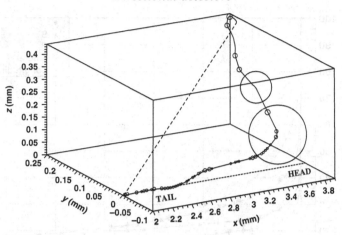

Fig. 22.8. An example simulated S recoil track at 100 keV in 40 torr CS_2 [1665].

Most notable are NEWAGE, studying CF_4 [1456], and MIMAC, studying CF_4, ^{4}He and ^{3}He [1688] (Section 22.10). In these cases there is always the possibility of reducing the absolute diffusion by using a shorter drift region or attempting higher electric fields, although this of course increases requirement for readout and hence the cost if the total mass of target is to remain reasonable.

Turning now briefly to straggling, we show in Fig. 22.8 an example sulphur recoil track at 100 keV in 40 torr CS_2, simulated using SRIM [1665], and in Fig. 22.9 plots of the expected lateral and longitudinal straggling as a function of range for target ions F, S and C. It can be seen that for long-range events, >8 mm, the straggling is typically <20% of the total range (defined as the direct line distance between the start and end of the track). However, as might be expected, towards lower ranges (corresponding to energies of most interest for dark matter searches) straggling greatly increases, for instance reaching >40% for S recoils below 1 mm. This degree of straggling, over which there is no control, when folded with the expected distribution of velocity vectors for WIMPs in the Galaxy, provides the main intrinsic limit to the feasibility of directional detectors. It is also the main motivation for seeking the additional, more powerful, information that could come if the absolute sense of the track direction could be determined, so-called head–tail discrimination (see Section 22.9).

22.5 TPC gamma background rejection and energy threshold

Building on the gas physics, we turn now to the issue of background rejection capability and energy threshold. The potential intrinsic discrimination

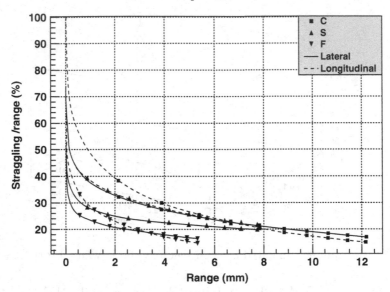

Fig. 22.9. Lateral (solid line) and longitudinal (dashed line) straggling/range as a function of range for fluorine ions in CF_4, sulphur and carbon ions in CS_2. Energy is marked up to 500 keV with an interval of 50 keV.

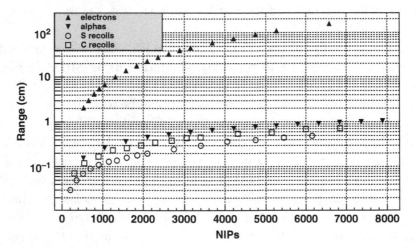

Fig. 22.10. Sulphur, carbon, alpha and electron range in CS_2 vs. ionization (NIPs) created as calculated by [1665].

power of the TPC concept is illustrated in Fig. 22.10. This shows simulations of the C, S and electron ranges in 40 torr CS_2 versus energy with straggling included but not diffusion. Comparing ions and electrons of the same energy we see that at all energies there is typically a factor of 10 difference in the range [1665]. This range difference itself provides a powerful discrimination

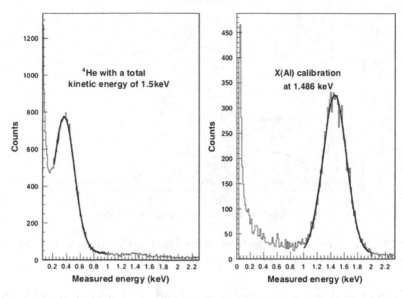

Fig. 22.11. Time of flight measurements performed with the MIMAC device comparing the spectrum from 1.5 keV ^{4}He ions with 1.5 keV X-rays [1687].

parameter. However, implied also is the associated difference in energy density along the track for different species (the $\mathrm{d}E/\mathrm{d}x$). It is the measurement of both parameters, range and $\mathrm{d}E/\mathrm{d}x$, through a position-sensitive energy measurement on each readout element (WMPC wire in DRIFT for instance), plus the potential for a low (below 10 keV) energy threshold, that allows exceptional discrimination against gamma background [1775]. The discrimination factor in DRIFT is measured to be $>10^6$ [472].

Concerning energy threshold, this is determined foremost by the intrinsic W-values (energy to create an ion pair) of the gas used, the avalanche gain or amplification process possible in the gas at the readout and the electronic noise. The W-values and associated so-called quench factors have been experimentally measured as a function of energy for C and S recoils down to 20 keV ($W = 34$ eV) and 30 keV ($W = 54$ eV) [1776] respectively and for F and He by the MIMAC group. These measurements are in broad agreement with theoretical estimates, although further studies are required to confirm the behaviour at lower energy. Such measurements are hard, generally attempted by using neutron beam techniques to induce the nuclear recoils [1862]. However, for gas directional detectors the MIMAC group have also pioneered use of ion beams, using differential pumping across thin foils separating the ion source from the target gas. Figure 22.11 shows for instance an exceptional measurement for He recoils [1687].

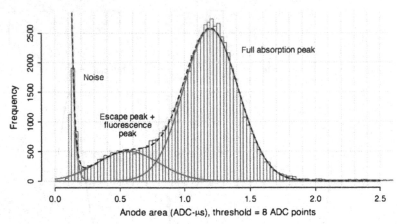

Fig. 22.12. The ^{55}Fe energy spectrum from DRIFT-IIb showing the escape peak and the full absorption peak at 5.9 keV (courtesy of D. Muna).

Based on the W-values and gains measured, it should be possible to reach sub-keV recoil thresholds in a directional gas experiment. Devices of realistic volume have not yet been optimized to achieve this in practice, but the DRIFT-IIb experiment has demonstrated the potential. Figure 22.12 shows an example ^{55}Fe spectrum taken from exposure of one-half of the $1\,\mathrm{m}^3$ DRIFT-II experiment. Although not taken in a self-triggering mode, and requiring some off-line noise reduction software, this plot does demonstrate that S-recoil thresholds of a few keV ($\sim$3.5 keV here) are feasible in a large volume directional detector [1488].

22.6 TPC neutron background rejection, solar neutrinos and radon

Although gamma (and alpha) rejection efficiency should be excellent in a TPC via $\mathrm{d}E/\mathrm{d}x$, track topology and range discrimination, the main background worry, in common with all dark matter detectors, remains high-energy neutrons as these can produce nuclear recoils potentially indistinguishable from those induced by WIMPs. The low density of a directional TPC target and the intrinsic insensitivity to electromagnetic (low $\mathrm{d}E/\mathrm{d}x$) events is actually a particular challenge here [489]. First, the long interaction path length for neutrons means their rejection by detection of multi-site scatter events is less feasible than in solid-state detectors, at least until very large active volumes are used. Second, muons, muon-induced secondary particles and electromagnetic secondaries coincident with neutrons are less likely to be detected and so could be unavailable as a means of neutron vetoing. Furthermore, back-scattering of neutrons from the internal walls of the vessel increases the total flux of neutrons inside the detector relative to the case of a denser target. As with all WIMP experiments, sufficient overburden can solve the muon neutron issue and massive low-Z shielding can deal with the rock neutrons. The greatest ultimate danger is thus from internal detector components via fission and (α, n) reactions for which shielding options are limited.

Extensive simulations have shown overall that material purification, careful design of internal passive neutron shielding and external muon vetoes can overcome the intrinsic neutron issues for a TPC, provided the depth is at least 1000 mwe [489]. However, a directional detector of course has the unique advantage that WIMP events should be correlated with Galactic motion (the sidereal day), while any terrestrial source of background, including neutron-induced recoils for instance from a local point source of activity, cannot be. This can allow specific identification of a neutron background and, in principle, an additional means of background reduction by statistical subtraction not available in other conventional experiments. The ease of operating with different and multiple target elements is a further route to neutron suppression here. For instance, in CS_2, WIMP interactions are more likely with sulphur than with carbon, owing to the higher A, whereas for a typical neutron background the rates are similar. It has been noted that coherent scattering from solar neutrinos could ultimately also be a background in a sensitive WIMP experiment [1465]. In principle, directional sensitivity again, uniquely, might be used to reject such background, since in this case events would be correlated with the direction of the Sun.

A final, specific particle background issue in all gaseous directional detectors is background from radon emanation by internal materials. The alpha-particle products from the radon decay chain are themselves easily rejected by energy and dE/dx cuts. However, certain of the daughter nuclei, notably ^{218}Po and ^{210}Pb, are produced charged and so become easily attached to surfaces at high voltage, in particular the cathode plane used to produce the necessary drift field. In the subsequent alpha decay it is possible for such nuclei to recoil from the attached surface back into the gas, with energy of $\sim$100 keV. If the coincident alpha is not detected, for instance because of absorption in the cathode plane, then this background recoil can easily mimic a WIMP event. Such recoils from all other surfaces, such as drift field rings or the readout plane, are easily removed by edge vetoes (as in DRIFT) or position reconstruction in the x–y plane. However, there can be no absolute z position information without an absolute event time trigger.

This background, known as radon progeny recoils (RPR), has been extensively studied by DRIFT [472] and new techniques developed to mitigate the effects. No straightforward means to veto events close to a high-voltage cathode plane have been implemented so far (but see Section 22.9). However, events produced close to the cathode, farthest away from the readout, will undergo the maximum pulse broadening due to diffusion, so this can be used as a basis for software discrimination cuts. Application of this, plus removal of radon-emanating components, gas flushing and etching of surfaces with nitric acid to remove ^{210}Pb, has proved highly successful in DRIFT, resulting in a reduction in background of a factor 3000 in the DRIFT-II module.

22.7 Electronic noise and other background

Operation of a low-pressure TPC in very low-background mode with unusual gases, high electric fields and sensitive electronics inevitably reveals classes of previously unobserved events with origin not associated with any particle interactions. Examples revealed, for instance, in DRIFT include so-called micro-sparks, ringers, pre-ionization events and others [472] yielding a typical raw uncut event rate of order 1 Hz. Care is needed to identify and understand these in order to correctly reject them in analysis. However, the slow drift times with CS_2 are an advantage here since such events have fast time structures and can be easily separated.

22.8 WIMP detection and directional sensitivity in practice

Turning now to sensitivity, we first note that for a directional detector this term has several meanings, for instance: (i) sensitivity for the detection of WIMPs regardless of directionality (in common with non-directional technologies), (ii) directional sensitivity, quantifiable in terms of the number of WIMP events required to show that a signal is Galactic (i.e. does not have an isotropic distribution in direction), and (iii) sensitivity to distinguishing between different WIMP velocity distributions. In each case there is the additional issue of whether the background is assumed to be zero. In the case of a non-zero background there is the further prospect of using the directional sensitivity to suppress background statistically, as a means of improving limits for instance.

Concerning (i), the powerful particle identification that can yield event-by-event, gamma, beta, alpha and RPR rejection, plus the expected intrinsically low detector backgrounds and low threshold, suggests good prospects. For example, we show in Fig. 22.13 a typical predicted WIMP spin-independent detection sensitivity for a 10 kg CS_2 TPC assuming zero background and 10 keV S-recoil threshold. This sensitivity is comparable to conventional technologies with the extra benefit, of course, of the additional directional sensitivity not shown here. However, recognizing that the volume of such a detector

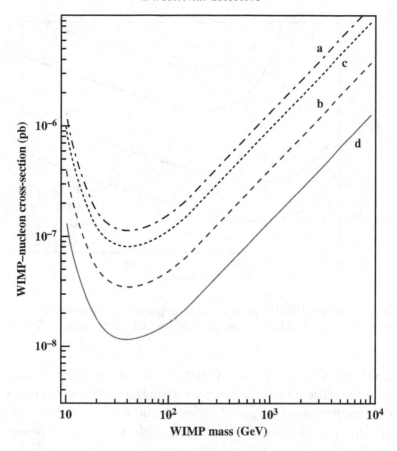

Fig. 22.13. Example predicted sensitivity limits due to neutron backgrounds in a small low-pressure gas TPC assuming one year of running with 100% efficiency above 10 keV and 100% rejection of gamma and alpha events. Dashed-dotted curve (a) 3.33 kg CS_2 detector with four neutrons detected; dashed curve (b) 3.33 kg with 0 events; dotted curve (c) 10 kg with 12 events detected; solid curve (d) 10 kg detector with 0 events [489].

is perceived to be a short-term issue for construction (for instance, DRIFT-II at 1 m^3 volume provides only ~170 g of target CS_2 at 40 Torr), there has been work to use CF_4 or He gas [1457; 1687; 1713]. The high sensitivity to spin-dependent interactions, owing to the high spin factors available, means competitive limits are feasible even in relatively small volumes (see Fig. 22.14). Furthermore, the flexibility of a gas target means that in principle the sensitivity to a non-directional signal can be increased simply by raising the pressure.

For directional sensitivity (items ii and iii) there are several approaches possible. Ideally, the objective would be full 3D reconstruction of tracks such that the most accurate fits can be made to build a distribution histogram of initial recoil directions in Galactic coordinates. This 'sky map' of recoil directions can then be compared with the distribution expected from the

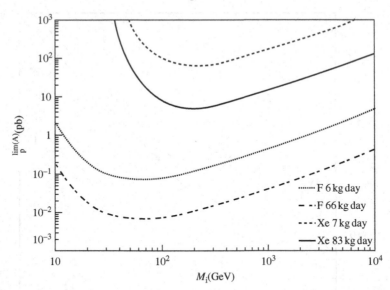

Fig. 22.14. Example WIMP–proton (spin-dependent) cross-section sensitivity curves as a function of WIMP mass for 0.05 bar CF_4:CS_2 and ^{129}Xe:CS_2 LPNI-TPCs.

assumed velocity distribution of WIMPs for a chosen galactic halo model. In this case the critical factor is the reconstruction pointing accuracy in 3D. In practice, for instance because of diffusion and because a detector is likely to have position resolution different along each axis, this response will be a complex function where the pointing accuracy depends critically on the absolute position of an event and its direction relative to the detector axis, as well as energy. Figure 22.15 shows results of an example pointing accuracy simulation for an ideal DRIFT-like detector (with all the gas physics in place but perfect position resolution).

In this response matrix, the pointing accuracy is defined as the probability that the direction of a recoil will be reconstructed to within 30° of the actual initial direction. The potential accuracy is surprisingly good, with, for example, >80% of events meeting this criterion above 40 keV with 25 cm drift distance. DRIFT has generated sky maps from data, while NEWAGE has done this and then used observation of an isotropic distribution as a means of statistically suppressing background to produce a limit [1457].

The only way to measure experimentally the directional sensitivity in practice is to use neutrons from a source, positioned or collimated to provide a crude directional beam. For this it has been found that ^{252}Cf neutrons mimic quite well the recoil energy spectrum and directional distribution expected from WIMPs [473]. Figure 22.16 shows an example sulphur recoil

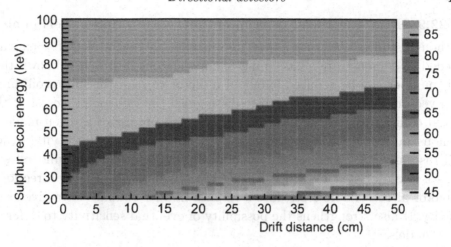

Fig. 22.15. Example pointing accuracy simulation for DRIFT-II. Shadings denote the probability that reconstruction reproduces the direction within 30 degrees of the known input initial direction.

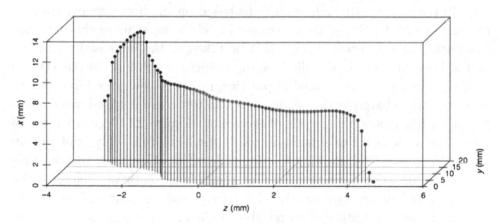

Fig. 22.16. Example 100 keV sulphur recoil reconstruction in 3D from DRIFT-II.

reconstruction in 3D from DRIFT-II taken during such a directed neutron run. While it is possible from such analysis to extract recoil directions as above, in practice a simpler approach, which does not necessarily severely reduce the sensitivity to a non-isotropic distribution of directions, is simply to use the 1D projection. DRIFT has demonstrated such basic 1D directional sensitivity from sulphur recoils down to 47 keV using a simple directed neutron analysis in which measurements of the three range components Δx, Δy and Δz were made for each event. The average magnitude of these components was found to increase for neutrons directed along the corresponding x, y or z axis. For instance, the normalized parameter $\Delta z/\Delta x$ yielded a magnitude of 14% at threshold increasing to >25% at high energy [474].

22.9 Head–tail recoil discrimination, theory and experiment

Although studies have shown that, in principle, only $\sim$100 WIMP events are needed to demonstrate a basic range component signature, it is known that this number can be reduced by about a factor of 10 if the full recoil direction vector can be reconstructed to distinguish the tail (point closest to the initial collision) from the head. Recently much progress has been made in demonstrating both theoretically and by experiment that such vital head–tail discrimination is possible, something previously not well understood. This is a major step towards realizing the possibility of a large directional detector at reasonable cost, since it implies smaller volumes for a given sensitivity. It also strengthens the possibility of eventual sensitivity to different halo models.

Original theoretical predictions for binary gases relevant to directionality were performed by Hitachi *et al.* [1108] but these did not account for the complexities introduced by ion straggling, diffusion and the readout geometry. In particular, although predictions based on the linear energy transfer calculations of the Bragg-like curves used might suggest more observed ionization at the tail, account must also be taken of the increased scattering at the head of a track, so-called ball-up, which, depending on the 1D, 2D or 3D readout projections used in practice, might be integrated in a way to produce more charge at the head instead. Later, more detailed simulations by [1665] did indeed reveal this phenomenon. Figure 22.17 gives examples from this work of a full simulation result for head–tail analysis of carbon recoils in 40 torr CS_2, showing the dramatic effect of different readout projections. This work demonstrated that the balance between charge at the head and tail varies significantly with energy and readout and in certain cases, for instance with carbon recoils <20 keV, can reverse so that more charge would be integrated at the head.

Some experimental evidence of the existence in practice of head–tail identification has come from Dujmic *et al.* [726], observing 1D projections of F recoils of relatively high energy (>200 keV) in 100–380 torr CF_4 using CCD readout in a small device (see Fig. 22.18). There was also earlier evidence from the DRIFT-I detector. The large 1 m^3 DRIFT-II detector at 40 Torr has achieved results using directed neutron studies of sulphur recoils down to the much lower energies of 1.5 keV amu^{-1} ($\sim$40 keV S-recoil) [473]. DRIFT-II is uniquely designed for head–tail studies by virtue of containing two TPCs identical in every way but for their oppositely directed drift fields. Comparison of data from this TPC pair, for a given directed neutron exposure, allows suppression of systematics, for instance from electronics.

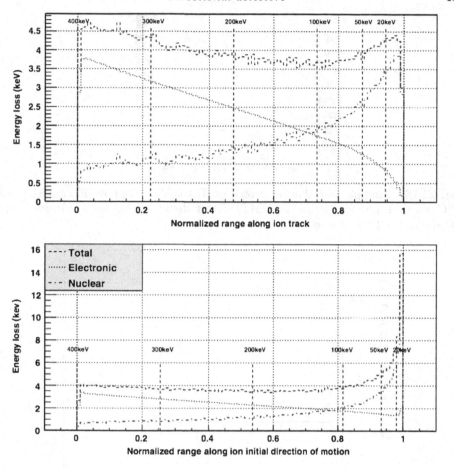

Fig. 22.17. Simulation result for head–tail analysis of carbon recoils in 40 torr CS_2 comparing the response for two different readout geometries: (top) normalized along the track; (bottom) along the ion initial direction.

Figure 22.19 shows results from a simple analysis that calculates the ratio of charge deposited in the first and second halves of tracks as projected (essentially) onto the z (time readout) axis, taking the ratio of this for optimal (z directed) neutrons with respect to anti-optimal (x or y directed) neutrons. This parameter shows a clear directional effect of 30% difference at high energy reducing to 14% at threshold.

22.10 Experimental status and readout technology

Having detailed the basis of directionality we give here a summary of the status and essential features of the main experiments under way (see Table 22.1) but with particular emphasis on the readout concepts used. This aspect

N. Spooner

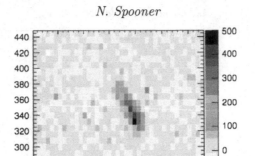

Fig. 22.18. Example scintillation profile of 313 keV nuclear recoil candidate in a ^{252}Cf exposure at 75 torr in the DMTPC test device. Neutrons are incident from the right, along the x axis.

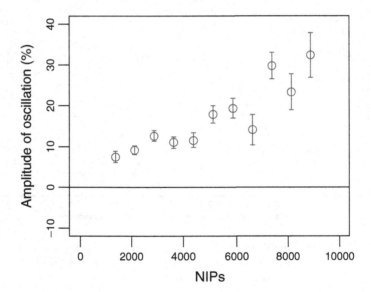

Fig. 22.19. Head–tail signal: oscillation amplitude vs. NIPs measured in the DRIFT-II detector [473].

provides perhaps the biggest technological challenge in the field since eventually very large areas of readout will be needed (see Section 22.11).

22.10.1 NEWAGE

The Japan-based NEWAGE team have pioneered the use in a gas TPC of micropixel chamber (µ-PIC) readout technology [1458], making use of new double-sided printed circuit board techniques with 100 µm thick polyimide substrate (see Fig. 22.20). Anode and cathode strips are formed on both sides of this with a pitch of 400 µm. An electric plating process is used to

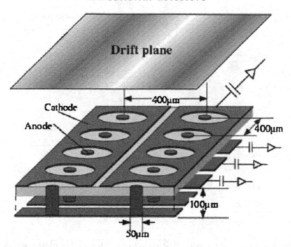

Fig. 22.20. Schematic structure of the micropixel chamber of NEWAGE [1458].

produce pixel anodes at the centre of cathode holes. For dark matter tests a μ-PIC plane with detection area 31×31 cm^2 has been used, with a 23×28 cm^2 gas electron multiplier added to provide a total gas gain of typically 2300 when operated with 150 torr CF$_4$, installed to read out a total TPC detection volume of $23 \times 28 \times 31$ cm^3 all housed in a stainless steel vessel.

This device has been successfully used to study prospects for directional WIMP–proton sensitivity with operation on the surface and at Kamioka mine. As with DRIFT, the readout provides an anisotropic response with two axes from the μ-PIC, with spatial resolution 800 μm, and the z axis from time information but with no absolute z position and, currently, no head–tail information. A gamma rejection factor of 1.5×10^{-4} has been measured with ^{137}Cs and a direction-dependent recoil detection efficiency map obtained with mean efficiency at 40%. Although operated with a relatively high threshold of ~ 100 keV$_{ee}$ and on the surface where neutrons provide a limiting background, it has been possible to demonstrate functionality, to produce a recoil direction all-sky map (see Fig. 22.21) and from this produce a crude cross-section limit of 1.36×10^4 pb at 100 GeV. Significant work is needed to reduce neutron backgrounds now, by relocation to Kamioka mine, and to improve the angular resolution.

22.10.2 MIMAC

The group of Santos *et al.* [1688] has also developed the use of pixelized anode readout and electronics but with avalanche amplification determined by use of a mesh based on the micromegas concept [410], rather than GEMs. Here,

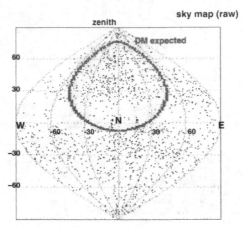

Fig. 22.21. An example directional sky map of the north sky derived by NEWAGE. The recoil directions of the fluorine and carbon nuclei are used. The thick line indicates the direction of the solar motion.

the gain region is defined by a wire mesh separated from the anode array by insulating stand-off rods of typically 50 μm. The original micromegas concept has been further developed for such applications to allow bulk fabrication of meshes up to typically ∼50 × 50 cm^2 that can be pre-formed with any chosen anode strip or pixel configuration. Whilst MIMAC plans a directional experiment, the main focus has been on fundamental studies of the relevant gas physics, particularly on measurement of quench factors and W-values, and the use of ^{3}He and CF$_4$ as a target. Use of ^{3}He, a spin-1/2 nucleus, is motivated by high sensitivity for spin-dependent interactions and for light WIMPs (<40 GeV). Helium-3 is also attractive owing to a relatively high quenching factor (QF), >70%; insensitivity to gammas, via the very low Compton cross-section; and the possibility of detecting neutron backgrounds via $n + ^3$He $\rightarrow p + ^3$H $+ 764$ keV.

In order to measure QFs, MIMAC has pioneered the use of an ion source linked to a dedicated micromegas TPC via a thin window, using differential pumping so that low-energy He or other ions can be fired directly into the target gas and the QF and W-values measured as a function of energy. Figure 22.11 shows an example response from 3.7 keV ^{3}He ions, indicating here that a recoil threshold below 1 keV looks feasible. Theoretical studies, based on the implied low mass sensitivity and using a 10 kg ^{3}He detector, show that a significant part of the SUSY parameter space not accessible by spin-dependent efforts could be explored [1688]. Such an experiment could run at two pressures, starting at 1–3 bar in a first phase to search for a signal, then at low pressure, 100–200 mbar, to extract directional information.

Again, CF_4 provides again an alternative target to mitigate the challenge of obtaining large quantities of ^{3}He.

22.10.3 DMTPC

The US DMTPC collaboration have revived the original idea of Masek and co-workers [1322] and the DRIFT collaboration of using light readout of a directional TPC via a CCD camera. The concept, here again based on CF_4 as the target in order to obtain spin-dependent sensitivities, uses meshes to define an avalanche region where electroluminescent proportional scintillation photons are produced, along with electrons. A system of optical lenses and CCD cameras can then be used to extract a 2D projected image. The average ionization energy in CF_4 is 54 eV so that typically 10^3 electrons are released for a 50 keV event. The group has built various prototypes of this design [727] demonstrating typical gas gains of 10^4–10^5 and an intrinsic spatial resolution of order 100 μm (discounting diffusion). In the avalanche about a third as many photons as electrons are produced. A highlight has been use of a small prototype to show head–tail discrimination with high-energy F recoils (see Section 22.9).

An advantage of this CCD technique is the relatively simple, and commercially available, means of reading many pixels, which also avoids the noise and stability issues often associated with charge readout. Design of the lens optics and light collection to minimize photon losses and maximize the area that can be covered by a single camera is critical here. A 10-litre detector has been built so far, designed for underground operation, and a multi-CCD version is envisaged to achieve a cubic metre volume (see Section 22.9). One issue requiring investigation is the possibility that additional degradation of position resolution is introduced during the electroluminescence photon production process in the electron avalanche.

22.10.4 DRIFT

DRIFT-IIa and b have successfully run underground since 2005, demonstrating stable room-temperature operation for long periods (months) in a variety of tests with and without neutron shielding. Calibrations are easily performed using internal ^{55}Fe (5.9 keV) gamma sources for gain and stability; external ^{60}Co for gamma rejection; and an external ^{252}Cf neutron source for nuclear recoil response. The latter has confirmed the high expected efficiency for recoil detection, typically 40% for S-recoils after all noise cuts [472]. Similarly the gamma rejection factor is found to be $>10^7$

N. Spooner

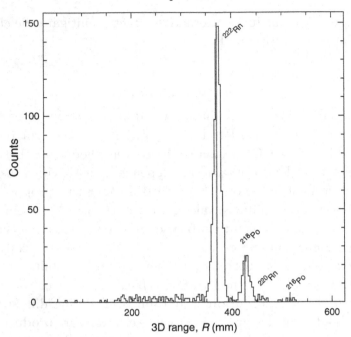

Fig. 22.22. Histogram of the 3D range R for GPCC alpha events in DRIFT-IIa selected to have $\Delta y < 60$ mm. The peaks are attributed to alpha decays of ^{222}Rn (at 371 mm), ^{218}Po (425 mm), ^{220}Rn (458 mm) and ^{216}Po (515 mm). Vertical lines indicate the mean peak positions. A low-level continuum is also visible below $R = 350$ mm, attributed to alpha-particles that traversed part of a cathode wire and subsequently escaped into the gas with degraded energy [475].

at 1000 NIPs (equivalent to 47 keV sulphur recoil). The energy threshold is currently limited by noise from the current electronics. Nevertheless, operating in triggerless mode, an excellent ^{55}Fe response is found (see Fig. 22.9) that reveals a potential threshold of about 2 keV for S recoils.

The particle identification power of DRIFT is illustrated in Fig. 22.22, which shows a spectrum of contained background alpha events in DRIFT-IIa [475]. The plot here is of a range spectrum, not an energy spectrum, obtained from reconstructing alpha tracks in 3D and extracting their 3D range in the gas, typically 30 cm for 5 MeV. Clear identification of various alpha species is possible. This reconstruction can be achieved also for S and C nuclear recoils in the low-energy region of interest for dark matter below 200 keV (see Fig. 22.16). However, at these low energies the expected range is <6 mm so reconstruction is compromised in the x direction by the crude MWPC wire spacing of 2 mm.

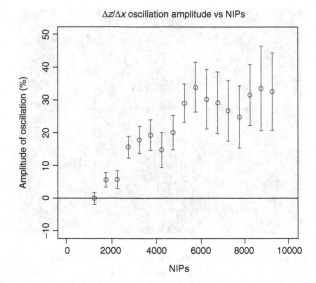

Fig. 22.23. The amplitude of the $\Delta z/\Delta x$ oscillation, expressed as a percentage of the average value, as a function of NIPs using a bin size of 500 NIPs. The data show a strong correlation with NIPs, or recoil energy, and also suggest that the effect disappears below the threshold of the detector, ~1000 NIPs.

So although 3D sky maps (theta, phi in Galactic coordinates) can be produced, an alternative means of extracting a directional signature has been used so far. Figure 22.23 shows the results of this 1D analysis [474]. Shielded dark matter runs of DRIFT-II underground at Boulby with all noise cuts in place and efficiencies understood revealed a remaining background of recoil-like events interpreted as RPRs (see Section 22.6). Cleaning procedures and radon reduction techniques have successfully reduced these by a factor of 1000 towards a background rate <1 event per day. Further R&D has demonstrated successful operation with CS_2/CF_4 mixtures and opened prospects for a scale-up array combining sensitivity to both spin-dependent and spin-independent WIMP couplings.

22.11 Scale-up and a future WIMP telescope

Despite the engineering challenge of scale implied by the low density of target necessary for a gaseous directional detector, it is already possible to produce conceptual designs for the ton-scale experiments necessary for sufficient sensitivity to produce a definitive directional discovery of spin-independent WIMPs, even at 10^{-10} pb. Such an experiment would need hundreds of square metres of readout plane. In Fig. 22.24 we introduce one such conceptual scale-up design, based on DRIFT principles using a

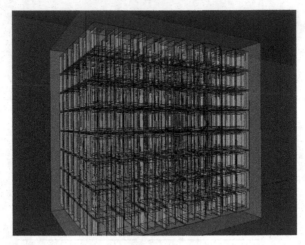

Fig. 22.24. Image showing the geometry of a possible multi-module DRIFT-III detector array. The detector vessels contain the MWPC and cathode support frames and are contained within 40 g cm^2 of CH$_2$ shielding with 10 g cm^{-2} of CH$_2$ between modules. The cavern size is $18 \times 20 \times 17\,\mathrm{m}^3$.

multi-module array. The concept here is mass production of many identical, low-cost TPC units of $2 \times 2 \times 2$ m each, all mounted within a large single vessel. The total volume required, for instance for a 1 ton target, is comparable to several particle physics experiments running underground, for instance MINOS, and much smaller than required for next-generation proton decay and neutrino physics experiments being proposed. Cavern excavation costs are well understood and so not necessarily a major cost driver. Furthermore, the relative simplicity of the technology, including room-temperature operation and minimal shielding requirements, is very favourable in comparison to the non-directional technologies.

There are nevertheless several generic design issues to address for scale-up construction. Of these, it is particularly important to establish: (i) that operation with sufficiently low levels of non-particle and neutron background can be achieved; (ii) that active background control through fiducialization of the drift volume can be achieved; and, most importantly, (iii) that a readout technology can be chosen to achieve very large areas with adequate spatial resolution, but at low cost per channel and without compromising internal background. Long-term underground operation of DRIFT-II is providing answers for item (i), except that further simulations are needed to establish the requirements for a massive external veto, perhaps using water cherenkov, depending on the site depth and internal intrinsic activity achievable. Development of full fiducialization, item (ii), via absolute z-axis measurement is critical, particularly to rejection of RPR events. Detection of positive ions at the high-voltage cathode is a promising route here, or, for the CCD option, provision of a pre-trigger through detection of primary scintillation light. It is unclear yet whether sufficient low-energy threshold can be achieved with that technique.

Item (iii) remains the biggest technical issue and also the greatest cost driver. For instance, development of finer readout spatial resolution to resolve shorter tracks could allow operation with higher gas pressure and hence with smaller detectors. This could reduce costs, but only if the electronics cost per channel is also lower, since more channels are likely needed. This has partly motivated development of the low-cost CCD optical technique (Section 22.10.2) and the various micropattern anode technologies using micromegas, GEMs and the μ-PIC technology of NEWAGE. For instance, the latter are developing a new 4×4 array scale-up of μ-PIC to make $1\,m^2$ [1457]. Advances in GEM and micromegas technology are allowing gain stage units of $1\,m^2$. There has also been demonstration of GEM and micromegas readout with CS_2 as a route to replacing MWPCs in DRIFT [1332]. For MWPCs themselves the pitch between wires is limited to 1–2 mm for mechanical and electrostatic reasons so cannot be improved, although interpolation does allow spatial resolutions potentially better than this by about a factor of 5 in one of the x–y directions.

Despite progress with micropattern readout in principle, a major issue for scale-up remains the signal feedthrough mechanics – simply the internal cabling and electronics necessary to transfer out of the vessel the signals from so many pixels without this introducing extra sources of radioactive background and dead space. In DRIFT this is solved by grouping MWPC wires (each 2 mm apart) so that every eighth wire is internally linked. This trick allows the whole $1\,m^3$, with x–y information from two TPCs, to be read out using only 36 channels, plus vetoes. It sacrifices absolute position measurement, introducing division of the detector into 16 mm wide sections that are degenerate such that although a recoil event (always <16 mm long) in any strip can be reconstructed in 3D, knowledge of absolute position (i.e. the particular strip hit) is lost. This is an acceptable sacrifice, reducing the number of channels by a factor of 50, since it is only the orientation of tracks that is important for directionality, not the absolute position. Mircopattern readout is likely to need to achieve similar simplifications. However, designs to achieve this without introducing unacceptable extra capacitative noise are a challenge.

The CCD technique does provide a powerful means to high-resolution 2D readout that avoids some of these issues. However, here there is the potential for added intrinsic background and additional volume associated with the

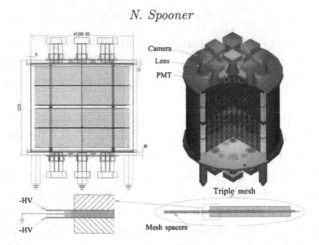

Fig. 22.25. Preliminary drawings of the 1 m³ DMTPC detector.

need to mount hundreds of optical systems and cameras within the experiment. Fiducialization and 3D reconstruction are also challenging with this technology since current CCDs are too slow to provide timing information (z axis) directly without adding a further charge collection grid and electronics. Nevertheless, conceptual designs are being considered. Figure 22.25 shows a single 1 m³ CCD-based scale-up module suggested by DMTPC [1713].

Finally, confirmation now that head–tail discrimination is feasible, something previously a major R&D goal, opens up new opportunities for scale-up. The extra information will clearly allow improved sensitivity and cost per unit volume in the 3D reconstruction concepts detailed here. However, the power of head–tail discrimination now opens a new, much simpler concept. It allows, in principle, operation with 1D readout only, simply observing a charge versus time signal from multiple-aligned single-anode planes used to determine statistically the average head–tail orientation of events. Such a scenario sacrifices the weaker track orientation information, but benefits from far less stringent electronics requirements.

23

Axion searches

Stephen Asztalos

At the time the Peccei–Quinn (PQ) solution [1555] to the strong CP problem [1553] was proposed it seemed perfectly reasonable to identify the PQ symmetry breaking scale with the electroweak scale. As the mass and the couplings scale inversely with the symmetry breaking scale (m_a, $g_{aii} \propto f_a^{-1}$), the expectation was that the mass of the axion would be hundreds of keV with couplings strong enough to be seen in accelerator and reactor experiments. The experimental situation quickly began to disfavour a massive axion as null results began to be amassed from a variety of particle and nuclear physics experiments. Tacitly acknowledging that there was no strong theoretical guidance for such a choice, new models soon appeared with arbitrarily large f_a, resulting in couplings so extraordinarily weak as to render the axion effectively invisible. Frustrating as this might have been to experimentalists, an unintended consequence of such a small coupling would be the prospect for the axion to be the dark matter pervading the Universe (since $\Omega_a \propto f_a^{7/6}$) [1600].

Still, it was unsatisfying that axion dark matter could, in principle, completely escape detection. Fortunately, in 1983 Pierre Sikivie made a proposal that could make 'invisible' axions 'visible' again [1277; 1754; 1755]. To overcome the dreadfully small couplings and excessively long decay natural lifetimes, a technique was borrowed from the gravitational wave community. Specifically, Sikivie proposed that axion decay into two photons could be stimulated with a high-Q oscillator. As such an enhancement in and of itself is far from sufficient to tease the 'invisible' axion into existence, it was further proposed that this resonant structure also reside in a strong magnetic field. With this arrangement virtual photons, forming the external field, supply one leg of the Primakoff interaction (Fig. 23.1), enhancing

Particle Dark Matter: Observations, Models and Searches, ed. Gianfranco Bertone. Published by Cambridge University Press. © Cambridge University Press 2010.

S. Asztalos

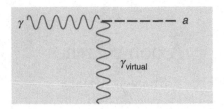

Fig. 23.1. The Primakoff diagram is the underpinning of most competitive axion search experiments. A virtual photon supplies one leg of the triangle diagram (suppressed) to produce a real photon or axion. A real photon from the decay of an axion with a mass ~1 μev lies in the microwave portion of the electromagnetic spectrum.

the exceedingly improbable case of decay into two real photons. Although it would be a decade or more before technology would catch up, the concept of externally providing one or more of the photons via the Primakoff process is the basis of all competitive axion searches today (e.g. ADMX, PVLAS, CAST and the various photon regeneration experiments – see below). This observation is not meant to detract from the numerous ingenious searches based on other couplings. Rather, the Primakoff effect gives experimentalists a handle largely absent in other axion decay channels. In the following sections various limits based on the $a\gamma\gamma$ channel will be described.

23.1 Constraints on axion properties

Before proceeding it is useful to recall the modest model dependence in the $a\gamma\gamma$ channel (Chapter 11). In the KSVZ model axions couple to hadrons, while in the DFSZ model the coupling is generalized to include leptons. Generally speaking, the KSVZ is the more optimistic of the two and hence is more accessible to search experiments. To simplify the following discussion we will restrict ourselves largely to limits on KSVZ-coupled axions.

By no means are all of these limits derived from data acquired in the laboratory. As will be seen, were it not for astrophysical limits the axion search space would be impossibly large. Only when cosmology, astrophysics and laboratory limits are combined can the context of present and future axion search experiments be understood. As befits a review of experimental axion searches, the discussion of astrophysical limits on the axion mass and couplings is necessarily circumspect. For a more in-depth discussion of the latter one may wish to consult the rather exhaustive literature on the role of the axion in astrophysics and cosmology [1265; 1621].

23.1.1 Astrophysical constraints

Globular clusters

Stellar objects like our Sun have lifetimes dictated by large photon opacities. By contrast, with their extremely small couplings axions would freely stream out of stellar interiors and modify stellar evolution. The net effect would be less time spent on the main sequence. In the particular case of globular clusters, axions emitted in the cores of helium burning HB stars would reduce their number with respect to that of red giants (HB progenitors – axion emission being highly suppressed in the degenerate helium cores of the latter.) The observed ratio of HB to red giant stars in clusters constrains the axion mass $m_a \leq 10\,\text{keV}$ [1622]. Because stellar evolution is well understood the derived constraint is considered very robust. The lack of evidence for highly penetrating particles in beam dump experiments at CERN [1073] and SLAC [365] rules out more massive axions.

Supernovae

As opposed to stellar evolution, the physics of supernovae (SNe) explosions is less well understood. However, the timescale for a SN explosion is not controversial: neutrino opacities are sufficiently high that neutrinos must 'evaporate' from the surface of the nascent Type II SN. Shortly after 1987a (a subluminous Type II event) exploded in the Large Magellanic Cloud it was pointed out that axion emission would have significantly limited the duration of the neutrino pulses observed in Kamiokande, IMB and Baksan should its mass lie in the range 1–$10\,\text{meV} < m_a < 10\,\text{eV}$ [481].

Cosmology

After consideration of various upper bounds to the axion mass we turn now to establishing a lower limit. Recall from Chapter 11 that various mechanisms, including vacuum realignment, give rise to a contribution from axions to the $\Omega_a \approx (0.1 \text{ to } 10)(\frac{\mu\text{eV}}{m_a})^{7/6}$. Precision measurements of cosmological parameters yields $\Omega_{\text{CDM}} \approx 0.22$ [276], implying a lower limit on the axion mass of order $0.5\,\mu\text{eV}$, though with considerable theoretical uncertainties.

When these various constraints are taken into account one is left with a viable axion search window spanning $0.5\mu\,\text{eV} < m_a < 1 - 10\,\text{meV}$; see Fig. 23.2.

23.1.2 Laboratory constraints

Even after astrophysical and cosmological constraints are taken into account, a daunting three or four decades of axion mass parameter search space

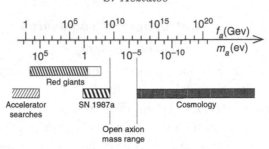

Fig. 23.2. The allowed axion mass range, bounded from below by the requirement that axions not overclose the Universe, and from above by null accelerator searches and stellar evolution arguments.

remain. Nonetheless, it is sufficiently constrained that experiments can be designed to target portions of the allowed region. To describe these experiments it is first instructive to introduce three additional diagrams relevant to the laboratory detection of axions (Fig. 23.3). The upper diagram (a) again is the Primakoff process. If the final state is an axion, the diagram describes polarization (rotation) experiments; a final state photon describes the RF cavity experiments. An extension of the Primakoff process is shown in Fig. 23.3(b). Where the intermediate axion is real, this describes the solar and photon regeneration experiments. The case where the intermediate axion is virtual describes birefringence of the vacuum. In polarization (ellipticity) experiments this latter process must compete with the pure QED process shown in Fig. 23.3(c). The reader is encouraged to consult these diagrams as the particular experiment is described below.

Polarization. Should the axion exist, its mere existence would impart both a rotation and ellipticity to polarized light. However, experiments based on subtle changes in the (presumably tiny) axion signal must be teased out from vastly larger classical and quantum-mechanical electrodynamic processes. While in theory it is possible to measure both ellipticity and rotation, in practice it is advantageous to focus on the latter as the magnitude of a second-order QED effect (vacuum birefringence, Fig. 23.3(c)) should swamp the former. (The contribution from QED to a rotation is negligible.) And unless great care is taken to remove residual gases, classical electrodynamic processes such as Faraday rotation and Cotton–Mouton-induced ellipticity will swamp the desired signal. Although a potential nuisance, these classical processes also can be useful calibration tools.

The basic approach is to inject linearly polarized light into an optical cavity within a magnet, with the electric field vector **E** at some non-zero angle φ with respect to the **B** field (optimally 45°). The probability for

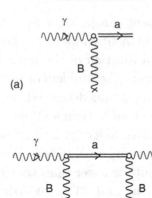

(a)

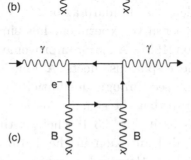

(b)

(c)

Fig. 23.3. (a) The Primakoff process describes the production of a real axion. Vacuum birefringence, which refers to a change in the index of refraction of the vacuum, can result from (b) a virtual axion or (c) higher-order QED effects.

single axion–photon conversion along a magnetic field can be obtained from classical electrodynamics modified in the presence of a massive scalar and is given by [1890]

$$\Pi = (\omega/4k_a)(g_{a\gamma\gamma}B_0 l)^2 F^2(q) \approx (1/4)(g_{a\gamma\gamma}B_0 l)^2, \qquad (23.1)$$

where l is the length of the magnet and B_0 its strength, and $F^2(q)$ is the Fourier transform of the magnetic field profile in z, normalized to unity for $q = 0$. As the axion is massive ($k_a = (\omega^2 - m_a^2)^{1/2}$) and the photon massless ($k_\gamma = \omega$), there is a momentum mismatch $q \equiv k_\gamma - k_a \approx m_a^2/2\omega$, for $m_a << \omega$ which defines the oscillation length in the problem. Insertion of an optical cavity increases the conversion probability by the number of passes N. The factor $g_{a\gamma\gamma}B_0 l^2$ is found in the conversion probabilities for all techniques premised on axion conversion along a magnetic field line. This same factor will be encountered again when discussing the solar helioscope and photon regeneration experiments.

For experiments seeking evidence of rotation a simple classical physical picture emerges. Recalling that the relevant Lagrangian is proportional to

$a\mathbf{E}_\omega \cdot \mathbf{B}$, only the $E_\parallel$ component can produce an axion. The parallel component is thus foreshortened by a tiny amount, whereas the perpendicular component is not, leading to a rotation of the linearly polarized light. The problem is somewhat analogous to the problem of neutrino mixing and oscillation. Not long after this approach was described, a Rochester–BNL–FNAL (RBF) collaboration performed an experiment with two superconducting dipoles end-to-end and an optical cavity, with the number of passes $N \sim 500$ [1726]. The demonstrated sensitivity to ellipticity, although impressive, was nonetheless ~ 4 orders of magnitude lower than that required to observe the predicted effect from QED alone and 13 or so orders of magnitude lower than that required to detect a standard axion.

A considerably more sensitive experiment has since been carried out by the PVLAS collaboration [1957]. A major improvement was the introduction of heterodyning to remove spurious effects. A 1 m, 5 T magnet was rotated (at $\omega_{\rm M}$) and the signal passed through an ellipticity modulator (at $\omega_{\rm SOM}$, for stress optic modulator) so that the signal (classical, QED or axion) should appear at $\omega_{\rm SOM} \pm 2\omega_{\rm M}$ as in Fig. 23.4. Their initial report of a rotation signal ($\sim 2 \times 10^{-7}$) implied an axion in the 1–1.5 meV mass range, and $g_{a\gamma\gamma} \sim 0.15$–$0.5 \times 10^6 \, \text{GeV}^{-1}$. How such an axion could escape detection in solar helioscopes or not modify HB stars was puzzling, as was the observation that the signal was orders of magnitude larger than the predicted QED ellipticity signal. A follow-up experiment especially designed to minimize possible sources of systematic error reported a null result [1958], resolving, at least for the time being, a potential showdown.

Photon regeneration. We have seen above how creative nature can be in generating spurious rotations and ellipticities. The photon regeneration approach ingeniously avoids these difficulties by requiring that a real axion pass through an opaque barrier and be regenerated by the inverse process on the other side. An advantage that the regeneration approach enjoys over the polarization approach is its scaling: the expected signal is largely independent of the background (e.g. dark current, radioactivity) in the regeneration scheme. The price to be paid is that the detection probability now scales as the fourth power of $g_{a\gamma\gamma}$ (presumably a very small number).

An early regeneration experiment was carried out using two dipoles (4.4 m long each, 3.7 T) and an argon ion laser (1.5 W, 200 traversals in an optical cavity). A limit of $g_{a\gamma\gamma} < 7.7 \times 10^{-7} \, \text{GeV}^{-1}$ was established, for $m_a < 10^{-3} \, \text{eV}$ [1661]. However, the suggestion of a signal at PVLAS spurred nearly half a dozen regeneration experiments by mid 2007. At least one of these has already confirmed the null result [1613], illustrating another

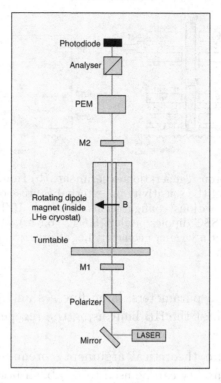

Fig. 23.4. A schematic of the PVLAS polarization experiment. Linearly polarized light passes through a 1 m long dipole magnet resting on a turntable. M1 and M2 comprise a Fabry–Perot optical resonator with an optical path of about 60 km. As the polarization direction is kept fixed, rotating the magnet gives rise to an output signal which is time-varying and can thus beat against a known carrier. Modulation techniques like this are common when measuring small signals in the presence of large backgrounds.

advantage of this approach, that it has been readily and inexpensively implemented. The OSQAR experiment (Optical Search for QED vacuum magnetic birefringence) serves as a representative example. An 18 W Ar$^+$ laser is incident on a polarizer before traversing two 9 T superconducting magnets borrowed from the LHC, separated by an optical barrier. On the detection end sits a liquid-nitrogen-cooled CCD with 1100 pixels with a dark count less than 1 count per minute. The lack of a signal has allowed the OSQAR collaboration to rule out the PVLAS result, although it continues to operate in a mode to detect the QED vacuum birefringence effect described above.

Recently, it has been proposed that by frequency coupling two optical cavities the detection probability can be enhanced by the finesse of a second optical cavity [1757]. As the conversion probability scales with the fourth

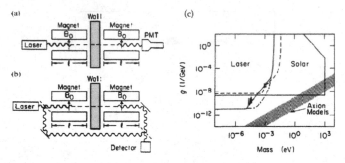

Fig. 23.5. (a) The photon regeneration experiment. (b) Homodyning largely eliminates the dependence of the sensitivity on external sources of noise. (c) Sensitivity of the experiment for various configurations. Dashed: 10 T, 10 m magnet, 1 kW optical laser. Solid: 30 SSC dipoles per leg (6.6 T, 16.6 m), 100 kW CO_2 laser. Dot-dash: Same as solid, but a wiggler geometry, i.e. alternating polarity of magnets to maximize $F^2(q)$ [1890].

power of experimental parameters, large finesses and vast laser power are required if one is to best the HB limit using the regeneration technique.

Solar. Although pure theoretical arguments premised on constraints in solar luminosity variability can be used to establish bounds on stellar axion production [1265], a direct detection experiment situated only 1 AU from an astrophysical object with a photon luminosity $\sim 10^{26}$ W seems irresistible.

Unlike the nearly monoenergetic axions in laser experiments, solar axions have a broad energy spectrum characteristic of the temperature in the solar core. The expected differential solar axion flux is given by

$$\frac{d\Phi_a}{dE_a} = g_{10}^2 3.821 \times 10^{10} \frac{(E_a/\text{keV})^3}{(e^{E_a/1.103 \text{ keV}} - 1)} \text{ cm}^{-2} \text{ s}^{-1} \text{ keV}^{-1}, \qquad (23.2)$$

reflecting an approximately thermal spectral production of axions in the solar interior from the Primakoff process (Fig. 23.6). The integrated solar flux at Earth is given by

$$\Phi_a = g_{10}^2 3.67 \times 10^{11} \text{ cm}^{-2} \text{ s}^{-1}, \qquad (23.3)$$

where $g_{10} \equiv g_{a\gamma\gamma} 10^{10}$ GeV. Using Eq. (23.2), the expected number of photons from converted solar axions that reach an X-ray detector is:

$$N_\gamma = \int \frac{d\Phi_a}{dE_a} P_{a\to\gamma} S T \, dE_a, \qquad (23.4)$$

where S is the magnet bore area (cm^2), T is the measurement time (s) and $P_{a\to\gamma}$ is again given by Eq. (23.1). Note further that the decay of keV axions

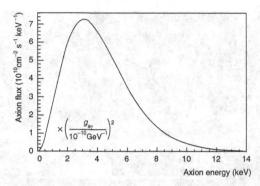

Fig. 23.6. The expected solar axion flux spectrum at the Earth. The mean axion kinetic energy is ∼4.2 keV. Figure from ref. [1891].

generates photons in the X-ray portion of the spectrum, so that an entirely new suite of devices is required for this application.

Solar detection of axions has reached considerable maturity. The pioneering implementation of the axion helioscope concept was performed at BNL [1315]. More recently, the Tokyo axion helioscope [1451] with $L = 2.3$ m and $B = 3.9$ T has provided new results [1151]. The third-generation CAST experiment at CERN is the first solar experiment to reach model sensitivities and will be described further. CAST improves on its predecessors by borrowing a powerful (9 T) superconducting test magnet from the Large Hadron Collider (LHC) experiment, combined with a series of sensitive X-ray detectors at either end of two parallel pipes of length $L = 9.26$ m and a cross-sectional area $S = 2 \times 14.5$ cm^2. During the first stage of operation of the CAST experiment, the magnetic field pipes were under high vacuum (below ∼10^{-6} mbar). The magnet is mounted on a moving platform, allowing for pointing at the Sun for 1.5 h at both sunrise and sunset throughout the whole year. The time the Sun is outside the field of view of the helioscope it is used for background measurements. Figure 23.7 gives a schematic view of the experimental set-up, showing the magnet, the platform to move it and the working principle.

At both ends of the magnet, three different detectors search for excess X-rays during solar tracking compared with the periods when the Sun is completely out of the field of view of CAST. These are: a conventional time-projection chamber (TPC), a smaller gaseous chamber of the micromegas type [953] and an X-ray mirror system with a charge coupled device [1824] at its focal plane 1.7 m downstream. Both the pn-CCD and the X-ray telescope are prototypes developed for X-ray astronomy [70]. An axion signal should appear as an excess above the background, which is mainly due to

Fig. 23.7. Schematic view of the CAST experimental set-up. The 9.26 m long LHC test magnet is mounted on a platform, allowing a movement of ±8° vertically and ±40° horizontally (courtesy of Thomas Sahne/CERN). Detectors are located at both ends of the magnet and exposed to axion-induced X-rays during ∼1.5 h at sunrise and ∼1.5 h at sunset.

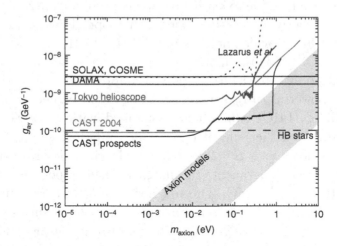

Fig. 23.8. Results from CAST data compared with other experimental/theoretical results [1972]. The shaded band represents typical axion models. Also shown is the envisaged future CAST performance (CAST, private communication).

environmental radioactivity and cosmic rays (mostly secondary neutrons and γ-rays).

The result with data taken with vacuum inside the magnetic pipes of the CAST magnet and all three detectors combined is shown in Fig. 23.8, along with results from previous helioscope experiments. The CAST result is for the first time below the best astrophysically derived limit

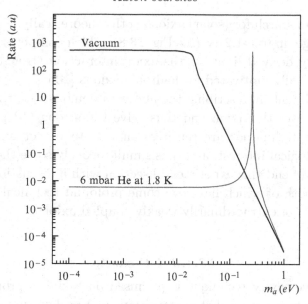

Fig. 23.9. Number of photons reaching an X-ray detector placed at the end of the CAST magnet as a function of the rest mass of the incoming solar axion. The upper line represents a vacuum inside the magnet, the lower one a fill gas – here helium at 6.0 mbar of pressure.

$g_{a\gamma\gamma} < 10^{-10}\,\mathrm{GeV^{-1}}$. Note this constraint is rest-mass-independent, since the axion–photon oscillation length exceeds the length of the magnet. However, for m_a above 0.02 eV and vacuum inside the magnetic pipes, not only does the signal strength diminish rapidly, but the spectral shape is also distorted (not shown). As with the photon regeneration experiments, this fall-off is due to increased momentum mismatch. In the case of CAST, coherence can be partially restored by filling the magnetic conversion region with a refractive gas, e.g. helium (or better, hydrogen, owing to its smaller photoabsorption). The fill gas in essence imparts a mass to the photon, thus slowing it down and minimizing the photon–axion momentum mismatch. For an appropriate gas pressure, coherence can thus be preserved, although for a narrow axion rest mass window the limit acquires a resonance-like line shape as shown in Fig. 23.9.

CAST Phase II is now taking data with helium gas inside the 1.8 K magnetic pipes: first with low pressure ^{4}He (up to ∼14 mbar, before it liquefies) and later with ^{3}He (up to ∼60 mbar or higher, owing to its higher vapour pressure). This will allow the solar axion searches to be extended up to ∼1.2 eV [1971]. This extension in axion rest mass sensitivity may be essential for a direct solar axion detection. In CAST, for the first time, a laboratory experiment can search for (solar) axions in the theoretically motivated axion rest mass range up to ~1.2 eV (see Fig. 23.8), which competes with the best astrophysically derived limit for the axion interaction strength, while it is also cosmologically motivated for hadronic axions [1068].

It was the intent in describing the above to demonstrate the diligence of efforts to discover the axion and thus solve the strong CP problem. Still, none of these techniques are sensitive enough to detect axions that are also of cosmological interest, and thus simultaneously solve the dark matter conundrum. We end this experimental section with a discussion of two such approaches, both of which have overcome profound technical challenges to allow detection of extraordinarily weakly coupled axions.

23.1.3 RF cavity

The microwave cavity technique is premised on Sikivie's proposal to resonantly stimulate axion-to-photon transitions in a background magnetic field of strength B_0. In practice, a high-Q cavity is placed in a large magnetic field and conversion is enormously enhanced by both the increased density of states around the final photon frequency and the virtual photons supplied by the external field. With photon regeneration conversion occurs only for axions travelling along the field lines; in the Sikivie approach overlap of the magnetic and electric field of the converted photon is dictated by the mode structure of the cavity. Consequently, the conversion probability does not depend on the length of the magnet, but rather on its volume. In a right cylindrical cavity one observes that maximum overlap with the external field of a solenoid magnet occurs when the cavity is tuned to the lowest TM$_{mln}$ mode. Resonant conversion of axions in our Galactic halo occurs when the cavity frequency equals m_A. The corresponding power on-resonance is given by [1755]

$$P_a = \left(\frac{\alpha}{\pi}\frac{g_{a\gamma\gamma}}{f_a}\right)^2 V\, B_0^2 \rho_a C \frac{1}{m_a} \mathrm{Min}(Q_\mathrm{L}, Q_a)$$

$$= 0.5 \times 10^{-26}\,\mathrm{watt} \left(\frac{V}{500\,\mathrm{litre}}\right)\left(\frac{B_0}{7\,\mathrm{tesla}}\right)^2 C\left(\frac{g_\gamma}{0.36}\right)^2$$

$$\times \left(\frac{\rho_a}{\frac{1}{2}\cdot 10^{-24}\frac{g}{\mathrm{cm}^3}}\right)\left(\frac{m_a}{2\pi(\mathrm{GHz})}\right)\mathrm{Min}(Q_\mathrm{L}, Q_a), \qquad (23.5)$$

where V is the volume of the cavity, B_0 is the magnetic field strength, Q_L is its loaded quality factor, $Q_a = 10^6$ is the 'quality factor' of the Galactic halo axion signal (i.e. the ratio of their energy to their energy spread), ρ_a is the density of Galactic halo axions on Earth and C is a mode-dependent form factor. The prefactor of 10^{-26} W sets the scale for the expected power.

At first, the temptation might be to conclude that the prefactor of $g_{a\gamma\gamma}^2$ makes this approach superior to the solar and regeneration experiments (with their prefactors of $g_{a\gamma\gamma}^4$). Unfortunately, the 'no-free-lunch' theorem makes its appearance in the source term: here the Galactic halo as opposed to the Sun or laser production. Consequently, even when combining the strongest magnets and highest-Q cavities available, the expected signal still is extraordinarily weak owing to the paucity of dark matter per volume of interstellar space ($\sim 0.45\,\mathrm{GeV\,cm^{-3}}$).

Any signal thus must compete with thermally generated backgrounds. The situation here would be hopeless were it not for the observation by Dicke that thermal noise can be averaged down over many realizations of the same experiment. The well-known Dicke radiometer equation specifies the time required to achieve a given signal-to-noise ratio (SNR) as

$$\mathrm{SNR} = \frac{P_a}{P_\mathrm{n}}\sqrt{Bt} = \frac{P_a}{k_\mathrm{B}T_\mathrm{n}}\sqrt{\frac{t}{B}}, \qquad (23.6)$$

where P_n is the average thermal power, T_n is the sum of the physical temperature of the cavity plus the noise temperature of the microwave receiver that detects the photons, t is time and B is the bandwidth. From Eq. (23.6), to detect a signal the cavity physical temperature and equivalent electronic noise temperature must be made as low as practicable.

Since the axion mass is unknown, a crucial element to many search experiments is tunability. Incremental changes in cavity frequency are achieved by moving metallic or dielectric tuning rods. For a fixed SNR, the maximum scan rate is given by

$$\frac{df}{dt} = \frac{12\,\mathrm{GHz}}{\mathrm{year}}\left(\frac{4}{\mathrm{SNR}}\right)^2\left(\frac{V}{500\,\mathrm{litre}}\right)^2\left(\frac{B_0}{7\,\mathrm{tesla}}\right)^4 C^2\left(\frac{g_\gamma}{0.36}\right)^4$$

$$\times\left(\frac{\rho_a}{\frac{1}{2}\cdot 10^{-24}\frac{g}{\mathrm{cm^3}}}\right)^2\left(\frac{3K}{T_\mathrm{n}}\right)^2\left(\frac{f}{\mathrm{GHz}}\right)^2\frac{Q_\mathrm{L}}{Q_a}. \qquad (23.7)$$

Equations (23.5) and (23.7) dictate the nature of the search strategies. If theoretical guidance existed as to the mass of the axion then the obvious strategy would be to tune B or t in Eq. (23.6) to improve the SNR continuously. In the absence of such guidance one can instead optimize some combination of T_n, B_0 and V. For reasons that will become clearer in subsequent sections, a great deal of effort has gone into reducing T_n and, hence, improving the scan rate.

23.1.4 Previous experiments

Shortly after the publication of the cavity haloscope proposal [1754], two small proof-of-principle experiments got under way, one led by a Rochester–Brookhaven–Fermilab (RBF) collaboration [631; 1946], and the other by a group at the University of Florida (UF) [1052]. It is a tribute to these two efforts that the design and operational strategy of the current Axion Dark Matter eXperiment (ADMX) in the United States, and to a lesser degree CARRACK in Japan, have carried over directly from their work. Although groundbreaking in their own way, amplifier technology would keep limits on $g_{a\gamma\gamma}$ from these experiments two to three orders of magnitude above KSVZ models.

23.1.5 The Axion Dark Matter eXperiment (ADMX)

A collaboration of US institutions (LLNL, University of Washington, University of Florida, UC Berkeley and NRAO) has built the next generation of microwave cavity experiment, whose scientific goal is to cover the (1–10 μeV) mass range with sensitivity to the KSVZ axion models. Called ADMX, the gain in sensitivity was predicated upon a scale-up of the active volume of the experiment, and projected improvements in the effective noise temperature of the microwave amplifier. The experiment began operation in 1996, and covered nearly an octave in frequency range 450–810 MHz (1.86–3.36 μeV), before shutting down in 2004 for an upgrade.

A schematic of the experiment is shown in Fig. 23.10. A single copper-plated stainless steel cavity 1 m long × 0.5 m diameter is inserted within the NbTi superconducting solenoidal magnet of 8.5 T central field. Radial translation of metal (copper) and dielectric (alumina) tuning rods within the cavity allows a roughly ±50% tuning range. A brief run with an array of four cavities was carried out to establish the feasibility of multicavity operation to reach higher frequencies [1238].

Power from the TM_{010} mode of the cavity is extracted by a capacitively coupled antenna and fed to a low-noise GaAs hybrid amplifier (HFET). The remainder of the electronics chain is an extremely stable and quiet version of that found in an undergraduate radioastronomy laboratory. Suffice it for here to describe briefly how a double heterodyne receiver operates. A GHz signal is shifted down to the intermediate frequency (IF), where filtering and amplification is practical. A second heterodyne operation mixes the signal down to near-audio frequencies, where digital sampling can be done without generating massive data streams. The sampling rate is such that the resultant bin size is well matched to quasi-thermalized axions. A single pass over a region

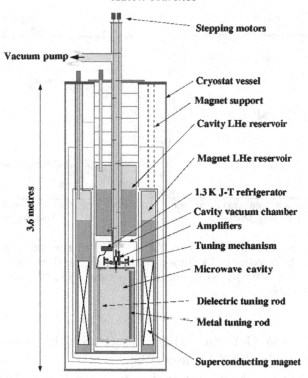

Fig. 23.10. Schematic of the Axion Dark Matter eXperiment (ADMX). Stepper motors at 300 K drive the tuning rods at 2 K via long G-10 shafts to minimize heat conduction. Separate LHe reservoirs cool the external magnet and cavity. Figure from ref. [1564].

of interest is rarely sufficient to acquire the requisite SNR. Multiple ($\sim$3) passes are typically required, although the process of co-adding samples can be repeated ad nauseam – or at least until a systematic noise floor is reached. In the case of ADMX, that resultant spectral sensitivity is $\sim$10^{-26} W; clearly overkill for an experiment that needs 'only' $\sim$10^{-23} W of sensitivity for discovery. Further, with three decades of axion mass to be covered one simply cannot afford to integrate a narrow region for 30 days. In practice, a trade-off is made between sensitivity and mass (frequency) coverage such that for each tuning position, power is integrated for $\sim$80 s. Candidates are encountered whenever the power in a single 125 Hz bin (or combined six bins) exceeds a threshold established by Monte Carlo techniques to give better than 90% confidence. All such candidates are examined by retuning the cavity to the candidate frequency and performing an extended integration. No candidates survived this exacting process, and as of May 2004 the limits shown in Fig. 23.11(a) were established at 90% confidence for the maximum likelihood

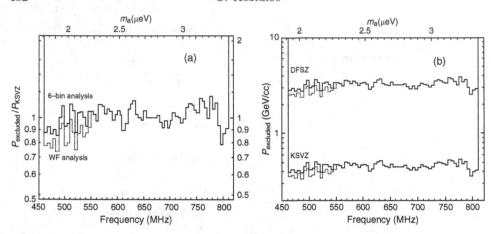

Fig. 23.11. Exclusion limits (90%) by ADMX for (a) the axion–photon coupling $g_{a\gamma\gamma}$, assuming the maximum likelihood estimate of the dark matter halo density [139]; (b) the halo density ρ_a, assuming KSVZ axions WF, Weiner filter; P, power. Figure from ref. [138].

halo density of 0.45 GeV cm^{-3} [139]. Figure 23.11(b) shows the excluded halo density assuming KSVZ axions [138].

A separate receiver chain resolves the cavity much more finely to look for axion dark matter that has (comparatively) recently fallen into the potential well of our Galaxy. This high-resolution channel has a very high SNR and thus holds great prospects for untangling mysteries of Galactic evolution should a signal ever be observed, although the models of infall are somewhat more speculative. Figure 23.12 shows results from this high-resolution addition to the ADMX data stream.

A path forward. From the radiometer Eq. (23.6) one may derive two scaling laws critical to the microwave cavity experiment. First, for a search conducted at a fixed value of $g_{a\gamma\gamma}$ the frequency scan rate goes as $(1/\nu)(d\nu/dt) \propto (B^4V^2)T_{\mathrm{s}}^{-2}$, where T_{s} is the total system noise temperature. Second, for a search conducted at a fixed $(1/\nu)(d\nu/dt)$, the power sensitivity goes as $g_{a\gamma\gamma}^2 \propto (B^2V)^{-1}T_{\mathrm{s}}$. With B near its theoretical maximum (at least for a magnet requiring a bore of 50 cm), reducing the system noise temperature (the sum of the physical plus amplifier temperatures) remains the most viable way to cover the parameter space. In the case of ADMX, there is still much room for improvement, as the theoretically irreducible noise of any classical device must obey the standard quantum limit, which at 1 GHz is $kT_{\mathrm{n}} \approx h\nu \sim 50\,\mathrm{mK}$.

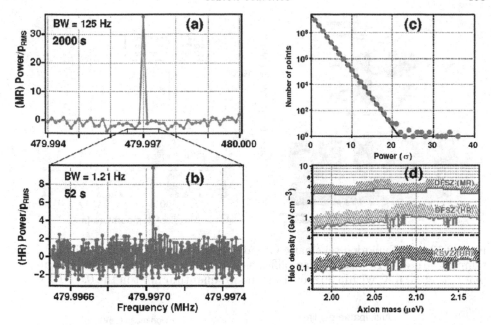

Fig. 23.12. ADMX high-resolution analysis. An environmental peak appears in both the medium-resolution (MR) spectrum (a), and high-resolution (HR) spectrum (b), with identical power when adjusted for integration time and bin width, BW. (c) The power fluctuation distribution of the high-resolution fast Fourier transform ($\Delta\nu = 0.019\,\mathrm{Hz}$). (d) Exclusion limits in ρ_a for the high-resolution analysis, assuming a dark-matter flow with a narrow velocity dispersion, $\leq 25\,\mathrm{m\ s^{-1}}$. Figure from ref. [725].

SQUIDs (Superconducting Quantum Interference Devices) have long been used as ultra-sensitive magnetometers and ultra-low noise amplifiers in the quasi-static regime, but until recently had never been able to be used as high-gain amplifiers above 100 MHz. Motivated by the microwave cavity axion experiment, in the late 1990s a series of breakthroughs in design have allowed the microstrip-couple SQUID amplifiers to approach the standard quantum limit in noise in the 0.5–1 GHz range (Fig. 23.13(b)), and the microstrip design has been demonstrated to work up to 3 GHz. The noise temperature of a SQUID can be approximated as 25% of the physical temperature, whereas the best GaAs noise temperature is of the order of 2 K (with little or no improvement expected with further cooling). (For a summary of the DC SQUID and a complete bibliography of the microstrip-coupled SQUID, see [435].)

After a 4-year construction hiatus (ending in 2008) ADMX resumed operation with a single SQUID amplifier replacing the balanced HFET design

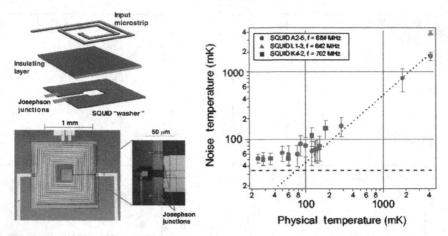

Fig. 23.13. (a) Exploded schematic view of a microstrip-coupled SQUID. (b) Noise temperature of a DC SQUID amplifier as a function of physical temperature. T_n plateaus at 50 mK, within 50% of the standard quantum limit in the 0.5 GHz range.

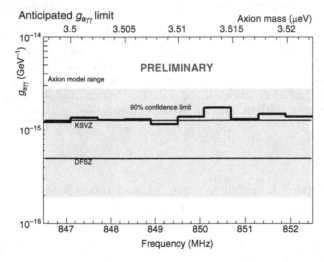

Fig. 23.14. Preliminary data from the Phase I ADMX upgrade demonstrating KSVZ sensitivity over a previously unexplored, albeit narrow, frequency range.

as the first stage amplification. The SQUID is located 1 m or so above the cavity top plate and the field in its vicinity is reduced to 50 nT or so through a combination of a bucking coil, μ-metal shields and Nb plating on the amplifier package. The SQUID and the cavity each have their own dedicated refrigeration, although both are fed by a common liquid helium reservoir. An initial run around 850 MHz achieved KSZV sensitivity in spite of higher-than-expected physical temperatures (Fig. 23.14). The intention is to run with the present configuration over the region $900 \leq f \leq 1000$ MHz before retrofitting the experiment with a dilution refrigerator. The total system noise temperature at that juncture should be of the order of 100–200 mK. This would enable the experiment to exclude DFSZ axions, even at densities much less than the maximum-likelihood halo estimate, while also permitting an appreciable improvement in the rate of frequency coverage. Microwave resonator and amplifier concepts need to be developed for the search to proceed to the 10–100 μeV range; even concepts are lacking for the third decade in open mass range, i.e. 100 μeV to 1 meV.

23.1.6 The Kyoto Axion Experiment (CARRACK)

The Axion Dark Matter experiment in the United States and its predecessors benefited from the steady improvements in noise temperature of conventional HFET amplifiers over years of development. Nonetheless, the upgrade of ADMX with d.c. SQUID amplifiers will ultimately run into a fundamental limiting noise temperature, the so-called standard quantum limit, $T_N \approx h\nu/k_B$, where k_B is the Boltzmann constant. A complementary approach exists which avoids the standard quantum limit at the expense of the loss of any phase information. (A photomultiplier tube operates on such a principle.) A novel experiment exploiting the particle nature of light has been developed in Kyoto. As with ADMX, axion conversion still occurs in a resonant cavity. However, detection is not done with a phase-sensitive amplifier, but instead a highly excited (Rydberg) atom. A properly prepared Rydberg atom has several salient features that enhance its attractiveness for this purpose: a large probability for absorbing electric dipole radiation between adjacent levels ($\sim n^2$ where n is the principal quantum number of the prepared state), a long lifetime and emission in the microwave region. To minimize false signals due to quantum tunnelling, electronic levels are then processed in a series of extremely sophisticated manipulations before the excited atom is preferentially extracted and counted [1830].

The implementation of the Rydberg-atom single-quantum detector in the microwave cavity experiment is shown in Fig. 23.15. Here, in addition to the conversion cavity, there is a second, detection, cavity coupled to it; the two are kept locked together in frequency. An atomic beam, e.g. ^{85}Kr, is doubly optically pumped to a suitable Rydberg level. The Kyoto group has measured blackbody photons in the CARRACK1 apparatus, as a function of physical temperature, and the agreement with theory is excellent (Fig. 23.16). The lowest temperature data point, 67 mK, is almost a factor of two lower than the standard quantum limit corresponding to the cavity

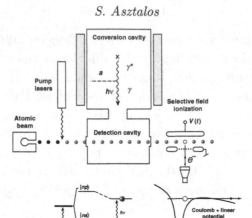

Fig. 23.15. Schematic of the microwave cavity axion experiment with a Rydberg-atom single-quantum detector for detection of the RF photons.

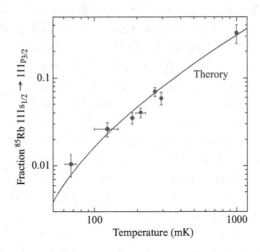

Fig. 23.16. Measurement of blackbody photons in the Kyoto CARRACK1 experiment. The probability of transition to the upper state from the prepared state $(111s_{1/2} \rightarrow 111p_{3/2})$ as it passes through the conversion cavity is compared with absolute (unnormalized) numbers from the theoretical calculation (unpublished).

frequency ($\sim$120 mK at 2527 MHz). This group has performed a search for axions for a range of about 10% in frequency with this cavity, but results are as yet unpublished. A larger version, CARRACK2, comparable in size to ADMX, has been constructed; it has been designed to be cooled to physical temperatures as low as 10–15 mK.

23.2 Conclusions

With present limits on the neutron electric dipole moment approaching $10^{-26} e$ cm [175] the strong CP problem continues to plague an otherwise extremely successful theory of quantum chromodynamics. Although the Standard Model has been modified in numerous ways to accommodate these data, perhaps none is as elegant as the PQ model. An immediate consequence of the broken PQ symmetry is the axion, a hypothetical entity with particle and astrophysical ramifications. The axion has proved a capable foe, successfully evading detection for more than 30 years. Nevertheless, the efforts described in this chapter are testimony to the progress that is being made.

It bears repeating that although the axion enjoys many couplings, it is the induced coupling to photons that has proved the most powerful. Nearly every major axion search in the past two decades has been based on the Primakoff effect by which a pseudoscalar can convert in an external electromagnetic field to a single real photon. Their unprecedented sensitivities derive from advances in modern superconducting magnet technology. On equal footing is the role that coherence plays in making possible axion–photon mixing experiments with large spatial volumes. Together, they represent Sikivie's key insight for detecting what otherwise might have forever remained the 'invisible' axion, and they are the driving force behind two of the most promising avenues: the microwave cavity search for dark-matter axions and the search for solar axions.

The overall experimental situation is summarized in Fig. 23.17. The cavity microwave searches, which rely on the assumption that axions constitute the dark matter in our halo, have already probed well into the band of plausible axion models. In Phase II ADMX will be equipped with the technology to push upwards of 10^{-5} eV or so, and firm concepts are in place (although not yet realized) to eventually reach upwards of 25 GHz or so. For the decade 10^{-4} to 10^{-3} eV, amplifier and cavity concepts do not yet exist. The solar axion experiments, by contrast, are broadband in mass coverage. The third-generation experiment, CAST, has equalled and is poised to better the astrophysical bounds, specifically those from horizontal branch stars. By using a low-Z gas of tunable pressure, the axion and photon dispersion-relations may be matched, allowing the search to push up into the $O(1$ eV$)$ mass range, and thus into the space of realistic models. One may claim that this region is already excluded, but theoretical limits established solely from SN1987a should be viewed with appropriate scepticism. A general rule of thumb is that experiments and direct observations should always be performed as a check of astrophysical inferences.

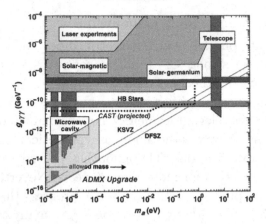

Fig. 23.17. Excluded $g_{a\gamma\gamma}$ vs. m_a with all experimental and observational constraints. The RF cavity and solar axion search experiments already possess sensitivities to discover KSVZ axions.

There is no concept that could feasibly push the sensitivity in $g_{a\gamma\gamma}$ down by more than an order of magnitude across the entire mass range. Pessimistically, an axion with a mass 0.1–10 meV and possessing standard couplings will remain undiscovered, although even here ideas have been proposed. As unhappy an idea as that may be, such an axion is too heavy to be the dark matter in the Universe. Optimistically, a lighter axion (< 0.1 meV) or one possessing unexpectedly stronger couplings to photons should be discovered, perhaps within a decade. Axionic dark matter could potentially reveal phenomenal detail about the formation and phase-space structure of our Galaxy, as the signal represents the total energy of axions and should thus manifest dramatic Doppler modulation, especially if there exists fine structure.

In conclusion, the axion has proven to be remarkably tenacious, in terms of both its ability to avoid detection and its persistence as an elegant resolution of the strong CP problem. Several detailed reviews have recently appeared specifically devoted to the RF microwave cavity technique [435] and, more generally, to searches for the 'invisible' axion [140; 1652].

Part V

Indirect detection and astrophysical constraints

24

Gamma-rays

Lars Bergström and Gianfranco Bertone

Dark matter particles can be searched for indirectly, i.e. through the detection of secondary particles produced by their annihilation or decay. Among these secondary products, gamma-rays are particularly interesting, as they travel in straight lines and practically unabsorbed in the local Universe. Recent advances in detection techniques, due to the great success of the new generation of air Cherenkov telescopes, and to the new data of the Fermi satellite, make the prospects for detecting gamma-rays from dark matter annihilation or decay particularly promising. We review in this chapter the theoretical prediction for the shape of the annihilation or decay spectrum, and the optimal targets for detection.

24.1 Annihilation

The idea that a detectable signature may be seen in gamma-rays created in WIMP annihilations has an old history (see, for example, the now strongly disfavoured very massive neutrino case in [1040]). It is especially important for the new large imaging air Cherenkov telescopes currently operating (e.g. HESS,[1] MAGIC,[2] CANGAROO[3] and VERITAS[4]) or being planned [460], and for the Fermi satellite[5] (formerly GLAST, the gamma-ray large-area space telescope), which was successfully launched in 2008, and which is now taking exciting new data (for pre-launch estimates of its dark matter detection potential, see [180]). There are several recent reviews covering this

---

[1] http://www.mpi-hd.mpg.de/hfm/HESS
[2] http://wwwmagic.mppmu.mpg.de
[3] http://icrhp9.icrr.u-tokyo.ac.jp
[4] http://veritas.sao.arizona.edu
[5] http://fermi.gsfc.nasa.gov/

Particle Dark Matter: Observations, Models and Searches, ed. Gianfranco Bertone. Published by Cambridge University Press. © Cambridge University Press 2010.

topic [290; 304; 337; 977; 1184]. Here we only point out some of the most important highlights.

The annihilation rate towards a direction making the angle ψ with respect to the Galactic centre is conveniently given by the factorized expression [304]

$$\Phi_\gamma(\psi) \simeq 0.94 \times 10^{-13} \left(\frac{N_\gamma\, v\sigma}{10^{-29}\ \mathrm{cm}^3\mathrm{s}^{-1}} \right) \left(\frac{100\,\mathrm{GeV}}{m_\chi} \right)^2$$
$$\times\, J(\psi)\ \mathrm{cm}^{-2}\ \mathrm{s}^{-1}\ \mathrm{sr}^{-1}, \qquad (24.1)$$

where we have defined the dimensionless function

$$J(\psi) = \frac{1}{8.5\,\mathrm{kpc}} \cdot \left(\frac{1}{0.3\,\mathrm{GeV\ cm}^{-3}} \right)^2 \int_{\mathrm{line\ of\ sight}} \rho^2(l)\, \mathrm{d}\,l(\psi), \qquad (24.2)$$

with $\rho(l)$ being the dark matter density along the line of sight $l(\psi)$. (Note the numerical factor in Eq. (24.1) differs by a factor of $1/2$ from that given in the original reference [304]; this takes into account the fact that the annihilating particles are identical, as is the case for supersymmetric neutralinos. See the footnote in connection to Eq. (21) of ref. [1880] for a detailed explanation.)

The particle physics factor $N_\gamma v\sigma$, which is the annihilation rate times the number of photons created per annihilation, can usually be rather accurately computed for a given dark matter candidate. One of the publicly available computer codes, DarkSUSY [980], for instance, uses PYTHIA [1766] to estimate the number of photons with continuum energy distribution created by mesons (mostly π^0) in the quark fragmentation for given Standard Model final states (see Chapter 16 for further information on publicly available computer codes). We show in Fig. 24.2 the spectra relative to different annihilation channels. Special calculations are needed for some of the higher-order QED corrections, which may be very important as we will see below.

There have been many improvements of the first, rather complete treatment in [304], and many promising new features keep appearing. Here we consider a few of these.

24.1.1 Large contributions at high energy from internal QED bremsstrahlung

This interesting phenomenon, originally proposed in [289], which one has to go beyond leading order to see, has recently been proposed as a method to detect an otherwise undetectable dark matter candidate [177]. It has also been applied to MSSM and mSUGRA models, and found to be very important [444], causing sometimes large boosts to the highest-energy end of the

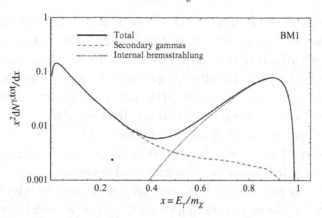

Fig. 24.1. The gamma-ray spectrum expected for one of the benchmark models (BM1) defined in [444]. Here $x = E_\gamma/m_\chi$, and the distribution without the bremsstrahlung is shown by the dashed line. There will also be a 2γ and a $Z\gamma$ contribution, not shown here.

gamma-ray spectrum. The idea, for fermion final states, is that a Majorana fermion (as many dark matter candidates are) suffers a helicity suppression for s-wave annihilation [973]. However, by emitting a photon from an internal charged leg, which only costs a factor of α_{em}/π, the helicity suppression may be avoided. The effect will be that these radiative corrections, instead of as usual being a per cent of the lowest-order process, may instead give enhancement factors of several thousand to a million times the suppressed lowest-order, low-velocity rate [289]. The resulting spectra will have a characteristic very sharp drop at the endpoint $E_\gamma = m_\chi$ of the gamma-ray spectrum; see Fig. 24.1. Interestingly, the very same process allows stringent constraints to be set on MeV DM candidates [240].

24.1.2 Annihilation into 2γ or $Z\gamma$

One feature that distinguishes dark matter annihilation into gamma-rays from the other indirect detection methods is the possibility that spectral features may reveal the mass of the WIMP. Such a 'smoking gun' signature of dark matter may be provided by the loop-induced two-photon annihilation [294; 301; 302; 304; 305; 418; 956; 1182; 1660]. Since the annihilating dark matter particles move with non-relativistic velocities, energy and momentum conservation gives two back-to-back photons with energy $E_\gamma = m_{WIMP}$ with an intrinsic energy spread of only $\sim 10^{-3}$, given mainly by the Doppler effect if there is a centre-of-mass motion of the annihilating pair. If such a gamma-ray line were observed, with one of the new generation of detectors having at least an energy resolution better than 10%, it would be a discovery essentially without uncertainties, owing to the lack of known background processes with the same peculiar properties. A similar gamma-ray peak is expected for the $Z\gamma$ final state [300; 411; 1879; 1885], but there the intrinsic width of the Z boson also has to be taken into account.

Since these processes are typically suppressed by $(\alpha/\pi)^2$ except in some interesting special cases (see below), it may quite often happen, unfortunately, that they fall below the experimental sensitivity.

Recently a class of models has been proposed [196] in which the 2γ and $Z\gamma$ processes may in fact be dominant [1043]. These models, which in some sense are more conventional than, for example, supersymmetric dark matter models, are so-called 'inert' Higgs models with just one extra Higgs doublet, protected by a Z_2 symmetry. Besides other interesting phenomenology, these models may also break electroweak symmetry radiatively [1065]. Although a large portion of the parameter space is excluded by LEP data, there are remaining portions which should show up in the phenomenology of LHC [1355].

24.1.3 *Sommerfeld enhancement and 'explosive' annihilation*

The expression found in [304] for annihilation into 2γ and $Z\gamma$ has a remarkable behaviour as the annihilating particles are either electroweak doublets (pure higgsinos) or triplets (pure gauginos). Namely, the cross-section tends to a constant value proportional to $1/m_W^2$ instead of $1/m_\chi^2$ as could be expected on dimensional grounds. This means that the unitarity limit [1162]

$$\sigma_{\text{unitarity}} < \frac{4\pi}{vm^2} \qquad (24.3)$$

will be violated at very high masses. This led Hisano, Matsumoto and Nojiri [1104] to investigate the behaviour of the amplitude near that limit. They discovered that including perturbatively higher-order corrections, they would get a value slightly higher than that found in [302] but, more importantly, unitarity was restored. A crucial step forward was then taken in ref. [1105], where they non-perturbatively summed up in the ladder approximation to all orders the attractive t-channel exchange diagrams and found a zero-energy bound state for some particular dark matter masses and typical Galactic velocities. The appearance of the bound state makes the cross-section increase by two to three orders of magnitude, compared with

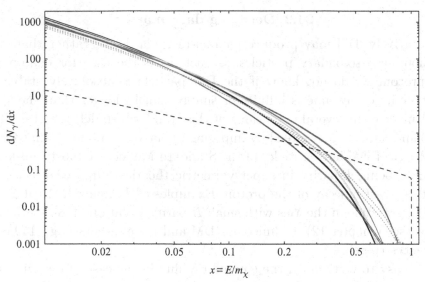

Fig. 24.2. Photon multiplicity for various final states. Solid lines show the rather soft and featureless spectra relative to the $u\bar{u}/d\bar{d}$, $s\bar{s}$, $c\bar{c}$, $b\bar{b}$, $t\bar{t}$, WW and ZZ channels (see ref. [332] for a detailed description of the spectra). The dashed line is the final-state radiation (FSR) spectrum from e^+e^- final states. All spectra are plotted for both $m_\chi = 500\,\mathrm{GeV}$ and $m_\chi = 1000\,\mathrm{GeV}$. From ref. [332].

that for velocities corresponding to the freeze-out temperature $T \sim m_\chi/20$ ($v \sim 0.3$–0.4). This phenomenon, discovered in [1105] and verified in [113; 557; 561] gives the possibility of very strong gamma-ray signals for particular masses (usually in the TeV region). It is analogous to what happens for positronium near bound state thresholds, as originally discussed by Sommerfeld in the early days of quantum mechanics. It is today called 'Sommerfeld enhancement' of annihilation rate, and may be a generic phenomenon. Of course, supersymmetric TeV particles interacting through Standard Model gauge bosons may have difficulty in giving the required relic density, unless one tolerates some fine-tuning, as is explicitly done in 'split SUSY' models [960]. The Sommerfeld enhancement is today extensively discussed in connection with the surprising new results on the high positron flux at high energies (see Chapter 26).

If Sommerfeld enhancement is active for positrons, one would expect signals in radio waves and gamma-rays [335]. See also [1963] for a discussion on the possible role of the inverse Compton process in determining the cause of the positron signal.

24.2 Decaying dark matter

Alternatively, DM may produce gamma-rays via decay, rather than annihilation, into secondary particles. In fact, as is the case for example for the proton, we do not know if the DM particle is absolutely stable, i.e. whether its decay time is infinite or simply (much) longer than the age of the Universe. In several extensions of the Standard Model, the stability of DM candidates is guaranteed by imposing by hand a symmetry that forbids the decay of DM into particles of the Standard Model, the most well-known example being R-parity in supersymmetric theories, imposed precisely to avoid too fast a decay of the proton. Examples of decaying DM candidates are the gravitino in theories with small R-parity violation [458], sterile neutrinos (see Chapter 12), technicolour DM and many others (e.g. [1496] and references therein).

It is easy to write down the gamma-ray flux in the case of decaying DM, in analogy to the case of annihilating DM

$$\frac{\mathrm{d}\phi}{\mathrm{d}E} = \frac{2}{m_\chi}\frac{\mathrm{d}N_\gamma}{\mathrm{d}E}\frac{1}{8\pi\tau_\chi}\int_{\mathrm{los}}\rho_\chi(\mathrm{l})\mathrm{dl}, \qquad (24.4)$$

where τ_χ is the decay time, and the integrand depends linearly on the density profile, unlike the case of annihilating DM, where it depends on the square of the density profile.

In the case of gravitinos with small R-parity violation [458], the model predicts a small photino–neutrino mixing $|U_{\tilde{\gamma}\nu}| = \mathcal{O}(10^{-8})$, leading to a decay of the gravitino into photon and neutrino with lifetime

$$\tau_{3/2}^{\text{2-body}} \simeq 3.8 \times 10^{27}\mathrm{s}\left(\frac{|U_{\tilde{\gamma}\nu}|}{10^{-8}}\right)^{-2}\left(\frac{m_{3/2}}{10\ \mathrm{GeV}}\right)^{-3}. \qquad (24.5)$$

The different dependence on ρ between Eqs. (24.2) and (24.4) leads to different detection strategies in the cases of annihilating and decaying DM. Annihilating DM is better searched for towards the Galactic centre, where the flux may be boosted by the presence of cusps. Decaying DM is instead better searched for towards the Galactic poles, where the signal is not much less than towards the centre, and the astrophysical background is much less intense. We note that the strategies of detection inevitably depend on the angular size of the detector. Sterile neutrinos searches at X-ray energies are performed with detectors with an angular resolution which is far better than current gamma-ray experiments. This implies that searching for point sources is much more promising in the former case, while the search for a diffuse emission is a better strategy for GeV energies and above [458].

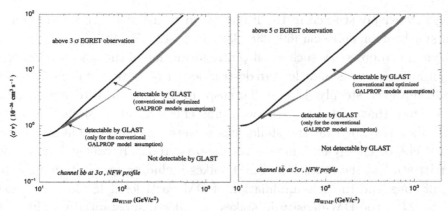

Fig. 24.3. Cross-sections $\langle \sigma v \rangle$ $(v \to 0)$ versus the WIMP mass m_{WIMP} for the $b\bar{b}$ annihilation channel. Left panel shows the result for 3σ significance, right panel shows the result for 5σ significance for 5 years of GLAST operation. The upper part of the plots corresponds to regions already excluded by the EGRET data around the Galactic centre and the lower part corresponds to regions not detectable by GLAST. The 'detectable by GLAST region' corresponds to models detectable by GLAST for both 'conventional' and 'optimized' astrophysical background. The shaded region represents models which can be detected only under the assumption of 'conventional' Galactic diffuse background. See text for more details.

24.3 Galactic centre

The Galactic centre has long been considered as the optimal target for indirect detection [286; 304; 341; 418; 513; 876; 1800].

The presence of an intense astrophysical background is in fact compensated for in this case by the strong enhancement of the annihilation signal due to the presence of the central cusp, as predicted by (naive) theoretical arguments and numerical simulations. We refer the interested reader to Chapter 2 for a discussion of the DM profiles derived from numerical simulations, and to Chapter 5 for a discussion on how astrophysical processes (mergers, presence of a black hole, gravitational scattering off stars) affect the DM profiles. Note in particular that the presence of a black hole at the Galactic centre inevitably leads to a modification of the DM profile. This process has been studied extensively in the context of stellar cusps around massive black holes in clusters of stars or at the centres of galaxies (see e.g. ref. [873] for an extended discussion and relevant references) and it has been subsequently applied to the distribution of DM at the centre of the Galaxy [985]. The consequent enhancement in the DM density around the central SMBH is usually referred to as a 'spike', in order to avoid confusion with DM 'cusps' at the centres of haloes in the cold dark matter model of structure formation. The case of annihilations from the DM spike at the centre of the Galaxy has been extensively studied in the literature in terms of neutrino, gamma-ray and synchrotron emission [69; 338; 339; 341; 342; 975; 985].

Dynamical processes such as off-centre formation of the seed black hole, or major merger events, may lead to destruction or reduction of the spike [1430; 1883]. However, steeply rising stellar cusps in the innermost regions of galaxies suggest that such processes were not effective, at least in the case of the Milky Way, or that the stellar cusps were regenerated via star formation [1449] or energy exchange between stars [1603]. Recently, Bertone and Merritt studied the evolution of DM spikes including gravitational scattering off stars and the self-annihilation of DM particles [338; 339]. As shown in Fig. 24.4, the DM density in spikes is, indeed, substantially reduced by these effects, but the enhancement of the annihilation signal could be still significant with respect to ordinary DM cusps.

The EGRET collaboration reported the detection of a point source (3EG J1746-2851) within 0.2 degrees from Sgr A* [1406], which triggered the interest of particle astrophysicists for its possible interpretation in terms of DM annihilations. However, a re-analysis based on photons with energies above 1 GeV has shown that the source appears to be slightly offset with respect to the Galactic centre [1120]. HESS observations have subsequently revealed a source of very high energy gamma-ray emission (HESS J1745-290) lying within $7'' \pm 14''_{\text{stat}} \pm 28''_{\text{syst}}$ from the supermassive black hole Sgr A*, and compatible with a point source of size less than $1.2'$ [37].

There have been many attempts to interpret gamma-ray data from the Galactic centre in terms of DM annihilations [340; 342; 513; 688; 1607; 1634; 1955]. However, the bad news is that the energy spectrum is well fitted by a power law $dN_\gamma/dE = \Phi_0 E^{-2.25\pm0.04}$ with $\Phi_0 = 2.25 \times 10^{-12}$ TeV cm^{-2} s^{-1}, over two decades in energy, as also confirmed by the MAGIC collaboration [45], which points towards a conventional astrophysics source. Furthermore the implied mass scale of the DM particle (well above 20 TeV, to be consistent with the observed spectrum) appears to be difficult to reconcile with the properties of commonly studied candidates. Nevertheless the Galactic centre remains an interesting target for the Fermi Large Area Telescope, since it will explore a range of energies below the relatively high threshold of HESS, where a DM signal could be hiding [180]. We show in Fig. 24.3 the expected reach of the Fermi LAT in the σ versus mass plane.

Aside from the central source, an interesting signature may arise from the analysis of the diffuse emission in an angular window around the Galactic centre. The size and shape of this 'annulus' of dark matter depend on the DM profile and on the shape and spectrum of the diffuse Galactic background [1727; 1811].

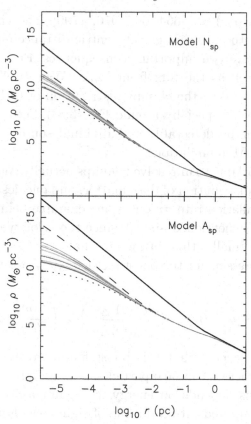

Fig. 24.4. Evolved DM density profiles at $\tau = 10$ (roughly 10^{10} yr) starting from two different initial DM profiles. Grey curves: benchmark models; dashed lines: $\sigma v = 0$; dotted curves: $\sigma v = 3 \times 10^{-26} \mathrm{cm}^3 \, \mathrm{s}^{-1}$, $m = 50$ GeV; thick line: initial DM density. From ref. [339].

24.4 Substructures

24.4.1 Subhaloes

As we have seen in Chapter 2, the presence of substructures is a generic prediction of the CDM paradigm. This is particularly important for indirect searches since any inhomogeneity in the distribution leads to a substantial increase of the annihilation flux. The properties of DM substructures have been discussed in Section 2.2.3. We recall here that the cumulative mass functions of subhaloes can be approximated by $M^{-\alpha}$, with $\alpha = 1.9$ to 2.0, normalized so that the mass in subhaloes larger than $10^{-6} M_{\mathrm{host}}$ is between about 5 and 15%; and that their spatial distribution within the host halo is independent of the mass threshold and it can be approximated as $n_{\text{sub},M0}(r) \propto \rho_{\text{DM}}(r) \times r$ [664; 667; 916; 1790]. The virial concentration of clumps varies with mass and galactocentric distance (see Fig. 2.6).

Substructures have two important consequences. First, the diffuse annihilation flux, as well as the flux from Milky Way satellites and external structures, is 'boosted' by the clumpy distribution of DM (see also the discussion in Chapter 3, especially around Eq. (3.8)). Second, the more luminous clumps might be detectable as individual sources, thus providing a smoking gun for DM annihilations.

The 'boost factor' due to unresolved clumps actually consists of two parts, one related to the distribution of dark matter, and one describing the dependence of the annihilation rate σv on v (one example being the Sommerfeld enhancement factor mentioned above), since the velocity dispersions of DM in DM clumps are smaller than in the host halo.

The general expression for the boost factor $B_{\Delta V; \sigma v}$ for indirect detections is thus

$$B_{\Delta V; \sigma v} = B_{\Delta V} \times B_{\sigma v} = \left(\frac{\langle n^2(r) \rangle_{\Delta V}}{\langle n_0^2(r) \rangle_{\Delta V}} \right) \times \left(\frac{\langle \sigma v \rangle_{v \simeq 10^{-3}}}{\langle \sigma v \rangle_{v \simeq 0.3}} \right). \qquad (24.6)$$

It is important to note that this boost factor involves an average over the volume ΔV where secondary particles (e.g. antiprotons and positrons) diffuse, and it thus depends on energy. For gamma-ray observations, the enhancement is computed within the line-of-sight cone. For sufficiently dense and nearby clumps, the boost factor may be so large as to bring the clump sufficiently above the diffuse background to make it detectable with gamma-ray telescopes.

The computation of the boost factor in realistic astrophysical and particle physics scenarios is a formidable task, which has so far only been partially addressed. In fact, the most recent numerical simulations suggest that large astrophysical boost factors are unlikely, and realistic estimates suggest that for gamma-rays it should be $\mathcal{O}(10)$ or less [664; 1790].

Even assuming a fiducial DM candidate, the prospects for detecting individual sources strongly depend on poorly constrained extrapolations of quantities such as the aforementioned virial concentration (as a function of mass and redshift) and mass function. In order to assess the detectability of substructures, one has to take into account that the very same process that provides the enhancement of individual clumps inevitably boosts at the same time the diffuse flux, which acts as a background for the annihilation signal in clumps. The relative importance of the various signals that contribute to the total annihilation flux is shown in Fig. 24.5. It is therefore difficult to make robust estimates of the number of detectable objects, which may vary

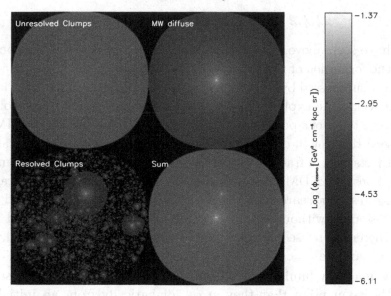

Fig. 24.5. Map of Φ_{cosmo} (proportional to the annihilation signal) for one of the subhalo models discussed in ref. [1578], in a cone of 50 degrees around the Galactic centre, as seen from the position of the Sun. Upper left: smooth subhalo contribution from unresolved haloes. Upper right: smooth contribution of the Milky Way halo. Lower left: contribution from resolved haloes. Lower right: sum of the three contributions. Figure taken from ref. [1578].

between $\mathcal{O}(100)$ and none; see ref. [1578] for 5 years of observation with the Fermi satellite. It is probably worth stressing that in the case of actual detection, this scenario would provide unambiguous evidence for DM, since all the detected substructures would appear as (possibly diffuse) gamma-ray sources with an identical spectrum.

Alternatively, one could search for substructures at other wavelengths, caused by the synchrotron and inverse Compton emission produced by secondary electrons and positrons that interact with the Galactic magnetic field or photons respectively, possibly leading to detectable signatures in X-rays, microwaves or radio wavelengths [184; 368; 384]. Furthermore, we note that the presence of substructures has important consequences for the determination of the angular power spectrum of anisotropies of the diffuse gamma-ray emission [87; 88; 874; 1749; 1750], which may provide a clear signature of DM annihilations even in the absence of clear signatures in the energy spectrum.

It may be anticipated that a deeper understanding of the properties of DM substructures will be one of the main problem areas of future indirect detection studies of dark matter.

24.4.2 Intermediate-mass black holes

As we have seen above, the strong enhancement of the annihilation signal due to the formation of the supermassive black hole at the Galactic centre is very likely suppressed by a combination of astrophysical processes. However, theories seeking to explain the formation of supermassive black holes predict the existence of a population of intermediate-mass black holes (IMBHs), which seed the growth of larger objects under appropriate circumstances. If these objects exist, a fraction of them would never experience major mergers, and the boosting of DM density may lead to observational gamma-ray annihilation rates to be searched for by Fermi [343]. These objects would appear as point sources without optical counterpart, and may be studied further by air Cherenkov telescopes once their location has been established, e.g. by the Fermi satellite.

The gamma-ray luminosity of these mini-spikes can be easily estimated under the assumption that they grow adiabatically from an initial NFW profile. The total gamma-ray flux in this case can be written as

$$
\Phi(E, D) = \Phi_0 \frac{\mathrm{d}N}{\mathrm{d}E} \left(\frac{\sigma v}{10^{-26} \mathrm{cm}^3\,\mathrm{s}^{-1}} \right) \left(\frac{m_\chi}{100\,\mathrm{GeV}} \right)^{-2}
$$
$$
\times \left(\frac{D}{\mathrm{kpc}} \right)^{-2} \left(\frac{\rho(r_{\mathrm{sp}})}{10^2\,\mathrm{GeV\ cm}^{-3}} \right)^2
$$
$$
\times \left(\frac{r_{\mathrm{sp}}}{\mathrm{pc}} \right)^{\frac{14}{3}} \left(\frac{r_{\mathrm{cut}}}{10^{-3}\,\mathrm{pc}} \right)^{-\frac{5}{3}}, \tag{24.7}
$$

where r_{sp} and r_{cut} are the outer and inner size of the mini-spike, and $\Phi_0 = 9 \times 10^{-10} \mathrm{cm}^{-2}\,\mathrm{s}^{-1}$.

If these objects exist, many of them should be detected as bright gamma-ray sources, even for a pessimistic choice of particle physics parameters (as shown in Fig. 24.6). Either strong constraints on a combination of the astrophysics and particle physics of this scenario, or an actual detection, should be possible within the first year of operation of Fermi.

Another potential smoking-gun signature of DM annihilations around IMBHs would be the detection of a distribution of point sources isotropically distributed in a $\sim 3°$ circle around the centre of M31. For a neutralino-like DM candidate with a mass $m_\chi = 150$ GeV, up to 20 sources would be detected with Fermi, while with air Cherenkov telescopes such as MAGIC and VERITAS, up to 10 sources might be detected, provided that the mass of neutralino is in the TeV range or above [875].

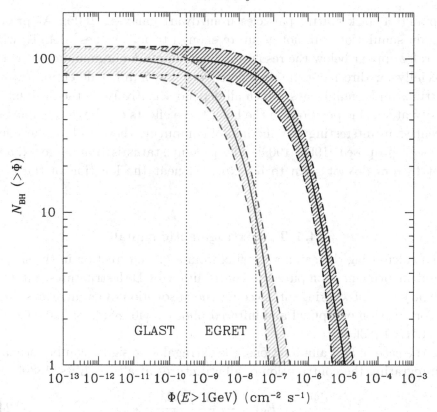

Fig. 24.6. IMBH integrated luminosity function, i.e. number of black holes producing a gamma-ray flux larger than a given flux, as a function of the flux, for our scenario B (i.e. for IMBHs with mass $\sim 10^5 M_\odot$). The upper (lower) line corresponds to $m_\chi = 100$ GeV, $\sigma v = 3 \times 10^{-26}$ cm^3 s^{-1} ($m_\chi = 1$ TeV, $\sigma v = 10^{-29}$ cm^3 s^{-1}). For each curve we also show the 1σ scatter among different realizations of Milky-Way-sized host DM haloes. The figure can be interpreted as the number of IMBHs that can be detected from experiments with point source sensitivity Φ (above 1 GeV), as a function of Φ. We show for comparison the 5σ point source sensitivity above 1 GeV of EGRET and GLAST (1 year). From ref. [343].

Finally, angular correlations of the diffuse extragalactic background could provide a tool to disentangle the signal induced by DM annihilation in mini-spikes from a conventional astrophysical component [1838].

24.4.3 Caustics and streams

The microscopic structure of the dark matter halo may also contain dark matter streams [882; 1505; 1756], created by late infall of dark matter particles on an existing halo (for an early, simplified model, see [1756]), or by tidal stripping of dark matter particles from dwarf galaxies [1505]. At present, N-body simulations are not accurate enough to follow these details, which generally appear below the resolution limit. It can be argued, however, that especially for direct dark matter detection, and indirect detection through neutrinos and gamma-rays, which all depend sensitively on the dark matter density at specific positions in the halo, these effects can be very important. Recently, an interesting new method of computing these and similar effects has been proposed [1908; 1909]. The present status is that large exhancement factors do not seem to be expected near the location of the Milky Way.

24.5 The extragalactic signal

When calculating the gamma-ray flux from a given direction in the sky, one should in principle compute the contribution of DM structures out to an arbitrary redshift, taking into account the absorption of gamma-rays due to pair production on optical and infrared photons [40; 88; 298; 600; 601; 662; 790; 1847; 1880].

In the case of DM annihilations, the extragalactic signal can be estimated by integrating the contribution of all structures at all redshifts [1880]

$$\left\langle \frac{\mathrm{d}\Phi}{\mathrm{d}E_0 \mathrm{d}\Omega} \right\rangle (E_0) = \frac{\sigma v}{8\pi} \frac{c}{H_0} \frac{\rho_{\mathrm{cr}}^2 \Omega_{0,m}^2}{m_\chi^2} \times \tag{24.8}$$

$$\int \mathrm{d}z(1+z)^3 \frac{\Delta^2(z)}{h(z)} \frac{\mathrm{d}N_\gamma(E_0(1+z))}{\mathrm{d}E} \mathrm{e}^{-\tau(z,E_0)},$$

with

$$\Delta^2(z) = \int \mathrm{d}M \frac{\nu(z,M)f(\nu(z,M))}{\sigma(M)} \left| \frac{\mathrm{d}\sigma}{\mathrm{d}M} \right| \Delta_M^2(z,M) \tag{24.9}$$

and

$$\Delta_M^2(z,M) = \int \mathrm{d}c' P(c(M,z),c') \frac{\Delta_{\mathrm{vir}}}{3} \frac{I_2(x_{\min},c')}{I_1^2(x_{\min},c')} (c')^3 \mathrm{d}c'. \tag{24.10}$$

$\Delta_M^2(z,M)$ represents the enhancement in the gamma-ray flux due to the presence of a DM halo with a mass M at a redshift z, while the integrals I_n are simply given by $I_n = (x_{\min}, x_{\max}) = \int g^n x^2 \mathrm{d}x$. The presence of substructures is taken into account by replacing Δ_M^2 in Eq. (24.10) with the following expression:

$$\Delta_M^2(M) \to (1 - f(M))^2 \Delta_M^2(M) + \tag{24.11}$$

$$+ \frac{1}{M} \int \mathrm{d}M_{\mathrm{s}} M_{\mathrm{s}} \frac{\mathrm{d}n}{\mathrm{d}M_{\mathrm{s}}} (M_{\mathrm{s}}, z) \Delta_{M_{\mathrm{s}}}^2(z, M_{\mathrm{s}}).$$

We refer the interested reader to [874; 1880] for further details on the technical aspects of this calculation, and limit ourselves to noticing here that the extragalactic flux is typically below the diffuse Galactic emission (dominated, outside the few inner degrees, by Galactic substructures), at least for typical choices of the properties of DM at small scales [874].

For the case of decaying DM, the calculation is more straightforward (see e.g. ref. [333] and references therein)

$$
\frac{\mathrm{d}\phi}{\mathrm{d}E} = A_{\mathrm{eg}} \frac{2}{m_{\mathrm{DM}}} \left(1 + \kappa \left(\frac{2E}{m_{\mathrm{DM}}} \right)^3 \right)^{-1/2} \left(\frac{2E}{m_{\mathrm{DM}}} \right)^{1/2} \Theta \left(1 - \frac{2E}{m_{\mathrm{DM}}} \right),
$$
(24.12)

with $\kappa = \Omega_\Lambda / \Omega_M \simeq 3$, and

$$
A_{\mathrm{eg}} = \frac{\Omega_{\mathrm{DM}} \rho_{\mathrm{c}}}{4\pi \tau_{\mathrm{DM}} m_{\mathrm{DM}} H_0 \Omega_M^{1/2}} = 10^{-7} \ (\mathrm{cm}^2 \ \mathrm{s} \ \mathrm{str})^{-1} \left(\frac{\tau_{\mathrm{DM}}}{10^{27} \ \mathrm{s}} \right)^{-1} \left(\frac{m_{\mathrm{DM}}}{10 \ \mathrm{GeV}} \right)^{-1}.
$$
(24.13)

In this case too, the Galactic signal typically dominates over the extragalactic one [333].

This is an area where observations by Fermi will be crucial but, because of the needed subtraction of the galactic diffuse foreground, this is one of the most difficult measurements, and will probably take several years to gather significant data. If coming air Cherenkov telescope arrays have the possibility of operating in a survey mode, this could also open up new regions of sensitivity. The eventual existence of a spectral signature, such as a line (somewhat asymmetrical and smeared because of the intergalactic absorption) or a broader feature predicted by some models with large radiative corrections mentioned above, could be crucial.

24.6 Connection between antimatter and gamma-ray signal

Recently, it has been argued that the spectral distortion, the so-called 'GeV anomaly', of EGRET's measurements of Galactic diffuse gamma-rays can be explained by supersymmetric WIMP annihilation [626]. This is done by making a very particular model of the distribution of dark matter, which is largely concentrated in the disk, and in addition has two very massive rings of dark matter in the plane of the disk. It is claimed that by using this dark matter distribution, and adjusting the annihilation rate by a rather large 'boost factor', the measured gamma-ray angular and energy distributions can be reproduced with a canonical supersymmetric WIMP of mass below 100 GeV.

It has been suggested that the GeV excess of EGRET might be an experimental artifact [1799], and in fact the preliminary data from Fermi (presented at several conferences, but not published yet) show no evidence of it. In any case, it has been shown that the proposal of de Boer *et al.* [629] has the problem of a severe overproduction of antiprotons [296]. In a similar way, an excess in the positron flux recently claimed by the PAMELA and ATIC collaborations [24; 515] is severely constrained by gamma-ray (and radio) observations (see e.g. [291; 335] and references therein), thus proving the strong interplay between various indirect detection strategies, and the need for a multi-wavelength approach (see Chapter 27 for a more detailed discussion).

25

High-energy neutrinos from WIMP annihilations in the Sun

Francis Halzen and Dan Hooper

25.1 Searching for dark matter with neutrinos

As the Solar System moves through the halo of the Milky Way, WIMPs become swept up by the Sun. Although dark matter particles interact only weakly, they occasionally scatter elastically with nuclei in the Sun and lose enough momentum to become gravitationally bound. Over the lifetime of the Sun, a sufficient density of WIMPs can accumulate in its centre so that an equilibrium is established between their capture and annihilation rates. The annihilation products of these WIMPs include neutrinos, which escape the Sun with minimal absorption, and thus potentially constitute an indirect signature of dark matter. Such neutrinos can be generated through the decays of heavy quarks, gauge bosons and other products of WIMP annihilation, and then proceed to travel to Earth where they can be efficiently identified using large volume neutrino detectors.

Compared with other dark matter detection techniques, indirect dark matter searches using neutrinos involve minimal astrophysical uncertainties. Although the capture rate of WIMPs onto the Sun depends on the local density and velocity distribution of dark matter (as do direct detection rates), the rate at which WIMPs annihilate is determined by the total number of WIMPs in the core of the Sun, which have accumulated over billions of years. As a consequence, any structure or other variations in the local dark matter density (subhaloes, streams, etc.) become averaged out. Furthermore, unlike indirect detection techniques using gamma-rays or charged particles, the rate of WIMP annihilations in the Sun does not depend on the distribution of dark matter throughout the Galaxy or on the properties of the Galactic magnetic or radiation fields. Thus, given a dark matter

Particle Dark Matter: Observations, Models and Searches, ed. Gianfranco Bertone. Published by Cambridge University Press. © Cambridge University Press 2010.

candidate, one can reliably predict the corresponding signal in a neutrino telescope. If not observed, the model is ruled out.

Extensions of the Standard Model which provide a solution to the hierarchy problem naturally yield viable candidates for dark matter. The neutralino in supersymmetric models, for example, has been intensively studied as a possible dark matter candidate. Detecting this particle has become the benchmark by which experiments are evaluated and mutually compared. Supersymmetric models allow for a large number of free parameters, however, leading to a large range of possible masses, couplings and other phenomenological properties.

One of the primary missions of neutrino telescopes is to search for an excess of high-energy neutrinos from the direction of the Sun, signalling the presence of annihilating dark matter. Neutrino telescopes detect the Cherenkov radiation from secondary particles produced in the interactions of high-energy neutrinos in highly transparent and well-shielded deep water or ice. They take advantage of the comparatively large cross-section of high-energy neutrinos and the long range of the muons produced. The instrumentation consists of a lattice of photomultipliers deployed in a shielded and optically clear medium that is transformed into a Cherenkov detector. As of 2008, the IceCube telescope is currently under construction and is taking data with a partial array of 2400 ten-inch photomultipliers positioned between 1500 and 2500 metres below the geographic South Pole. With the completion of the detector by 2010–2011, the instrumented volume will be doubled from 0.5 to $1 \, \text{km}^3$ [653]. Data taken with the first 22 strings of IceCube have resulted in a limit on an excess flux from the Sun. The limit improves on previous results of the Super-Kamiokande [651] and AMANDA [20] collaborations by factors of 3 to 5 for WIMPs heavier than approximately 250 GeV. An extension of the experiment dubbed DeepCore is under construction and will enable IceCube to place the strongest limits yet for WIMPs as light as 50 GeV. The Antares experiment is also designed to be sensitive to relatively light WIMPs [454]. For WIMPs lighter than approximately 50 GeV, the performance of Super-Kamiokande is unmatched by existing detectors.

In the remainder of this chapter, we describe the processes of capture and annihilation of WIMPs in the Sun, and predict the resulting flux of high-energy neutrinos. Before moving on to the details of this subject, however, we show a summary of the overall status of indirect dark matter detection using neutrinos in Fig. 25.1. This figure shows the current limits on the neutrino-induced muon flux from the direction of the Sun, as well as the projected sensitivity of IceCube. Also shown is a sampling of the supersymmetric

Solar neutralinos

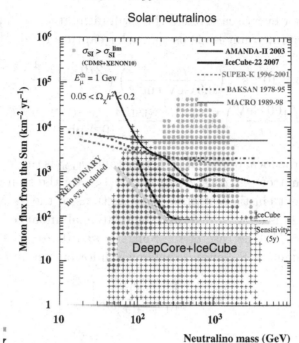

Fig. 25.1. The current status of indirect searches for dark matter using high-energy neutrinos. In addition to the various experimental constraints, the projected sensitivity of IceCube is shown, along with the flux of neutrino-induced muons predicted over the parameter space of the minimal supersymmetric Standard Model.

parameter space. In the later sections of this chapter, we will discuss the sensitivity of this technique to both neutralino and Kaluza–Klein dark matter candidates.

25.2 The capture and annihilation of WIMPs in the Sun

Beginning with a simple estimate, we expect WIMPs to be captured in the Sun at a rate approximately given by:

$$C^{\odot} \sim \phi_{\chi}(M_{\odot}/m_p)\sigma_{\chi p}, \tag{25.1}$$

where ϕ_{χ} is the flux of WIMPs in the Solar System, $M_{\odot}$ is the mass of the Sun and $\sigma_{\chi p}$ is the WIMP–proton elastic scattering cross-section. Reasonable estimates of the local distribution of WIMPs lead to a capture rate of $C^{\odot} \sim 10^{20}\,\mathrm{s}^{-1} \times (100\,\mathrm{GeV}/m_{\chi})\,(\sigma_{\chi p}/10^{-6}\,\mathrm{pb})$, where m_{χ} is the mass of the WIMP. This neglects, however, a number of potentially important factors, including the gravitational focusing of the WIMP flux towards the Sun, and the fact that not every scattered WIMP will ultimately be captured. Taking these effects into account leads us to a solar capture rate of [993]:

$$
C^\odot \approx 1.3 \times 10^{21}\,\mathrm{s}^{-1} \left(\frac{\rho_{\mathrm{local}}}{0.3\,\mathrm{GeV\ cm}^{-3}} \right) \left(\frac{270\,\mathrm{km\ s}^{-1}}{\bar{v}_{\mathrm{local}}} \right)
$$

$$
\times \left(\frac{100\,\mathrm{GeV}}{m_\chi} \right) \sum_i \left(\frac{A_i\,(\sigma_{\chi i,\mathrm{SD}} + \sigma_{\chi i,\mathrm{SI}})\,S(m_\chi/m_i)}{10^{-6}\,\mathrm{pb}} \right), \qquad (25.2)
$$

where ρ_{local} is the local dark matter density and $\bar{v}_{\mathrm{local}}$ is the local rms velocity of halo dark matter particles; $\sigma_{\chi i,\mathrm{SD}}$ and $\sigma_{\chi i,\mathrm{SI}}$ are the spin-dependent and spin-independent elastic scattering cross-sections of the WIMP with nuclei species i; and A_i is a factor denoting the relative abundance and form factor for each species. In the case of the Sun, $A_\mathrm{H} \approx 1.0$, $A_\mathrm{He} \approx 0.07$ and $A_\mathrm{O} \approx 0.0005$. The quantity S contains dynamical information and is given by:

$$
S(x) = \left[\frac{A(x)^{3/2}}{1 + A(x)^{3/2}} \right]^{2/3}, \qquad (25.3)
$$

where

$$
A(x) = \frac{3}{2} \frac{x}{(x-1)^2} \left(\frac{v_{\mathrm{esc}}}{\bar{v}_{\mathrm{local}}} \right)^2, \qquad (25.4)
$$

and $v_{\mathrm{esc}} \approx 1156\,\mathrm{km\ s}^{-1}$ is the escape velocity of the Sun. Notice that for WIMPs much heavier than the target nuclei, $S \propto 1/m_\chi$, leading the capture rate to be suppressed by two factors of the WIMP mass. In this case ($m_\chi \gtrsim 30\,\mathrm{GeV}$), we can write the capture rate as:

$$
C^\odot \approx 3.35 \times 10^{20}\,\mathrm{s}^{-1} \left(\frac{\rho_{\mathrm{local}}}{0.3\,\mathrm{GeV\ cm}^{-3}} \right) \left(\frac{270\,\mathrm{km\ s}^{-1}}{\bar{v}_{\mathrm{local}}} \right)^3 \left(\frac{100\,\mathrm{GeV}}{m_\chi} \right)^2
$$

$$
\times \left(\frac{\sigma_{\chi\mathrm{H,SD}} + \sigma_{\chi\mathrm{H,SI}} + 0.07\,\sigma_{\chi\mathrm{He,SI}} + 0.0005\,S(m_\chi/m_\mathrm{O})\,\sigma_{\chi\mathrm{O,SI}}}{10^{-6}\,\mathrm{pb}} \right).
$$

$$
(25.5)
$$

If the capture rate and annihilation cross-sections are sufficiently large, equilibrium will be reached between these processes. For $N(t)$ WIMPs in the Sun, the rate of change of this quantity is given by

$$
\dot{N}(t) = C^\odot - A^\odot N(t)^2 - E^\odot N, \qquad (25.6)
$$

where $C^\odot$ is the capture rate described above, $A^\odot$ is the annihilation cross-section times the relative WIMP velocity per volume and $E^\odot$ is the inverse time for a WIMP to escape the Sun via evaporation. Evaporation is highly suppressed for WIMPs heavier than a few GeV [994; 1019]. The annihilation cross-section $A^\odot$ can be approximated by

$$A^\odot = \frac{\langle \sigma v \rangle}{V_{\text{eff}}}, \tag{25.7}$$

where V_{eff} is the effective volume of the core of the Sun determined roughly by matching the core temperature with the gravitational potential energy of a single WIMP at the core radius and is given by [994; 1019]

$$V_{\text{eff}} = 5.7 \times 10^{27} \, \text{cm}^3 \left(\frac{100 \, \text{GeV}}{m_\chi} \right)^{3/2}. \tag{25.8}$$

Neglecting evaporation, the present WIMP annihilation rate is given by

$$\Gamma = \frac{1}{2} A^\odot N(t_\odot)^2 = \frac{1}{2} C^\odot \tanh^2 \left(\sqrt{C^\odot A^\odot} \, t_\odot \right), \tag{25.9}$$

where $t_\odot \approx 4.5$ billion years is the age of the Solar System. The annihilation rate is maximized when it reaches equilibrium with the capture rate. This occurs when

$$\sqrt{C^\odot A^\odot} t_\odot \gg 1. \tag{25.10}$$

If this condition is met, the final annihilation rate (and corresponding neutrino flux and event rate) is determined entirely by the capture rate and has no further dependence on the dark matter particle's annihilation cross-section.

25.3 The neutrino spectrum

Through their annihilations, WIMPs can generate neutrinos through a variety of channels. Annihilations to heavy quarks, tau leptons, gauge bosons and/or Higgs bosons can each generate energetic neutrinos in their subsequent decays [1183]. In some models, WIMPs can also annihilate directly to neutrino–antineutrino pairs. Annihilations to light quarks or muons, however, do not contribute to the high-energy neutrino spectrum, as these particles come to rest in the solar medium before decaying.

Neglecting the effects of oscillations and interactions with the solar medium, the spectrum of neutrinos from WIMP annihilations to a final state, $X\bar{X}$, is given by:

$$\frac{\text{d}N_\nu}{\text{d}E_\nu} = \frac{1}{2} \int_{E_\nu/\gamma(1+\beta)}^{E_\nu/\gamma(1-\beta)} \frac{1}{\gamma\beta} \frac{\text{d}E'}{E'} \left(\frac{\text{d}N_\nu}{\text{d}E_\nu} \right)_{X\bar{X}}^{\text{rest}}, \tag{25.11}$$

where $\gamma = m_\chi/m_X$, $\beta = \sqrt{1 - \gamma^{-2}}$ and $(\mathrm{d}N_\nu/\mathrm{d}E_\nu)^{\text{rest}}_{XX}$ is the spectrum of neutrinos produced in the decay of an X at rest.

As an example, consider the simple case of WIMPs annihilating to a pair of Z bosons, $\chi\chi \to ZZ$. In this case, the most energetic neutrinos are produced by the direct decay of a Z to a neutrino–antineutrino pair, for which

$$\left(\frac{\mathrm{d}N_\nu}{\mathrm{d}E_\nu}\right)^{\text{rest}}_{ZZ} = 2\Gamma_{Z\to\nu\bar\nu}\,\delta(E_\nu - m_Z/2), \qquad (25.12)$$

which leads to

$$\left(\frac{\mathrm{d}N_\nu}{\mathrm{d}E_\nu}\right)_{ZZ} = \frac{2\,\Gamma_{Z\to\nu\bar\nu}}{m_Z\,\gamma\,\beta}, \quad \text{for} \quad m_\chi(1-\beta)/2 < E_\nu < m_\chi(1+\beta)/2,$$
$$(25.13)$$

where $\Gamma_{Z\to\nu\bar\nu} \approx 0.067$ is the branching fraction to neutrino pairs (per flavour). In addition, Z bosons also produce neutrinos through their decays to $\tau^+\tau^-$, $b\bar b$ and $c\bar c$. An expression similar to Eq. (25.13) can be written down for the case of WIMP annihilations to W^+W^-, followed by $W^\pm \to l^\pm\nu$.

Tau leptons produce neutrinos through a variety of channels, including through the semi-leptonic decays $\tau \to \mu\nu\nu$, $e\nu\nu$ and the hadronic decays $\tau \to \pi\nu$, $K\nu$, $\pi\pi\nu$ and $\pi\pi\pi\nu$. Top quarks decay to a $W^\pm$ and a bottom quark nearly 100% of the time, each of which can generate neutrinos in their subsequent decay. For bottom and charm quarks, only the semi-leptonic decays contribute to the neutrino spectrum (with the exception of the neutrinos resulting from taus and c quarks produced in decays of b quarks). In the case of b and c quark decays, the process of hadronization reduces the fraction of energy that is transferred to the resulting neutrinos and other decay products. To account for such effects fully and accurately, programs such as PYTHIA are often used.

Once produced in the Sun's core, neutrinos then propagate through the solar medium and to the Earth. Over this journey, they can potentially be absorbed, lose energy and/or change flavour. In particular, charged current interactions of electron and muon neutrinos in the Sun lead to their absorption. The probability of absorption taking place can be estimated by $1 - \exp(-E_\nu/E_{\text{abs}})$, where E_{abs} is approximately 130 GeV for electron or muon neutrinos and 200 GeV for electron or muon antineutrinos. Absorption, therefore, only plays an important role for the case of relatively heavy WIMPs.

The effect of charged current interactions on tau neutrinos in the Sun is somewhat more complicated. The tau leptons produced in such interactions quickly decay and thus regenerate the absorbed tau neutrino, albeit with a reduced energy. Neutral current interactions of all three neutrino flavours similarly reduce the neutrinos' energy without depleting their number.

Vacuum oscillations lead to the full mixing of muon and tau neutrinos over their propagation to the Earth, making the observed muon neutrino spectrum (which is the relevant one for detection in neutrino telescopes) effectively the average of the muon and tau flavours prior to mixing. Electron neutrinos can also oscillate into muon flavour through matter effects in the Sun (the MSW effect). Electron antineutrinos can generally be neglected, as their oscillations to muon or tau flavours are highly suppressed [1323].

Programs such as DarkSUSY [981], which include effects such as hadronization, absorption, regeneration and oscillations, are very useful in making detailed predictions for the neutrino spectrum resulting from WIMP annihilations in the Sun.

25.4 Neutrino telescopes

Once they reach Earth, neutrinos can potentially be detected in large volume neutrino telescopes. Neglecting oscillations and solar absorption, the muon neutrino spectrum at the Earth resulting from WIMP annihilations in the Sun is given by:

$$\frac{dN_{\nu_\mu}}{dE_{\nu_\mu}} = \frac{C_\odot F_{\mathrm{Eq}}}{4\pi D_{\mathrm{ES}}^2} \left(\frac{dN_{\nu_\mu}}{dE_{\nu_\mu}}\right)^{\mathrm{Inj}}, \tag{25.14}$$

where $C_\odot$ is the WIMP capture rate in the Sun, F_{Eq} is the non-equilibrium suppression factor (~ 1 for capture–annihilation equilibrium), D_{ES} is the Earth–Sun distance and $\left(\frac{dN_{\nu_\mu}}{dE_{\nu_\mu}}\right)^{\mathrm{Inj}}$, where Inj stands for injection, is the muon neutrino spectrum from the Sun per WIMP annihilating.

Muon neutrinos can produce muons through charged current interactions with ice or water nuclei inside or near the detector volume of a high-energy neutrino telescope. The rate of neutrino-induced muons observed in a high-energy neutrino telescope is estimated by:

$$\begin{aligned} N_{\mathrm{events}} \approx &\int\int \frac{dN_{\nu_\mu}}{dE_{\nu_\mu}} \frac{d\sigma_\nu}{dy}(E_{\nu_\mu}, y) \left[R_\mu(E_\mu) + L\right] A_{\mathrm{eff}}\, dE_{\nu_\mu}\, dy \\ &+ \int\int \frac{dN_{\bar\nu_\mu}}{dE_{\bar\nu_\mu}} \frac{d\sigma_{\bar\nu}}{dy}(E_{\bar\nu_\mu}, y) \left[R_\mu(E_\mu) + L\right] A_{\mathrm{eff}}\, dE_{\bar\nu_\mu}\, dy, \end{aligned}$$

$$\tag{25.15}$$

where σ_ν ($\sigma_{\bar\nu}$) is the neutrino–nucleon (antineutrino–nucleon) charged current interaction cross-section, $(1-y)$ is the fraction of neutrino/antineutrino energy which goes into the muon, A_{eff} is the effective area of the detector, $R_\mu(E_\mu)$ is the distance a muon of energy $(1 - y)\, E_\nu$ travels before falling below the muon energy threshold of the experiment (ranging from approximately 1 to 100 GeV), called the muon range, and L is the depth of the detector volume. The muon range in water or ice is approximately given by:

$$R_\mu(E_\mu) \approx 2.4\,\text{km} \times \ln\left[\frac{2.0 + 0.0042\, E_\mu(\text{GeV})}{2.0 + 0.0042\, E_\mu^{\text{thr}}(\text{GeV})}\right], \qquad (25.16)$$

where E_μ^{thr} is the muon energy threshold of the experiment.

When completed, the IceCube experiment will possess a full square kilometre of effective area and kilometre depth, and will be sensitive to muons above approximately 50 GeV. The DeepCore extension of IceCube will be sensitive down to 10 GeV. The Super-Kamiokande detector, in contrast, has 10^{-3} times the effective area of IceCube and a depth of only 36.2 m [651]. For low-mass WIMPs, however, Super-Kamiokande benefits over large volume detectors such as IceCube by being sensitive to muons with as little energy as ~ 1 GeV.

The spectrum and flux of neutrinos generated in WIMP annihilations depend on the annihilation modes that dominate, and thus are model-dependent. As long as the majority of annihilations proceed to final states such as $b\bar{b}$, $t\bar{t}$, $\tau^+\tau^-$, W^+W^-, ZZ or some combination of Higgs and gauge bosons, however, the variation between different final states is not dramatic. In Fig. 25.2, we plot the event rate in a kilometre-scale neutrino telescope such as IceCube as a function of the WIMP's effective elastic scattering cross-section for four possible annihilation modes. The effective elastic scattering cross-section used here is defined as $\sigma_{\text{eff}} = \sigma_{\chi\text{H,SD}} + \sigma_{\chi\text{H,SI}} + 0.07\,\sigma_{\chi\text{He,SI}} + 0.0005\,S(m_\chi/m_0)\,\sigma_{\chi0,\text{SI}}$ (see Eq. (25.5)).

The elastic scattering cross-section of a WIMP is constrained by the absence of a positive signal in direct detection experiments. Currently, the strongest limits on the WIMP–nucleon spin-independent elastic scattering cross-section have been made by the CDMS [39] and XENON [90] experiments. These results exclude spin-independent cross-sections larger than approximately 5×10^{-8} pb for a 25–100 GeV WIMP or 2×10^{-7} pb $(m_\chi/500\text{ GeV})$ for a heavier WIMP.

With these results in mind, consider as an example a 300 GeV WIMP with an elastic scattering cross-section with nucleons which is largely spin-independent. With a cross-section near the CDMS bound, say 1×10^{-7} pb, we can determine from Fig. 25.2 the corresponding rates in a kilometre-scale neutrino telescope, such as IceCube. Sadly, we find that this cross-section yields less than 1 event per year for annihilations to $b\bar{b}$, about 2 events

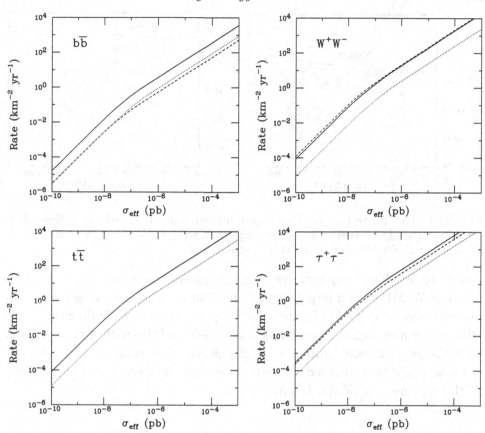

Fig. 25.2. The event rate in a kilometre-scale neutrino telescope such as IceCube as a function of the WIMP's effective elastic scattering cross-section in the Sun for a variety of annihilation modes. The effective elastic scattering cross-section is defined as $\sigma_{\text{eff}} = \sigma_{\chi\text{H,SD}} + \sigma_{\chi\text{H,SI}} + 0.07\,\sigma_{\chi\text{He,SI}} + 0.0005\,S(m_\chi/m_0)\,\sigma_{\chi0,\text{SI}}$ (see Eq. (25.5)). The dashed, solid and dotted lines correspond to WIMPs of mass 100, 300 and 1000 GeV, respectively. A 50 GeV muon energy threshold and an annihilation cross-section of 3×10^{-26} cm^3 s^{-1} have been adopted.

per year for annihilations to W^+W^- or $t\bar{t}$ and about 8 per year for annihilations to $\tau^+\tau^-$, none of which are sufficient to be identified over the background of atmospheric neutrinos by IceCube or other planned experiments (see Fig. 25.1). Clearly, WIMPs that scatter with nucleons mostly through spin-independent interactions are not likely to be detected with IceCube or other planned neutrino telescopes.

The same conclusion is not reached for the case of spin-dependent scattering, however. The strongest bounds on the WIMP–proton spin-dependent cross-section have been placed by the COUPP [251] and KIMS [1318] collaborations, but these constraints are approximately 7 orders of magnitude

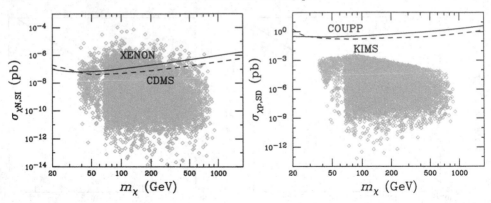

Fig. 25.3. The lightest neutralino's spin-independent (left) and spin-dependent (right) scattering cross-sections for a range of MSSM parameters. Also shown are the current limits from direct detection experiments.

less stringent than those corresponding to spin-independent couplings. As a result, a WIMP with a largely spin-dependent scattering cross-section with protons may be capable of generating large event rates in high-energy neutrino telescopes. Again considering a 300 GeV WIMP with a cross-section near the experimental limit, Fig. 25.2 suggests that rates as high as $\sim 10^6$ per year could be generated if purely spin-dependent scattering contributes to the capture rate of WIMPs in the Sun.

25.5 The case of neutralino dark matter

The lightest neutralino in R-parity conserving supersymmetric models is arguably the best motivated and certainly the most studied candiate for dark matter. The elastic scattering and annihilation cross-sections of the lightest neutralino depend on its various couplings, and on the mass spectrum of the Higgs bosons and superpartners. These couplings, in turn, depend on the neutralino's composition; in the minimal supersymmetric Standard Model (MSSM), the lightest neutralino can be any mixture of the bino, neutral wino and two neutral Higgsinos.

Spin-dependent, axial-vector, scattering of neutralinos with the quarks or gluons within a nucleon is made possible through the t-channel exchange of a Z-boson, or the s-channel exchange of a squark. Spin-independent scattering occurs at the tree level through s-channel squark exchange and t-channel Higgs exchange, and at the one-loop level through diagrams involving a loop of quarks and/or squarks.

The cross-sections for these processes can vary dramatically depending on the details of the supersymmetric model under consideration. It is clear from

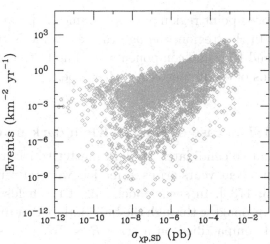

Fig. 25.4. The rate of events at a kilometre-scale neutrino telescope such as IceCube from dark matter annihilations in the Sun, as a function of the WIMP's spin-dependent elastic scattering cross-section. Each point shown is beyond the reach of present and near-future direct detection experiments.

Fig. 25.3, however, that the neutralino's spin-dependent cross-section can in some cases be considerably larger than its corresponding spin-independent interaction, which is desirable for the purposes of indirect detection. In particular, very large spin-dependent cross-sections ($\sigma_{\mathrm{SD}} \gtrsim 10^{-3}\,\mathrm{pb}$) are possible even in models with very small spin-independent scattering rates. Such a model would go easily undetected in all planned direct detection experiments, while still generating of the order of $\sim$1000 events per year at IceCube.

In Fig. 25.4, we demonstrate this by plotting the rate in a kilometre-scale neutrino telescope from WIMP annihilation in the Sun versus the WIMP's spin-dependent cross-section with protons. In this figure, each point shown is beyond the reach of present (as of 2008) and near-future direct detection experiments. We thus conclude that neutralinos may be observable by IceCube while remaining beyond the reach of current and next generation direct detection experiments.

A neutralino which has a large spin-dependent cross-section generally has a sizable coupling to the Z, and thus has a large Higgsino component. Writing the lightest neutralino as a superposition of the bino, wino and Higgsino states, $\chi^0 = f_B \tilde{B} + f_W \tilde{W} + f_{H_1} \tilde{H}_1 + f_{H_2} \tilde{H}_2$, the spin-dependent scattering cross-section through the exchange of a Z is proportional to the square of the quantity $|f_{H_1}|^2 - |f_{H_2}|^2$. Neutralinos with a few per cent Higgsino fraction or more are likely to be within the reach of IceCube [295; 1061].

This makes the focus point region of supersymmetric parameter space especially promising. In this region, the lightest neutralino is typically a strong mixture of bino and Higgsino components, leading to the prediction of hundreds or thousands of events per year at IceCube.

25.6 The case of Kaluza–Klein dark matter

One of the alternative candidates for dark matter which has received quite some attention in recent years arises in models with universal extra dimensions (UED) [533; 1730]. In such models, all of the fields of the Standard Model are able to propagate through the bulk of the extra-dimensional space, which is flat and compactified on a scale $R \sim \text{TeV}^{-1}$. As a consequence, each Standard Model particle is accompanied by a tower of Kaluza–Klein (KK) states. Such states appear as heavy versions of their Standard Model counterparts, with their extra mass being the result of momentum in the compactified dimension(s) of space.

The lightest KK particle (LKP) in this scenario can be naturally stable, in a way analogous to the way that the lightest superpartner is stabilized in R-parity conserving models of supersymmetry. In the case of one universal extra dimension, the most likely choice for the LKP is the first KK excitation of the hypercharge gauge boson, $B^{(1)}$. The thermal relic density of such a state can match the observed dark matter density over a range of masses; a few TeV $\gtrsim m_{B^{(1)}} \gtrsim 500$ GeV [478; 1268].

The range of elastic scattering cross-sections predicted for the LKP are quite challenging to reach with direct detection experiments, but are more favourable for detection using neutrino telescopes. The spin-independent LKP–nucleon cross-section, which is generated through the exchange of KK quarks and the Higgs boson, is rather small and typically falls within the range of 10^{-9} to 10^{-12} pb [1729], well beyond the sensitivity of current or upcoming direct detection experiments. The spin-dependent scattering cross-section for the LKP with a proton, however, is considerably larger and is given by [1729]

$$\sigma_{H,\text{SD}} = \frac{g_1^4\, m_p^2}{648\pi m_{B^{(1)}}^4 r_q^2}(4\Delta_u^p + \Delta_d^p + \Delta_s^p)^2$$

$$\approx 4.4 \times 10^{-6}\,\text{pb}\left(\frac{800\,\text{GeV}}{m_{B^{(1)}}}\right)^4\left(\frac{0.1}{r_q}\right)^2, \qquad (25.17)$$

where $r_q \equiv (m_{q^{(1)}} - m_{B^{(1)}})/m_{B^{(1)}}$ is fractional shift of the KK quark masses over the LKP mass, which is expected to be of the order of 10%. The Δs

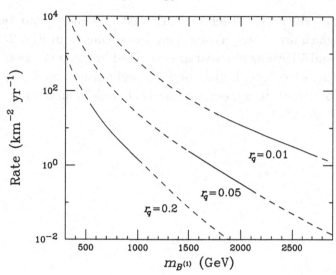

Fig. 25.5. The event rate in a kilometre-scale neutrino telescope as a function of the LKP mass [1122]. The three lines correspond to fractional mass splittings of the KK quarks of 20%, 5% and 1%. The solid sections of these lines reflect the approximate range in which it is possible to generate the observed thermal relic abundance. A 50 GeV muon energy threshold has been used.

parameterize the fraction of spin carried by each variety of quark within the proton.

In addition to this somewhat large spin-dependent scattering cross-section, the annihilation products of the LKP are very favourable for the purposes of generating observable neutrinos. In sharp contrast to neutralinos, approximately 60% of LKP annihilations generate a pair of charged leptons (20% to each type). Although most of the remaining 40% of LKP annihilations produce up-type quarks, about 4% generate neutrino pairs. The neutrino and tau lepton final states each contribute substantially to the event rate in a neutrino telescope.

The event rates in a kilometre-scale neutrino telescope from KK dark matter annihilating in the Sun [1122] are estimated in Fig. 25.5. There are competing effects which contribute to these results. In particular, a small mass splitting between the LKP and KK quarks yields a large spin-dependent elastic scattering cross-section, as seen in Eq. (25.17). On the other hand, KK quarks which are not much heavier than the LKP contribute to the freeze-out process and increase the range of LKP masses in which the thermal abundance matches the observed dark matter density. Over the range of masses for which the observed dark matter density can be thermally generated, which are shown as the solid line segments in Fig. 25.5, between roughly 0.5 and 50 events per year are expected in a kilometre-scale neutrino telescope such as IceCube. If non-thermal mechanisms, such as decays of KK gravitons, contribute to generating the LKP relic abundance, much larger rates may be possible.

26

Indirect dark matter detection with cosmic antimatter

Pierre Salati, Fiorenza Donato and Nicolao Fornengo

26.1 Production of antimatter in the Galaxy

The indirect detection of particle dark matter (DM) is based on the search for anomalous components in cosmic rays (CRs) due to the annihilation of DM pairs in the Galactic halo, on the top of the standard astrophysical production. These additional exotic components are potentially detectable at Earth as spectral distortions for the various cosmic radiations:

$$\chi + \chi \rightarrow q\bar{q}, W^+W^- \ldots \rightarrow \bar{p}, \bar{D}, e^+ \gamma, \nu. \tag{26.1}$$

Detection of the DM annihilation products has motivated the spectacular development of several new experimental techniques. They range from detectors on balloons or in space for the study of antimatter and gamma-rays, to large-area cosmic-ray and gamma-ray detectors on the ground, to neutrino telescopes underground for the study of the neutrino component. In the following, we will discuss in detail the antimatter component of DM indirect searches, namely antiprotons, antideuterons and positrons.

26.2 Propagation of antinuclei in the Galaxy

Whatever the mechanism responsible for their production, charged cosmic rays subsequently propagate through the Galactic magnetic field and are deflected by its irregularities: the Alfvén waves. In the regime where the magnetic turbulence is strong – which is the case for the Milky Way – Monte Carlo simulations [492] indicate that it is similar to space diffusion with a coefficient:

$$K(E) = K_0 \beta (\mathcal{R}/1 \text{ GV})^\delta, \tag{26.2}$$

Particle Dark Matter: Observations, Models and Searches, ed. Gianfranco Bertone. Published by Cambridge University Press. © Cambridge University Press 2010.

which increases as a power law with the rigidity $\mathcal{R} = pc/Ze$ of the particle, where p denotes momentum. In addition, because the scattering centres drift inside the Milky Way with a velocity $V_a \sim 20$ to 100 km s^{-1}, a second-order Fermi mechanism is responsible for some mild diffusive reacceleration. Its coefficient K_{EE} depends on the particle velocity β and total energy E and is related to the space diffusion coefficient $K(E)$ through:

$$K_{EE} = \frac{2}{9} V_a^2 \frac{E^2 \beta^4}{K(E)}. \qquad (26.3)$$

Finally, Galactic convection wipes cosmic rays away from the disk with a velocity $V_C \sim 5$ to 15 km s^{-1}. We can assume steady state for the various populations of particles and write the master equation for the space and energy distribution function $\psi = dn/dE$ as [285]:

$$\partial_z \left(V_C \, \psi \right) - K \, \Delta \psi + \partial_E \left\{ b^{\text{loss}}(E) \, \psi - K_{EE}(E) \, \partial_E \psi \right\} = q \left(\mathbf{x}, E \right). \qquad (26.4)$$

This equation applies to any charged species – nuclei, protons, antiprotons or positrons – as long as the rates for production q and energy loss $b^{\text{loss}}(E)$ are properly accounted for.

The solution of the master equation (26.4) has been investigated in depth, and several different techniques lead to very similar fluxes at the Earth [1402; 1822]. One possibility is a completely numerical solution. This is the way followed in the Galprop model [1819; 1823], which, thanks to a realistic gas distribution, can calculate gamma-rays in addition to charged species [1589]. A completely different approach to the CR transport through the diffusive volume relies on the calculation of the Green function of Eq. (26.4), which describes the probability for a CR that is produced at a location $\mathbf{x}$ with energy E_S to be detected at the Earth with a degraded energy E. The Green function is then integrated over the diffusive volume and the energy range. This is the approach chosen to propagate positrons [644; 645; 1483].

The Bessel expansion method, based on the cylindrical symmetry of the DH and on approximate values for the ISM (not relevant for charged CR propagation), permits a 2D fully analytical model. Numerical solution is required only for the diffusion in energy space. This is the model developed in [705; 1401; 1402; 1403] and detailed more extensively in the rest of this chapter. Very similar results for the propagation of stable primary and secondary nuclei have been obtained with the modified weighted slab technique [1178].

According to this approach the region of the Galaxy inside which cosmic rays diffuse – the so-called diffusive halo (DH) – is pictured as a thick disk which matches the circular structure of the Milky Way as shown in Fig. 26.1.

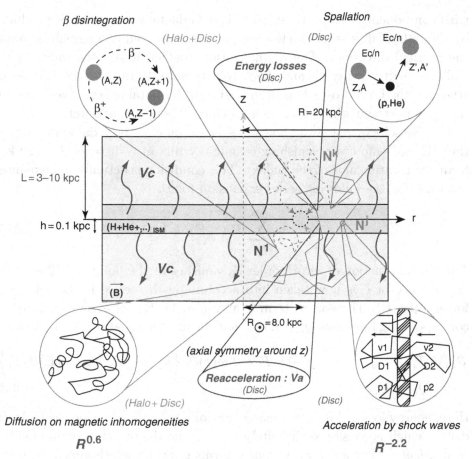

Fig. 26.1. Schematic edge-on view of the Milky Way as seen by a cosmic-ray physicist. The stellar and gaseous disk is sandwiched between two thick layers which contain turbulent magnetic fields. After having been accelerated by supernova-driven shock waves or produced by DM species annihilating in the Galactic halo, cosmic rays diffuse on magnetic inhomogeneities and are wiped away by a Galactic wind with velocity V_C. They can lose energy and are also mildly subject to diffusive reacceleration. The former process is by far the dominant one in the case of electrons and positrons. This diagram has been borrowed from the review [1402].

The Galactic disk of stars and gas, where primary cosmic rays are accelerated, lies in the middle. It extends radially 20 kpc from the centre and has a half-thickness h of 100 pc. Confinement layers where cosmic rays are trapped by diffusion lie above and beneath this thin disk of gas. The intergalactic medium starts at the vertical boundaries $z = \pm L$ as well as beyond a radius of $r = R \equiv 20$ kpc. The half-thickness L of the diffusive halo is not known, and reasonable values range from 1 to 15 kpc. The diffusion coefficient K is the same everywhere whereas the convective velocity is exclusively vertical with component $V_C(z) = V_C \, \text{sign}(z)$. This Galactic wind, which is produced by the bulk of disk stars like the Sun, drifts away from its progenitors along the vertical directions. The normalization coefficient K_0, the index δ, the Galactic drift velocity V_C and the Alfvén velocity V_a may be determined by studying the boron-to-carbon ratio (B/C), which is quite sensitive to cosmic ray transport and which may be used efficiently as a constraint.

The Bessel expansion method takes advantage of the axial symmetry of the DH and enforces a vanishing cosmic-ray flux at a distance $R = 20$ kpc from the rotation axis of the Galaxy. This condition is actually implemented naturally by the following series expansion for ψ:

$$\psi(r, z, E) = \sum_{i=1}^{+\infty} P_i(z, E) \, J_0(\alpha_i \, r/R). \tag{26.5}$$

The Bessel function of zeroth order J_0 vanishes at the points α_i. The radial dependence of ψ is now taken into account by the set of its Bessel transforms $P_i(z, E)$. The source term q may also be Bessel expanded into the corresponding functions $Q_i(z, E)$ so that the master equation (26.4) becomes

$$\partial_z(V_C P_i) - K\partial_z^2 P_i + K\left\{\frac{\alpha_i}{R}\right\}^2 P_i + 2h\delta(z)\partial_E\left\{b^{\text{loss}}(E)P_i - K_{EE}(E)\partial_E P_i\right\}$$
$$= Q_i(z, E). \tag{26.6}$$

Here, energy loss and diffusive reacceleration are confined inside the Galactic disk – which is considered infinitely thin, hence the presence of an effective term $2h\,\delta(z)$. The form of the source terms $Q_i(z, E)$ which appear in equation (26.6) depends on the nature of the cosmic-ray particle.

26.3 Antiprotons in cosmic rays

In the case of antiprotons, the following mechanisms can in principle contribute to the source term in the transport equation (26.4):

- The spallation of high-energy primary nuclei impinging on the atoms of the interstellar medium inside the Galactic disk produces secondary antiprotons.
- The annihilation of DM candidate particles throughout the Milky Way halo generates primary antiprotons. Notice that WIMP annihilations take place all over the diffusive halo.
- Tertiary antiprotons result from the inelastic and non-annihilating interactions with a nucleon at rest. The energy transfer may be sufficient to excite it as a Δ resonance. This mechanism redistributes antiprotons toward

lower energies and flattens their spectrum [297]. This yields the source term:

$$q_{\bar{p}}^{\text{ter}}(r, E_{\bar{p}}) = \int_{E_{\bar{p}}}^{+\infty} \frac{\mathrm{d}\sigma_{\bar{p}}\,\mathrm{H} \to \bar{p}\,X}{\mathrm{d}E_{\bar{p}}}(E_{\bar{p}}' \to E_{\bar{p}})\, n_{\mathrm{H}}\, \beta_{\bar{p}}'\, \psi_{\bar{p}}(r, E_{\bar{p}}')\, \mathrm{d}E_{\bar{p}}'$$
$$- \sigma_{\bar{p}}\,\mathrm{H} \to \bar{p}\,X(E_{\bar{p}})\, n_{\mathrm{H}}\, \beta_{\bar{p}}\, \psi_{\bar{p}}(r, E_{\bar{p}}), \tag{26.7}$$

where the inelastic and non-annihilating differential cross-section in this expression can be approximated by:

$$\frac{\mathrm{d}\sigma_{\bar{p}}\,\mathrm{H} \to \bar{p}\,X}{\mathrm{d}E_{\bar{p}}} = \frac{\sigma_{\bar{p}}\,\mathrm{H} \to \bar{p}\,X}{T_{\bar{p}}'}. \tag{26.8}$$

The initial antiproton kinetic energy is denoted by $T_{\bar{p}}'$. In order to take into account elastic scatterings on helium, one simply has to replace the hydrogen density by $n_{\mathrm{H}} + 4^{2/3}\, n_{\mathrm{He}}$.

- Antiprotons may also annihilate on interstellar H and He. This leads to a negative source term $-\Gamma_{\bar{p}}^{\text{ann}}\,\psi$, where the annihilation rate $\Gamma_{\bar{p}}^{\text{ann}}$ is defined as

$$\Gamma_{\bar{p}}^{\text{ann}} = \sigma_{\bar{p}\,\mathrm{H}}^{\text{ann}}\, \beta_{\bar{p}}\, n_{\mathrm{H}} + \sigma_{\bar{p}\,\mathrm{He}}^{\text{ann}}\, \beta_{\bar{p}}\, n_{\mathrm{He}}. \tag{26.9}$$

The annihilation cross-section $\sigma_{\bar{p}\,\mathrm{H}}^{\text{ann}}$ can be borrowed from [1835; 1836] and multiplied by a factor of $4^{2/3} \sim 2.5$, taking into account the higher geometric cross-section, to get $\sigma_{\bar{p}\,\mathrm{He}}^{\text{ann}}$. The average hydrogen n_{H} and helium n_{He} densities in the Galactic disk are respectively set equal to 0.9 and $0.1\,\mathrm{cm}^{-3}$.

26.3.1 Secondary antiprotons

Secondary antiprotons are produced by CR spallation on the interstellar medium (ISM). They represent the background when searching for small contributions coming from exotic sources, such as signals from DM annihilation.

The rate for the production of secondary antiprotons takes the following form:

$$q_{\bar{p}}^{\text{sec}}(r, E_{\bar{p}}) = \int_{E_{p,\alpha}^0}^{+\infty} n_{\mathrm{H},\alpha} \times \beta_{p,\alpha}\, \psi_{p,\alpha}(r, E_{p,\alpha}) \times \mathrm{d}E_{p,\alpha} \times \frac{\mathrm{d}\sigma}{\mathrm{d}E_{\bar{p}}}(E_{p,\alpha} \to E_{\bar{p}}),$$
$$\tag{26.10}$$

for interactions between cosmic-ray protons and alpha-particles, and hydrogen and helium nuclei in the ISM.

Equation (26.4) may be solved according to the method outlined in Appendix B of [704] for an antiproton source located only in the Galactic disk, as is the case for secondary antiprotons.

Some approximations may be at hand: in particular, setting the energy loss rate b^{loss} and the energy diffusion coefficient K_{EE} equal to zero does not sizably affect the solution of the diffusion equation.

As for the cosmic-ray proton and helium fluxes, they can be borrowed from [704], where a fit to high-energy (>20 GeV/n) data is proposed. A more recent parameterization of interstellar (IS) fluxes has been derived in [703].

Once we have the IS fluxes of antiprotons at the Sun's position in the Galaxy, we have to further propagate them inside the heliosphere, where the cosmic-ray particles which eventually reach the Earth are affected by the presence of the solar wind. We model the effect of solar modulation by adopting the force field approximation of the full transport equation [1567]. In this model, the top-of-atmosphere (TOA) flux for a cosmic species Φ^{TOA} is obtained as:

$$\frac{\Phi^{\mathrm{TOA}}(E^{\mathrm{TOA}})}{\Phi^{\mathrm{IS}}(E^{\mathrm{IS}})} = \left(\frac{p^{\mathrm{TOA}}}{p^{\mathrm{IS}}}\right)^2, \tag{26.11}$$

where E and p denote the total energies and momenta of interstellar and TOA antiprotons, which are related by the energy shift:

$$E^{\mathrm{TOA}} = E^{\mathrm{IS}} - \phi, \tag{26.12}$$

where the parameter ϕ is determined by fits to cosmic-ray data. A value $\phi = 500$ MV is indicative of periods of minimal solar activity.

The secondary IS $\bar{p}$ flux is displayed in the left panel of Fig. 26.2 along with the data demodulated according to the force-field prescription. We use either the DTUNUC [704] $\bar{p}$ production cross-sections (solid line) or those discussed in refs. [445; 730] (dashed line). The differences between the two curves illustrate the uncertainty related to the production cross-sections, as emphasized in [704], where a careful and conservative analysis within the DTUNUC simulation settled a nuclear uncertainty of $\sim 25\%$ in the energy range $0.1 - 100$ GeV. The conclusion is similar here, although the two sets of cross-sections differ mostly at low energy. In the right panel, along with the demodulated $\bar{p}/p$ data, we show the curves bounding the propagation uncertainty on the $\bar{p}$ calculation based on either the DTUNUC [704] $\bar{p}$ production cross-sections (solid lines) or those borrowed from [445] (dashed lines). The uncertainty arising from propagation is comparable to the nuclear one [704]. The detailed calculation of that secondary component [1403] has required the determination of the propagation–diffusion parameters that are consistent

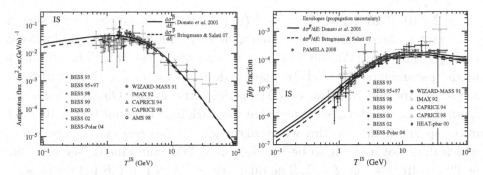

Fig. 26.2. Left panel: interstellar (IS) antiproton flux for the B/C best fit model and two parameterizations of the production cross-section. Right panel: propagation uncertainty envelopes of the IS $\bar{p}/p$ ratio for the same production cross-sections as in the left panel. All data are demodulated using the force-field approximation: AMS 98 [34], IMAX 92 [1454], CAPRICE 94 [375], WIZARD-MASS 91 [227], CAPRICE 98 [376], BESS 93 [1462], BESS 95+97 [1535], BESS 98 [1370], BESS 99 and 2000 [134], BESS 2002 [1059], BESS Polar [17], WIZARD-MASS 1 [1113], IIEAT-$\bar{p}$ [239] and PAMELA [23]

with the B/C data [1401]. By varying those parameters over the entire range allowed by the cosmic-ray nuclei measurements, the theoretical uncertainty on the antiproton secondary flux has been found to be 9% from 100 MeV to 1 GeV. It reaches a maximum of 24% at 10 GeV and decreases to 10% at 100 GeV.

From Fig. 26.2, it is manifest that the secondary contribution alone explains experimental data in the whole energetic range. It is not necessary to invoke an additional component to the standard astrophysical one.

26.3.2 Antiprotons from DM annihilation

The antiproton signal from annihilating DM particles leads to a primary component directly produced throughout the DH. The differential rate of production per unit volume and time is a function of space coordinates and antiproton kinetic energy $T_{\bar{p}}$. It is defined as:

$$q_{\bar{p}}^{\mathrm{DM}}(r, z, T_{\bar{p}}) = \xi^2 \langle \sigma v \rangle \, g(T_{\bar{p}}) \left(\frac{\rho_\chi(r, z)}{m_\chi} \right)^2, \qquad (26.13)$$

where $\langle \sigma v \rangle$ denotes the average over the Galactic velocity distribution function of the WIMP pair annihilation cross-section σ_{ann} multiplied by the relative velocity v. For a relic CDM particle able to explain the observed amount of cosmological dark matter [729; 1100; 1266] its value falls in the range 2×10^{-26} to 3×10^{-26} cm^3 s^{-1} (unless special situations occur, such as dominant p-wave annihilation, dominant co-annihilations or modified cosmology). For any DM candidate, σ_{ann} is nevertheless calculated from the model parameters. The WIMP mass is denoted by σ_{ann} and $\rho_\chi(r, z)$ is the mass distribution function of DM particles inside the Galactic halo. The quantity ξ parameterizes the fact that the dark halo may not be totally made of the species under scrutiny when this candidate possesses a relic abundance which does not allow it to be the dominant DM component (see e.g. [402] or [387]). In this case $\xi < 1$. The quantity $g(T_{\bar{p}})$ in Eq. (26.13) denotes the antiproton differential spectrum per annihilation event, defined as:

$$
\frac{\mathrm{d}N_{\bar{p}}}{\mathrm{d}E_{\bar{p}}} = \sum_{\mathrm{F},h} B_{\chi h}^{(\mathrm{F})} \frac{\mathrm{d}N_{\bar{p}}^h}{\mathrm{d}E_{\bar{p}}} \, .
\tag{26.14}
$$

The annihilation into a quark or a gluon h is realized through the various final states F with branching ratios $B_{\chi h}^{(\mathrm{F})}$. Quarks or gluons may in fact be directly produced when a WIMP pair annihilates, or they may alternatively result from the intermediate production of Higgs bosons or gauge bosons. Each quark or gluon h then generates jets whose subsequent fragmentation and hadronization yield an antiproton energy spectrum $\mathrm{d}N_{\bar{p}}^h/\mathrm{d}E_{\bar{p}}$. The single production spectra are usually evaluated within Monte Carlo simulations of electroweak annihilation events [1764]. For details on the calculation of pure state $\mathrm{d}N_{\bar{p}}^h/\mathrm{d}E_{\bar{p}}$ and of $g(T_{\bar{p}})$ in a minimal supersymmetric extension of the Standard Model (MSSM), see the Appendix of ref. [700].

The distribution of DM inside galaxies is a much-debated issue. Different analyses of rotational curves observed for several types of galaxies strongly favour a cored dark matter distribution, flattened towards the central regions (ref. [706] and references therein). On the other side, many collisionless cosmological N-body simulations in ΛCDM models are in good agreement among themselves [1083], but for the very central regions some resolution issues remain open. It has been recently stressed that asymptotic slopes may not be reached at all at small scales [1002; 1426; 1503; 1643; 1810]. However, it is not clear whether the central cusp is steepened or flattened when the baryonic distribution is taken into account (e.g. [1366; 1393]). A detailed discussion may be found in Chapters 2, 3 and 5 of this volume. We wish to stress that the DM distribution is a crucial ingredient for the indirect signal into gamma-rays or neutrinos, while for antimatter it has been shown to be less relevant [646; 700], since antimatter diffuses and therefore the signal at Earth is more local, less dependent on the inner structure of the Galaxy than for gamma-rays and neutrinos.

Table 26.1. *Parameters in Eq. (26.15) for different halo models.*

Halo model	α	β	γ	ρ_s ($10^6\,M_\odot$ kpc^{-3})	r_s (kpc)
Cored isothermal [165]	2	2	0	7.90	4
NFW 97 [1502]	1	3	1	5.38	21.75
Moore 04 [666]	1	3	1.16	2.54	32.62

The scale radius r_s and density ρ_s are strongly correlated with the virial mass of the Galaxy [744] and the values are borrowed from [876] for the Milky Way. When the DM distribution is cuspy ($\gamma \geq 1$) the divergence at the Galactic centre is smoothed according to the prescription of [445].

To be definite, we will consider here a spherical DM Galactic distribution with a dependence on galactocentric distance r parameterized by:

$$\rho(r) = \rho_s \left(\frac{r_s}{r}\right)^\gamma \left\{ 1 + \left(\frac{r}{r_s}\right)^\alpha \right\}^{(\gamma-\beta)/\alpha}. \tag{26.15}$$

The different profiles of Table 26.1 basically span the whole range of reasonable halo models with respect to indirect dark matter detection prospects.

The solution of the transport equation for a generic source term $q^{\mathrm{prim}}(r, z, E)$ within the Bessel expansion has been derived in [220]:

$$N_i^{\bar{p},\mathrm{prim}}(z) = \exp\left(\frac{V_c(|z|-L)}{2K}\right) \frac{y_i(L)}{A_i \sinh(S_i L/2)}$$

$$\left[\cosh(S_i z/2) + \frac{(V_c + 2h\Gamma_{\bar{p}}^{ine})}{KS_i A_i} \sinh(S_i z/2)\right] - \frac{y_i(z)}{KS_i},$$

$$\tag{26.16}$$

where:

$$y_i(z) = 2 \int_0^z \exp\left(\frac{V_c}{2K}(z-z')\right) \sinh\left(\frac{S_i}{2}(z-z')\right) q_i^{\mathrm{prim}}(z')\mathrm{d}z' \tag{26.17}$$

and the quantities S_i and A_i are defined as:

$$S_i \equiv \left\{\frac{V_c^2}{K^2} + 4\frac{\zeta_i^2}{R^2}\right\}^{1/2} \quad \text{and} \quad A_i(E) \equiv 2\,h\,\Gamma_{\bar{p}}^{ine} + V_c + K\,S_i \coth\left\{\frac{S_i L}{2}\right\}. \tag{26.18}$$

In particular, at $z = 0$ where fluxes are measured, we have:

$$N_i^{\bar{p},\mathrm{prim}}(0) = \exp\left(\frac{-V_c L}{2K}\right) \frac{y_i(L)}{A_i \sinh(S_i L/2)}. \tag{26.19}$$

Table 26.2. *Typical combinations of diffusion parameters that are*
compatible with the B/C analysis [1401]. As shown in [700], these
propagation models correspond respectively to minimal, medium and
maximal primary antiproton fluxes.

Case	δ	K_0 (kpc^2 Myr^{-1})	L (kpc)	V_C (km s^{-1})	V_a (km s^{-1})
MIN	0.85	0.0016	1	13.5	22.4
MED	0.70	0.0112	4	12	52.9
MAX	0.46	0.0765	15	5	117.6

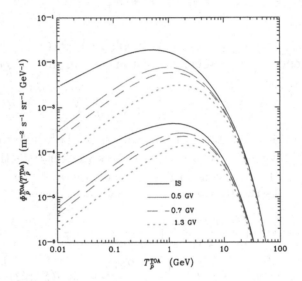

Fig. 26.3. Top-of-atmosphere antiproton fluxes as a function of the antiproton kinetic energy for the $m_\chi = 100$ GeV reference case. The upper (lower) set of curves refer to the maximal (minimal) set of astrophysical parameters. Solid curves show the interstellar fluxes. Broken curves show the effect of solar modulation at different periods of solar activity: $\phi = 500$ MV (long dashed), $\phi = 700$ MV (short dashed), $\phi = 1300$ MV (dotted).

The three propagation models featured in Table 26.2 have been drawn from [700]. The MED configuration provides the best fit to the B/C measurements whereas the MIN and MAX models point respectively to the minimal and maximal allowed antiproton fluxes which can be produced by WIMP annihilation.

Figure 26.3 shows the TOA antiproton fluxes for a reference WIMP with $m_\chi = 100$ GeV, 2.3×10^{-26} cm^3 s^{-1}, pure $\bar{b}b$ annihilation final state and for the maximal and minimal sets of astrophysical parameters. The variation of the astrophysical parameters induces a much larger uncertainty on the

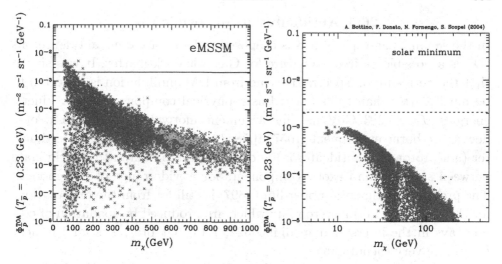

Fig. 26.4. Left: The scatter plot shows the antiproton flux at solar minimum from neutralino annihilation calculated at $T_{\bar{p}} = 0.23$ GeV, as a function of the neutralino mass for a generic scan in a low-energy MSSM and for the MED set of astrophysical parameters [699]. Crosses refer to cosmologically dominant neutralinos ($0.05 \leq \Omega_\chi h^2 \leq 0.3$); dots refer to subdominant relic neutralinos ($\Omega h^2 < 0.05$). Right: The same as in the left panel, but calculated for a scan in a supersymmetric framework where gaugino non-universality is not assumed, and therefore lighter neutralinos are present [398]. The astrophysical parameters are set at the MIN case, and solar modulation is at solar minimum.

primary than on the secondary flux: in the first case, the uncertainty reaches two orders of magnitude for energies $T_{\bar{p}} \lesssim 1$ GeV, while in the second case it never exceeds 25%. The figure shows that solar modulation has the effect of depleting the low-energy tail of the antiproton flux. The effect is clearly more pronounced for periods of high solar activity, when the solar wind is stronger.

Predictions for antiprotons in various realizations of supersymmetric theories have been performed (see for example [161; 398; 401; 699; 736; 847]).

In Fig. 26.4 we show the predicted antiproton flux at kinetic energiy $T_{\bar{p}} = 0.23$ GeV, for two different supersymmetric models: the left panel [699] refers to a low-energy realization of the MSSM, while the right panel [699] stands for a scan of a supersymmetric scheme where gaugino non-universality is not assumed and therefore light neutralinos are present [397; 409]. In both cases the region below the horizontal line denotes the amount of antiprotons, in excess of the secondary component, which can be accommodated at $T_{\bar{p}} = 0.23$ GeV in order not to exceed the observed flux, as measured by BESS. All the points of the scatter plot that lie below the horizontal black line are therefore compatible with observations.

26.4 Antideuterons in cosmic rays

In the seminal paper [701], it was proposed to look for cosmic antideuterons ($\bar{\text{D}}$) as a possible indirect signature for Galactic dark matter. It was shown that the antideuteron spectra deriving from DM annihilation is expected to be much flatter than the standard astrophysical component at low kinetic energies, $T_{\bar{\text{D}}} \lesssim 2\text{--}3$ GeV/n. This argument motivated the proposal of a new space-borne experiment [1057; 1058; 1478] looking for cosmic antimatter (antiproton and antideuteron) and having the potential to discriminate between standard and exotic components for a wide range of DM models. The present experimental upper limit [897] is still far from the expectations for the secondary antideuteron flux which are produced by spallation of cosmic rays in the interstellar medium [522; 730], but prospects for the near future are very encouraging.

Once the astrophysical framework for the transport of (anti)nuclei is set, the calculation of the antideuteron flux rests on the $\bar{\text{D}}$ specificities regarding the source term (i.e. primary or secondary) and its nuclear interactions (right-hand side of Eq. (26.4)). The tertiary term (see Eq. (26.7)) requires specific inelastic non-annihilating cross-sections which are detailed in the Appendix of ref. [698].

The production of cosmic antideuterons is based on the fusion process of a $\bar{p}$ and $\bar{n}$ pair. One simple but powerful treatment of the fusion of two or more nucleons is based on the so-called coalescence model, which, despite its simplicity, is able to reproduce remarkably well the available data on light nuclei and antinuclei production in different kinds of collisions. In the coalescence model, the momentum distribution of the (anti)deuteron is proportional to the product of the (anti)proton and (anti)neutron momentum distribution [482; 1712]. That function depends on the difference $\Delta_{\mathbf{k}}$ between (anti)nucleon momenta. It is strongly peaked around $\Delta_{\mathbf{k}} \simeq \mathbf{0}$ (compare the minimum energy to form a $\bar{\text{D}}$, i.e. $4m_p$, with the binding energy ~ 2.2 MeV), so that

$$\mathbf{k}_{\bar{p}} \simeq \mathbf{k}_{\bar{n}} \simeq \frac{\mathbf{k}_{\bar{\text{D}}}}{2} \ . \tag{26.20}$$

The $\bar{\text{D}}$ density in momentum space is thus written as the $\bar{p}$ density times the probability of finding an $\bar{n}$ within a sphere of radius p_0 around $\mathbf{k}_{\bar{p}}$ (see for example ref. [1042]):

$$\gamma \frac{\mathrm{d}\mathcal{N}_{\bar{\text{D}}}}{\mathrm{d}\mathbf{k}_{\bar{\text{D}}}} = \frac{4\pi}{3} p_0^3 \cdot \gamma \frac{\mathrm{d}\mathcal{N}_{\bar{p}}}{\mathrm{d}\mathbf{k}_{\bar{p}}} \cdot \gamma \frac{\mathrm{d}\mathcal{N}_{\bar{n}}}{\mathrm{d}\mathbf{k}_{\bar{n}}} \ . \tag{26.21}$$

The coalescence momentum p_0 is a free parameter constrained by data on hadronic production. In [730], a large set of data is used, including many

pA reactions (see their Table I and references therein). This leads to an estimate of $p_0 = 79$ MeV. At LEP energies, (anti)deuteron production occurs through e^+e^- annihilations into $q\bar{q}$ pairs, an electroweak mechanism similar to the $\bar{D}$ production in DM annihilation reactions. Based on theoretical arguments, it has been argued [1042] that the antideuteron yields in e^+e^- reactions should be smaller than in hadronic reactions. However, the ALEPH collaboration [1700] has found that this theoretical prediction (see Fig. 5 in ALEPH paper) underestimates their measured $\bar{D}$ inclusive cross-section. They derive (see their Fig. 6) a value $B_2 = 3.3 \pm 0.4 \pm 0.1 \times 10^{-3}$ GeV^2 at the Z resonance, which translates into $p_0 = 71.8 \pm 3.6$ MeV, very close to the $p_0 = 79$ MeV derived for the hadronic production. From a rough estimate of the fusion process, one can estimate that the antideuteron fluxes are a factor of 10^4 lower than the antiproton ones. We will see in the following, and by comparing the results of the previous section, that this result is recovered for both the primary and the secondary $\bar{D}$ fluxes at Earth.

26.4.1 Secondary antideuterons

The secondary $\bar{D}$ flux is the sum of the six contributions corresponding to p, He and $\bar{p}$ cosmic-ray fluxes impinging on H and He IS gas (other reactions are negligible [730]). Contributions to the $\bar{D}$ flux from $\bar{p} + $ H and $\bar{p} + $ He reactions are evaluated using the $\bar{p}$ flux calculated in the same run. The production cross-sections for these specific processes are those given in ref. [730]. The solution to the propagation equation has the same expression as for secondary antiprotons [698].

The different contributions to the total secondary antideuteron flux, calculated for the best fit propagation configuration (the MED one in Table 26.2), are shown in the left panel of Fig. 26.5. As expected, the dominant production channel is the one from p–H collisions, followed by the one from cosmic protons on IS helium (p–He). As shown in ref. [730], the $\bar{p} + $ H channel is dominant at low energies, and negligible beyond a few GeV/n. The effects of energy losses, reacceleration and tertiaries add up to replenish the low-energy tail. The maximum of the total flux reaches the value of 2×10^{-7} particles (m^2 s sr GeV/n)$^{-1}$ at 3–4 GeV/n. At 100 MeV/n it is decreased by an order of magnitude, thus preserving an interesting window for possible exotic contributions characterized by a flatter spectrum (see next section).

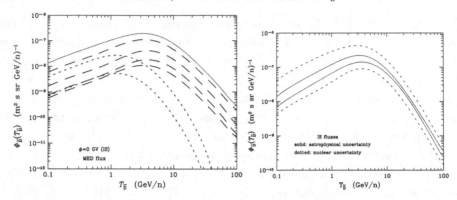

Fig. 26.5. Left panel: Contribution of all nuclear channels to the $\bar{D}$ secondary flux. Dashed lines, from top to bottom, refer to: $p + H$, $p + He$, $He+H$, $He+He$. Dotted lines, from top to bottom, stand for: $\bar{p} + H$, $\bar{p} + He$. Solid line: sum of all the components. Right panel: Dominant uncertainties on the interstellar secondary $\bar{D}$ flux. Solid lines: propagation uncertainty band. Dotted lines: nuclear uncertainty band.

For the determination of the propagation uncertainties, one can calculate the secondary antideuteron flux for all the propagation parameter combinations providing an acceptable fit to stable nuclei [1401]. The resulting envelope for the secondary antideuteron flux is presented in Fig. 26.5(left). The solid lines delimit the uncertainty band due to the degeneracy of the propagation parameters: at energies below 1–2 GeV/n, the uncertainty is 40–50% around the average flux, while at 10 GeV/n it decreases to $\sim 15\%$. This behaviour is analogous to that obtained for $\bar{p}$ [704] and is easily understood. The degenerate transport parameters combine to give the same grammage in order to reproduce the B/C ratio. Indeed, the grammage crossed by C to produce the secondary species B is also crossed by p and He to produce the secondary $\bar{p}$ and $\bar{D}$. In short, a similar propagation history associated with a well-constrained B/C ratio explains the small uncertainty. The possible nuclear uncertainty can arise from two different sources. The first one is directly related to the elementary production process and has been estimated to be conservatively $\pm 50\%$ [698]. Second, there is the uncertainty in the coalescence momentum p_0. Using an independent model (i.e. different from the coalescence scheme) for $\bar{D}$ production, ref. [730] found that, conservatively, the $\bar{D}$ background was certainly no more than twice the flux calculated with $p_0 = 79$ MeV. The dotted lines in Fig. 26.5 (right panel) take into account the sum of all the possible uncertainties of nuclear source [698]. At the lowest energies the flux is uncertain by almost one order of magnitude, at 100 GeV/n by a factor of 4.

26.4.2 Antideuterons from DM annihilation

The source term for primary $\bar{\mathrm{D}}$ has the same form as the one for primary antiprotons in Eq. (26.13) ($\bar{p} \to \bar{\mathrm{D}}$). The production of antideuterons from the pair-annihilation of supersymmetric dark matter particles in the halo of our Galaxy was proposed in [701], and subsequently discussed in [162] also for universal extra-dimension, Kaluza–Klein and warped extra-dimension dark matter models. In ref. [698] primary antideuterons in different super-symmetric scenarios have been calculated in the full 2D propagation model, with a thorough estimation of all the possible uncertainty sources. As previously discussed (see Section 26.4) the production of a $\bar{\mathrm{D}}$ relies on the availability of a $\bar{p}$–$\bar{n}$ pair in a single DM annihilation. As in ref. [701], we assume that the probability of forming an antiproton (or an antineutron) with momentum $\mathbf{k}_{\bar{p}}$ ($\mathbf{k}_{\bar{n}}$) is essentially isotropic:

$$\frac{\mathrm{d}N_{\bar{p}}}{\mathrm{d}F_{\bar{p}}}(\chi + \chi \to \bar{p} + \cdots) = 4\pi\, k_{\bar{p}}\, E_{\bar{p}}\, \mathcal{F}_{\bar{p}}(\sqrt{s} = 2m, E_{\bar{p}}) . \tag{26.22}$$

Applying the factorization-coalescence scheme discussed above leads to the antideuteron differential multiplicity

$$\frac{\mathrm{d}N_{\bar{\mathrm{D}}}}{\mathrm{d}E_{\bar{\mathrm{D}}}} = \left(\frac{4\,p_0^3}{3\,k_{\bar{\mathrm{D}}}}\right) \cdot \left(\frac{m_{\bar{\mathrm{D}}}}{m_{\bar{p}}^2}\right) \cdot \sum_{\mathrm{F},\mathrm{h}} B_{\chi\mathrm{h}}^{(\mathrm{F})} \left\{\frac{\mathrm{d}N_{\bar{p}}^{\mathrm{h}}}{\mathrm{d}E_{\bar{p}}}\left(E_{\bar{p}} = \frac{E_{\bar{\mathrm{D}}}}{2}\right)\right\}^2 . \tag{26.23}$$

We assume, as discussed in Section 26.4, that the same value of the coalescence momentum $p_0 = 79$ MeV holds as for hadronic reactions.

The solution to the diffusion equation is provided by Eq. (26.19) where the specific cross-sections have been reviewed, for instance, in ref. [698]. In the left panel of Fig. 26.6 the secondary $\bar{\mathrm{D}}$ flux for the median configuration of Table 26.2 is plotted alongside the primary flux from $m_\chi = 50$ GeV, calculated for the maximal, median and minimal propagation scenarios. The present BESS upper limit on the (negative) antideuteron search [897] is at a level of 2×10^{-4} (m^2 s sr GeV/n)$^{-1}$. We also plot the estimated sensitivities of the gaseous antiparticle spectrometer GAPS on a long-duration balloon flight (LDB) and an ultra-long duration balloon mission (ULDB) [1057; 1058; 1258], and of AMS-02 for three years of data taking. The prospects for exploring a part of the region where DM annihilation is mostly expected (i.e. the low-energy tail) are very promising. If one of these experiments measures at least one antideuteron, it will be a clear signal of an exotic contribution to the cosmic antideuterons.

The discrimination power between primary and secondary $\bar{\mathrm{D}}$ flux may be deduced from the right panel of Fig. 26.6. The ratio of the primary to

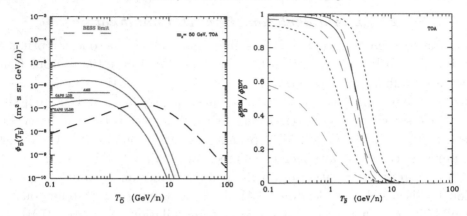

Fig. 26.6. Left panel: TOA primary (solid lines) and secondary (dashed line) antideuteron fluxes, modulated at solar minimum. The signal is derived for a m_χ=50 GeV WIMP and for the three propagation models of Table 26.2. The secondary flux is shown for the median propagation model. The upper dashed horizontal line shows the current BESS upper limit on the search for cosmic antideuterons. The three horizontal solid lines are the estimated sensitivities for (from top to bottom): AMS-02, GAPS on long (LDB) and ultra-long (ULDB) duration balloon flights [1057; 1058; 1258]. Right panel: Ratio of the primary to total (signal+background) TOA antideuteron flux. Solid curve refers to a WIMP mass of $m_\chi = 50$ GeV and for the MED propagation parameters. Dotted lines show the MAX (upper) and MIN (lower) cases. Dashed lines refer to the MED propagation parameters and different masses, which are (from top to bottom): m_χ=10, 100, 500 GeV.

total TOA $\bar{\rm D}$ flux is plotted as a function of the kinetic energy per nucleon, for the three representative propagation models and different WIMP masses (the annihilation cross-section is again fixed at the reference value). This ratio stays higher than 0.7 for $T_{\bar{\rm D}} < 1$ GeV/n except for $m_\chi = 500$ GeV. For propagation models with $L \gtrsim 4$ kpc – which is a very reasonable expectation – this ratio is at least 0.9 for masses below 100 GeV. Increasing the WIMP mass, we must descend to lower energies in order to maximize the primary-to-secondary ratio. However, for an $m_\chi = 500$ GeV WIMP we still have a 50–60% of DM contribution in the 0.1–0.5 GeV/n range. Of course, the evaluation of the theoretical uncertainties presented in this chapter must be kept in mind while comparing with real data. Figure 26.6 clearly states that the antideuteron indirect DM detection technique is probably the most powerful one for low and intermediate WIMP-mass haloes.

The uncertainties due to propagation are similar to the ones quoted for primary antiprotons [698]. At the lowest energies of hundreds of MeV/n the total uncertainty reaches almost two orders of magnitude, while at energies above 1 GeV/n it is about a factor of 30.

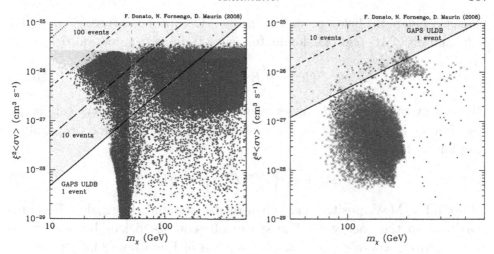

Fig. 26.7. Left: GAPS ultra-long duration balloon flight (ULDB) reach compared with predictions for neutralino dark matter in low-energy supersymmetric models, shown in the plane of effective annihilation cross-section $\xi^2 \langle \sigma_{ann} v \rangle_0$ vs. neutralino mass m_χ [698]. The solid, long-dashed and short-dashed lines show our estimate for the capability of GAPS ULDB of measuring 1, 10 and 100 events, respectively, for the MED propagation model. The scatter plot reports the quantity $\xi^2 \langle \sigma_{ann} v \rangle_0$ calculated in a low-energy MSSM (for masses above the vertical dashed line) and in non-universal gaugino models which predict low-mass neutralinos [397; 409]. Crosses refer to cosmologically dominant neutralinos, while dots stand for subdominant neutralinos. Faint points in the upper left corner are excluded by antiproton searches. Right: The same as in the left panel, but for SUGRA scheme with non-universality in the Higgs sector [698].

Examples for theoretical calculations for neutralino dark matter are shown in Fig. 26.7, where the reaching capabilities of the GAPS detector are derived and shown in terms of the neutralino annihilation parameters. The shaded band denotes the region which will be covered by GAPS with an ultra-long duration flight: sensitivity to neutralinos from a mass of a few GeV up to a few hundreds of GeV is at hand and it will cover a large portion of the supersymmetric parameter space for models with gaugino non-universality [397; 409] or low-energy MSSM [698].

26.5 Positrons in cosmic rays

In the case of positrons and electrons, the master equation (26.4) describing the propagation of cosmic rays throughout the DH is dominated by space diffusion and energy losses. Above a few GeV, synchrotron radiation in the

Galactic magnetic fields as well as inverse Compton scattering on stellar light and on CMB photons dominate, hence the positron loss rate

$$b^{\text{loss}}(E) = \left\langle \dot{E} \right\rangle = -\frac{E^2}{E_0 \tau_E} \, . \tag{26.24}$$

The energy of reference E_0 is set equal to 1 GeV while the typical energy loss time τ_E is of order 10^{16} s. The master equation for positron propagation simplifies into

$$-K \, \Delta\psi + \partial_E \left\{ b^{\text{loss}}(E) \, \psi \right\} = q\left(\mathbf{x}, E\right). \tag{26.25}$$

Above a few MeV, positrons are ultra-relativistic and the rigidity $\mathcal{R}$ is proportional to the energy E. The space diffusion coefficient boils down to $K(\epsilon) = K_0 \, \epsilon^\delta$, where $\epsilon = E/E_0$. The solution of Equation (26.25) proposed by [179; 464] is based on replacing the energy E by the pseudo-time

$$\tilde{t}(E) = \tau_E \left\{ v(E) = \frac{\epsilon^{\delta-1}}{1-\delta} \right\}, \tag{26.26}$$

and leads to the well-known heat equation

$$\frac{\partial\tilde{\psi}}{\partial\tilde{t}} - K_0 \, \Delta\tilde{\psi} = \tilde{q}\left(\mathbf{x}, \tilde{t}\right), \tag{26.27}$$

where the space and energy positron density is now given by $\tilde{\psi} = \epsilon^2 \, \psi$ whereas the positron production rate has become $\tilde{q} = \epsilon^{2-\delta} \, q$. Because ϵ is dimensionless, both $\tilde{\psi}$ and $\tilde{q}$ have the same dimensions as before. In this formalism, the energy losses which positrons experience are described by an evolution of their density with respect to the pseudo-time $\tilde{t}$. Equipped with these notations, we can write the cosmic-ray positron density as the convolution

$$\psi_{e^+}(\mathbf{x}, E) = \int_{E_S=E}^{E_S=+\infty} \mathrm{d}E_S \int_{\text{DH}} \mathrm{d}^3\mathbf{x}_S \; G_{e^+}(\mathbf{x}, E \leftarrow \mathbf{x}_S, E_S) \, q_{e^+}(\mathbf{x}_S, E_S). \tag{26.28}$$

The positron propagator $G_{e^+}(\mathbf{x}, E \leftarrow \mathbf{x}_S, E_S)$ is defined as the probability for a particle injected at $\mathbf{x}_S$ with the energy E_S to reach the location $\mathbf{x}$ with the degraded energy $E \leq E_S$. It is proportional to the Green function $\tilde{G}$ of the heat equation (26.27) through

$$G_{e^+}(\mathbf{x}, E \leftarrow \mathbf{x}_S, E_S) = \frac{\tau_E}{E_0 \, \epsilon^2} \, \tilde{G}\left(\mathbf{x}, \tilde{t} \leftarrow \mathbf{x}_S, \tilde{t}_S\right), \tag{26.29}$$

where the connection between the energy E and pseudo-time $\tilde{t}$ is given by relation (26.26).

To build up our intuition, it is useful to derive the heat Green function $\tilde{G}$ connecting a source at $\mathbf{x}_S$ to the Earth in the simple case of an infinite

DH. In this limit, there are no vertical boundaries and we get the Gaussian distribution

$$\tilde{G}\left(\mathbf{x}_{\odot}, \tilde{t} \leftarrow \mathbf{x}_S, \tilde{t}_S\right) = \left\{\frac{1}{4\pi K_0 \tilde{\tau}}\right\}^{3/2} \exp\left\{-\frac{r_{\oplus}^2}{4 K_0 \tilde{\tau}}\right\}, \qquad (26.30)$$

where $\tilde{\tau} = \tilde{t} - \tilde{t}_S$ is the typical duration over which the positron energy decreases from E_S to E. That timescale also includes information on the diffusion process. The distance between the source at $\mathbf{x}_S$ and the Earth is denoted by $r_{\oplus}$. The concept of positron horizon is based on the Gaussian distribution (26.30), which is roughly constant within a sphere of radius

$$\lambda_{\mathrm{D}} = \sqrt{4K_0\tilde{\tau}}, \qquad (26.31)$$

and decreases sharply outside. So does the positron Green function G_{e+}. The positron sphere – whose centre is at the Earth where the observer stands – actually delineates the region of the diffusive halo from which positrons predominantly originate. The typical diffusion length λ_{D} gauges how far particles produced at the energy E_S travel before being detected with the energy E. It encodes at the same time the energy loss process and the diffusion throughout the magnetic fields of the Galaxy. A rapid inspection of Eq. (26.26) shows that λ_{D} increases as the detected energy E decreases, except for energies E_S at the source very close to E. The positron sphere is indeed fairly small at high energies, say above $\sim 100\,\mathrm{GeV}$, whereas it spreads over several kiloparsecs below 10 GeV. In the case where $E_S = 100$ GeV for instance, λ_{D} exceeds 3 kpc below an energy E of $\sim 8\,\mathrm{GeV}$.

The diffusive halo inside which cosmic rays propagate before escaping into the intergalactic medium is actually finite and $\tilde{G}$ should account for that effect. In spite of the boundaries at $r = R \equiv 20$ kpc, we can decide that cosmic-ray diffusion is not limited along the radial direction and that it operates as if it took place inside an infinite horizontal slab with half-thickness L. Sources located beyond the Galactic radius R are disregarded since the convolution (26.28) is performed only over the DH. Because their energy is rapidly degraded as they propagate, positrons are produced close to where they are observed. Neglecting the effect of radial boundaries on the propagator G_{e+} turns out to be a fair approximation [645] because positrons do not originate from far away on average. The effects of the radial boundaries down at the Earth are not significant insofar as cosmic rays tend to leak above and beneath the diffusive halo at $z = \pm L$ instead of travelling a long distance along the Galactic plane.

The infinite slab hypothesis allows the radial and vertical directions to be disentangled and the reduced propagator $\tilde{G}$ to be expressed as

$$\tilde{G}\left(\mathbf{x}, \tilde{t} \leftarrow \mathbf{x}_S, \tilde{t}_S\right) = \frac{1}{4\pi K_0 \tilde{\tau}} \exp\left\{-\frac{r^2}{4K_0\tilde{\tau}}\right\} \tilde{V}\left(z, \tilde{t} \leftarrow z_S, \tilde{t}_S\right), \quad (26.32)$$

where the radial distance between the source at $\mathbf{x}_S$ and the point $\mathbf{x}$ of observation is now defined as

$$r = \left\{(x - x_S)^2 + (y - y_S)^2\right\}^{1/2}. \quad (26.33)$$

Should the half-thickness L be very large, expression (26.32) would boil down to the Gaussian distribution (26.30) and the vertical propagator $\tilde{V}$ would be given by the 1D solution $\mathcal{V}_{1D}$ of the heat equation (26.27)

$$\tilde{V}\left(z, \tilde{t} \leftarrow z_S, \tilde{t}_S\right) \equiv \mathcal{V}_{1D}\left(z, \tilde{t} \leftarrow z_S, \tilde{t}_S\right) = \frac{1}{\sqrt{4\pi K_0 \tilde{\tau}}} \exp\left\{-\frac{(z - z_S)^2}{4K_0\tilde{\tau}}\right\}. \quad (26.34)$$

But the diffusive halo has a finite vertical extent. We need to implement the corresponding boundary conditions and impose that the positron density vanishes at $z = \pm L$.

(i) A first approach relies on the method of the so-called electrical images and has been discussed in [179]. Any point-like source inside the slab is associated to the infinite series of its multiple images through the boundaries at $z = \pm L$ which act as mirrors. The nth image is located at

$$z_n = 2Ln + (-1)^n z_S, \quad (26.35)$$

and has a positive or negative contribution depending on whether n is an even or odd number. When the diffusion time $\tilde{\tau}$ is small, the 1D solution (26.34) is quite a good approximation. The relevant parameter is actually

$$\zeta = \frac{L^2}{4K_0\tilde{\tau}} \equiv \frac{L^2}{\lambda_D^2}, \quad (26.36)$$

and in the regime where it is much larger than 1, the propagation is insensitive to the vertical boundaries. On the contrary, when ζ is much smaller than 1, a large number of images need to be taken into account in the sum

$$\tilde{V}\left(z, \tilde{t} \leftarrow z_S, \tilde{t}_S\right) = \sum_{n=-\infty}^{+\infty} (-1)^n \, \mathcal{V}_{1D}\left(z, \tilde{t} \leftarrow z_n, \tilde{t}_S\right), \quad (26.37)$$

and convergence may be a problem.

(ii) It is fortunate that a quite different approach is possible in that case. The 1D diffusion equation (26.27) actually looks like the Schrödinger equation – although in imaginary time – that accounts for the behaviour of a particle inside an infinitely deep 1D potential well which extends from $z = -L$ to $z = +L$. The eigenfunctions of the associated Hamiltonian are both even

$$\varphi_n(z) = \sin\{k_n (L - |z|)\} \qquad (26.38)$$

and odd

$$\varphi'_n(z) = \sin\{k'_n (L - z)\} \qquad (26.39)$$

functions of the vertical coordinate z. The wavevectors k_n and k'_n are respectively defined as

$$k_n = \left(n - \frac{1}{2}\right) \frac{\pi}{L} \text{ (even)} \quad \text{and} \quad k'_n = n \frac{\pi}{L} \text{ (odd)}. \qquad (26.40)$$

The vertical propagator may be expanded as the series

$$\tilde{V}\left(z, \tilde{t} \leftarrow z_S, \tilde{t}_S\right) = \sum_{n=1}^{+\infty} \frac{1}{L} \left\{ e^{-\lambda_n \tilde{\tau}} \varphi_n(z_S) \varphi_n(z) + e^{-\lambda'_n \tilde{\tau}} \varphi'_n(z_S) \varphi'_n(z) \right\},$$

$$(26.41)$$

where the time constants λ_n and λ'_n are respectively equal to $K_0 k_n{}^2$ and $K_0 k'_n{}^2$. In the regime where ζ is much smaller than 1, for very large values of the diffusion time $\tilde{\tau}$, just a few eigenfunctions need to be considered in order for the sum (26.41) to converge. A close examination of these various expressions for $\tilde{G}$ indicate that the energies E and E_S always come into play through the diffusion length λ_D so that the positron propagator may be written as

$$G_{e^+}(\mathbf{x}, E \leftarrow \mathbf{x}_S, E_S) = \frac{\tau_E}{E_0 \, \epsilon^2} \tilde{G}\left(\mathbf{x} \leftarrow \mathbf{x}_S; \lambda_D\right). \qquad (26.42)$$

26.5.1 Secondary positrons

Like for antiprotons, a background of secondary positrons is produced by the spallation of the interstellar medium by impinging high-energy particles. In that respect, the Milky Way looks like a giant accelerator where cosmic rays play the role of the beam whereas the Galactic disk and its gas behave as the target. The dominant mechanism is the collision of protons with hydrogen atoms at rest producing charged pions $\pi^\pm$ which decay into muons $\mu^\pm$. The latter are also unstable and eventually lead to electrons and positrons through the chain

$$p + H \longrightarrow X + \pi^{\pm}$$
$$\pi^{\pm} \longrightarrow \nu_{\mu} + \mu^{\pm}$$
$$\mu^{\pm} \longrightarrow \nu_{\mu} + \nu_e + e^{\pm}.$$

Below $\sim 3\,\text{GeV}$, one of the protons is predominantly excited to a Δ resonance which subsequently decays into either a neutral or a charged pion. The former species produces gamma-rays whereas the latter particle decays into positrons. Above $\sim 7\,\text{GeV}$, pion production is well described in the framework of the scaling model. Various parameterizations are given in the literature [152; 1837] for the Lorentz invariant (LI) cross-section $E_\pi\,\mathrm{d}^3\sigma/\mathrm{d}^3p_\pi$. Positrons may also be produced through kaons, although this channel is rare. The positron production cross-sections of these processes have been carefully computed in the appendices of [1483]. Notice also that useful parametric expressions for the yield and spectra of the stable secondary species produced in p–p collisions have been derived from experimental data and summarized in [1198].

Cosmic-ray protons with energy E_p induce a production of positrons per hydrogen atom with a rate

$$\mathrm{d}\Gamma^{\text{sec}}_{e+}(E_e) = \frac{\mathrm{d}\sigma}{\mathrm{d}E_e}(E_p \to E_e) \times \beta_p \times \{\mathrm{d}n_p \equiv \psi_p(E_p) \times dE_p\}. \quad (26.43)$$

This leads to the source term

$$q^{\text{sec}}_{e+}(\mathbf{x}, E_e) = 4\pi n_{\mathrm{H}}(\mathbf{x}) \int \Phi_p(\mathbf{x}, E_p) \times \mathrm{d}E_p \times \frac{\mathrm{d}\sigma}{\mathrm{d}E_e}(E_p \to E_e). \quad (26.44)$$

That relation can be generalized in order to incorporate cosmic-ray helium nuclei as well as interstellar helium. The gas of the Galactic plane is generally assumed to be homogeneously spread. Because positrons detected at the Earth originate mostly from the solar neighbourhood, we can safely disregard the space dependence of the proton and helium fluxes. Making use then of the measurements [704; 1746] in relation (26.44), we can express the secondary positron flux as

$$\Phi^{\text{sec}}_{e+}(\odot, \epsilon \equiv E_e/E_0) = \frac{\beta_{e+}}{4\pi} \times \frac{\tau_E}{\epsilon^2} \times \int_\epsilon^{+\infty} \mathrm{d}\epsilon_S \times \tilde{I}(\lambda_{\mathrm{D}}) \times q^{\text{sec}}_{e+}(\odot, \epsilon_S). \quad (26.45)$$

The integral $\tilde{I}$ is the convolution of the reduced positron Green function $\tilde{G}$ over the Galactic disk alone

$$\tilde{I}(\lambda_{\mathrm{D}}) = \int_{\text{disk}} \mathrm{d}^3\mathbf{x}_S \ \tilde{G}(\mathbf{x}_\odot \leftarrow \mathbf{x}_S; \lambda_{\mathrm{D}}), \quad (26.46)$$

and depends only on λ_{D} as discussed above.

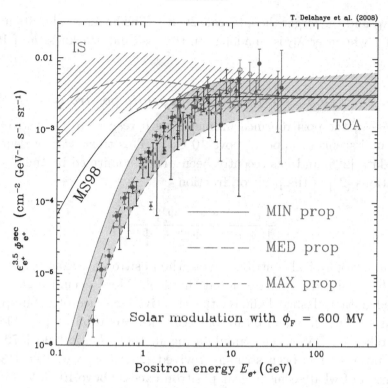

Fig. 26.8. Secondary positron flux as a function of the positron energy. The hatched band corresponds to the cosmic-ray propagation uncertainty on the interstellar prediction whereas the shaded strip refers to the top-of-atmosphere fluxes. The long-dashed curves feature our reference model MED with a differential production cross-section borrowed from [1198] and the most recent measurements by BESS [1746] of the cosmic-ray proton and helium fluxes. The MIN, MED and MAX propagation parameters are displayed in Table 26.2. The observations by CAPRICE [374] (squares), HEAT [226] (triangles), MASS [1023] (open circles) and AMS [35; 49] (dots) are also indicated.

This method has been recently used [644] to derive the positron flux featured in Fig. 26.8. At 1 GeV, the width of the IS uncertainty strip corresponds to an increase by a factor of ∼6 between the smallest and the largest positron fluxes allowed by the B/C constraint. That factor decreases down to 3.9 at 10 GeV and reaches 2.9 at 100 GeV. Once modulated with a Fisk potential ϕ_F of 600 MV, the hatched region is transformed into the shaded TOA band. Quite surprisingly, the MIN model (solid curve) corresponds now to the maximal secondary positron flux whereas the MAX configuration (short-dashed line) yields the minimal prediction. At fixed energies ϵ and ϵ_S, the smaller the diffusion coefficient K_0, the smaller the diffusion length λ_D and the larger the integral (26.46). The latter reaches its maximal value of 1 whenever λ_D is smaller than the disk half-thickness h of 100 pc.

26.5.2 DM signals in cosmic positrons

An excess of the positron measurements with respect to the astrophysical background seems to appear above 10 GeV. This trend was present in the HEAT data [226] and has recently been nicely confirmed by the PAMELA observations [24] of the positron fraction

$$\frac{e^+}{e^+ + e^-} \equiv \frac{\Phi_{e^+}^{\text{tot}}}{\Phi_{e^+}^{\text{tot}} + \Phi_{e^-}^{\text{tot}}} \, . \tag{26.47}$$

Different astrophysical contributions to the positron fraction in the 10 GeV region have already been explored in [226]. More accurate and energy extended data will shed light on the effective presence of a bump in the positron absolute flux and on its physical interpretation. A possible explanation relies on DM species annihilating in the galactic halo [179; 1128]. This hypothesis has been strongly revived [292; 559] with the PAMELA observation of what seems a clear positron excess above 10 GeV.

Such an interpretation, although very exciting, is at some point limited by the uncertainties in the halo structure and in the cosmic-ray propagation modelling. Equations (26.28) and (26.42) can be combined to yield the positron flux generated by the WIMP annihilations taking place within the Milky Way diffusive halo

$$\Phi_{e^+}^{\text{DM}}(\odot, \epsilon \equiv E_e/E_0) = \mathcal{F} \times \frac{\tau_E}{\epsilon^2} \times \int_{\epsilon}^{m_\chi/E_0} d\epsilon_S \, g(\epsilon_S) \, \tilde{I}_{\text{DM}}(\lambda_D). \tag{26.48}$$

The information related to particle physics has been factored out in

$$\mathcal{F} = \frac{\beta}{4\pi} \, \xi^2 \, \langle \sigma_{\text{ann}} v \rangle \left\{ \frac{\rho_\odot}{m_\chi} \right\}^2 . \tag{26.49}$$

The energy distribution $g(\epsilon_S)$ describes the positron spectrum at the source and depends on the details of the WIMP annihilation mechanism. The halo integral $\tilde{I}_{\text{DM}}$ is the convolution of the reduced positron propagator $\tilde{G}$ with the square of the DM Galactic density

$$\tilde{I}_{\text{DM}}(\lambda_D) = \int_{\text{DH}} d^3 \mathbf{x}_S \, \tilde{G}(\mathbf{x}_\odot \leftarrow \mathbf{x}_S; \lambda_D) \left\{ \frac{\rho_\chi(\mathbf{x}_S)}{\rho_\odot} \right\}^2 . \tag{26.50}$$

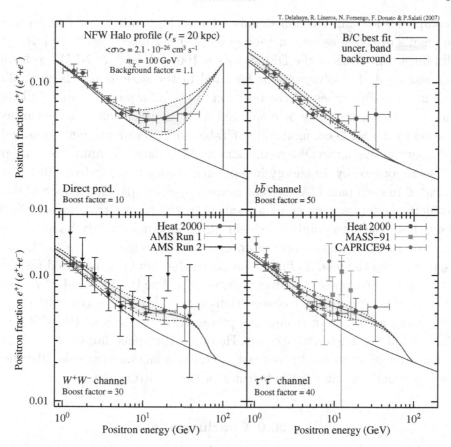

Fig. 26.9. The positron fraction $e^+/(e^- + e^+)$ is plotted versus the positron energy E for a 100 GeV WIMP in the case of an NFW DM profile. The four panels refer to different annihilation final states : direct e^+e^- production (top left), $b\bar{b}$ (top right), W^+W^- (bottom left) and $\tau^+\tau^-$ (bottom right). In each panel, the thin solid line stands for the positron background borrowed from [1483] and parameterized by [179]. The thick solid curve refers to the total positron flux where the signal is propagated with the best-fit choice of the astrophysical parameters, that is, the configuration MED of Table 26.2. The pale shaded area features the total uncertainty band arising from cosmic-ray propagation. The different models found in [1401] to be compatible with the B/C ratio all yield a positron fraction which is enclosed inside this shaded strip. Experimental data from HEAT [226], AMS [35; 49], CAPRICE [374] and MASS [1023] are also presented for comparison. Figure adapted from [645].

In Fig. 26.9, the positron fraction (26.47) is presented as a function of the positron energy E. The total positron flux at the Earth

$$\Phi_{e^+}^{\text{tot}} = \Phi_{e^+}^{\text{DM}} + \Phi_{e^+}^{\text{sec}} \qquad (26.51)$$

encompasses the annihilation signal and a background component for which the results of [1483] as parameterized by [179] have been used – see the thin solid lines. The mass of the DM species is 100 GeV and an NFW profile has been assumed. The observations featured in the various panels are indications of a possible excess of the positron fraction for energies above 10 GeV. Those measurements may be compared with the thick solid curves that correspond to the MED configuration of Table 26.2. In order to get a reasonable agreement between the DM predictions and the data, the annihilation signal has been boosted by an energy-independent factor ranging from 10 to 50 as indicated in each panel. As is clear in the upper left panel, the case of direct production offers a very good agreement with the positron excess. Notice how well all the data points lie within the shaded uncertainty band. A boost factor of 10 is enough to obtain an excellent agreement between the measurements and the median flux. A smaller value would be required for a flux at the upper envelope of the uncertainty strip. The W^+W^- and $\tau^+\tau^-$ channels may also reproduce the observations reasonably well, especially once the uncertainty arising from cosmic-ray propagation throughout the Milky Way diffusive halo is taken into account. However, they need larger boost factors of the order of 30 to 40. In contrast, softer production channels, like the $b\bar{b}$ case, are unable to match the features of the positron bump.

26.6 Conclusions

Crucial information on the astronomical dark matter pervading the Galaxy will be provided by comparing the positron [24] and antiproton [23] observations. An excess seems to be present in the positron flux but does not show up in the antiproton signal. This may lead to strong constraints on the properties of the DM species as mentioned, for instance, in [559; 703]. Pure leptonic models are a priori favoured since even a small branching ratio to quarks or gauge bosons would translate into an overproduction of antiprotons which is not seen. Care needs to be taken, however, as antiprotons could be rehabilitated depending on how cosmic-ray transport is modelled. The forthcoming measurements will be crucial to ascertain the presence of WIMPs in outer space and to study their nature.

27

Multi-wavelength studies

Stefano Profumo and Piero Ullio

27.1 Introduction

The class of weakly interacting massive particles (WIMPs) is the leading category of particle dark matter candidates: on the one hand, the mechanism of thermal freeze-out of a stable WIMP χ leads to a non relativistic relic population whose relative matter density can be approximated as $\Omega_\chi h^2 \approx 3 \times 10^{-27}$ cm^3 s$^{-1}/\langle \sigma_{\text{Ann}} v \rangle$, where h is the Hubble constant in units of 100 km s^{-1} Mpc^{-1}, Ω_χ is the ratio of the χ density over the critical density and $\langle \sigma_{\text{Ann}} v \rangle$ is the thermally averaged χ pair-annihilation cross-section. If the new physics connected to the WIMP is at the electroweak scale, m_{EW}, one can estimate $\langle \sigma_{\text{Ann}} v \rangle \approx \alpha_{\text{EW}}^2/m_{\text{EW}}^2 \approx 10^{-25}$ cm^3 s^{-1}, giving a thermal leftover χ population that lies in the same ballpark as the dark matter abundance inferred from cosmic microwave background anisotropies, large-scale structure and other astronomical observations [1266]. On the other hand, numerous motivated particle physics extensions to the Standard Model encompass a stable WIMP, including supersymmetry [1184] and universal extra dimensions (UED) [1126], in virtue of unbroken discrete symmetries (R-parity in the case of supersymmetry, and Kaluza–Klein parity in the case of UED).

Since WIMPs were once kept in thermal equilibrium by pair annihilations into Standard Model particles and inverse WIMP pair production processes, even in today's cold Universe, occasionally, WIMPs can pair annihilate, giving rise to energetic, stable 'ordinary' Standard Model particles. The rate of WIMP pair annihilation depends on the number density of dark matter in a given location and on the pair-annihilation rate $\langle \sigma_{\text{Ann}} v \rangle$. The spectrum and particle content of the final state products of a pair-annihilation

Particle Dark Matter: Observations, Models and Searches, ed. Gianfranco Bertone. Published by Cambridge University Press. © Cambridge University Press 2010.

event are, instead, determined once the microscopic details of the specific WIMP model are specified. Given a WIMP χ, the production rate of a final state Standard Model particle species $f = \gamma,\ e^{\pm}, \bar{p},\ \nu, \dots$, emitted in a pair-annihilation event in a location $\mathbf{x}$ with dark matter density $\rho_{\mathrm{DM}}(\mathbf{x})$, reads

$$Q(p, \mathbf{x}) \equiv \frac{\rho_{\mathrm{DM}}^2(\mathbf{x})}{2m_\chi} \langle \sigma_{\mathrm{Ann}} v \rangle \frac{\mathrm{d}N_f}{\mathrm{d}p}. \tag{27.1}$$

In the formula above, $\mathrm{d}N_f$ is the number of f particles with modulus of the momenta between p and $p + \mathrm{d}p$ resulting from the WIMP pair-annihilation event. Numerous studies have investigated the indirect detection of WIMPs through signatures in the spectrum of cosmic-ray positrons and antiprotons, or through the emission of energetic gamma-rays. With the recent successful launch of the space-based antimatter detector PAMELA [1576] and of the Fermi gamma-ray space telescopes [180], dark matter indirect detection faces new opportunities, with a forthcoming dramatic increase in the quality and quantity of available data.

The pair-annihilation of WIMPs yields gamma-rays from the two-photon decay of neutral pions, and, in the high-energy end of the spectrum, from internal bremsstrahlung from charged particle final states. Subdominant one-loop annihilation modes give rise to the additional emission of monochromatic photons: the direct pair-annihilation into two photons, for instance, yields two gamma-rays with $E_\gamma = m_\chi$. Photons from prompt emission are generally in the gamma-ray band, where absorption is negligible, hence the induced fluxes or intensities can be straightforwardly derived by summing contributions along the line of sight.

At lower frequencies, a conspicuous non-thermal population of energetic electrons and positrons results from the decays of charged pions produced by the hadronization of strongly interacting particles in the final state, as well as from the decays of gauge bosons, Higgs bosons and charged leptons. This non-thermal $e^{\pm}$ population loses energy and produces secondary radiation through several processes: synchrotron in the presence of magnetic fields, inverse Compton scattering off starlight and cosmic microwave background photons, and bremsstrahlung in the presence of ionized gas. This radiation can actually cover the whole electromagnetic spectrum between the radio and the gamma-ray band.

The computation of the multi-wavelength emissions from WIMP-induced energetic $e^{\pm}$ involves several steps: (i) assessment of the spectrum and production rate of electrons and positrons from dark matter pair annihilations (that follows from Eq. (27.1) with $f = e^{\pm}$); (ii) computation of the effects of propagation and energy losses, possibly leading to a steady state $e^{\pm}$ config-uration; (iii) computation of the actual emissions from the mentioned equi-librium configuration; (iv) evaluation of eventual absorption of the emitted radiation along the line of sight to derive fluxes and intensities for a local observer. We describe items (i) and (ii) in Section 27.1.1, while we devote Section 27.1.2 to the calculation of the secondary emission from dark mat-ter annihilation-induced $e^{\pm}$. Section 27.2 then applies the general outlined framework to the case of galaxy clusters, specifically Coma (Section 27.2.1), 1E 0657-56 (the so-called Bullet cluster, 27.2.2) and Ophiuchus (27.2.3). The following Section 27.3 studies the other extreme of galactic scales: dwarf galaxies. In particular, we focus on the case of the dwarf spheroidal (dSph) galaxy Draco in 27.3.1 and on constraints on dark matter particle models from X-ray observations of dSph galaxies Fornax, Ursa Minor and Carina in 27.3.2. Section 27.4 specializes to the case of our own Milky Way, and more specifically on the Galactic centre (27.4.1), on dark Galactic mini-haloes (27.4.2) and on the dark matter annihilation origin of the so-called WMAP haze (27.4.3). Finally, Section 27.5 addresses radio constraints from cosmo-logical dark matter annihilation at all redshifts, and Section 27.6 concludes and provides an overview.

27.1.1 The electron–positron equilibrium number densities

Except for WIMPs that annihilate mainly into light leptons, the bulk of elec-trons and positrons from WIMP annihilations are secondaries from interme-diate steps in which charged pions are produced and decay along the chain: $\pi^{\pm} \rightarrow \mu^{\pm}\nu_{\mu}(\bar{\nu}_{\mu})$, with $\mu^{\pm} \rightarrow e^{\pm} + \bar{\nu}_{\mu}(\nu_{\mu}) + \nu_{e}(\bar{\nu}_{e})$. These are twin processes with respect to the production of neutral pions and their $\pi^{0} \rightarrow \gamma\gamma$ decay, and therefore, at the level of the production rates, the yield into electron and positrons in Eq. (27.1) is at the same level as the gamma-ray yield and has analogous spectral shape. In principle one should take into account other stable yields, such as protons, which can be confined in the astrophysical environment in which WIMP annihilation is considered, and subsequently give rise to further secondaries (such as pions in proton collisions with the ambient interstellar material); their multiplicity in WIMP annihilation is, however, much smaller than for electrons and positrons and, therefore, they can be safely neglected.

After emission, electrons and positrons go through a random walk in the turbulent and regular components of the ambient magnetic field, los-ing energy by radiative emissions, as well as losing and/or gaining energy in the convective or advective effects associated with plasma outflows or inflows. The sum of these effects is usually described in terms of a transport equation of the form:

$$\frac{\partial n_e(\mathbf{x}, p, t)}{\partial t} = Q_e(\mathbf{r}, p, t) + \nabla \cdot (D_{xx} \nabla n_e) + \frac{\partial}{\partial p} p^2 D_{pp} \frac{\partial}{\partial p} \frac{1}{p^2} n_e$$

$$+ \frac{\partial}{\partial p}(\dot{p} \, n_e) - \nabla \cdot (\mathbf{V} n_e) + \frac{\partial}{\partial p} \left[\frac{p}{3} (\nabla \cdot \mathbf{V}) n_e \right], \quad (27.2)$$

where $n_e(\mathbf{r}, p, t)$ is electron/positron number density per unit particle momentum p, D_{xx} is the spatial diffusion coefficient, diffusive re-acceleration is described as diffusion in momentum space and is determined by the coefficient D_{pp}, $\dot{p} \equiv dp/dt$ is the momentum gain or loss rate, and $\mathbf{V}$ is the convection or advection velocity. Except for when considering the effect of individual dark matter substructures orbiting within a given astrophysical object, the WIMP source function can be regarded as time-independent, and the emission rate is large enough to reach equilibrium, so that Eq. (27.2) can be solved in the steady state approximation, i.e. setting the left-hand side of Eq. (27.2) to zero. Simplifying ansätze are usually adopted for the morphology of the source (namely spherical symmetry when considering galaxy clusters, dwarf satellites and the Galactic centre region, axial symmetry when considering the whole Milky Way) and the spatial dependence of the parameters in the model; another useful approximation is to consider the limit of free escape to occur at the boundary of the diffusion region, i.e. impose $n_e = 0$ outside the region where it is estimated that the propagation equation can be applied. We discuss briefly below how relevant the different terms in the transport equation are for the astrophysical sources that we will consider.

Energy loss term

The radiative losses affecting the $e^+ - e^-$ propagation are inverse Compton scattering on CMB and starlight, synchrotron emission, bremsstrahlung, ionization and Coulomb scattering. The term $\dot{p}$ in Eq. (27.2) can be written as:

$$\dot{p} = \dot{p}_{\text{IC}}(p) + \dot{p}_{\text{syn}}(p) + \dot{p}_{\text{brem}}(p) + \dot{p}_{\text{ion}}(p) + \dot{p}_{\text{Coul}}(p)$$

$$\simeq \dot{p}_{\text{IC}}(p) + \dot{p}_{\text{syn}}(p)$$

$$\simeq b_{\text{IC}}^0 \frac{U_{\text{ph}}}{1 \, \text{eV cm}^{-3}} \left(\frac{p}{1 \, \text{GeV}} \right)^2 + b_{\text{syn}}^0 \left(\frac{B}{1 \, \mu\text{G}} \right)^2 B_\mu^2 \left(\frac{p}{1 \, \text{GeV}} \right)^2, \quad (27.3)$$

where we have indicated that all cases of interest to us lie in the relativistic regime, where either inverse Compton or synchrotron energy loss processes dominate, and the coefficients $b_{\text{IC}}^0 \simeq 0.76$ and $b_{\text{syn}}^0 \simeq 0.025$ are found, respectively, for a background photon energy density U_{ph} of $1 \, \text{eV cm}^{-3}$ (the CMB energy density is about $0.25 \, \text{eV cm}^{-3}$, while the starlight energy density in the Milky Way is about $0.6\,\mathrm{eV\,cm^{-3}}$ locally and $8\,\mathrm{eV\,cm^{-3}}$ in the Galactic centre region [1589]) and a radiating magnetic field B of $1\,\mu\mathrm{G}$ strength (say, within one order of magnitude of what is expected in galaxy clusters, dwarf galaxies or the local portion of the Milky Way, while in the Galactic centre region magnetic fields should be much stronger, at the $1\,\mathrm{G}$ level or even higher).

Spatial diffusion and re-acceleration

There is no definite prediction for the diffusion coefficient $D_{xx}(p)$. Its spectral shape α and normalization depend on unknown variables describing turbulence, namely the amplitude of the random magnetic field and the scale and the spectrum of the turbulence. For example, one finds that in the case of Bohm diffusion $\alpha = 1$, while for Kolmogorov and Kraichnan diffusion, the scaling in energy is, respectively, $\alpha = 1/3$ and $\alpha = 1/2$. On top of this, there is the uncertainty on whether and how to include the re-acceleration term $D_{pp}(p)$, which is in general predicted to be related to $D_{xx}(p)$. When considering the case for diffusion of cosmic rays within the Milky Way one can rely on the comparison between the model and a rather vast and variegated set of observables, such as the ratio of local measurements of abundances of secondary versus primary cosmic-ray species, and global maps of photon emissivities in different energy bands. A nearly self-consistent picture (see further discussion below) can be obtained for, say, a diffusion coefficient at 1 GeV of the order of $\sim 10^{28}\mathrm{cm^2\,s^{-1}}$, $\alpha \sim 0.6$ and no re-acceleration, or $\alpha = 1/3$ and including re-acceleration (see for example [1822]).

Assuming, for simplicity, that re-acceleration can be neglected, to obtain a feeling for whether diffusion is relevant, we can guess, on dimensional grounds, that the diffusion length scale should be of the order of the square root of the diffusion coefficient times the timescale relevant in the transport equation, i.e. $\lambda_{xx} \sim \sqrt{D_{xx}\tau_\mathrm{loss}}$, with τ_loss the timescale for the energy loss associated to radiative processes, namely $\tau_\mathrm{loss} \sim p/\dot{p}$. This guess gives a result that is not too far from the diffusion length scale one obtains in one of the limits in which the transport equation can be solved analytically, namely for spatially constant diffusion coefficient and energy loss term, and neglecting re-acceleration and convection [179]: $\lambda_{xx}(p) = 2\sqrt{v(p) - v(p')}$, where we introduced the variable v, in place of the momentum p of the radiating electron or positron, through the double change of variables:

$$v = \int_{u_{\min}}^{u} \mathrm{d}\tilde{u}\,D(\tilde{u}) \quad \text{and} \quad u = \int_{p}^{p_{\max}} \frac{\mathrm{d}\tilde{p}}{\dot{p}(\tilde{p})} \qquad (27.4)$$

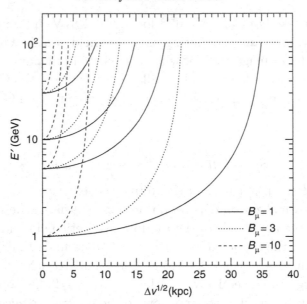

Fig. 27.1. The figure shows the distance $(\Delta v)^{1/2}$ which, on average, an electron covers while losing energy from its energy at emission E' and the energy when it interacts E, for a few values of E: 30 GeV, 10 GeV, 5 GeV and 1 GeV, and for a few values of the magnetic field (in μG); we are focusing on a WIMP of mass 100 GeV, hence cutting $E' < 100$ GeV.

and p' is the positron or electron momentum at emission. As a sample case, we plot in Fig. 27.1 the square root of $\Delta v \equiv v(p) - v(p')$, for a few values of radiating and emission energies, for a few sample values of the ambient magnetic field in units of μG B_μ and with diffusion coefficient within an extrapolation which should apply to galaxy clusters (see e.g. [367; 565]). Length scales of a few (tens of) kpc should be compared to the typical scales over which the dark matter density profiles vary or to the length scale corresponding, for a given object distance, to the angular resolution of the detector searching for the induced signal. In galaxy clusters the scale parameter of the dark matter halo is generally much larger, and diffusion turns out to be relevant only for dark matter profiles which are singular towards their centre, on angular portions on the sky which are however small compared with the detector resolution; it follows that spatial diffusion is a marginal effect in galaxy clusters, and can be safely neglected for most practical purposes in this case [567]. In the limit of negligible spatial diffusion (as well as negligible re-acceleration and convection), the equilibrium number density n_e reduces to:

$$n_e(\mathbf{x}, p) = \frac{1}{\dot{p}(p)} \int_p^{M_\chi} \mathrm{d}p'\, Q_e(\mathbf{x}, p'). \tag{27.5}$$

On the other hand, in the case of Milky Way dwarf satellites (whose size is a few kpc), and for estimates of the induced signals in the local neighbourhood or even extended portions of the Milky Way (with Milky Way diffusion region extending only a few kpc above the Galactic plane, up to possibly 10 kpc or so [1821]), spatial propagation of electrons and positrons is an important effect and should be appropriately modelled (see for example [179; 567]).

Convection or advection effects

In the Galactic centre region the energy loss timescale decreases dramatically, since synchrotron radiation losses are boosted by the very large magnetic field; still, one cannot simply refer to Eq. (27.5). The last two terms on the right-hand side of Eq. (27.2) need to be included to model, in the innermost part of the Galaxy, the accretion flow of gas onto the central black hole [69; 1416]. In the limit of radial free fall $\mathbf{V}$ is oriented towards the black hole and scales like $\sim c\sqrt{R_{\mathrm{BH}}/r}$ (with R_{BH} the gravitational radius of the black hole, and r the radial distance) up to the accretion radius which is about $R_{\mathrm{acc}} \sim 0.04$ pc [1416]. A particle propagating in such accretion flow gains momentum since it feels an adiabatic compression in the direction of the black hole. In Fig. 27.2 we show the interplay between synchrotron loss and adiabatic heating, plotting electron/positron trajectories in the plane of radial distance versus particle momentum, in the limit of equipartition for the magnetic field in the central region of the Galaxy, i.e. assuming the magnetic energy completely balances the kinetic pressure [69; 1416]. The synchrotron loss dominates at high energies, while the advection gain takes over at low energies; electrons and positrons accumulate on the trajectory separating the two regimes (black curve in the figure). Because, when approaching the BH, the scaling in radius of the synchrotron loss is faster than the advection gain, $\dot{p}_{\mathrm{syn}} \propto r^{-5/2}$ versus $\dot{p}_{\mathrm{adv}} \propto r^{-3/2}$, the advection-dominated region becomes smaller and smaller and disappears for radii very close to the BH horizon [1634]. As stated above, in the region with $r > R_{\mathrm{acc}}$ we neglect the advection and thus the trajectories are just horizontal lines.

A convection effect, due to the outflow of primary cosmic rays from the Galactic disk, where sources are located, could be relevant also for the propagation of particles in the whole Milky Way diffusion region [1821]; wind velocities are expected, however, to be much smaller and their effect much less dramatic.

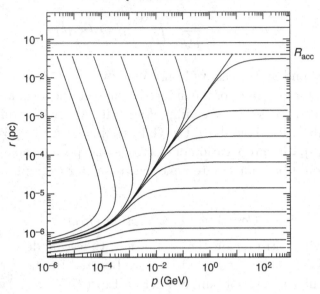

Fig. 27.2. Electron/positron trajectories in the plane of radius versus momentum in the case of magnetic field for the Galactic centre region satisfying the equipartition condition. Far from the turning points, synchrotron loss is dominant in horizontal trajectories, while adiabatic heating takes over in the near-vertical trajectories. The black solid line represents the curve along which the $e^+ - e^-$ accumulate since the two effects balance each other. The dotted line is the accretion radius $R_{\mathrm{acc}} = 0.04\,\mathrm{pc}$, where advection is assumed to stop.

27.1.2 Multi-frequency emissivities and intensities

The solution of Eq. (27.2) provides the e^+ or e^- number density n_e in the stationary limit. For a radiative process i, with associated power P_i, the photon emissivity at frequency ν is obtained by folding n_e with the power [1663]:

$$j_i(\mathbf{x}, \nu) = 2 \int_{m_e}^{M_\chi} \mathrm{d}E\, P_i(\mathbf{x}, E, \nu)\, n_e(\mathbf{x}, E), \qquad (27.6)$$

where the factor 2 takes into account electrons and positrons. The corresponding intensity I_i, as measured by a detector, follows from solution of the differential equation:

$$\frac{\mathrm{d}I_i(\nu, s)}{\mathrm{d}s} = -\alpha(\nu, s)\, I_i(\nu, s, \tilde{\theta}) + \frac{j_i(\nu, s)}{4\pi}, \qquad (27.7)$$

i.e. within the increment $\mathrm{d}s$ along a line of sight of observation, there is a gain in intensity $j_i/(4\pi)\,\mathrm{d}s$, while a decrease $\alpha\, I_i\, \mathrm{d}s$ could be due to absorption.

Radio emission

At radio frequencies, the DM-induced emission is dominated by the synchrotron radiation. The power for synchrotron emission takes the form [1663]:

$$P_{\text{syn}}(\mathbf{x}, E, \nu) = \frac{\sqrt{3}\, e^3}{m_e c^2}\, B(\mathbf{x}) F(\nu/\nu_c)\,, \qquad (27.8)$$

where m_e is the electron mass, the critical synchrotron frequency is defined as $\nu_c \equiv 3/(4\,\pi) \cdot c\, e/(m_e c^2)^3 B(\mathbf{x}) E^2$, and $F(t) \equiv t \int_t^\infty dz K_{5/3}(z)$ is the standard function setting the spectral behaviour of synchrotron radiation. Synchrotron self-absorption can be a relevant effect; indeed one finds that, when considering the region around the Galactic centre, the synchrotron luminosity is so large that low-frequency intensities are significantly changed by absorption [69; 341; 975]; in principle, synchrotron self-Compton effects could be sizable as well, but they turn out to be subdominant [69]. For all other sources, absorption can be safely neglected.

It is instructive to compare the synchrotron to the gamma-ray luminosity under a few simplifying assumptions. Consider the limit of radiative losses as the dominant term in the transport equation, so that the electron/positron equilibrium number density is given by Eq. (27.5). Suppose also that the energy loss rate is dominated by synchrotron emission, i.e. $\dot{p} \simeq \dot{p}_{\text{syn}}$, and that it is sufficient to implement the monochromatic approximation for the synchrotron power, i.e. assume $F(\nu/\nu_c) \sim \delta(\nu/\nu_c - 0.29)$ [1663]. In the monochromatic approximation there is a one-to-one correspondence between the energy of the radiating electron (peak energy in the power) and the frequency of the emitted photon, that is $E_{\text{peak}} = \nu^{1/2}(0.29\, B\, c_0)^{-1/2}$ with $c_0 = 3/(4\,\pi) \cdot c\, e/(m_e c^2)^3$, or, introducing values for numerical constants, the peak energy in GeV is $\widehat{E}_{\text{peak}} \simeq 0.463\, \widehat{\nu}^{1/2} \widehat{B}^{-1/2}$, with $\widehat{\nu}$ the frequency in MHz and $\widehat{B}$ the magnetic field in μG. Under this set of approximations, the induced synchrotron luminosity becomes:

$$\nu L_\nu^{\text{syn}} = \frac{9\sqrt{3}}{4} \frac{\sigma v}{M_\chi^2} \int d^3 x \rho(\mathbf{x})^2 E_{\text{peak}}\, Y_e(E_{\text{peak}}), \qquad (27.9)$$

where we have defined $Y_e(E) = \int_E^{M_\chi} dE' dN_e/dE'$. Analogously, the induced gamma-ray luminosity is:

$$\nu L_\nu^\gamma = 2\pi \frac{\sigma v}{M_\chi^2} \int d^3 x \rho(\mathbf{x})^2 E^2 \frac{dN_\gamma}{dE}\,. \qquad (27.10)$$

Having already stressed that electron/positron yield from WIMP annihilation is at the same level as the gamma-ray yield, it follows that the radio and gamma-ray luminosities are also expected to be roughly comparable.

From the UV to the gamma-ray band

For large magnetic field, such as for the Galactic centre, synchrotron emission may extend to the UV and (possibly) the X-ray band. The emission through inverse Compton (IC) scattering of the ultra-relativistic electrons from WIMP annihilations on cosmic microwave and starlight background photons gives rise to a spectrum of photons stretching from below the extreme ultraviolet up to the soft gamma-ray band, peaking in the soft X-ray energy band. The inverse Compton power is given by:

$$P_{IC}(\mathbf{x}, E, \nu) = c\, h\nu \int d\epsilon\, \frac{dn_\gamma}{d\epsilon}(\epsilon, \mathbf{x})\, \sigma(\epsilon, \nu, E), \qquad (27.11)$$

where ϵ is the energy of the target photons, $dn_\gamma/d\epsilon$ is their differential energy spectrum and σ is the Klein–Nishina cross-section. Additional radiative emission is expected from the process of non-thermal bremsstrahlung, i.e. the emission of gamma-ray photons in the deflection of the charged particles by gas electrostatic potential, as well as via ionization and Coulomb scattering; in general, these processes give subdominant contributions with respect to synchrotron and inverse Compton emission.

At gamma-ray frequencies, radiative emission matches the hard gamma-ray component which, as already mentioned, arises from prompt emission in WIMP pair annihilations. We present in Fig. 27.3 an example of the multi-wavelength spectrum from a 100 GeV WIMP annihilating in $b\bar{b}$ in the dwarf spheroidal galaxy Draco, with the details of the origin of radiation at various frequencies (from ref. [568]).

27.2 The multi-wavelength approach and galaxy clusters

Analyses of multi-wavelength signals from dark matter have been discussed in the context of galaxy clusters for a few case studies, including the Coma cluster [567], the Bullet cluster [566] and the Ophiuchus cluster [1609]. We give below a summary of the results of these studies.

27.2.1 The Coma cluster

In ref. [567] there is a thorough analysis of the transport and diffusion properties of WIMP annihilation products in the Coma cluster of galaxies,

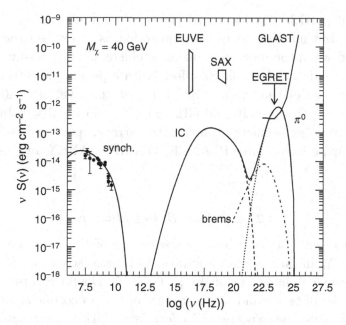

Fig. 27.3. The multi-wavelength spectrum from a 100 GeV WIMP annihilating in $b\bar{b}$ in the dwarf spheroidal galaxy Draco. The WIMP pair annihilation rate has been tuned to give a gamma-ray signal at the level of the EGRET limit on the gamma-ray emission.

investigating the resulting multi-frequency signals, from radio to gamma-ray frequencies. Other relevant astrophysical effects of WIMP annihilations are also discussed, including the DM-induced Sunyaev–Zel'dovich effect and intra-cluster gas heating. Colafrancesco *et al.* [567] find that Coma radio-halo data (the spectrum and the surface brightness) can be nicely fitted by the WIMP-induced signal for certain particle physics models and under reasonable assumptions for the structure of the intra-cluster magnetic field. Fitting the radio data and moving to higher frequencies, they found out that the multi-frequency spectral energy distributions are typically dim at EUV and X-ray frequencies (with respect to the data), but show a non-negligible gamma-ray emission, depending on the amplitude of the Coma magnetic field. A simultaneous fit to the radio, EUV and hard X-ray data is not possible without violating the gamma-ray EGRET upper limit. However, the inverse Compton emission can reproduce both the spectrum and the spatial distribution of the EUV emission observed in Coma, provided that a quite small average magnetic field $B \sim 0.15$ μG is assumed. Such a low value of the field is also able to make the radio data and the hard X-ray data of Coma consistent within a synchrotron/IC model for their origins.

The best-fit particle physics model yields substantial heating of the intra-cluster gas, but not sufficient energy injection as to explain the quenching of cooling flows in the innermost region of clusters. The best-fit ($b\bar{b}$) WIMP model also yields a detectable SZ effect (with a peculiar spectrum very different from that of the thermal SZ effect) at the level of ~ 40 to 10 μK in the frequency range ~ 10–200 GHz, which could be observable with the next generation high-sensitivity bolometric arrays, space and balloon-borne microwave experiments, like PLANCK, OLIMPO, APEX and ALMA.

27.2.2 The Bullet cluster

The SZ_{DM} effect is an inevitable consequence of the presence and of the nature of DM in large-scale structures. Colafrancesco et al. [566] showed that microwave observations of the 1E 0657-556 cluster can provide crucial probes for the presence and for the nature of DM in cosmic structures. The authors calculate the expected SZ effect from dark matter annihilation in the main mass concentrations of the cluster, and estimate the sources of contamination, confusion and bias to assess its significance. It is found that SZ observations at $\nu \approx 223$ GHz can resolve the SZ_{DM} signal both spatially and spectrally and isolate it from the other SZ signals, and mainly from the thermal SZ effect which is null at $\nu \sim 220$–223 GHz for the case of 1E 0657-556. SZ observations of the SZ_{DM} effect with sub-arcmin resolution and sub-microkelvin sensitivity of 1E 0657-556 are therefore crucial, and maybe unique, to find direct astrophysical probes of the existence and of the nature of Dark Matter, or to set strong experimental limits.

27.2.3 The Ophiuchus cluster

Profumo [1609] investigates a scenario where the recently discovered non-thermal hard X-ray emission from the Ophiuchus cluster [732] originates from inverse Compton scattering of energetic electrons and positrons produced in WIMP dark matter pair annihilations. This scenario is compatible with all observational information on the cluster, and it can account for both the X-ray and the radio emission, provided the average magnetic field is of the order of 0.1 μG. It is also shown that Fermi-LAT will conclusively test the dark matter annihilation hypothesis. Depending on the particle dark matter model, the LAT might even detect the monochromatic line produced by dark matter pair annihilation into two photons.

27.3 The multi-wavelength approach and dwarf galaxies

A class of low-background targets for multi-wavelength searches for WIMP dark matter annihilation is nearby dwarf galaxies. The general set-up and an application to the specific case of the Draco dSph galaxy were discussed in [568], while constraints on WIMP models from XMM observations of the three satellite dSph galaxies Fornax, Carina and Ursa Minor have been discussed in [1172].

27.3.1 *The case of Draco*

After investigating mass models for Draco in the light of available observational data, ref. [568] models the dark matter density profile, taking advantage of numerical simulations of hierarchical structure formation. The gamma-ray and electron/positron yield expected for WIMP models is then analysed. Unlike in larger dark matter structures – such as galaxy clusters – spatial diffusion plays an important role here. While Draco would appear as a point-like gamma-ray source, synchrotron emission from electrons and positrons produced by WIMP annihilations features a spatially extended structure. Depending on the cosmic-ray propagation set-up and the size of the magnetic fields, the search for a diffuse radio emission from Draco can be a more sensitive indirect dark matter probe than gamma-rays. In some cases, an extended radio emission could be detectable from Draco even if no gamma-ray source is identified by Fermi-LAT or by atmospheric Cherenkov telescopes, making this technique the most promising search for dark matter signatures from the class of objects under consideration, i.e. nearby dwarf spheroidal galaxies. Reference [568] also points out that available data are consistent with the presence of a black hole at the centre of Draco: if this is indeed the case, very significant enhancements of the rates for gamma-rays and other emissions related to dark matter annihilations are expected.

27.3.2 *X-ray constraints from local dwarfs*

Local dSph galaxies are an ideal environment for particle dark matter searches with X-rays. Reference [1172] describes how X-rays are produced as secondary radiation in inverse Compton scattering off cosmic microwave background photons of electrons and positrons resulting from particle dark matter annihilation. The resulting spectrum is only mildly dependent on the details of the particle dark matter model (the dark matter mass and the dominant final state into which it pair-annihilates), and it is, generically, hard (spectral index smaller than ~ 1.5). The normalization of the emission depends on (i) the particle dark matter pair annihilation rate, (ii) the diffusion set-up and (iii) the dark matter density distribution. For reasonable choices of these three a-priori unknown inputs of the problem, the X-ray emission is potentially within reach of current X-ray detectors. Interestingly enough, the shape of the spectral energy distribution indicates that for dSph galaxies X-rays have a sensitivity to indirect dark matter detection comparable to, if not better than, gamma rays.

Jeltema and Profumo [1172] use XMM-Newton archival data on three Local Group dSph galaxies, Ursa Minor, Fornax and Carina, to search for the diffuse X-ray emission expected from dark matter annihilation. They study the optimal energy and radial range to search for this type of emission, and conclude that for XMM-Newton and for the dSph galaxies under investigation these correspond to an energy band between 0.5 and 8 keV and to a radius of around 6′. No significant signal over background is found, and this, in turn, is turned into constraints on particle dark matter models. The best constraints result from both the Fornax and the Ursa Minor observations, while data from Carina result in bounds that are a factor of a few weaker. Ursa Minor features the largest dark matter density, making it the best candidate target, but it has the shortest usable XMM exposure.

In determining the impact on particle dark matter searches of our X-ray constraints, ref. [1172] points out the uncertainties resulting from the modelling of cosmic-ray diffusion processes, and from the dark matter distribution. In particular, including dark matter substructures can boost our constraints significantly. The bounds obtained are expressed in terms of the dominant dark matter annihilation final states. For those final states relevant for specific dark matter models, such as supersymmetry, the constraints on the mass versus pair annihilation plane are very similar. In the most conservative set-up, only models with rather large annihilation cross-sections are excluded. Assuming a smaller diffusion coefficient, or factoring in the effect of dark matter substructures, our constraints fall well within the interesting region where the supersymmetric dark matter can be a thermal relic from the early Universe. Also, limits can be set on particular supersymmetric dark matter scenarios, such as Wino or Higgsino lightest neutralino dark matter.

An important result of the analysis is that even assuming a conservative diffusion set-up, the sensitivities of X-rays and of gamma-rays to particle dark matter annihilation in dSph galaxies are comparable. This fact has twofold implications: first, if longer observations of dSph galaxies were carried out with existing telescopes, it is possible that the first astronomical signature of particle dark matter annihilation would come from X-rays. Second, should a signature be detected in the future with gamma-ray telescopes, it would be extremely important to confirm the nature of the signal via X-ray observations.

Future X-ray telescopes, such as Constellation-X and XEUS, will also have greatly increased effective areas with respect to current instruments. Using the currently available projections for the effective areas and backgrounds of these telescopes it is estimated that the X-ray limits (0.5–8 keV band) placed by a 100 ks observation of Ursa Minor with Constellation-X or XEUS would improve over the limits placed in this chapter by factors of roughly 35 and 70, respectively. Thus even for a conservative diffusion model, the future generation of X-ray telescopes will place constraints on dark matter annihilation from dwarfs similar to those from Fermi-LAT, and stronger constraints at particle masses below a few hundred GeV. In addition, a signal from Fermi could be confirmed with X-ray observations.

In short, X-rays can play an important role in exploring the nature of particle dark matter and in pinpointing its properties. This role is complementary, but not subsidiary, to searches with gamma-rays. Very exciting results at both frequencies might be just around the corner.

27.4 The multi-wavelength approach and the Milky Way

Galactic sources, including the central regions of the Galaxy as well as Galactic DM clumps, are interesting targets for multi-wavelength searches for WIMP annihilations. We summarize here results on the innermost Galactic centre region from ref. [1634], dark matter 'mini-haloes' [184] and the dark matter interpretation of the so-called WMAP 'haze' [1121].

27.4.1 The Galactic centre

Regis and Ullio [1634] present a systematic, self-consistent study of the multi-wavelength emission due to WIMP pair annihilations in the Galactic centre region. High-energy electrons and positrons emitted in a region with large magnetic fields give rise to synchrotron emission covering radio frequencies up to, possibly, the X-ray band. The authors discuss spectral and angular features, and sketch the correlations among signals in the different energy bands. The critical assumptions for this region stem from uncertainties in the DM source functions, regarding both WIMP models and DM distributions, and from the modelling of propagation for electrons and positrons, and the assumptions on magnetic field profiles. Radio to millimetre synchrotron emission is essentially independent of the shape of the magnetic field in the innermost region of the Galaxy, while at shorter wavelengths, i.e. in the infrared and, especially, the X-ray band, a different choice for the magnetic field may greatly change predictions. Radio signals have in general very large angular sizes, larger than the typical size for the source function and hence of the gamma-ray signals. The size of the region of synchrotron X-ray emissivity shrinks dramatically going to larger frequencies, smaller WIMP masses or softer annihilation channels.

The luminosity of the WIMP source at the different frequencies, and especially comparing the radio with the gamma-ray band, is essentially at a comparable level, with luminosity ratios depending rather weakly on WIMP mass and annihilation channels. This is interesting, since the GC astrophysical source Sgr A*, an unusual source, certainly very different from typical Galactic or extragalactic compact sources associated to black holes, has a very low luminosity over the whole spectrum, at a level at which it is plausible that a WIMP-induced component may be relevant. Indeed, while none of the fluxes detected in the direction of the Galactic centre has spectral or angular features typical of a DM source, still all data sets contribute to place significant constraints on the WIMP parameter space. Although the gamma-ray band is the regime in which it is most straightforward to make the connection between a given dark matter model and the induced signal (hence it is also the regime on which most previous analyses have concentrated), it does not seem to be the energy range with the best signal to background ratios. In the case of large magnetic fields close to the Galactic centre, X-ray data can give much tighter constraints. Radio and near-infrared measurements, which are less model-dependent, tend to be more constraining as well.

27.4.2 Galactic mini-haloes

Baltz and Wai [184] calculate the spectrum and spatial extent of diffuse emission from the charged particle products of dark matter annihilations in Galactic satellites that are currently within the diffusion zone, namely within a few kpc of the stellar disk. For satellites moving with typical Galactic halo velocities of 300 km s^{-1}, the crossing time of the diffusion zone is of the same order as the diffusion time, and thus an inherently time-dependent treatment is required. For annihilation sources, e.g. Galactic satellites at typical distances of 10 kpc, the diffuse emission in both inverse Compton and synchrotron extends over roughly 300 square degrees. At least in terms of the number of photons, the diffuse inverse Compton emission might be detectable by Fermi-LAT, assuming bright enough annihilation sources. The spatial extent of the emission, however, makes its detection problematic.

27.4.3 The WMAP 'haze'

Hooper *et al.* [1121] show that the observed features of the WMAP haze match the expected signal produced through the synchrotron emission of dark matter annihilation products for a model with a cusped halo profile scaling as $\rho(r) \propto r^{-1.2}$ in the inner kpc, and an annihilation cross-section of $\sim 3 \times 10^{-26}$ cm^3 s^{-1}. A wide range of annihilation modes are consistent with the synchrotron spectrum, and no boost factors are required. The properties required of a WIMP to generate the haze are precisely those anticipated for the most theoretically attractive particle dark matter candidates. If the haze is generated through dark matter annihilations, this will have very interesting implications for Fermi. If the $\rho(r) \propto r^{-1.2}$ slope of the halo profile continues to the inner Galaxy, the gamma-ray flux from the Galactic centre will be observable by Fermi, so long as the WIMP is lighter than several hundred GeV, in spite of the presence of the observed HESS source in the region.

27.5 Radio observations

We summarize here the findings of two studies, [384] and [1964], that addressed the detection of a radio emission from WIMP annihilation.

Borriello *et al.* [384] calculate the secondary WIMP-induced radio emission from the Galactic halo as well as from its expected substructures, comparing it with the measured diffuse radio background. They employ a multi-frequency approach using data in the relevant frequency range 100 MHz to 100 GHz, as well as the WMAP haze data at 23 GHz. The derived constraints are of the order $\langle \sigma v \rangle = 10^{-24}$ cm^3 s^{-1} for a DM mass $m_\chi = 100$ GeV but they depend strongly on the astrophysical uncertainties, in particular on the assumptions on the Galactic magnetic field model. The signal from single bright clumps offers only weak signals because of diffusion effects which spread the electrons over large areas, diluting the radio signal. The diffuse signal from the halo and the unresolved clumps is instead relevant and can be compared to the radio astrophysical background to derive constraints on the DM mass and annihilation cross-section. Conservative constraints are at the level of $\langle \sigma v \rangle \sim 10^{-23}$ cm^{3}s^{-1} for a DM mass $m_\chi = 100$ GeV from the WMAP haze at 23 GHz. However, depending on the astrophysical uncertainties, constraints as strong as $\langle \sigma v \rangle \sim 10^{-25}$ cm^3 s^{-1} can be achieved.

Zhang and Sigl [1964] calculate the intensity and angular power spectrum of the cosmological background of synchrotron emission from WIMP annihilations into electron–positron pairs, and compare this background with the intensity and anisotropy of astrophysical and cosmological radio backgrounds, such as from normal galaxies, radio-galaxies, galaxy cluster accretion shocks, the cosmic microwave background and Galactic foregrounds. Under modest assumptions for the dark matter clustering, they find that around 2 GHz, the average intensity and fluctuations of the radio background at sub-degree scales allow them to probe dark matter masses $\gtrsim 100\,\text{GeV}$ and annihilation cross-sections not far from the natural values $\langle \sigma v \rangle \sim 3 \times 10^{-26}\,\text{cm}^3\,\text{s}^{-1}$. The angular power spectrum of the signal from dark matter annihilation tends to be flatter than that from astrophysical radio backgrounds. Furthermore, radio source counts have comparable constraining power. Such signatures are interesting especially for future radio detectors such as SKA.

27.6 Conclusions and overview

If dark matter particles pair annihilate into stable Standard Model particles, a population of energetic, non-thermal electrons and positrons is necessarily produced. The secondary radiation resulting from the energy losses of this non-standard population (including synchrotron and inverse Compton up-scattering of background radiation photons) in turn produces a peculiar multi-wavelength spectrum, extending from radio to gamma-ray frequencies. In some cases, constraints from radio or X-ray observations of Galactic or extragalactic targets place constraints on dark matter models which are more stringent than those resulting from gamma-ray data. Reversing this statement, multi-wavelength observations will complement gamma-ray observations as probes of particle dark matter, since the expected luminosities at different frequencies are generically comparable. The indirect search for dark matter with astronomical observations at various frequencies is therefore a crucial and indispensable element in the quest for the fundamental nature of dark matter.

28

Particle dark matter and Big Bang nucleosynthesis

Karsten Jedamzik and Maxim Pospelov

28.1 Introduction

In the late 1940s and throughout the 1950s a number of visionary scientists including Alpher, Fermi, Follin, Gamow, Hayashi, Herman and Turkevich attempted to explain nuclear abundance patterns observed in the nearby Universe, such as the peculiar high helium mass fraction $Y_p \approx 0.25$. This initially speculative work on an era of nucleosynthesis (element formation) in an expanding Universe at very high temperature $T \sim 10^9 \, \text{K}$ developed slowly but steadily over the coming decades into what is now known as the Standard Model of Big Bang nucleosynthesis (BBN). The idea that the Universe may have undergone a very hot and dense early phase was triggered by the observations of Hubble, in the 1920s, of the recession velocity of galaxies being proportional to their inferred distance from the Milky Way, which were most elegantly explained by a Universe in expansion. The 'expanding hot Big Bang' idea received further support from the observation by Penzias and Wilson in 1965 of the cosmic microwave background radiation (CMBR), believed to be the left-over radiation of the early Universe. Detailed observational and theoretical studies of BBN as well as the CMBR and the Hubble flow have developed into the main pillars on which present-day cosmology rests.

BBN takes place between eras with (CMBR) temperatures $T \simeq 3 \, \text{MeV}$ and $T \simeq 10 \, \text{keV}$, in the cosmic time window $t \simeq 0.1\text{--}10^4 \, \text{s}$, and may be characterized as a freeze-out from nuclear statistical equilibrium of a cosmic plasma at very low ($\sim 10^{-9}$) baryon-to-photon number ratio (cf. Section 28.2), conditions which are not encountered in stars. It produces the bulk of ^4He and ^2H (D), as well as good fractions of ^3He and ^7Li observed in the current

Particle Dark Matter: Observations, Models and Searches, ed. Gianfranco Bertone. Published by Cambridge University Press. © Cambridge University Press 2010.

Universe, whereas all other elements are believed to be produced either by stars or by cosmic rays. In its standard version it assumes a Universe expanding according to the laws of general relativity, at a given homogeneously distributed baryon-to-photon ratio η_b, with only Standard Model particle degrees of freedom excited, with negligible $\mu_l \ll T$ lepton chemical potentials, and in the absence of any significant perturbations from primordial black holes, decaying particles, etc. By a detailed comparison of observationally inferred abundances (cf. Section 28.3) with those theoretically predicted, fairly precise constraints or conclusions about the cosmic conditions during the BBN era may thus be derived. BBN has been instrumental, for example, in constraining the contribution of extra 'degrees of freedom' excited in the early Universe to the total energy density, such as predicted in many models of particle physics beyond the Standard Model. Such contributions may lead to an enhanced expansion rate at $T \sim 1\,\mathrm{MeV}$ implying an increased $^4\mathrm{He}$ mass fraction. It is now known that aside from baryons and other subdominant components, not much more than the already known relativistic degrees of freedom (i.e. photons, γ; electrons and positrons, $e^{\pm}$; and three left-handed neutrinos, ν) could have been present during the BBN era. BBN is also capable of constraining very sensitively any non-thermal perturbations as induced, for example, by the residual annihilation of weak-scale dark matter particles (Section 28.5), or by the decay of relic particles (Section 28.7) and the possible concomitant production of dark matter. Moreover, the sheer presence of negatively charged or strongly interacting weak mass-scale particles during BBN (Section 28.6) may lead to dramatic shifts in yields of light elements through the catalytic phenomena. BBN may therefore constrain properties and production mechanisms of dark matter particles, and this chapter aims at revealing this connection.

It is possible that the biggest contribution of BBN towards understanding the dark matter enigma has already been made. Before the advent of precise estimates of the fractional contribution of baryons to the present critical density, $\Omega_b \approx (0.02273 \pm 0.00062)/h^2$, where h is the Hubble constant in units $100\,\mathrm{km\ s^{-1}Mpc^{-1}}$, by detailed observations and interpretations of the anisotropies in the CMBR [729], BBN was the only comparatively precise way to estimate Ω_b. As it was not clear if the 'missing' dark matter was simply in the form of brown dwarfs, white dwarfs, black holes (formed from baryons) and/or $T \sim 10^6\,\mathrm{K}$ hot gas, various attempts to reconcile a BBN era at large $\Omega_b \sim 1$ with the observationally inferred light element abundances were made. These included, for example, BBN in a baryon-inhomogeneous environment, left over possibly owing to a first-order QCD phase transition at $T \approx 100\,\mathrm{MeV}$, or BBN with late-decaying particles, such as the supersymmetric gravitino (for reviews, see [1154; 1376; 1690]). Only continuous theoretical efforts of this sort, and their constant 'failure' to account for large Ω_b, gave way to the notion that the dark matter must be in the form of 'exotic', non-baryonic material, such as a new fundamental particle as investigated in the present book.

28.2 Standard BBN theory

Standard BBN (SBBN) theory is well understood and described in detail in many modern cosmology textbooks. The essence of SBBN is represented by a set of Boltzmann equations that may be written in the following schematic form:

$$\frac{\mathrm{d}Y_i}{\mathrm{d}t} = -H(T)T\frac{\mathrm{d}Y_i}{\mathrm{d}T} = \sum(\Gamma_{ij}Y_j + \Gamma_{ikl}Y_kY_l + \cdots),\qquad(28.1)$$

where $Y_i = n_i/s$ are the time t- (or temperature T-) dependent ratios between number density n_i and entropy density s of light elements $i = {}^1\mathrm{H}$, n, D, ${}^4\mathrm{He}$, etc.; the $\Gamma_{ij...}$ are generalized rates for element interconversion and decay that can be estimated by experiments and/or theoretical calculations, and $H(T)$ is the temperature-dependent Hubble expansion rate. The system of equations (28.1) assumes thermal equilibrium, i.e. Maxwell–Boltzmann distributions for nuclei, which is an excellent approximation maintained by frequent interactions with the numerous gamma-rays and $e^{\pm}$ in the plasma. The initial conditions for this set of equations are well specified: for temperatures much in excess of the neutron–proton mass difference, neutron and proton abundances are equal and related to the baryon-to-entropy ratio, $Y_{\mathrm{neutron}} \simeq Y_{\mathrm{proton}} \simeq \frac{1}{2}\,n_{\mathrm{baryon}}/s$, while the abundance of all other elements is essentially zero. At tempartures relevant to BBN, the baryonic contribution to the Hubble rate is minuscule, and $H(T)$ is given by the standard radiation-domination formula:

$$H(T) = T^2 \times \left(\frac{8\pi^3 g_* G_N}{90}\right)^{1/2},\quad \text{where } g_* = g_{\mathrm{boson}} + \frac{7}{8}\,g_{\mathrm{fermion}},\qquad(28.2)$$

with g denoting the excited relativistic degrees of freedom. This expression needs to be interpolated in a known way across the brief epoch of the electron–positron annihilation, after which the photons become slightly hotter than neutrinos and $H(T) \simeq T_9^2/(178\ \mathrm{s})$, where T_9 is the photon temperature in units of $10^9\ \mathrm{K}$. A number of well-developed integration routines that go back to an important work of Wagoner, Fowler and Hoyle [1912] allow one to solve the BBN system of equations numerically and obtain the freeze-out values of the light elements. A qualitative 'computer-free' insight to these solutions can be found in, for example, ref. [1486].

In a nutshell, SBBN may be described as follows. After all weak rates have fallen below the Hubble expansion rate, the neutron-to-proton ratio freezes to $\sim 1/6$, subject to a slow further decrease to $\sim 1/7$ by $T_9 \simeq 0.85$ via neutron decay and out-of-equilibrium weak conversion. At this point, to a good approximation, all neutrons available will be incorporated into ^{4}He, since it is the light element with the highest binding energy per nucleon. Synthesis of ^{4}He, and all other elements, has to await the presence of significant amounts of D (the 'deuterium bottleneck'). This occurs rather late, at $T_9 \simeq 0.85$, since at higher T_9 the fragile D is rapidly photodisintegrated by the multitude of CMBR photons. At $T_9 \lesssim 0.85$ the fairly complete nuclear burning of all D then results in only trace amounts $O(10^{-5})$ of D (and ^{3}He) being left over after SBBN has ended. Elements with nucleon number $A > 4$ are produced in even smaller quantities, owing to appreciable Coulomb barrier suppression at such low T_9, resulting in only $O(10^{-10})$ of ^{7}Li, and even lower abundances of other isotopes. SBBN terminates because of the combination of a lack of free neutrons and the importance of Coulomb barriers at low T. In the following a few more details are given.

$O(0.1)$ abundances: ^{4}He

The ^{4}He mass fraction Y_p is dependent on the timing of major BBN events, such as the neutron-to-proton freeze-out at $T \simeq 0.7$ MeV, post-freeze-out neutron depletion before the deuterium bottleneck, and the position of this bottleneck itself as a function of temperature. Consequently, Y_p is dependent on such well-measured quantities as Newton's constant, the neutron–proton mass difference, neutron lifetime and deuterium binding energy, and to a much lesser degree on less precisely known values for the nuclear reaction rates. This sensitivity to the timing of the BBN events makes ^{4}He an important probe of the Hubble expansion rate, and of all possible additional non-standard contributions that could modify it. The SBBN predicts Y_p with an impressive precision, $Y_p = 0.2486 \pm 0.0002$, where we use the most recent evaluation [604].

$O(10^{-5})$ abundances: D and ^{3}He

Deuterium and ^{3}He BBN predictions are more sensitive both to nuclear physics and to η_b input. Reactions involving these elements are well measured, and with the current WMAP input SBBN is capable of making fairly precise predictions of these abundances: D/H $= 2.49 \pm 0.17 \times 10^{-5}$; ^{3}He/H $= (1.00 \pm 0.07) \times 10^{-5}$.

$O(10^{-10})$ *abundances:* ^7Li

Among all observable BBN abundances, ^{7}Li is the most sensitive to the η_b and nuclear physics inputs. The actual observable that BBN predicts is the combined abundance of ^{7}Li and ^{7}Be, as later in the course of the cosmological evolution ^{7}Be is transformed into ^{7}Li via electron capture. At the CMBR-measured value of the baryon-to-photon ratio η_b, more than 90% of primordial lithium is produced in the form of ^{7}Be in the radiative capture process ^{4}He + ^{3}He → ^{7}Be + γ. As the rate for this process per ^{3}He nucleus is much slower than the Hubble rate, the output of ^{7}Be is almost linearly dependent on the corresponding S-factor for this reaction. With recent improvement in its experimental determination [451; 1045; 1495], the current $\sim$15% accuracy prediction for ^{7}Be+ ^{7}Li stands at $5.24^{+0.71}_{-0.67} \times 10^{-10}$ [604].

$O(10^{-14})$ *and lower abundances:* ^6Li *and* $A \geq 9$ *elements*

Lithium-6 is formed in the BBN reaction

$$^4\text{He} + \text{D} \rightarrow {}^6\text{Li} + \gamma, \quad Q = 1.47\,\text{McV}, \tag{28.3}$$

which at BBN temperatures is about four orders of magnitude suppressed relative to other radiative capture reactions such as ^{4}He + ^{3}H → ^{7}Li + γ, and seven or eight orders of magnitude suppressed relative to other photonless nuclear rates. The extra suppression is in a way accidental: it comes from the same charge to mass ratio for ^{4}He and D, which inhibits the E1 transition, making this radiative capture extremely inefficient. This results in $O(10^{-14})$ level prediction for primordial ^{6}Li which is well below the detection capabilities. Heavier elements with $A \geq 9$ such as ^{9}Be, ^{10}B and ^{11}B are never made in any significant quantities in the SBBN framework, and the main reason for that is the absence of stable $A = 8$ nuclei, as ^{8}Be is underbound by 92 keV and decays to two α.

28.3 Observed light element abundances

In the following, we will briefly discuss the observationally inferred light elements, element by element. Here the discussion will also include isotopes, or isotope ratios, such as ^{6}Li, ^{9}Be, and ^{3}He/^{2}H, which are not always considered in SBBN but are very useful to constrain deviations from SBBN.

28.3.1 ^{4}He

The primordial ^{4}He/H ratio is inferred from observations of hydrogen- and helium-emission lines in extragalactic low-metallicity HII-regions and compact blue galaxies, illuminated by young star clusters. Two particular groups have performed such analysis for years now, with their most recent results $Y_p \approx 0.2477 \pm 0.0029$ [1561] and $Y_p \approx 0.2516 \pm 0.0011$ [1159]. These estimates are (surprisingly) considerably larger than earlier estimates by both groups (0.239 and 0.242, respectively), explained in large parts by a new estimate for HeI emissivities [1588]. Other differences with respect to older studies, and/or between the two new studies themselves, are the adopted rates for collisional excitation of H- (He-) emission lines, corrections for a temperature structure in these galaxies ('temperature variations'), corrections for the presence of neutral ^{4}He ('icf – ionization correction factor') and as corrections for troughs in the stellar spectra at the position of the ^{4}He- (H-) emission lines ('underlying stellar absorption'). All of these may have an impact on a level of 1% or more. This, as well as the comparatively large change from earlier estimates (coincidentally going into the direction of agreement with the SBBN prediction of $Y_p \approx 0.248$), implies that a conservative estimate $Y_p \approx 0.249 \pm 0.009$ [1532] (see also $Y_p \approx 0.250 \pm 0.004$) [899] of the error bar is more appropriate when constraining perturbations of SBBN.

28.3.2 D

For the observational determination of primordial D/H ratios, high-resolution observations of low-metallicity quasar absorption line systems (QALS) are used (cf. [477; 590; 1329; 1534; 1573]). QALS are clouds of partially neutral gas which fall on the line of sight between the observer and a high-redshift quasar. The neutral component in these clouds yields absorption features, for example, at the redshifted position of the Lyman-α wavelength. For the very rare QALS of sufficiently simple velocity structure, one may compare the absorption at the Lyman-α position of H with that of D (shifted by 81 km s^{-1}) to infer a D/H ratio. Here the low metallicity of these QALS is conducive to the belief that stellar D destruction in such clouds is negligible. Currently there exist only about six to eight QALS with D/H determinations. When averaged they yield typically 2.68–2.82×10^{-5} [851; 1534; 1573; 1803] for the central value, with inferred statistical 1σ error bars of 0.2–0.3×10^{-5}, comparing favourably to the SBBN prediction of $2.49 \pm 0.17 \times 10^{-5}$ [604] at the WMAP inferred η_b. Nevertheless, as an important cautionary remark, the various inferred D/H ratios in QALS show a spread considerably larger than that expected from the above quoted error bars only. This is usually a sign of the existence of unknown systematic errors. Until these systematics are better understood, primordial values as high as D/H $\approx 4 \times 10^{-5}$ should not be considered to be ruled out.

28.3.3 $^3He/D$

Observational determinations of ^{3}He/H-ratios are possible within our Galaxy, which is chemically evolved. The chemical evolution of ^{3}He is, however, rather involved, with ^{3}He known to be produced in some stars and destroyed in others. Furthermore, any D entering stars will be converted to ^{3}He by proton burning. The net effect of all this is an observed approximate constancy of $(D+^3He)/H \approx 3.6 \pm 0.5 \times 10^{-5}$ [923] over the past few billion years in our Galaxy. Whereas the relation of Galactic observed ^{3}He/H ratios to the primordial one is obscure, the ratio of ^{3}He/D as observed in the presolar nebulae is invaluable in constraining perturbations of SBBN. This ratio $0.83^{+0.53}_{-0.25}$ [923] (where the error bars are obtained when using the independent 2σ ranges of ^{3}He/H and D/H) provides a firm upper limit on the primordial ^{3}He/D [1752]. This is because ^{3}He may be either produced or destroyed in stars, while D is always destroyed, such that the cosmological ^{3}He/D ratio may only grow in time.

28.3.4 ^7Li

Ratios of ^{7}Li/H may be inferred from observations of absorption lines (such as the 6708A doublet) in the atmospheres of low-metallicity Galactic halo stars. When this is done for stars at low metallicity [Z], ^{7}Li/H ratios show a well-known anomaly (with respect to other elements), i.e. ^{7}Li/H ratios are constant over a wide range of (low) [Z] and some range of temperature (the 'Spite plateau'). As most elements are produced by stars and/or cosmic rays, which themselves produce metallicity, the ^{7}Li Spite plateau is believed to be an indication of a primordial origin of this isotope. This interpretation is strengthened by the absence of any observed scatter in the ^{7}Li abundance for such stars. There have been several observational determinations of the ^{7}Li abundance on the Spite plateau. Most of them fall in the range ^{7}Li/H $\approx$ 1–2×10^{-10} such as $1.23^{+0.68}_{-0.32} \times 10^{-10}$ [1133; 1662] and 1.1–1.5×10^{-10} [136], with some being somewhat higher, such as $2.19 \pm 0.28 \times 10^{-10}$ [381]. Here differences may be due to differing methods of atmospheric temperature estimation. These values should be compared with the SBBN prediction $5.24^{+0.71}_{-0.67} \times 10^{-10}$ [604] (with 1σ error estimates), clearly indicating a conflict which is often referred to as the 'lithium problem'. It is essentially ruled out that this problem be solved by either an erroneous atmospheric temperature determination, or significant changes in the rates at which ^{7}Li is produced or destroyed in SBBN. There remain only two viable possibilities of a resolution to this statistically significant $(4$–$5)\sigma$ problem. First, it is conceivable that atmospheric ^{7}Li has been partially destroyed in such stars owing to nuclear burning in the stellar interior. Although far from understood, one may indeed construct (currently ad hoc) models that deplete ^{7}Li in such stars, reducing it by about half while respecting all other observations [1271; 1640]. Second, it is possible that the lithium problem points directly towards physics beyond the SBBN model, possibly connected to the production of the dark matter (cf. Section 28.7).

28.3.5 ^{6}Li and ^{9}Be

The isotope of ^{6}Li is usually not associated with BBN, as its standard BBN production ^{6}Li/H $\sim 10^{-14}$ is very low. However, the smallest deviations from SBBN may lead to important cosmological ^{6}Li abundances. It is therefore interesting that the existence of ^{6}Li has been claimed in about 10 low-metallicity stars [136], with, nevertheless, each of these observations only at the 2–4σ statistical significance level. Asplund et al. infer an average of ^{6}Li/^{7}Li ≈ 0.044 (corresponding to ^{6}Li/H $\approx 6 \times 10^{-12}$) for their star sample [136], whereas Cayrel et al. infer ^{6}Li/^{7}Li $\approx 0.052 \pm 0.019$ for the star HD84937 [497]. Such claims, if true, would be of great interest, as the inferred ^{6}Li/^{7}Li in very low metallicity stars is exceedingly hard to explain by cosmic-ray production [1599], although in situ production in stellar flares may be conceivable [1843]. Moreover, the ^{6}Li observations seem to be consistent with a plateau structure at low metallicity as expected when originating from BBN. However, recent work [496] has cast a significant shadow over the claim of elevated ^{6}Li/^{7}Li ratios at low [Z]. Similar to ^{7}Li, ^{6}Li is inferred from observations of atmospheric stellar absorption features. Unlike in the case of D and H in QALS, the absorption lines of ^{7}Li and ^{6}Li are always blended together. The ^{6}Li/^{7}Li ratios may therefore be obtained only by observations of a minute asymmetry in the 6708 line. Such asymmetries could be due to ^{6}Li, but may also be due to asymmetric convective motions in the stellar atmospheres. The analysis in Cayrel et al. [496] prefers the latter explanation.

Unlike the case of ^{6}Li, detection of ^{9}Be in many stars at low metallicities is not controversial. Observations of ^{9}Be [373; 1605] are far above the $O(10^{-18})$ SBBN prediction, and exhibit linear correlation with oxygen, clearly indicating its secondary (spallation) origin [852]. The lowest level of detected ^{9}Be/H is a few times 10^{-14}, which translates into the limit on the primordial fraction at 2×10^{-13} [1592], assuming no significant depletion of ^{9}Be in stellar atmospheres.

28.4 Cascade nucleosynthesis from energy injection

The possibility that BBN may be significantly perturbed by the presence of energetic, non-thermal SM particles in the plasma first received detailed attention in the 1980s [672; 766; 1328; 1637]. Although much of the pioneering work had been done, the first fully realistic calculations of coupled thermal nuclear reactions and non-energetic phenomena have only recently been presented [603; 1166; 1213; 1214]. Energetic particles may be injected as products of the decay or annihilation of relic non-SM particles, or perhaps via more exotic mechanisms such as evaporation of primordial black holes or supersymmetric Q-balls. The injected energetic photons (γ), electrons/positrons ($e^{\pm}$), neutrinos (ν), muons ($\mu^{\pm}$), pions (π), nucleons and antinucleons (N and $\bar{N}$) and gauge bosons (Z and $W^{\pm}$), etc. may be considered as the 'cosmic rays' of the early Universe. In contrast to their present-day counterparts, and with the exception of neutrinos, these early cosmic rays thermalize rapidly within a small fraction of the Hubble time $H^{-1}(T)$ for all cosmic temperatures above $T \sim 1\,\mathrm{eV}$. This, of course, happens only after all unstable species (π, μ, Z and $W^{\pm}$) have decayed leaving only γ, $e^{\pm}$, ν and N. Many of the changes in BBN light-element production occur during the course of this thermalization. One often distinguishes between hadronically (π, N and $\bar{N}$) and electromagnetically (γ, $e^{\pm}$) interacting particles, mainly because the former may change BBN yields at times as early as $\tau \gtrsim 0.1\,\mathrm{s}$ (i.e. $T \lesssim 3\,\mathrm{MeV}$), whereas the latter only have impact for $\tau \gtrsim 10^5\,\mathrm{s}$ (i.e. $T \lesssim 3\,\mathrm{keV}$). In the following we summarize the most important interactions and outline the impact of such particles on BBN. For hadronically interacting particles these effects include the following.

(i) $\pi^{\pm}$s may cause charge exchange, i.e. $\pi^- + p \rightarrow \pi^0 + n$ between $1\,\mathrm{MeV} \gtrsim T \gtrsim 300\,\mathrm{keV}$ thereby creating extra neutrons after n/p freeze-out and increasing the helium mass fraction Y_{p}.

(ii) Antinucleons $\bar{N}$ injected in the primordial plasma preferentially annihilate on protons, thereby raising the effective n/p ratio and increasing Y_{p}.

(iii) At higher temperatures, neutrons completely thermalize through magnetic moment scattering on $e^{\pm}$ ($T \gtrsim 80\,\mathrm{keV}$), whereas protons do so through Coulomb interactions with $e^{\pm}$ and Thomson scattering off CMBR photons ($T \gtrsim 20\,\mathrm{keV}$). Any extra neutrons at $T \sim 40$ keV may lead to an important depletion of $^7\mathrm{Be}$.

(iv) At lower temperatures, both energetic neutrons and protons may spall $^4\mathrm{He}$, e.g. $n + {}^4\mathrm{He} \rightarrow {}^3\mathrm{H} + p + n + (\pi\mathrm{s})$, or $n + {}^4\mathrm{He} \rightarrow \mathrm{D} + p + 2n + (\pi\mathrm{s})$. Both reactions are important as they may either increase the $^2\mathrm{H}$ abundance or lead to $^6\mathrm{Li}$ formation via the secondary non-thermal reactions of energetic $^3\mathrm{H}(^3\mathrm{He})$ on ambient alpha particles.

The main features of electromagnetic injection are as follows.

(i)   Energetic photons may pair-produce on CMBR photons, i.e. $\gamma + \gamma_{\mathrm{CMBR}} \rightarrow e^- + e^+$ as long as their energy is above the threshold $E_C \approx m_e^2/22T$ for this process. The created energetic $e^\pm$ in turn undergo inverse Compton scattering, i.e. $e^\pm + \gamma_{\mathrm{CMBR}} \rightarrow e^\pm + \gamma$, to produce further photons. Interactions with CMBR photons completely dominate interactions with matter owing to the exceedingly small cosmic baryon-to-photon ratio η.

(ii) Only when $E_\gamma \lesssim E_C$ do interactions with matter become important. These include Bether–Heitler pair production $\gamma + p(^4\mathrm{He}) \rightarrow p(^4\mathrm{He}) + e^+ + e^-$ and Compton scattering $\gamma + e^- \rightarrow \gamma + e^-$ off plasma electrons, as well as photo-disintegration (see below).

(iii) A small fraction of photons with $E_\gamma \lesssim E_C$ may photodisintegrate first D at $T \lesssim 3\,\mathrm{keV}$, when E_C becomes larger than $E_b^{\mathrm{D}} \approx 2.2\,\mathrm{MeV}$, the D binding energy, and later $^4\mathrm{He}$, at $T \lesssim 0.3\,\mathrm{keV}$ since $E_b^{^4\mathrm{He}} \approx 19.8\,\mathrm{MeV}$. Such processes may cause first D destruction and, later, D and $^3\mathrm{He}$ production, and more importantly $^3\mathrm{He}/\mathrm{D}$ overproduction. They may also lead to $^6\mathrm{Li}$ production.

These interactions are also displayed in the chart of Fig. 28.1.

In the context of non-thermal energy injection related to particle dark matter, there are two very important processes that have profound impact on $^6\mathrm{Li}$ and $^7\mathrm{Li}$ abundances and deserve further comments. Energetic $^3\mathrm{He}$ and $^3\mathrm{H}$ produced via electromagnetic or hadronic energy injection (i.e. via spallation or photodisintegration) provide the possibility of efficient production of $^6\mathrm{Li}$ via the non-thermal nuclear reactions on thermal $^4\mathrm{He}$:

$$^3\mathrm{H} + {}^4\mathrm{He} \rightarrow {}^6\mathrm{Li} + n, \quad Q = -4.78\,\mathrm{MeV};$$
$$^3\mathrm{He} + {}^4\mathrm{He} \rightarrow {}^6\mathrm{Li} + p, \quad Q = -4.02\,\mathrm{MeV}. \tag{28.4}$$

For energies of projectiles $\sim$10 MeV, the cross-sections for these non-thermal processes are of the order of 100 mb, and indeed 10^7 times larger than the SBBN cross-section for producing $^6\mathrm{Li}$ (28.3). This enhancement figure under-lines the $^6\mathrm{Li}$ sensitivity to non-thermal BBN, and makes it an important probe of energy injection mechanisms in the early Universe.

Another important aspect of the non-thermal BBN is the possibility of alleviating the tension between the Spite plateau value and the predicted abundance of $^7\mathrm{Li}$ – that is, 'solve the $^7\mathrm{Li}$ problem'. To achieve that, the energy injection should occur in the temperature interval $60\,\mathrm{keV} \gtrsim T \gtrsim 30\,\mathrm{keV}$, i.e. during or just after $^7\mathrm{Be}$ synthesis. The essence of this mechanism consists in the injection of 10^{-5} or more neutrons per baryon which will enhance $^7\mathrm{Be} \rightarrow {}^7\mathrm{Li}$ interconversion followed by the p-destruction of $^7\mathrm{Li}$ via the thermal reaction sequence [1166]:

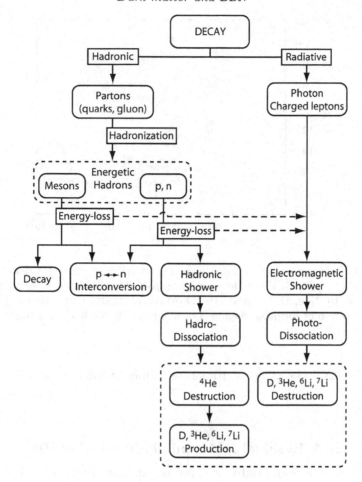

Fig. 28.1. Diagram illustrating the effects of relic particle decay on the plasma, from ref. [1213].

$$n + {}^7\text{Be} \to p + {}^7\text{Li}; \quad p + {}^7\text{Li} \to {}^4\text{He} + {}^4\text{He}. \tag{28.5}$$

Note that this is the same mechanism that depletes ^{7}Be in SBBN, but with elevated neutron concentration due to the hadronic energy injection. This mechanism of depleting ^{7}Be is tightly constrained by the deuterium abundance, as extra neutrons could easily overproduce D.

Figure 28.2 summarizes constraints on abundance versus lifetime of relic decaying particles. It is convenient to measure the abundance in terms of $\Omega_X h^2$, a present-day fraction of total energy density if these particles were to remain stable. It is seen that constraints become increasingly stringent when the lifetime τ_X increases, implying also that, under generic circumstances, the production of dark matter X (with $\Omega_X h^2 \sim 0.1$) by the decay of a parent

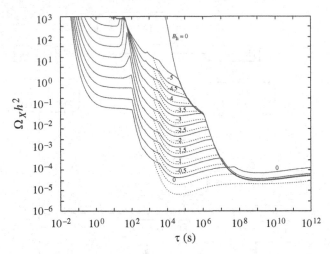

Fig. 28.2. Constraints on the abundance $\Omega_X h^2$ of relic particles decaying at τ_X assuming $M_X = 100\,\text{GeV}$ for the particle mass. The limits are given, from early to late times, by ^{4}He, D, ^{6}Li and ^{3}He/D overproduction, respectively. The various lines are for different $\log_{10} B_h$, as labelled, where B_h is the hadronic branching ratio. From ref. [1168].

particle $Y \to X + \cdots$ at $\tau_X \gg 10^3\,\text{s}$ is extremely problematic, if it is possible at all.

28.5 Residual DM annihilation during BBN

Many dark matter candidates X may be ultimately visible in our Galaxy because of the cosmic rays they inject induced by residual XX self-annihilations. This is, for example, the case for supersymmetric neutralinos, provided that their annihilation products can be distinguished from astrophysical backgrounds. Residual annihilation events in the early Universe may also be of importance as they may lead to cosmologically significant ^{6}Li abundances [1167] induced by the non-thermal nuclear reaction discussed in Section 28.4. Given an X annihilation rate $\langle \sigma v \rangle$ and X density n_X one may determine the approximate fraction f_X of X particles which annihilate in the early Universe at temperature T

$$ f_X \approx \frac{1}{n_X} \frac{\mathrm{d}n_X}{\mathrm{d}t} \Delta t_H \approx \langle \sigma v \rangle \frac{n_X}{s} \, s H^{-1}, \qquad (28.6) $$

where $\Delta t_H \approx H^{-1} = (90 M_{pl}^2 / \pi^2 g T^4)^{1/2}$ is the characteristic Hubble time at T, g is the appropriate particle statistical weight, $s = 4\pi^2/90 \, g T^3$ is radiation entropy and $\langle ... \rangle$ denotes a thermal average, which can be taken once the velocity dependence of σv is specified. Many scenarios for the production of dark matter envisage a stable self-annihilating particle, typically a weakly interacting massive particle or WIMP, whose final asymptotic abundance is given by its annihilation rate. Straightforward considerations of thermal WIMP freeze-out require the annihilation rate at $T_f^{\text{th}} \simeq 0.05 m_X$ to be $\langle \sigma v \rangle_f^{\text{th}} \approx 1\,\text{pb} \times c = 3 \times 10^{-26}\,\text{cm}^3\,\text{s}^{-1}$ if the X particle is to be the dominant component of dark matter, $\Omega_X h^2 \approx 0.1$. Less straightforward but still plausible scenarios that include the non-thermal production of dark matter, e.g. via evaporation of Q-balls and/or decay of relic particles $Y \rightarrow X + \cdots$ with subsequent X self-annihilation, may require $\langle \sigma v \rangle_f$ well in excess of $(\sigma v)_f^{\text{th}}$ for $\Omega_X h^2 \approx 0.1$. We concentrate on the WIMP example, and parameterize the velocity dependence of $\langle \sigma v \rangle$ by $\langle \sigma v \rangle = (\sigma v)_0 S(v)$, with the chosen normalization $(\sigma v)_0 = \langle \sigma v \rangle_f$. We are interested in finding the fraction of annihilating WIMP particles at $T < 10\,\text{keV}$, a temperature scale below which ^{6}Li is no longer susceptible to nuclear burning (destruction). Exploiting (28.6) at the freeze-out temperature T_f, where $f_X(T_f) \approx 1$, as well as at an arbitrary other $T \ll T_f$, we obtain

$$f_X(T) \approx \left(\frac{g(T)}{g(T_f)} \right)^{1/2} \left(\frac{T}{T_f} \right) \frac{\langle S(v) \rangle_T}{\langle S(v) \rangle_{T_f}} \qquad (28.7)$$

for $f_X \ll 1$. Several generic options are possible for the temperature scaling of the S-ratio in (28.7). If the s-wave annihilation is mediated by short-distance physics and occurs away from sharp narrow resonances, $\langle S(v) \rangle_T = \langle S(v) \rangle_{T_f} = 1$, where the second equality is due to our chosen normalization. Using this conservative assumption and Eq. (28.7), for a WIMP of mass $M_X = 100\,\text{GeV}$, so that $g(T)/g(T_f) \simeq 0.1$, one finds that only a small fraction, $f_X \approx 6 \times 10^{-7}$, of X particles has a chance to annihilate at $T \simeq 10\,\text{keV}$ and below. Nevertheless, even this tiny fraction is sufficient to produce a ^{6}Li abundance of ^{6}Li/H $\approx 1.6 \times 10^{-12}$.

Existence of an attractive Coulomb-like force of some strength α' in the WIMP sector may lead to a significant enhancement of annihilation at low temperatures/velocities, possibly leading to a much higher yield of ^{6}Li. In this case the Sommerfeld-like scaling $\sigma v \sim (\pi \alpha'/v)[1 - \exp(-\pi \alpha'/v)]^{-1}$ enhances the annihilation at small $v \lesssim \pi \alpha'$, i.e. $\langle S(v) \rangle_T \simeq \pi \alpha'/v$. This leads to a $\sim T^{-1/2}$ scaling of $\langle S(v) \rangle_T$ in (28.7) when X particles are still in thermal equilibrium with the plasma. After they have dropped out of thermal equilibrium, $\langle S(v) \rangle_T$ falls even more rapidly as $\sim T^{-1}$, with the net effect that weak-mass-scale X particles usually have much smaller velocities at the end of BBN than in the Milky Way. Similarly, the presence of narrow

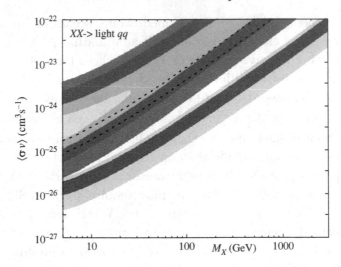

Fig. 28.3. Dark matter annihilation rate versus dark matter mass. The lower thin band shows parameters where ^{6}Li due to residual dark matter annihilation may account for the ^{6}Li abundance as inferred in HD84937 (^{6}Li/^{7}Li $\approx$ 0.014–0.09 at 2σ), whereas the upper thick region shows where ^{7}Li is efficiently destroyed (e.g. ^{7}Li/H $< 2 \times 10^{-10}$).

resonances just above the XX annihilation threshold may greatly boost the annihilation at low energies. Both mechanisms of enhancing the annihilation have been widely discussed in an attempt to link some cosmic-ray anomalies to dark matter annihilation, as for example an elevated positron fraction $e^+/(e^- + e^+)$ observed by the PAMELA instrument [24].

Keeping the annihilation rate as a free parameter, Fig. 28.3 shows dark matter parameters which lead to the production of a ^{6}Li/^{7}Li ratio as claimed to be observed in the star HD84937 at 1σ and 2σ (dark and light regions, respectively, in lower thin band). Here a completely hadronic $XX \rightarrow q\bar{q}$ annihilation has been employed, an assumption which could be further confronted with the constraints on antiproton fluxes in our Galaxy. Electromagnetic annihilations, as often implied in recent dark-matter interpretations of PAMELA [24], may also lead to ^{6}Li, but only annihilations below $T \lesssim 0.3\,\mathrm{keV}$ may do so efficiently. Note that the figure already implicitly assumes a factor $\sim$3–4 stellar destruction (i.e. the abundance is a quarter to a third what it would have been) of ^{7}Li (and ^{6}Li) to solve the lithium problem. It is therefore found that weak-scale mass dark matter particles, if fairly light, and if annihilating into hadronically interacting particles, may account for all of the observed ^{6}Li in HD84937. The figure also shows in the upper thick band the dark matter parameters that would lead to significant

^{7}Li destruction due to residual dark matter annihilations, with the nuclear destruction mechanism described in the previous section. Since those regions are much above the ^{6}Li band, the possibility of depletion in ^{7}Li by a half or more is severely constrained by ^{6}Li overproduction. It is intriguing to realize that primordial ^{6}Li production by residual (hadronic) dark matter annihilations dominates standard BBN ^{6}Li production for annihilation rates as small as $\langle \sigma v \rangle \sim 10^{-27} \mathrm{cm}^3 \, \mathrm{s}^{-1}$, well below $\langle \sigma v \rangle_f^{\mathrm{th}}$. It is thus possible that in the most primeval gas clouds and in the oldest stars the bulk of ^{6}Li is due to dark matter annihilations. Unfortunately, ^{6}Li abundances as low as ^{6}Li/H $\lesssim 10^{-12}$ are difficult to observe.

28.6 Catalysed BBN (CBBN)

The idea of particle physics catalysis of nuclear reactions goes back to the 1950s, and muon-catalysed fusion has been a subject of active theoretical and experimental research in nuclear physics. In recent years there has been interest in a possibility of nuclear catalysis by hypothetical negatively charged particles that live long enough to participate in nuclear reactions at the BBN time. The essence of the idea is very simple: a negatively charged massive particle that we call X^- gets into a bound state with the nucleus of mass m_N and charge Z, forming a large compound nucleus with the charge $Z - 1$, mass $M_X + m_N$, and binding energy in the $O(0.1\text{--}1)$ MeV range. Once the bound state is formed, the Coulomb barrier is reduced, signalling a higher 'reactivity' of the compound nucleus with other nuclei. But the most important effects of catalysis prove to be new reaction channels which may open up and avoid SBBN-suppressed production mechanisms [1591], such as in Eq. (28.3), thus clearing a path to synthesis of elements such as ^{6}Li and ^{9}Be. Although in this chapter we discuss catalysis by negatively charged heavy relics, this is not the only option for CBBN. For example, strongly interacting relics may also participate and catalyse certain nuclear reactions.

Although the connection between dark matter and CBBN is not immediate – after all, the dark matter may not be charged – it is possible that dark matter particles do have a relatively long-lived charged counterpart. One example of this kind is supersymmetry with the lightest supersymmetric particle (LSP) the gravitino and the next-to-LSP (NLSP) a charged slepton, to be examined in the next section. In that case the decay of the NLSP is tremendously delayed by the small size of the gravitino–lepton–slepton coupling $\sim G_N^{1/2}$. Another example in the same vein is the nearly degenerate stau–neutralino system, in which case the longevity of the charged stau

Table 28.1. *Properties of the bound states.*

Bound state	a_0(fm)	$|E_b|$(keV)	T_0(keV)
pX^-	29	25	0.6
$^4\mathrm{He}X^-$	3.63	346	8.2
$^7\mathrm{Be}X^-$	1.03	1350	32
$^8\mathrm{Be}X^-$	0.91	1430	34

Bohr radius $a_0 = 1/(Z\alpha m_N)$, binding energies E_b calculated for realistic charge radii and 'photo-dissociation decoupling' temperatures T_0.

against the decay to the dark matter neutralino is ensured as long as the mass splitting of the stau–neutralino system is below 100 MeV. Both the gravitino and neutralino in these two examples represent viable dark matter candidates. A very important aspect of CBBN is that the abundance of charged particles before they start decaying is given by their annihilation rate at freeze-out. In most of the models their abundance is easily calculable, and if no special mechanisms are introduced to boost the annihilation rate, the abundance of charged particles per nucleon is not small, and in the typical ballpark of $Y_X \sim (0.001\text{--}0.1) \times m_X/\mathrm{TeV}$.

Properties of the bound states

For light nuclei participating in BBN, we can assume that the reduced mass of the nucleus–X^- system is well approximated by the nuclear mass, so that the binding energy is given by $Z^2\alpha^2 m_A/2$ when the Bohr orbit is larger than the nuclear radius. It turns out that this is a bad approximation for all nuclei heavier than $A = 4$, and the effect of the finite nuclear charge radius has to be taken into account. In Table 28.1 we give the binding energies, as well as the recombination temperature, defined as the temperature at which the photodissociation rate of bound states becomes smaller than the Hubble expansion rate. Below these temperatures bound states are practically stable, and the most important benchmark temperatures for the CBBN are $T \sim 30,\ 8,\ 0.5$ keV, when $(^7\mathrm{Be}X^-)$, $(^4\mathrm{He}X^-)$ and (pX^-) can be formed without efficient suppression by the photodissociation processes. It is important to emphasize that these properties of the bound states are generic to any CBBN realization: i.e. they are completely determined by the charge of X^- and electromagnetic properties of nuclei, and thus are applicable to SUSY or non-SUSY models alike. It is also important to note that the $(^8\mathrm{Be}X^-)$ compound nucleus is stable, which may open the path to synthesis of $A > 8$ elements in CBBN.

Catalysis at 30 keV: suppression of ^7Be

When the Universe cools to temperatures of 30 keV, the abundances of deuterium, ^{3}He, ^{4}He, ^{7}Be and ^{7}Li are already close to their freeze-out values, although several nuclear processes remain faster than the Hubble rate. At these temperatures, a negatively charged relic can get into bound states with ^{7}Be and form a $(^7\text{Be}X^-)$ composite object. Once this object is formed, some new destruction mechanisms for ^{7}Be appear. For models with weak currents connecting nearly mass-degenerate X^-–X^0 states, a very fast internal conversion is followed by the p-destruction of ^{7}Li:

$$(^7\text{Be}X^-) \to {}^7\text{Li} + X^0; \quad {}^7\text{Li} + p \to 2\alpha. \tag{28.8}$$

When $X^- \to X^0$ is energetically disallowed, the destruction of ^{7}Be can be achieved via the following chain:

$$(^7\text{Be}X^-) + p \to (^8\text{B}X^-) + \gamma : \quad (^8\text{B}X^-) \to (^8\text{Be}X^-) + e^+\nu, \tag{28.9}$$

which is much enhanced by the atomic resonances in the $(^7\text{Be}X^-)$ system [359].

The rates for both mechanisms may be faster than the Hubble rate, possibly leading to a sizable suppression of ^{7}Be abundance if $(^7\text{Be}X^-)$ bound states are efficiently forming. In other words, $(^7\text{Be}X^-)$ serves as a bottleneck for the CBBN depletion of ^{7}Be. The recombination rate per ^{7}Be nucleus leading to $(^7\text{Be}X^-)$ is given by the product of recombination cross-section and the concentration of X^-particles. It can be easily shown that for $Y_X < 0.01$ the recombination rate is too slow to lead to a significant depletion of ^{7}Be. Detailed calculations of recombination rate and numerical analyses of the CBBN at 30 keV [359; 1290] find that the suppression of ^{7}Be by a factor of 2 is possible for $Y_X \geq 0.1$ if only mechanism (28.9) is operative, and for $Y_X \geq 0.02$ if the internal conversion (28.8) is allowed.

Catalysis at 8 keV: enhancement of ^6Li and ^9Be

As the Universe continues to cool below 10 keV, an efficient formation of $(^4\text{He}X^-)$ bound states becomes possible. With the reasonable assumption of $Y_X < Y_{\text{He}}$ the rate of formation of bound states per X^- particle is given by the recombination cross-section and the concentration of the helium nuclei. Numerical analysis of recombination reveals that at $T \simeq 5$ keV about 50% of available X^- particles will be in bound states with ^{4}He [1591].

As soon as $(^4\text{He}X^-)$ is formed, new reaction channels open up. In particular, a photonless thermal production of ^{6}Li becomes possible

$$(^4\text{He}X^-) + \text{D} \to {}^6\text{Li} + X^-; \quad Q \simeq 1.13\,\text{MeV}, \tag{28.10}$$

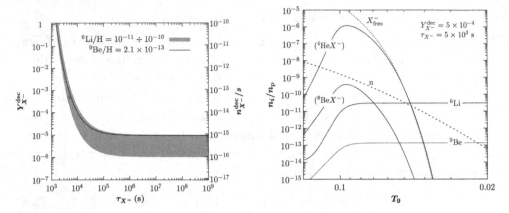

Fig. 28.4. Left panel shows CBBN constraints on the abundance vs. lifetime of X^-. The cross in the shaded area corresponds to a point in the parameter space for which the temporal development of ^{6}Li and ^{9}Be is shown in the right panel, following ref. [1592].

which exceeds the SBBN production rate by about six orders of magnitude. The production of ^{9}Be may also be catalysed, possibly by many orders of magnitude relative to the SBBN values, with the following thermal nuclear chain [1590]:

$$(^4\mathrm{He}X^-) + {}^4\mathrm{He} \to (^8\mathrm{Be}X^-) + \gamma; \quad (^8\mathrm{Be}X^-) + n \to {}^9\mathrm{Be} + X^-. \quad (28.11)$$

Both reactions at these energies are dominated by the resonant contributions, although the efficiency of the second process in (28.11) is not fully understood.

   Current estimates or calculations of the CBBN rates are used to determine the generic constraints on lifetimes/abundances of charged particles. The essence of these limits is displayed in Fig. 28.4, which shows that for typical X^- abundances the lifetime of the charged particles would have to be limited to a few thousand seconds! This is the main conclusion to be learned from CBBN. Note that while, to lowest order, non-thermal BBN is sensitive to the energy density of decaying particles, the CBBN processes are controlled by the number density of X^-, which underlines the complementary character of these constraints. In some models, where both catalysis and cascade nucleosynthesis occur, catalysis dominates cascade production of ^{6}Li for all particles with hadronic branching ratio $B_h \lesssim 10^{-2}$ [1169], whereas ^{7}Li destruction is usually dominated by cascade effects unless $B_h \lesssim 10^{-4}$. Beryllium-9 production, on the other hand, is conceivable only through catalysis.

Catalysis below 1 keV and nuclear uncertainties

Finally we comment on the possibility of (pX^-) catalysis of nuclear reactions, discussed in refs. [671; 1170]. Although it is conceivable that the absence of the Coulomb barrier for this compound nucleus may lead to significant changes of SBBN/CBBN predictions, in practice it turns out that in most cases (pX^-)-related mechanisms are of secondary importance. The large radius and shallow binding of this system leads to a fast charge-exchange reaction on helium, $(pX^-) + {}^4\mathrm{He} \rightarrow ({}^4\mathrm{He}X^-) + p$, that reduces the abundance of (pX^-) below 10^{-6} relative to hydrogen, as long as $Y_{X^-} \lesssim Y_{{}^4\mathrm{He}}$, making further reactions inconsequential for any observable element [1592]. In the less likely case, $Y_{X^-} \gtrsim Y_{{}^4\mathrm{He}}$, significant late-time processing due to (pX^-) bound states may still occur. Such late-time BBN, nevertheless, typically leads to observationally unacceptable final BBN yields.

Unlike in the SBBN case and even in cascade nucleosynthesis that uses mostly measured nuclear reaction rates, CBBN rates cannot be measured in the laboratory, and significant nuclear theory input for the calculation of the reaction rates is required. However, since the X^- participates only in electromagnetic interactions, such calculations are feasible, and dedicated nuclear theory studies [1062] in this direction have already commenced. The reaction rates for some CBBN processes, such as (28.9) and (28.10) are already known within an accuracy of a factor of 2, and the detailed calculations for the ${}^9\mathrm{Be}$ synthesis are under way [1199].

28.7 DM production during BBN: NLSP→LSP example

Dark matter particles may be produced by the decay of relic parent particles Y during BBN. Examples, well studied by different groups, include the production of gravitino-LSP dark matter by NLSP decays (often charged sleptons or neutralinos) or production of neutralino dark matter by heavier gravitinos. Other conceivable possibilities include the production of superweakly interacting Kaluza–Klein dark matter, and more generally the cascade decays to any superweakly interacting dark matter candidates. In the case of charged NLSP decays, both non-thermal and CBBN processes must be accounted for. In the framework of gravitino-LSP/stau-NLSP the lifetime of the charged slepton in the limit of $m_{\tilde{G}} \ll M_{\mathrm{NLSP}}$ is given by

$$\tau_{\mathrm{NLSP}} \approx 2.4 \times 10^4\,\mathrm{s} \times \left(\frac{M_{\mathrm{NLSP}}}{300\mathrm{GeV}}\right)^{-5} \left(\frac{m_{\tilde{G}}}{10\mathrm{GeV}}\right)^2, \tag{28.12}$$

where M_{NLSP} and $m_{\tilde{G}}$ denote NLSP and gravitino mass, respectively.

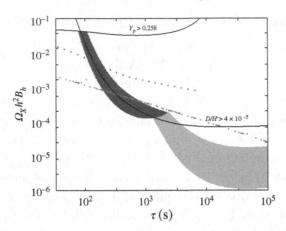

Fig. 28.5. Parameter space in the relic decaying particle abundance times hadronic branching ratio B_h, i.e. $\Omega_Y h^2 B_h$, and lifetime τ_Y plane, where ^{7}Li is significantly reduced (shaded region between 10^2 and 10^3 seconds) and ^{6}Li is efficiently produced (shaded region for $\tau > 10^3$ seconds) or both (the overlapping region at 10^3 seconds). See text for further details. From ref. [173].

It becomes much more difficult with increasing τ_Y to obtain observational consistency with inferred primordial abundances (cf. Fig. 28.2). BBN therefore plays an important role in constraining such scenarios (see Sections 28.4 and 28.6). However, BBN may not only constrain, but also favour particular scenarios, if current discrepancies with ^{6}Li and ^{7}Li abundances are to be taken seriously. Both trends, the reduction of ^{7}Li and the production of ^{6}Li via mechanisms described in Sections 28.4 and 28.6, are seen in Fig. 28.5. There the shaded region between 10^2 and 10^3 seconds shows decaying particle parameter space resulting in more than a factor of 2 suppressed ^{7}Li abundance relative to the SBBN prediction, and the shaded area at $>10^3$ seconds shows regions where significant ^{6}Li production ($0.015 \lesssim {}^6$Li/^{7}Li $\lesssim 0.3$) occurs. In the overlap of these areas, both effects may be achieved simultaneously [1166]. Figure 28.5 also shows the prediction of supersymmetric scenarios with the gravitino-LSP, for some representative values of other supersymmetric mass parameters. In particular, the lower plotted row of dots shows predictions of stau NLSPs with gravitino LSPs of mass $m_{\tilde{G}} = 50\,\text{GeV}$ within the so-called constrained minimal supersymmetric SM (CMSSM), whereas the upper row of dots shows the case of neutralino NLSPs decaying into $m_{\tilde{G}} = 100\,\text{MeV}$ gravitino LSPs within the gauge-mediated supersymmetry breaking scenario. It is seen that both scenarios naturally cross the region of '^{7}Li destruction'. The assumption underlying these models is a thermal freeze-out abundance of the NLSP. Since this typically leads to NLSP abundances, $10^{-3} \lesssim \Omega_{\text{NLSP}} \lesssim 10^3$, and taking into account that gravitino energy density due to NLSP decays is $\Omega_{\tilde{G}} = \Omega_{\text{NLSP}} \left(m_{\tilde{G}}/M_{\text{NLSP}} \right)$, the resulting $\Omega_{\tilde{G}}$

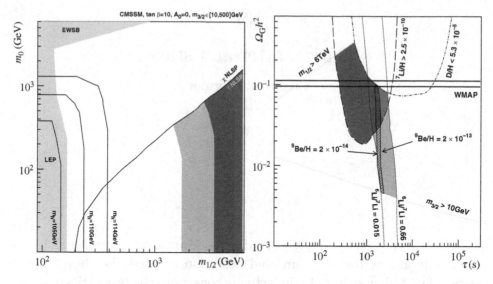

Fig. 28.6. Parameter space in the CMSSM which may impact the primordial ^{7}Li and/or ^{6}Li abundances. Left: Unifying scalar soft mass m_0 versus gaugino soft mass $m_{1/2}$. The three shaded areas to the right show, from left to right, parameter space of significant ^{6}Li production, significant ^{6}Li production as well as observationally favoured ^{7}Li destruction, and only ^{7}Li destruction, respectively. Right: Gravitino abundance $\Omega_{\tilde{G}}h^2$ versus NLSP lifetime τ for the same models as shown in the left panel. From ref. [173].

produced in such scenarios may come close to the observed dark matter density. This is particularly the case for heavy gravitinos $m_{\tilde{G}} \sim 100\,\mathrm{GeV}$ in the CMSSM, for which more-detailed results are shown in Fig. 28.6. It is intriguing, and perhaps purely coincidental, that when resolving the tension between observed and predicted ^{7}Li abundance by staus decaying into gravitinos, the resulting gravitino abundance may account for all the dark matter. For stau decay times $\tau \approx 10^3\,\mathrm{s}$ it is furthermore possible to synthesize a primordial ^{6}Li abundance as claimed to be observed in low-metallicity stars. Moreover, although less certain, the same parameter space could also lead to an important ^{9}Be abundance due to catalytic effects (cf. Section 28.6), as indicated by the cross-hatched region in Fig. 28.6. Finally, since produced by decays, gravitino dark matter in such scenarios is significantly warm, with free-streaming velocities of the same order as those of a $m \approx 3\,\mathrm{keV}$ early freezing-out relic particle, which has important implications for the small-scale structures in the present-day Universe. It is therefore not impossible that, some time in the future, anomalies in the primordial light elements may have been understood as signs of the dark matter. Nevertheless, independent verification by particle accelerators such as the LHC is required. Unfortunately, scenarios as presented in Fig. 28.6 require staus of mass $m_{\tilde{\tau}} \gtrsim 1\,\mathrm{TeV}$, too heavy to be produced at the LHC.

29

Dark matter and stars

Gianfranco Bertone

If dark matter (DM) actually pervades the Universe, it is natural to ask whether it affects the structure and evolution of stars. We have seen in Section 11.1.3 that stellar physics actually constrains the properties of axions. In this chapter, we focus on weakly interacting massive particles (WIMPs), a class of DM candidates that has been extensively discussed in this book, and that includes the most popular DM candidates, such as the neutralino arising in supersymmetric scenarios, and the $B_0^{(1)}$ in theories with universal extra dimensions.

Despite their weak interactions, WIMPs may lead to macroscopic effects in astrophysical objects. In fact, DM particles travelling through planets or stars can lose enough kinetic energy in collisions with ordinary matter to remain gravitationally bound to them – in other words, they can be 'captured', and sink to the gravitational centre of the star or planet. Although the total mass collected in realistic astrophysical environments amounts to a tiny fraction of the host celestial bodies, the cloud of WIMPs may still influence the hosts' physical properties through the injection of energy owing to the interaction of secondary particles, produced in DM self-annihilation, with the surrounding medium. As we have seen for example in Chapter 7, the self-annihilation rate is in fact proportional to the square of the DM density, which can reach extremely high values at the centre of astrophysical objects. We discuss here the details of the capture and annihilation of DM in astrophysical objects, and review the consequences of these processes on their physical properties.

The first studies of the effects of DM particles on stars date back to the 1980s. Press and Spergel have in fact studied the effect of DM particles on the energy transport in the solar core [1602; 1785]. Their results matched

Particle Dark Matter: Observations, Models and Searches, ed. Gianfranco Bertone. Published by Cambridge University Press. © Cambridge University Press 2010.

those obtained by Gilliland and Faulkner, which were unpublished when the papers of Spergel and Press appeared, although their results had been cited by Steigman *et al.* in a paper published in 1978 [1805]. Shortly after refs. [1602] and [1785] appeared, Faulkner and Gilliland published their earlier results [819], and they subsequently co-authored with Press and Spergel a more refined analysis of the problem [950]. The main aim of these studies (see also [1638]) was to solve in terms of WIMPs the (in)famous 'solar neutrino problem', i.e. the discrepancy between the predicted and observed flux of ∼MeV neutrinos from the Sun, which instead found a satisfactory solution in 1992, with the discovery of neutrino oscillations [898]. Following this pioneering work, however, many authors further pursued this line of research, and tried to detect the effects of DM particles on the Sun and other main-sequence stars, compact objects and planets. We review here the results so far obtained, including the constraints on the DM properties and the prospects for detecting DM effects on the structure and evolution of astrophysical objects.

29.1 DM capture and annihilation in stars

Given a Maxwellian distribution of DM particles, the number of particles captured per unit time by a star moving at velocity v_* with respect to them is [994]

$$C_\chi = 4\pi \int_0^{R_*} \mathrm{d}r\, r^2 \frac{\mathrm{d}C(r)}{\mathrm{d}V} \tag{29.1}$$

with

$$\frac{\mathrm{d}C(r)}{\mathrm{d}V} = \left(\frac{6}{\pi}\right)^{1/2} \sigma_{\chi,N} \frac{\rho_i(r)}{M_i} \frac{\rho_\chi}{m_\chi} \frac{v^2(r)}{\bar{v}^2} \frac{\bar{v}}{2\eta A^2} \tag{29.2}$$

$$\times \left\{ \left(A_+ A_- - \frac{1}{2} \right) [\chi(-\eta, \eta) - \chi(A_-, A_+)] \right.$$

$$\left. + \frac{1}{2} A_+ \mathrm{e}^{-A_-^2} - \frac{1}{2} A_- \mathrm{e}^{-A_+^2} - \frac{1}{2} \eta \mathrm{e}^{-\eta^2} \right\}$$

$$A^2 = \frac{3v^2(r)\mu}{2\bar{v}^2 \mu_-^2}, \qquad A_\pm = A \pm \eta, \qquad \eta^2 = \frac{3v_*^2}{2\bar{v}^2}$$

$$\chi(a, b) = \frac{\sqrt{\pi}}{2} [\mathrm{Erf}(b) - \mathrm{Erf}(a)] = \int_a^b \mathrm{d}y\, \mathrm{e}^{-y^2}$$

$$\mu_- = (\mu_i - 1)/2, \qquad \mu_i = m_\chi/M_i,$$

where $\rho_i(r)$ is the mass density profile of a given chemical element in the interior of the star and M_i refers to its atomic mass, while ρ_χ, m_χ and $\bar{v}$ are respectively the WIMP mass and the WIMP density and velocity dispersion at the star position. The velocity of the star is assumed to be equal to $\bar{v}$, giving therefore $\eta = \sqrt{3/2}$. The radial escape velocity profile depends on $M(r)$, i.e. the mass enclosed within a radius r

$$v^2(r) = 2 \int_r^\infty GM(r')/r'^2 \mathrm{d}r'. \tag{29.3}$$

The WIMP scattering cross-section off nuclei, $\sigma_{\chi,N}$, is constrained by direct detection experiments and for a WIMP mass of 100 GeV the current upper limits are $\sigma_p^{\mathrm{SI}} = 10^{-43}$ cm^2 and $\sigma_p^{\mathrm{SD}} = 10^{-38}$ cm^2 respectively for spin-independent and spin-dependent WIMP interactions off a proton (see Chapter 17). We adopt these upper limits here as reference values, but the capture rate can be easily rescaled for other scattering cross-sections by using Eq. (29.3). The spin-independent interactions with nucleons inside nuclei add up coherently giving an enhancement factor A^4 with respect to the interaction with a single nucleon

$$\sigma_{\chi,N}^{\mathrm{SI}} = A^4 \sigma_p^{\mathrm{SI}}, \tag{29.4}$$

where A is the mass number. There is no such enhancement for the spin-dependent interactions. Once captured, WIMPs are redistributed in the interior of the star, reaching, in a characteristic time τ_{th}, a thermal distribution [1019]:

$$n_\chi(r) = n_0 e^{\frac{-r^2}{r_\chi^2}}, \quad r_\chi = \sqrt{\frac{3kT_\mathrm{c}}{2\pi G \rho_\mathrm{c} m_\chi}}, \tag{29.5}$$

where T_c and ρ_c refer to the core temperature and density. The number of scattering events needed for DM particles to thermalize with the nuclei in the star is of order m_χ/M_H, where H is hydrogen, so an upper limit on the thermalization time can be obtained as

$$\tau_{\mathrm{th}} = \frac{m_\chi}{M_\mathrm{H}} \frac{1}{\sigma_p^{\mathrm{SD}} \bar{n}_\mathrm{H} \bar{v}}, \tag{29.6}$$

where $\bar{n}_\mathrm{H}$ is the average density of the star.

The WIMPs luminosity is simply

$$L_\chi(r) = 4\pi(\sigma v) m_\chi c^2 n_\chi^2(r). \tag{29.7}$$

For the annihilation cross-section times relative velocity (σv), we assume the value 3×10^{-26} cm^2, as appropriate for a thermal WIMP (see Chapter 7), but note that the total WIMP luminosity at equilibrium does not depend on this quantity, as it does not depend on m_χ. In fact, after a time

$$\tau_\chi = \left(\frac{C_\chi(\sigma v)}{\pi^{3/2} r_\chi^3} \right)^{-1/2}, \tag{29.8}$$

an equilibrium between capture and annihilation is established, and the total WIMP luminosity is simply given by $L_\chi = C_\chi m_\chi c^2$. Since $C_\chi \propto n \equiv \rho_\chi / m_\chi$, then L_χ does not depend on m_χ at equilibrium.

Aside from direct energy injection, WIMPs modify stellar interiors by contributing to energy transport. To quantify this effect, it is customary to introduce the 'Knudsen number'

$$K(t) = \frac{l_\chi(0, t)}{r_\chi(t)}, \tag{29.9}$$

where $l_\chi(r, t)$ is the WIMP mean free path in the star

$$l_\chi(r, t) \equiv \left[\sum_i \sigma_i n_i(r, t) \right]^{-1}. \tag{29.10}$$

The Knudsen number indicates whether the WIMPs transport energy locally $(K < 1)$, in which case the energy transport matches the case of conductive transport by a gas of massive particles, or non-locally $(K > 1)$, in which case the treatment of energy transport is much more complicated. Given the space constraints, we refer the interested reader to ref. [1714], for useful analytic formulas and references to the existing literature.

29.2 The Earth and other planets

It is natural to ask whether DM capture and annihilation can affect the physical properties of the most constrained astrophysical object: the Earth. The first studies have focused on the details of capture and consequent annihilation of DM particles, especially in view of the possibility of detecting neutrino fluxes [880; 910; 994; 995; 998]. However, in most cases the predicted fluxes are very low for common DM candidates, and in any case smaller than the flux from the Sun (see Chapter 25).

Interesting constraints can instead be derived through the analysis of the Earth's internal heat flow. In fact, geologists have extensively studied the Earth's internal heat, by drilling boreholes (as deep as 12 km) into the ground. Measured temperature gradients are typically between 10 and $50 \, \text{K} \, \text{km}^{-1}$, but it is clear that they cannot hold for lower depths (otherwise

all the rock in the mantle would be molten, in contradiction of seismic measurements). Current estimates actually place temperature gradients deep
inside Earth between 0.6 and 0.8 $\mathrm{K\,km^{-1}}$; the heat flux is then calculated
by multiplying these temperature gradients by the thermal conductivity
of the relevant material. Over 20 200 borehole measurements, the average
measured heat flux is 0.087 ± 0.002 $\mathrm{W\,m^{-2}}$, which upon integration over
the surface of the Earth leads to a heat flow of 44.2 ± 1 TW. The decay
of radioactive elements is known to contribute to the heating, but there
is a residual 50%, thus about 20 TW that remains unaccounted for (see
ref. [1368] and references therein). We can use this flux as an upper limit to
the energy deposited by WIMPs inside the Earth.

The energy injected by WIMPs at the centre of the Earth can be calculated
with Eq. (29.7), but to give a rough estimate of the total number of particles
captured, we can work out the limiting case in which the scattering cross-
section is so high that the Earth captures all the particles passing through it.
The flux of DM crossing the Earth is given by $n_\chi v_\chi$, where n_χ is the DM number density and v is the incoming DM velocity. The capture rate is then found
by multiplying by Earth's geometric cross-section, $\sigma_\oplus = 4\pi R_\oplus^2 \simeq 5.1 \times 10^{18}$
$\mathrm{cm^2}$. For $v_\chi = 270$ $\mathrm{km\,s^{-1}}$, we find that the maximal capture rate is

$$C_{\chi,\oplus}^{\mathrm{max}} = 4.1 \times 10^{25} \left(\frac{\mathrm{GeV}}{m_\chi} \right) \mathrm{s^{-1}}, \qquad (29.11)$$

which leads to an injection of energy equal to $L_{\chi,\oplus}^{\mathrm{max}} = C_{\chi,\oplus}^{\mathrm{max}} m_\chi c^2 \approx$
1.6×10^3 TW.

A more refined treatment of the capture rate allows the WIMP luminosity to be determined for any value of the scattering cross-section, and
the comparison with the measured heat flow allows interesting constraints
to be set on the parameters of the DM particles. The resulting curve is
shown in Fig. 29.1, as the lower boundary in the shaded exclusion region.
At small values of the DM mass, the lower bound of the excluded region
has its minimum at the mass of the experimental target. Below the lower
edge, capture is not efficient (less than 90%) and there is no heating constraint. For masses larger than $m_\chi \approx 10^{10}$ GeV, the value of the annihilation
cross-section that guarantees equilibrium between capture and annihilation
exceeds the maximum value allowed by the unitarity bound [1018]. Finally,
models above the upper edge of the shaded region drift to the centre of the
Earth on a timescale larger than 1 Gyr, and they are thus excluded by this
analysis.

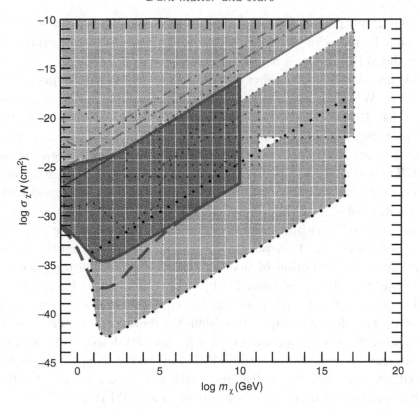

Fig. 29.1. Inside the heavily shaded region, dark matter annihilations would over-heat the Earth. Figure taken from ref. [1368].

29.3 Main-sequence stars

Shortly after the pioneering studies discussed in the introduction, a number of papers appeared on the possible effects of 'cosmions', a light version of WIMPs which was fashionable back then, on stars. Helioseismology was found to provide interesting constraints [612; 820; 949], and this technique was actually revived recently in the context of WIMPs [1350; 1351], although a careful analysis of the systematic and statistical errors associated with the reconstruction of the sound speed profile in the innermost regions of the Sun has shown that the prospects for detecting or constraining DM with this technique are not particularly promising [406]. Alternatively, since WIMPs tend to decrease the temperature at the centre of the Sun, and since the neutrino flux depends strongly on the central temperature (for the ^{8}B neutrino flux, e.g. $\phi_{\rm B} \propto T_{\rm c}^{20}$), one can exclude models that lead to a decrease of the solar neutrino flux larger than the $\sim 20\%$ experimental uncertainty on the measured flux (see the introduction for a discussion of the original motivation of earlier studies and ref. [1349] for a more recent discussion in terms of WIMPs). However, this leads to detectable effects only if the WIMP mass is below ∼30 GeV [406].

Aside from the Sun, a number of authors have discussed the perturbations induced by WIMPs on the structure of generic stars [415; 416; 417; 634; 635; 1670; 1784]. Because of space constraints, we will not review here all the relevant literature but limit ourselves to discussing the most recent results and citing the most recent papers, where references to original literature can be found. The old analytic and perturbative approaches have been superseded by modern codes of stellar evolution that compute in a self-consistent way the evolution of stars in the presence of WIMPs, taking into account in a detailed fashion the capture rate over the various nuclei in the star and the energy deposition and transport in the stellar interior. Recently, a systematic study of the evolution of main-sequence stars in the presence of DM has been obtained in refs. [806; 1714], where the authors have shown how the Herstzprung–Russell diagram (i.e. the equilibrium configurations in the luminosity vs. effective temperature plane) is modified for stars of different masses, evolving in environments with different DM density. In Fig. 29.2, we show an example where the total WIMP luminosity is equal to the energy injected per unit time by nuclear reactions, which corresponds to the case of a star evolving in an environment of very high DM density.

Let us summarize the effect of WIMPs on main-sequence stars:

- For $L_\chi << L_{nuc}$, stars evolve as normal.
- For $L_\chi \approx L_{nuc}$, owing to the negative heat capacity of stars, the central temperature and density drop, with consequent reduction of nuclear burning (this is incidentally the reason why T_c, and thus the solar neutrino flux, decrease).
- For $L_\chi >> L_{nuc}$, the stellar core expands and cools drastically, shutting down nuclear burning.

For stars supported by WIMPs annihilations, the time to consume the hydrogen core is lengthened, and the main-sequence lifetime is extended. The increase in main-sequence lifetime is most prominent at low Z, as we will see in the section dedicated to (zero-metallicity) Pop III stars. Needless to say, the most interesting effects occur for extreme values of the DM density, which can occur, in the Galaxy, only in the innermost regions, especially if a strong DM overdensity extends around the supermassive black hole lying at the Galactic centre (see Chapters 5 and 24).

In ref. [1714] it was shown that in this region, velocity dispersions are large as well (since we are essentially dealing with Keplerian velocities around

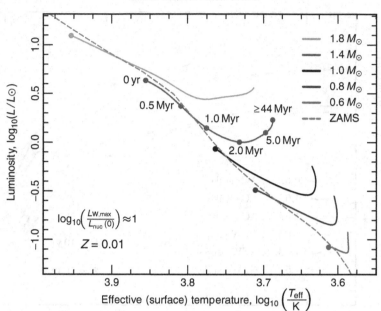

Fig. 29.2. Evolutionary tracks of stars of different mass in the HR diagram. Filled, unlabelled circles indicate the starting points of tracks, whilst labelled ones give indicative ages during the evolution of $1.4M_\odot$ stars. Tracks have been halted when the star exhausts the supply of hydrogen in its core or reaches the current age of the Universe. Stars with a greater luminosity contribution from WIMPs push further up the Hayashi track and spend longer there before returning to the main sequence. Stars that come to be entirely dominated by WIMP annihilation (bottom right) evolve quickly back up the Hayashi track and halt, holding their position in the HR diagram well beyond the age of the Universe. From ref. [1714].

the supermassive BH), which tends to compensate the increase of the capture rate due to the large density. The only cases where interesting effects may realistically be observed are stars with large eccentricities, as shown in Fig. 29.3.

29.4 Compact objects

As we have seen, the capture of DM onto stars is proportional to the number of nucleons in the star times the escape velocity, so compact objects such as white dwarfs or neutron stars are ideal targets for searches aimed at detecting the effects of DM accretion. Both classes of degenerate compact objects are unfortunately also usually rather hot, so that any increase in temperature due to the accretion of WIMPs would be difficult to detect, unless they evolve in regions with very high DM density. One example is the

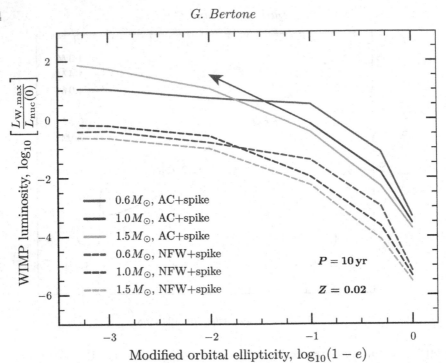

Fig. 29.3. WIMP luminosities achievable by stars on 10-year orbits about the GC. On realistic orbits, annihilation can provide up to 100 times the power of nuclear fusion. If the Galactic halo has undergone adiabatic contraction (AC+spike), break-even between fusion and annihilation energy occurs in stars on any orbit with an eccentricity $e \gtrsim 0.9$, for all masses $M_\star \lesssim 1.5\,M_\odot$. If not (NFW+spike), annihilation begins to rival nuclear fusion in stars of a solar mass or less on orbits with $e \gtrsim 0.99$. These curves have been obtained by calculating the capture rates on such orbits near the GC, applying small boosts due to the non-Gaussian distribution of WIMP velocities, and converting to the maximum WIMP-to-nuclear burning ratios expected during a star's evolution. The arrow indicates that the $1\,M_\odot$, AC+spike curve is expected to continue upwards, but there is no reliable way to convert capture rates to WIMP luminosities in this region because it is beyond the range of parameters we considered in our grid of stellar evolutionary models. From ref. [1714].

Galactic centre, as for main-sequence stars. Moskalenko and Wai have in fact shown that for a white dwarf orbiting close enough to an adiabatically grown supermassive black hole (SMBH, see Chapter 5), the WIMPs luminosity may be comparable to or even exceed the stellar luminosity of the star, thus predicting the existence of unusual stars, called WIMP burners, in the vicinity of an SMBH [1484]. The capture rate is shown in Fig. 29.4 for oxygen white dwarfs (WD). Note that for a WD with mass $(M \gtrsim M_\odot)$ and/or for nuclei with atomic number $A_n \gg 1$, almost all WIMPs crossing the star

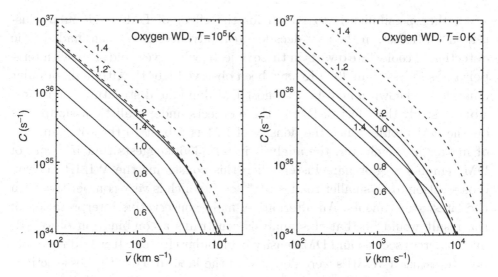

Fig. 29.4. WIMP capture rate by oxygen white dwarfs (WD) ($A_n = 16$) vs. velocity dispersion for $\rho_\chi = \rho_\chi^{\text{max}}$. The left panel shows calculations for WDs without hydrogen envelope and $T = 100\,000$ K. Right panel corresponds to the zero-temperature approximation. The solid lines show the capture rate calculated using the modified cross-section; the dashed lines are calculated for $\sigma_0' = \sigma_0$. The labels show WD mass in $M_\odot$ units. From ref. [1484].

will be captured. The luminosity of the star purely due to the accretion and subsequent annihilation of DM is at equilibrium [336]

$$
\begin{aligned}
L &= \left(\frac{8}{3\pi}\right)^{1/2} \rho_\chi \frac{3 v_{\text{esc}}^2}{2\bar{v}} \sigma_n^{\text{SI}} \sum_i \frac{M_*}{m_p} \frac{x_i}{A_i} A_i^4 \\
&= 3 \times 10^{27} \left(\frac{\rho_{\text{DM}}}{50 M_\odot\,\text{pc}^{-3}}\right) \left(\frac{2 M_*}{M_\odot}\right)^2 \left(\frac{\sigma_n^{\text{SI}}}{10^{-44}\,\text{cm}^2}\right) \left(\frac{10^9\text{cm}}{R_*}\right)\,\text{erg s}^{-1},
\end{aligned}
\tag{29.12}
$$

which can be regarded as the minimum expected luminosity of a star in an environment of weakly interacting, self-annihilating DM of density ρ_{DM} and velocity $\bar{v}$.

The observational consequence of this scenario would be the existence of a concentration of very hot WDs at the Galactic centre, which could be searched for by using the spectra of known hot WDs as templates for spectroscopic analysis of WDs at the Galactic centre, although only a limited part of the near-IR band can be used because of absorption in the interstellar medium.

Another possibility is to search for the effect of DM in globular clusters [336]. In fact, in the past decade, great advances have been made in the detection of cool white dwarfs in these objects, where very old stars with temperatures of less than 4000 K have been observed [1641]. Although globular clusters are known not to be the most DM-dominated structures, it is possible to study the effect of DM in these objects under realistic assumptions on the DM profile. Since the density of DM is robust to the total amount of matter in the cluster, the analysis in ref. [336] suggests that if a core of DM remains in globular clusters then this constrains the WIMP nucleon cross-section to be smaller than $\sim 10^{-44}$ cm^2, which is very competitive with the latest experiments. An alternative, more conservative interpretation of the result would be that this analysis constrains a combination of WIMP nucleon cross-section and DM density in globular clusters. It would therefore be interesting if WIMPs were detected in the laboratory with cross-sections close to this level of 10^{-44} cm^2, as it could open a new branch of DM astronomy using white dwarfs [1274]. In the case of neutron stars, it is a rather straightforward exercise to calculate the heating of the star, with consequent increase of the surface temperature, due to the energy injected by annihilating DM particles. For neutron stars and WIMP–nucleon cross-sections close to the experimental limit, the surface area of the star is larger than the sum of the cross-sections for the individual nuclei. As discussed in ref. [336], unlike a nucleus with a large atomic number, a medium with a constant number density of nucleons gives rise to no resonant enhancement, although the presence of rod- and sheet-like structures in the neutron matter may give rise to an order of magnitude enhancement as they can for the neutrino–nucleon cross-section [1132].

Recent calculations [336; 1274] have shown that at the solar radius the heating due to DM is less than the minimum temperature one might expect for a neutron star which started at some high temperature 13 billion years ago. So unless one can find a way to measure the temperature of cool neutron stars towards the Galactic centre, in which case one would be able to place constraints on the heating due to DM by looking at the non-observation of black body temperature e.g. from pulsars, it will be difficult to use this argument to constrain DM properties. Fortunately, the build up of DM in neutron stars leads to other interesting effects [336] (see also ref. [997] for the case of charged DM).

The timescale for DM inside the star to thermalize with the background neutrons can be estimated: DM particles falling on a neutron star will be semi-relativistic and therefore need to lose an amount of kinetic energy roughly equivalent to their own mass through collisions with nuclei before they reach equilibrium. We can derive from Eq. (29.6) the minimum value of the DM nucleon cross-section such that the thermalization timescale is less than a million years: $\sigma_{\chi n} > 10^{-60}(m_\chi/m_{\text{nuc}})$ cm^2. Once thermalized, the thermal radius within which the majority of thermalized DM particles will be located is given by (29.5), which, inserting typical values for a neutron star, becomes

$$r_{\text{th}} \sim 64\,\text{cm} \left(\frac{T}{10^5\text{K}}\right)^{1/2} \left(\frac{10^{14}\,\text{g}\,\text{cm}^{-3}}{\rho_c}\right)^{1/2} \left(\frac{100\,\text{GeV}}{m_\chi}\right)^{1/2}, \qquad (29.13)$$

where we have assumed that the phase-space density of DM particles is low enough that Maxwell–Boltzmann statistics are still valid, an issue we will return to below. The number of particles required to reach equilibrium between annihilation and capture is then given by

$$N_{\text{eq}} = 5 \times 10^{30} \left(\frac{\Gamma_c}{10^{29}\,\text{s}^{-1}}\right)^{1/2} \left(\frac{T}{10^5\,\text{K}}\right)^{3/4}$$

$$\times \left(\frac{10^{15}\,\text{g}\,\text{cm}^{-3}}{\rho_c}\right)^{3/4} \left(\frac{100\,\text{GeV}}{m_\chi}\right)^{3/4} \left(\frac{10^{-26}\,\text{cm}^3\,\text{s}^{-1}}{\langle\sigma_{\text{ann}}v\rangle}\right)^{1/2}, \qquad (29.14)$$

where $\sigma_{\text{ann}}v = 10^{-26}$ cm^3 s^{-1} is the appropriate value for a thermal relic. The capture rate of 10^{29} s^{-1} corresponds to a rather large density of DM, but one which could be feasibly found at the Galactic centre for 100 GeV WIMPs.

As the density of DM rises in the core of the neutron star, there will come a point at which the self-gravity of the DM core is greater than the gravity due to the baryonic matter within the same volume. The number of WIMPs N_{SG} at the moment when this occurs is such that $4\pi r_{\text{th}}^3 \rho_c/3 = N_{\text{SG}} m_\chi$. This means that if $N_{\text{eq}} \geq N_{\text{SG}}$ the core of DM will become self-gravitating. For a given DM particle mass, this will occur when there are

$$N_{\text{SG}} = 6 \times 10^{40} \left(\frac{T}{10^5\,\text{K}}\right)^{3/2} \left(\frac{10^{14}\,\text{g}\,\text{cm}^{-3}}{\rho_c}\right)^{1/2} \left(\frac{100\,\text{GeV}}{m_\chi}\right)^{5/2} \qquad (29.15)$$

DM particles in the star. The region of the parameter space where this happens is shown in Fig. 29.5. Once in a self-gravitating configuration, the WIMPs will continue to increase in number. There is, however, a limit beyond which the degeneracy pressure cannot support gravity any longer. The critical number of particles to trigger gravitational collapse (the Chandrasekhar limit) is

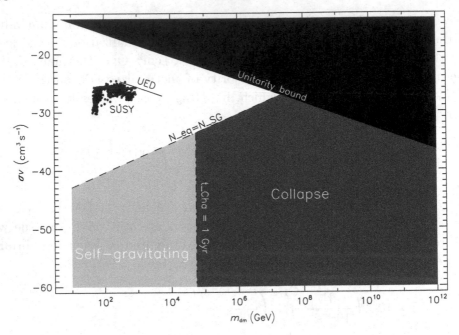

Fig. 29.5. Different outcomes of the accumulation of DM inside a neutron star, in the DM mass vs. annihilation cross-section plane, for a capture rate of $\Gamma_c = (100\,\text{GeV}/m_\chi)10^{29}\,\text{s}^{-1}$. In the top left corner we show the results of a scan of the supersymmetric parameter space, where neutralino models are compatible with accelerator and cosmological constraints (as obtained with DarkSUSY [980]). The solid line is relative to viable DM models in universal extra dimensions (see text for further details). In the shaded region below the dashed line, DM particles become self-gravitating before equilibrium between capture and annihilation is reached. For models on the right of the dotted vertical line, particles reach the critical mass for gravitational collapse in less than 1 Gyr. These models can probably be ruled out, since they lead either to large injection of energy in the core of the neutron star, or to gravitational collapse to a black hole, rapidly destroying the star.

$$N_{\text{Cha}} \sim \left(\frac{M_{\text{Pl}}}{m_\chi}\right)^3 \sim 10^{51}\left(\frac{100\,\text{GeV}}{m_\chi}\right)^3, \qquad (29.16)$$

which for a 100 GeV DM particle is roughly equal to the mass of a planet like Mars. For a heavy DM candidate, with mass $m_\chi = 10^{12}$ GeV, the Chandrasekhar limit is much less, of the order of 1000 tonnes. If that is the case, the accretion of matter onto the newly formed mini-BH would rapidly destroy the star over very short timescales (see ref. [997] for a discussion of the growth of the mini black hole, and consequent destruction of the neutron star). We have summarized the different behaviours in Fig. 29.5, where we have assumed a large DM accretion rate of $\Gamma_c = (100\,\text{GeV}/m_\chi)10^{29}\,\text{s}^{-1}$.

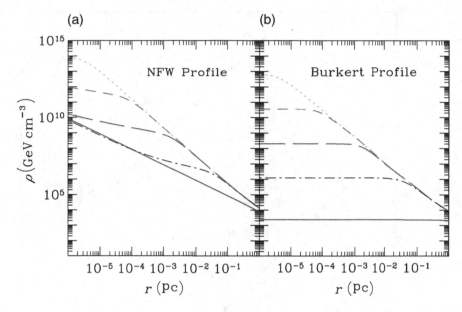

Fig. 29.6. Adiabatically contracted DM profiles for (a) an initial NFW profile and (b) an initial Burkert profile, for $M_{\mathrm{vir}} = 10^6 M_\odot$, $c_{\mathrm{vir}} = 10$ and $z = 19$. The solid lines show the initial profile. Black dot-dash lines correspond to a baryonic core density of $10^7\,\mathrm{cm}^{-3}$, long-dashed lines to $10^{10}\,\mathrm{cm}^{-3}$, dashed lines to $10^{13}\,\mathrm{cm}^{-3}$ and dotted lines to $n \sim 10^{16}\,\mathrm{cm}^{-3}$. From ref. [1787].

29.5 Pop III stars

Alternatively, one may focus on the first stars, which are thought to form from gas collapsing at the centre of 10^6–$10^8 M_\odot$ DM haloes at redshift $z \lesssim 10$–30 (see e.g. refs. [449; 1646] for recent reviews). The consequences of DM annihilations in Pop III stars were first investigated by Spolyar, Freese and Gondolo [1787], who have shown that the energy released by WIMP annihilations in these mini-haloes, during the formation of a protostar, may exceed any cooling mechanism, thus leading to a new phase of stellar evolution, which consists in a star supported only by DM annihilations, also called a 'Dark Star'. The reason that this can happen is shown in Fig. 29.6. The DM distribution of the mini-halo hosting the star responds to the evolving gravitational potential as the baryons cool and collapse (this is the so-called adiabatic contraction of the DM profile), as shown in Fig. 29.6 for two different initial profiles.

The DM heating due to the coupling of the annihilation radiation in the adiabatically compressed profile with the baryons must overcome the cooling mechanisms at work in the protostellar object, the dominant being H_2 cooling. Setting the heating rate equal to the cooling rate gives the critical

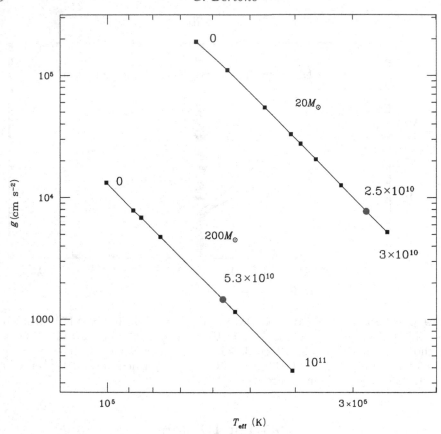

Fig. 29.7. Zero-age main sequence (ZAMS) positions of 20 and $200M_\odot$ Pop III stars in the g vs. T_{eff} plane for different DM densities (labels in units of GeV cm^{-3}). Larger circles correspond to the critical WIMP density (see text). Adapted from ref. [1840].

temperature $T_c(n)$, as a function of the ambient density n. If the star forms exactly at the centre of the halo, as should be the case if one believes current numerical simulations, Spolyar *et al.* have shown that no matter what the WIMP mass and H_2 fraction are, the heating always overcomes the cooling, meaning that, if these processes are actually at work, DM heating dominates inside the core and leads to the formation of a 'dark star', supported by DM annihilations instead of nuclear burning. The pioneering work of Spolyar *et al.* has triggered rather intense activity on these objects. The formation of proto-stars was found to be significantly delayed [884; 1153]. It was also shown in refs. [885; 1152] that the annihilation of DM particles captured by the star (as opposed to the case of annihilations in the adiabatic compressed profile), owing to scattering off the stellar nuclei, can lead to an energy injection that overwhelms nuclear reactions. This is particularly important because this means that the star can be fuelled not only by the WIMPs that happen to be in the adiabatically compressed profile, but also by the stars in the mini-halo whose orbits happen to cross the star or protostar. In practice, the entire mini-halo becomes a reservoir of energy for the star. In ref. [1152] the stellar evolution was followed to the main-sequence phase, as was also done subsequently by Yoon *et al.* [1953] and by Taoso *et al.* [1840].

There are some obvious astrophysical consequences of such a modified scenario for Pop III star formation. Primordial star formation is necessarily modified, with consequences, among other things, for the reionization history of the Universe (see e.g. refs. [1497; 1703]).

It has also been argued that dark stars may remain frozen in a non-evolving configuration, provided that their DM reservoir is big enough. The observational consequences are particularly interesting, since this leads to the prediction of a population of objects at redshift zero, and thus for example in our Galaxy, with very peculiar properties [1840]. In Fig. 29.7, for instance, we show the effective temperature and gravity acceleration at the surface of these frozen Pop III stars, kept in the H-burning phase, for different DM densities. Frozen stars would thus appear to be much bigger and to have much lower surface temperatures than normal stars.

Acknowledgements

This chapter is partially based on work done in collaboration with Malcolm Fairbairn, Georges Meynet and Marco Taoso. It is a pleasure to thank Fabio Iocco and Marco Taoso for useful comments.

References

[1] G. Aad *et al.* The ATLAS experiment at the CERN Large Hadron Collider, *JINST*, **3**:S08003, 2008.

[2] C. E. Aalseth *et al.* Experimental constraints on a dark matter origin for the DAMA annual modulation effect, *Phys. Rev. Lett.*, **101**:251301, 2008, arXiv:0807.0879 [astro-ph].

[3] T. Aaltonen *et al.* Inclusive search for squark and gluino production in ppbar collisions at $\sqrt{s} = 1.96$ TeV, 2008, arXiv:0811.2512 [hep-ex].

[4] T. Aaltonen *et al.* Search for pair production of scalar top quarks decaying to a τ lepton and a b quark in $p\bar{p}$ collisions at $\sqrt{s} = 1.96$ TeV, *Phys. Rev. Lett.*, **101**:071802, 2008, arXiv:0802.3887 [hep-ex].

[5] T. Aaltonen *et al.* Search for supersymmetry in $p\bar{p}$ collisions at $\sqrt{s} = 1.96$-TeV using the trilepton signature of chargino-neutralino production, *Phys. Rev. Lett.*, **101**:251801, 2008, arXiv:0808.2446 [hep-ex].

[6] M. Aaronson. Accurate radial velocities for carbon stars in Draco and Ursa Minor – The first hint of a dwarf spheroidal mass-to-light ratio, *Astrophys. J.*, **266**:L11–L15, 1983.

[7] K. Abazajian, G. M. Fuller, and M. Patel. Sterile neutrino hot, warm, and cold dark matter, *Phys. Rev.*, D**64**:023501, 2001, astro-ph/0101524.

[8] K. Abazajian, G. M. Fuller, and W. H. Tucker. Direct detection of warm dark matter in the X-ray, *Astrophys. J.*, **562**:593–604, 2001, astro-ph/0106002.

[9] V. M. Abazov *et al.* Search for neutral Higgs bosons at high $\tan\beta$ in the b (h/H/A)→b tau tau channel, 2008, arXiv:0811.0024 [hep-ex].

[10] V. M. Abazov *et al.* Search for squarks and gluinos in events with jets and missing transverse energy using 2.1 fb^{-1} of $p\bar{p}$ collision data at $\sqrt{(s)} = 1.96$ TeV, *Phys. Lett.*, B**660**:449–457, 2008, arXiv:0712.3805 [hep-ex].

[11] V. M. Abazov *et al.* Search for the lightest scalar top quark in events with two leptons in ppbar collisions at $\sqrt{(s)}$ = 1.96 TeV, 2008, arXiv:0811.0459 [hep-ex].

[12] B. Abbott *et al.* Search for the trilepton signature from the associated production of SUSY $\tilde{\chi}_1^\pm \tilde{\chi}_2^0$ gauginos, *Phys. Rev. Lett.*, **80**:1591–1596, 1998, hep-ex/9705015.

[13] L. F. Abbott and P. Sikivie. A cosmological bound on the invisible axion, *Phys. Lett.*, B**120**:133–136, 1983.

[14] L. F. Abbott and M. B. Wise. Large-scale anisotropy of the microwave background and the amplitude of energy density fluctuations in the early universe, *Astrophys. J.*, **282**:L47–L50, 1984.

[15] K. Abe. The XMASS experiment, *J. Phys. Conf. Ser.*, **120**:042022, 2008.

[16] K. Abe *et al.* Distillation of liquid xenon to remove krypton, 2008, arXiv:0809.4413 [physics.ins-det].

[17] K. Abe *et al.* Measurement of cosmic-ray low-energy antiproton spectrum with the first BESS-Polar Antarctic flight, *ArXiv e-prints*, May 2008.

[18] A. Abulencia *et al.* Search for neutral MSSM Higgs bosons decaying to tau pairs in $p\bar{p}$ collisions at $\sqrt{s}$ = 1.96 TeV, *Phys. Rev. Lett.*, **96**:011802, 2006, hep-ex/0508051.

[19] E. Accomando, R. L. Arnowitt, B. Dutta, and Y. Santoso. Neutralino proton cross sections in supergravity models, *Nucl. Phys.*, B**585**:124–142, 2000, hep-ph/0001019.

[20] M. Ackermann *et al.* Limits to the muon flux from neutralino annihilations in the sun with the AMANDA detector, *Astropart. Phys.*, **24**:459–466, 2006, astro-ph/0508518.

[21] S. L. Adler. Axial vector vertex in spinor electrodynamics, *Phys. Rev.*, **177**:2426–2438, 1969.

[22] R. Adolphi *et al.* The CMS experiment at the CERN LHC, *JINST*, **3**:S08004, 2008.

[23] O. Adriani *et al.* A new measurement of the antiproton-to-proton flux ratio up to 100 GeV in the cosmic radiation, 2008, arXiv:0810.4994 [astro-ph].

[24] O. Adriani *et al.* Observation of an anomalous positron abundance in the cosmic radiation, 2008, arXiv:0810.4995 [astro-ph].

[25] N. Afshordi, R. Mohayaee, and E. Bertschinger. Hierarchical phase space structure of dark matter haloes: Tidal debris, caustics, and dark matter annihilation, 2008, arXiv:0811.1582 [astro-ph].

[26] K. Agashe, A. Belyaev, T. Krupovnickas, G. Perez, and J. Virzi. LHC signals from warped extra dimensions, *Phys. Rev.*, D**77**:015003, 2008, hep-ph/0612015.

[27] K. Agashe, R. Contino, and R. Sundrum. Top compositeness and precision unification, *Phys. Rev. Lett.*, **95**:171804, 2005, hep-ph/0502222.

[28] K. Agashe, A. Delgado, M. J. May, and R. Sundrum. RS1, custodial isospin and precision tests, *JHEP*, **08**:050, 2003, hep-ph/0308036.

[29] K. Agashe, N. G. Deshpande, and G. H. Wu. Universal extra dimensions and b → s gamma, *Phys. Lett.*, B**514**:309–314, 2001, hep-ph/0105084.

[30] K. Agashe *et al.* LHC signals for warped electroweak neutral gauge bosons, *Phys. Rev.*, D**76**:115015, 2007, arXiv:0709.0007 [hep-ph].

[31] K. Agashe, A. Falkowski, I. Low, and G. Servant. KK parity in warped extra dimension, *JHEP*, **04**:027, 2008, arXiv:0712.2455 [hep-ph].

[32] K. Agashe and G. Servant. Warped unification, proton stability and dark matter, *Phys. Rev. Lett.*, **93**:231805, 2004, hep-ph/0403143.

[33] K. Agashe and G. Servant. Baryon number in warped GUTs: Model building and (dark matter related) phenomenology, *JCAP*, **0502**:002, 2005, hep-ph/0411254.

[34] M. Aguilar *et al.* The Alpha Magnetic Spectrometer (AMS) on the International Space Station: Part I – results from the test flight on the space shuttle, *Phys. Rep.*, **366**:331–405, August 2002.

[35] M. Aguilar *et al.* Cosmic-ray positron fraction measurement from 1-GeV to 30-GeV with AMS-01, *Phys. Lett.*, B**646**:145–154, 2007, astro-ph/0703154.

[36] J. A. Aguilar-Saavedra *et al.* Supersymmetry parameter analysis: SPA convention and project, *Eur. Phys. J.*, C**46**:43–60, 2006, hep-ph/0511344.

[37] F. Aharonian *et al.* Very high energy gamma rays from the direction of Sagittarius A*, *Astron. Astrophys.*, **425**:L13–L17, 2004, astro-ph/0408145.

[38] M. Ahlers, J. Kersten, and A. Ringwald. Long-lived staus at neutrino telescopes, *JCAP*, **0607**:005, 2006, hep-ph/0604188.

[39] Z. Ahmed *et al.* Search for weakly interacting massive particles with the first five-tower data from the Cryogenic Dark Matter Search at the Soudan Underground Laboratory, *Phys. Rev. Lett.*, **102**:011301, 2009, arXiv:0802.3530 [astro-ph].

[40] E.-J. Ahn, G. Bertone, and D. Merritt. Impact of astrophysical processes on the gamma-ray background from dark matter annihilations, *Phys. Rev.*, D**76**:023517, 2007, astro-ph/0703236.

[41] D. S. Akerib *et al.* Exclusion limits on the WIMP nucleon cross-section from the first run of the Cryogenic Dark Matter Search in the Soudan underground lab, *Phys. Rev.*, D**72**:052009, 2005, astro-ph/0507190.

[42] E. K. Akhmedov, V. A. Rubakov, and A. Yu. Smirnov. Baryogenesis via neutrino oscillations, *Phys. Rev. Lett.*, **81**:1359–1362, 1998, hep-ph/9803255.

[43] D. Akimov *et al.* Measurements of scintillation efficiency and pulse-shape for low energy recoils in liquid xenon, *Phys. Lett.*, B**524**:245–251, 2002, hep-ex/0106042.

[44] D. Yu. Akimov *et al.* The ZEPLIN-III dark matter detector: Instrument design, manufacture and commissioning, *Astropart. Phys.*, **27**:46–60, 2007, astro-ph/0605500.

[45] J. Albert *et al.* Observation of gamma rays from the galactic center with the MAGIC telescope, *Astrophys. J.*, **638**:L101–L104, 2006, astro-ph/0512469.

[46] I. Albuquerque, G. Burdman, and Z. Chacko. Neutrino telescopes as a direct probe of supersymmetry breaking, *Phys. Rev. Lett.*, **92**:221802, 2004, hep-ph/0312197.

[47] I. F. M. Albuquerque, G. Burdman, and Z. Chacko. Direct detection of supersymmetric particles in neutrino telescopes, *Phys. Rev.*, D**75**:035006, 2007, hep-ph/0605120.

[48] I. F. M. Albuquerque, G. Burdman, C. A. Krenke, and B. Nosratpour. Direct detection of Kaluza–Klein particles in neutrino telescopes, *Phys. Rev.*, D**78**:015010, 2008, arXiv:0803.3479 [hep-ph].

[49] J. Alcaraz *et al.* Leptons in near Earth orbit, *Phys. Lett.*, B**484**:10–22, 2000.

[50] M. S. Alenazi and P. Gondolo. Directional recoil rates for WIMP direct detection, *Phys. Rev.*, D**77**:043532, 2008, arXiv:0712.0053 [astro-ph].

[51] R. Allahverdi and M. Drees. Production of massive stable particles in inflaton decay, *Phys. Rev. Lett.*, **89**:091302, 2002, hep-ph/0203118.

[52] R. Allahverdi and M. Drees. Thermalization after inflation and production of massive stable particles, *Phys. Rev.*, D**66**:063513, 2002, hep-ph/0205246.

[53] B. Allanach *et al.* SUSY Les Houches Accord 2, 2008, arXiv:0801.0045 [hep-ph].

[54] B. Allanach, S. Kraml, and W. Porod. Comparison of SUSY mass spectrum calculations, 2002, hep-ph/0207314.

[55] B. C. Allanach. SOFTSUSY: A C++ program for calculating supersymmetric spectra, *Comput. Phys. Commun.*, **143**:305–331, 2002, hep-ph/0104145.

[56] B. C. Allanach. SUSY predictions and SUSY tools at the LHC, 2008, arXiv:0805.2088 [hep-ph].

[57] B. C. Allanach, G. Belanger, F. Boudjema, and A. Pukhov. Requirements on collider data to match the precision of WMAP on supersymmetric dark matter, *JHEP*, **12**:020, 2004, hep-ph/0410091.

[58] B. C. Allanach, K. Cranmer, C. G. Lester, and A. M. Weber. Natural priors, CMSSM fits and LHC weather forecasts, *JHEP*, **08**:023, 2007, arXiv:0705.0487 [hep-ph].

[59] B. C. Allanach et al. The Snowmass points and slopes: Benchmarks for SUSY searches, 2002, hep-ph/0202233.

[60] B. C. Allanach and D. Hooper. Panglossian prospects for detecting neutralino dark matter in light of natural priors, 2008, arXiv:0806.1923 [hep-ph].

[61] B. C. Allanach and C. G. Lester. Multi-dimensional mSUGRA likelihood maps, *Phys. Rev.*, **D73**:015013, 2006, hep-ph/0507283.

[62] B. C. Allanach, C. G. Lester, M. A. Parker, and B. R. Webber. Measuring sparticle masses in non-universal string inspired models at the LHC, *JHEP*, **09**:004, 2000, hep-ph/0007009.

[63] B. C. Allanach, C. G. Lester, and A. M. Weber. The dark side of mSUGRA, *JHEP*, **12**:065, 2006, hep-ph/0609295.

[64] B. Allgood et al. The shape of dark matter halos: Dependence on mass, redshift, radius, and formation, *Mon. Not. Roy. Astron. Soc.*, **367**:1781–1796, 2006, astro-ph/0508497.

[65] G. J. Alner et al. The DRIFT-I dark matter detector at Boulby: Design, installation and operation, *Nucl. Instrum. Meth.*, A**535**:644–655, 2004.

[66] G. J. Alner et al. First limits on nuclear recoil events from the ZEPLIN I galactic dark matter detector, *Astropart. Phys.*, **23**:444–462, 2005.

[67] G. J. Alner et al. The DRIFT-II dark matter detector: Design and commissioning, *Nucl. Instrum. Meth.*, A**555**:173–183, 2005.

[68] G. J. Alner et al. First limits on WIMP nuclear recoil signals in ZEPLIN-II: A two phase xenon detector for dark matter detection, *Astropart. Phys.*, **28**:287–302, 2007, astro-ph/0701858.

[69] R. Aloisio, P. Blasi, and A. V. Olinto. Neutralino annihilation at the galactic center revisited, *JCAP*, **0405**:007, 2004, astro-ph/0402588.

[70] J. Altmann, W. J. Egle, U. Bingel et al. Mirror system for the German X-ray satellite ABRIXAS: I. Flight mirror fabrication, integration, and testing. In R. B. Hoover and A. B. Walker, eds., volume 3444 of *Society of Photo-Optical Instrumentation Engineers (SPIE) Conference Series*, 350–358, November 1998.

[71] B. Altunkaynak, M. Holmes, and B. D. Nelson. Solving the LHC inverse problem with dark matter observations, *JHEP*, **10**:013, 2008, arXiv:0804.2899 [hep-ph].

[72] L. Alvarez-Gaume, M. Claudson, and M. B. Wise. Low-energy supersymmetry, *Nucl. Phys.*, **B207**:96, 1982.

[73] L. Alvarez-Gaume, J. Polchinski, and M. B. Wise. Minimal low-energy supergravity, *Nucl. Phys.*, **B221**:495, 1983.

[74] A. Alves, O. Eboli, and T. Plehn. It's a gluino, *Phys. Rev.*, **D74**:095010, 2006, hep-ph/0605067.

[75] J. Alwall, S. de Visscher, and F. Maltoni. QCD radiation in the production of heavy colored particles at the LHC, 2008, arXiv:0810.5350 [hep-ph].

[76] J. Alwall *et al.* MadGraph/MadEvent v4: The new Web generation, *JHEP*, **09**:028, 2007, arXiv:0706.2334 [hep-ph].

[77] J. Alwall, M.-P. Le, M. Lisanti, and J. G. Wacker. Searching for gluinos at the Tevatron, *Phys. Rev. Lett.*, **B666**:34–37, 2008, arXiv:0803.0019 [hep-ph].

[78] J. Alwall, D. Rainwater, and T. Plehn. Same-sign charginos and Majorana neutralinos at the LHC, *Phys. Rev.*, **D76**:055006, 2007, arXiv:0706.0536 [hep-ph].

[79] U. Amaldi, W. de Boer, and H. Furstenau. Comparison of grand unified theories with electroweak and strong coupling constants measured at LEP, *Phys. Lett.*, **B260**:447, 1991.

[80] J. Amare, B. Beltran, S. Cebrian *et al.* Light yield of undoped sapphire at low temperature under particle excitation. *Appl. Phys. Lett.*, **87**(26):264102, 2005.

[81] J. Amare *et al.* Recent developments on scintillating bolometers for WIMP searches: ROSEBUD status, *J. Phys. Conf. Ser.*, **39**:133–135, 2006.

[82] J. Amare *et al.* Scintillation of sapphire under particle excitation at low temperature, *J. Phys. Conf. Ser.*, **39**:200, 2006.

[83] M. Ambrosio *et al.* Limits on dark matter WIMPs using upward-going muons in the MACRO detector, *Phys. Rev.*, **D60**:082002, 1999, hep-ex/9812020.

[84] C. Amsler *et al.* Luminescence quenching of the triplet excimer state by air traces in gaseous argon, *JINST*, **3**:P02001, 2008, arXiv:0708.2621 [physics.ins-det].

[85] C. Amsler *et al.* Review of particle physics, *Phys. Lett.*, **B667**:1, 2008.

[86] S. Ando, J. F. Beacom, S. Profumo, and D. Rainwater. Probing new physics with long-lived charged particles produced by atmospheric

and astrophysical neutrinos, *JCAP*, **0804**:029, 2008, arXiv:0711.2908 [hep-ph].

[87] S. Ando and E. Komatsu. Anisotropy of the cosmic gamma-ray background from dark matter annihilation, *Phys. Rev.*, D**73**:023521, 2006, astro-ph/0512217.

[88] S. Ando, E. Komatsu, T. Narumoto, and T. Totani. Dark matter annihilation or unresolved astrophysical sources? Anisotropy probe of the origin of cosmic gamma-ray background, *Phys. Rev.*, D**75**:063519, 2007, astro-ph/0612467.

[89] S. Andreas, T. Hambye, and M. H. G. Tytgat. WIMP dark matter, Higgs exchange and DAMA, *JCAP*, **0810**:034, 2008, arXiv:0808.0255 [hep-ph].

[90] J. Angle *et al.* First results from the XENON10 dark matter experiment at the Gran Sasso National Laboratory, *Phys. Rev. Lett.*, **100**:021303, 2008, arXiv:0706.0039 [astro-ph].

[91] J. Angle *et al.* Limits on spin-dependent WIMP-nucleon cross-sections from the XENON10 experiment, *Phys. Rev. Lett.*, **101**:091301, 2008, arXiv:0805.2939 [astro-ph].

[92] G. Angloher *et al.* Limits on WIMP dark matter using sapphire cryogenic detectors, *Astropart. Phys.*, **18**:43–55, 2002.

[93] G. Angloher *et al.* Limits on WIMP dark matter using scintillating CaWO-4 cryogenic detectors with active background suppression, *Astropart. Phys.*, **23**:325–339, 2005, astro-ph/0408006.

[94] G. W. Angus, B. Famaey, and D. A. Buote. X-ray group and cluster mass profiles in MOND: unexplained mass on the group scale, *Mon. Not. Roy. Astron. Soc.*, **387**:1470–1480, July 2008, 0709.0108.

[95] G. W. Angus, B. Famaey, and H.-S. Zhao. Can MOND take a bullet? Analytical comparisons of three versions of MOND beyond spherical symmetry, *Mon. Not. Roy. Astron. Soc.*, **371**:138, 2006, astro-ph/0606216.

[96] G. W. Angus, H.-Y. Shan, H.-S. Zhao, and B. Famaey. On the law of gravity, the mass of neutrinos and the proof of dark matter, *Astrophys. J.*, **654**:L13–L16, 2007, astro-ph/0609125.

[97] A. Anisimov, Y. Bartocci, and F. L. Bezrukov. Inflaton mass in the νMSM inflation, 2008, arXiv:0809.1097 [hep-ph].

[98] C. Ankenbrandt *et al.* Muon Collider Task Force Report. 2007. FERMILAB-TM-2399-APC.

[99] I. Antoniadis, N. Arkani-Hamed, S. Dimopoulos, and G. R. Dvali. New dimensions at a millimeter to a Fermi and superstrings at a TeV, *Phys. Lett.*, B**436**:257–263, 1998, hep-ph/9804398.

[100] T. Appelquist, H.-C. Cheng, and B. A. Dobrescu. Bounds on universal extra dimensions, *Phys. Rev.*, D**64**:035002, 2001, hep-ph/0012100.

[101] T. Appelquist and H.-U. Yee. Universal extra dimensions and the Higgs boson mass, *Phys. Rev.*, D**67**:055002, 2003, hep-ph/0211023.

[102] E. Aprile, A. Bolotnikov, D. Chen, and R. Mukherjee. W value in liquid krypton, *Phys. Rev. A*, **48**:1313–1318, August 1993.

[103] E. Aprile *et al.* Scintillation response of liquid xenon to low energy nuclear recoils, *Phys. Rev.*, D**72**:072006, 2005, astro-ph/0503621.

[104] E. Aprile *et al.* Simultaneous measurement of ionization and scintillation from nuclear recoils in liquid xenon as target for a dark matter experiment, *Phys. Rev. Lett.*, **97**:081302, 2006, astro-ph/0601552.

[105] E. Aprile *et al.* New measurement of the relative scintillation efficiency of xenon nuclear recoils below 10 keV, 2008, arXiv:0810.0274 [astro-ph].

[106] E. Aprile *et al.* The XENON10 dark matter search experiment, *submitted to Phys. Rev. D*, 2008.

[107] A. Arbey and F. Mahmoudi. SUSY constraints from relic density: high sensitivity to pre-BBN expansion rate, *Phys. Lett.*, B**669**:46–51, 2008, arXiv:0803.0741 [hep-ph].

[108] C. Arina, F. Bazzocchi, N. Fornengo, J. C. Romao, and J. W. F. Valle. Minimal supergravity sneutrino dark matter and inverse seesaw neutrino masses, *Phys. Rev. Lett.*, **101**:161802, 2008, arXiv:0806.3225 [hep-ph].

[109] C. Arina and N. Fornengo. Sneutrino cold dark matter, a new analysis: relic abundance and detection rates, *JHEP*, **11**:029, 2007, arXiv:0709.4477 [hep-ph].

[110] K. Arisaka *et al.* XAX: a multi-ton, multi-target detection system for dark matter, double beta decay and pp solar neutrinos, 2008, arXiv:0808.3968 [astro-ph].

[111] N. Arkani-Hamed, S. Dimopoulos, and G. R. Dvali. The hierarchy problem and new dimensions at a millimeter, *Phys. Lett.*, B**429**:263–272, 1998, hep-ph/9803315.

[112] N. Arkani-Hamed *et al.* MARMOSET: The path from LHC data to the new standard model via on-shell effective theories, 2007, hep-ph/0703088.

[113] N. Arkani-Hamed, D. P. Finkbeiner, T. Slatyer, and N. Weiner. A theory of dark matter, 2008, arXiv:0810.0713 [hep-ph].

[114] N. Arkani-Hamed, L. J. Hall, H. Murayama, D. Tucker-Smith, and N. Weiner. Neutrino masses at v**(3/2), 2000, hep-ph/0007001.

[115] N. Arkani-Hamed, L. J. Hall, H. Murayama, D. Tucker-Smith, and N. Weiner. Small neutrino masses from supersymmetry breaking, *Phys. Rev.*, D**64**:115011, 2001, hep-ph/0006312.

[116] N. Arkani-Hamed, G. L. Kane, J. Thaler, and L. T. Wang. Supersymmetry and the LHC inverse problem, 2006, hep-ph/0512190.

[117] N. Arkani-Hamed, M. Porrati, and L. Randall. Holography and phenomenology, *JHEP*, **08**:017, 2001, hep-th/0012148.

[118] N. Arkani-Hamed and N. Weiner. LHC signals for a superunified theory of dark matter, 2008, arXiv:0810.0714 [hep-ph].

[119] F. Arneodo *et al.* Scintillation efficiency of nuclear recoil in liquid xenon, *Nucl. Instrum. Meth.*, A**449**:147–157, 2000.

[120] R. L. Arnowitt, A. H. Chamseddine, and P. Nath. Masses of superpartners of quarks, leptons, and gauge mesons in supergravity grand unified theories, *Phys. Rev. Lett.*, **50**:232, 1983.

[121] R. L. Arnowitt and B. Dutta. Dark matter detection rates in SUGRA models, 2001, hep-ph/0112157.

[122] R. L. Arnowitt and B. Dutta. Dark matter, muon g-2 and other accelerator constraints, 2002, hep-ph/0211417.

[123] R. L. Arnowitt and B. Dutta. SUSY dark matter: Closing the parameter space, 2002, hep-ph/0210339.

[124] R. L. Arnowitt, B. Dutta, T. Kamon, N. Kolev, and D. A. Toback. Detection of SUSY in the stau-neutralino coannihilation region at the LHC, *Phys. Lett.*, B**639**:46–53, 2006, hep-ph/0603128.

[125] R. L. Arnowitt *et al.* Indirect measurements of the stau – neutralino 1(0) mass difference and mSUGRA in the co-annihilation region of mSUGRA models at the LHC, *Phys. Lett.*, B**649**:73–82, 2007.

[126] R. L. Arnowitt *et al.* Determining the dark matter relic density in the mSUGRA stau–neutralino co-annihilation region at the LHC, *Phys. Rev. Lett.*, **100**:231802, 2008, arXiv:0802.2968 [hep-ph].

[127] R. L. Arnowitt and P. Nath. Annual modulation signature for the direct detection of Milky Way wimps and supergravity models, *Phys. Rev.*, D**60**:044002, 1999, hep-ph/9902237.

[128] S. Arrenberg, L. Baudis, K. Kong, K. T. Matchev, and J. Yoo. Kaluza–Klein dark matter: Direct detection vis-a-vis LHC, 2008, arXiv:0805.4210 [hep-ph].

[129] T. Asaka, S. Blanchet, and M. Shaposhnikov. The nuMSM, dark matter and neutrino masses, *Phys. Lett.*, B**631**:151–156, 2005, hep-ph/0503065.

[130] T. Asaka, M. Laine, and M. Shaposhnikov. On the hadronic contribution to sterile neutrino production, *JHEP*, **06**:053, 2006, hep-ph/0605209.

[131] T. Asaka, M. Laine, and M. Shaposhnikov. Lightest sterile neutrino abundance within the nuMSM, *JHEP*, **01**:091, 2007, hep-ph/0612182.

[132] T. Asaka and M. Shaposhnikov. The nuMSM, dark matter and baryon asymmetry of the universe, *Phys. Lett.*, **B620**:17–26, 2005, hep-ph/0505013.

[133] T. Asaka, M. Shaposhnikov, and A. Kusenko. Opening a new window for warm dark matter, *Phys. Lett.*, **B638**:401–406, 2006, hep-ph/0602150.

[134] Y. Asaoka *et al.* Measurements of cosmic-ray low-energy antiproton and proton spectra in a transient period of solar field reversal, *Phys. Rev. Lett.*, **88**(5):051101, 2002.

[135] Y. Ascasibar, Y. Hoffman, and S. Gottlöber. Secondary infall and dark matter haloes, *Mon. Not. Roy. Astron. Soc.*, **376**:393–404, 2007.

[136] M. Asplund, D. L. Lambert, P. E. Nissen, F. Primas, and V. V. Smith. Lithium isotopic abundances in metal-poor halo stars, *Astrophys. J.*, **644**:229–259, 2006.

[137] J. Astrom *et al.* Fracture processes studied in CRESST, *Nucl. Instrum. Meth.*, **A559**:754–756, 2006.

[138] S. J. Asztalos *et al.* Experimental constraints on the axion dark matter halo density, *Astrophys. J.*, **571**:L27–L30, 2002, astro-ph/0104200.

[139] S. J. Asztalos *et al.* An improved RF cavity search for halo axions, *Phys. Rev.*, **D69**:011101, 2004, astro-ph/0310042.

[140] S. J. Asztalos, L. J. Rosenberg, K. van Bibber, P. Sikivie, and K. Zioutas. Searches for astrophysical and cosmological axions, *Ann. Rev. Nucl. Part. Sci.*, **56**:293–326, 2006.

[141] C. Athanasiou, C. G. Lester, J. M. Smillie, and B. R. Webber. Distinguishing spins in decay chains at the Large Hadron Collider, *JHEP*, **08**:055, 2006, hep-ph/0605286.

[142] ATLAS Collaboration. ATLAS detector and physics performance. Technical design report, 1999. Vol. 2. CERN-LHCC-99-15.

[143] ATLAS Collaboration. ATLAS detector and physics performance. Technical design report, 1999. CERN-LHCC-99-14 and 15.

[144] ATLAS Collaboration. Detector and physics performance. Technical design report, 1999. CERN-LHCC-1999-015.

[145] ATLAS Collaboration. Expected performance of the ATLAS Experiment, detector, trigger and physics, 2008. CERN-OPEN-2008-020.

[146] F. T. Avignone III, P. S. Barbeau, and J. I. Collar. Comments on new limits on spin-independent couplings of low-mass WIMP dark matter with a germanium detector at a threshold of 200 eV, 2008, arXiv:0806.1341 [hep-ex].

[147] M. Axenides, R. H. Brandenberger, and M. S. Turner. Development of axion perturbations in an axion dominated universe, *Phys. Lett.*, **B126**:178, 1983.

[148] G. Azuelos *et al.* Exploring little Higgs models with ATLAS at the LHC, *Eur. Phys. J.*, **C39S2**:13–24, 2005, hep-ph/0402037.

[149] H. Bachacou, I. Hinchliffe, and F. E. Paige. Measurements of masses in SUGRA models at LHC, *Phys. Rev.*, **D62**:015009, 2000, hep-ph/9907518.

[150] D. J. Bacon, D. M. Goldberg, B. T. P. Rowe, and A. N. Taylor. Weak gravitational flexion, *Mon. Not. Roy. Astron. Soc.*, **365**:414–428, January 2006.

[151] D. J. Bacon, A. R. Refregier, and R. S. Ellis. Detection of weak gravitational lensing by large-scale structure, *Mon. Not. Roy. Astron. Soc.*, **318**:625–640, October 2000.

[152] G. D. Badhwar, R. L. Golden, and S. A. Stephens. Analytic representation of the proton–proton and proton–nucleus cross-sections and its application to the sea-level spectrum and charge ratio of muons, *Phys. Rev.*, **D15**:820–831, February 1977.

[153] S. Baek, D. G. Cerdeno, Y. G. Kim, P. Ko, and C. Munoz. Direct detection of neutralino dark matter in supergravity, *JHEP*, **06**:017, 2005, hep-ph/0505019.

[154] H. Baer and C. Balazs. χ^2 analysis of the minimal supergravity model including WMAP, $g_\mu - 2$ and $b \rightarrow s\gamma$ constraints, *JCAP*, **0305**:006, 2003, hep-ph/0303114.

[155] H. Baer, C. Balazs, A. Belyaev, and J. O'Farrill. Direct detection of dark matter in supersymmetric models, *JCAP*, **0309**:007, 2003, hep-ph/0305191.

[156] H. Baer and M. Brhlik. Cosmological relic density from minimal supergravity with implications for collider physics, *Phys. Rev.*, **D53**:597, 1996, hep-ph/9508321.

[157] H. Baer and M. Brhlik. Neutralino dark matter in minimal supergravity: Direct detection vs. collider searches, *Phys. Rev.*, **D57**:567, 1998, hep-ph/9706509.

[158] H. Baer *et al.* Yukawa unified supersymmetric SO(10) model: Cosmology, rare decays and collider searches, *Phys. Rev.*, D**63**:015007, 2001, hep-ph/0005027.

[159] H. Baer *et al.* Updated constraints on the minimal supergravity model, *JHEP*, **07**:050, 2002, hep-ph/0205325.

[160] H. Baer, T. Krupovnickas, S. Profumo, and P. Ullio. Model independent approach to focus point supersymmetry: From dark matter to collider searches, *JHEP*, **10**:020, 2005, hep-ph/0507282.

[161] H. Baer, A. Mustafayev, S. Profumo, A. Belyaev, and X. Tata. Direct, indirect and collider detection of neutralino dark matter in SUSY models with non-universal Higgs masses, *JHEP*, **07**:065, 2005, hep-ph/0504001.

[162] H. Baer and S. Profumo. Low energy antideuterons: Shedding light on dark matter, *JCAP*, **0512**:008, 2005, astro-ph/0510722.

[163] J. Bagnasco, M. Dine, and S. D. Thomas. Detecting technibaryon dark matter, *Phys. Lett.*, B**320**:99–104, 1994, hep-ph/9310290.

[164] J. N. Bahcall, M. Schmidt, and R. M. Soneira. The Galactic Spheroid, *Astrophys. J.*, **265**:730, 1983.

[165] J. N. Bahcall and R. M. Soneira. The Universe at faint magnitudes. 2. Models for the predicted star counts, *Astrophys. J. Suppl.*, **44**:73–110, 1980.

[166] J. N. Bahcall and R. A. Wolf. Star distribution around a massive black hole in a globular cluster, *Astrophys. J.*, **209**:214–232, October 1976.

[167] J. N. Bahcall and R. A. Wolf. The star distribution around a massive black hole in a globular cluster. II. Unequal star masses, *Astrophys. J.*, **216**:883–907, September 1977.

[168] M. Bahr *et al.* Herwig++ Physics and Manual, *Eur. Phys. J.*, C**58**:639–707, 2008, arXiv:0803.0883 [hep-ph].

[169] Y. Bai. Mini Little Higgs and dark matter, *Phys. Lett.*, B**666**:332–335, 2008, arXiv:0801.1662 [hep-ph].

[170] Y. Bai and Z. Han. A Unified dark matter model in sUED, 2008, arXiv:0811.0387 [hep-ph].

[171] J. Bailin *et al.* Internal alignment of the halos of disk galaxies in cosmological hydrodynamic simulations, *Astrophys. J.*, **627**:L17–L20, 2005, astro-ph/0505523.

[172] J. Bailin and M. Steinmetz. Internal and external alignment of the shapes and angular momenta of LCDM halos, *Astrophys. J.*, **627**:647–665, 2005, astro-ph/0408163.

[173] S. Bailly, K. Jedamzik, and G. Moultaka. Gravitino dark matter and the cosmic lithium abundances, 2008, arXiv:0812.0788 [hep-ph].

[174] G. Bakale, U. Sowada, and W. F. Schmidt. Effect of an electric field on electron attachment to sulfur hexafluoride, nitrous oxide, and molecular oxygen in liquid argon and xenon, *J. Phys. Chem.*, **80**:2556–2559, 1976.

[175] C. A. Baker *et al.* An improved experimental limit on the electric dipole moment of the neutron, *Phys. Rev. Lett.*, **97**:131801, 2006, hep-ex/0602020.

[176] E. A. Baltz, M. Battaglia, M. E. Peskin, and T. Wizansky. Determination of dark matter properties at high-energy colliders, *Phys. Rev.*, D**74**:103521, 2006, hep-ph/0602187.

[177] E. A. Baltz and L. Bergstrom. Detection of leptonic dark matter, *Phys. Rev.*, D**67**:043516, 2003, hep-ph/0211325.

[178] E. A. Baltz, C. Briot, P. Salati, R. Taillet, and J. Silk. Detection of neutralino annihilation photons from external galaxies, *Phys. Rev.*, D**61**(2):023514, January 2000.

[179] E. A. Baltz and J. Edsjo. Positron propagation and fluxes from neutralino annihilation in the halo, *Phys. Rev.*, D**59**:023511, 1999, astro-ph/9808243.

[180] E. A. Baltz *et al.* Pre-launch estimates for GLAST sensitivity to dark matter annihilation signals, *JCAP*, **0807**:013, 2008, arXiv:0806.2911 [astro-ph].

[181] E. A. Baltz and P. Gondolo. Improved constraints on supersymmetric dark matter from muon g-2, *Phys. Rev.*, D**67**:063503, 2003, astro-ph/0207673.

[182] E. A. Baltz and P. Gondolo. Markov chain Monte Carlo exploration of minimal supergravity with implications for dark matter, *JHEP*, **10**:052, 2004, hep-ph/0407039.

[183] E. A. Baltz and D. Hooper. Kaluza–Klein dark matter, electrons and gamma ray telescopes, *JCAP*, **0507**:001, 2005, hep-ph/0411053.

[184] E. A. Baltz and L. Wai. Diffuse inverse Compton and synchrotron emission from dark matter annihilations in galactic satellites, *Phys. Rev.*, D**70**:023512, 2004, astro-ph/0403528.

[185] S. R. Bandler *et al.* Recoil direction sensitivity in a superfluid helium particle detector, *Proc. Workshop on the Identification of dark matter*, N. J. C. Spooner, ed., World Scientific, 469–474, 1997.

[186] M. Bando, T. Kugo, T. Noguchi, and K. Yoshioka. Brane fluctuation and suppression of Kaluza–Klein mode couplings, *Phys. Rev. Lett.*, **83**:3601–3604, 1999, hep-ph/9906549.

[187] R. Barate *et al.* Search for the standard model Higgs boson at LEP, *Phys. Lett.*, B565:61, 2003, hep-ex/0306033.

[188] P. S. Barbeau, J. I. Collar, and O. Tench. Large-mass ultra-low noise germanium detectors: Performance and applications in neutrino and astroparticle physics, *JCAP*, 0709:009, 2007, nucl-ex/0701012.

[189] P. S. Barbeau, J. I. Collar, and P. M. Whaley. Design and characterization of a neutron calibration facility for the study of sub-keV nuclear recoils, *Nucl. Instrum. Meth.*, A574:385–391, 2007, nucl-ex/0701011.

[190] E. Barberio *et al.* Averages of b-hadron properties at the end of 2005, 2006, hep-ex/0603003.

[191] R. Barbieri and A. Dolgov. Bounds on sterile-neutrinos from nucleosynthesis, *Phys. Lett.*, B237:440, 1990.

[192] R. Barbieri and A. Dolgov. Neutrino oscillations in the early universe, *Nucl. Phys.*, B349:743–753, 1991.

[193] R. Barbieri, S. Ferrara, and C. A. Savoy. Gauge models with spontaneously broken local supersymmetry, *Phys. Lett.*, B119:343, 1982.

[194] R. Barbieri, M. Frigeni, and G. F. Giudice. Dark matter neutralinos in supergravity theories, *Nucl. Phys.*, B313:725, 1989.

[195] R. Barbieri, T. Gregoire, and L. J. Hall. Mirror world at the Large Hadron Collider, 2005, hep-ph/0509242.

[196] R. Barbieri, L. J. Hall, and V. S. Rychkov. Improved naturalness with a heavy Higgs: An alternative road to LHC physics, *Phys. Rev.*, D74:015007, 2006, hep-ph/0603188.

[197] R. Barbieri and A. Strumia. What is the limit on the Higgs mass?, *Phys. Lett.*, B462:144–149, 1999, hep-ph/9905281.

[198] R. Barbieri and A. Strumia. The 'LEP paradox', 2000, hep-ph/0007265.

[199] S. Bardeau, G. Soucail, J.-P. Kneib, *et al.* A CFH12k lensing survey of X-ray luminous galaxy clusters. II. Weak lensing analysis and global correlations, *Astron. Astrophys.*, 470:449–466, August 2007.

[200] J. M. Bardeen, P. J. Steinhardt, and M. S. Turner. Spontaneous creation of almost scale-free density perturbations in an inflationary universe, *Phys. Rev.*, D28:679, 1983.

[201] W. A. Bardeen and S.-H. H. Tye. Current algebra applied to properties of the light Higgs boson, *Phys. Lett.*, B74:229, 1978.

[202] G. Barenboim and J. D. Lykken. Minimal noncanonical cosmologies, *JHEP*, 07:016, 2006, astro-ph/0604528.

[203] G. Barenboim and J. D. Lykken. Quintessence, inflation and baryogenesis from a single pseudo-Nambu–Goldstone boson, *JHEP*, **10**:032, 2007, arXiv:0707.3999 [astro-ph].

[204] V. Barger *et al.* Recoil detection of the lightest neutralino in MSSM singlet extensions, *Phys. Rev.*, D**75**:115002, 2007, hep-ph/0702036.

[205] V. Barger, W.-Y. Keung, and G. Shaughnessy. Spin dependence of dark matter scattering, *Phys. Rev.*, D**78**:056007, 2008, arXiv:0806.1962 [hep-ph].

[206] V. Barger, P. Langacker, M. McCaskey, M. J. Ramsey-Musolf, and G. Shaughnessy. LHC phenomenology of an extended standard model with a real scalar singlet, *Phys. Rev.*, D**77**:035005, 2008, arXiv:0706.4311 [hep-ph].

[207] V. D. Barger *et al.* CP-violating phases in SUSY, electric dipole moments, and linear colliders, *Phys. Rev.*, D**64**:056007, 2001, hep-ph/0101106.

[208] V. D. Barger and C. Kao. Relic density of neutralino dark matter in supergravity models, *Phys. Rev.*, D**57**:3131, 1998, hep-ph/9704403.

[209] V. D. Barger and C. Kao. Implications of new CMB data for neutralino dark matter, *Phys. Lett.*, B**518**:117, 2001, hep-ph/0106189.

[210] V. D. Barger, R. J. N. Phillips, and S. Sarkar. Remarks on the KARMEN anomaly, *Phys. Lett.*, B**352**:365–371, 1995, hep-ph/9503295.

[211] M. Barkovich, J. C. D'Olivo, and R. Montemayor. Active–sterile neutrino oscillations and pulsar kicks, *Phys. Rev.*, D**70**:043005, 2004, hep-ph/0402259.

[212] N. Baro, F. Boudjema, and A. Semenov. Automatised full one-loop renormalisation of the MSSM I: The Higgs sector, the issue of tan(beta) and gauge invariance, 2008, arXiv:0807.4668 [hep-ph].

[213] N. Baro, F. Boudjema, and A. Semenov. Full one-loop corrections to the relic density in the MSSM: A few examples, *Phys. Lett.*, B**660**:550–560, 2008, arXiv:0710.1821 [hep-ph].

[214] A. Barr, C. Lester, and P. Stephens. m(T2): The truth behind the glamour, *J. Phys.*, G**29**:2343–2363, 2003, hep-ph/0304226.

[215] A. J. Barr. Using lepton charge asymmetry to investigate the spin of supersymmetric particles at the LHC, *Phys. Lett.*, B**596**:205–212, 2004, hep-ph/0405052.

[216] A. J. Barr. Measuring slepton spin at the LHC, *JHEP*, **02**:042, 2006, hep-ph/0511115.

[217] A. J. Barr, B. Gripaios, and C. G. Lester. Weighing WIMPs with kinks at colliders: Invisible particle mass measurements from Endpoints, *JHEP*, **02**:014, 2008, arXiv:0711.4008 [hep-ph].

[218] A. J. Barr, A. Pinder, and M. Serna. Precision determination of invisible-particle masses at the CERN LHC: II, 2008, arXiv:0811.2138 [hep-ph].

[219] S. M. Barr, R. S. Chivukula, and E. Farhi. Electroweak fermion number violation and the production of stable particles in the early universe, *Phys. Lett.*, B**241**:387–391, 1990.

[220] A. Barrau, G. Boudoul, F. Donato, D. Maurin, P. Salati, and R. Taillet. Antiprotons from primordial black holes, *Astron. Astrophys.*, **388**:676–687, June 2002.

[221] A. Barrau *et al.* Kaluza–Klein dark matter and galactic antiprotons, *Phys. Rev.*, D**72**:063507, 2005, astro-ph/0506389.

[222] J. D. Barrow. Massive particles as a probe of the early universe, *Nucl. Phys.*, B**208**:501–508, 1982.

[223] M. Bartelmann. Arcs from a universal dark-matter halo profile. *Astron. Astrophys.*, **313**:697–702, September 1996.

[224] M. Bartelmann, R. Narayan, S. Seitz, and P. Schneider. Maximum-likelihood Cluster Reconstruction, *Astrophys. J.*, **464**:L115+, June 1996.

[225] M. Bartelmann and P. Schneider. Weak gravitational lensing, *Phys. Rep.*, **340**:291–472, January 2001.

[226] S. W. Barwick, J. J. Beatty, A. Bhattacharyya, *et al.* (HEAT Collaboration). Measurements of the cosmic-ray positron fraction from 1 to 50 GeV, *Astrophys. J.*, **482**:L191, 1997.

[227] G. Basini *et al.* The flux of cosmic ray antiprotons from 3.7 to 24 GeV, *26th International Cosmic Ray Conference, Salt Lake City*, **3**:77, August 1999.

[228] G. Battaglia, A. Helmi, E. Tolstoy, and M. Irwin. The mass content of the Sculptor dwarf spheroidal galaxy, *ArXiv e-prints*, July 2008, 0807.2334.

[229] M. Battaglia, A. Datta, A. De Roeck, K. Kong, and K. T. Matchev. Contrasting supersymmetry and universal extra dimensions at the CLIC multi-TeV $e^+ e^-$ collider, *JHEP*, **07**:033, 2005, hep-ph/0502041.

[230] M. Battaglia, A. K. Datta, A. De Roeck, K. Kong, and K. T. Matchev. Contrasting supersymmetry and universal extra dimensions at colliders, 2005, hep-ph/0507284.

[231] M. Battaglia, N. Kelley, and B. Hooberman. A study of $e^+e^- \rightarrow H^0 A^0$ production and the constraint on dark matter density, *Phys. Rev.*, D**78**:015021, 2008, arXiv:0805.1506 [hep-ex].

[232] M. Battaglia, A. De Roeck, J. R. Ellis and D. Schulte, eds. Physics at the CLIC multi-TeV linear collider: Report of the CLIC Physics Working Group, 2004, CERN-2004-005.

[233] R. A. Battye and E. P. S. Shellard. Axion string constraints, *Phys. Rev. Lett.*, **73**:2954–2957, 1994, astro-ph/9403018.

[234] R. A. Battye and E. P. S. Shellard. Global string radiation, *Nucl. Phys.*, B**423**:260–304, 1994, astro-ph/9311017.

[235] H. Baumgardt, J. Makino, and T. Ebisuzaki. Massive black holes in star clusters. II. Realistic cluster models, *Astrophys. J.*, **613**:1143–1156, October 2004.

[236] I. Bavykina *et al.* Development of cryogenic phonon detectors based on $CaMoO_4$ and $ZnWO_4$ scintillating crystals for direct dark matter search experiments, 2008, arXiv:0811.1786 [astro-ph].

[237] G. L. Bayatian *et al.* CMS physics: Technical design report, 2006, CERN-LHCC-2006-001.

[238] G. L. Bayatian *et al.* CMS technical design report, volume II: Physics performance, *J. Phys.*, G**34**:995–1579, 2007.

[239] A. S. Beach *et al.* Measurement of the cosmic-ray antiproton-to-proton abundance ratio between 4 and 50 GeV, *Phys. Rev. Lett.*, **87**:271101, December 2001.

[240] J. F. Beacom, N. F. Bell, and G. Bertone. Gamma-ray constraint on Galactic positron production by MeV dark matter, *Phys. Rev. Lett.*, **94**:171301, 2005, astro-ph/0409403.

[241] R. Bean, J. Dunkley, and E. Pierpaoli. Constraining isocurvature initial conditions with WMAP 3-year data, *Phys. Rev.*, D**74**:063503, 2006, astro-ph/0606685.

[242] J. J. Beatty *et al.* New measurement of the cosmic-ray positron fraction from 5-GeV to 15-GeV, *Phys. Rev. Lett.*, **93**:241102, 2004, astro-ph/0412230.

[243] P. Bechtle, K. Desch, W. Porod, and P. Wienemann. Determination of MSSM parameters from LHC and ILC observables in a global fit, *Eur. Phys. J.*, C**46**:533–544, 2006, hep-ph/0511006.

[244] P. Bechtle, K. Desch, and P. Wienemann. Fittino, a program for determining MSSM parameters from collider observables using an iterative method, *Comput. Phys. Commun.*, **174**:47–70, 2006, hep-ph/0412012.

[245] W. Beenakker *et al.* The production of charginos/neutralinos and sleptons at hadron colliders, *Phys. Rev. Lett.*, **83**:3780–3783, 1999, hep-ph/9906298.

[246] W. Beenakker, R. Hopker, and M. Spira. PROSPINO: A program for the PROduction of Supersymmetric Particles In Next-to-leading Order QCD, 1996, hep-ph/9611232.

[247] W. Beenakker, R. Hopker, M. Spira, and P. M. Zerwas. Squark and gluino production at hadron colliders, *Nucl. Phys.*, B**492**:51–103, 1997, hep-ph/9610490.

[248] W. Beenakker, M. Kramer, T. Plehn, M. Spira, and P. M. Zerwas. Stop production at hadron colliders, *Nucl. Phys.*, B**515**:3–14, 1998, hep-ph/9710451.

[249] M. C. Begelman, R. D. Blandford, and M. J. Rees. Massive black hole binaries in active galactic nuclei, *Nature*, **287**:307–309, September 1980.

[250] K. G. Begeman, A. H. Broeils, and R. H. Sanders. Extended rotation curves of spiral galaxies: Dark haloes and modified dynamics, *Mon. Not. Roy. Astron. Soc.*, **249**:523, 1991.

[251] E. Behnke *et al.* Improved spin-dependent WIMP limits from a bubble chamber, *Science*, **319**:933–936, 2008, arXiv:0804.2886 [astro-ph].

[252] J. Bekenstein and M. Milgrom. Does the missing mass problem signal the breakdown of Newtonian gravity?, *Astrophys. J.*, **286**:7–14, 1984.

[253] J. D. Bekenstein. The missing light puzzle: a hint about gravitation? In A. Coley, C. Dyer, and T. Tupper, eds., *Proceedings of the 2nd Canadian Conference on General Relativity and Relativistic Astrophysics*, 68–104, 1988.

[254] J. D. Bekenstein. Relativistic gravitation theory for the modified newtonian dynamics paradigm. *Phys. Rev. D*, **70**:083509, 2004.

[255] J. D. Bekenstein. Modified gravity vs dark matter: Relativistic theory for MOND, *PoS*, JHW2004:012, 2005, astro-ph/0412652.

[256] J. D. Bekenstein. The modified Newtonian dynamics – MOND – and its implications for new physics, *Contem. Phys.*, **47**:387–403, 2006.

[257] J. D. Bekenstein and E. Sagi. Do Newton's G and Milgrom's a_0 vary with cosmological epoch?, *Phys. Rev.*, D**77**:103512, 2008, arXiv:0802.1526 [astro-ph].

[258] J. D. Bekenstein and R. H. Sanders. A primer to relativistic MOND theory. In G. A. Mamon, F. Combes, C. Deffayet, and B. Fort, eds., *Mass Profiles and Shapes of Cosmological Structures*, volume 20 of *EAS Publications Series*, 225–230, 2006.

[259] G. Belanger, F. Boudjema, A. Cottrant, A. Pukhov, and A. Semenov. WMAP constraints on SUGRA models with non-universal gaugino

masses and prospects for direct detection, *Nucl. Phys.*, B**706**:411–454, 2005, hep-ph/0407218.

[260] G. Belanger, F. Boudjema, A. Pukhov, and A. Semenov. micrOMEGAs: A program for calculating the relic density in the MSSM, *Comput. Phys. Commun.*, **149**:103–120, 2002, hep-ph/0112278.

[261] G. Belanger, F. Boudjema, A. Pukhov, and A. Semenov. micrOMEGAs: Version 1.3, *Comput. Phys. Commun.*, **174**:577–604, 2006, hep-ph/0405253.

[262] G. Belanger, F. Boudjema, A. Pukhov, and A. Semenov. micrOMEGAs2.0: A program to calculate the relic density of dark matter in a generic model, *Comput. Phys. Commun.*, **176**:367–382, 2007, hep-ph/0607059.

[263] G. Belanger, F. Boudjema, A. Pukhov, and A. Semenov. Dark matter direct detection rate in a generic model with micrOMEGAs2.1, 2008, arXiv:0803.2360 [hep-ph].

[264] G. Belanger, E. Nezri, and A. Pukhov. Discriminating dark matter candidates using direct detection, 2008, arXiv:0810.1362 [hep-ph].

[265] G. Belanger, A. Pukhov, and G. Servant. Dirac Neutrino Dark Matter, *JCAP*, **0801**:009, 2008, arXiv:0706.0526 [hep-ph].

[266] A. A. Belavin, A. M. Polyakov, A. S. Shvarts, and Yu. S. Tyupkin. Pseudoparticle solutions of the Yang–Mills equations, *Phys. Lett.*, B**59**:85–87, 1975.

[267] E. F. Bell *et al.* The accretion origin of the Milky Way's stellar halo, *Astrophys. J.*, **680**:295–311, 2008, arXiv:0706.0004 [astro-ph].

[268] J. S. Bell and R. Jackiw. A PCAC puzzle: pi0 → gamma gamma in the sigma model, *Nuovo Cim.*, A**60**:47–61, 1969.

[269] P. Belli, R. Cerulli, N. Fornengo, and S. Scopel. Effect of the galactic halo modeling on the DAMA/NaI annual modulation result: An extended analysis of the data for WIMPs with a purely spin-independent coupling, *Phys. Rev.*, D**66**:043503, 2002, hep-ph/0203242.

[270] P. Belli *et al.* Extending the DAMA annual-modulation region by inclusion of the uncertainties in astrophysical velocities, *Phys. Rev.*, D**61**:023512, 2000, hep-ph/9903501.

[271] V. Belokurov *et al.* An orphan in the field of streams, *Astrophys. J.*, **658**:337–344, 2007, astro-ph/0605705.

[272] V. Belokurov *et al.* Cats and dogs, hair and a hero: A quintet of new Milky Way companions, *Astrophys. J.*, **654**:897–906, 2007, astro-ph/0608448.

[273] M. Beltran, J. Garcia-Bellido, and J. Lesgourgues. Isocurvature bounds on axions revisited, *Phys. Rev.*, **D75**:103507, 2007, hep-ph/0606107.

[274] P. Benetti *et al.* First results from a dark matter search with liquid argon at 87-K in the Gran Sasso underground laboratory, *Astropart. Phys.*, **28**:495–507, 2008, astro-ph/0701286.

[275] J. Benjamin *et al.* Cosmological constraints from the 100-deg^2 weak-lensing survey, *Mon. Not. Roy. Astron. Soc.*, **381**:702–712, October 2007.

[276] C. Bennett *et al.* First year Wilkinson Microwave Anisotropy Probe (WMAP) observations: Foreground emission, *Astrophys. J. Suppl.*, **148**:97, 2003, astro-ph/0302208.

[277] G. W. Bennett *et al.* Final report of the muon E821 anomalous magnetic moment measurement at BNL, *Phys. Rev.*, **D73**:072003, 2006, hep-ex/0602035.

[278] A. Benoit *et al.* Improved exclusion limits from the EDELWEISS WIMP search, *Phys. Lett.*, **B545**:43–49, 2002, astro-ph/0206271.

[279] A. Benoit *et al.* Sensitivity of the EDELWEISS WIMP search to spin-dependent interactions, *Phys. Lett.*, **B616**:25–30, 2005, astro-ph/0412061.

[280] A. Benoit *et al.* Measurement of the response of heat-and-ionization germanium detectors to nuclear recoils, *Nucl. Instrum. Meth.*, **A577**:558–568, 2007, astro-ph/0607502.

[281] M. C. Bento, O. Bertolami, R. Rosenfeld, and L. Teodoro. Self-interacting dark matter and invisibly decaying Higgs, *Phys. Rev.*, **D62**:041302, 2000, astro-ph/0003350.

[282] Z. Berezhiani. Mirror world and its cosmological consequences, *Int. J. Mod. Phys.*, **A19**:3775–3806, 2004, hep-ph/0312335.

[283] Z. Berezhiani. Through the Looking-Glass: Alice's Adventures in Mirror World, 2005, hep-ph/0508233.

[284] Z. Berezhiani and A. Lepidi. Cosmological bounds on the 'millicharges' of mirror particles, 2008, arXiv:0810.1317 [hep-ph].

[285] V. S. Berezinskii, S. V. Bulanov, V. A. Dogiel, and V. S. Ptuskin. *Astrophysics of Cosmic Rays*. V. L. Ginsburg, ed., North-Holland, 1990.

[286] V. Berezinsky, A. Bottino, and G. Mignola. High-energy gamma radiation from the galactic center due to neutralino annihilation, *Phys. Lett.*, **B325**:136–142, 1994, hep-ph/9402215.

[287] V. Berezinsky *et al.* Neutralino dark matter in supersymmetric models with non-universal scalar mass terms, *Astropart. Phys.*, **5**:1–26, 1996, hep-ph/9508249.

[288] V. S. Berezinsky, A. V. Gurevich, and K. P. Zybin. Distribution of dark matter in the Galaxy and the lower limits for the masses of supersymmetric particles, *Phys. Lett.*, **B294**:221–228, November 1992.

[289] L. Bergstrom. Radiative processes in dark matter photino annihilation, *Phys. Lett.*, **B225**:372, 1989.

[290] L. Bergstrom. Non-baryonic dark matter: Observational evidence and detection methods, *Rept. Prog. Phys.*, **63**:793, 2000, hep-ph/0002126.

[291] L. Bergstrom, G. Bertone, T. Bringmann, J. Edsjo, and M. Taoso. Gamma-ray and radio constraints of high positron rate dark matter models annihilating into new light particles, 2008, arXiv:0812.3895 [astro-ph].

[292] L. Bergstrom, T. Bringmann, and J. Edsjo. New positron spectral features from supersymmetric dark matter – a way to explain the PAMELA data?, 2008, arXiv:0808.3725 [astro-ph].

[293] L. Bergstrom, T. Bringmann, M. Eriksson, and M. Gustafsson. Gamma rays from Kaluza–Klein dark matter, *Phys. Rev. Lett.*, **94**:131301, 2005, astro-ph/0410359.

[294] L. Bergstrom, T. Bringmann, M. Eriksson, and M. Gustafsson. Two photon annihilation of Kaluza–Klein dark matter, *JCAP*, **0504**:004, 2005, hep-ph/0412001.

[295] L. Bergstrom, J. Edsjo, and P. Gondolo. Indirect detection of dark matter in km-size neutrino telescopes, *Phys. Rev.*, **D58**:103519, 1998, hep-ph/9806293.

[296] L. Bergstrom, J. Edsjo, M. Gustafsson, and P. Salati. Is the dark matter interpretation of the EGRET gamma excess compatible with antiproton measurements?, *JCAP*, **0605**:006, 2006, astro-ph/0602632.

[297] L. Bergstrom, J. Edsjo, and P. Ullio. Cosmic antiprotons as a probe for supersymmetric dark matter?, *Astrophys. J.*, **526**:215–235, 1999, astro-ph/9902012.

[298] L. Bergstrom, J. Edsjo, and P. Ullio. Spectral gamma-ray signatures of cosmological dark matter annihilations, *Phys. Rev. Lett.*, **87**:251301, 2001, astro-ph/0105048.

[299] L. Bergstrom and D. Hooper. Dark matter and gamma-rays from Draco: MAGIC, GLAST and CACTUS, *Phys. Rev.*, D**73**:063510, 2006, hep-ph/0512317.

[300] L. Bergstrom and J. Kaplan. Gamma-ray lines from TeV dark matter, *Astropart. Phys.*, **2**:261–268, 1994, hep-ph/9403239.

[301] L. Bergstrom and H. Snellman. Observable monochromatic photons from cosmic photino annihilation, *Phys. Rev.*, D**37**:3737–3741, 1988.

[302] L. Bergstrom and P. Ullio. Full one-loop calculation of neutralino annihilation into two photons, *Nucl. Phys.*, B**504**:27–44, 1997, hep-ph/9706232.

[303] L. Bergström, P. Ullio, and J. H. Buckley. Observability of gamma rays from dark matter neutralino annihilations in the Milky Way halo, *Astropart. Phys.*, **9**:137–162, 1998.

[304] L. Bergstrom, P. Ullio, and J. H. Buckley. Observability of gamma rays from dark matter neutralino annihilations in the Milky Way halo, *Astropart. Phys.*, **9**:137–162, 1998, astro-ph/9712318.

[305] Z. Bern, P. Gondolo, and M. Perelstein. Neutralino annihilation into two photons, *Phys. Lett.*, B**411**:86–96, 1997, hep-ph/9706538.

[306] R. Bernabei. Competitiveness of a very low radioactive ton scintillator for particle dark matter search. In *The Identification of Dark Matter*, World Scientific, 574, 1997.

[307] R. Bernabei, P. Belli, A. Incicchitti, and D. Prosperi. Liquid noble gases for dark matter searches: a synoptic survey, 2008, arXiv: 0806.0011 [astro-ph].

[308] R. Bernabei and A. Bottino. DAMA sheds light on dark-matter particles, *Nature*, **449**:24, 2007.

[309] R. Bernabei *et al.* A multiton low activity NaI(Tl) detector for underground physics, *Astropart. Phys.*, **4**:45–54, 1995.

[310] R. Bernabei *et al.* Searching for WIMPs by the annual modulation signature, *Phys. Lett.*, B**424**:195–201, 1998.

[311] R. Bernabei *et al.* Investigation on possible diurnal effects induced by dark matter particles, *Nuovo Cim.*, **112**A:1541–1552, 1999.

[312] R. Bernabei *et al.* On a further search for a yearly modulation of the rate in particle dark matter direct search, *Phys. Lett.*, B**450**:448–455, 1999.

[313] R. Bernabei *et al.* Performances of the about 100 kg NaI(Tl) set-up of the DAMA experiment at Gran Sasso, *Il Nuovo Cim.*, A**112**:545–575, 1999.

[314] R. Bernabei *et al.* On the investigation of possible systematics in WIMP annual modulation search, *Eur. Phys. J.*, C**18**:283–292, 2000.

[315] R. Bernabei *et al.* Search for WIMP annual modulation signature: Results from DAMA / NaI-3 and DAMA / NaI-4 and the global combined analysis, *Phys. Lett.*, B**480**:23–31, 2000.

[316] R. Bernabei *et al.* Investigating the DAMA annual modulation data in a mixed coupling framework, *Phys. Lett.*, B**509**:197–203, 2001.

[317] R. Bernabei *et al.* Light response of a pure liquid xenon scintillator irradiated with 2.5-MeV neutrons, *Eur. Phys. J. direct*, C**3**:11, 2001.

[318] R. Bernabei *et al.* Investigating the DAMA annual modulation data in the framework of inelastic dark matter, *Eur. Phys. J.*, C**23**:61–64, 2002.

[319] R. Bernabei *et al.* Dark matter search, *Riv. Nuovo Cim.*, **26**N1:1–73, 2003, astro-ph/0307403.

[320] R. Bernabei *et al.* Dark matter particles in the galactic halo: Results and implications from DAMA/NaI, *Int. J. Mod. Phys.*, D**13**:2127–2160, 2004, astro-ph/0501412.

[321] R. Bernabei *et al.* Investigating halo substructures with annual modulation signature, *Eur. Phys. J.*, C**47**:263–271, 2006, astro-ph/0604303.

[322] R. Bernabei *et al.* Investigating pseudoscalar and scalar dark matter, *Int. J. Mod. Phys.*, A**21**:1445–1470, 2006, astro-ph/0511262.

[323] R. Bernabei *et al.* On electromagnetic contributions in WIMP quests, *Int. J. Mod. Phys.*, A**22**:3155–3168, 2007, arXiv:0706.1421 [astro-ph].

[324] R. Bernabei *et al.* First results from DAMA/LIBRA and the combined results with DAMA/NaI, *Eur. Phys. J.*, C**56**:333–355, 2008, arXiv:0804.2741 [astro-ph].

[325] R. Bernabei *et al.* Investigating electron interacting dark matter, *Phys. Rev.*, D**77**:023506, 2008, arXiv:0712.0562 [astro-ph].

[326] R. Bernabei *et al.* Investigation on light dark matter, *Mod. Phys. Lett.*, A**23**:2125–2140, 2008, arXiv:0802.4336 [astro-ph].

[327] R. Bernabei *et al.* Possible implications of the channeling effect in NaI(Tl) crystals, *Eur. Phys. J.*, C**53**:205–213, 2008, arXiv:0710.0288 [astro-ph].

[328] R. Bernabei *et al.* The DAMA/LIBRA apparatus, *Nucl. Instrum. Meth.*, A**592**:297–315, 2008, arXiv:0804.2738 [astro-ph].

[329] F. Bernardeau, L. van Waerbeke, and Y. Mellier. Weak lensing statistics as a probe of {OMEGA} and power spectrum. *Astron. Astrophys.*, **322**:1–18, June 1997.

[330] V. Bertin, E. Nezri, and J. Orloff. Neutralino dark matter beyond CMSSM universality, *JHEP*, **02**:046, 2003, hep-ph/0210034.

[331] G. Bertone. Dark matter: the connection with gamma-ray astrophysics, *Astrophys. Space Sci.*, **309**:505–515, June 2007.

[332] G. Bertone, T. Bringmann, R. Rando, G. Busetto, and A. Morselli. GLAST sensitivity to point sources of dark matter annihilation, 2006, astro-ph/0612387.

[333] G. Bertone, W. Buchmuller, L. Covi, and A. Ibarra. Gamma-rays from decaying dark matter, *JCAP*, **0711**:003, 2007, arXiv:0709.2299 [astro-ph].

[334] G. Bertone, D. G. Cerdeno, J. I. Collar, and B. C. Odom. WIMP identification through a combined measurement of axial and scalar couplings, *Phys. Rev. Lett.*, **99**:151301, 2007, arXiv:0705.2502 [astro-ph].

[335] G. Bertone, M. Cirelli, A. Strumia, and M. Taoso. Gamma-ray and radio tests of the e+e- excess from DM annihilations, 2008, arXiv:0811.3744 [astro-ph].

[336] G. Bertone and M. Fairbairn. Compact stars as dark matter probes, *Phys. Rev.*, D**77**:043515, 2008, arXiv:0709.1485 [astro-ph].

[337] G. Bertone, D. Hooper, and J. Silk. Particle dark matter: Evidence, candidates and constraints, *Phys. Rep.*, **405**:279–390, 2005, hep-ph/0404175.

[338] G. Bertone and D. Merritt. Dark matter dynamics and indirect detection, *Mod. Phys. Lett.*, A**20**:1021–1036, 2005, astro-ph/0504422.

[339] G. Bertone and D. Merritt. Time-dependent models for dark matter at the Galactic center, *Phys. Rev.*, D**72**:103502, 2005, astro-ph/0501555.

[340] G. Bertone, G. Servant, and G. Sigl. Indirect detection of Kaluza-Klein dark matter, *Phys. Rev.*, D**68**:044008, 2003, hep-ph/0211342.

[341] G. Bertone, G. Sigl, and J. Silk. Astrophysical limits on massive dark matter, *Mon. Not. Roy. Astron. Soc.*, **326**:799–804, 2001, astro-ph/0101134.

[342] G. Bertone, G. Sigl, and J. Silk. Annihilation radiation from a dark matter spike at the Galactic Centre, *Mon. Not. Roy. Astron. Soc.*, **337**:98, 2002, astro-ph/0203488.

[343] G. Bertone, A. R. Zentner, and J. Silk. A new signature of dark matter annihilations: Gamma-rays from intermediate-mass black holes, *Phys. Rev.*, D**72**:103517, 2005, astro-ph/0509565.

[344] E. Bertschinger. Self-similar secondary infall and accretion in an Einstein–de Sitter universe, *Astrophys. J. Suppl.*, **58**:39–65, May 1985.

[345] E. Bertschinger. The effects of cold dark matter decoupling and pair annihilation on cosmological perturbations. *Phys. Rev.*, D**74**:063509, 2006.

[346] E. Bertschinger. Multiscale gaussian random fields for cosmological simulations, *Astrophys. J. Suppl.*, **137**:1, 2001, astro-ph/0103301.

[347] F. L. Bezrukov. nuMSM predictions for neutrinoless double beta decay, *Phys. Rev.*, D**72**:071303, 2005, hep-ph/0505247.

[348] F. L. Bezrukov and M. Shaposhnikov. Searching for dark matter sterile neutrino in laboratory, *Phys. Rev.*, D**75**:053005, 2007, hep-ph/0611352.

[349] F. L. Bezrukov and M. Shaposhnikov. The Standard Model Higgs boson as the inflaton, *Phys. Lett.*, B**659**:703–706, 2008, arXiv:0710.3755 [hep-th].

[350] B. Bhattacherjee and A. Kundu. The International Linear Collider as a Kaluza–Klein factory, *Phys. Lett.*, B**627**:137–144, 2005, hep-ph/0508170.

[351] X.-J. Bi, M.-z. Li, and X.-m. Zhang. Quintessino as dark matter, *Phys. Rev.*, D**69**:123521, 2004, hep-ph/0308218.

[352] X.-J. Bi, J.-X. Wang, C. Zhang, and X.-m. Zhang. Phenomenology of quintessino dark matter, *Phys. Rev.*, D**70**:123512, 2004, hep-ph/0404263.

[353] P. L. Biermann and A. Kusenko. Relic keV sterile neutrinos and reionization, *Phys. Rev. Lett.*, **96**:091301, 2006, astro-ph/0601004.

[354] N. Bilic, R. J. Lindebaum, G. B. Tupper, and R. D. Viollier. On the formation of degenerate heavy neutrino stars, *Phys. Lett.*, B**515**:105–110, 2001, astro-ph/0106209.

[355] P. Binetruy, G. Girardi, and P. Salati. Constraints on a system of two neutral fermions from cosmology, *Nucl. Phys.*, B**237**:285, 1984.

[356] P. Binetruy, Y. Mambrini, and E. Nezri. Direct and indirect detection of dark matter in heterotic orbifold models, *Astropart. Phys.*, **22**:1–18, 2004, hep-ph/0312155.

[357] J. Binney and A. Knebe. Two-body relaxation in cosmological simulations, *Mon. Not. Roy. Astron. Soc.*, **333**:378, 2002, astro-ph/0105183.

[358] J. Binney and S. Tremaine. *Galactic Dynamics*, Princeton University Press, 2008.

[359] C. Bird, K. Koopmans, and M. Pospelov. Primordial lithium abundance in catalyzed Big Bang nucleosynthesis, *Phys. Rev.*, D**78**:083010, 2008, hep-ph/0703096.

[360] A. Birkedal, K. T. Matchev, M. Perelstein, and A. Spray. Robust gamma ray signature of WIMP dark matter, 2005, hep-ph/0507194.

[361] A. Birkedal, A. Noble, M. Perelstein, and A. Spray. Little Higgs dark matter, *Phys. Rev.*, D**74**:035002, 2006, hep-ph/0603077.

[362] A. Birkedal-Hansen and B. D. Nelson. Relic neutralino densities and detection rates with nonuniversal gaugino masses, *Phys. Rev.*, D**67**:095006, 2003, hep-ph/0211071.

[363] N. D. Birrell and P. C. W. Davies. *Quantum Fields in Curved Space*, Cambridge University Press, 1982.

[364] M. Bisset, R. Lu, and N. Kersting. Improving SUSY spectrum determinations at the LHC with Wedgebox and hidden threshold techniques, 2008, arXiv:0806.2492 [hep-ph].

[365] J. D. Bjorken *et al.* Search for neutral metastable penetrating particles produced in the SLAC beam dump, *Phys. Rev.*, D**38**:3375, 1988.

[366] G. A. Blair, W. Porod, and P. M. Zerwas. Reconstructing supersymmetric theories at high energy scales, *Phys. Rev.*, D**63**:017703, 2001, hep-ph/0007107.

[367] P. Blasi and S. Colafrancesco. Cosmic rays, radio halos and nonthermal X-ray emission in clusters of galaxies, *Astropart. Phys.*, **122**:169–183, 1999, astro-ph/9905122.

[368] P. Blasi, A. V. Olinto, and C. Tyler. Detecting WIMPS in the microwave sky, *Astropart. Phys.*, **18**:649–662, 2003, astro-ph/0202049.

[369] M. Blennow, J. Edsjo, and T. Ohlsson. Neutrinos from WIMP annihilations using a full three-flavor Monte Carlo, *JCAP*, **0801**:021, 2008, arXiv:0709.3898 [hep-ph].

[370] G. R. Blumenthal, S. M. Faber, R. Flores, and J. R. Primack. Contraction of dark matter galactic halos due to baryonic infall, *Astrophys. J.*, **301**:27–34, February 1986.

[371] C. Bobin *et al.* Towards measurements of recoils below 4-keV with the *ROSEBUD* experiment, *Nucl. Phys. Proc. Suppl.*, **70**:90–95, 1999.

[372] P. Bode, J. P. Ostriker, and N. Turok. Halo formation in warm dark matter models, *Astrophys. J.*, **556**:93–107, 2001, astro-ph/0010389.

[373] A. M. Boesgaard and M. C. Novicki. Beryllium in disk and halo stars – evidence for a beryllium dispersion in old stars, *Astrophys. J.*, **641**:1122–1130, 2006, astro-ph/0512317.

[374] M. Boezio, P. Carlson, T. Francke, *et al.* The cosmic-ray electron and positron spectra measured at 1 AU during solar minimum activity, *Astrophys. J.*, **532**:653–669, 2000.

[375] M. Boezio *et al.* The cosmic-ray antiproton flux between 0.62 and 3.19 G eV measured near solar minimum activity, *Astrophys. J.*, **487**:415, 1997.

[376] M. Boezio *et al.* The cosmic-ray antiproton flux between 3 and 49 GeV, *Astrophys. J.*, **561**:787–799, 2001.

[377] T. Böker, S. Laine, R. P. van der Marel, *et al.* A Hubble Space Telescope census of nuclear star clusters in late-type spiral galaxies. I. Observations and image analysis, *Astron. J.*, **123**:1389–1410, 2002.

[378] A. S. Bolton, S. Burles, L. V. E. Koopmans, *et al.* The Sloan Lens ACS Survey. V. The full ACS strong-lens sample, *Astrophys. J.*, **682**:964–984, August 2008, 0805.1931.

[379] A. S. Bolton, S. Burles, L. V. E. Koopmans, T. Treu, and L. A. Moustakas. The Sloan Lens ACS Survey. I. A large spectroscopically selected sample of massive early-type lens galaxies, *Astrophys. J.*, **638**:703–724, 2006.

[380] M. Bolz, A. Brandenburg, and W. Buchmuller. Thermal production of gravitinos, *Nucl. Phys.*, B**606**:518–544, 2001, hep-ph/0012052.

[381] P. Bonifacio *et al.* The lithium content of the globular cluster NGC 6397, 2002, astro-ph/0204332.

[382] H. Bonnet, Y. Mellier, and B. Fort. First detection of a gravitational weak shear at the periphery of CL 0024+1654, *Astrophys. J.*, **427**:L83–L86, June 1994.

[383] E. Boos *et al.* CompHEP 4.4: Automatic computations from Lagrangians to events, *Nucl. Instrum. Meth.*, A**534**:250–259, 2004, hep-ph/0403113.

[384] E. Borriello, A. Cuoco, and G. Miele. Radio constraints on dark matter annihilation in the galactic halo and its substructures, *Phys. Rev.*, D**79**:023518, 2009, arXiv:0809.2990 [astro-ph].

[385] F. Borzumati, T. Bringmann, and P. Ullio. Dark matter from late decays and the small-scale structure problems, *Phys. Rev.*, D**77**:063514, 2008, hep-ph/0701007.

[386] F. Borzumati and Y. Nomura. Low-scale see-saw mechanisms for light neutrinos, *Phys. Rev.*, D**64**:053005, 2001, hep-ph/0007018.

[387] A. Bottino, V. de Alfaro, N. Fornengo, G. Mignola, and S. Scopel. A new investigation about neutralino dark matter: Relic density and detection rates, *Astropart. Phys.*, **1**:61–76, 1992.

[388] A. Bottino, V. de Alfaro, N. Fornengo, S. Mignola, and S. Scopel. On the neutralino as dark matter candidate. 2. Direct detection, *Astropart. Phys.*, **2**:77–90, 1994, hep-ph/9309219.

[389] A. Bottino, F. Donato, N. Fornengo, and S. Scopel. Pinning down neutralino properties from a possible modulation signal in WIMP direct search, *Phys. Lett.*, B**423**:109–117, 1998, hep-ph/9709292.

[390] A. Bottino, F. Donato, N. Fornengo, and S. Scopel. Combining the data of annual modulation effect in WIMP direct detection with measurements of WIMP indirect searches, *Astropart. Phys.*, **10**:203–210, 1999, hep-ph/9809239.

[391] A. Bottino, F. Donato, N. Fornengo, and S. Scopel. Compatibility of the new DAMA/NaI data on an annual modulation effect in WIMP direct search with a relic neutralino in supergravity schemes, *Phys. Rev.*, D**59**:095004, 1999, hep-ph/9808459.

[392] A. Bottino, F. Donato, N. Fornengo, and S. Scopel. Neutralino properties in the light of a further indication of an annual modulation effect in WIMP direct search, *Phys. Rev.*, D**59**:095003, 1999, hep-ph/9808456.

[393] A. Bottino, F. Donato, N. Fornengo, and S. Scopel. Further investigation of a relic neutralino as a possible origin of an annual-modulation effect in WIMP direct search, *Phys. Rev.*, D**62**:056006, 2000, hep-ph/0001309.

[394] A. Bottino, F. Donato, N. Fornengo, and S. Scopel. Implications for relic neutralinos of the theoretical uncertainties in the neutralino nucleon cross-section, *Astropart. Phys.*, **13**:215–225, 2000, hep-ph/9909228.

[395] A. Bottino, F. Donato, N. Fornengo, and S. Scopel. Probing the supersymmetric parameter space by WIMP direct detection, *Phys. Rev.*, D**63**:125003, 2001, hep-ph/0010203.

[396] A. Bottino, F. Donato, N. Fornengo, and S. Scopel. Size of the neutralino nucleon cross-section in the light of a new determination of the pion nucleon sigma term, *Astropart. Phys.*, **18**:205–211, 2002, hep-ph/0111229.

[397] A. Bottino, F. Donato, N. Fornengo, and S. Scopel. Lower bound on the neutralino mass from new data on CMB and implications for relic neutralinos, *Phys. Rev.*, D**68**:043506, 2003, hep-ph/0304080.

[398] A. Bottino, F. Donato, N. Fornengo, and S. Scopel. Indirect signals from light neutralinos in supersymmetric models without gaugino mass unification, *Phys. Rev.*, D**70**:015005, 2004, hep-ph/0401186.

[399] A. Bottino, F. Donato, N. Fornengo, and S. Scopel. Light neutralinos and WIMP direct searches, *Phys. Rev.*, D**69**:037302, 2004, hep-ph/0307303.

[400] A. Bottino, F. Donato, N. Fornengo, and S. Scopel. Do current WIMP direct measurements constrain light relic neutralinos?, *Phys. Rev.*, D**72**:083521, 2005, hep-ph/0508270.

[401] A. Bottino, F. Donato, N. Fornengo, and S. Scopel. Interpreting the recent results on direct search for dark matter particles in terms of relic neutralino, *Phys. Rev.*, D**78**:083520, 2008, arXiv:0806.4099 [hep-ph].

[402] A. Bottino, F. Donato, N. Fornengo, and S. Scopel. Upper bounds on signals due to WIMP self-annihilation: comments on the case of the synchrotron radiation from the galactic center and the WMAP haze, *Phys. Rev.*, D**77**:127301, 2008, arXiv:0802.0714 [hep-ph].

[403] A. Bottino, F. Donato, N. Fornengo, and S. Scopel. Zooming in on light relic neutralinos by direct detection and measurements of galactic antimatter, *Phys. Rev.*, D**77**:015002, 2008, arXiv:0710.0553 [hep-ph].

[404] A. Bottino *et al.* Direct versus indirect searches for neutralino dark matter, *Mod. Phys. Lett.*, A**7**:733–748, 1992.

[405] A. Bottino *et al.* Exploring the supersymmetric parameter space by direct search for WIMPs, *Phys. Lett.*, B**402**:113–121, 1997, hep-ph/9612451.

[406] A. Bottino *et al.* Does solar physics provide constraints to weakly interacting massive particles?, *Phys. Rev.*, D**66**:053005, 2002, hep-ph/0206211.

[407] A. Bottino, N. Fornengo, G. Polesello, and S. Scopel. Light neutralinos at LHC in cosmologically-inspired scenarios: new benchmarks in the search for supersymmetry, *Phys. Rev.*, D**77**:115026, 2008, arXiv:0801.3334 [hep-ph].

[408] A. Bottino, N. Fornengo, and S. Scopel. Implications of a possible 115-GeV supersymmetric Higgs boson on detection and cosmological abundance of relic neutralinos, *Nucl. Phys.*, B**608**:461–474, 2001, hep-ph/0012377.

[409] A. Bottino, N. Fornengo, and S. Scopel. Light relic neutralinos, *Phys. Rev.*, D**67**:063519, 2003, hep-ph/0212379.

[410] J. Bouchez *et al.* Bulk micromegas detectors for large TPC applications, *Nucl. Instrum. Meth.*, A**574**:425–432, 2007.

[411] F. Boudjema, A. Semenov, and D. Temes. Self-annihilation of the neutralino dark matter into two photons or a Z and a photon in the MSSM, *Phys. Rev.*, D**72**:055024, 2005, hep-ph/0507127.

[412] F. Boudjema, A. Semenov, and D. Temes. SUSY dark matter: Loops and precision from particle physics, *Nucl. Phys. Proc. Suppl.*, **157**:172–178, 2006.

[413] M. Boulay *et al.* DEAP/CLEAN experiment at SNOLAB, 2008. Talk at IDM08, Stockholm, August 2008.

[414] M. G. Boulay and A. Hime. Technique for direct detection of weakly interacting massive particles using scintillation time discrimination in liquid argon, *Astropart. Phys.*, **25**:179–182, 2006.

[415] A. Bouquet, J. Kaplan, and F. Martin. Weakly interacting massive particles and stellar structure, *Astron. Astrophys.*, **222**:103–116, September 1989.

[416] A. Bouquet and P. Salati. Dark matter and the suppression of stellar core convection, *Astrophys. J.*, **346**:284–288, November 1989.

[417] A. Bouquet and P. Salati. Life and death of cosmions in stars, *Astron. Astrophys.*, **217**:270–282, June 1989.

[418] A. Bouquet, P. Salati, and J. Silk. Gamma-ray lines as a probe for a cold dark matter halo, *Phys. Rev.*, D**40**:3168, 1989.

[419] R. Bousso and J. Polchinski. Quantization of four-form fluxes and dynamical neutralization of the cosmological constant, *JHEP*, **06**:006, 2000, hep-th/0004134.

[420] R. J. Bouwens, G. D. Illingworth, M. Franx, and H. Ford. $z \approx 7$–10 galaxies in the HUDF and GOODS fields, and their UV luminosity functions, 2008, arXiv:0803.0548 [astro-ph].

[421] M. Bowen, Y. Cui, and J. D. Wells. Narrow trans-TeV Higgs bosons and H − h h decays: Two LHC search paths for a hidden sector Higgs boson, *JHEP*, **03**:036, 2007, hep-ph/0701035.

[422] A. Boyarsky, J. W. den Herder, A. Neronov, and O. Ruchayskiy. Search for the light dark matter with an X-ray spectrometer, *Astropart. Phys.*, **28**:303–311, 2007, astro-ph/0612219.

[423] A. Boyarsky, D. Iakubovskyi, O. Ruchayskiy, and V. Savchenko. Constraints on decaying dark matter from XMM-Newton observations of M31, *Mon. Not. Roy. Astron. Soc.*, **387**:1361, 2008, arXiv:0709.2301 [astro-ph].

[424] A. Boyarsky, J. Lesgourgues, O. Ruchayskiy, and M. Viel. Lyman-α constraints on warm and on warm-plus-cold dark matter models, 2008, arXiv:0812.0010 [astro-ph].

[425] A. Boyarsky, D. Malyshev, A. Neronov, and O. Ruchayskiy. Constraining DM properties with SPI, 2007, arXiv:0710.4922 [astro-ph].

[426] A. Boyarsky, A. Neronov, O. Ruchayskiy, and M. Shaposhnikov. Restrictions on parameters of sterile neutrino dark matter from observations of galaxy clusters, *Phys. Rev.*, D**74**:103506, 2006, astro-ph/0603368.

[427] A. Boyarsky, A. Neronov, O. Ruchayskiy, and M. Shaposhnikov. The masses of active neutrinos in the nuMSM from X-ray astronomy, *JETP Lett.*, **83**:133–135, 2006, hep-ph/0601098.

[428] A. Boyarsky, A. Neronov, O. Ruchayskiy, M. Shaposhnikov, and I. Tkachev. How to find a dark matter sterile neutrino?, *Phys. Rev. Lett.*, **97**:261302, 2006, astro-ph/0603660.

[429] A. Boyarsky and O. Ruchayskiy. Bounds on light dark matter, 2008, arXiv:0811.2385 [astro-ph].

[430] A. Boyarsky, O. Ruchayskiy, and D. Iakubovskyi. A lower bound on the mass of dark matter particles, 2008, arXiv:0808.3902 [hep-ph].

[431] M. Boylan-Kolchin, C.-P. Ma, and E. Quataert. Core formation in galactic nuclei due to recoiling black holes, *Astrophys. J.*, **613**:L37–L40, September 2004.

[432] R. Brada and M. Milgrom. Stability of disk galaxies in the modified dynamics, *Astrophys. J.*, **519**:590–598, 1999.

[433] M. Bradač, S. W. Allen, T. Treu, *et al.* Revealing the properties of dark matter in the merging cluster MACS J0025.4-1222, *Astrophys. J.*, **687**:959–967, 2008, 0806.2320.

[434] M. Bradač, D. Clowe, A. H. Gonzalez, *et al.* Strong and weak lensing united. III. Measuring the mass distribution of the merging galaxy cluster 1ES 0657-558, *Astrophys. J.*, **652**:937–947, 2006.

[435] R. Bradley *et al.* Microwave cavity searches for dark-matter axions, *Rev. Mod. Phys.*, **75**:777–817, 2003.

[436] T. G. Brainerd, R. D. Blandford, and I. Smail. Weak gravitational lensing by galaxies, *Astrophys. J.*, **466**:623, 1996.

[437] C. B. Bratton *et al.* Angular distribution of events from SN1987A, *Phys. Rev.*, D**37**:3361, 1988.

[438] J. Brau *et al.*, eds. International Linear Collider reference design report. 1: Executive summary. 2: Physics at the ILC. 3: Accelerator. 4: Detectors, 2007, ILC-REPORT-2007-001.

[439] M. Bravin *et al.* Simultaneous measurement of phonons and scintillation light for active background rejection in the CRESST experiment, *Nucl. Instrum. Meth.*, A**444**:323–326, 2000.

[440] M. Brhlik and L. Roszkowski. WIMP velocity impact on direct dark matter searches, *Phys. Lett.*, B**464**:303–310, 1999, hep-ph/9903468.

[441] S. Bridle, J. Shawe-Taylor, A. Amara, *et al.* Handbook for the GREAT08 Challenge: An image analysis competition for cosmological lensing, 2008, arXiv:0802.1214.

[442] A. Brignole, L. E. Ibanez, and C. Munoz. Soft supersymmetry-breaking terms from supergravity and superstring models, 1997, hep-ph/9707209.

[443] T. Bringmann. High-energetic cosmic antiprotons from Kaluza–Klein dark matter, *JCAP*, **0508**:006, 2005, astro-ph/0506219.

[444] T. Bringmann, L. Bergstrom, and J. Edsjö. New gamma-ray contributions to supersymmetric dark matter annihilation, *JHEP*, **01**:049, 2008, arXiv:0710.3169 [hep-ph].

[445] T. Bringmann and P. Salati. Galactic antiproton spectrum at high energies: Background expectation versus exotic contributions, *Phys. Rev. D*, **75**:083006, 2007.

[446] T. Broadhurst, N. Benítez, D. Coe, *et al.* Strong-lensing analysis of A1689 from Deep Advanced Camera Images, *Astrophys. J.*, **621**:53–88, 2005.

[447] T. Broadhurst, M. Takada, K. Umetsu *et al.* The surprisingly steep mass profile of A1689, from a lensing analysis of Subaru images, *Astrophys. J.*, **619**:L143 L146, 2005.

[448] T. Broadhurst, K. Umetsu, E. Medezinski, M. Oguri, and Y. Rephaeli. Comparison of cluster lensing profiles with ΛCDM Predictions, *Astrophys. J.*, **685**:L9–L12, 2008, arXiv:0805.2617.

[449] V. Bromm and R. B. Larson. The first stars, *Ann. Rev. Astron. Astrophys.*, **42**:79–118, 2004, astro-ph/0311019.

[450] A. Broniatowski *et al.* Cryogenic Ge detectors with interleaved electrodes: Design and modeling, *J. Low Temp. Phys.*, **151**:830–834, 2008.

[451] T. A. D. Brown *et al.* The ^{3}He + ^{4}He → ^{7}Be astrophysical S-factor, *Phys. Rev.*, **C76**:055801, 2007, arXiv:0710.1279 [nucl-ex].

[452] T. Bruch, J. Read, L. Baudis, and G. Lake. Detecting the Milky Way's dark disk, 2008, arXiv:0804.2896 [astro-ph].

[453] P. Brun. Towards a new tool for the indirect detection of dark matter: Building of a SuSy spectrum generator based on micrOMEGAs, 2006, astro-ph/0603387.

[454] J. Brunner. Status of the ANTARES project, *Nucl. Phys. Proc. Suppl.*, **145**:323–326, 2005.

[455] O. Buchmueller *et al.* Prediction for the lightest Higgs boson mass in the CMSSM using indirect experimental constraints, *Phys. Lett.*, **B657**:87–94, 2007, arXiv:0707.3447 [hep-ph].

[456] O. Buchmueller *et al.* Predictions for supersymmetric particle masses in the CMSSM using indirect experimental and cosmological constraints, *JHEP*, **09**:117, 2008, arXiv:0808.4128 [hep-ph].

[457] W. Buchmuller. Baryogenesis – 40 years later, 2007, arXiv:0710.5857 [hep-ph].

[458] W. Buchmuller, L. Covi, K. Hamaguchi, A. Ibarra, and T. Yanagida. Gravitino dark matter in R-parity breaking vacua, *JHEP*, **03**:037, 2007, hep-ph/0702184.

[459] W. Buchmuller, K. Hamaguchi, M. Ratz, and T. Yanagida. Supergravity at colliders, *Phys. Lett.*, B**588**:90–98, 2004, hep-ph/0402179.

[460] J. Buckley *et al.* Section on prospects for dark matter detection of the *White Paper on the Status and Future of Ground-Based TeV Gamma-Ray Astronomy*, 2008, arXiv:0812.0795 [astro-ph].

[461] M. R. Buckley, S. Y. Choi, K. Mawatari, and H. Murayama. Determining spin through quantum azimuthal-angle correlations, 2008, arXiv:0811.3030 [hep-ph].

[462] M. R. Buckley, B. Heinemann, W. Klemm, and H. Murayama. Quantum interference effects among helicities at LEP-II and Tevatron, *Phys. Rev.*, D**77**:113017, 2008, arXiv:0804.0476 [hep-ph].

[463] M. R. Buckley, H. Murayama, W. Klemm, and V. Rentala. Discriminating spin through quantum interference, *Phys. Rev.*, D**78**:014028, 2008, arXiv:0711.0364 [hep-ph].

[464] S. V. Bulanov and V. A. Dogel. The influence of the energy dependence of the diffusion coefficient on the spectrum of the electron component of cosmic rays and the radio background radiation of the Galaxy, *Astrophys. Space Sci.*, **29**:305–318, August 1974.

[465] J. S. Bullock *et al.* Profiles of dark haloes: evolution, scatter, and environment, *Mon. Not. Roy. Astron. Soc.*, **321**:559–575, 2001, astro-ph/9908159.

[466] J. S. Bullock and K. V. Johnston. Tracing galaxy formation with stellar halos I: Methods, *Astrophys. J.*, **635**:931–949, 2005, astro-ph/0506467.

[467] J. S. Bullock, A. V. Kravtsov, and D. H. Weinberg. Hierarchical galaxy formation and substructure in the Galaxy's stellar halo, *Astrophys. J.*, **548**:33–46, 2001, astro-ph/0007295.

[468] A. J. Buras, A. Poschenrieder, M. Spranger, and A. Weiler. The impact of universal extra dimensions on B → X/s gamma, B → X/s gluon, B → X/s mu+ mu-, K(L) → pi0 e+ e-, and epsilon'/epsilon, *Nucl. Phys.*, B**678**:455–490, 2004, hep-ph/0306158.

[469] A. J. Buras, M. Spranger, and A. Weiler. The impact of universal extra dimensions on the unitarity triangle and rare K and B decays, *Nucl. Phys.*, B**660**:225–268, 2003, hep-ph/0212143.

[470] G. Burdman, B. A. Dobrescu, and E. Ponton. Resonances from two universal extra dimensions, *Phys. Rev.*, D**74**:075008, 2006, hep-ph/0601186.

[471] C. P. Burgess, M. Pospelov, and T. ter Veldhuis. The minimal model of nonbaryonic dark matter: A singlet scalar, *Nucl. Phys.*, B**619**:709–728, 2001, hep-ph/0011335.

[472] S. Burgos *et al.* First results from the DRIFT-IIa dark matter detector, *Astropart. Phys.*, **28**:409–421, 2007, arXiv:0707.1488 [hep-ex].

[473] S. Burgos *et al.* First measurement of the Head-Tail directional nuclear recoil signature at energies relevant to WIMP dark matter searches, *Astropart. Phys.*, 2008, arXiv:0809.1831 [astro-ph].

[474] S. Burgos *et al.* Measurement of the Range Component Directional Signature in a DRIFT-II Detector using 252Cf Neutrons, *Nucl. Instrum. Meth.*, 2008, arXiv:0807.3969 [hep-ex].

[475] S. Burgos *et al.* Track reconstruction and performance of DRIFT directional dark matter detectors using alpha particles, *Nucl. Instrum. Meth.*, A**584**:114–128, 2008, arXiv:0707.1758 [physics.ins-det].

[476] A. Burkert. The structure of dark matter halos in dwarf galaxies, *Astrophys. J. Lett.*, **447**:L25, 1995.

[477] S. Burles and D. Tytler. The deuterium abundance towards Q1937–1009, *Astrophys. J.*, **499**:699, 1998, astro-ph/9712108.

[478] F. Burnell and G. D. Kribs. The abundance of Kaluza-Klein dark matter with coannihilation, *Phys. Rev.*, D**73**:015001, 2006, hep-ph/0509118.

[479] M. Burns, K. Kong, K. T. Matchev, and M. Park. A general method for model-independent measurements of particle spins, couplings and mixing angles in cascade decays with missing energy at hadron colliders, *JHEP*, **10**:081, 2008, arXiv:0808.2472 [hep-ph].

[480] M. Burns, K. Kong, K. T. Matchev, and M. Park. Using subsystem MT2 for complete mass determinations in decay chains with missing energy at hadron colliders, 2008, arXiv:0810.5576 [hep-ph].

[481] A. Burrows, M. Ted Ressell, and M. S. Turner. Axions and SN1987A: Axion trapping, *Phys. Rev.*, D**42**:3297–3309, 1990.

[482] S. T. Butler and C. A. Pearson. Deuterons from high-energy proton bombardment of matter, *Phys. Rev.*, **129**:836–842, 1963.

[483] R. A. Cabanac, C. Alard, M. Dantel-Fort, *et al.* The CFHTLS strong lensing legacy survey. I. Survey overview and T0002 release sample, *Astron. Astrophys.*, **461**:813–821, 2007.

[484] J. A. R. Caldwell and J. P. Ostriker. The mass distribution within our Galaxy: A three component model, *Astrophys. J.*, **251**:61–87, 1981.

[485] C. G. Jr Callan, R. F. Dashen, and D. J. Gross. The structure of the gauge theory vacuum, *Phys. Lett.*, **B63**:334–340, 1976.

[486] M. Campanelli, C. O. Lousto, Y. Zlochower, and D. Merritt. Maximum gravitational recoil, *Phys. Rev. Lett.*, **98**:231102, 2007.

[487] B. Canadas, D. G. Cerdeno, C. Munoz, and S. Panda. Stau detection at neutrino telescopes in scenarios with supersymmetric dark matter, 2008, arXiv:0812.1067 [hep-ph].

[488] M. Carena, A. D. Medina, B. Panes, N. R. Shah, and C. E. M. Wagner. Collider phenomenology of gauge-Higgs unification scenarios in warped extra dimensions, *Phys. Rev.*, **D77**:076003, 2008, arXiv:0712.0095 [hep-ph].

[489] M. J. Carson *et al.* Simulations of neutron background in a time projection chamber relevant to dark matter searches, *Nucl. Instrum. Meth.*, **A546**:509–522, 2005, hep-ex/0503017.

[490] J. A. Casas, J. R. Espinosa, and I. Hidalgo. Expectations for LHC from naturalness: Modified vs. SM Higgs sector, *Nucl. Phys.*, **B777**:226–252, 2007, hep-ph/0607279.

[491] J. A. Casas, J. R. Espinosa, and M. Quiros. Standard Model stability bounds for new physics within LHC reach, *Phys. Lett.*, **B382**:374–382, 1996, hep-ph/9603227.

[492] F. Casse, M. Lemoine, and G. Pelletier. Transport of cosmic rays in chaotic magnetic fields, *Phys. Rev.*, **D65**:023002, 2002, astro-ph/0109223.

[493] M. Casse, J. Paul, G. Bertone, and G. Sigl. Gamma rays from the Galactic bulge and large extra dimensions, *Phys. Rev. Lett.*, **92**:111102, 2004, hep-ph/0309173.

[494] R. Catena, N. Fornengo, A. Masiero, M. Pietroni, and F. Rosati. Dark matter relic abundance and scalar-tensor dark energy, *Phys. Rev.*, **D70**:063519, 2004, astro-ph/0403614.

[495] R. Catena, N. Fornengo, A. Masiero, M. Pietroni, and M. Schelke. Enlarging mSUGRA parameter space by decreasing pre-BBN Hubble rate in scalar-tensor cosmologies, *JHEP*, **10**:003, 2008, arXiv:0712.3173 [hep-ph].

[496] R. Cayrel *et al.* Line shift, line asymmetry, and the ^{6}Li/^{7}Li isotopic ratio determination, 2007, arXiv:0708.3819 [astro-ph].

[497] R. Cayrel, M. Spite, F. Spite, E. Vangioni-Flam, M. Casse, and J. Audouze. New high S/N observations of the Li-6/Li-7 blend in HD

84937 and two other metal-poor stars, *Astron. Astrophys.*, **343**:923, 1999, arXiv:9901205 [astro-ph].

[498] R. Cayrel, M. Steffen, P. Bonifacio, H.-G. Ludwig, and E. Caffau. Overview of the lithium problem in metal-poor stars and new results on ^{6}Li, 2008, arXiv:0810.4290 [astro-ph].

[499] J. A. R. Cembranos, A. Dobado, and A. L. Maroto. Cosmological and astrophysical limits on brane fluctuations, *Phys. Rev.*, D**68**:103505, 2003, hep-ph/0307062.

[500] J. A. R. Cembranos, A. Dobado, and A. Lopez Maroto. Brane-world dark matter, *Phys. Rev. Lett.*, **90**:241301, 2003, hep-ph/0302041.

[501] J. A. R. Cembranos, J. L. Feng, A. Rajaraman, and F. Takayama. Superwimp solutions to small scale structure problems. *Phys. Rev. Lett.*, **95**:181301, 2005.

[502] R. Cen and J. P. Ostriker. Where are the baryons?, *Astrophys. J.*, **514**:1, 1999, astro-ph/9806281.

[503] D. G. Cerdeño, E. Gabrielli, D. E. Lopez-Fogliani, C. Munoz, and A. M. Teixeira. Phenomenological viability of neutralino dark matter in the NMSSM, *JCAP*, **0706**:008, 2007, hep-ph/0701271.

[504] D. G. Cerdeño, C. Hugonie, D. E. Lopez-Fogliani, C. Munoz, and A. M. Teixeira. Theoretical predictions for the direct detection of neutralino dark matter in the NMSSM, *JHEP*, **12**:048, 2004, hep-ph/0408102.

[505] D. G. Cerdeño, C. Muñoz, and O. Seto. Right-handed sneutrino as thermal dark matter, 2008, arXiv:0807.3029 [hep-ph].

[506] D. Cerdeno and A. Green. Direct detection of WIMPs, *This Volume*, 2009.

[507] D. G. Cerdeno, K.-Y. Choi, K. Jedamzik, L. Roszkowski, and R. Ruiz de Austri. Gravitino dark matter in the CMSSM with improved constraints from BBN, *JCAP*, **0606**:005, 2006, hep-ph/0509275.

[508] D. G. Cerdeno *et al.* Determination of the string scale in D-brane scenarios and dark matter implications, *Nucl. Phys.*, B**603**:231–258, 2001, hep-ph/0102270.

[509] D. G. Cerdeno, E. Gabrielli, M. E. Gomez, and C. Munoz. Neutralino nucleon cross section and charge and colour breaking constraints, *JHEP*, **06**:030, 2003, hep-ph/0304115.

[510] D. G. Cerdeno, E. Gabrielli, and C. Munoz. Experimental constraints on the neutralino nucleon cross section, 2002, hep-ph/0204271.

[511] D. G. Cerdeno, T. Kobayashi, and C. Munoz. Prospects for the direct detection of neutralino dark matter in orbifold scenarios, *JHEP*, **01**:009, 2008, arXiv:0709.0858 [hep-ph].

[512] D. G. Cerdeno and C. Munoz. Neutralino dark matter in supergravity theories with non-universal scalar and gaugino masses, *JHEP*, **10**:015, 2004, hep-ph/0405057.

[513] A. Cesarini, F. Fucito, A. Lionetto, A. Morselli, and P. Ullio. The galactic center as a dark matter gamma-ray source, *Astropart. Phys.*, **21**:267–285, 2004, astro-ph/0305075.

[514] Z. Chacko, H.-S. Goh, and R. Harnik. The twin Higgs: Natural electroweak breaking from mirror symmetry, *Phys. Rev. Lett.*, **96**:231802, 2006, hep-ph/0506256.

[515] J. Chang *et al.* An excess of cosmic ray electrons at energies of 300.800 GeV, *Nature*, **456**:362–365, 2008.

[516] S. Chang, C. Hagmann, and P. Sikivie. Studies of the motion and decay of axion walls bounded by strings, *Phys. Rev.*, D**59**:023505, 1999, hep-ph/9807374.

[517] S. Chang, J. Hisano, H. Nakano, N. Okada, and M. Yamaguchi. Bulk standard model in the Randall-Sundrum background, *Phys. Rev.*, D**62**:084025, 2000, hep-ph/9912498.

[518] S. Chang, G. D. Kribs, D. Tucker-Smith, and N. Weiner. Inelastic dark matter in light of DAMA/LIBRA, 2008, arXiv:0807.2250 [hep-ph].

[519] S. Chang, A. Pierce, and N. Weiner. Using the energy spectrum at DAMA/LIBRA to probe light dark matter, 2008, arXiv:0808.0196 [hep-ph].

[520] S. C. Chapman *et al.* The kinematic footprints of five stellar streams in Andromeda's halo, 2008, arXiv:0808.0755 [astro-ph].

[521] G. Chardin. Dark matter direct detection using cryogenic detectors, 2004, astro-ph/0411503.

[522] P. Chardonnet, J. Orloff, and P. Salati. The production of antimatter in our galaxy, *Phys. Lett. B*, **409**:313–320, 1997.

[523] J. Charles, A. Hocker, H. Lacker, F. R. Le Diberder, and S. T'Jampens. Bayesian statistics at work: The troublesome extraction of the CKM phase alpha, 2006, hep-ph/0607246.

[524] U. Chattopadhyay, A. Corsetti, and P. Nath. Supersymmetric dark matter and Yukawa unification, *Phys. Rev.*, D**66**:035003, 2002, hep-ph/0201001.

[525] U. Chattopadhyay, A. Corsetti, and P. Nath. WMAP constraints, SUSY dark matter and implications for the direct detection of SUSY, *Phys. Rev.*, D**68**:035005, 2003, hep-ph/0303201.

[526] U. Chattopadhyay and P. Nath. Interpreting the new Brookhaven g(mu)-2 result, *Phys. Rev.*, D**66**:093001, 2002, hep-ph/0208012.

[527] U. Chattopadhyay and P. Nath. Modular invariant soft breaking, WMAP, dark matter and sparticle mass limits, *Phys. Rev.*, D**70**:096009, 2004, hep-ph/0405157.

[528] U. Chattopadhyay and D. P. Roy. Higgsino dark matter in a SUGRA model with nonuniversal gaugino masses, *Phys. Rev.*, D**68**:033010, 2003, hep-ph/0304108.

[529] D.-M. Chen. Strong lensing probability in TeVeS (tensor vector scalar) theory, *J. Cosmol. Astropart. Phys.*, **1**:6, 2008.

[530] D.-M. Chen and H.-S. Zhao. Strong lensing probability for testing TeVeS theory, *Astrophys. J.*, **650**:L9–L12, 2006.

[531] X. Chen and S.-H. H. Tye. Heating in brane inflation and hidden dark matter, *JCAP*, **0606**:011, 2006, hep-th/0602136.

[532] H.-C. Cheng, D. Engelhardt, J. F. Gunion, Z. Han, and B. McElrath. Accurate mass determinations in decay chains with missing energy, *Phys. Rev. Lett.*, **100**:252001, 2008, arXiv:0802.4290 [hep-ph].

[533] H.-C. Cheng, J. L. Feng, and K. T. Matchev. Kaluza–Klein dark matter, *Phys. Rev. Lett.*, **89**:211301, 2002, hep-ph/0207125.

[534] H.-C. Cheng, J. F. Gunion, Z. Han, G. Marandella, and B. McElrath. Mass determination in SUSY-like events with missing energy, *JHEP*, **12**:076, 2007, arXiv:0707.0030 [hep-ph].

[535] H.-C. Cheng and I. Low. TeV symmetry and the little hierarchy problem, *JHEP*, **09**:051, 2003, hep-ph/0308199.

[536] H.-C. Cheng and I. Low. Little hierarchy, little Higgses, and a little symmetry, *JHEP*, **08**:061, 2004, hep-ph/0405243.

[537] H.-C. Cheng, K. T. Matchev, and M. Schmaltz. Bosonic supersymmetry? Getting fooled at the LHC, *Phys. Rev.*, D**66**:056006, 2002, hep-ph/0205314.

[538] H.-C. Cheng, K. T. Matchev, and M. Schmaltz. Radiative corrections to Kaluza-Klein masses, *Phys. Rev.*, D**66**:036005, 2002, hep-ph/0204342.

[539] H.-Y. Chiu. Symmetry between particle and anti-particle populations in the universe, *Phys. Rev. Lett.*, **17**:712, 1966.

[540] M.-C. Chiu, C.-M. Ko, and Y. Tian. Theoretical aspects of gravitational lensing in TeVeS, *Astrophys. J.*, **636**:565–574, 2006.

[541] R. S. Chivukula and T. P. Walker. Technicolor cosmology, *Nucl. Phys.*, B**329**:445, 1990.

[542] W. S. Cho, K. Choi, Y. G. Kim, and C. B. Park. Gluino stransverse mass, *Phys. Rev. Lett.*, **100**:171801, 2008, arXiv:0709.0288 [hep-ph].

[543] W. S. Cho, K. Choi, Y. G. Kim, and C. B. Park. Measuring superparticle masses at hadron collider using the transverse mass kink, *JHEP*, **02**:035, 2008, arXiv:0711.4526 [hep-ph].

[544] J.-H. Choi, M. D. Weinberg, and N. Katz. The dynamics of satellite disruption in cold dark matter haloes, 2008, arXiv:0812.0009 [astro-ph].

[545] K. Choi, A. Falkowski, H. P. Nilles, and M. Olechowski. Soft supersymmetry breaking in KKLT flux compactification, *Nucl. Phys.*, B**718**:113, 2005, hep-th/0503216.

[546] K. Choi, K. S. Jeong, and K. Okumura. Phenomenology of mixed modulus-anomaly mediation in fluxed string compactifications and brane models, *JHEP*, **09**:039, 2005, hep-ph/0504037.

[547] N. D. Christensen and C. Duhr. FeynRules – Feynman rules made easy, 2008, arXiv:0806.4194 [hep-ph].

[548] D. J. H. Chung, P. Crotty, E. W. Kolb, and A. Riotto. On the gravitational production of superheavy dark matter, *Phys. Rev.*, D**64**:043503, 2001, hep-ph/0104100.

[549] D. J. H. Chung, E. W. Kolb, and A. Riotto. Nonthermal supermassive dark matter, *Phys. Rev. Lett.*, **81**:4048–4051, 1998, hep-ph/9805473.

[550] D. J. H. Chung, E. W. Kolb, and A. Riotto. Production of massive particles during reheating, *Phys. Rev.*, D**60**:063504, 1999, hep-ph/9809453.

[551] D. J. H. Chung, E. W. Kolb, and A. Riotto. Superheavy dark matter, *Phys. Rev.*, D**59**:023501, 1999, hep-ph/9802238.

[552] D. J. H. Chung, A. Notari, and A. Riotto. Minimal theoretical uncertainties in inflationary predictions, *JCAP*, **0310**:012, 2003, hep-ph/0305074.

[553] P. Ciarcelluti and R. Foot. Early Universe cosmology in the light of the mirror dark matter interpretation of the DAMA/Libra signal, 2008, arXiv:0809.4438 [astro-ph].

[554] P. Ciarcelluti and A. Lepidi. Thermodynamics of the early Universe with mirror dark matter, 2008, arXiv:0809.0677 [astro-ph].

[555] L. Ciotti and J. Binney. Two-body relaxation in modified Newtonian dynamics, *Mon. Not. Roy. Astron. Soc.*, **351**:285, 2004, astro-ph/0403020.

[556] M. Cirelli *et al.* Spectra of neutrinos from dark matter annihilations, *Nucl. Phys.*, B**727**:99–138, 2005, hep-ph/0506298. http://www.marcocirelli.net/DMnu.html.

[557] M. Cirelli, N. Fornengo, and A. Strumia. Minimal dark matter, *Nucl. Phys.*, B**753**:178–194, 2006, hep-ph/0512090.

[558] M. Cirelli, R. Franceschini, and A. Strumia. Minimal dark matter predictions for galactic positrons, anti-protons, photons, *Nucl. Phys.*, B**800**:204–220, 2008, arXiv:0802.3378 [hep-ph].

[559] M. Cirelli, M. Kadastik, M. Raidal, and A. Strumia. Model-independent implications of the e+, e-, anti-proton cosmic ray spectra on properties of dark matter, 2008, arXiv:0809.2409 [hep-ph].

[560] M. Cirelli and A. Strumia. Minimal dark matter predictions and the PAMELA positron excess, 2008, arXiv:0808.3867 [astro-ph].

[561] M. Cirelli, A. Strumia, and M. Tamburini. Cosmology and astrophysics of minimal dark matter, *Nucl. Phys.*, B**787**:152–175, 2007, arXiv:0706.4071 [hep-ph].

[562] D. Clowe, M. Bradač, A. H. Gonzalez, *et al.* A direct empirical proof of the existence of dark matter, *Astrophys. J.*, **648**:L109–L113, 2006, astro-ph/0608407.

[563] D. Clowe, A. Gonzalez, and M. Markevitch. Weak-lensing mass reconstruction of the interacting cluster 1E 0657–558: Direct evidence for the existence of dark matter, *Astrophys. J.*, **604**:596–603, 2004.

[564] A. G. Cohen, T. S. Roy, and M. Schmaltz. Hidden sector renormalization of MSSM scalar masses, *JHEP*, **02**:027, 2007, hep-ph/0612100.

[565] S. Colafrancesco and P. Blasi. Clusters of galaxies and the diffuse gamma-ray background, *Astropart. Phys.*, **9**:227–246, 1998, astro-ph/9804262.

[566] S. Colafrancesco, P. de Bernardis, S. Masi, G. Polenta, and P. Ullio. Direct probes of dark matter in the cluster 1ES0657–556 through microwave observations, 2007, astro-ph/0702568.

[567] S. Colafrancesco, S. Profumo, and P. Ullio. Multi-frequency analysis of neutralino dark matter annihilations in the Coma cluster, *Astron. Astrophys.*, **455**:21, 2006, astro-ph/0507575.

[568] S. Colafrancesco, S. Profumo, and P. Ullio. Detecting dark matter WIMPs in the Draco dwarf: a multi-wavelength perspective, *Phys. Rev.*, D**75**:023513, 2007, astro-ph/0607073.

[569] P. Colin, A. Klypin, and A. V. Kravtsov. Velocity bias in a LCDM model, *Astrophys. J.*, **539**:561–569, 2000, astro-ph/9907337.

[570] LEP Collaboration. A combination of preliminary electroweak measurements and constraints on the standard model, 2003, hep-ex/0312023.

[571] J. I. Collar and F. T. Avignone. Diurnal modulation effects in cold dark matter experiments, *Phys. Lett.*, B**275**:181–185, 1992.

[572] J. M. Comerford and P. Natarajan. The observed concentration-mass relation for galaxy clusters, *Mon. Not. Roy. Astron. Soc.*, **379**:190–200, July 2007.

[573] C. R. Contaldi, T. Wiseman, and B. Withers. TeVeS gets caught on caustics, *Phys. Rev. D*, **78**:044034, 2008, arXiv:0802.1215.

[574] R. Contino, Y. Nomura, and A. Pomarol. Higgs as a holographic pseudo-Goldstone boson, *Nucl. Phys.*, B**671**:148–174, 2003, hep-ph/0306259.

[575] R. Contino and G. Servant. Discovering the top partners at the LHC using same-sign dilepton final states, *JHEP*, **06**:026, 2008, arXiv:0801.1679 [hep-ph].

[576] C. J. Copi, J. Heo, and L. M. Krauss. Directional sensitivity, WIMP detection, and the galactic halo, *Phys. Lett.*, B**461**:43–48, 1999, hep-ph/9904499.

[577] C. J. Copi and L. M. Krauss. Angular signatures for galactic halo WIMP scattering in direct detectors: Prospects and challenges, *Phys. Rev.*, D**63**:043507, 2001, astro-ph/0009467.

[578] C. J. Copi and L. M. Krauss. Comparing WIMP interaction rate detectors with annual modulation detectors, *Phys. Rev.*, D**67**:103507, 2003, astro-ph/0208010.

[579] C. J. Copi, L. M. Krauss, D. Simmons-Duffin, and S. R. Stroiney. Assessing alternatives for directional detection of a WIMP halo, *Phys. Rev.*, D**75**:023514, 2007, astro-ph/0508649.

[580] C. Coppi *et al.* Quenching factor measurement for CaWO-4 by neutron scattering, *Nucl. Instrum. Meth.*, A**559**:396–398, 2006.

[581] A. Corsetti and P. Nath. Gaugino mass nonuniversality and dark matter in SUGRA, strings and D brane models, *Phys. Rev.*, D**64**:125010, 2001, hep-ph/0003186.

[582] N. Cosme, L. Lopez Honorez, and M. H. G. Tytgat. Leptogenesis and dark matter related?, *Phys. Rev.*, D**72**:043505, 2005, hep-ph/0506320.

[583] P. Côté, L. Ferrarese, A. Jordán, *et al.* The ACS Fornax Cluster Survey. II. The central brightness profiles of early-type galaxies: A characteristic radius on nuclear scales and the transition from

central luminosity deficit to excess, *Astrophys. J.*, **671**:1456–1465, 2007, arXiv:0711.1358.

[584] P. Côté, S. Piatek, L. Ferrarese, *et al.* The ACS Virgo Cluster Survey. VIII. The nuclei of early-type galaxies, *Astrophys. J. Suppl.*, **165**:57–94, 2006.

[585] L. Covi, H.-B. Kim, J. E. Kim, and L. Roszkowski. Axinos as dark matter, *JHEP*, **05**:033, 2001, hep-ph/0101009.

[586] L. Covi, J. E. Kim, and L. Roszkowski. Axinos as cold dark matter, *Phys. Rev. Lett.*, **82**:4180–4183, 1999, hep-ph/9905212.

[587] H. R. Crane. CO_2-CS_2 Geiger counter, *Rev. Sci. Instrum.*, **32**:953, 1961.

[588] E. Cremmer *et al.* SuperHiggs effect in supergravity with general scalar interactions, *Phys. Lett.*, **B79**:231, 1978.

[589] E. Cremmer *et al.* Spontaneous symmetry breaking and Higgs effect in supergravity without cosmological constant, *Nucl. Phys.*, **B147**:105, 1979.

[590] N. H. M. Crighton, J. K. Webb, A. Ortiz-Gill, and A. Fernandez-Soto. D/H in a new Lyman limit quasar absorption system at $z = 3.256$ towards PKS1937–1009, *Mon. Not. Roy. Astron. Soc.*, **355**:1042, 2004, astro-ph/0403512.

[591] P. Crotty, J. Garcia-Bellido, J. Lesgourgues, and A. Riazuelo. Bounds on isocurvature perturbations from CMB and LSS data, *Phys. Rev. Lett.*, **91**:171301, 2003, astro-ph/0306286.

[592] C. S. Kochanck. The mass of the Milky Way, *Astrophys. J.*, **457**:228, 1996.

[593] C. Csaki, M. L. Graesser, and G. D. Kribs. Radion dynamics and electroweak physics, *Phys. Rev.*, **D63**:065002, 2001, hep-th/0008151.

[594] C. Csaki, C. Grojean, J. Hubisz, Y. Shirman, and J. Terning. Fermions on an interval: Quark and lepton masses without a Higgs, *Phys. Rev.*, **D70**:015012, 2004, hep-ph/0310355.

[595] C. Csaki, J. Heinonen, and M. Perelstein. Testing gluino spin with three-body decays, *JHEP*, **10**:107, 2007, arXiv:0707.0014 [hep-ph].

[596] C. Csaki, J. Hubisz, and S. J. Lee. Radion phenomenology in realistic warped space models, *Phys. Rev.*, **D76**:125015, 2007, arXiv:0705.3844 [hep-ph].

[597] K. M. Cudworth. The local galactic escape velocity revisited: Improved proper motions for critical stars, *Astron. J.*, **99**:590, 1990.

[598] A. J. Cuesta, F. Prada, A. Klypin, and M. Moles. The virialized mass of dark matter haloes, 2007, arXiv:0710.5520 [astro-ph].

[599] D. Cumberbatch *et al.* Solving the cosmic lithium problems with primordial late-decaying particles, *Phys. Rev.*, D**76**:123005, 2007, arXiv:0708.0095 [astro-ph].

[600] A. Cuoco, J. Brandbyge, S. Hannestad, T. Haugboelle, and G. Miele. Angular signatures of annihilating dark matter in the cosmic gamma-ray background, *Phys. Rev.*, D**77**:123518, 2008, arXiv:0710.4136 [astro-ph].

[601] A. Cuoco *et al.* The signature of large scale structures on the very high energy gamma-ray sky, *JCAP*, **0704**:013, 2007, astro-ph/0612559.

[602] R. H. Cyburt, J. R. Ellis, B. D. Fields, and K. A. Olive. Updated nucleosynthesis constraints on unstable relic particles, *Phys. Rev.*, D**67**:103521, 2003, astro-ph/0211258.

[603] R. H. Cyburt, J. R. Ellis, B. D. Fields, K. A. Olive, and V. C. Spanos. Bound-state effects on light-element abundances in gravitino dark matter scenarios, *JCAP*, **0611**:014, 2006, astro-ph/0608562.

[604] R. H. Cyburt, B. D. Fields, and K. A. Olive. A bitter pill: The primordial lithium problem worsens, 2008, arXiv:0808.2818 [astro-ph].

[605] E. S. Cypriano, L. J. Sodré, J.-P. Kneib, and L. E. Campusano. Weak-lensing mass distributions for 24 X-ray Abell Clusters, *Astrophys. J.*, **613**:95–108, 2004.

[606] A. Czarnecki and W. J. Marciano. The muon anomalous magnetic moment: A harbinger for 'new physics', *Phys. Rev.*, D**64**:013014, 2001.

[607] A. Dabholkar and J. M. Quashnock. Pinning down the axion, *Nucl. Phys.*, B**333**:815, 1990.

[608] H. Dahle, S. Hannestad, and J. Sommer-Larsen. The density profile of cluster-scale dark matter halos, *Astrophys. J.*, **588**:L73–L76, 2003.

[609] J. J. Dalcanton, C. R. Canizares, A. Granados, C. C. Steidel, and J. T. Stocke. Observational limits on Omega in stars, brown dwarfs, and stellar remnants from gravitational microlensing, *Astrophys. J.*, **424**:550–568, 1994.

[610] J. J. Dalcanton and C. J. Hogan. Halo cores and phase space densities: Observational constraints on dark matter physics and structure formation, *Astrophys. J.*, **561**:35–45, 2001, astro-ph/0004381.

[611] T. Damour and B. Pichon. Big Bang nucleosynthesis and tensor-scalar gravity, *Phys. Rev.*, D**59**:123502, 1999, astro-ph/9807176.

[612] W. Dappen, R. L. Gilliland, and J. Christensen-Dalsgaard. Weakly interacting massive particles, solar neutrinos, and solar oscillations, *Nature*, **321**:229–231, 1986.

[613] A. Datta, G. L. Kane, and M. Toharia. Is it SUSY?, 2005, hep-ph/0510204.

[614] A. Datta, K. Kong, and K. T. Matchev. Discrimination of super-symmetry and universal extra dimensions at hadron colliders, *Phys. Rev.*, D**72**:096006, 2005, hep-ph/0509246.

[615] R. Dave, D. N. Spergel, P. J. Steinhardt, and B. D. Wandelt. Halo properties in cosmological simulations of self-interacting cold dark matter, *Astrophys. J.*, **547**:574–589, 2001, astro-ph/0006218.

[616] M. Davier, A. Hoecker, B. Malaescu, C. Z. Yuoin, and Z. Zhang. Reevaluation of the hadronic contribution to the muon magnetic anomaly using new $e^+ e^- \to pi^+ pi^-$ cross section data from BABAR, arXiv:0908.4300 [hep-ph].

[617] R. L. Davis. Goldstone bosons in string models of galaxy formation, *Phys. Rev.*, D**32**:3172, 1985.

[618] R. L. Davis. Cosmic axions from cosmic strings, *Phys. Lett.*, B**180**:225, 1986.

[619] R. L. Davis and E. P. S. Shellard. Do axions need inflation?, *Nucl. Phys.*, B**324**:167, 1989.

[620] H. Davoudiasl, J. L. Hewett, and T. G. Rizzo. Bulk gauge fields in the Randall-Sundrum model, *Phys. Lett.*, B**473**:43–49, 2000, hep-ph/9911262.

[621] H. Davoudiasl, T. G. Rizzo, and A. Soni. On direct verification of warped hierarchy-and-flavor models, *Phys. Rev.*, D**77**:036001, 2008, arXiv:0710.2078 [hep-ph].

[622] R. Ruiz de Austri, R. Trotta, and L. Roszkowski. A Markov chain Monte Carlo analysis of the CMSSM, *JHEP*, **05**:002, 2006, hep-ph/0602028.

[623] W. J. G. de Blok. CDM in LSB galaxies: Toward the optimal halo profile. In S. Ryder, D. Pisano, M. Walker, and K. Freeman, eds., *IAU Symposium*, 69, 2004.

[624] W. J. G. de Blok. Halo mass profiles and low surface brightness galaxy rotation curves, *Astrophys. J.*, **634**:227–238, 2005.

[625] W. J. G. de Blok and A. Bosma. High-resolution rotation curves of low surface brightness galaxies, *Astron. Astrophys.*, **385**:816–846, 2002.

[626] W. de Boer, C. Sander, V. Zhukov, A. V. Gladyshev, and D. I. Kazakov. EGRET excess of diffuse galactic gamma rays as tracer of dark matter, *Astron. Astrophys.*, **444**:51, 2005, astro-ph/0508617.

[627] A. de Gouvea. See-saw energy scale and the LSND anomaly, *Phys. Rev.*, D**72**:033005, 2005, hep-ph/0501039.

[628] A. de Gouvea, J. Jenkins, and N. Vasudevan. Neutrino phenomenology of very low-energy seesaws, *Phys. Rev.*, D**75**:013003, 2007, hep-ph/0608147.

[629] G. De Lucia *et al.* Substructures in cold dark matter halos, *Mon. Not. Roy. Astron. Soc.*, **348**:333, 2004, astro-ph/0306205.

[630] G. De Lucia and A. Helmi. The Galaxy and its stellar halo: insights on their formation from a hybrid cosmological approach, 2008, arXiv:0804.2465 [astro-ph].

[631] S. De Panfilis *et al.* Limits on the abundance and coupling of cosmic axions at 4.5 μeV $< m_a <$ 5.0 μeV, *Phys. Rev. Lett.*, **59**:839, 1987.

[632] A. De Roeck *et al.* Supersymmetric benchmarks with non-universal scalar masses or gravitino dark matter, *Eur. Phys. J.*, C**49**:1041–1066, 2007, hep-ph/0508198.

[633] G. de Vaucouleurs. Recherches sur les Nebuleuses Extragalactiques, *Ann. Astrophys.*, **11**:247, January 1948.

[634] D. Dearborn, G. Raffelt, P. Salati, J. Silk, and A. Bouquet. Dark matter and the age of globular clusters, *Nature*, **343**:347–+, January 1990.

[635] D. Dearborn, G. Raffelt, P. Salati, J. Silk, and A. Bouquet. Dark matter and thermal pulses in horizontal-branch stars, *Astrophys. J.*, **354**:568–582, May 1990.

[636] D. S. P. Dearborn, D. N. Schramm, and G. Steigman. Astrophysical constraints on the couplings of axions, majorons, and familons, *Phys. Rev. Lett.*, **56**:26, 1986.

[637] V. P. Debattista *et al.* The causes of halo shape changes induced by cooling baryons: Disks versus substructures, 2007, arXiv:0707.0737 [astro-ph].

[638] X. Defay *et al.* Cryogenic Ge detectors for dark matter search: Surface event rejection with ionization signals, *J. Low Temp. Phys.*, **151**:896–901, 2008.

[639] T. A. DeGrand, T. W. Kephart, and T. J. Weiler. Invisible axions and the QCD phase transition in the early universe, *Phys. Rev.*, D**33**:910, 1986.

[640] G. Degrassi, P. Gambino, and G. F. Giudice. B $\rightarrow X_s$ γ in supersymmetry: Large contributions beyond the leading order, *JHEP*, **12**:009, 2000, hep-ph/0009337.

[641] G. Degrassi, P. Gambino, and P. Slavich. SusyBSG: a fortran code for BR[$B \rightarrow X_s\gamma$] in the MSSM with Minimal Flavor Violation, *Comput. Phys. Commun.*, **179**:759–771, 2008, arXiv:0712.3265 [hep-ph].

[642] G. Degrassi, S. Heinemeyer, W. Hollik, P. Slavich, and G. Weiglein. Towards high-precision predictions for the MSSM Higgs sector, *Eur. Phys. J.*, C**28**:133–143, 2003, hep-ph/0212020.

[643] W. Dehnen and J. Binney. Local stellar kinematics from Hipparcos data, *Mon. Not. Roy. Astron. Soc.*, **298**:387–394, 1998, astro-ph/9710077.

[644] T. Delahaye *et al.* Galactic secondary positron flux at the Earth. *Astron. Astrophys.*, **501**:821–833, 2009, arXiv:0809.5268 [astro-ph].

[645] T. Delahaye, R. Lineros, F. Donato, N. Fornengo, and P. Salati. Positrons from dark matter annihilation in the galactic halo: Theoretical uncertainties, *Phys. Rev.*, D**77**:063527, 2008, 0712.2312.

[646] T. Delahaye, R. Lineros, F. Donato, N. Fornengo, and P. Salati. Positrons from dark matter annihilation in the galactic halo: Theoretical uncertainties, *Phys. Rev.*, D**77**:063527, 2008, arXiv:0712.2312 [astro-ph].

[647] C. Dennis, M. Karagoz Unel, G. Servant, and J. Tseng. Multi-W events at LHC from a warped extra dimension with custodial symmetry, hep-ph/0701158.

[648] F. Deppisch and A. Pilaftsis. Thermal right-handed sneutrino dark matter in the F_D-term model of hybrid inflation, *JHEP*, **10**:080, 2008, arXiv:0808.0490 [hep-ph].

[649] F. D'Eramo. Dark matter and Higgs boson physics, *Phys. Rev.*, D**76**:083522, 2007, arXiv:0705.4493 [hep-ph].

[650] R. Dermisek, S. Raby, L. Roszkowski, and R. Ruiz De Austri. Dark matter and B/s → mu+ mu- with minimal SO(10) soft SUSY breaking, *JHEP*, **04**:037, 2003, hep-ph/0304101.

[651] S. Desai *et al.* Search for dark matter WIMPs using upward through-going muons in Super-Kamiokande, *Phys. Rev.*, D**70**:083523, 2004, hep-ex/0404025.

[652] N. G. Deshpande and E. Ma. Pattern of symmetry breaking with two Higgs doublets, *Phys. Rev.*, D**18**:2574, 1978.

[653] T. DeYoung. Status of the IceCube neutrino telescope, *Int. J. Mod. Phys.*, A**20**:3160–3162, 2005.

[654] J. L. Diaz-Cruz. Holographic dark matter and Higgs, *Phys. Rev. Lett.*, **100**:221802, 2008, arXiv:0711.0488 [hep-ph].

[655] J. L. Diaz-Cruz, J. R. Ellis, K. A. Olive, and Y. Santoso. On the feasibility of a stop NLSP in gravitino dark matter scenarios, *JHEP*, **05**:003, 2007, hep-ph/0701229.

[656] L. M. Diaz-Rivera, L. Samushia, and B. Ratra. Inflation and accelerated expansion tensor-vector-scalar cosmological solutions, *Phys. Rev. D*, **73**:083503, 2006.

[657] D. A. Dicus, E. W. Kolb, and V. L. Teplitz. Cosmological upper bound on heavy neutrino lifetimes, *Phys. Rev. Lett.*, **39**:168, 1977.

[658] D. A. Dicus, E. W. Kolb, V. L. Teplitz, and R. V. Wagoner. Astrophysical bounds on the masses of axions and Higgs particles, *Phys. Rev.*, D**18**:1829, 1978.

[659] D. A. Dicus, E. W. Kolb, V. L. Teplitz, and R. V. Wagoner. Astrophysical bounds on very low mass axions, *Phys. Rev.*, D**22**:839, 1980.

[660] J. Diemand *et al.* Clumps and streams in the local dark matter distribution, *Nature*, **454**:735–738, 2008, arXiv:0805.1244 [astro-ph].

[661] J. Diemand and M. Kuhlen. Infall caustics in dark matter halos?, *Astrophys. J.*, **680**:L25–L28, 2008, arXiv:0804.4185 [astro-ph].

[662] J. Diemand, M. Kuhlen, and P. Madau. Early supersymmetric cold dark matter substructure, *Astrophys. J.*, **649**:1–13, 2006, astro-ph/0603250.

[663] J. Diemand, M. Kuhlen, and P. Madau. Dark matter substructure and gamma-ray annihilation in the Milky Way halo, *Astrophys. J.*, **657**:262, 2007, astro-ph/0611370.

[664] J. Diemand, M. Kuhlen, and P. Madau. Formation and evolution of galaxy dark matter halos and their substructure, *Astrophys. J.*, **667**:859, 2007, astro-ph/0703337.

[665] J. Diemand, P. Madau, and B. Moore. The distribution and kinematics of early high-sigma peaks in present-day haloes: implications for rare objects and old stellar populations, *Mon. Not. Roy. Astron. Soc.*, **364**:367–383, 2005, astro-ph/0506615.

[666] J. Diemand, B. Moore, and J. Stadel. Convergence and scatter of cluster density profiles, *Mon. Not. Roy. Astron. Soc.*, **353**:624–632, September 2004, astro-ph/0402267.

[667] J. Diemand, B. Moore, and J. Stadel. Velocity and spatial biases in CDM subhalo distributions, *Mon. Not. Roy. Astron. Soc.*, **352**:535, 2004, astro-ph/0402160.

[668] J. Diemand, B. Moore, and J. Stadel. Earth-mass dark-matter haloes as the first structures in the early universe, *Nature*, **433**:389–391, 2005, astro-ph/0501589.

[669] J. Diemand, B. Moore, J. Stadel, and S. Kazantzidis. Two body relaxation in CDM simulations, *Mon. Not. Roy. Astron. Soc.*, **348**:977, 2004, astro-ph/0304549.

[670] J. Diemand, M. Zemp, B. Moore, J. Stadel, and M. Carollo. Cusps in CDM halos: The density profile of a billion particle halo, *Mon. Not. Roy. Astron. Soc.*, **364**:665, 2005, astro-ph/0504215.

[671] S. Dimopoulos, D. Eichler, R. Esmailzadeh, and G. D. Starkman. Getting a charge out of dark matter, *Phys. Rev.*, **D41**:2388, 1990.

[672] S. Dimopoulos, R. Esmailzadeh, L. J. Hall, and G. D. Starkman. Is the universe closed by baryons? Nucleosynthesis with a late decaying massive particle, *Astrophys. J.*, **330**:545, 1988.

[673] S. Dimopoulos and S. Raby. Supercolor, *Nucl. Phys.*, **B192**:353, 1981.

[674] M. Dine and W. Fischler. The not-so-harmless axion, *Phys. Lett.*, **B120**:137–141, 1983.

[675] M. Dine, W. Fischler, and M. Srednicki. A simple solution to the strong CP problem with a harmless axion, *Phys. Lett.*, **B104**:199, 1981.

[676] M. Dine, W. Fischler, and M. Srednicki. Supersymmetric technicolor, *Nucl. Phys.*, **B189**:575–593, 1981.

[677] M. Dine, A. E. Nelson, Y. Nir, and Y. Shirman. New tools for low-energy dynamical supersymmetry breaking, *Phys. Rev.*, **D53**:2658–2669, 1996, hep-ph/9507378.

[678] M. Dine, A. E. Nelson, and Y. Shirman. Low-energy dynamical supersymmetry breaking simplified, *Phys. Rev.*, **D51**:1362–1370, 1995, hep-ph/9408384.

[679] A. Djouadi, M. Drees, and J. L. Kneur. Constraints on the minimal supergravity model and prospects for SUSY particle production at future linear e+ e- colliders, *JHEP*, **08**:055, 2001, hep-ph/0107316.

[680] A. Djouadi, J. Kalinowski, and M. Spira. HDECAY: A program for Higgs boson decays in the standard model and its supersymmetric extension, *Comput. Phys. Commun.*, **108**:56–74, 1998, hep-ph/9704448.

[681] A. Djouadi, J.-L. Kneur, and G. Moultaka. SuSpect: A Fortran code for the supersymmetric and Higgs particle spectrum in the MSSM, *Comput. Phys. Commun.*, **176**:426–455, 2007, hep-ph/0211331.

[682] A. Djouadi, G. Moreau, and R. K. Singh. Kaluza–Klein excitations of gauge bosons at the LHC, *Nucl. Phys.*, **B797**:1–26, 2008, arXiv:0706.4191 [hep-ph].

[683] A. Djouadi, M. M. Muhlleitner, and M. Spira. Decays of supersymmetric particles: The program SUSY-HIT (SUspect-SdecaY-Hdecay-InTerface), *Acta Phys. Polon.*, **B38**:635–644, 2007, hep-ph/0609292.

[684] B. A. Dobrescu, D. Hooper, K. Kong, and R. Mahbubani. Spinless photon dark matter from two universal extra dimensions, *JCAP*, **0710**:012, 2007, arXiv:0706.3409 [hep-ph].

[685] B. A. Dobrescu, K. Kong, and R. Mahbubani. Leptons and photons at the LHC: Cascades through spinless adjoints, *JHEP*, **07**:006, 2007, hep-ph/0703231.

[686] B. A. Dobrescu and E. Ponton. Chiral compactification on a square, *JHEP*, **03**:071, 2004, hep-th/0401032.

[687] S. Dodelson. *Modern Cosmology*, Academic Press, 2003.

[688] S. Dodelson, D. Hooper, and P. D. Serpico. Extracting the gamma ray signal from dark matter annihilation in the Galactic center region, *Phys. Rev.*, D**77**:063512, 2008, arXiv:0711.4621 [astro-ph].

[689] S. Dodelson and M. Liguori. Can cosmic structure form without dark matter?, *Phys. Rev. Lett.*, **97**:231301, 2006.

[690] S. Dodelson and L. M. Widrow. Sterile-neutrinos as dark matter, *Phys. Rev. Lett.*, **72**:17–20, 1994, hep-ph/9303287.

[691] T. Doke. *Radiat. Phys.*, **1**:24, 1977.

[692] T. Doke *et al. Nucl. Instr. Methods*, A **269**:291, 1988.

[693] A. D. Dolgov and S. H. Hansen. Massive sterile neutrinos as warm dark matter, *Astropart. Phys.*, **16**:339–344, 2002, hep-ph/0009083.

[694] A. D. Dolgov, K. Kohri, O. Seto, and J. Yokoyama. Stabilizing dilaton and baryogenesis, *Phys. Rev.*, D**67**:103515, 2003, hep-ph/0210223.

[695] D. Dominici, B. Grzadkowski, J. F. Gunion, and M. Toharia. The scalar sector of the Randall–Sundrum model, *Nucl. Phys.*, B**671**:243–292, 2003, hep-ph/0206192.

[696] D. Dominici, B. Grzadkowski, J. F. Gunion, and M. Toharia. Higgs-boson interactions within the Randall–Sundrum model, *Acta Phys. Polon.*, B**33**:2507–2522, 2002, hep-ph/0206197.

[697] G. Domokos. Dark matter or modified dynamics?, Preprint JHU-TIPAC-96009, 1996, ArXiv e-prints.

[698] F. Donato, N. Fornengo, and D. Maurin. Antideuteron fluxes from dark matter annihilation in diffusion models, *Phys. Rev.*, D**78**:043506, 2008, arXiv:0803.2640 [hep-ph].

[699] F. Donato, N. Fornengo, D. Maurin, and P. Salati. Antiprotons in cosmic rays from neutralino annihilation, 2004, astro-ph/0306207.

[700] F. Donato, N. Fornengo, D. Maurin, P. Salati, and R. Taillet. Antiprotons in cosmic rays from neutralino annihilation, *Phys. Rev.*, D**69**:063501, 2004.

[701] F. Donato, N. Fornengo, and P. Salati. Antideuterons as a signature of supersymmetric dark matter, *Phys. Rev.*, D62:043003, 2000, hep-ph/9904481.

[702] F. Donato, N. Fornengo, and S. Scopel. Effects of galactic dark halo rotation on WIMP direct detection, *Astropart. Phys.*, 9:247–260, 1998, hep-ph/9803295.

[703] F. Donato, D. Maurin, P. Brun, T. Delahaye, and P. Salati. Constraints on WIMP dark matter from the high energy PAMELA $\bar{p}/p$ data, 2008, arXiv:0810.5292 [astro-ph].

[704] F. Donato, D. Maurin, P. Salati, A. Barrau, G. Boudoul, and R. Taillet. Antiprotons from spallations of cosmic rays on interstellar matter, *Astrophys. J.*, 563:172–184, 2001.

[705] F. Donato, D. Maurin, and R. Taillet. β–radioactive cosmic rays in a diffusion model: test for a local bubble?, *Astron. Astrophys.*, 381:539–559, 2002, astro-ph/0108079.

[706] F. Donato and P. Salucci. Cores of dark matter halos correlate with disk scale lengths, *Mon. Not. Roy. Astron. Soc.*, 353:L17–L22, 2004, astro-ph/0403206.

[707] T. W. Donnelly, S. J. Freedman, R. S. Lytel, R. D. Peccei, and M. Schwartz. Do axions exist?, *Phys. Rev.*, D18:1607, 1978.

[708] O. Doré, M. Martig, Y. Mellier, *et al.* Testing Gravity with the CFHTLS-Wide Cosmic Shear Survey and SDSS LRGs, 2007, arXiv:0712.1599.

[709] M. R. Douglas. The statistics of string/M theory vacua, *JHEP*, 05:046, 2003, hep-th/0303194.

[710] M. Drees, H. Iminniyaz, and M. Kakizaki. Abundance of cosmological relics in low-temperature scenarios, *Phys. Rev.*, D73:123502, 2006, hep-ph/0603165.

[711] M. Drees, H. Iminniyaz, and M. Kakizaki. Constraints on the very early universe from thermal WIMP dark matter, *Phys. Rev.*, D76:103524, 2007, arXiv:0704.1590 [hep-ph].

[712] M. Drees, Y. G. Kim, T. Kobayashi, and M. M. Nojiri. Direct detection of neutralino dark matter and the anomalous dipole moment of the muon, *Phys. Rev.*, D63:115009, 2001, hep-ph/0011359.

[713] M. Drees and M. Nojiri. Neutralino–nucleon scattering revisited, *Phys. Rev.*, D48:3483–3501, 1993, hep-ph/9307208.

[714] M. Drees and M. M. Nojiri. New contributions to coherent neutralino – nucleus scattering, *Phys. Rev.*, D47:4226–4232, 1993, hep-ph/9210272.

[715] M. Drees and M. M. Nojiri. The neutralino relic density in minimal $N = 1$ supergravity, *Phys. Rev.*, D**47**:376, 1993, hep-ph/9207234.

[716] M. Drees, M. M. Nojiri, D. P. Roy, and Y. Yamada. Light Higgsino dark matter, *Phys. Rev.*, D**56**:276–290, 1997, hep-ph/9701219.

[717] M. Drees and C.-L. Shan. Model-independent determination of the WIMP mass from direct dark matter detection data, *JCAP*, **0806**:012, 2008, arXiv:0803.4477 [hep-ph].

[718] M. Drees and X. Tata. Signals for heavy exotics at hadron colliders and supercolliders, *Phys. Lett.*, B**252**:695–702, 1990.

[719] A. K. Drukier, K. Freese, and D. N. Spergel. Detecting cold dark matter candidates, *Phys. Rev.*, D**33**:3495–3508, 1986.

[720] J. Dubinski, I. Berentzen, and I. Shlosman. Anatomy of the bar instability in cuspy dark matter halos, 2008, arXiv:0810.4925 [astro-ph].

[721] J. Dubinski and R. G. Carlberg. The structure of cold dark matter halos, *Astrophys. J.*, **378**:496–503, 1991.

[722] G. Duda, A. Kemper, and P. Gondolo. Model independent form factors for spin independent neutralino nucleon scattering from elastic electron scattering data, *JCAP*, **0704**:012, 2007, hep-ph/0608035.

[723] E. Dudas, S. Lavignac, and J. Parmentier. A light neutralino in hybrid models of supersymmetry breaking, *Nucl. Phys.*, B**808**:237–259, 2009, arXiv:0808.0562 [hep-ph].

[724] S. Duffau *et al.* Spectroscopy of QUEST RR Lyrae Variables: the new Virgo Stellar Stream, *Astrophys. J.*, **636**:L97–L100, 2006, astro-ph/0510589.

[725] L. Duffy, P. Sikivie, D. B. Tanner, *et al.* Results of a search for cold flows of dark matter axions, *Phys. Rev. Lett.*, **95**:091304, 2005.

[726] D. Dujmic *et al.* Charge amplification concepts for direction-sensitive dark matter detectors, *Astropart. Phys.*, **30**:58–64, 2008, arXiv:0804.4827 [astro-ph].

[727] D. Dujmic *et al.* DMTPC-10L: Direction-sensitive dark matter detector prototype, 2008, arXiv:0810.2769 [physics.ins-det].

[728] J. Dunkley, M. Bucher, P. G. Ferreira, K. Moodley, and C. Skordis. Fast and reliable MCMC for cosmological parameter estimation, *Mon. Not. Roy. Astron. Soc.*, **356**:925–936, 2005, astro-ph/0405462.

[729] J. Dunkley *et al.* Five-Year Wilkinson Microwave Anisotropy Probe (WMAP) observations: Likelihoods and parameters from the WMAP data, 2008, arXiv:0803.0586 [astro-ph].

[730] R. Duperray *et al.* Flux of light antimatter nuclei near earth, induced by cosmic rays in the galaxy and in the atmosphere, *Phys. Rev.*, D**71**:083013, 2005, astro-ph/0503544.

[731] A. A. Dutton, F. C. van den Bosch, A. Dekel, and S. Courteau. A revised model for the formation of disk galaxies: Quiet history, low spin, and dark-halo expansion, *Astrophys. J.*, **654**:27–52, 2006, astro-ph/0604553.

[732] D. Eckert, N. Produit, S. Paltani, A. Neronov, and T. J.-L. Courvoisier. INTEGRAL discovery of non-thermal hard X-ray emission from the Ophiuchus cluster, 2007, arXiv:0712.2326 [astro-ph].

[733] A. S. Eddington. The distribution of stars in globular clusters, *Mon. Not. Roy. Astron. Soc.*, **76**:572, 1916.

[734] J. Edsjö. WimpSim – a general WIMP annihilation and neutrino propagation code for WIMP annihilations in the Sun/Earth, 2008, http://www.physto.se/~edsjo/wimpsim/.

[735] J. Edsjö and P. Gondolo. Neutralino relic density including coannihilations, *Phys. Rev.*, **D56**:1879–1894, 1997, hep-ph/9704361.

[736] J. Edsjö, M. Schelke, and P. Ullio. Direct versus indirect detection in mSUGRA with self-consistent halo models, *JCAP*, **0409**:004, 2004, astro-ph/0405414.

[737] O. J. Eggen, D. Lynden-Bell, and A. R. Sandage. Evidence from the motions of old stars that the Galaxy collapsed, *Astrophys. J.*, **136**:748, 1962.

[738] R. Eichler *et al.* Limits for shortlived neutral particles emitted mu+ or pi+ decay, *Phys. Lett.*, **B175**:101, 1986.

[739] S. Eidelman *et al.* Review of particle physics, *Phys. Lett.*, **B592**:1, 2004.

[740] J. Einasto. Influence of the atmospheric and instrumental dispersion on the brightness distribution in a galaxy, *Trudy Inst. Astrofiz. Alma-Ata*, **5**:87, 1965.

[741] J. Einasto. On constructing models of stellar systems. V. The binomial model, *Publications of the Tartuskoj Astrofizica Observatory*, **36**:414–441, 1968.

[742] J. Einasto. Dark matter, 2009, arXiv:0901.0632 [astro-ph.CO].

[743] J. Einasto and U. Haud. Galactic models with massive corona. I. Method. II. Galaxy, *Astron. Astrophys.*, **223**:89–106, 1989.

[744] V. R. Eke, J. F. Navarro, and M. Steinmetz. The power spectrum dependence of dark matter halo concentrations, *Astrophys. J.*, **554**:114–125, June 2001.

[745] A. A. El-Zant, Y. Hoffman, J. Primack, F. Combes, and I. Shlosman. Flat-cored dark matter in cuspy clusters of galaxies, *Astrophys. J.*, **607**:L75–L78, 2004, astro-ph/0309412.

[746] P. J. Elahi, R. J. Thacker, L. M. Widrow, and E. Scannapieco. Sub-haloes in scale-free cosmologies, 2008, arXiv:0811.0206 [astro-ph].

[747] J. R. Ellis. Beyond the standard model for hillwalkers, 1998, hep-ph/9812235.

[748] J. R. Ellis, T. Falk, G. Ganis, and K. A. Olive. Supersymmetric dark matter in the light of LEP and the Tevatron Collider, *Phys. Rev.*, D**62**:075010, 2000, hep-ph/0004169.

[749] J. R. Ellis, T. Falk, G. Ganis, K. A. Olive, and M. Schmitt. Charginos and neutralinos in the light of radiative corrections: Sealing the fate of Higgsino dark matter, *Phys. Rev.*, D**58**:095002, 1998, hep-ph/9801445.

[750] J. R. Ellis, T. Falk, G. Ganis, K. A. Olive, and M. Srednicki. The CMSSM parameter space at large tan beta, *Phys. Lett.*, B**510**:236, 2001, hep-ph/0102098.

[751] J. R. Ellis, T. Falk, K. A. Olive, and Y. Santoso. Exploration of the MSSM with non-universal Higgs masses, *Nucl. Phys.*, B**652**:259, 2003, hep-ph/0210205.

[752] J. R. Ellis, T. Falk, K. A. Olive, and M. Schmitt. Supersymmetric dark matter in the light of LEP 1.5, *Phys. Lett.*, B**388**:97, 1996, hep-ph/9607292.

[753] J. R. Ellis, T. Falk, K. A. Olive, and M. Schmitt. Constraints on neutralino dark matter from LEP2 and cosmology, *Phys. Lett.*, B**413**:355, 1997, hep-ph/9705444.

[754] J. R. Ellis, A. Ferstl, and K. A. Olive. Re-evaluation of the elastic scattering of supersymmetric dark matter, *Phys. Lett.*, B**481**: 304–314, 2000, hep-ph/0001005.

[755] J. R. Ellis, A. Ferstl, K. A. Olive, and Y. Santoso. Direct detection of dark matter in the MSSM with non-universal Higgs masses, *Phys. Rev.*, D**67**:123502, 2003, hep-ph/0302032.

[756] J. R. Ellis and R. A. Flores. Elastic supersymmetric relic – nucleus scattering revisited, *Phys. Lett.*, B**263**:259–266, 1991.

[757] J. R. Ellis, R. A. Flores, and J. D. Lewin. Rates for inelastic nuclear excitation by dark matter particles, *Phys. Lett.*, B**212**:375, 1988.

[758] J. R. Ellis and M. K. Gaillard. Strong and weak CP violation, *Nucl. Phys.*, B**150**:141, 1979.

[759] J. R. Ellis, G. B. Gelmini, J. L. Lopez, D. V. Nanopoulos, and S. Sarkar. Astrophysical constraints on massive unstable neutral relic particles, *Nucl. Phys.*, B**373**:399–437, 1992.

[760] J. R. Ellis, J. S. Hagelin, D. V. Nanopoulos, K. A. Olive, and M. Srednicki. Supersymmetric relics from the big bang, *Nucl. Phys.*, B**238**:453, 1984.

[761] J. R. Ellis, J. S. Hagelin, D. V. Nanopoulos, and K. Tamvakis. Weak symmetry breaking by radiative corrections in broken supergravity, *Phys. Lett.*, B**125**:275, 1983.

[762] J. R. Ellis, S. Heinemeyer, K. A. Olive, A. M. Weber, and G. Weiglein. The supersymmetric parameter space in light of B-physics observables and electroweak precision data, *JHEP*, **08**:083, 2007, arXiv:0706.0652 [hep-ph].

[763] J. R. Ellis, S. Kelley, and D. V. Nanopoulos. Precision LEP data, supersymmetric GUTs and string unification, *Phys. Lett.*, B**249**:441, 1990.

[764] J. R. Ellis, S. Kelley, and D. V. Nanopoulos. Probing the desert using gauge coupling unification, *Phys. Lett.*, B**260**:131, 1991.

[765] J. R. Ellis, J. E. Kim, and D. V. Nanopoulos. Cosmological gravitino regeneration and decay, *Phys. Lett.*, B**145**:181, 1984.

[766] J. R. Ellis, D. V. Nanopoulos, and S. Sarkar. The cosmology of decaying gravitinos, *Nucl. Phys.*, B**259**:175, 1985.

[767] J. R. Ellis, D. V. Nanopoulos, and K. Tamvakis. Grand unification in simple supergravity, *Phys. Lett.*, B**121**:123, 1983.

[768] J. R. Ellis and K. A. Olive. Constraints on light particles from supernova SN1987a, *Phys. Lett.*, B**193**:525, 1987.

[769] J. R. Ellis, K. A. Olive, and P. Sandick. What if supersymmetry breaking appears below the GUT scale?, *Phys. Lett.*, B**642**:389, 2006, hep-ph/0607002.

[770] J. R. Ellis, K. A. Olive, and P. Sandick. Phenomenology of GUT-less supersymmetry breaking, *JHEP*, **06**:079, 2007, arXiv:0704.3446 [hep-ph].

[771] J. R. Ellis, K. A. Olive, and P. Sandick. Sparticle discovery potentials in the CMSSM and GUT-less supersymmetry-breaking scenarios, *JHEP*, **08**:013, 2008, arXiv:0801.1651 [hep-ph].

[772] J. R. Ellis, K. A. Olive, and Y. Santoso. Constraining supersymmetry, *New J. Phys.*, **4**:32, 2002, hep-ph/0202110.

[773] J. R. Ellis, K. A. Olive, and Y. Santoso. The MSSM parameter space with non-universal Higgs masses, *Phys. Lett.*, B**539**:107, 2002, hep-ph/0204192.

[774] J. R. Ellis, K. A. Olive, and Y. Santoso. Sneutrino NLSP scenarios in the NUHM with gravitino dark matter, *JHEP*, **10**:005, 2008, arXiv:0807.3736 [hep-ph].

[775] J. R. Ellis, K. A. Olive, Y. Santoso, and V. C. Spanos. Phenomeno-logical constraints on patterns of supersymmetry breaking, *Phys. Lett.*, **B573**:162, 2003, hep-ph/0305212.

[776] J. R. Ellis, K. A. Olive, Y. Santoso, and V. C. Spanos. Supersymmetric dark matter in light of WMAP, *Phys. Lett.*, **B565**:176, 2003, hep-ph/0303043.

[777] J. R. Ellis, K. A. Olive, Y. Santoso, and V. C. Spanos. Gravitino dark matter in the CMSSM, *Phys. Lett.*, **B588**:7, 2004, hep-ph/0312262.

[778] J. R. Ellis, K. A. Olive, Y. Santoso, and V. C. Spanos. Likelihood analysis of the CMSSM parameter space, *Phys. Rev.*, **D69**:095004, 2004, hep-ph/0310356.

[779] J. R. Ellis, K. A. Olive, Y. Santoso, and V. C. Spanos. Prospects for sparticle discovery in variants of the MSSM, *Phys. Lett.*, **B603**:51, 2004, hep-ph/0408118.

[780] J. R. Ellis, K. A. Olive, Y. Santoso, and V. C. Spanos. Very constrained minimal supersymmetric standard models, *Phys. Rev.*, **D70**:055005, 2004, hep-ph/0405110.

[781] J. R. Ellis, K. A. Olive, Y. Santoso, and V. C. Spanos. Update on the direct detection of supersymmetric dark matter, *Phys. Rev.*, **D71**:095007, 2005, hep-ph/0502001.

[782] J. R. Ellis, K. A. Olive, Y. Santoso, and V. C. Spanos. On B/s $\rightarrow$ mu+ mu- and cold dark matter scattering in the MSSM with non-universal Higgs masses, *JHEP*, **05**:063, 2006, hep-ph/0603136.

[783] J. R. Ellis, K. A. Olive, and C. Savage. Hadronic uncertainties in the elastic scattering of supersymmetric dark matter, *Phys. Rev.*, **D77**:065026, 2008, arXiv:0801.3656 [hep-ph].

[784] J. R. Ellis, K. A. Olive, and E. Vangioni. Effects of unstable particles on light-element abundances: Lithium versus deuterium and He-3, *Phys. Lett.*, **B619**:30, 2005, astro-ph/0503023.

[785] J. R. Ellis, A. R. Raklev, and O. K. Oye. Gravitino dark matter scenarios with massive metastable charged sparticles at the LHC, *JHEP*, **10**:061, 2006, hep-ph/0607261.

[786] J. R. Ellis, G. Ridolfi, and F. Zwirner. On radiative corrections to supersymmetric Higgs boson masses and their implications for LEP searches, *Phys. Lett.*, **B262**:477, 1991.

[787] J. R. Ellis, G. Ridolfi, and F. Zwirner. Radiative corrections to the masses of supersymmetric Higgs bosons, *Phys. Lett.*, **B257**:83, 1991.

[788] J. R. Ellis and D. Ross. A light Higgs boson would invite supersymmetry, *Phys. Lett.*, **B506**:331, 2001, hep-ph/0012067.

[789] U. Ellwanger, J. F. Gunion, and C. Hugonie. NMHDECAY: A Fortran code for the Higgs masses, couplings and decay widths in the NMSSM, *JHEP*, **02**:066, 2005, hep-ph/0406215.

[790] D. Elsaesser and K. Mannheim. Supersymmetric dark matter and the extragalactic gamma ray background, *Phys. Rev. Lett.*, **94**:171302, 2005, astro-ph/0405235.

[791] R. Enberg, P. J. Fox, L. J. Hall, A. Y. Papaioannou, and M. Papucci. LHC and dark matter signals of improved naturalness, *JHEP*, **11**:014, 2007, arXiv:0706.0918 [hep-ph].

[792] M. Endo and F. Takahashi. Non-thermal production of dark matter from late-decaying scalar field at intermediate scale, *Phys. Rev.*, D**74**:063502, 2006, hep-ph/0606075.

[793] J. Engel. Nuclear form-factors for the scattering of weakly interacting massive particles, *Phys. Lett.*, B**264**:114–119, 1991.

[794] J. Engel, S. Pittel, and P. Vogel. Nuclear physics of dark matter detection, *Int. J. Mod. Phys.*, E1:1–37, 1992.

[795] K. Enqvist, K. Kainulainen, and J. Maalampi. Cosmic abundances of very heavy neutrinos, *Nucl. Phys.*, B**317**:647–664, 1989.

[796] K. Enqvist, K. Kainulainen, and J. Maalampi. Neutrino asymmetry and oscillations in the early universe, *Phys. Lett.*, B**244**:186–190, 1990.

[797] K. Enqvist, K. Kainulainen, and J. Maalampi. Resonant neutrino transitions and nucleosynthesis, *Phys. Lett.*, B**249**:531–534, 1990.

[798] EROS-2 Collaboration. Limits on the Machos from EROS-2, *Nucl. Phys. B Proc. Suppl.*, **173**:40–43, 2007.

[799] J. R. Espinosa, T. Konstandin, J. M. No, and M. Quiros. Some cosmological implications of hidden sectors, *Phys. Rev.*, D**78**:123528, 2008, arXiv:0809.3215 [hep-ph].

[800] J. R. Espinosa and M. Quiros. Novel effects in electroweak breaking from a hidden sector, *Phys. Rev.*, D**76**:076004, 2007, hep-ph/0701145.

[801] R. Essig. Direct detection of non-chiral dark matter, *Phys. Rev.*, D**78**:015004, 2008, arXiv:0710.1668 [hep-ph].

[802] N. W. Evans, C. M. Carollo, and P. T. de Zeeuw. Triaxial haloes and particle dark matter detection, *Mon. Not. Roy. Astron. Soc.*, **318**:1131, 2000, astro-ph/0008156.

[803] N. W. Evans, F. Ferrer, and S. Sarkar. A 'Baedecker' for the dark matter annihilation signal, *Phys. Rev.*, D**69**:123501, 2004, astro-ph/0311145.

[804] G. Fahlman, N. Kaiser, G. Squires, and D. Woods. Dark matter in MS 1224 from distortion of background galaxies, *Astrophys. J.*, **437**:56–62, December 1994.

[805] M. Fairbairn and T. Schwetz. Spin-independent elastic WIMP scattering and the DAMA annual modulation signal, 2008, arXiv:0808.0704 [hep-ph].

[806] M. Fairbairn, P. Scott, and J. Edsjö. The zero age main sequence of WIMP burners, *Phys. Rev.*, D**77**:047301, 2008, arXiv:0710.3396 [astro-ph].

[807] T. Falk, K. A. Olive, and M. Srednicki. Heavy sneutrinos as dark matter, *Phys. Lett.*, B**339**:248, 1994, hep-ph/9409270.

[808] A. Falkowski, O. Lebedev, and Y. Mambrini. SUSY phenomenology of KKLT flux compactifications, *JHEP*, **11**:034, 2005, hep-ph/0507110.

[809] A. Faltenbacher and J. Diemand. Velocity distributions in clusters of galaxies, *Mon. Not. Roy. Astron. Soc.*, **369**:1698–1702, 2006, astro-ph/0602197.

[810] A. Faltenbacher *et al.* Spatial and kinematic alignments between central and satellite halos, 2007, arXiv:0706.0262 [astro-ph].

[811] A. Faltenbacher *et al.* Three different types of galaxy alignment within dark matter halos, 2007, arXiv:0704.0674 [astro-ph].

[812] B. Famaey, G. W. Angus, G. Gentile, H.-Y. Shan, and H. S. Zhao. The wedding of modified dynamics and non-exotic dark matter in galaxy clusters, arXiv:0706.1279, 2007, 0706.1279.

[813] D. S. M. Fantin, M. R. Merrifield, and A. M. Green. Modelling ultra-fine structure in dark matter halos, *Mon. Not. Roy. Astron. Soc.*, **390**:1055, 2008, arXiv:0808.1050 [astro-ph].

[814] A. E. Faraggi and M. Pospelov. Self-interacting dark matter from the hidden heterotic-string sector, *Astropart. Phys.*, **16**:451–461, 2002, hep-ph/0008223.

[815] D. Fargion, R. Konoplich, M. Grossi, and M. Khlopov. On the heavy relic neutrino–galactic gamma halo connection, 1998, astro-ph/9902327.

[816] G. R. Farrar and P. Fayet. Phenomenology of the production, decay, and detection of new hadronic states associated with supersymmetry, *Phys. Lett.*, B**76**:575, 1978.

[817] G. R. Farrar and G. Zaharijas. Dark matter and the baryon asymmetry of the universe, 2004, hep-ph/0406281.

[818] G. R. Farrar and G. Zaharijas. Dark matter and the baryon asymmetry, *Phys. Rev. Lett.*, **96**:041302, 2006, hep-ph/0510079.

[819] J. Faulkner and R. L. Gilliland. Weakly interacting, massive particles and the solar neutrino flux, *Astrophys. J.*, **299**:994–1000, 1985.

[820] J. Faulkner, D. O. Gough, and M. N. Vahia. Weakly interacting massive particles and solar oscillations, *Nature*, **321**:226–229, May 1986.

[821] C. Faure, J.-P. Kneib, G. Covone *et al.* First catalog of strong lens candidates in the COSMOS field, *Astrophys. J. Suppl.*, **176**:19–38, 2008, arXiv:0802.2174.

[822] P. Fayet. Supersymmetry and weak, electromagnetic and strong interactions, *Phys. Lett.*, **B64**:159, 1976.

[823] P. Fayet. Spontaneously broken supersymmetric theories of weak, electromagnetic and strong interactions, *Phys. Lett.*, **B69**:489, 1977.

[824] P. Fayet. Relations between the masses of the superpartners of leptons and quarks, the Goldstino couplings and the neutral currents, *Phys. Lett.*, **B84**:416, 1979.

[825] P. Fayet and S. Ferrara. Supersymmetry, *Phys. Rep.*, **32**:249, 1977.

[826] M. Feast and P. Whitelock. Galactic kinematics of Cepheids from HIPPARCOS proper motions, *Mon. Not. Roy. Astron. Soc.*, **291**:683, 1997.

[827] M. Feix, C. Fedeli, and M. Bartelmann. Asymmetric gravitational lenses in TeVeS and application to the bullet cluster, *Astron. Astrophys.*, **480**:313–325, 2008, 0707.0790.

[828] J. L. Feng and D. E. Finnell. Squark mass determination at the next generation of linear e^+e^- colliders, *Phys. Rev.*, **D49**:2369–2381, 1994, hep-ph/9310211.

[829] J. L. Feng and J. Kumar. Dark-matter particles without weak-scale masses or weak interactions, *Phys. Rev. Lett.*, **101**:231301, 2008, arXiv:0803.4196 [hep-ph].

[830] J. L. Feng, J. Kumar, J. Learned, and L. E. Strigari. Testing the dark matter interpretation of the DAMA/LIBRA result with Super-Kamiokande, 2008, arXiv:0808.4151 [hep-ph].

[831] J. L. Feng, J. Kumar, and L. E. Strigari. Explaining the DAMA signal with WIMPless dark matter, 2008, arXiv:0806.3746 [hep-ph].

[832] J. L. Feng, K. T. Matchev, and T. Moroi. Focus points and naturalness in supersymmetry, *Phys. Rev.*, **D61**:075005, 2000, hep-ph/9909334.

[833] J. L. Feng, K. T. Matchev, and F. Wilczek. Neutralino dark matter in focus point supersymmetry, *Phys. Lett.*, **B482**:388–399, 2000, hep-ph/0004043.

[834] J. L. Feng and T. Moroi. Tevatron signatures of long-lived charged sleptons in gauge-mediated supersymmetry breaking models, *Phys. Rev.*, D**58**:035001, 1998, hep-ph/9712499.

[835] J. L. Feng and M. E. Peskin. Selectron studies at e^-e^- and e^+e^- colliders, *Phys. Rev.*, D**64**:115002, 2001, hep-ph/0105100.

[836] J. L. Feng, A. Rajaraman, and F. Takayama. Graviton cosmology in universal extra dimensions, *Phys. Rev.*, D**68**:085018, 2003, hep-ph/0307375.

[837] J. L. Feng, A. Rajaraman, and F. Takayama. Superweakly-interacting massive particles, *Phys. Rev. Lett.*, **91**:011302, 2003, hep-ph/0302215.

[838] J. L. Feng, A. Rajaraman, and F. Takayama. SuperWIMP dark matter signals from the early universe, *Phys. Rev.*, D**68**:063504, 2003, hep-ph/0306024.

[839] J. L. Feng, A. Rajaraman, and F. Takayama. Probing gravitational interactions of elementary particles, *Int. J. Mod. Phys.*, D**13**:2355–2359, 2004, hep-th/0405248.

[840] J. L. Feng and B. T. Smith. Slepton trapping at the Large Hadron and International Linear Colliders, *Phys. Rev.*, D**71**:015004, 2005, hep-ph/0409278.

[841] J. L. Feng, S. Su, and F. Takayama. Supergravity with a gravitino LSP, *Phys. Rev.*, D**70**:075019, 2004, hep-ph/0404231.

[842] J. L. Feng, S. Su, and F. Takayama. Lower limit on dark matter production at the Large Hadron Collider, *Phys. Rev. Lett.*, **96**:151802, 2006, hep-ph/0503117.

[843] J. L. Feng, S.-f. Su, and F. Takayama. SuperWIMP gravitino dark matter from slepton and sneutrino decays, *Phys. Rev.*, D**70**:063514, 2004, hep-ph/0404198.

[844] J. L. Feng, H. Tu, and H.-B. Yu. Thermal relics in hidden sectors, *JCAP*, 0810:043, 2008, arXiv:0808.2318 [hep-ph].

[845] L. Ferrarese and H. Ford. Supermassive black holes in galactic nuclei: Past, present and future research, *Space Sci. Rev.*, **116**:523–624, 2005.

[846] P. M. Ferreira, I. Jack, and D. R. T. Jones. The three-loop SSM beta-functions, *Phys. Lett.*, B**387**:80–86, 1996, hep-ph/9605440.

[847] F. Ferrer, L. M. Krauss, and S. Profumo. Indirect detection of light neutralino dark matter in the NMSSM, *Phys. Rev.*, D**74**:115007, 2006, hep-ph/0609257.

[848] I. Ferreras, P. Saha, and S. Burles. Unveiling dark haloes in lensing galaxies, *Mon. Not. Roy. Astron. Soc.*, **383**:857–863, 2008, arXiv:0710.3159.

[849] I. Ferreras, P. Saha, and L. L. R. Williams. Stellar and total mass in early-type lensing galaxies, *Astrophys. J.*, **623**:L5–L8, April 2005.

[850] I. Ferreras, M. Sakellariadou, and M. F. Yusaf. Necessity of dark matter in modified Newtonian dynamics within galactic scales, *Phys. Rev. Lett.*, **100**:031302, 2008, arXiv:0709.3189.

[851] B. Fields and S. Sarkar. Big-bang nucleosynthesis (PDG minireview), 2006, astro-ph/0601514.

[852] B. D. Fields, K. A. Olive, and E. Vangioni-Flam. Implications of a new temperature scale for halo dwarfs on LiBeB and chemical evolution, *Astrophys. J.*, **623**:1083–1091, 2005, astro-ph/0411728.

[853] J. A. Fillmore and P. Goldreich. Self-similar gravitational collapse in an expanding universe, *Astrophys. J.*, **281**:1–8, 1984.

[854] D. P. Finkbeiner, T. Slatyer, and N. Weiner. Nuclear scattering of dark matter coupled to a new light scalar, 2008, arXiv:0810.0722 [hep-ph].

[855] S. Fiorucci *et al.* Identification of backgrounds in the EDELWEISS-I dark matter search experiment, *Astropart. Phys.*, **28**:143–153, 2007, astro-ph/0610821.

[856] A. L. Fitzpatrick, J. Kaplan, L. Randall, and L.-T. Wang. Searching for the Kaluza-Klein graviton in bulk RS models, *JHEP*, **09**:013, 2007, hep-ph/0701150.

[857] D. J. Fixsen *et al.* The Cosmic Microwave Background Spectrum from the full COBE/FIRAS data set, *Astrophys. J.*, **473**:576, 1996, astro-ph/9605054.

[858] T. Flacke, A. Menon, and D. J. Phalen. Non-minimal universal extra dimensions, 2008, arXiv:0811.1598 [hep-ph].

[859] H. Flores *et al.* 3D spectroscopy with VLT/GIRAFFE. I. The true Tully Fisher relationship at $z \simeq 0.6$, *Astron. Astrophys.*, **445**: 107–118, 2006.

[860] R. Foot. Experimental implications of mirror matter-type dark matter, *Int. J. Mod. Phys.*, **A19**:3807–3818, 2004, astro-ph/0309330.

[861] R. Foot. Exploring the mirror matter interpretation of the DAMA experiment: Has the dark matter problem been solved?, 2004, astro-ph/0403043.

[862] R. Foot. Implications of the DAMA and CRESST experiments for mirror matter-type dark matter, *Phys. Rev.*, **D69**:036001, 2004, hep-ph/0308254.

[863] R. Foot. Mirror matter-type dark matter, *Int. J. Mod. Phys.*, D**13**:2161–2192, 2004, astro-ph/0407623.

[864] R. Foot. Reconciling the positive DAMA annual modulation signal with the negative results of the CDMS II experiment, *Mod. Phys. Lett.*, A**19**:1841–1846, 2004, astro-ph/0405362.

[865] R. Foot. Implications of the DAMA/NaI and CDMS experiments for mirror matter-type dark matter, *Phys. Rev.*, D**74**:023514, 2006, astro-ph/0510705.

[866] R. Foot. Mirror dark matter, *Int. J. Mod. Phys.*, A**22**:4951–4957, 2007, arXiv:0706.2694 [hep-ph].

[867] R. Foot. Mirror dark matter and the new DAMA/LIBRA results: A simple explanation for a beautiful experiment, *Phys. Rev.*, D**78**:043529, 2008, arXiv:0804.4518 [hep-ph].

[868] R. Foot, G. C. Joshi, H. Lew, and R. R. Volkas. Charge quantization in the standard model and some of its extensions, *Mod. Phys. Lett.*, A**5**:2721–2732, 1990.

[869] R. Foot and Z. K. Silagadze. Supernova explosions, 511-keV photons, gamma ray bursts and mirror matter, *Int. J. Mod. Phys.*, D**14**:143–152, 2005, astro-ph/0404515.

[870] LEP Higgs Working Group for Higgs boson searches. Search for neutral Higgs bosons at LEP. Paper submitted to ICHEP04, Beijing, LHWG-NOTE-2004-01, ALEPH-2004-008, DELPHI-2004-042, L3-NOTE-2820, OPAL-TN-744, http://lephiggs.web.cern.ch/LEPHIGGS/papers/August2004_MSSM/index.html.

[871] C. Adam-Bourdarios for the SFitter collaboration. Extrapolation to the high scale at LHC and ILC. Talk at GDR-SUSY, 2008.

[872] D. A. Forbes, P. Lasky, A. W. Graham, and L. Spitler. Uniting old stellar systems: from globular clusters to giant ellipticals, *Mon. Not. Roy. Astron. Soc.*, **389**:1924–1936, 2008, arXiv:0806.1090.

[873] M. Fornasa and G. Bertone. Black holes as dark matter annihilation boosters, *Int. J. Mod. Phys.*, D**17**:1125–1157, 2008, arXiv:0711.3148 [astro-ph].

[874] M. Fornasa, L. Pieri, G. Bertone, and E. Branchini. Anisotropy probe of galactic and extra-galactic dark matter annihilations, 2009, arXiv:0901.2921 [astro-ph].

[875] M. Fornasa, M. Taoso, and G. Bertone. Gamma-rays from dark matter mini-spikes in M31, *Phys. Rev.*, D**76**:043517, 2007, astro-ph/0703757.

[876] N. Fornengo, L. Pieri, and S. Scopel. Neutralino annihilation into gamma-rays in the Milky Way and in external galaxies, *Phys. Rev.*, D**70**:103529, 2004, hep-ph/0407342.

[877] N. Fornengo, A. Riotto, and S. Scopel. Supersymmetric dark matter and the reheating temperature of the universe, *Phys. Rev.*, D**67**:023514, 2003, hep-ph/0208072.

[878] B. Fort, O. Le Fevre, F. Hammer, and M. Cailloux. An arc system with a radial gravitational image in the cluster MS 2137-23, *Astrophys. J.*, **399**:L125–L127, 1992.

[879] M. Frank *et al.* The Higgs boson masses and mixings of the complex MSSM in the Feynman-diagrammatic approach, *JHEP*, **02**:047, 2007, hep-ph/0611326.

[880] K. Freese. Can scalar neutrinos or masive Dirac neutrinos be the missing mass?, *Phys. Lett.*, B**167**:295–300, 1986.

[881] K. Freese, J. A. Frieman, and A. Gould. Signal modulation in cold dark matter detection, *Phys. Rev.*, D**37**:3388, 1988.

[882] K. Freese, P. Gondolo, and H. J. Newberg. Detectability of weakly interacting massive particles in the Sagittarius dwarf tidal stream, *Phys. Rev.*, D**71**:043516, 2005, astro-ph/0309279.

[883] K. Freese, P. Gondolo, H. J. Newberg, and M. Lewis. The effects of the Sagittarius dwarf tidal stream on dark matter detectors, *Phys. Rev. Lett.*, **92**:111301, 2004, astro-ph/0310334.

[884] K. Freese, P. Gondolo, J. A. Sellwood, and D. Spolyar. Dark matter densities during the formation of the first stars and in dark stars, 2008, arXiv:0805.3540 [astro-ph].

[885] K. Freese, D. Spolyar, and A. Aguirre. Dark matter capture in the first star: a power source and a limit on stellar mass, 2008, arXiv:0802.1724 [astro-ph].

[886] A. Freitas and K. Kong. Two universal extra dimensions and spinless photons at the ILC, *JHEP*, **02**:068, 2008, arXiv:0711.4124 [hep-ph].

[887] A. Freitas, A. von Manteuffel, and P. M. Zerwas. Slepton production at e^+e^- and e^-e^- linear colliders. (Addendum), *Eur. Phys. J.*, C**40**:435–445, 2005, hep-ph/0408341.

[888] C. S. Frenk, S. D. M. White, G. Efstathiou, and M. Davis. Cold dark matter, the structure of galactic haloes and the origin of the Hubble sequence, *Nature*, **317**:595–597, October 1985.

[889] J. A. Frieman and G. F. Giudice. Cosmologically benign gravitinos at the weak scale, *Phys. Lett.*, B**224**:125, 1989.

[890] J. A. Frieman and G. F. Giudice. Cosmic technicolor nuggets, *Nucl. Phys.*, B**355**:162–191, 1991.

[891] T. Fritzsche and W. Hollik. One-loop calculations for SUSY processes, *Nucl. Phys. Proc. Suppl.*, **135**:102–106, 2004, hep-ph/0407095.

[892] L. Fu, E. Semboloni, H. Hoekstra *et al.* Very weak lensing in the CFHTLS wide: cosmology from cosmic shear in the linear regime, *Astron. Astrophys.*, **479**:9–25, 2008, arXiv:0712.0884.

[893] M. Fujii and K. Hamaguchi. Non-thermal dark matter via Affleck-Dine baryogenesis and its detection possibility, *Phys. Rev.*, **D66**:083501, 2002, hep-ph/0205044.

[894] M. Fujii and M. Ibe. Neutralino dark matter from MSSM flat directions in light of WMAP result, *Phys. Rev.*, **D69**:035006, 2004, hep-ph/0308118.

[895] J. Fujimoto *et al.* GRACE/SUSY: Automatic generation of tree amplitudes in the minimal supersymmetric standard model, *Comput. Phys. Commun.*, **153**:106–134, 2003, hep-ph/0208036.

[896] J. Fujimoto *et al.* Two-body and three-body decays of charginos in one-loop order in the MSSM, *Phys. Rev.*, **D75**:113002, 2007.

[897] H. Fuke, T. Maeno, K. Abe, *et al.* Search for cosmic-ray antideuterons, *Phys. Rev. Lett.*, **95**:081101, 2005.

[898] Y. Fukuda *et al.* Evidence for oscillation of atmospheric neutrinos, *Phys. Rev. Lett.*, **81**:1562–1567, 1998, hep-ex/9807003.

[899] M. Fukugita and M. Kawasaki. Primordial helium abundance: A reanalysis of the Izotov–Thuan spectroscopic sample, *Astrophys. J.*, **646**:691, 2006.

[900] M. Fukugita and T. Yanagida. Baryogenesis without grand unification, *Phys. Lett.*, **B174**:45, 1986.

[901] T. Fukushige, A. Kawai, and J. Makino. Structure of dark matter halos from hierarchical clustering. III. Shallowing of the inner cusp, *Astrophys. J.*, **606**:625–634, 2004, astro-ph/0306203.

[902] T. Fukushige and J. Makino. Structure of dark matter halos from hierarchical clustering, *Astrophys. J.*, **557**:533, 2001, astro-ph/0008104.

[903] T. Fukushige and J. Makino. Structure of dark matter halos from hierarchical clustering: II. Universality and self-similarity in cluster-sized halos, 2001, astro-ph/0108014.

[904] G. M. Fuller, A. Kusenko, I. Mocioiu, and S. Pascoli. Pulsar kicks from a dark-matter sterile neutrino, *Phys. Rev.*, **D68**:103002, 2003, astro-ph/0307267.

[905] S. A. Fulling. Remarks on positive frequency and Hamiltonians in expanding universes, *Gen. Rel. Grav.*, **10**:807–824, 1979.

[906] S. A. Fulling. Aspects of quantum field theory in curved space-time, *London Math. Soc. Student Texts*, **17**:1–315, 1989.

[907] S. R. Furlanetto and A. Loeb. Constraining the collisional nature of the dark matter through observations of gravitational wakes, *Astrophys. J.*, **565**:854–866, 2002.

[908] E. Gabrielli, S. Khalil, C. Munoz, and E. Torrente-Lujan. Initial scales, supersymmetric dark matter and variations of neutralino nucleon cross sections, *Phys. Rev.*, **D63**:025008, 2001, hep-ph/0006266.

[909] T. K. Gaisser, G. Steigman, and S. Tilav. Limits on cold dark matter candidates from deep underground detectors, *Phys. Rev.*, **D34**:2206, 1986.

[910] T. K. Gaisser, G. Steigman, and S. Tilav. Limits on cold-dark-matter candidates from deep underground detectors, *Phys. Rev. D*, **34**:2206–2222, 1986.

[911] R. J. Gaitskell. Direct detection of dark matter, *Ann. Rev. Nucl. Part. Sci.*, **54**:315–359, 2004.

[912] R. J. Gaitskell *et al.* The design of a cryogenic dark matter detector based on the detection of the recoil direction of target nuclei, *Nucl. Instrum. Meth.*, **A370**:162–164, 1996.

[913] C. Galbiati and R. Purtschert. Discovery of underground argon with low level of radioactive ^{39}Ar and possible applications to WIMP dark matter detectors, *J. Phys. Conf. Ser.*, **120**:042015, 2008, arXiv:0712.0381 [astro-ph].

[914] A. K. Ganguly, P. Jain, S. Mandal, and S. Stokes. Self interacting dark matter in the Solar System, *Phys. Rev.*, **D76**:025026, 2007, hep-ph/0611006.

[915] L. Gao and T. Theuns. Lighting the Universe with filaments, *Science*, **317**:1527, 2007, arXiv:0709.2165 [astro-ph].

[916] L. Gao, S. D. M. White, A. Jenkins, F. Stoehr, and V. Springel. The subhalo populations of LCDM dark haloes, *Mon. Not. Roy. Astron. Soc.*, **355**:819, 2004, astro-ph/0404589.

[917] B. Garbrecht, C. Pallis, and A. Pilaftsis. Anatomy of F(D)-term hybrid inflation, *JHEP*, **12**:038, 2006, hep-ph/0605264.

[918] E. I. Gates, G. Gyuk, and M. S. Turner. The local halo density, *Astrophys. J.*, **449**:L123–L126, 1995, astro-ph/9505039.

[919] R. Gavazzi, B. Fort, Y. Mellier, R. Pelló, and M. Dantel-Fort. A radial mass profile analysis of the lensing cluster MS 2137.3-2353, *Astron. Astrophys.*, **403**:11–27, 2003.

[920] R. Gavazzi, T. Treu, J. D. Rhodes, *et al.* The Sloan Lens ACS Survey. IV. The mass density profile of early-type galaxies out to 100 effective radii, *Astrophys. J.*, **667**:176–190, 2007.

[921] C. G. R. Geddes *et al.* High-quality electron beams from a laser wakefield accelerator using plasma-channel guiding, *Nature*, **431**:538–541, 2004.

[922] M. Geha *et al.* The least luminous galaxy: Spectroscopy of the Milky Way satellite Segue 1, 2008, arXiv:0809.2781 [astro-ph].

[923] J. Geiss and G. Gloeckler. Linking primordial to solar and galactic composition, *Space Sci. Rev.*, **130**:5, 2007.

[924] M. Gell-Mann, P. Ramond, and R. Slansky. Complex spinors and unified theories, 1980, Print-80-0576 (CERN).

[925] G. Gelmini and P. Gondolo. WIMP annual modulation with opposite phase in late-infall halo models, *Phys. Rev.*, D**64**:023504, 2001, hep-ph/0012315.

[926] G. Gelmini, P. Gondolo, A. Soldatenko, and C. E. Yaguna. The effect of a late decaying scalar on the neutralino relic density, *Phys. Rev.*, D**74**:083514, 2006, hep-ph/0605016.

[927] G. Gelmini, E. Osoba, S. Palomares-Ruiz, and S. Pascoli. MeV sterile neutrinos in low reheating temperature cosmological scenarios, *JCAP*, **0810**:029, 2008, arXiv:0803.2735 [astro-ph].

[928] G. Gelmini, S. Palomares-Ruiz, and S. Pascoli. Low reheating temperature and the visible sterile neutrino, *Phys. Rev. Lett.*, **93**:081302, 2004, astro-ph/0403323.

[929] G. Gelmini and C. E. Yaguna. Constraints on minimal SUSY models with warm dark matter neutralinos, *Phys. Lett.*, B**643**:241–245, 2006, hep-ph/0607012.

[930] G. B. Gelmini and P. Gondolo. Neutralino with the right cold dark matter abundance in (almost) any supersymmetric model, *Phys. Rev.*, D**74**:023510, 2006, hep-ph/0602230.

[931] G. B. Gelmini and P. Gondolo. Ultra-cold WIMPs: relics of non-standard pre-BBN cosmologies, *JCAP*, **0810**:002, 2008, arXiv:0803.2349 [astro-ph].

[932] G. B. Gelmini, P. Gondolo, and E. Roulet. Neutralino dark matter searches, *Nucl. Phys.*, B**351**:623–644, 1991.

[933] G. Gentile, A. Burkert, P. Salucci, U. Klein, and F. Walter. The dwarf galaxy DDO 47 as a dark matter laboratory: Testing cusps hiding in triaxial halos, *Astrophys. J. Lett.*, **634**:L145–L148, 2005.

[934] R. Genzel *et al.* The rapid formation of a large rotating disk galaxy three billion years after the Big Bang, *Nature*, **442**:786–789, 2006.

[935] S. S. Gershtein and Ya. B. Zeldovich. Rest mass of muonic neutrino and cosmology, *JETP Lett.*, **4**:120–122, 1966.

[936] T. Gherghetta, G. F. Giudice, and J. D. Wells. Phenomenological consequences of supersymmetry with anomaly-induced masses, *Nucl. Phys.*, **B559**:27–47, 1999, hep-ph/9904378.

[937] T. Gherghetta and A. Pomarol. Bulk fields and supersymmetry in a slice of AdS, *Nucl. Phys.*, **B586**:141–162, 2000, hep-ph/0003129.

[938] A. M. Ghez, S. Salim, N. N. Weinberg *et al.* Measuring distance and properties of the Milky Way's central supermassive black hole with stellar orbits, 2008, arXiv:0808.2870.

[939] S. Ghigna *et al.* Density profiles and substructure of Dark matter halos: converging results at ultra-high numerical resolution, *Astrophys. J.*, **544**:616, 2000, astro-ph/9910166.

[940] S. Ghigna, B. Moore, F. Governato, G. Lake, T. Quinn, and J. Stadel. Dark matter halos within clusters, *Mon. Not. Roy. Astron. Soc.*, **300**:146–162, 1998, astro-ph/9801192.

[941] K. Ghosh. Probing two Universal Extra Dimension model with leptons and photons at the LHC and ILC, 2008, arXiv:0809.1827 [hep-ph].

[942] K. Ghosh and A. Datta. Phenomenology of spinless adjoints in two Universal Extra Dimensions, *Nucl. Phys.*, **B800**:109–126, 2008, arXiv:0801.0943 [hep-ph].

[943] K. Ghosh and A. Datta. Probing two Universal Extra Dimensions at International Linear Collider, *Phys. Lett.*, **B665**:369–373, 2008, arXiv:0802.2162 [hep-ph].

[944] D. Giannios. Spherically symmetric, static spacetimes in a tensor-vector-scalar theory. *Phys. Rev. D*, **71**:103511, 2005.

[945] W. R. Gilks, S. Richardson, and D. J. Spegelhalter, eds., *Markov Chain Monte Carlo in Practice*, Chapman and Hall, 1996.

[946] S. P. D. Gill, A. Knebe, and B. K. Gibson. The evolution of substructure III: the outskirts of clusters, *Mon. Not. Roy. Astron. Soc.*, **356**:1327–1332, 2005, astro-ph/0404427.

[947] S. Gillessen, F. Eisenhauer, S. Trippe *et al.* Monitoring stellar orbits around the Massive Black Hole in the Galactic Center, 2008, arXiv:0810.4674.

[948] L. Gilliland *et al.* Solar models with energy transport by weakly interacting particles, *Astrophys. J.*, **306**:703, 1986.

[949] R. L. Gilliland and W. Dappen. Oscillations in solar models with weakly interacting massive particles, *Astrophys. J.*, **324**:1153–1157, January 1988.

[950] R. L. Gilliland, J. Faulkner, W. H. Press, and D. N. Spergel. Solar models with energy transport by weakly interacting particles, *Astrophys. J.*, **306**:703–709, 1986.

[951] G. Gilmore *et al.* Observed properties of dark matter: dynamical studies of dSph galaxies, *Nucl. Phys. Proc. Suppl.*, **173**:15–18, 2007, astro-ph/0608528.

[952] G. Gilmore, M. I. Wilkinson, R. F. G. Wyse *et al.* The observed properties of dark matter on small spatial scales, *Astrophys. J.*, **663**: 948–959, 2007.

[953] Y. Giomataris, P. Rebourgeard, J. P. Robert, and G. Charpak. MICROMEGAS: A high-granularity position-sensitive gaseous detector for high particle-flux environments, *Nucl. Instrum. Meth.*, **A376**:29–35, 1996.

[954] L. Girardello and M. T. Grisaru. Soft breaking of supersymmetry, *Nucl. Phys.*, **B194**:65, 1982.

[955] J. Gironnet, V. B. Mikhailik, H. Kraus, P. de Marcillac, and N. Coron. Scintillation studies of (BGO) down to a temperature of 6-K, *Nucl. Instrum. Meth.*, **A594**:358–361, 2008.

[956] G. F. Giudice and K. Griest. Rate for annihilation of galactic dark matter into two photons, *Phys. Rev.*, **D40**:2549, 1989.

[957] G. F. Giudice, E. W. Kolb, and A. Riotto. Largest temperature of the radiation era and its cosmological implications, *Phys. Rev.*, **D64**:023508, 2001, hep-ph/0005123.

[958] G. F. Giudice, E. W. Kolb, A. Riotto, D. V. Semikoz, and I. I. Tkachev. Standard model neutrinos as warm dark matter, *Phys. Rev.*, **D64**:043512, 2001, hep-ph/0012317.

[959] G. F. Giudice, R. Rattazzi, and J. D. Wells. Graviscalars from higher-dimensional metrics and curvature-Higgs mixing, *Nucl. Phys.*, **B595**:250–276, 2001, hep-ph/0002178.

[960] G. F. Giudice and A. Romanino. Split supersymmetry, *Nucl. Phys.*, **B699**:65–89, 2004, hep-ph/0406088.

[961] C. Giunti, C. W. Kim, and U. W. Lee. Running coupling constants and grand unification models, *Mod. Phys. Lett.*, **A6**:1745, 1991.

[962] B. K. Gjelsten, D. J. Miller, and P. Osland. Measurement of SUSY masses via cascade decays for SPS 1a, *JHEP*, **12**:003, 2004, hep-ph/0410303.

[963] B. K. Gjelsten, D. J. Miller, and P. Osland. Measurement of the gluino mass via cascade decays for SPS 1a, *JHEP*, **06**:015, 2005, hep-ph/0501033.

[964] S. L. Glashow. Partial symmetries of weak interactions, *Nucl. Phys.*, **22**:579–588, 1961.

[965] T. Gleisberg *et al.* Event generation with SHERPA 1.1, 2008, arXiv:0811.4622 [hep-ph].

[966] O. Y. Gnedin, A. V. Kravtsov, A. A. Klypin, and D. Nagai. Response of dark matter halos to condensation of baryons: cosmological simulations and improved adiabatic contraction model, *Astrophys. J.*, **616**:16–26, 2004, astro-ph/0406247.

[967] O. Y. Gnedin and H. Zhao. Maximum feedback and dark matter profiles of dwarf galaxies, *Mon. Not. Roy. Astron. Soc.*, **333**:299, 2002, astro-ph/0108108.

[968] T. Goerdt, O. Y. Gnedin, B. Moore, J. Diemand, and J. Stadel. The survival and disruption of CDM micro-haloes: implications for direct and indirect detection experiments, *Mon. Not. Roy. Astron. Soc.*, **375**:191–198, 2007, astro-ph/0608495.

[969] T. Goerdt, B. Moore, J. I. Read, J. Stadel, and M. Zemp. Does the Fornax dwarf spheroidal have a central cusp or core? *Mon. Not. Roy. Astron. Soc.*, **368**:1073–1077, May 2006.

[970] T. Goerdt, J. I. Read, B. Moore, and J. Stadel. Core creation in galaxies and haloes via sinking massive objects: application to binary nuclei, 2008, arXiv:0806.1951 [astro-ph].

[971] I. Gogoladze and C. Macesanu. Precision electroweak constraints on universal extra dimensions revisited, *Phys. Rev.*, D**74**:093012, 2006, hep-ph/0605207.

[972] J. L. Goity, W. J. Kossler, and M. Sher. Production, collection and utilization of very longlived heavy charged leptons, *Phys. Rev.*, D**48**:5437–5439, 1993, hep-ph/9305244.

[973] H. Goldberg. Constraint on the photino mass from cosmology, *Phys. Rev. Lett.*, **50**:1419, 1983.

[974] M. E. Gomez and J. D. Vergados. Cold dark matter detection in SUSY models at large tan(beta), *Phys. Lett.*, B**512**:252–260, 2001, hep-ph/0012020.

[975] P. Gondolo. Either neutralino dark matter or cuspy dark halos, *Phys. Lett.*, B**494**:181–186, 2000, hep-ph/0002226.

[976] P. Gondolo. Recoil momentum spectrum in directional dark matter detectors, *Phys. Rev.*, D**66**:103513, 2002, hep-ph/0209110.

[977] P. Gondolo. Non-baryonic dark matter, *NATO Sci. Ser. II*, **187**: 279–333, 2005, astro-ph/0403064.

[978] P. Gondolo, J. Edsjö, L. Bergstrom, P. Ullio, and E. A. Baltz. Dark-SUSY: A numerical package for dark matter calculations in the MSSM, 2000, astro-ph/0012234.

[979] P. Gondolo *et al.* DarkSUSY: A numerical package for supersymmetric dark matter calculations, 2002, astro-ph/0211238.

[980] P. Gondolo *et al.* DarkSUSY: Computing supersymmetric dark matter properties numerically, *JCAP*, **0407**:008, 2004, astro-ph/0406204.

[981] P. Gondolo *et al.* DarkSUSY 4.00 neutralino dark matter made easy, *New Astron. Rev.*, **49**:149–151, 2005.

[982] P. Gondolo and G. Gelmini. Cosmic abundances of stable particles: Improved analysis, *Nucl. Phys.*, B**360**:145–179, 1991.

[983] P. Gondolo and G. Gelmini. Compatibility of DAMA dark matter detection with other searches, *Phys. Rev.*, D**71**:123520, 2005, hep-ph/0504010.

[984] P. Gondolo and G. Raffelt. Solar neutrino limit on the axion-like interpretation of the DAMA signal, 2008, arXiv:0807.2926 [astro-ph].

[985] P. Gondolo and J. Silk. Dark matter annihilation at the galactic center, *Phys. Rev. Lett.*, **83**:1719–1722, 1999, astro-ph/9906391.

[986] Y. Gong, T.-J. Zhang, T. Lan, and X.-L. Chen. Dark energy and neutrino mass constraints from weak lensing, supernova, and relative galaxy ages, 2008, arXiv:0810.3572.

[987] A. H. Gonzalez, D. Zaritsky, and A. I. Zabludoff. A census of baryons in galaxy clusters and groups, 2007, arXiv:0705.1726 [astro-ph].

[988] M. W. Goodman and E. Witten. Detectability of certain dark-matter candidates, *Phys. Rev.*, D**31**:3059, 1985.

[989] D. Gorbunov, A. Khmelnitsky, and V. Rubakov. Constraining sterile neutrino dark matter by phase-space density observations, *JCAP*, **0810**:041, 2008, arXiv:0808.3910 [hep-ph].

[990] D. Gorbunov and M. Shaposhnikov. How to find neutral leptons of the nuMSM?, *JHEP*, **10**:015, 2007, arXiv:0705.1729 [hep-ph].

[991] T. Goto, K. Kawagoe, and M. M. Nojiri. Study of the slepton non-universality at the CERN Large Hadron Collider, *Phys. Rev.*, D**70**:075016, 2004, hep-ph/0406317.

[992] J. R. I. Gott. On the formation of elliptical galaxies, *Astrophys. J.*, **201**:296–310, 1975.

[993] A. Gould. Cosmological density of WIMPs from solar and terrestrial annihilations, IASSNS-AST-91-34.

[994] A. Gould. Resonant enhancements in WIMP capture by the Earth, *Astrophys. J.*, **321**:571, 1987.

[995] A. Gould. Direct and indirect capture of WIMPs by the Earth, *Astrophys. J.*, **328**:919–939, 1988.

[996] A. Gould. Gravitational diffusion of solar system WIMPs, *Astrophys. J.*, **368**:610–615, 1991.

[997] A. Gould, B. T. Draine, R. W. Romani, and S. Nussinov. Neutron stars: Graveyard of charged dark matter, *Phys. Lett.*, B**238**:337, 1990.

[998] A. Gould, J. A. Frieman, and K. Freese. Probing the Earth with WIMPs, *Phys. Rev.*, D**39**:1029, 1989.

[999] F. Governato *et al.* Forming disk galaxies in lambda CDM simulations, *Mon. Not. Roy. Astron. Soc.*, **374**:1479–1494, 2007, astro-ph/0602351.

[1000] F. Governato *et al.* Forming a large disk galaxy from a $z < 1$ major merger, 2008, arXiv:0812.0379 [astro-ph].

[1001] A. W. Graham. Core depletion from coalescing supermassive black holes, *Astrophys. J. Lett.*, **613**:L33–L36, 2004.

[1002] A. W. Graham, D. Merritt, B. Moore, J. Diemand, and B. Terzić. Empirical models for dark matter halos. II. Inner profile slopes, dynamical profiles, and $rho/sigma^3$, *Astron. J.*, **132**:2701–2710, 2006, astro-ph/0608613.

[1003] A. W. Graham, D. Merritt, B. Moore, J. Diemand, and B. Terzić. Empirical models for dark matter halos. III. The Kormendy Relation and the $\log rho_e$– $\log R_e$ relation, *Astron. J.*, **132**:2711–2716, December 2006.

[1004] L. Grandi. WARP: an argon double phase technique for dark matter search, Ph.D. Thesis, University of Pavia, 2005.

[1005] R. Gray *et al.* Measuring mass and cross section parameters at a focus point region, 2005, hep-ex/0507008.

[1006] E. K. Grebel, J. S. Gallagher, and D. Harbeck. The progenitors of dwarf spheroidal galaxies, *Astron. J.*, **125**:1926, 2003, astro-ph/0301025.

[1007] A. M Green. A potential WIMP signature for the caustic ring halo model, *Phys. Rev.*, D**63**:103003, 2001, astro-ph/0012393.

[1008] A. M. Green. The WIMP annual modulation signal and non-standard halo models, *Phys. Rev.*, D**63**:043005, 2001, astro-ph/0008318.

[1009] A. M. Green. Effect of halo modelling on WIMP exclusion limits, *Phys. Rev.*, D**66**:083003, 2002, astro-ph/0207366.

[1010] A. M. Green. Effect of realistic astrophysical inputs on the phase and shape of the WIMP annual modulation signal, *Phys. Rev.*, D**68**:023004, 2003, astro-ph/0304446.

[1011] A. M. Green. Determining the WIMP mass using direct detection experiments, *JCAP*, **0708**:022, 2007, hep-ph/0703217.

[1012] A. M. Green. Determining the WIMP mass from a single direct detection experiment, a more detailed study, *JCAP*, **0807**:005, 2008, arXiv:0805.1704 [hep-ph].

[1013] A. M. Green, S. Hofmann, and D. J. Schwarz. The power spectrum of SUSY-CDM on sub-galactic scales, *Mon. Not. Roy. Astron. Soc.*, **353**:L23, 2004, astro-ph/0309621.

[1014] A. M. Green and B. Morgan. Optimizing WIMP directional detectors, *Astropart. Phys.*, **27**:142–149, 2007, astro-ph/0609115.

[1015] R. M. Green. *Spherical Astronomy*, Cambridge University Press, 1985.

[1016] K. Griest. Calculations of rates for direct detection of neutralino dark matter, *Phys. Rev. Lett.*, **61**:666–669, 1988.

[1017] K. Griest. Cross sections, relic abundance, and detection rates for neutralino dark matter, *Phys. Rev.*, **D38**:2357, 1988.

[1018] K. Griest and M. Kamionkowski. Unitarity limits on the mass and radius of dark matter particles, *Phys. Rev. Lett.*, **64**:615, 1990.

[1019] K. Griest and D. Seckel. Cosmic asymmetry, neutrinos and the Sun, *Nucl. Phys.*, **B283**:681, 1987.

[1020] K. Griest and D. Seckel. Three exceptions in the calculation of relic abundances, *Phys. Rev.*, **D43**:3191, 1991.

[1021] K. Griest and J. Silk. Prospects fade for neutrino cold dark matter, *Nature*, **343**:26–27, 1990.

[1022] C. J. Grillmair. Four new stellar debris streams in the Galactic Halo, 2008, arXiv:0811.3965 [astro-ph].

[1023] C. Grimani, S. A. Stephens, F. S. Cafagna, *et al.* Measurements of the absolute energy spectra of cosmic-ray positrons and electrons above 7 GeV, *Astron. Astrophys.*, **392**:287–294, 2002.

[1024] D. Grin, G. Covone, J.-P. Kneib, M. Kamionkowski, A. Blain, and E. Jullo. Telescope search for decaying relic axions, *Phys. Rev. D*, **75**:105018, 2007.

[1025] D. Grin, T. L. Smith, and M. Kamionkowski. Axion constraints in non-standard thermal histories, *Phys. Rev.*, **D77**:085020, 2008, arXiv:0711.1352 [astro-ph].

[1026] D. J. Gross, R. D. Pisarski, and L. G. Yaffe. QCD and instantons at finite temperature, *Rev. Mod. Phys.*, **53**:43, 1981.

[1027] Y. Grossman and M. Neubert. Neutrino masses and mixings in non-factorizable geometry, *Phys. Lett.*, **B474**:361–371, 2000, hep-ph/9912408.

[1028] Joint LEP 2 Supersymmetry Working Group. Combined LEP chargino results, up to 208 GeV, http://lepsusy.web.cern.ch/ lepsusy/www/inos_ Moriond01/charginos_pub.html.

[1029] Joint LEP 2 Supersymmetry Working Group. Combined LEP selectron/smuon/stau results, 183–208 GeV, http://lepsusy.web. cern.ch/lepsusy/www/sleptons_ summer02/slep_2002.html.

[1030] LEP Electroweak Working Group. http://lepewwg.web.cern.ch/ LEPEWWG/.

[1031] LEP2 SUSY Working Group. Combined LEP selectron/smuon/stau results, 183–208 GeV, 2004, LEPSUSYWG/04-01.1.

[1032] Tevatron Electroweak Working Group. http://tevewwg.fnal.gov/.

[1033] M. Guchait, F. Mahmoudi, and K. Sridhar. Tevatron constraint on the Kaluza-Klein gluon of the bulk Randall–Sundrum model, *JHEP*, **05**:103, 2007, hep-ph/0703060.

[1034] M. Guchait, F. Mahmoudi, and K. Sridhar. Associated production of a Kaluza-Klein excitation of a gluon with a t t(bar) pair at the LHC, *Phys. Lett.*, **B666**:347–351, 2008, arXiv:0710.2234 [hep-ph].

[1035] S. B. Gudnason, C. Kouvaris, and F. Sannino. Dark matter from new technicolor theories, *Phys. Rev.*, **D74**:095008, 2006, hep-ph/0608055.

[1036] S. B. Gudnason, C. Kouvaris, and F. Sannino. Towards working technicolor: Effective theories and dark matter, *Phys. Rev.*, **D73**:115003, 2006, hep-ph/0603014.

[1037] J. F. Gunion, D. Hooper, and B. McElrath. Light neutralino dark matter in the NMSSM, *Phys. Rev.*, **D73**:015011, 2006, hep-ph/0509024.

[1038] J. F. Gunion, M. Toharia, and J. D. Wells. Precision electroweak data and the mixed radion–Higgs sector of warped extra dimensions, *Phys. Lett.*, **B585**:295–306, 2004, hep-ph/0311219.

[1039] J. E. Gunn. Massive galactic halos. I. Formation and evolution, *Astrophys. J.*, **218**:592–598, 1977.

[1040] J. E. Gunn, B. W. Lee, I. Lerche, D. N. Schramm, and G. Steigman. Some astrophysical consequences of the existence of a heavy stable neutral lepton, *Astrophys. J.*, **223**:1015–1031, 1978.

[1041] W.-L. Guo, L.-M. Wang, Y.-L. Wu, Y.-F. Zhou, and C. Zhuang. Gauge-singlet dark matter in a left-right symmetric model with spontaneous CP violation, 2008, arXiv:0811.2556 [hep-ph].

[1042] G. Gustafson and J. Hakkinen. Deuteron production in e+ e- annihilation, *Z. Phys.*, **C61**:683–688, 1994.

[1043] M. Gustafsson, E. Lundstrom, L. Bergstrom, and J. Edsjö. Significant gamma lines from inert Higgs dark matter, *Phys. Rev. Lett.*, **99**:041301, 2007, astro-ph/0703512.

[1044] A. H. Guth and S. Y. Pi. Fluctuations in the new inflationary universe, *Phys. Rev. Lett.*, **49**:1110–1113, 1982.

[1045] Gy. Gyurky *et al.* ^{3}He(alpha,gamma)^{7}Be cross section at low energies, *Phys. Rev.*, **C75**:035805, 2007, nucl-ex/0702003.

[1046] H. E. Haber and R. Hempfling. Can the mass of the lightest Higgs boson of the minimal supersymmetric model be larger than m(Z)?, *Phys. Rev. Lett.*, **66**:1815, 1991.

[1047] H. E. Haber, R. Hempfling, and A. H. Hoang. Approximating the radiatively corrected Higgs mass in the minimal supersymmetric model, *Z. Phys.*, **C75**:539–554, 1997, hep-ph/9609331.

[1048] H. E. Haber and G. L. Kane. The search for supersymmetry: probing physics beyond the Standard Model, *Phys. Rep.*, **117**:75, 1985.

[1049] A. Habig. An indirect search for WIMPs with Super-Kamiokande, 2001, hep-ex/0106024.

[1050] K. Hagiwara, A. D. Martin, D. Nomura, and T. Teubner. Improved predictions for $g-2$ of the muon and $\alpha_{qed}(m_z^2)$, *Phys. Lett.*, **B649**:173–179, 2007.

[1051] C. Hagmann, S. Chang, and P. Sikivie. Axion radiation from strings, *Phys. Rev.*, **D63**:125018, 2001, hep-ph/0012361.

[1052] C. Hagmann and P. Sikivie. Computer simulations of the motion and decay of global strings, *Nucl. Phys.*, **B363**:247–280, 1991.

[1053] T. Hahn. Automatic loop calculations with FeynArts, FormCalc, and LoopTools, *Nucl. Phys. Proc. Suppl.*, **89**:231–236, 2000, hep-ph/0005029.

[1054] T. Hahn. Generating Feynman diagrams and amplitudes with FeynArts 3, *Comput. Phys. Commun.*, **140**:418–431, 2001, hep-ph/0012260.

[1055] T. Hahn and M. Perez-Victoria. Automatized one-loop calculations in four and D dimensions, *Comput. Phys. Commun.*, **118**:153–165, 1999, hep-ph/9807565.

[1056] T. Hahn and C. Schappacher. The implementation of the minimal supersymmetric standard model in FeynArts and FormCalc, *Comput. Phys. Commun.*, **143**:54–68, 2002, hep-ph/0105349.

[1057] C. J. Hailey, T. Aramaki, W. W. Craig, *et al.* Accelerator testing of the general antiparticle spectrometer; a novel approach to indirect dark matter detection, *J. Cosmol. Astropart. Phys.*, **1**:7, 2006.

[1058] C. J. Hailey *et al.* Development of the gaseous antiparticle spectrometer for space-based antimatter detection, *Nucl. Instrum. Meth. Phys. Res. B*, **214**:122–125, 2004.

[1059] S. Haino *et al.* Measurement of cosmic-ray antiproton spectrum with BESS-2002, **3**:13, 2005.

[1060] J. Hall and P. Gondolo. Stellar orbit constraints on neutralino annihilation at the galactic center, *Phys. Rev. D*, **74**:063511, 2006.

[1061] F. Halzen and D. Hooper. Prospects for detecting dark matter with neutrino telescopes in light of recent results from direct detection experiments, *Phys. Rev.*, **D73**:123507, 2006, hep-ph/0510048.

[1062] K. Hamaguchi, T. Hatsuda, M. Kamimura, Y. Kino, and T. T. Yanagida. Stau-catalyzed Li-6 production in big-bang nucleosynthesis, *Phys. Lett.*, **B650**:268, 2007, hep-ph/0702274.

[1063] K. Hamaguchi, Y. Kuno, T. Nakaya, and M. M. Nojiri. A study of late decaying charged particles at future colliders, *Phys. Rev.*, **D70**:115007, 2004, hep-ph/0409248.

[1064] T. Hambye and K. Riesselmann. Matching conditions and Higgs mass upper bounds revisited, *Phys. Rev.*, **D55**:7255–7262, 1997, hep-ph/9610272.

[1065] T. Hambye and M. H. G. Tytgat. Electroweak symmetry breaking induced by dark matter, *Phys. Lett.*, **B659**:651–655, 2008, arXiv:0707.0633 [hep-ph].

[1066] F. Hammer, I. M. Gioia, E. J. Shaya, P. Teyssandier, O. Le Fevre, and G. A. Luppino. Detailed lensing properties of the MS 2137-2353 core and reconstruction of sources from Hubble Space Telescope imagery, *Astrophys. J.*, **491**:477, 1997.

[1067] S. Hannestad. What is the lowest possible reheating temperature?, *Phys. Rev.*, **D70**:043506, 2004, astro-ph/0403291.

[1068] S. Hannestad, A. Mirizzi, and G. Raffelt. New cosmological mass limit on thermal relic axions, *JCAP*, **0507**:002, 2005, hep-ph/0504059.

[1069] S. Hannestad and G. G. Raffelt. Stringent neutron-star limits on large extra dimensions, *Phys. Rev. Lett.*, **88**:071301, 2002, hep-ph/0110067.

[1070] S. H. Hansen, J. Lesgourgues, S. Pastor, and J. Silk. Closing the window on warm dark matter, *Mon. Not. Roy. Astron. Soc.*, **333**:544–546, 2002, astro-ph/0106108.

[1071] S. H. Hansen, B. Moore, and J. Stadel. A universal density slope – velocity anisotropy relation, 2005, astro-ph/0509799.

[1072] S. H. Hansen, B. Moore, M. Zemp, and J. Stadel. A universal velocity distribution of relaxed collisionless structures, *JCAP*, **0601**:014, 2006, astro-ph/0505420.

[1073] T. Hansl *et al.* Results of a beam dump experiment at the CERN SPS Neutrino Facility, *Phys. Lett.*, B**74**:139, 1978.

[1074] J. G. Hao and R. Akhoury. Can relativistic MOND theory resolve both the dark matter and dark energy paradigms?, arXiv:astro-ph/0504130, 2005.

[1075] D. Harari and P. Sikivie. On the evolution of global strings in the early universe, *Phys. Lett.*, B**195**:361, 1987.

[1076] B. V. Harling and A. Hebecker. Sequestered dark matter, *JHEP*, **05**:031, 2008, arXiv:0801.4015 [hep-ph].

[1077] R. Harnik and G. D. Kribs. An effective theory of dirac dark matter, 2008, arXiv:0810.5557 [hep-ph].

[1078] T. Haruyama *et al. Cryocoolers*, **13**:689, 2005.

[1079] F. Hasenbalg *et al.* Cold dark matter identification: Diurnal modulation revisited, *Phys. Rev.*, D**55**:7350–7355, 1997, astro-ph/9702165.

[1080] S. W. Hawking. The development of irregularities in a single bubble inflationary universe, *Phys. Lett.*, B**115**:295, 1982.

[1081] S. Heinemeyer, W. Hollik, and G. Weiglein. The masses of the neutral CP-even Higgs bosons in the MSSM: Accurate analysis at the two-loop level, *Eur. Phys. J.*, C**9**:343, 1999, hep-ph/9812472.

[1082] S. Heinemeyer, W. Hollik, and G. Weiglein. FeynHiggs: a program for the calculation of the masses of the neutral CP-even Higgs bosons in the MSSM, *Comput. Phys. Commun.*, **124**:76, 2000, hep-ph/9812320.

[1083] K. Heitmann, Z. Lukić, P. Fasel, *et al.* The cosmic code comparison project, *Comput. Sci. Disc.*, **1**:015003, 2008, 0706.1270.

[1084] R. H. Helm. Inelastic and elastic scattering of 187-MeV electrons from selected even-even nuclei, *Phys. Rev.*, **104**:1466–1475, 1956.

[1085] A. Helmi, S. D. M. White, and V. Springel. The phase-space structure of a dark-matter halo: Implications for dark-matter direct detection experiments, *Phys. Rev.*, D**66**:063502, 2002, astro-ph/0201289.

[1086] T. K. Hemmick *et al.* A search for anomalously heavy isotopes of low-Z nuclei, *Phys. Rev.*, D**41**:2074, 1990.

[1087] J. F. Hennawi, N. Dalal, P. Bode, and J. P. Ostriker. Characterizing the cluster lens population, *Astrophys. J.*, **654**:714–730, 2007.

[1088] J. F. Hennawi, M. D. Gladders, M. Oguri, *et al.* A new survey for giant arcs, *Astron. J.*, **135**:664–681, 2008.

[1089] M. Hénon. L'évolution initiale d'un amas sphérique, *Ann. Astrophys.*, **27**:83, 1964.

[1090] L. Hernquist. An analytical model for spherical galaxies and bulges, *Astrophys. J.*, **356**:359–364, 1990.

[1091] J. L. Hewett and T. G. Rizzo. Shifts in the properties of the Higgs boson from radion mixing, *JHEP*, **08**:028, 2003, hep-ph/0202155.

[1092] J. N. Hewitt, E. L. Turner, D. P. Schneider, B. F. Burke, and G. I. Langston. Unusual radio source MG1131+0456 – A possible Einstein ring, *Nature*, **333**:537–540, June 1988.

[1093] C. Heymans, L. Van Waerbeke, D. Bacon, *et al.* The shear testing programme. I. Weak lensing analysis of simulated ground-based observations, *Mon. Not. Roy. Astron. Soc.*, **368**:1323–1339, 2006.

[1094] J. Hidaka and G. M. Fuller. Dark matter sterile neutrinos in stellar collapse: Alteration of energy/lepton number transport and a mechanism for supernova explosion enhancement, *Phys. Rev.*, **D74**:125015, 2006, astro-ph/0609425.

[1095] C. T. Hill and G. G. Ross. Models and new phenomenological implications of a class of pseudogoldstone bosons, *Nucl. Phys.*, **B311**:253, 1988.

[1096] I. Hinchliffe and F. E. Paige. Measurements in SUGRA models with large tan(beta) at LHC, *Phys. Rev.*, **D61**:095011, 2000, hep-ph/9907519.

[1097] I. Hinchliffe, F. E. Paige, M. D. Shapiro, J. Soderqvist, and W. Yao. Precision SUSY measurements at LHC, *Phys. Rev.*, **D55**:5520–5540, 1997, hep-ph/9610544.

[1098] M. Hindmarsh. Axions and the QCD phase transition, *Phys. Rev.*, **D45**:1130–1138, 1992.

[1099] M. Hindmarsh and O. Philipsen. WIMP dark matter and the QCD equation of state, *Phys. Rev.*, **D71**:087302, 2005, hep-ph/0501232.

[1100] G. Hinshaw *et al.* Five-year Wilkinson Microwave Anisotropy Probe (WMAP) observations: Data processing, sky maps, & basic results, 2008, arXiv:0803.0732 [astro-ph].

[1101] K. S. Hirata *et al.* Observation in the Kamiokande-II detector of the neutrino burst from supernova SN 1987a, *Phys. Rev.*, **D38**:448–458, 1988.

[1102] J. Hisano, K. Kohri, and M. M. Nojiri. Neutralino warm dark matter, *Phys. Lett.*, **B505**:169–176, 2001, hep-ph/0011216.

[1103] J. Hisano, S. Matsumoto, M. Nagai, O. Saito, and M. Senami. Non-perturbative effect on thermal relic abundance of dark matter, *Phys. Lett.*, **B646**:34–38, 2007, hep-ph/0610249.

[1104] J. Hisano, S. Matsumoto, and M. M. Nojiri. Unitarity and higher-order corrections in neutralino dark matter annihilation into two photons, *Phys. Rev.*, **D67**:075014, 2003, hep-ph/0212022.

[1105] J. Hisano, S. Matsumoto, and M. M. Nojiri. Explosive dark matter annihilation, *Phys. Rev. Lett.*, **92**:031303, 2004, hep-ph/0307216.

[1106] A. Hitachi. Properties of liquid rare gas scintillation for WIMP searches. Prepared for *IDM 2004: 5th International Workshop on the Identification of Dark Matter*, Edinburgh, UK, 6–10 September 2004.

[1107] A. Hitachi. Properties of liquid xenon scintillation for dark matter searches, *Astropart. Phys.*, **24**:247–256, 2005.

[1108] A. Hitachi. Bragg-like curve for dark matter searches: binary gases, 2008, arXiv:0804.1191.

[1109] A. Hitachi *et al.* Effect of ionization density on the time dependence of luminescence from liquid argon and xenon, *Phys. Rev.*, **B27**: 5279–5285, 1983.

[1110] A. Hocker, H. Lacker, S. Laplace, and F. Le Diberder. A new approach to a global fit of the CKM matrix, *Eur. Phys. J.*, **C21**: 225–259, 2001, hep-ph/0104062.

[1111] H. Hoekstra and B. Jain. Weak gravitational lensing and its cosmological applications, *Ann. Rev. Nucl. Part. Sci.*, **58**:99–123, 2008, arXiv:0805.0139.

[1112] H. Hoekstra, H. K. C. Yee, and M. D. Gladders. Properties of galaxy dark matter halos from weak lensing, *Astrophys. J.*, **606**:67–77, 2004.

[1113] M. Hof *et al.* Measurement of cosmic-ray antiprotons from 3.7 to 19 GeV, *Astrophys. J.*, **467**:L33, 1996.

[1114] S. Hofmann, D. J. Schwarz, and H. Stoecker. Damping scales of neutralino cold dark matter, *Phys. Rev.*, **D64**:083507, 2001.

[1115] W. Hofmann and the Hess Collaboration. The High Energy Stereoscopic System (HESS) project. In B. L. Dingus, M. H. Salamon, and D. B. Kieda, eds., volume 515 of *American Institute of Physics Conference Series*, 500, 2000.

[1116] C. J. Hogan and J. J. Dalcanton. New dark matter physics: Clues from halo structure, *Phys. Rev.*, **D62**:063511, 2000, astro-ph/0002330.

[1117] C. J. Hogan and M. J. Rees. Axion miniclusters, *Phys. Lett.*, **B205**:228–230, 1988.

[1118] J. Hong *et al.* The scintillation efficiency of carbon and hydrogen recoils in an organic liquid scintillator for dark matter searches, *Astropart. Phys.*, **16**:333–338, 2002.

[1119] D. Hooper, I. de la Calle Perez, J. Silk, F. Ferrer, and S. Sarkar. Have atmospheric Cerenkov telescopes observed dark matter?, *J. Cosmol. Astropart. Phys.*, **9**:2, 2004.

[1120] D. Hooper and B. L. Dingus. Limits on supersymmetric dark matter from EGRET observations of the galactic center region, *Phys. Rev.*, D**70**:113007, 2004, astro-ph/0210617.

[1121] D. Hooper, D. P. Finkbeiner, and G. Dobler. Evidence of dark matter annihilations in the WMAP haze, *Phys. Rev.*, D**76**:083012, 2007, arXiv:0705.3655 [astro-ph].

[1122] D. Hooper and G. D. Kribs. Probing Kaluza–Klein dark matter with neutrino telescopes, *Phys. Rev.*, D**67**:055003, 2003, hep-ph/0208261.

[1123] D. Hooper and G. D. Kribs. Kaluza–Klein dark matter and the positron excess, *Phys. Rev.*, D**70**:115004, 2004, hep-ph/0406026.

[1124] D. Hooper, J. March-Russell, and S. M. West. Asymmetric sneutrino dark matter and the Omega(b)/Omega(DM) puzzle, *Phys. Lett.*, B**605**:228–236, 2005, hep-ph/0410114.

[1125] D. Hooper, F. Petriello, K. M. Zurek, and M. Kamionkowski. The new DAMA dark-matter window and energetic-neutrino searches, 2008, arXiv:0808.2464 [hep-ph].

[1126] D. Hooper and S. Profumo. Dark matter and collider phenomenology of universal extra dimensions, *Phys. Rept.*, **453**:29–115, 2007, hep-ph/0701197.

[1127] D. Hooper and G. Servant. Indirect detection of Dirac right-handed neutrino dark matter, *Astropart. Phys.*, **24**:231–246, 2005, hep-ph/0502247.

[1128] D. Hooper and J. Silk. Searching for dark matter with future cosmic positron experiments, *Phys. Rev.*, D**71**:083503, 2005, hep-ph/0409104.

[1129] D. Hooper and L.-T. Wang. Direct and indirect detection of neutralino dark matter in selected supersymmetry breaking scenarios, *Phys. Rev.*, D**69**:035001, 2004, hep-ph/0309036.

[1130] D. Hooper and K. M. Zurek. Natural supersymmetric model with MeV dark matter, *Phys. Rev.*, D**77**:087302, 2008, arXiv:0801.3686 [hep-ph].

[1131] D. Horns. TeV γ-radiation from dark matter annihilation in the Galactic center, *Phys. Lett. B*, **607**:225–232, 2005.

[1132] C. J. Horowitz, M. A. Perez-Garcia, J. Carriere, D. K. Berry, and J. Piekarewicz. Nonuniform neutron-rich matter and coherent neutrino scattering, *Phys. Rev.*, C**70**:065806, 2004, astro-ph/0409296.

[1133] A. Hosford, S. G. Ryan, A. E. Garcia Perez, J. E. Norris, and K. A. Olive. Lithium abundances of halo dwarfs based on excitation temperature. I. LTE, 2008, arXiv:0811.2506 [astro-ph].

[1134] O. Host and S. H. Hansen. What it takes to measure a fundamental difference between dark matter and baryons: the halo velocity anisotropy, *JCAP*, **0706**:016, 2007, arXiv:0704.2909 [astro-ph].

[1135] K. Hsieh, R. N. Mohapatra, and S. Nasri. Dark matter in universal extra dimension models: Kaluza–Klein photon and right-handed neutrino admixture, *Phys. Rev.*, D**74**:066004, 2006, hep-ph/0604154.

[1136] W. Hu and J. Silk. Thermalization constraints and spectral distortions for massive unstable relic particles, *Phys. Rev. Lett.*, **70**:2661–2664, 1993.

[1137] P. Huang, N. Kersting, and H. H. Yang. Hidden thresholds: A technique for reconstructing new physics masses at hadron colliders, 2008, arXiv:0802.0022 [hep-ph].

[1138] Y. Huang, M. H. Reno, I. Sarcevic, and J. Uscinski. Weak interactions of supersymmetric staus at high energies, *Phys. Rev.*, D**74**:115009, 2006, hep-ph/0607216.

[1139] M. E. Huber, B. Cabrera, M. A. Taber, and R. D. Gardner. Limit on the flux of cosmic ray magnetic monopoles from operation of an eight loop superconducting detector, *Phys. Rev. Lett.*, **64**:835–838, 1990.

[1140] J. Hubisz, J. Lykken, M. Pierini, and M. Spiropulu. Missing energy look-alikes with 100 pb^{-1} at the LHC, *Phys. Rev.*, D**78**:075008, 2008, arXiv:0805.2398 [hep-ph].

[1141] C. Hugonie, G. Belanger, and A. Pukhov. Dark matter in the constrained NMSSM, *JCAP*, **0711**:009, 2007, arXiv:0707.0628 [hep-ph].

[1142] S. D. Hunter *et al.* EGRET observations of the diffuse gamma-ray emission from the galactic plane, *Astrophys. J.*, **481**:205–240, 1997.

[1143] A. Huss, B. Jain, and M. Steinmetz. How universal are the density profiles of dark halos?, *Astrophys. J.*, **517**:64–69, 1999.

[1144] P. Hut. Limits on masses and number of neutral weakly interacting particles, *Phys. Lett.*, B**69**:85, 1977.

[1145] L. E. Ibanez. Locally supersymmetric SU(5) grand unification, *Phys. Lett.*, B**118**:73, 1982.

[1146] L. E. Ibanez and G. G. Ross. SU(2)$_L$ x U(1) symmetry breaking as a radiative effect of supersymmetry breaking in GUTs, *Phys. Lett.*, B**110**:215, 1982.

[1147] R. Ibata *et al.* The haunted halos of Andromeda and Triangulum: A panorama of galaxy formation in action, 2007, arXiv:0704.1318 [astro-ph].

[1148] R. Ibata, M. Irwin, G. F. Lewis, and A. Stolte. Galactic halo substructure in the Sloan Digital Sky Survey: The ancient tidal stream from the Sagittarius dwarf galaxy, *Astrophys. J*, **547**:L133–L136, 2001.

[1149] K. Ichiki, M. Takada, and T. Takahashi. Constraints on neutrino masses from weak lensing, 2008, arXiv:0810.4921.

[1150] A. S. Ilyin, K. P. Zybin, and A. V. Gurevich. Dark matter in galaxies and the growth of giant black holes, *Sov. J. Exp. Theor. Phys.*, **98**:1–13, 2004.

[1151] Y. Inoue *et al.* Search for sub-electronvolt solar axions using coherent conversion of axions into photons in magnetic field and gas helium, *Phys. Lett.*, B**536**:18–23, 2002, astro-ph/0204388.

[1152] F. Iocco. Dark matter capture and annihilation over the first stars: Preliminary estimates, *Astrophys. J.*, **677**:L1, 2008, arXiv:0802.0941 [astro-ph].

[1153] F. Iocco *et al.* Dark matter annihilation effects on the first stars, 2008, arXiv:0805.4016 [astro-ph].

[1154] F. Iocco, G. Mangano, G. Miele, O. Pisanti, and P. D. Serpico. Primordial nucleosynthesis: From precision cosmology to fundamental physics, 2008, arXiv:0809.0631 [astro-ph].

[1155] I. G. Irastorza *et al.* Present status of IGEX dark matter search at Canfranc Underground Laboratory, 2002, astro-ph/0211535.

[1156] M. Irwin and D. Hatzidimitriou. Structural parameters for the Galactic dwarf spheroidals, *Mon. Not. Roy. Astron. Soc.*, **277**:1354–1378, 1995.

[1157] C. Isaila *et al.* Scintillation light detectors with Neganov-Luke amplification, *Nucl. Instrum. Meth.*, A**559**:399–401, 2006.

[1158] T. Ishiyama, T. Fukushige, and J. Makino. Variation of the subhalo abundance in dark matter halos, 2008, arXiv:0812.0683 [astro-ph].

[1159] Y. I. Izotov, T. X. Thuan, and G. Stasinska. The primordial abundance of 4He: a self-consistent empirical analysis of systematic effects in a large sample of low-metallicity HII regions, *Astrophys. J.*, **662**:15–38, 2007, astro-ph/0702072.

[1160] I. Jack, D. R. T. Jones, and A. F. Kord. Snowmass benchmark points and three-loop running, *Ann. Phys.*, **316**:213–233, 2005, hep-ph/0408128.

[1161] R. Jackiw and C. Rebbi. Vacuum periodicity in a Yang–Mills quantum theory, *Phys. Rev. Lett.*, **37**:172–175, 1976.

[1162] M. Jacob and G. C. Wick. On the general theory of collisions for particles with spin, *Ann. Phys.*, **7**:404–428, 1959.

[1163] H.-T. Janka, W. Keil, G. Raffelt, and D. Seckel. Nucleon spin fluctuations and the supernova emission of neutrinos and axions, *Phys. Rev. Lett.*, **76**:2621–2624, 1996, astro-ph/9507023.

[1164] J. R. Jardel and J. A. Sellwood. Halo density reduction by baryonic settling?, 2008, arXiv:0808.3449 [astro-ph].

[1165] M. Jarvis, G. M. Bernstein, P. Fischer, *et al.* Weak-lensing results from the 75 square degree Cerro Tololo Inter-American Observatory survey, *Astron. J.*, **125**:1014–1032, 2003.

[1166] K. Jedamzik. Did something decay, evaporate, or annihilate during Big Bang nucleosynthesis?, *Phys. Rev.*, **D70**:063524, 2004, astro-ph/0402344.

[1167] K. Jedamzik. Neutralinos and Big Bang nucleosynthesis, *Phys. Rev.*, **D70**:083510, 2004, astro-ph/0405583.

[1168] K. Jedamzik. Big Bang nucleosynthesis constraints on hadronically and electromagnetically decaying relic particles, *Phys. Rev. D*, **74**:103509, 2006.

[1169] K. Jedamzik. Bounds on long-lived charged massive particles from Big Bang nucleosynthesis, *JCAP*, **0803**:008, 2008, arXiv:0710.5153 [hep-ph].

[1170] K. Jedamzik. The cosmic ^{6}Li and ^{7}Li problems and BBN with long-lived charged massive particles, *Phys. Rev.*, **D77**:063524, 2008, arXiv:0707.2070 [astro-ph].

[1171] M.-J. Jee, H. C. Ford, G. D. Illingworth, *et al.* Discovery of a ringlike dark matter structure in the core of the galaxy cluster Cl 0024+17, *Astrophys. J.*, **661**:728–749, 2007, arXiv:0705.2171.

[1172] T. E. Jeltema and S. Profumo. Searching for dark matter with X-ray observations of local dwarf galaxies, *Astrophys. J.*, **686**:1045–1055, 2008, 0805.1054.

[1173] Y. P. Jing and Y. Suto. Triaxial modeling of halo density profiles with high-resolution N-body simulations, *Astrophys. J.*, **574**:538, 2002, astro-ph/0202064.

[1174] D. E. Johnston, E. S. Sheldon, R. H. Wechsler, *et al.* Cross-correlation weak lensing of SDSS galaxy clusters II: Cluster density profiles and the mass–richness relation, 2007, arXiv:0709.1159.

[1175] K. V. Johnston. A prescription for building the Milky Way's halo from disrupted satellites, *Astrophys. J.*, **495**:297, 1998.

[1176] K. V. Johnston, L. Hernquist, and M. Bolte. Fossil signatures of ancient accretion events in the halo, *Astrophys. J*, **465**:278, 1996.

[1177] K. V. Johnston, D. N. Spergel, and C. Haydn. How lumpy is the Milky Way's dark matter halo?, 2001, astro-ph/0111196.

[1178] F. C. Jones, A. Lukasiak, V. Ptuskin, and W. Webber. The modified weighted slab technique: Models and results, 2000, astro-ph/0007293.

[1179] A. Juillard *et al*. Development of Ge/NbSi detectors for EDELWEISS-II with identification of near-surface events, *Nucl. Instrum. Meth.*, A**559**:393–395, 2006.

[1180] E. Jullo, J.-P. Kneib, M. Limousin, Á. Elíasdóttir, P. J. Marshall, and T. Verdugo. A Bayesian approach to strong lensing modelling of galaxy clusters, *New J. Phys.*, **9**:447, 2007, 0706.0048.

[1181] G. Jungman. Neutdriver, 2000, http://t8web.lanl.gov/people/jungman/neut-package.html.

[1182] G. Jungman and M. Kamionkowski. Gamma-rays from neutralino annihilation, *Phys. Rev.*, D**51**:3121–3124, 1995, hep-ph/9501365.

[1183] G. Jungman and M. Kamionkowski. Neutrinos from particle decay in the Sun and Earth, *Phys. Rev.*, D**51**:328–340, 1995, hep-ph/9407351.

[1184] G. Jungman, M. Kamionkowski, and K. Griest. Supersymmetric dark matter, *Phys. Rep.*, **267**:195–373, 1996, hep-ph/9506380.

[1185] M. Juric *et al.* The Milky Way tomography with SDSS. 1. Stellar number density distribution, *Astrophys. J.*, **673**:864–914, 2008, astro-ph/0510520.

[1186] R. Juszkiewicz, J. Silk, and A. Stebbins. Constraints on cosmologically regenerated gravitinos, *Phys. Lett.*, B**158**:463–467, 1985.

[1187] M. M. Kado and C. G. Tully. The searches for Higgs bosons at LEP, *Ann. Rev. Nucl. Part. Sci.*, **52**:65–113, 2002.

[1188] K. Kadota. Sterile neutrino dark matter in warped extra dimensions, *Phys. Rev.*, D**77**:063509, 2008, arXiv:0711.1570 [hep-ph].

[1189] K. Kainulainen, K. Tuominen, and J. Virkajarvi. The WIMP of a minimal technicolor theory, *Phys. Rev.*, D**75**:085003, 2007, hep-ph/0612247.

[1190] N. Kaiser and G. Squires. Mapping the dark matter with weak gravitational lensing, *Astrophys. J.*, **404**:441–450, 1993.

[1191] N. Kaiser, G. Wilson, and G. A. Luppino. Large-scale cosmic shear measurements, 2007, arXiv:astro-ph/0003338.

[1192] K. Kajantie, M. Laine, K. Rummukainen, and Y. Schroder. The pressure of hot QCD up to g**6 ln(1/g), *Phys. Rev.*, D**67**:105008, 2003, hep-ph/0211321.

[1193] M. Kakizaki, S. Matsumoto, Y. Sato, and M. Senami. Significant effects of second KK particles on LKP dark matter physics, *Phys. Rev.*, D**71**:123522, 2005, hep-ph/0502059.

[1194] M. Kakizaki, S. Matsumoto, Y. Sato, and M. Senami. Relic abundance of LKP dark matter in UED model including effects of second KK resonances, *Nucl. Phys.*, B**735**:84–95, 2006, hep-ph/0508283.

[1195] M. Kakizaki, S. Matsumoto, and M. Senami. Relic abundance of dark matter in the minimal universal extra dimension model, *Phys. Rev.*, D**74**:023504, 2006, hep-ph/0605280.

[1196] N. Kallivayalil, R. P. van der Marel, and C. Alcock. Is the SMC bound to the LMC? The HST proper motion of the SMC, *Astrophys. J.*, **652**:1213–1229, 2006, astro-ph/0606240.

[1197] T. Kaluza. On the problem of unity in physics, *Sitzungsber. Preuss. Akad. Wiss. Berlin (Math. Phys.)*, **1921**:966–972, 1921.

[1198] T. Kamae, N. Karlsson, T. Mizuno, T. Abe, and T. Koi. Parameterization of γ, $e^{+/-}$, and neutrino spectra produced by p–p interaction in astronomical environments, *Astrophys. J.*, **647**:692–708, 2006.

[1199] M. Kamimura, Y. Kino, and E. Hiyama. Big-Bang nucleosynthesis reactions catalyzed by a long-lived negatively-charged leptonic particle, 2008, arXiv:0809.4772 [nucl-th].

[1200] M. Kamionkowski and A. Kinkhabwala. Galactic halo models and particle dark matter detection, *Phys. Rev.*, D**57**:3256–3263, 1998, hep-ph/9710337.

[1201] M. Kamionkowski and S. M. Koushiappas. Galactic substructure and direct detection of dark matter, *Phys. Rev.*, D**77**:103509, 2008, arXiv:0801.3269 [astro-ph].

[1202] M. Kamionkowski and S. Profumo. Early annihilation and diffuse backgrounds in models of weakly interacting massive particles in which the cross section for pair annihilation is enhanced by $1/v$, *Phys. Rev. Lett.*, **101**:261301, 2008, arXiv:0810.3233 [astro-ph].

[1203] M. Kamionkowski and M. S. Turner. Thermal relics: do we know their abundances?, *Phys. Rev.*, D**42**:3310–3320, 1990.

[1204] G. L. Kane, A. A. Petrov, J. Shao, and L.-T. Wang. Initial determination of the spins of the gluino and squarks at LHC, 2008, arXiv:0805.1397 [hep-ph].

[1205] D. B. Kaplan. A single explanation for both the baryon and dark matter densities, *Phys. Rev. Lett.*, **68**:741–743, 1992.

[1206] D. B. Kaplan and K. M. Zurek. Exotic axions, *Phys. Rev. Lett.*, **96**:041301, 2006, hep-ph/0507236.

[1207] M. Kaplinghat. Dark matter from early decays, *Phys. Rev.*, D**72**:063510, 2005.

[1208] M. Kaplinghat and L. E. Strigari. Proper motion of Milky Way dwarf spheroidals from line-of-sight velocities, 2008, arXiv:0805.0795 [astro-ph].

[1209] N. Katz and J. E. Gunn. Dissipational galaxy formation. I. Effects of gasdynamics, *Astrophys. J.*, **377**:365–381, 1991.

[1210] G. Kauffmann and M. Haehnelt. A unified model for the evolution of galaxies and quasars, *Mon. Not. Roy. Astron. Soc.*, **311**:576–588, 2000.

[1211] L. Kaufmann and A. Rubbia. The ArDM project: A direct detection experiment, based on liquid argon, for the search of dark matter, *Nucl. Phys. Proc. Suppl.*, **173**:141–143, 2007.

[1212] K. Kawagoe, M. M. Nojiri, and G. Polesello. A new SUSY mass reconstruction method at the CERN LHC, *Phys. Rev.*, D**71**:035008, 2005, hep-ph/0410160.

[1213] M. Kawasaki, K. Kohri, and T. Moroi. Big-Bang nucleosynthesis and hadronic decay of long-lived massive particles, *Phys. Rev. D*, **71**:083502, 2005.

[1214] M. Kawasaki, K. Kohri, and T. Moroi. Hadronic decay of late-decaying particles and big-bang nucleosynthesis, *Phys. Lett.*, B**625**:7–12, 2005.

[1215] M. Kawasaki, T. Moroi, and T. Yanagida. Constraint on the reheating temperature from the decay of the Polonyi field, *Phys. Lett.*, B**370**:52–58, 1996, hep-ph/9509399.

[1216] S. Kazantzidis, J. S. Bullock, A. R. Zentner, A. V. Kravtsov, and L. A. Moustakas. Cold dark matter substructure and galactic disks. I. Morphological signatures of hierarchical satellite accretion, *Astrophys. J.*, **688**:254–276, 2008, 0708.1949.

[1217] S. Kazantzidis *et al.* Density profiles of cold dark matter substructure: Implications for the missing satellites problem, *Astrophys. J.*, **608**:663–3679, 2004, astro-ph/0312194.

[1218] S. Kazantzidis *et al.* The effect of gas cooling on the shapes of dark matter halos, *Astrophys. J.*, **611**:L73–L76, 2004, astro-ph/0405189.

[1219] S. Kazantzidis, J. Magorrian, and B. Moore. Generating equilibrium dark matter halos: Inadequacies of the local Maxwellian approximation, *Astrophys. J.*, **601**:37–46, 2004, astro-ph/0309517.

[1220] C. R. Keeton, C. S. Kochanek, and U. Seljak. Shear and ellipticity in gravitational lenses, *Astrophys. J.*, **482**:604, 1997.

[1221] W. Keil *et al.* A fresh look at axions and SN 1987A, *Phys. Rev.*, D**56**:2419–2432, 1997, astro-ph/9612222.

[1222] D. Keres, N. Katz, D. H. Weinberg, and R. Dave. How do galaxies get their gas?, *Mon. Not. Roy. Astron. Soc.*, **363**:2–28, 2005, astro-ph/0407095.

[1223] F. J. Kerr and D. Lynden-Bell. Review of galactic constants, *Mon. Not. Roy. Astron. Soc.*, **221**:1023, 1986.

[1224] N. Kersting. On measuring split-SUSY gaugino masses at the LHC, 2008, arXiv:0806.4238 [hep-ph].

[1225] S. Khalil, C. Munoz, and E. Torrente-Lujan. Relic neutralino density in scenarios with intermediate unification scale, *New J. Phys.*, **4**:27, 2002, hep-ph/0202139.

[1226] M. Yu. Khlopov and C. Kouvaris. Composite dark matter from a model with composite Higgs boson, *Phys. Rev.*, D**78**:065040, 2008, arXiv:0806.1191 [astro-ph].

[1227] M. Yu. Khlopov and A. D. Linde. Is it easy to save the gravitino?, *Phys. Lett.*, B**138**:265–268, 1984.

[1228] V. Khotilovich, R. L. Arnowitt, B. Dutta, and T. Kamon. The stau neutralino co-annihilation region at an International Linear Collider, *Phys. Lett.*, B**618**:182–192, 2005, hep-ph/0503165.

[1229] M. Kilbinger, K. Benabed, J. Guy, *et al.* Dark energy constraints and correlations with systematics from CFHTLS weak lensing, SNLS supernovae Ia and WMAP5, 2008, arXiv:0810.5129.

[1230] W. Kilian, T. Ohl, and J. Reuter. WHIZARD: Simulating multi-particle processes at LHC and ILC, 2007, arXiv:0708.4233 [hep-ph].

[1231] C. Kilic, L.-T. Wang, and I. Yavin. On the existence of angular correlations in decays with heavy matter partners, *JHEP*, **05**:052, 2007, hep-ph/0703085.

[1232] J. E. Kim. Weak interaction singlet and strong CP invariance, *Phys. Rev. Lett.*, **43**:103, 1979.

[1233] J. E. Kim. Light pseudoscalars, particle physics and cosmology, *Phys. Rep.*, **150**:1–177, 1987.

[1234] S. S. Kim, D. F. Figer, and M. Morris. Dynamical friction on galactic center star clusters with an intermediate-mass black hole, *Astrophys. J. Lett.*, **607**:L123–L126, 2004.

[1235] Y. D. Kim. The status of XMASS experiment, *Phys. Atom. Nucl.*, **69**:1970–1974, 2006.

[1236] Y. G. Kim and K. Y. Lee. The minimal model of fermionic dark matter, *Phys. Rev.*, D**75**:115012, 2007, hep-ph/0611069.

[1237] Y. G. Kim, T. Nihei, L. Roszkowski, and R. Ruiz de Austri. Upper and lower limits on neutralino WIMP mass and spin-independent scattering cross section, and impact of new (g- 2)(mu) measurement, *JHEP*, **12**:034, 2002, hep-ph/0208069.

[1238] D. S. Kinion. First results from a multiple microwave cavity search for dark matter axions, PhD thesis, University of California, Davis, 2001. UMI-30-19020.

[1239] R. Kinnunen, S. Lehti, F. Moortgat, A. Nikitenko, and M. Spira. Measurement of the h/a → tau tau cross section and possible constraints on tan beta, *Eur. Phys. J.*, **C40N5**:23–32, 2005.

[1240] E. N. Kirby, J. D. Simon, M. Geha, P. Guhathakurta, and A. Frebel. Uncovering extremely metal-poor stars in the Milky Way's ultra-faint dwarf spheroidal satellite galaxies, 2008, arXiv:0807.1925 [astro-ph].

[1241] R. Kitano and I. Low. Dark matter from baryon asymmetry, *Phys. Rev.*, **D71**:023510, 2005, hep-ph/0411133.

[1242] R. Kitano, H. Murayama, and M. Ratz. Unified origin of baryons and dark matter, *Phys. Lett.*, **B669**:145–149, 2008, arXiv:0807.4313 [hep-ph].

[1243] R. Kitano and Y. Nomura. Supersymmetry, naturalness, and signatures at the LHC, *Phys. Rev.*, **D73**:095004, 2006, hep-ph/0602096.

[1244] O. Klein. Quantum theory and five-dimensional theory of relativity, *Z. Phys.*, **37**:895–906, 1926.

[1245] J. Kleyna, M. I. Wilkinson, N. W. Evans, G. Gilmore, and C. Frayn. Dark matter in dwarf spheroidals. II. Observations and modelling of Draco, *Mon. Not. Roy. Astron. Soc.*, **330**:792–806, 2002.

[1246] J. T. Kleyna, M. I. Wilkinson, G. Gilmore, and N. W. Evans. A dynamical fossil in the Ursa Minor dwarf spheroidal galaxy, *Astrophys. J. Lett.*, **588**:L21–L24, 2003, astro-ph/0304093.

[1247] A. Klypin, A. V. Kravtsov, J. S. Bullock, and J. R. Primack. Resolving the structure of cold dark matter halos, *Astrophys. J.*, **554**:903–915, 2001, astro-ph/0006343.

[1248] A. Klypin, O. Valenzuela, P. Colin, and T. Quinn. Dynamics of barred galaxies: Effects of disk height, 2008, arXiv:0808.3422 [astro-ph].

[1249] A. Klypin, H. Zhao, and R. S. Somerville. LCDM-based models for the Milky Way and M31 I: Dynamical models, *Astrophys. J.*, **573**:597–613, 2002, astro-ph/0110390.

[1250] A. A. Klypin, A. V. Kravtsov, O. Valenzuela, and F. Prada. Where are the missing galactic satellites?, *Astrophys. J.*, **522**:82–92, 1999, astro-ph/9901240.

[1251] A. Knebe, H. Yahagi, H. Kase, G. Lewis, and B. K. Gibson. The radial alignment of dark matter subhalos: from simulations to observations, 2008, arXiv:0805.1823 [astro-ph].

[1252] J.-P. Kneib, P. Hudelot, R. S. Ellis, *et al.* A wide-field Hubble Space Telescope study of the cluster Cl 0024+1654 at $z = 0.4$. II. The cluster mass distribution, *Astrophys. J.*, **598**:804–817, 2003.

[1253] J. P. Kneib, Y. Mellier, B. Fort, and G. Mathez. The distribution of dark matter in distant cluster lenses – Modelling A:370, *Astron. Astrophys.*, **273**:367, 1993.

[1254] J. L. Kneur and N. Sahoury. Bottom-up reconstruction scenarios for (un)constrained MSSM parameters at the LHC, 2008, arXiv:0808.0144 [hep-ph].

[1255] M. Kobayashi and T. Maskawa. CP violation in the renormalizable theory of weak interaction, *Prog. Theor. Phys.*, **49**:652–657, 1973.

[1256] C. S. Kochanek. Evidence for dark matter in MG 1654+134, *Astrophys. J.*, **445**:559–577, 1995.

[1257] C. S. Kochanek and N. Dalal. Tests for substructure in gravitational lenses, *Astrophys. J.*, **610**:69–79, 2004.

[1258] J. E. Koglin. Antideuterons as an indirect dark matter signature: design and preparation for a balloon-born GAPS experiment, *JOP:Conference Series, TAUP 2007 Proceedings*, in press.

[1259] K. Kohri, T. Moroi, and A. Yotsuyanagi. Big-bang nucleosynthesis with unstable gravitino and upper bound on the reheating temperature, *Phys. Rev.*, D**73**:123511, 2006, hep-ph/0507245.

[1260] K. Kohri and Y. Santoso. Cosmological scenario of stop NLSP with gravitino LSP and the cosmic lithium problem, 2008, arXiv:0811.1119 [hep-ph].

[1261] E. W. Kolb and K. A. Olive. The Lee–Weinberg bound revisited, *Phys. Rev.*, D**33**:1202, 1986.

[1262] E. W. Kolb, G. Servant, and T. M. P. Tait. The radionactive universe, *JCAP*, **0307**:008, 2003, hep-ph/0306159.

[1263] E. W. Kolb and I. I. Tkachev. Axion miniclusters and Bose stars, *Phys. Rev. Lett.*, **71**:3051–3054, 1993, hep-ph/9303313.

[1264] E. W. Kolb and I. I. Tkachev. Femtolensing and picolensing by axion miniclusters, *Astrophys. J.*, **460**:L25–L28, 1996, astro-ph/9510043.

[1265] E. W. Kolb and M. S. Turner. *The Early Universe*, Frontiers in Physics, Addison-Wesley, 1988, 1990.

[1266] E. Komatsu *et al.* Five-year Wilkinson Microwave Anisotropy Probe (WMAP) observations: Cosmological interpretation, *Astrophys. J. Suppl.*, **180**:330–376, 2008.

[1267] P. Konar, K. Kong, and K. T. Matchev. $\sqrt{s}_{m}in$: a global inclusive variable for determining the mass scale of new physics in events with missing energy at hadron colliders, 2008, arXiv:0812.1042 [hep-ph].

[1268] K. Kong and K. T. Matchev. Precise calculation of the relic density of Kaluza–Klein dark matter in universal extra dimensions, *JHEP*, **01**:038, 2006, hep-ph/0509119.

[1269] L. V. E. Koopmans, T. Treu, A. S. Bolton, S. Burles, and L. A. Moustakas. The Sloan Lens ACS Survey. III. The structure and formation of early-type galaxies and their evolution since $z \approx 1$, *Astrophys. J.*, **649**:599–615, 2006.

[1270] S. Koposov *et al.* The luminosity function of the Milky Way satellites, *Astrophys. J.*, **686**:279–291, 2008, arXiv:0706.2687 [astro-ph].

[1271] A. J. Korn *et al.* A probable stellar solution to the cosmological lithium discrepancy, *Nature*, **442**:657–659, 2006.

[1272] K. Kosack *et al.* TeV gamma-ray observations of the Galactic Center, *Astrophys. J. Lett.*, **608**:L97–L100, June 2004.

[1273] C. Kouvaris. The dark side of strong coupled theories, *Phys. Rev.*, D**78**:075024, 2008, arXiv:0807.3124 [hep-ph].

[1274] C. Kouvaris. WIMP annihilation and cooling of neutron stars, *Phys. Rev.*, D**77**:023006, 2008, arXiv:0708.2362 [astro-ph].

[1275] H. Kraus *et al.* EURECA: The European future of dark matter searches with cryogenic detectors, *Nucl. Phys. Proc. Suppl.*, **173**: 168–171, 2007.

[1276] F. Krauss, R. Kuhn, and G. Soff. AMEGIC++ 1.0: A matrix element generator in C++, *JHEP*, **02**:044, 2002, hep-ph/0109036.

[1277] L. Krauss, J. Moody, F. Wilczek, and D. E. Morris. Calculations for cosmic axion detection, *Phys. Rev. Lett.*, **55**:1797, 1985.

[1278] L. M. Krauss. New constraints on ino masses from cosmology. 1. Supersymmetric inos, *Nucl. Phys.*, B**227**:556, 1983.

[1279] A. V. Kravtsov *et al.* The dark side of the halo occupation distribution, *Astrophys. J.*, **609**:35–49, 2004, astro-ph/0308519.

[1280] A. V. Kravtsov, O. Y. Gnedin, and A. A. Klypin. The tumultuous lives of galactic dwarfs and the missing satellites problem, *Astrophys. J.*, **609**:482–497, 2004, astro-ph/0401088.

[1281] A. V. Kravtsov and A. A. Klypin. Origin and evolution of halo bias in linear and non-linear regimes, *Astrophys. J.*, **520**:437–453, 1999, astro-ph/9812311.

[1282] J. M. Kubo, S. S. Allam, J. Annis, *et al.* The Sloan Bright Arcs Survey: Six strongly lensed galaxies at $z = 0.4$–1.4, 2008, arXiv:0812.3934.

[1283] M. Hishida, S. Kubota, and J. Ruan. *Jpn. J. Phys.*, C **11**:2645, 1978.

[1284] M. Kuhlen, J. Diemand, and P. Madau. The shapes, orientation, and alignment of Galactic dark matter subhalos, 2007, arXiv:0705.2037 [astro-ph].

[1285] M. Kuhlen, J. Diemand, and P. Madau. The dark matter annihilation signal from galactic substructure: Predictions for GLAST, *Astrophys. J.*, **686**:262–278, 2008, arXiv:0805.4416 [astro-ph].

[1286] M. Kuhlen, L. E. Strigari, A. R. Zentner, J. S. Bullock, and J. R. Primack. Dark energy and dark matter halos, *Mon. Not. Roy. Astron. Soc.*, **357**:387–400, 2005, astro-ph/0402210.

[1287] A. Kulesza and L. Motyka. Threshold resummation for squark-antisquark and gluino-pair production at the LHC, 2008, arXiv:0807.2405 [hep-ph].

[1288] P. Kumar. Neutrino masses, baryon asymmetry, dark matter and the moduli problem – A complete framework, 2008, arXiv:0809.2610 [hep-ph].

[1289] A. Kurylov and M. Kamionkowski. Generalized analysis of weakly-interacting massive particle searches, *Phys. Rev.*, D**69**:063503, 2004, hep-ph/0307185.

[1290] M. Kusakabe, T. Kajino, R. N. Boyd, T. Yoshida, and G. J. Mathews. The X^- solution to the ^{6}Li and ^{7}Li Big Bang nucleosynthesis problems, 2007, arXiv:0711.3858 [astro-ph].

[1291] A. Kusenko. Sterile neutrinos, dark matter, and the pulsar velocities in models with a Higgs singlet, *Phys. Rev. Lett.*, **97**:241301, 2006, hep-ph/0609081.

[1292] A. Kusenko and G. Segre. Neutral current induced neutrino oscillations in a supernova, *Phys. Lett.*, B**396**:197–200, 1997, hep-ph/9701311.

[1293] R. Kuzio de Naray, S. S. McGaugh, W. J. G. de Blok, and A. Bosma. High resolution optical velocity fields of 11 low surface brightness galaxies, *Astrophys. J. Suppl.*, **165**:461–479, 2006, astro-ph/0604576.

[1294] V. Kuzmin and I. Tkachev. Ultra-high energy cosmic rays, super-heavy long-living particles, and matter creation after inflation, *JETP Lett.*, **68**:271–275, 1998, hep-ph/9802304.

[1295] V. Kuzmin and I. Tkachev. Matter creation via vacuum fluctuations in the early universe and observed ultra-high energy cosmic ray events, *Phys. Rev.*, D**59**:123006, 1999, hep-ph/9809547.

[1296] V. A. Kuzmin. Simultaneous solution to baryogenesis and dark-matter problems, *Phys. Part. Nucl.*, **29**:257–265, 1998, hep-ph/9701269.

[1297] V. A. Kuzmin, V. A. Rubakov, and M. E. Shaposhnikov. On the anomalous electroweak baryon number nonconservation in the early universe, *Phys. Lett.*, B155:36, 1985.

[1298] R. Lafaye, T. Plehn, M. Rauch, and D. Zerwas. Measuring supersymmetry, *Eur. Phys. J.*, C54:617–644, 2008, arXiv:0709.3985 [hep-ph].

[1299] R. Lafaye, T. Plehn, and D. Zerwas. SFITTER: SUSY parameter analysis at LHC and LC, 2004, hep-ph/0404282.

[1300] M. Laffranchi and A. Rubbia. The ArDM project: A liquid argon TPC for dark matter detection, 2007, hep-ph/0702080.

[1301] A. B. Lahanas and D. V. Nanopoulos. WMAPing out supersymmetric dark matter and phenomenology, *Phys. Lett.*, B568:55, 2003, hep-ph/0303130.

[1302] A. B. Lahanas, D. V. Nanopoulos, and V. C. Spanos. Neutralino dark matter elastic scattering in a flat and accelerating universe, *Mod. Phys. Lett.*, A16:1229, 2001, hep-ph/0009065.

[1303] A. B. Lahanas and V. C. Spanos. Implications of the pseudo-scalar Higgs boson in determining the neutralino dark matter, *Eur. Phys. J.*, C23:185, 2002, hep-ph/0106345.

[1304] M. Laine and Y. Schroder. Quark mass thresholds in QCD thermodynamics, *Phys. Rev.*, D73:085009, 2006, hep-ph/0603048.

[1305] M. Laine and M. Shaposhnikov. Sterile neutrino dark matter as a consequence of nuMSM-induced lepton asymmetry, *JCAP*, 0806:031, 2008, arXiv:0804.4543 [hep-ph].

[1306] P. Langacker and M.-x. Luo. Implications of precision electroweak experiments for m_t, ρ^0, $\sin^2 \theta_W$ and grand unification, *Phys. Rev.*, D44:817, 1991.

[1307] G. I. Langston, D. P. Schneider, S. Conner, *et al.* MG 1654+1346 – an Einstein Ring image of a quasar radio lobe, *Astron. J.*, **97**: 1283–1290, 1989.

[1308] M. Lattanzi and J. I. Silk. Can the WIMP annihilation boost factor be boosted by the Sommerfeld enhancement?, 2008, arXiv:0812.0360 [astro-ph].

[1309] T. R. Lauer, S. M. Faber, E. A. Ajhar, C. J. Grillmair, and P. A. Scowen. M32 +/- 1, *Astron. J.*, **116**:2263–2286, 1998.

[1310] T. R. Lauer, S. M. Faber, D. Richstone, *et al.* The masses of nuclear black holes in luminous elliptical galaxies and implications for the space density of the most massive black holes, *Astrophys. J.*, **662**:808–834, 2007.

[1311] J. Lavalle, Q. Yuan, D. Maurin, and X. J. Bi. Full calculation of clumpiness boost factors for antimatter cosmic rays in the light of ΛCDM *N*-body simulation results, 2007, arXiv:0709.3634 [astro-ph].

[1312] T. B. Lawson. Development of a WIMP dark matter detector with direction sensitivity, Ph.D. Thesis, University of Sheffield, UK, 2002.

[1313] G. Lazarides, C. Panagiotakopoulos, and Q. Shafi. Relaxing the cosmological bound on axions, *Phys. Lett.*, B**192**:323, 1987.

[1314] G. Lazarides, R. K. Schaefer, D. Seckel, and Q. Shafi. Dilution of cosmological axions by entropy production, *Nucl. Phys.*, B**346**:193–212, 1990.

[1315] D. M. Lazarus *et al.* A search for solar axions, *Phys. Rev. Lett.*, **69**:2333–2336, 1992.

[1316] V. N. Lebedenko *et al.* Result from the first science run of the ZEPLIN-III dark matter search experiment, 2008, arXiv:0812.1150 [astro-ph].

[1317] B. W. Lee and S. Weinberg. Cosmological lower bound on heavy-neutrino masses, *Phys. Rev. Lett.*, **39**:165, 1977.

[1318] H. S. Lee *et al.* Limits on WIMP-nucleon cross section with CsI(Tl) crystal detectors, *Phys. Rev. Lett.*, **99**:091301, 2007, arXiv:0704.0423 [astro-ph].

[1319] H.-S. Lee, K. T. Matchev, and S. Nasri. Revival of the thermal sneutrino dark matter, *Phys. Rev.*, D**76**:041302, 2007, hep-ph/0702223.

[1320] J. S. Lee *et al.* CPsuperH: A computational tool for Higgs phenomenology in the minimal supersymmetric standard model with explicit CP violation, *Comput. Phys. Commun.*, **156**:283–317, 2004, hep-ph/0307377.

[1321] T. D. Lee and C.-N. Yang. Question of parity conservation in weak interactions, *Phys. Rev.*, **104**:254–258, 1956.

[1322] M. J. Lehner, K. N. Buckland, and G. E. Masek. Electron diffusion in a low pressure methane detector for particle dark matter, *Astropart. Phys.*, **8**:43–50, 1997.

[1323] R. Lehnert and T. J. Weiler. Neutrino flavor ratios as diagnostic of solar WIMP annihilation, *Phys. Rev.*, D**77**:125004, 2008, arXiv:0708.1035 [hep-ph].

[1324] P. J. T. Leonard and S. Tremaine. The local Galactic escape speed, *Astrophys. J.*, **353**:486, 1990.

[1325] C. Lester and A. Barr. MTGEN: Mass scale measurements in pair-production at colliders, *JHEP*, **12**:102, 2007, arXiv:0708.1028 [hep-ph].

[1326] C. G. Lester. Constrained invariant mass distributions in cascade decays: The shape of the '$m(q_{ll})$-threshold' and similar distributions, *Phys. Lett.*, **B655**:39–44, 2007, hep-ph/0603171.

[1327] C. G. Lester and D. J. Summers. Measuring masses of semi-invisibly decaying particles pair produced at hadron colliders, *Phys. Lett.*, **B463**:99–103, 1999, hep-ph/9906349.

[1328] Yu. L. Levitan, I. M. Sobol, M. Yu. Khlopov, and V. M. Chechetkin. Production of light elements by cascades from energetic antiprotons in the early universe and problems of nuclear cosmoarchaeology, *Sov. J. Nucl. Phys.*, **47**:109–115, 1988.

[1329] S. A. Levshakov, M. Dessauges-Zavadsky, S. D'Odorico, and P. Molaro. Molecular hydrogen, deuterium and metal abundances in the damped Ly-alpha system at $z = 3.025$ toward QSO 0347-3819 s, 2001, astro-ph/0105529.

[1330] J. D. Lewin and P. F. Smith. Review of mathematics, numerical factors, and corrections for dark matter experiments based on elastic nuclear recoil, *Astropart. Phys.*, **6**:87–112, 1996.

[1331] H. Li, J. Liu, J.-Q. Xia, *et al.* Constraining cosmological parameters with observational data including weak lensing effects, 2008, arXiv:0812.1672.

[1332] P. K. Lightfoot, N. J. C. Spooner, T. B. Lawson, S. Aune, and I. Giomataris. First operation of bulk micromegas in low pressure negative ion drift gas mixtures for dark matter searches, *Astropart. Phys.*, **27**:490–499, 2007.

[1333] B. Lillie, L. Randall, and L.-T. Wang. The Bulk RS KK-gluon at the LHC, *JHEP*, **09**:074, 2007, hep-ph/0701166.

[1334] B. Lillie, J. Shu, and T. M. P. Tait. Kaluza–Klein gluons as a diagnostic of warped models, *Phys. Rev.*, **D76**:115016, 2007, arXiv:0706.3960 [hep-ph].

[1335] M. Limousin, J. Richard, J.-P. Kneib, *et al.* Strong lensing in Abell 1703: constraints on the slope of the inner dark matter distribution, *Astron. Astrophys.*, **489**:23–35, 2008, 0802.4292.

[1336] C. Lin. A search for universal extra dimensions in the multi- lepton channel from proton antiproton collisions at $\sqrt{s} = 1.8$-TeV, Fermilab-thesis-2005-69.

[1337] S. T. Lin *et al.* New limits on spin-independent couplings of low-mass WIMP dark matter with a germanium detector at a threshold of 200 eV, 2007, arXiv:0712.1645 [hep-ex].

[1338] S.-T. Lin, H. T. Wong, and for the TEXONO Collaboration. Limits on low-mass WIMP dark matter with an ultra-low-energy germanium detector at 220 eV threshold, 2008, arXiv:0810.3504 [astro-ph].

[1339] W. B. Lin, D. H. Huang, X. Zhang, and R. H. Brandenberger. Nonthermal production of WIMPs and the sub-galactic structure of the universe, *Phys. Rev. Lett.*, **86**:954, 2001, astro-ph/0009003.

[1340] A. D. Linde. A new inflationary universe scenario: A possible solution of the horizon, flatness, homogeneity, isotropy and primordial monopole problems, *Phys. Lett.*, **B108**:389–393, 1982.

[1341] A. D. Linde. Generation of isothermal density perturbations in the inflationary universe, *Phys. Lett.*, **B158**:375–380, 1985.

[1342] A. D. Linde. Axions in inflationary cosmology, *Phys. Lett.*, **B259**: 38–47, 1991.

[1343] A. D. Linde and D. H. Lyth. Axionic domain wall production during inflation, *Phys. Lett.*, **B246**:353–358, 1990.

[1344] J. Lindhard *et al. K. Dan. Viderask. Selsk., Math. Fys. Medd.*, **10**:10, 1968.

[1345] F.-S. Ling, P. Sikivie, and S. Wick. Diurnal and annual modulation of cold dark matter signals, *Phys. Rev.*, **D70**:123503, 2004, astro-ph/0405231.

[1346] W. H. Lippincott *et al.* Scintillation time dependence and pulse shape discrimination in liquid argon, *Phys. Rev.*, **C78**:035801, 2008, arXiv:0801.1531 [nucl-ex].

[1347] M. Lisanti and J. G. Wacker. Unification and dark matter in a minimal scalar extension of the standard model, 2007, arXiv:0704.2816 [hep-ph].

[1348] A. Loeb and M. Zaldarriaga. The small-scale power spectrum of cold dark matter, *Phys. Rev.*, **D71**:103520, 2005.

[1349] I. Lopes and J. Silk. Solar neutrinos: Probing the quasi-isothermal solar core produced by SUSY dark matter particles, *Phys. Rev. Lett.*, **88**:151303, 2002, astro-ph/0112390.

[1350] I. P. Lopes, G. Bertone, and J. Silk. Solar seismic model as a new constraint on supersymmetric dark matter, *Mon. Not. Roy. Astron. Soc.*, **337**:1179–1184, 2002, astro-ph/0205066.

[1351] I. P. Lopes, J. Silk, and S. H. Hansen. Helioseismology as a new constraint on SUSY dark matter, *Mon. Not. Roy. Astron. Soc.*, **331**:361, 2002, astro-ph/0111530.

[1352] L. Lopez Honorez, E. Nezri, J. F. Oliver, and M. H. G. Tytgat. The inert doublet model: An archetype for dark matter, *JCAP*, **0702**:028, 2007, hep-ph/0612275.

[1353] I. Low. T parity and the littlest Higgs, *JHEP*, **10**:067, 2004, hep-ph/0409025.

[1354] J. Lundberg and J. Edsjö. WIMP diffusion in the solar system including solar depletion and its effect on earth capture rates, *Phys. Rev.*, D**69**:123505, 2004, astro-ph/0401113.

[1355] E. Lundstrom, M. Gustafsson, and J. Edsjö. The inert doublet model and LEP II limits, 2008, arXiv:0810.3924 [hep-ph].

[1356] R. Lynds and V. Petrosian. Giant luminous arcs in galaxy clusters, *Bull. Am. Astron. Soc.*, **18**:1014, 1986.

[1357] D. H. Lyth. A limit on the inflationary energy density from axion isocurvature fluctuations, *Phys. Lett.*, B**236**:408, 1990.

[1358] D. H. Lyth. Axions and inflation: Sitting in the vacuum, *Phys. Rev.*, D**45**:3394–3404, 1992.

[1359] D. H. Lyth and E. D. Stewart. Constraining the inflationary energy scale from axion cosmology, *Phys. Lett.*, B**283**:189–193, 1992.

[1360] D. H. Lyth and E. D. Stewart. Thermal inflation and the moduli problem, *Phys. Rev.*, D**53**:1784–1798, 1996, hep ph/9510204.

[1361] E. Ma. Common origin of neutrino mass, dark matter, and baryogenesis, *Mod. Phys. Lett.*, A**21**:1777–1782, 2006, hep-ph/0605180.

[1362] E. Ma. Verifiable radiative seesaw mechanism of neutrino mass and dark matter, *Phys. Rev.*, D**73**:077301, 2006, hep-ph/0601225.

[1363] A. V. Macciò, A. A. Dutton, and F. C. van den Bosch. Concentration, spin and shape of dark matter haloes as a function of the cosmological model: WMAP1, WMAP3 and WMAP5 results, 2008, arXiv:0805.1926 [astro-ph].

[1364] A. V. Macciò *et al.* Concentration, spin and shape of dark matter haloes: Scatter and the dependence on mass and environment, *Mon. Not. Roy. Astron. Soc.*, **378**:55–71, 2007, astro-ph/0608157.

[1365] A. V. Macciò, X. Kang, and B. Moore. Central mass and luminosity of Milky Way satellites in the LCDM model, 2008, arXiv:0810.1734 [astro-ph].

[1366] A. V. Macciò, B. Moore, J. Stadel, and J. Diemand. Radial distribution and strong lensing statistics of satellite galaxies and substructure using high-resolution ΛCDM hydrodynamical simulations, *Mon. Not. Roy. Astron. Soc.*, **366**:1529–1538, 2006.

[1367] C. Macesanu, C. D. McMullen, and S. Nandi. Collider implications of universal extra dimensions, *Phys. Rev.*, D**66**:015009, 2002, hep-ph/0201300.

[1368] G. D. Mack, J. F. Beacom, and G. Bertone. Towards closing the window on strongly interacting dark matter: Far-reaching constraints from Earth's heat flow, *Phys. Rev.*, D**76**:043523, 2007, arXiv:0705.4298 [astro-ph].

[1369] P. Madau, J. Diemand, and M. Kuhlen. Dark matter subhalos and the dwarf satellites of the Milky Way, *Astrophys. J.*, **679**:1260, 2008, arXiv:0802.2265 [astro-ph].

[1370] T. Maeno *et al.* Successive measurements of cosmic-ray antiproton spectrum in a positive phase of the solar cycle, *Astropart. Phys.*, **16**:121–128, 2001.

[1371] R. Mahbubani and L. Senatore. The minimal model for dark matter and unification, *Phys. Rev.*, D**73**:043510, 2006, hep-ph/0510064.

[1372] A. Mahdavi, H. Hoekstra, A. Babul, D. D. Balam, and P. L. Capak. A dark core in Abell 520, *Astrophys. J.*, **668**:806–814, 2007, 0706.3048.

[1373] F. Mahmoudi. SuperIso: A program for calculating the isospin asymmetry of $B \to K^\star\gamma$ in the MSSM, *Comput. Phys. Commun.*, **178**:745–754, 2008, arXiv:0710.2067 [hep-ph].

[1374] L. Maiani. All you need to know about the Higgs boson. In *Proceedings, Gif-sur-Yvette Summer School on Particle Physics*, 1979, 1–52.

[1375] S. R. Majewski, M. F. Skrutskie, M. D. Weinberg, and J. C. Ostheimer. A two micron all sky survey view of the Sagittarius dwarf galaxy. I. Morphology of the Sagittarius core and tidal arms, *Astrophys. J.*, **599**:1082–1115, 2003.

[1376] R. A. Malaney and G. J. Mathews. Probing the early universe: A review of primordial nucleosynthesis beyond the standard Big Bang, *Phys. Rep.*, **229**:145–219, 1993.

[1377] J. M. Maldacena. The large N limit of superconformal field theories and supergravity, *Adv. Theor. Math. Phys.*, **2**:231–252, 1998, hep-th/9711200.

[1378] Y. Mambrini, C. Muñoz, E. Nezri, and F. Prada. Adiabatic compression and indirect detection of supersymmetric dark matter, *J. Cosmol. Astropart. Phys.*, **1**:10, 2006.

[1379] R. Mandelbaum, C. M. Hirata, T. Broderick, U. Seljak, and J. Brinkmann. Ellipticity of dark matter haloes with galaxy–galaxy weak lensing, *Mon. Not. Roy. Astron. Soc.*, **370**:1008–1024, 2006.

[1380] R. Mandelbaum, U. Seljak, R. J. Cool, *et al.* Density profiles of galaxy groups and clusters from SDSS galaxy-galaxy weak lensing, *Mon. Not. Roy. Astron. Soc.*, **372**:758–776, 2006.

[1381] R. Mandelbaum, U. Seljak, and C. M. Hirata. A halo mass–concentration relation from weak lensing, *J. Cosmol. Astropart. Phys.*, **8**:6, 2008, arXiv:0805.2552.

[1382] R. Mandelbaum, G. van de Ven, and C. R. Keeton. Galaxy density profiles and shapes – II. Selection biases in strong lensing surveys, 2008, arXiv:0808.2497.

[1383] M. L. Mangano. Standard Model backgrounds to supersymmetry searches, 2008, arXiv:0809.1567 [hep-ph].

[1384] R. Maoli, L. Van Waerbeke, Y. Mellier *et al.* Cosmic shear analysis in 50 uncorrelated VLT fields. Implications for $Omega_0$, $Sigma_8$, *Astron. Astrophys.*, **368**:766–775, 2001.

[1385] M. Mapelli, A. Ferrara, and E. Pierpaoli. Impact of dark matter decays and annihilations on reionization, *Mon. Not. Roy. Astron. Soc.*, **369**:1719–1724, 2006, astro-ph/0603237.

[1386] J. March-Russell, S. M. West, D. Cumberbatch, and D. Hooper. Heavy dark matter through the Higgs portal, *JHEP*, **07**:058, 2008, arXiv:0801.3440 [hep-ph].

[1387] J. D. March-Russell and S. M. West. WIMPonium and boost factors for indirect dark matter detection, 2008, arXiv:0812.0559 [astro-ph].

[1388] D. Marchesini *et al.* The evolution of the stellar mass function of galaxies from $z = 4.0$ and the first comprehensive analysis of its uncertainties: Evidence for mass-dependent evolution, 2008, arXiv:0811.1773 [astro-ph].

[1389] M. Markevitch, A. H. Gonzalez, L. David, *et al.* A textbook example of a bow shock in the merging galaxy cluster 1E 0657-56, *Astrophys. J.*, **567**:L27–L31, 2002.

[1390] N. F. Martin, J. T. A. de Jong, and H.-W. Rix. A comprehensive maximum likelihood analysis of the structural properties of faint Milky Way satellites, *Astrophys. J.*, **684**:1075–1092, 2008, 0805.2945.

[1391] S. P. Martin. A supersymmetry primer, 1997, hep-ph/9709356.

[1392] C. J. Martoff *et al.* Prototype direction-sensitive solid state detector for dark matter, *Phys. Rev. Lett.*, **76**:4882–4885, 1996.

[1393] S. Mashchenko, H. M. P. Couchman, and J. Wadsley. Cosmological puzzle resolved by stellar feedback in high redshift galaxies, *Nature*, **442**:539, 2006, astro-ph/0605672.

[1394] S. Mashchenko, A. Sills, and H. M. Couchman. Constraining global properties of the Draco dwarf spheroidal galaxy, *Astrophys. J.*, **640**:252–269, 2006.

[1395] S. Mashchenko, J. Wadsley, and H. M. P. Couchman. Stellar feedback in dwarf galaxy formation, *Science*, **319**:174, 2008, arXiv:0711.4803.

[1396] R. Massey, J. Rhodes, A. Leauthaud, *et al.* COSMOS: Three-dimensional weak lensing and the growth of structure, *Astrophys. J. Suppl.*, **172**:239–253, 2007.

[1397] M. Mateo, E. W. Olszewski, C. Pryor, D. L. Welch, and P. Fischer. The Carina dwarf spheroidal galaxy – How dark is it?, *Astron. J*, **105**:510–526, 1993.

[1398] M. L. Mateo. Dwarf galaxies of the Local Group, *Ann. Rev. Astron. Astrophys.*, **36**:435–506, 1998, astro-ph/9810070.

[1399] S. Matsumoto, T. Moroi, and K. Tobe. Testing the Littlest Higgs model with T-parity at the Large Hadron Collider, *Phys. Rev.*, D**78**:055018, 2008, arXiv:0806.3837 [hep-ph].

[1400] S. Matsumoto and M. Senami. Efficient coannihilation process through strong Higgs self-coupling in LKP dark matter annihilation, *Phys. Lett.*, B**633**:671–674, 2006, hep-ph/0512003.

[1401] D. Maurin, F. Donato, R. Taillet, and P. Salati. Cosmic rays below $Z = 30$ in a diffusion model: New constraints on propagation parameters, *Astrophys. J.*, **555**:585–596, 2001, astro-ph/0101231.

[1402] D. Maurin *et al.* Galactic cosmic ray nuclei as a tool for astroparticle physics, 2002, astro-ph/0212111.

[1403] D. Maurin, R. Taillet, and F. Donato. New results on source and diffusion spectral features of Galactic cosmic rays: I- B/C ratio, *Astron. Astrophys.*, **394**:1039–1056, 2002, astro-ph/0206286.

[1404] L. Mayer, F. Governato, and T. Kaufmann. The formation of disk galaxies in computer simulations, 2008, arXiv:0801.3845 [astro-ph].

[1405] L. Mayer and B. Moore. The baryonic mass–velocity relation: clues to feedback processes during structure formation and the cosmic baryon inventory, 2003, astro-ph/0309500.

[1406] H. A. Mayer-Hasselwander, D. L. Bertsch, B. L. Dingus, *et al.* High-energy gamma-ray emission from the Galactic Center, *Astron. Astrophys.*, **335**:161–172, 1998.

[1407] J. McDonald. WIMP densities in decaying particle dominated cosmology, *Phys. Rev.*, D**43**:1063–1068, 1991.

[1408] J. McDonald. Gauge singlet scalars as cold dark matter, *Phys. Rev.*, D**50**:3637–3649, 1994, hep-ph/0702143.

[1409] S. McGaugh. Some systematic properties of rotation curves. In G. A. Mamon, F. Combes, C. Deffayet, and B. Fort, eds., *Mass Profiles and Shapes of Cosmological Structures*, volume 20 of *EAS Publications Series*, 69–76, 2006.

[1410] S. S. McGaugh. The baryonic Tully–Fisher relation of galaxies with extended rotation curves and the stellar mass of rotating galaxies, *Astrophys. J.*, **632**:859–871, 2005, astro-ph/0506750.

[1411] S. S. McGaugh and W. J. G. de Blok. Testing the hypothesis of modified dynamics with low surface brightness galaxies and other evidence, *Astrophys. J.*, **499**:66–81, 1998, astro-ph/9801102.

[1412] D. N. McKinsey. CLEAN: A self-shielding detector for characterizing the low energy solar neutrino spectrum, 2000. Prepared for International Workshop on Low Energy Solar Neutrinos, Tokyo, Japan, 4–5 December 2000.

[1413] D. N. McKinsey. The Mini-CLEAN experiment, *Nucl. Phys. Proc. Suppl.*, **173**:152–155, 2007.

[1414] D. N. McKinsey and K. J. Coakley. Neutrino detection with CLEAN, *Astropart. Phys.*, **22**:355–368, 2005, astro-ph/0402007.

[1415] P. Meade and M. Reece. Top partners at the LHC: Spin and mass measurement, *Phys. Rev.*, D**74**:015010, 2006, hep-ph/0601124.

[1416] F. Melia, M. Wardle, P. Heimberg, M. Wardle, and P. Heimberg. On the stability of neutron star winds, 1992, STEWARD-1035.

[1417] Y. Mellier. Probing the universe with weak lensing, *Ann. Rev. Astron. Astrophys.*, **37**:127–189, 1999.

[1418] Y. Mellier, B. Fort, and J.-P. Kneib. The dark matter distribution in MS 2137-23 from the modeling of the multiple arc systems, *Astrophys. J.*, **407**:33–45, 1993.

[1419] M. Meneghetti, N. Yoshida, M. Bartelmann, *et al.* Giant cluster arcs as a constraint on the scattering cross-section of dark matter, *Mon. Not. Roy. Astron. Soc.*, **325**:435–442, 2001.

[1420] D Merritt. Spherical stellar systems with spheroidal velocity distributions, *Mon. Not. Roy. Astron. Soc.*, **90**:1027, 1984.

[1421] D. Merritt. Dynamical mapping of hot stellar systems, *Astrophys. J.*, **413**:79–94, 1993.

[1422] D. Merritt. Evolution of the dark matter distribution at the galactic center, *Phys. Rev. Lett.*, **92**:201304, 2004.

[1423] D. Merritt. Mass deficits, stalling radii, and the merger histories of elliptical galaxies, *Astrophys. J.*, **648**:976–986, 2006.

[1424] D. Merritt. Evolution of nuclear star clusters, 2008, arXiv:0802.3186.

[1425] D. Merritt and L. Ferrarese. Relationship of black holes to bulges. In *ASP Conf. Ser. 249: The Central Kiloparsec of Starbursts and AGN: The La Palma Connection*, 335, 2001.

[1426] D. Merritt, A. W. Graham, B. Moore, J. Diemand, and B. Terzić. Empirical models for dark matter halos. I. Nonparametric

700   *References*

construction of density profiles and comparison with parametric models, *Astron. J.*, **132**:2685–2700, 2006.

[1427] D. Merritt, S. Harfst, and G. Bertone. Collisionally regenerated dark matter structures in galactic nuclei, *Phys. Rev. D*, **75**:043517, 2007.

[1428] D. Merritt, S. Mikkola, and A. Szell. Long-term evolution of massive black hole binaries. III. Binary evolution in collisional nuclei, *Astrophys. J.*, **671**:53–72, 2007, arXiv:0705.2745.

[1429] D. Merritt, M. Milosavljević, M. Favata, S. A. Hughes, and D. E. Holz. Consequences of gravitational radiation recoil, *Astrophys. J. Lett.*, **607**:L9–L12, 2004.

[1430] D. Merritt, M. Milosavljević, L. Verde, and R. Jimenez. Dark matter spikes and annihilation radiation from the galactic center, *Phys. Rev. Lett.*, **88**:191301, 2002, astro-ph/0201376.

[1431] D. Merritt, J. F. Navarro, A. Ludlow, and A. Jenkins. A universal density profile for dark and luminous matter?, *Astrophys. J. Lett.*, **624**:L85–L88, 2005.

[1432] D. Merritt and A. Szell. Dynamical cusp regeneration, *Astrophys. J.*, **648**:890–899, 2006.

[1433] A. Meza, J. F. Navarro, M. G. Abadi, and M. Steinmetz. Accretion relicts in the Galactic disk: omegaCen dwarf debris in the solar neighbourhood, *Mon. Not. Roy. Astron. Soc.*, **359**:93–103, 2005, astro-ph/0408567.

[1434] M. Milgrom. A modification of the Newtonian dynamics as a possible alternative to the hidden mass hypothesis, *Astrophys. J.*, **270**:365–370, 1983.

[1435] M. Milgrom. A modification of the Newtonian dynamics: Implications for galaxies, *Astrophys. J.*, **270**:371–383, 1983.

[1436] M. Milgrom. A modification of the Newtonian dynamics: implications for galaxy systems, *Astrophys. J.*, **270**:384–389, 1983.

[1437] M. Milgrom. Solutions for the modified Newtonian dynamics field equation, *Astrophys. J.*, **302**:617–625, 1986.

[1438] M. Milgrom. On stability of galactic disks in the modified dynamics and the distribution of their mean surface brightness, *Astrophys. J.*, **338**:121–127, 1989.

[1439] M. Milgrom. Dynamics with a nonstandard inertia-acceleration relation: an alternative to dark matter in galactic systems, *Ann. Phys.*, **229**:384–415, 1994.

[1440] M. Milgrom. Modified dynamics predictions agree with observations of the H kinematics in faint dwarf galaxies contrary to the conclusions of Lo, Sargent, and Young, *Astrophys. J.*, **429**:540–544, 1994.

[1441] M. Milgrom. Do modified Newtonian dynamics follow from the cold dark matter paradigm?, *Astrophys. J.*, **571**:L81–L83, 2002.

[1442] M. Milgrom. MOND as modified inertia. In G. A. Mamon, F. Combes, C. Deffayet, and B. Fort, eds., *Mass Profiles and Shapes of Cosmological Structures*, volume 20 of *EAS Publications Series*, 217–224, 2006.

[1443] M. Milgrom. Marriage à-la-MOND: Baryonic dark matter in galaxy clusters and the cooling flow puzzle, *New Astron. Rev.*, **51**:906–915, 2008, arXiv:0712.4203.

[1444] M. Milgrom. The MOND paradigm, 2008, arXiv:0801.3133.

[1445] M. Milgrom and R. H. Sanders. Rings and shells of "Dark Matter" as MOND Artifacts, *Astrophys. J.*, **678**:131–143, 2008, arXiv:0709.2561.

[1446] D. J. Miller, P. Osland, and A. R. Raklev. Invariant mass distributions in cascade decays, *JHEP*, **03**:034, 2006, hep-ph/0510356.

[1447] J. P. Miller, E. de Rafael, and B. L. Roberts. Muon g-2: Review of theory and experiment, *Rep. Prog. Phys.*, **70**:795, 2007.

[1448] M. C. Miller and E. J. M. Colbert. Intermediate-mass black holes, *Int. J. Mod. Phys. D*, **13**:1–64, 2004.

[1449] M. Milosavljević and A. Loeb. The link between warm molecular disks in maser nuclei and star formation near the black hole at the Galactic Center, *Astrophys. J.*, **604**:L45, 2004, astro-ph/0401221.

[1450] P. Minkowski. mu → e gamma at a rate of one out of 1-billion muon decays?, *Phys. Lett.*, B**67**:421, 1977.

[1451] M. Minowa *et al.* The Tokyo axion helioscope experiment, *Nucl. Phys. Proc. Suppl.*, **72**:171–175, 1999, hep-ex/9806015.

[1452] J. Miralda-Escudé. Gravitational lensing by a cluster of galaxies and the central cD galaxy: Measuring the mass profile, *Astrophys. J.*, **438**:514–526, 1995.

[1453] J. Miralda-Escudé. A test of the collisional dark matter hypothesis from cluster lensing, *Astrophys. J.*, **564**:60–64, 2002.

[1454] J. W. Mitchell *et al.* Measurement of 0.25-3.2 GeV antiprotons in the cosmic radiation, *Phys. Rev. Lett.*, **76**:3057–3060, 1996.

[1455] S. Mitra. Has DAMA detected self-interacting dark matter?, *Phys. Rev.*, D**71**:121302, 2005, astro-ph/0409121.

[1456] K. Miuchi *et al.* Performance and applications of a mu-TPC, *Nucl. Instrum. Meth.*, A**535**:236–241, 2004.

[1457] K. Miuchi *et al.* Direction-sensitive dark matter search results in a surface laboratory, *Phys. Lett.*, B**654**:58–64, 2007, arXiv:0708.2579 [astro-ph].

[1458] K. Miuchi *et al.* Performance of a time-projection-chamber with a large-area micro-pixel-chamber readout, *Nucl. Instrum. Meth.*, A**576**:43–46, 2007, physics/0701085.

[1459] M. Miyajima, T. Takahashi, S. Konno *et al.* Average energy expended per ion pair in liquid argon, *Phys. Rev. A*, **9**:1438, 1974.

[1460] J. Miyamoto *et al.* GEM operation in negative ion drift gas mixtures, *Nucl. Instrum. Meth.*, A**526**:409–412, 2004.

[1461] R. N. Mohapatra and G. Senjanovic. Neutrino mass and spontaneous parity nonconservation, *Phys. Rev. Lett.*, **44**:912, 1980.

[1462] A. Moiseev *et al.* Cosmic-ray antiproton flux in the energy range from 200 to 600 MeV, *Astrophys. J.*, **474**:479, 1997.

[1463] A. A. Moiseev. Gamma-ray Large Area Space Telescope: Mission overview, *Nucl. Instrum. Meth. Phys. Res. A*, **588**:41–47, 2008.

[1464] K. Molikawa, M. Hattori, J.-P. Kneib, and K. Yamashita. The giant luminous arc statistics. II. Spherical lens models based on ROSAT HRI data, *Astron. Astrophys.*, **351**:413–432, 1999.

[1465] J. Monroe and P. Fisher. Neutrino backgrounds to dark matter searches, *Phys. Rev.*, D**76**:033007, 2007, arXiv:0706.3019 [astro-ph].

[1466] B. Moore, J. Diemand, P. Madau, M. Zemp, and J. Stadel. Globular clusters, satellite galaxies and stellar haloes from early dark matter peaks, *Mon. Not. Roy. Astron. Soc.*, **368**:563–570, 2006, astro-ph/0510370.

[1467] B. Moore, J. Diemand, and J. Stadel. On the age–radius relation and orbital history of cluster galaxies, 2004, astro-ph/0406615.

[1468] B. Moore *et al.* Dark matter substructure in galactic halos, 1999, astro-ph/9907411.

[1469] B. Moore *et al.* Dark matter substructure within galactic halos, *Astrophys. J.*, **524**:L19–L22, 1999.

[1470] B. Moore *et al.* Dark matter in Draco and the Local Group: Implications for direct detection experiments, *Phys. Rev.*, D**64**:063508, 2001, astro-ph/0106271.

[1471] B. Moore, F. Governato, T. R. Quinn, J. Stadel, and G. Lake. Resolving the structure of cold dark matter halos, *Astrophys. J. Lett.*, **499**:L5, 1998, astro-ph/9709051.

[1472] B. Moore, N. Katz, and G. Lake. On the destruction and overmerging of dark halos in dissipationless N body simulations, 1995, astro-ph/9503088.

[1473] B. Moore, S. Kazantzidis, J. Diemand, and J. Stadel. The origin and tidal evolution of cuspy triaxial haloes, 2003, astro-ph/0310660.

[1474] B. Moore, T. R. Quinn, F. Governato, J. Stadel, and G. Lake. Cold collapse and the core catastrophe, *Mon. Not. Roy. Astron. Soc.*, **310**:1147–1152, 1999, astro-ph/9903164.

[1475] A. Morales *et al.* Improved constraints on WIMPs from the International Germanium Experiment IGEX, *Phys. Lett.*, B**532**:8–14, 2002, hep-ex/0110061.

[1476] B. Morgan and A. M. Green. Directional statistics for WIMP direct detection. II: 2-d read-out, *Phys. Rev.*, D**72**:123501, 2005, astro-ph/0508134.

[1477] B. Morgan, A. M. Green, and N. J. C. Spooner. Directional statistics for WIMP direct detection, *Phys. Rev.*, D**71**:103507, 2005, astro-ph/0408047.

[1478] K. Mori, C. J. Hailey, E. A. Baltz, *et al.* A novel antimatter detector based on X-ray deexcitation of exotic atoms, *Astrophys. J.*, **566**:604–616, 2002.

[1479] T. Moroi, H. Murayama, and M. Yamaguchi. Cosmological constraints on the light stable gravitino, *Phys. Lett.*, B**303**:289–294, 1993.

[1480] T. Moroi and L. Randall. Wino cold dark matter from anomaly-mediated SUSY breaking, *Nucl. Phys.*, B**570**:455–472, 2000, hep-ph/9906527.

[1481] T. Moroi, M. Yamaguchi, and T. Yanagida. On the solution to the Polonyi problem with 0 (10-TeV) gravitino mass in supergravity, *Phys. Lett.*, B**342**:105–110, 1995, hep-ph/9409367.

[1482] D. J. Mortlock and E. L. Turner. Gravitational lensing in modified Newtonian dynamics, *Mon. Not. Roy. Astron. Soc.*, **327**:557–566, 2001.

[1483] I. V. Moskalenko and A. W. Strong. Production and propagation of cosmic-ray positrons and electrons, *Astrophys. J.*, **493**:694, 1998.

[1484] I. V. Moskalenko and L. L. Wai. Dark matter burners, *Astrophys. J.*, **659**:L29–L32, 2007, astro-ph/0702654.

[1485] M. Muhlleitner, A. Djouadi, and Y. Mambrini. SDECAY: A Fortran code for the decays of the supersymmetric particles in the MSSM, *Comput. Phys. Commun.*, **168**:46–70, 2005, hep-ph/0311167.

[1486] V. F. Mukhanov. Nucleosynthesis without a computer, *Int. J. Theor. Phys.*, **43**:669–693, 2004, astro-ph/0303073.

[1487] V. F. Mukhanov and G. V. Chibisov. Quantum fluctuation and non-singular universe. (In Russian), *JETP Lett.*, **33**:532–535, 1981.

[1488] D. Muna *et al.* Low energy electron and nuclear recoil thresholds in the DRIFT-II negative ion TPC for dark matter searches, *Nucl. Instrum. Meth.*, in press.

[1489] C. Munoz. Dark matter detection in the light of recent experimental results, *Int. J. Mod. Phys.*, A19:3093, 2004, hep-ph/0309346.

[1490] D. Munshi, P. Valageas, L. van Waerbeke, and A. Heavens. Cosmology with weak lensing surveys, *Phys. Rep.*, **462**:67–121, 2008.

[1491] T. Naka *et al.* Development of emulsion track expansion techniques for optical-microscopy-observation of low-velocity ion tracks with ranges beyond optical resolution limit, *Nucl. Instrum. Meth.*, A**581**:761–764, 2007.

[1492] T. Nakajima and M. Morikawa. An interpretation of flat density cores of clusters of galaxies by degeneracy pressure of fermionic dark matter: A case study of A1689, *Astrophys. J.*, **655**:135–143, 2007.

[1493] D. V. Nanopoulos, K. A. Olive, and M. Srednicki. After primordial inflation, *Phys. Lett.*, B**127**:30, 1983.

[1494] C. R. Nappi and B. A. Ovrut. Supersymmetric extension of the $SU(3) \times SU(2) \times U(1)$ model, *Phys. Lett.*, B**113**:175, 1982.

[1495] B. S. Nara Singh, M. Hass, Y. Nir-El, and G. Haquin. A new precision measurement of the $^3He(^4He,gamma)^7Be$ cross section, *Phys. Rev. Lett.*, **93**:262503, 2004, nucl-ex/0407017.

[1496] E. Nardi, F. Sannino, and A. Strumia. Decaying dark matter can explain the electron/positron excesses, 2008, arXiv:0811.4153 [hep-ph].

[1497] A. Natarajan, J. C. Tan, and B. W. O'Shea. Dark matter annihilation and primordial star formation, 2008, arXiv:0807.3769 [astro-ph].

[1498] P. Natarajan, A. Loeb, J.-P. Kneib, and I. Smail. Constraints on the collisional nature of the dark matter from gravitational lensing in the cluster A2218, *Astrophys. J.*, **580**:L17–L20, 2002.

[1499] P. Natarajan and H. Zhao. MOND plus classical neutrinos are not enough for cluster lensing, *Mon. Not. Roy. Astron. Soc.*, **389**:250–256, 2008, 0806.3080.

[1500] P. Nath and R. L. Arnowitt. Non-universal soft SUSY breaking and dark matter, *Phys. Rev.*, D**56**:2820–2832, 1997, hep-ph/9701301.

[1501] J. F. Navarro, C. S. Frenk, and S. D. M. White, *Astrophys. J.*, **462**:563–575, 1996, astro-ph/9508025.

[1502] J. F. Navarro, C. S. Frenk, and S. D. M. White. A universal density profile from hierarchical clustering, *Astrophys. J.*, **490**:493–508, 1997, astro-ph/9611107.

[1503] J. F. Navarro, E. Hayashi, C. Power, *et al.* The inner structure of LambdaCDM halos III: Universality and asymptotic slopes, *Mon. Not. Roy. Astron. Soc.*, **349**:1039–1051, 2004, astro-ph/0311231.

[1504] J. F. Navarro, A. Ludlow, V. Springel, *et al.* The diversity and similarity of cold dark matter halos, 2008, arXiv:0810.1522 [astro-ph].

[1505] H. J. Newberg *et al.* Sagittarius tidal debris 90 kpc from the Galactic Center, *Astrophys. J.*, **596**:L191–L194, 2003, astro-ph/0309162.

[1506] H. J. Newberg, B. Yanny, N. Cole, *et al.* The overdensity in Virgo, Sagittarius debris, and the asymmetric spheroid, *Astrophys. J*, **668**:221–235, 2007, 0706.3391.

[1507] E. Nezri, M. H. G. Tytgat, and G. Vertongen. Positrons and antiprotons from inert doublet model dark matter, 2009, arXiv:0901.2556 [hep-ph].

[1508] F. Nicastro, S. Mathur, and M. Elvis. Missing baryons and the warm-hot intergalactic medium, 2007, arXiv:0712.2375 [astro-ph].

[1509] H. P. Nilles. Supersymmetry, supergravity and particle physics, *Phys. Rep.*, **110**:1, 1984.

[1510] J. Ninkovic *et al.* New technique for the measurement of the scintillation efficiency of nuclear recoils, *Nucl. Instrum. Meth.*, **A564**:567–578, 2006, astro-ph/0604094.

[1511] C. Nipoti, L. Ciotti, J. Binney, and P. Londrillo. Dynamical friction in modified Newtonian dynamics, *Mon. Not. Roy. Astron. Soc.*, **386**:2194–2198, 2008, 0802.1122.

[1512] C. Nipoti, P. Londrillo, and L. Ciotti. Dissipationless collapses in modified Newtonian dynamics, *Astrophys. J.*, **660**:256–266, 2007.

[1513] C. Nipoti, P. Londrillo, and L. Ciotti. Galaxy merging in modified Newtonian dynamics, *Mon. Not. Roy. Astron. Soc.*, **381**:L104–L108, 2007, 0705.4633.

[1514] A. Nisati, S. Petrarca, and G. Salvini. On the possible detection of massive stable exotic particles at the LHC, *Mod. Phys. Lett.*, **A12**:2213–2222, 1997, hep-ph/9707376.

[1515] M. M. Nojiri, G. Polesello, and D. R. Tovey. Proposal for a new reconstruction technique for SUSY processes at the LHC, 2003, hep-ph/0312317.

[1516] M. M. Nojiri, G. Polesello, and D. R. Tovey. Constraining dark matter in the MSSM at the LHC, *JHEP*, **03**:063, 2006, hep-ph/0512204.

[1517] M. M. Nojiri, G. Polesello, and D. R. Tovey. A hybrid method for determining SUSY particle masses at the LHC with fully identified cascade decays, *JHEP*, **05**:014, 2008, arXiv:0712.2718 [hep-ph].

[1518] M. M. Nojiri, K. Sakurai, Y. Shimizu, and M. Takeuchi. Handling jets + missing E_T channel using inclusive mT2, *JHEP*, **10**:100, 2008, arXiv:0808.1094 [hep-ph].

[1519] M. M. Nojiri, Y. Shimizu, S. Okada, and K. Kawagoe. Inclusive transverse mass analysis for squark and gluino mass determination, *JHEP*, **06**:035, 2008, arXiv:0802.2412 [hep-ph].

[1520] Y. Nomura. Supersymmetric unification in warped space, 2004, hep-ph/0410348.

[1521] S. Nussinov. Technocosmology: Could a technibaryon excess provide a 'natural' missing mass candidate?, *Phys. Lett.*, **B165**:55, 1985.

[1522] Donal O'Connell, M. J. Ramsey-Musolf, and M. B. Wise. Minimal extension of the standard model scalar sector, *Phys. Rev.*, **D75**:037701, 2007, hep-ph/0611014.

[1523] M. Oguri and R. D. Blandford. What is the largest Einstein radius in the universe?, *Mon. Not. Roy. Astron. Soc.*, **392**:930–944, 2009, 0808.0192.

[1524] H. Ohki *et al.* Nucleon sigma term and strange quark content from lattice QCD with exact chiral symmetry, *Phys. Rev.*, **D78**:054502, 2008, arXiv:0806.4744 [hep-lat].

[1525] T. Ohnuki, D. P. Snowden-Ifft, and C. J. Martoff. Measurement of carbon disulfide anion diffusion in a TPC, *Nucl. Instrum. Meth.*, **A463**:142–148, 2001, physics/0004006.

[1526] V. K. Oikonomou, J. D. Vergados, and Ch. C. Moustakidis. Direct detection of dark matter-rates for various WIMPs, *Nucl. Phys.*, **B773**:19–42, 2007, hep-ph/0612293.

[1527] N. Okada and O. Seto. Gravitino dark matter from increased thermal relic particles, *Phys. Rev.*, **D77**:123505, 2008, arXiv: 0710.0449 [hep-ph].

[1528] Y. Okada, M. Yamaguchi, and T. Yanagida. Renormalization group analysis on the Higgs mass in the softly broken supersymmetric standard model, *Phys. Lett.*, **B262**:54, 1991.

[1529] Y. Okada, Ma. Yamaguchi, and T. Yanagida. Upper bound of the lightest Higgs boson mass in the minimal supersymmetric standard model, *Prog. Theor. Phys.*, **85**:1, 1991.

[1530] K. A. Olive. Introduction to supersymmetry: Astrophysical and phenomenological constraints, 1999, hep-ph/9911307.

[1531] K. A. Olive, D. N. Schramm, and G. Steigman. Limits on new superweakly interacting particles from primordial nucleosynthesis, *Nucl. Phys.*, **B180**:497, 1981.

[1532] K. A. Olive and E. D. Skillman. A realistic determination of the error on the primordial helium abundance: Steps toward non-parametric nebular helium abundances, *Astrophys. J.*, **617**:29, 2004, astro-ph/0405588.

[1533] W. Oller, H. Eberl, and W. Majerotto. Full one-loop corrections to neutralino pair production in e+ e- annihilation, *Phys. Lett.*, B**590**:273–283, 2004, hep-ph/0402134.

[1534] J. M. O'Meara *et al.* The deuterium to hydrogen abundance ratio towards the QSO SDSS1558-0031, *Astrophys. J.*, **649**:L61–L66, 2006, astro-ph/0608302.

[1535] S. Orito *et al.* Precision measurement of cosmic-ray antiproton spectrum, *Phys. Rev. Lett.*, **84**:1078–1081, 2000.

[1536] L. P. Osipkov. Spherical stellar systems with spheroidal velocity distributions, *Pis'ma Astron.*, **55**:77, 1979.

[1537] K. Ozone. Liquid xenon scintillation detector for the new $mu \to e\gamma$ search experiment, Ph.D. Thesis, University of Tokyo, 2005.

[1538] H. Pagels and J. R. Primack. Supersymmetry, cosmology and new TeV physics, *Phys. Rev. Lett.*, **48**:223, 1982.

[1539] F. E. Paige, S. D. Protopopescu, H. Baer, and X. Tata. ISAJET 7.69: A Monte Carlo event generator for p p, anti-p p, and e+ e- reactions, 2003, hep-ph/0312045.

[1540] P. B. Pal and L. Wolfenstein. Radiative decays of massive neutrinos, *Phys. Rev.*, D**25**:766, 1982.

[1541] A. Palazzo, D. Cumberbatch, A. Slosar, and J. Silk. Sterile neutrinos as subdominant warm dark matter, *Phys. Rev.*, D**76**:103511, 2007, arXiv:0707.1495 [astro-ph].

[1542] C. Pallis. b – tau unification and sfermion mass non-universality, *Nucl. Phys.*, B**678**:398–426, 2004, hep-ph/0304047.

[1543] C. Pallis. Massive particle decay and cold dark matter abundance, *Astropart. Phys.*, **21**:689–702, 2004, hep-ph/0402033.

[1544] G. Panico, E. Ponton, J. Santiago, and M. Serone. Dark matter and electroweak symmetry breaking in models with warped extra dimensions, *Phys. Rev.*, D**77**:115012, 2008, arXiv:0801.1645 [hep-ph].

[1545] G. Panico, M. Serone, and A. Wulzer. Electroweak symmetry breaking and precision tests with a fifth dimension, *Nucl. Phys.*, B**762**:189–211, 2007, hep-ph/0605292.

[1546] L. C. Parker, H. Hoekstra, M. J. Hudson, L. van Waerbeke, and Y. Mellier. The masses and shapes of dark matter halos from galaxy-galaxy lensing in the CFHT legacy survey, *Astrophys. J.*, **669**:21–31, 2007, 0707.1698.

[1547] M. Passera, W. J. Marciano, and A. Sirlin. The muon g-2 and the bounds on the Higgs boson mass, *Phys. Rev.*, D**78**:013009, 2008, arXiv:0804.1142 [hep-ph].

[1548] B. Patt and F. Wilczek. Higgs-field portal into hidden sectors, 2006, hep-ph/0605188.

[1549] M. M. Pavan, I. I. Strakovsky, R. L. Workman, and R. A. Arndt. The pion nucleon Sigma term is definitely large: Results from a GWU analysis of pi N scattering data, *PiN Newslett.*, **16**:110–115, 2002, hep-ph/0111066.

[1550] J. A. Peacock *et al.* A measurement of the cosmological mass density from clustering in the 2dF Galaxy Redshift Survey, *Nature*, **410**:169–173, 2001, astro-ph/0103143.

[1551] C. D. Peak. Advanced crystal scintillators for dark matter detection, Ph.D. Thesis, University of Sheffield, UK, 2000.

[1552] R. D. Peccei. Summary of the Beijing workshop on Weak Interactions and CP Violation (Invited talk), August 22–26, 1989.

[1553] R. D. Peccei. The strong CP problem and axions, *Lect. Notes Phys.*, **741**:3–17, 2008, hep-ph/0607268.

[1554] R. D. Peccei and H. R. Quinn. Constraints imposed by CP conservation in the presence of instantons, *Phys. Rev.*, D**16**:1791–1797, 1977.

[1555] R. D. Peccei and H. R. Quinn. CP conservation in the presence of instantons, *Phys. Rev. Lett.*, **38**:1440–1443, 1977.

[1556] P. J. E. Peebles. Structure of the Coma cluster of galaxies, *Astron. J.*, **75**:13, 1970.

[1557] P. J. E. Peebles. Star distribution near a collapsed object, *Astrophys. J.*, **178**:371–376, 1972.

[1558] P. J. E. Peebles. Large-scale background temperature and mass fluctuations due to scale-invariant primeval perturbations, *Astrophys. J.*, **263**:L1–L5, 1982.

[1559] P. J. E. Peebles. The sequence of cosmogony and the nature of primeval departures from homogeneity, *Astrophys. J.*, **274**:1–6, 1983.

[1560] P. Peiffer, T. Pollmann, S. Schonert, A. Smolnikov, and S. Vasiliev. Pulse shape analysis of scintillation signals from pure and xenon-doped liquid argon for radioactive background identification, *JINST*, **3**:P08007, 2008.

[1561] M. Peimbert, V. Luridiana, and A. Peimbert. Revised primordial helium abundance based on new atomic data, 2007, astro-ph/0701580.

[1562] H. V. Peiris *et al.* First year Wilkinson Microwave Anisotropy Probe (WMAP) observations: Implications for inflation, *Astrophys. J. Suppl.*, **148**:213, 2003, astro-ph/0302225.

[1563] J. Penarrubia, A. McConnachie, and A. Babul. On the formation of extended galactic disks by tidally disrupted dwarf galaxies, *Astrophys. J.*, **650**:L33–L36, 2006, astro-ph/0606101.

[1564] H. Peng *et al.* Cryogenic cavity detector for a large-scale cold dark-matter axion search, *Nucl. Instrum. Meth.*, A**444**:569–583, 2000.

[1565] M. J. Pereira, G. L. Bryan, and S. P. D. Gill. Radial alignment in simulated clusters, 2007, arXiv:0707.1702 [astro-ph].

[1566] M. Perelstein and A. Spray. Indirect detection of little Higgs dark matter, *Phys. Rev.*, D**75**:083519, 2007, hep-ph/0610357.

[1567] J. S. Perko. Solar modulation of galactic antiprotons, *Astron. Astrophys.*, **184**:119–121, 1987.

[1568] M. E. Peskin. Supersymmetry in elementary particle physics, 2008, arXiv:0801.1928 [hep-ph].

[1569] M. E. Peskin and T. Takeuchi. Estimation of oblique electroweak corrections, *Phys. Rev.*, D**46**:381–409, 1992.

[1570] A. H. G. Peter and S. Tremaine. Dynamics of WIMPs in the solar system and implications for detection, 2008, arXiv:0806.2133 [astro-ph].

[1571] K. Petraki and A. Kusenko. Dark-matter sterile neutrinos in models with a gauge singlet in the Higgs sector, *Phys. Rev.*, D**77**:065014, 2008, arXiv:0711.4646 [hep-ph].

[1572] F. Petriello and K. M. Zurek. DAMA and WIMP dark matter, *JHEP*, **09**:047, 2008, arXiv:0806.3989 [hep-ph].

[1573] M. Pettini, B. J. Zych, M. T. Murphy, A. Lewis, and C. C. Steidel. Deuterium abundance in the most metal-poor damped Lyman alpha system: Converging on Omega baryons, 2008, arXiv:0805.0594 [astro-ph].

[1574] N. Phinney. SLC final performance and lessons, *eConf*, C00082:MO102, 2000, physics/0010008.

[1575] S.-Y. Pi. Inflation without tears, *Phys. Rev. Lett.*, **52**:1725–1728, 1984.

[1576] P. Picozza *et al.* PAMELA: A payload for antimatter matter exploration and light-nuclei astrophysics, *Astropart. Phys.*, **27**:296–315, 2007, astro-ph/0608697.

[1577] D. M. Pierce, J. A. Bagger, K. T. Matchev, and R.-J. Zhang. Precision corrections in the minimal supersymmetric standard model, *Nucl. Phys.*, B**491**:3–67, 1997, hep-ph/9606211.

[1578] L. Pieri, G. Bertone, and E Branchini. Dark matter annihilation in substructures revised, *Mon. Not. Roy. Astron. Soc.*, **384**:1627, 2008, arXiv:0706.2101 [astro-ph].

[1579] T. Plehn, D. Rainwater, and P. Skands. Squark and gluino production with jets, *Phys. Lett.*, **B645**:217–221, 2007, hep-ph/0510144.

[1580] T. Plehn and T. M. P. Tait. Seeking sgluons, 2008, arXiv:0810.3919 [hep-ph].

[1581] E. Pointecouteau. The mass missing problem in clusters: dark matter or modified dynamics?, arXiv:astro-ph/0607142, 2006.

[1582] G. Polesello. Prospects for SUSY at the LHC, *Proceedings of the SUSY 2004 conference Tsukuba 2004*, KEK Proceedings 2004-12, 2004.

[1583] J. Polonyi. Generalization of the massive scalar multiplet coupling to the supergravity, 1978, Hungary Central Inst. Res. KFKI-77-93.

[1584] A. Pomarol. Gauge bosons in a five-dimensional theory with localized gravity, *Phys. Lett.*, **B486**:153–157, 2000, hep-ph/9911294.

[1585] A. Pomarol and A. Wulzer. Stable skyrmions from extra dimensions, *JHEP*, **03**:051–051, 2008, arXiv:0712.3276 [hep-th].

[1586] E. Ponton and L. Wang. Radiative effects on the chiral square, *JHEP*, **11**:018, 2006, hep-ph/0512304.

[1587] W. Porod. SPheno, a program for calculating supersymmetric spectra, SUSY particle decays and SUSY particle production at e^+e^- colliders, *Comput. Phys. Commun.*, **153**:275–315, 2003, hep-ph/0301101.

[1588] R. L. Porter, G. J. Ferland, and K. B. MacAdam. He I emission in the Orion Nebula and implications for primordial helium abundance, *Astrophys. J.*, **657**:327–337, 2007, astro-ph/0611579.

[1589] T. A. Porter, I. V. Moskalenko, A. W. Strong, E. Orlando, and L. Bouchet. Inverse Compton origin of the hard X-ray and soft gamma-ray emission from the Galactic Ridge, *Astrophys. J.*, **682**:400–407, 2008, arXiv:0804.1774 [astro-ph].

[1590] M. Pospelov. Bridging the primordial $A = 8$ divide with catalyzed Big Bang nucleosynthesis, 2007, arXiv:0712.0647 [hep-ph].

[1591] M. Pospelov. Particle physics catalysis of thermal big bang nucleosynthesis, *Phys. Rev. Lett.*, **98**:231301, 2007, hep-ph/0605215.

[1592] M. Pospelov, J. Pradler, and F. D. Steffen. Constraints on supersymmetric models from catalytic primordial nucleosynthesis of beryllium, *JCAP*, **0811**:020, 2008, arXiv:0807.4287 [hep-ph].

[1593] M. Pospelov and A. Ritz. Astrophysical signatures of secluded dark matter, 2008, arXiv:0810.1502 [hep-ph].

[1594] M. Pospelov, A. Ritz, and M. B. Voloshin. Bosonic super-WIMPs as keV-scale dark matter, 2008, arXiv:0807.3279 [hep-ph].

[1595] M. Pospelov, A. Ritz, and M. B. Voloshin. Secluded WIMP Dark Matter, *Phys. Lett.*, B**662**:53–61, 2008, arXiv:0711.4866 [hep-ph].

[1596] F. Prada *et al.* How far do they go? The outer structure of dark matter halos, *Astrophys. J.*, **645**:1001–1011, 2006, astro-ph/0506432.

[1597] F. Prada, A. Klypin, J. Flix, M. Martínez, and E. Simonneau. Dark matter annihilation in the Milky Way galaxy: Effects of baryonic compression, *Phys. Rev. Lett.*, **93**:241301, 2004.

[1598] F. Prada, A. A. Klypin, E. Simonneau, *et al.* How far do they go? The outer structure of galactic dark matter halos, *Astrophys. J.*, **645**:1001–1011, 2006.

[1599] N. Prantzos. Li6 in the early galaxy: Energetics, evolution and stellar depletion, 2005, astro-ph/0510122.

[1600] J. Preskill, M. B. Wise, and F. Wilczek. Cosmology of the invisible axion, *Phys. Lett.*, B**120**:127–132, 1983.

[1601] W. H. Press and P. Schechter. Formation of galaxies and clusters of galaxies by self-similar gravitational condensation, *Astrophys. J.*, **187**:425–438, 1974.

[1602] W. H. Press and D. N. Spergel. Capture by the sun of a galactic population of weakly interacting, massive particles, *Astrophys. J.*, **296**:679–684, 1985.

[1603] M. Preto, D. Merritt, and R. Spurzem. N-body growth of a Bahcall–Wolf cusp around a black hole, *Astrophys. J.*, **613**:L109, 2004, astro-ph/0406324.

[1604] J. R. Primack, D. Seckel, and B. Sadoulet. Detection of cosmic dark matter, *Ann. Rev. Nucl. Part. Sci.*, **38**:751–807, 1988.

[1605] F. Primas, M. Asplund, P. E. Nissen, and V. Hill. The beryllium abundance in the very metal-poor halo star G 64-12 from VLT/UVES observations, 2000, astro-ph/0009482.

[1606] B. J. Pritzl, K. A. Venn, and M. J. Irwin. A comparison of elemental abundance ratios in globular clusters, field stars, and dwarf spheroidal galaxies, *Astron. J.*, **130**:2140–2165, 2005, astro-ph/0506238.

[1607] S. Profumo. TeV γ-rays and the largest masses and annihilation cross sections of neutralino dark matter, *Phys. Rev. D*, **72**:103521, 2005.

[1608] S. Profumo. Hunting the lightest lightest neutralinos, *Phys. Rev.*, D**78**:023507, 2008, arXiv:0806.2150 [hep-ph].

[1609] S. Profumo. Non-thermal X-rays from the Ophiuchus cluster and dark matter annihilation, *Phys. Rev.*, D**77**:103510, 2008, arXiv:0801.0740 [astro-ph].

[1610] S. Profumo and M. Kamionkowski. Dark matter and the CACTUS gamma-ray excess from Draco, *JCAP*, **0603**:003, 2006, astro-ph/0601249.

[1611] S. Profumo, K. Sigurdson, and M. Kamionkowski. What mass are the smallest protohalos?, *Phys. Rev. Lett.*, **97**:031301, 2006.

[1612] S. Profumo and P. Ullio. SUSY dark matter and quintessence, *JCAP*, **0311**:006, 2003, hep-ph/0309220.

[1613] P. Pugnat *et al.* First results from the OSQAR photon regeneration experiment: No light shining through a wall, *Phys. Rev.*, D**78**:092003, 2008, arXiv:0712.3362 [hep-ex].

[1614] A. Pukhov. CalcHEP 3.2: MSSM, structure functions, event generation, batchs, and generation of matrix elements for other packages, 2004, hep-ph/0412191.

[1615] C. W. Purcell, J. S. Bullock, and A. R. Zentner. Shredded galaxies as the source of diffuse intrahalo light on varying scales, 2007, astro-ph/0703004.

[1616] M. E. Putman, J. Grcevich, and J. E. G. Peek. Fuel for galaxy disks, 2008, arXiv:0803.3069 [astro-ph].

[1617] G. D. Quinlan. The dynamical evolution of massive black hole binaries I. Hardening in a fixed stellar background, *New Astron.*, **1**:35–56, 1996.

[1618] P. J. Quinn, J. K. Salmon, and W. H. Zurek. Primordial density fluctuations and the structure of galactic haloes, *Nature*, **322**:329–335, 1986.

[1619] G. Raffelt and D. Seckel. Bounds on exotic particle interactions from SN 1987a, *Phys. Rev. Lett.*, **60**:1793, 1988.

[1620] G. G. Raffelt. Astrophysical methods to constrain axions and other novel particle phenomena, *Phys. Rep.*, **198**:1–113, 1990.

[1621] G. G. Raffelt, *Stars as Laboratories for Fundamental Physics: The Astrophysics of Neutrinos, Axions, and Other Weakly Interacting Particles*, University of Chicago Press, 1996.

[1622] G. G. Raffelt. Astrophysical axion bounds: An update, 1997, astro-ph/9707268.

[1623] G. G. Raffelt and D. S. P. Dearborn. Bounds on hadronic axions from stellar evolution, *Phys. Rev.*, D**36**:2211, 1987.

[1624] K. Rajagopal, M. S. Turner, and F. Wilczek. Cosmological implications of axinos, *Nucl. Phys.*, B**358**:447–470, 1991.

[1625] L. Randall and R. Sundrum. A large mass hierarchy from a small extra dimension, *Phys. Rev. Lett.*, **83**:3370–3373, 1999, hep-ph/9905221.

[1626] L. Randall and R. Sundrum. An alternative to compactification, *Phys. Rev. Lett.*, **83**:4690–4693, 1999, hep-th/9906064.

[1627] S. W. Randall, M. Markevitch, D. Clowe, A. H. Gonzalez, and M. Bradač. Constraints on the self-interaction cross section of dark matter from numerical simulations of the merging galaxy cluster 1E 0657-56, *Astrophys. J.*, **679**:1173–1180, 2008, 0704.0261.

[1628] A. P. Rasmussen *et al.* On the putative detection of z0 X-ray absorption features in the spectrum of Markarian 421, *Astrophys. J.*, **656**:129–138, 2007, astro-ph/0604515.

[1629] R. Rattazzi and A. Zaffaroni. Comments on the holographic picture of the Randall–Sundrum model, *JHEP*, **04**:021, 2001, hep-th/0012248.

[1630] J. I. Read, G. Lake, O. Agertz, and V. P. Debattista. Thin, thick and dark discs in LCDM, 2008, arXiv:0803.2714 [astro-ph].

[1631] I. H. Redmount and M. J. Rees. Gravitational-radiation rocket effects and galactic structure, *Comments Astrophys.*, **14**:165, 1989.

[1632] D. Reed *et al.* Dark matter subhaloes in numerical simulations, *Mon. Not. Roy. Astron. Soc.*, **359**:1537–1548, 2005, astro-ph/0406034.

[1633] A. Refregier. Weak gravitational lensing by large-scale structure, *Ann. Rev. Astron. Astrophys.*, **41**:645–668, 2003.

[1634] M. Regis and P. Ullio. Multi-wavelength signals of dark matter annihilations at the Galactic center, *Phys. Rev.*, D**78**:043505, 2008, arXiv:0802.0234 [hep-ph].

[1635] C. L. Reichardt *et al.* High resolution CMB power spectrum from the complete ACBAR data set, 2008, arXiv:0801.1491 [astro-ph].

[1636] M. J. Reid. The distance to the center of the galaxy, *Ann. Rev. Astron. Astrophys.*, **31**:345–372, 1993.

[1637] M. H. Reno and D. Seckel. Primordial nucleosynthesis: The effects of injecting hadrons, *Phys. Rev.*, D**37**:3441, 1988.

[1638] A. Renzini. Effects of cosmions in the sun and in globular cluster stars, *Astron. Astrophys.*, **171**:121, 1987.

[1639] J. Rich, D. Lloyd Owen, and M. Spiro. Experimental particle physics without accelerators, *Phys. Rep.*, **151**:239, 1987.

[1640] O. Richard, G. Michaud, and J. Richer. Implications of wmap observations on li abundance and stellar evolution models. *Astrophys. J.*, **619**:538–548, 2005.

[1641] H. B. Richer *et al.* White dwarfs in globular clusters: HST observations of M4, *Astrophys. J.*, **484**:741–760, 1997, astro-ph/9702169.

[1642] M. C. Richter, G. B. Tupper, and R. D. Viollier. A symbiotic scenario for the rapid formation of supermassive black holes, *JCAP*, **0612**:015, 2006, astro-ph/0611552.

[1643] M. Ricotti, A. Pontzen, and M. Viel. Is the concentration of dark matter halos at virialization universal?, *Astrophys. J. Lett.*, **663**:L53–L56, 2007, 0706.0856.

[1644] S. Riemann. Z′ signals from Kaluza–Klein dark matter, 2005, hep-ph/0508136.

[1645] S. Riemer-Sorensen, S. H. Hansen, and K. Pedersen. Sterile neutrinos in the Milky Way: Observational constraints, *Astrophys. J.*, **644**:L33–L36, 2006, astro-ph/0603661.

[1646] E. Ripamonti and T. Abel. The formation of primordial luminous objects, 2005, astro-ph/0507130.

[1647] E. Ripamonti, M. Mapelli, and A. Ferrara. Intergalactic medium heating by dark matter, *Mon. Not. Roy. Astron. Soc.*, **374**:1067–1077, 2007, astro-ph/0606482.

[1648] E. Ripamonti, M. Mapelli, and A. Ferrara. The impact of dark matter decays and annihilations on the formation of the first structures, *Mon. Not. Roy. Astron. Soc.*, **375**:1399–1408, 2007, astro-ph/0606483.

[1649] T. G. Rizzo. Probes of universal extra dimensions at colliders, *Phys. Rev.*, D**64**:095010, 2001, hep-ph/0106336.

[1650] T. G. Rizzo. Radion couplings to bulk fields in the Randall-Sundrum model, *JHEP*, **06**:056, 2002, hep-ph/0205242.

[1651] L. Rolandi and W. Blum. *Particle Detection with Drift Chambers*, Springer-Verlag, 1994.

[1652] L. J. Rosenberg and K. A. van Bibber. Searches for invisible axions, *Phys. Rep.*, **325**:1–39, 2000.

[1653] L. Roszkowski, R. R. de Austri, J. Silk, and R. Trotta. On prospects for dark matter indirect detection in the constrained MSSM, 2007, arXiv:0707.0622 [astro-ph].

[1654] L. Roszkowski, R. Ruiz de Austri, and K.-Y. Choi. Gravitino dark matter in the CMSSM and implications for leptogenesis and the LHC, *JHEP*, **08**:080, 2005, hep-ph/0408227.

[1655] L. Roszkowski, R. Ruiz de Austri, and T. Nihei. New cosmological and experimental constraints on the CMSSM, *JHEP*, **08**:024, 2001, hep-ph/0106334.

[1656] L. Roszkowski, R. Ruiz de Austri, and R. Trotta. Implications for the Constrained MSSM from a new prediction for b to s gamma, *JHEP*, **07**:075, 2007, arXiv:0705.2012 [hep-ph].

[1657] S. Roth *et al.* Cryogenic composite detectors for the dark matter experiments CRESST and EURECA, 2008, arXiv:0810.0423 [astro-ph].

[1658] A. Rubbia. ArDM: A ton-scale liquid argon experiment for direct detection of dark matter in the universe, *J. Phys. Conf. Ser.*, **39**:129–132, 2006, hep-ph/0510320.

[1659] O. Ruchayskiy. Restrictions on sterile neutrino parameters from astrophysical observations, 2007, arXiv:0704.3215 [astro-ph].

[1660] S. Rudaz. On the annihilation of heavy neutral fermion pairs into monochromatic gamma-rays and its astrophysical implications, *Phys. Rev.*, D**39**:3549, 1989.

[1661] G. Ruoso *et al.* Limits on light scalar and pseudoscalar particles from a photon regeneration experiment, *Z. Phys.*, C**56**:505–508, 1992.

[1662] S. G. Ryan, J. E. Norris, and T. C. Beers. The Spite Lithium plateau: ultra-thin but post-primordial, *Astrophys. J.*, **523**:654–677, 1999, astro-ph/9903059.

[1663] G. B. Rybicki and A. P. Lightman. *Radiative Processes in Astrophysics*, Wiley-Interscience, 1979.

[1664] V. S. Rychkov and A. Strumia. Thermal production of gravitinos, *Phys. Rev.*, D**75**:075011, 2007, hep-ph/0701104.

[1665] Burgos S. *et al.* SRIM simulation studies of the head-tail effect in gases relevant to dark matter searches, *Nucl. Instrum. Meth.*, in press.

[1666] R. K. Sachs and A. M. Wolfe. Perturbations of a cosmological model and angular variations of the microwave background, *Astrophys. J.*, **147**:73–90, 1967.

[1667] E. Sagi and J. D. Bekenstein. Black holes in the TeVeS theory of gravity and their thermodynamics, *Phys. Rev.*, D**77**:024010, 2008, arXiv:0708.2639 [gr-qc].

[1668] A. Salam. Weak and electromagnetic interactions, 1968. Originally printed in *Svartholm: Elementary Particle Theory*, Proceedings of the Nobel Symposium held 1968 at Lerum, Sweden, Stockholm 1968, 367–377.

[1669] P. Salati. Quintessence and the relic density of neutralinos, *Phys. Lett.*, B**571**:121–131, 2003, astro-ph/0207396.

[1670] P. Salati and J. Silk. A stellar probe of dark matter annihilation in galactic nuclei, *Astrophs. J.*, **338**:24–31, 1989.

[1671] L. V. Sales, J. F. Navarro, M. G. Abadi, and M. Steinmetz. Cosmic ménage à trois: The origin of satellite galaxies on extreme orbits, *Mon. Not. Roy. Astron. Soc.*, **379**:1475–1483, 2007, arXiv:0704.1773 [astro-ph].

[1672] P. Salucci. Review for IAU Symposium 244, Dark Galaxies and Lost Baryons, 2007, arXiv:0707.4370 [astro-ph].

[1673] P. Salucci and A. Burkert. Dark matter scaling relations, *Astrophys. J. Lett.*, **537**:L9–L12, 2000.

[1674] F. J. Sánchez-Salcedo, J. Reyes-Iturbide, and X. Hernandez. An extensive study of dynamical friction in dwarf galaxies: the role of stars, dark matter, halo profiles and MOND, *Mon. Not. Roy. Astron. Soc.*, **370**:1829–1840, 2006.

[1675] R. H. Sanders. A stratified framework for scalar-tensor theories of modified dynamics, *Astrophys. J.*, **480**:492, 1997.

[1676] R. H. Sanders. Clusters of galaxies with modified Newtonian dynamics, *Mon. Not. Roy. Astron. Soc.*, **342**:901–908, 2003.

[1677] R. H. Sanders. A tensor-vector-scalar framework for modified dynamics and cosmic dark matter, *Mon. Not. Roy. Astron. Soc.*, **363**:459, 2005, astro-ph/0502222.

[1678] R. H. Sanders. Solar system constraints on multi-field theories of modified dynamics, *Mon. Not. Roy. Astron. Soc.*, **370**:1519–1528, 2006, astro-ph/0602161.

[1679] R. H. Sanders. Neutrinos as cluster dark matter, *Mon. Not. Roy. Astron. Soc.*, **380**:331–338, 2007.

[1680] R. H. Sanders and K. G. Begeman. Modified dynamics – MOND – as a dark halo, *Mon. Not. Roy. Astron. Soc.*, **266**:360, 1994.

[1681] R. H. Sanders and S. S. McGaugh. Modified Newtonian dynamics as an alternative to dark matter, *Ann. Rev. Astron. Astrophys.*, **40**:263–317, 2002, astro-ph/0204521.

[1682] R. H. Sanders and E. Noordermeer. Confrontation of MOND with the rotation curves of early-type disc galaxies, *Mon. Not. Roy. Astron. Soc.*, **379**:702–710, 2007, astro-ph/0703352.

[1683] V. Sanglard *et al.* Final results of the EDELWEISS-I dark matter search with cryogenic heat-and-ionization Ge detectors, *Phys. Rev.*, **D71**:122002, 2005, astro-ph/0503265.

[1684] F. Sannino and K. Tuominen. Techniorientifold, *Phys. Rev.*, **D71**:051901, 2005, hep-ph/0405209.

[1685] D. I. Santiago, D. Kalligas, and R. V. Wagoner. Scalar-tensor cosmologies and their late time evolution, *Phys. Rev.*, **D58**:124005, 1998, gr-qc/9805044.

[1686] D. Santos *et al.* A project of a new detector for direct dark matter search: MACHe3, 2000, astro-ph/0005332.

[1687] D. Santos, O. Guillaudin, T. Lamy, F. Mayet, and E. Moulin. MIMAC-He3: Micro-TPC matrix of chambers of He3, 6th International Workshop on the Identification of Dark Matter, 2007, astro-ph/0701230.

[1688] D. Santos, O. Guillaudin, Th. Lamy, F. Mayet, and E. Moulin. MIMAC: A micro-TPC matrix of chambers for direct detection of Wimps, *J. Phys. Conf. Ser.*, **65**:012012, 2007, astro-ph/0703310.

[1689] Y. Santoso. Gravitino dark matter with stop as the NLSP, 2007, arXiv:0709.3952 [hep-ph].

[1690] S. Sarkar. Big Bang nucleosynthesis and physics beyond the standard model, *Rep. Prog. Phys.*, **59**:1493–1610, 1996, hep-ph/9602260.

[1691] W. C. Saslaw, M. J. Valtonen, and S. J. Aarseth. The gravitational slingshot and the structure of extragalactic radio sources, *Astrophys. J.*, **190**:253–270, 1974.

[1692] K. Sato and M. Kobayashi. Cosmological constraints on the mass and the number of heavy lepton neutrinos, *Prog. Theor. Phys.*, **58**:1775, 1977.

[1693] K. Sato and H. Sato. Higgs meson emission from a star and a constraint on its mass, *Prog. Theor. Phys.*, **54**:1564–1565, 1975.

[1694] C. Savage, K. Freese, and P. Gondolo. Annual modulation of dark matter in the presence of streams, *Phys. Rev.*, **D74**:043531, 2006, astro-ph/0607121.

[1695] C. Savage, G. Gelmini, P. Gondolo, and K. Freese. Compatibility of DAMA/LIBRA dark matter detection with other searches, 2008, arXiv:0808.3607 [astro-ph].

[1696] C. Savage, P. Gondolo, and K. Freese. Can WIMP spin dependent couplings explain DAMA data, in light of null results from other experiments?, *Phys. Rev.*, **D70**:123513, 2004, astro-ph/0408346.

[1697] W. Scandale and F. Zimmermann. Scenarios for sLHC and vLHC, *Nucl. Phys. Proc. Suppl.*, **177–178**:207–211, 2008.

[1698] R. Scarpa, G. Marconi, and R. Gilmozzi. Globular clusters as a test for gravity in the weak acceleration regime, *AIP Conf. Proc.*, **822**:102–104, 2006, astro-ph/0601581.

[1699] R. Schabinger and J. D. Wells. A minimal spontaneously broken hidden sector and its impact on Higgs boson physics at the Large Hadron Collider, *Phys. Rev.*, **D72**:093007, 2005, hep-ph/0509209.

[1700] S. Schael *et al.* Deuteron and anti-deuteron production in e^+e^- collisions at the Z resonance, *Phys. Lett. B*, **639**:192–201, 2006.

[1701] R. J. Scherrer and M. S. Turner. On the relic, cosmic abundance of stable weakly interacting massive particles, *Phys. Rev.*, D**33**:1585, 1986.

[1702] C. Schimd, J.-P. Uzan, and A. Riazuelo. Weak lensing in scalar-tensor theories of gravity, *Phys. Rev. D*, **71**:083512, 2005.

[1703] D. R. G. Schleicher, R. Banerjee, and R. S. Klessen. Dark stars: Implications and constraints from cosmic reionization and the extra-galactic X-ray background, 2008, arXiv:0809.1519 [astro-ph].

[1704] M. Schmaltz and D. Tucker-Smith. Little Higgs review, *Ann. Rev. Nucl. Part. Sci.*, **55**:229–270, 2005, hep-ph/0502182.

[1705] F. Schmidt. Weak lensing probes of modified gravity, *Phys. Rev. D*, **78**:043002, 2008, arXiv:0805.4812.

[1706] F. Schmidt, M. Liguori, and S. Dodelson. Galaxy-CMB cross-correlation as a probe of alternative models of gravity, *Phys. Rev. D*, **76**:083518, 2007, arXiv:0706.1775.

[1707] P. Schneider, J. Ehlers, and E. E. Falco, *Gravitational Lenses*, Springer-Verlag, 1992.

[1708] P. Schneider and X. Er. Weak lensing goes bananas: What flexion really measures, *Astron. Astrophys.*, **485**:363–376, 2008, 0709.1003.

[1709] P. Schneider, C. S. Kochanek, and J. Wambsganss, *Gravitational Lensing: Strong, Weak and Micro*, Saas-Fee Advanced Courses, volume 33, Springer-Verlag, 2006.

[1710] R. Schödel, A. Eckart, T. Alexander, *et al.* The structure of the nuclear stellar cluster of the Milky Way, *Astron. Astrophys.*, **469**:125–146, 2007.

[1711] R. Schödel, D. Merritt, and A. Eckart. The nuclear star cluster of the Milky Way, *J. Phys. Conf. Ser.*, **131**:012044, 2008, arXiv:0810.0204.

[1712] A. Schwarzschild and Č. Zupančič. Production of tritons, deuterons, nucleons, and mesons by 30-GeV protons on A1, Be, and Fe targets, *Phys. Rev.*, **129**:854–862, 1963.

[1713] G. Sciolla. Directional detection of dark matter, *Mod. Phys. Lett. A*, 2008, arXiv:0811.2764 [astro-ph].

[1714] P. Scott, M. Fairbairn, and J. Edsjö. Dark stars at the Galactic centre – the main sequence, 2008, arXiv:0809.1871 [astro-ph].

[1715] G. M. Seabroke *et al.* Is the sky falling? Searching for stellar streams in the local Milky Way disc in the CORAVEL and RAVE surveys, 2007, arXiv:0709.4219 [astro-ph].

[1716] L. Searle and R. Zinn. Compositions of halo clusters and the formation of the galactic halo, *Astrophys. J*, **225**:357–379, 1978.

[1717] D. Seckel and M. S. Turner. Isothermal density perturbations in an axion dominated inflationary universe, *Phys. Rev.*, **D32**:3178, 1985.

[1718] W. Seidel *et al.* The CRESST experiment. Prepared for International Workshop on Aspects of Dark Matter in Astrophysics and Particle Physics, Heidelberg, Germany, 16–20 September 1996.

[1719] C. Seitz and P. Schneider. Steps towards nonlinear cluster inversion through gravitational distortions. III. Including a redshift distribution of the sources, *Astron. Astrophys.*, **318**:687–699, 1997.

[1720] S. Seitz and P. Schneider. Cluster lens reconstruction using only observed local data: an improved finite-field inversion technique, *Astron. Astrophys.*, **305**:383, 1996.

[1721] U. Seljak, A. Makarov, P. McDonald, and H. Trac. Can sterile neutrinos be the dark matter?, *Phys. Rev. Lett.*, **97**:191303, 2006, astro-ph/0602430.

[1722] J. A. Sellwood. Bar-Halo friction in galaxies. III. Halo density changes, *Astrophys. J.*, **679**:379–396, 2008.

[1723] J. A. Sellwood and S. S. McGaugh. The compression of dark matter halos by baryonic infall, *Astrophys. J.*, **634**:70–76, 2005, astro-ph/0507589.

[1724] M. Selvi and for the LVD Collaboration. The LVD Core Facility: a study of LVD as muon veto and active shielding for dark matter experiments, 2008, arXiv:0811.2884 [hep-ex].

[1725] A. Semenov. LanHEP: A package for automatic generation of Feynman rules from the Lagrangian, *Comput. Phys. Commun.*, **115**:124–139, 1998.

[1726] Y. Semertzidis *et al.* Limits on the production of light scalar and pseudoscalar particles, *Phys. Rev. Lett.*, **64**:2988–2991, 1990.

[1727] P. D. Serpico and G. Zaharijas. Optimal angular window for observing Dark Matter annihilation from the Galactic Center region: the case of γ^- ray lines, *Astropart. Phys.*, **29**:380–385, 2008, arXiv:0802.3245 [astro-ph].

[1728] J. L. Sérsic, *Atlas de Galaxias Australes*, Cordoba, Argentina: Observatorio Astronomico, 1968.

[1729] G. Servant and T. M. P. Tait. Elastic scattering and direct detection of Kaluza–Klein dark matter, *New J. Phys.*, **4**:99, 2002, hep-ph/0209262.

[1730] G. Servant and T. M. P. Tait. Is the lightest Kaluza–Klein particle a viable dark matter candidate?, *Nucl. Phys.*, **B650**:391–419, 2003, hep-ph/0206071.

[1731] A. Seth, M. Agüeros, D. Lee, and A. Basu-Zych. The coincidence of nuclear star clusters and active galactic nuclei, *Astrophys. J.*, **678**:116–130, 2008, arXiv:0801.0439.

[1732] H.-Y. Shan, M. Feix, B. Famaey, and H.-S. Zhao. An analytic model for non-spherical lenses in covariant Modified Newtonian dynamics, *Mon. Not. Roy. Astron. Soc.*, **387**:1303–1312, 2008, arXiv0804.2668.

[1733] H. Shapley. A stellar system of a new type, *Harvard College Obs. Bull.*, **908**:1–11, 1938.

[1734] M. Shaposhnikov. A possible symmetry of the nuMSM, *Nucl. Phys.*, **B763**:49–59, 2007, hep-ph/0605047.

[1735] M. Shaposhnikov. Sterile neutrinos in cosmology and how to find them in the lab, 2008, arXiv:0809.2028 [hep-ph].

[1736] M. Shaposhnikov. The nuMSM, leptonic asymmetries, and properties of singlet fermions, *JHEP*, **08**:008, 2008, arXiv:0804.4542 [hep-ph].

[1737] M. Shaposhnikov and I. Tkachev. The nuMSM, inflation, and dark matter, *Phys. Lett.*, **B639**:414–417, 2006, hep-ph/0604236.

[1738] M. Shaposhnikov and D. Zenhausern. Quantum scale invariance, cosmological constant and hierarchy problem, 2008, arXiv:0809.3406 [hep-th].

[1739] M. Shaposhnikov and D. Zenhausern. Scale invariance, unimodular gravity and dark energy, 2008, arXiv:0809.3395 [hep-th].

[1740] S. Sharma and M. Steinmetz. Multi-dimensional density estimation and phase space structure of dark matter halos, 2005, astro-ph/0507550.

[1741] E. P. S. Shellard. Cosmic string interactions, *Nucl. Phys.*, **B283**: 624–656, 1987.

[1742] W. Shepherd, T. M. P. Tait, and G. Zaharijas. WIMPonium, 2009, arXiv:0901.2125 [hep-ph].

[1743] R. K. Sheth and G. Tormen. An Excursion set model of hierarchical clustering: Ellipsoidal collapse and the moving barrier, *Mon. Not. Roy. Astron. Soc.*, **329**:61, 2002, astro-ph/0105113.

[1744] X.-D. Shi and G. M. Fuller. A new dark matter candidate: Non-thermal sterile neutrinos, *Phys. Rev. Lett.*, **82**:2832–2835, 1999, astro-ph/9810076.

[1745] M. A. Shifman, A. I. Vainshtein, and V. I. Zakharov. Can confinement ensure natural CP invariance of strong interactions?, *Nucl. Phys.*, **B166**:493, 1980.

[1746] Y. Shikaze *et al.* Measurements of 0.2 20 GeV/n cosmic-ray proton and helium spectra from 1997 through 2002 with the BESS spectrometer, *Astropart. Phys.*, **28**:154–167, 2007.

[1747] I. Shlosman, J. Frank, and M. C. Begelman. Bars within bars – A mechanism for fuelling active galactic nuclei, *Nature*, **338**:45–47, 1989.

[1748] T. Shutt *et al.* A solution to the dead-layer problem in ionization and phonon-based dark matter detectors, *Nucl. Instrum. Meth.*, **A444**:340–344, 2000.

[1749] J. M. Siegal-Gaskins. Revealing dark matter substructure with anisotropies in the diffuse gamma-ray background, *JCAP*, **0810**:040, 2008, arXiv:0807.1328 [astro-ph].

[1750] J. M. Siegal-Gaskins and V. Pavlidou. Robust identification of isotropic diffuse gamma rays from Galactic dark matter, 2009, arXiv:0901.3776 [astro-ph.HE].

[1751] J. M. Siegal-Gaskins and M. Valluri. Signatures of LCDM substructure in tidal debris, *Astrophys. J.*, **681**:40, 2008, arXiv:0710.0385 [astro-ph].

[1752] G. Sigl, K. Jedamzik, D. N. Schramm, and V. S. Berezinsky. Helium photodesintegration and nucleosynthesis: Implications for topological defects, high energy cosmic rays, and massive black holes, *Phys. Rev.*, D**52**:6682–6693, 1995, astro-ph/9503094.

[1753] P. Sikivie. Of axions, domain walls and the early universe, *Phys. Rev. Lett.*, **48**:1156–1159, 1982.

[1754] P. Sikivie. Experimental tests of the "invisible" axion, *Phys. Rev. Lett.*, **51**:1415, 1983.

[1755] P. Sikivie. Detection rates for 'invisible' axion searches, *Phys. Rev.*, D**32**:2988, 1985.

[1756] P. Sikivie. Caustic rings of dark matter, *Phys. Lett.*, B**432**:139–144, 1998, astro-ph/9705038.

[1757] P. Sikivie, D. B. Tanner, and K. van Bibber. Resonantly enhanced axion - photon regeneration, *Phys. Rev. Lett.*, **98**:172002, 2007, hep-ph/0701198.

[1758] J. Silk. The Dark Side of the Universe, *Int. J. Mod. Phys. A*, **17**:167–179, 2002.

[1759] J. Silk and M. Srednicki. Cosmic-ray antiprotons as a probe of a photino-dominated universe, *Phys. Rev. Lett.*, **53**:624, 1984.

[1760] V. Silveira and A. Zee. Scalar phantoms, *Phys. Lett.*, B**161**:136, 1985.

[1761] V. Simha and G. Steigman. Constraining the universal lepton asymmetry, *JCAP*, **0808**:011, 2008, arXiv:0806.0179 [hep-ph].

[1762] J. D. Simon, A. D. Bolatto, A. Leroy, L. Blitz, and E. L. Gates. High-resolution measurements of the halos of four dark matter-dominated galaxies: Deviations from a universal density profile, *Astrophys. J.*, **621**:757–776, 2005, astro-ph/0412035.

[1763] J. D. Simon and M. Geha. The kinematics of the ultra-faint Milky Way satellites: Solving the missing satellite problem, *Astrophys. J.*, **670**:313–331, 2007, arXiv:0706.0516 [astro-ph].

[1764] T. Sjostrand *et al.* High-energy-physics event generation with PYTHIA 6.1, *Comput. Phys. Commun.*, **135**:238–259, 2001, hep-ph/0010017.

[1765] T. Sjostrand, S. Mrenna, and P. Skands. PYTHIA 6.4 physics and manual, *JHEP*, **05**:026, 2006, hep-ph/0603175.

[1766] T. Sjostrand, S. Mrenna, and P. Skands. A brief introduction to PYTHIA 8.1, *Comput. Phys. Commun.*, **178**:852–867, 2008, arXiv:0710.3820 [hep-ph].

[1767] P. Skands *et al.* SUSY Les Houches accord: Interfacing SUSY spectrum calculators, decay packages, and event generators, *JHEP*, **07**:036, 2004, hep-ph/0311123.

[1768] C. Skordis. Tensor-vector-scalar cosmology: Covariant formalism for the background evolution and linear perturbation theory, *Phys. Rev. D*, **74**:103513, 2006.

[1769] C. Skordis, D. F. Mota, P. G. Ferreira, and C. Boehm. Large scale structure in Bekenstein's theory of relativistic modified Newtonian dynamics, *Phys. Rev. Lett.*, **96**:011301, 2006.

[1770] J. M. Smillie. Spin correlations in decay chains involving W bosons, *Eur. Phys. J.*, C**51**:933–943, 2007, hep-ph/0609296.

[1771] J. M. Smillie and B. R. Webber. Distinguishing spins in supersymmetric and universal extra dimension models at the Large Hadron Collider, *JHEP*, **10**:069, 2005, hep-ph/0507170.

[1772] M. C. Smith *et al.* The RAVE Survey: Constraining the local galactic escape speed, *Mon. Not. Roy. Astron. Soc.*, **379**:755–772, 2007, astro-ph/0611671.

[1773] P. F. Smith. Terrestrial searches for new stable particles, *Contemp. Phys.*, **29**:159, 1988.

[1774] P. F. Smith and J. D. Lewin. Dark matter detection, *Phys. Rep.*, **187**:203, 1990.

[1775] D. P. Snowden-Ifft, C. J. Martoff, and J. M. Burwell. Low pressure negative ion drift chamber for dark matter search, *Phys. Rev.*, D**61**:101301, 2000, astro-ph/9904064.

[1776] D. P. Snowden-Ifft, T. Ohnuki, E. S. Rykoff, and C. J. Martoff. Neutron recoils in the drift detector, *Nucl. Instrum. Meth.*, A**498**:155–164, 2003.

[1777] D. P. Snowden-Ifft and A. J. Westphal. Unique signature of dark matter in ancient mica, *Phys. Rev. Lett.*, **78**:1628–1631, 1997, astro-ph/9701215.

[1778] Y. Sofue and V. Rubin. Rotation curves of spiral galaxies, *Ann. Rev. Astron. Astrophys.*, **39**:137–174, 2001, astro-ph/0010594.

[1779] J. Sommer-Larsen and A. Dolgov. Formation of disk galaxies: Warm dark matter and the angular momentum problem, *Astrophys. J.*, **551**:608–623, 2001, astro-ph/9912166.

[1780] G. Soucail, B. Fort, Y. Mellier, and J. P. Picat. A blue ring-like structure, in the center of the A 370 cluster of galaxies, *Astron. Astrophys.*, **172**:L14–L16, 1987.

[1781] G. Soucail, Y. Mellier, B. Fort, G. Mathez, and M. Cailloux. The giant arc in A 370 – Spectroscopic evidence for gravitational lensing from a source at $Z = 0.724$, *Astron. Astrophys.*, **191**:L19–L21, 1988.

[1782] K. Spekkens, R. Giovanelli, and M. P. Haynes. The cusp/core problem in galactic halos: Long-slit spectra for a large dwarf galaxy sample, *Astron. J.*, **129**:2119–2137, 2005.

[1783] D. N. Spergel. Motion of the earth and the detection of weakly interacting massive particles, *Phys. Rev. D*, **37**:1353–1355, 1988.

[1784] D. N. Spergel and J. Faulkner. Weakly interacting, massive particles in horizontal-branch stars, *Astrophys. J. Lett.*, **331**:L21–L24, 1988.

[1785] D. N. Spergel and W. H. Press. Effect of hypothetical, weakly interacting, massive particles on energy transport in the solar interior, *Astrophys. J.*, **294**:663–673, 1985.

[1786] L. Spitzer, *Dynamical Evolution of Globular Clusters*, Princeton University Press, 1987.

[1787] D. Spolyar, K. Freese, and P. Gondolo. Dark matter and the first stars: a new phase of stellar evolution, *Phys. Rev. Lett.*, **100**:051101, 2008, arXiv:0705.0521 [astro-ph].

[1788] N. J. C. Spooner *et al.* Measurements of carbon recoil scintillation efficiency and anisotropy in stilbene for WIMP searches with directional sensitivity. In N. J. C. Spooner, ed., *Proceedings of Workshop on Identification of Dark Matter*, World Scientific, 481–486, 1997.

[1789] N. J. C. Spooner, D. R. Tovey, C. D. Peak, and J. W. Roberts. Demonstration of nuclear recoil discrimination using recoil range in a mixed CaF-2 + liquid scintillator gel detector for dark matter searches, *Astropart. Phys.*, **8**:13–19, 1997.

[1790] V. Springel *et al.* The Aquarius Project: the subhalos of galactic halos, 2008, arXiv:0809.0898 [astro-ph].

[1791] V. Springel, C. S. Frenk, and S. D. M. White. The large-scale structure of the Universe, *Nature*, **440**, 2006, astro-ph/0604561.

[1792] M. Srednicki and R. Watkins. Coherent couplings of neutralinos to nuclei from squark mixing, *Phys. Lett.*, B**225**:140, 1989.

[1793] M. Srednicki, R. Watkins, and K. A. Olive. Calculations of relic densities in the early universe, *Nucl. Phys.*, B**310**:693, 1988.

[1794] J. Stadel *et al.* Quantifying the heart of darkness with GHALO – a multi-billion particle simulation of our galactic halo, 2008, arXiv:0808.2981 [astro-ph].

[1795] G. D. Starkman and D. N. Spergel. Proposed new technique for detecting supersymmetric dark matter, *Phys. Rev. Lett.*, **74**:2623–2625, 1995.

[1796] A. A. Starobinsky. Dynamics of phase transition in the new inflationary universe scenario and generation of perturbations, *Phys. Lett.*, B**117**:175–178, 1982.

[1797] J. Stasielak, P. L. Biermann, and A. Kusenko. Thermal evolution of the primordial clouds in warm dark matter models with keV sterile neutrinos, *Astrophys. J.*, **654**:290–303, 2007, astro-ph/0606435.

[1798] F. W. Stecker. Gamma ray constraints on dark matter reconsidered, *Phys. Lett.*, B**201**:529–532, 1988.

[1799] F. W. Stecker, S. D. Hunter, and D. A. Kniffen. The likely cause of the EGRET GeV anomaly and its implications, *Astropart. Phys.*, **29**:25–29, 2008, arXiv:0705.4311 [astro-ph].

[1800] F. W. Stecker and A. J. Tylka. Spectra, fluxes and observability of gamma-rays from dark matter annihilation in the galaxy, *Astrophys. J.*, **343**:169, 1989.

[1801] F. D. Steffen. Gravitino dark matter and cosmological constraints, *JCAP*, **0609**:001, 2006, hep-ph/0605306.

[1802] G. Steigman. Cosmology confronts particle physics, *Ann. Rev. Nucl. Part. Sci.*, **29**:313–338, 1979.

[1803] G. Steigman. Primordial nucleosynthesis in the precision cosmology era, *Ann. Rev. Nucl. Part. Sci.*, **57**:463–491, 2007, arXiv:0712.1100 [astro-ph].

[1804] G. Steigman, K. A. Olive, and D. N. Schramm. Cosmological constraints on superweak particles, *Phys. Rev. Lett.*, **43**:239, 1979.

[1805] G. Steigman and M. S. Turner. Cosmological constraints on the properties of weakly interacting massive particles, *Nucl. Phys.*, B**253**:375–386, 1985.

[1806] P. J. Steinhardt and M. S. Turner. Saving the invisible axion, *Phys. Lett.*, B**129**:51, 1983.

[1807] K. R. Stewart, J. S. Bullock, R. H. Wechsler, A. H. Maller, and A. R. Zentner. Merger histories of galaxy halos and implications for disk survival, *Astrophys. J.*, **683**:597, 2008, arXiv:0711.5027 [astro-ph].

[1808] D. Stiff and L. M. Widrow. Fine structure of dark matter halos and its effect on terrestrial detection experiments, *Phys. Rev. Lett.*, **90**:211301, 2003, astro-ph/0301301.

[1809] D. Stockinger. The muon magnetic moment and supersymmetry, *J. Phys.*, G**34**:R45–R92, 2007, hep-ph/0609168.

[1810] F. Stoehr. Circular velocity profiles of dark matter haloes, *Mon. Not. Roy. Astron. Soc.*, **365**:147–152, 2006.

[1811] F. Stoehr, S. D. M. White, V. Springel, G. Tormen, and N. Yoshida. Dark matter annihilation in the Milky Way's halo, *Mon. Not. Roy. Astron. Soc.*, **345**:1313, 2003, astro-ph/0307026.

[1812] F. Stoehr, S. D. M. White, G. Tormen, and V. Springel. The Milky Way's satellite population in a LambdaCDM universe, *Mon. Not. Roy. Astron. Soc.*, **335**:L84–L88, 2002, astro-ph/0203342.

[1813] L. E. Strigari, J. S. Bullock, and M. Kaplinghat. Determining the nature of dark matter with astrometry, *Astrophys. J.*, **657**:L1–L4, 2007, astro-ph/0701581.

[1814] L. E. Strigari *et al.* A large dark matter core in the Fornax dwarf spheroidal galaxy?, *Astrophys. J.*, **652**:306–312, 2006, astro-ph/0603775.

[1815] L. E. Strigari *et al.* Redefining the missing satellites problem, 2007, arXiv:0704.1817 [astro-ph].

[1816] L. E. Strigari *et al.* A common mass scale for satellite galaxies of the Milky Way, *Nature*, **454**:1096–1097, 2008, arXiv:0808.3772 [astro-ph].

[1817] L. E. Strigari, S. M. Koushiappas, J. S. Bullock, and M. Kaplinghat. Precise constraints on the dark matter content of Milky Way dwarf galaxies for gamma-ray experiments, *Phys. Rev.*, D**75**:083526, 2007, astro-ph/0611925.

[1818] L. E. Strigari, S. M. Koushiappas, J. S. Bullock, *et al.* The most dark-matter-dominated galaxies: Predicted gamma-ray signals from the faintest Milky Way dwarfs, *Astrophys. J.*, **678**:614–620, 2008, arXiv:0709.1510.

[1819] A. W. Strong and I. V. Moskalenko. Propagation of cosmic-ray nucleons in the Galaxy, *Astrophys. J.*, **509**:212–228, 1998.

[1820] A. W. Strong and I. V. Moskalenko. Models for Galactic cosmic-ray propagation, *Adv. Space Res.*, **27**:717–726, 2001, astro-ph/0101068.

[1821] A. W. Strong and I. V. Moskalenko. New developments in the GAL-PROP CR propagation model, 2001, astro-ph/0106504.

[1822] A. W. Strong, I. V. Moskalenko, and V. S. Ptuskin. Cosmic-ray propagation and interactions in the Galaxy, *Ann. Rev. Nucl. Part. Sci.*, **57**:285–327, 2007, astro-ph/0701517.

[1823] A. W. Strong, I. V. Moskalenko, and O. Reimer. Diffuse Galactic continuum gamma rays. A model compatible with EGRET data and cosmic-ray measurements, *Astrophys. J.*, **613**:962–976, 2004, astro-ph/0406254.

[1824] L. Struder *et al.* The European Photon Imaging Camera on XMM-Newton: The pn-CCD camera, *Astron. Astrophys.*, **365**:L18–L26, 2001.

[1825] A. Strumia and F. Vissani. Neutrino masses and mixings and..., 2006, hep-ph/0606054.

[1826] L. Susskind. The anthropic landscape of string theory, 2003, hep-th/0302219.

[1827] G. 't Hooft. Computation of the quantum effects due to a four-dimensional pseudoparticle, *Phys. Rev.*, D14:3432–3450, 1976.

[1828] G. 't Hooft. Symmetry breaking through Bell–Jackiw anomalies, *Phys. Rev. Lett.*, **37**:8–11, 1976.

[1829] G. 't Hooft *et al.* (eds.). *Recent Developments in Gauge Theories. Proceedings Nato Advanced Study Institute*, Cargese, France, August 26 – September 8, 1979. Nato Advanced Study Institutes Series B, Physics, 1980.

[1830] M. Tada, Y. Kishimoto, M. Shibata, *et al.* Manipulating ionization path in a Stark map: Stringent schemes for the selective field ionization in highly excited Rb Rydberg [rapid communication], *Phys. Lett. A*, **303**:285–291, 2002.

[1831] O. Tajima *et al.* Search for invisible decay of the Upsilon(1S), *Phys. Rev. Lett.*, **98**:132001, 2007, hep-ex/0611041.

[1832] R. Takahashi and T. Chiba. Weak lensing of galaxy clusters in modified Newtonian dynamics, *Astrophys. J.*, **671**:45–52, December 2007.

[1833] T. Takahashi *et al. Phys. Rev.*, A12:1771, 1975.

[1834] T. Tamaki. Post-Newtonian parameters in the tensor-vector-scalar theory, *Phys. Rev.*, D77:124020, 2008, arXiv:0803.4309 [gr-qc].

[1835] L. C. Tan and L. K. Ng. Parametrization of anti-p invariant cross-section in p p collisions using a new scaling variable, *Phys. Rev.*, D26:1179–1182, 1982.

[1836] L. C. Tan and L. K. Ng. Calculation of the equilibrium antiproton spectrum, *J. Phys. G Nucl. Phys.*, **9**:227–242, 1983.

[1837] L. C. Tan and L. K. Ng. Parametrization of hadron inclusive cross-sections in p p collisions extended to very low-energies, *J. Phys.*, G**9**:1289–1308, 1983.

[1838] M. Taoso, S. Ando, G. Bertone, and S. Profumo. Angular correlations in the cosmic gamma-ray background from dark matter annihilation around intermediate-mass black holes, 2008, arXiv:0811.4493 [astro-ph].

[1839] M. Taoso, G. Bertone, and A. Masiero. Dark matter candidates: A ten-point test, *JCAP*, **0803**:022, 2008, arXiv:0711.4996 [astro-ph].

[1840] M. Taoso, G. Bertone, G. Meynet, and S. Ekstrom. Dark matter annihilations in Pop III stars, 2008, arXiv:0806.2681 [astro-ph].

[1841] A. Tasitsiomi. The state of the cold dark matter models on galactic and subgalactic scales, *Int. J. Mod. Phys. D*, **12**:1157–1196, 2003.

[1842] A. Tasitsiomi, A. V. Kravtsov, S. Gottlober, and A. A. Klypin. Density profiles of LCDM clusters, *Astrophys. J.*, **607**:125–139, 2004, astro-ph/0311062.

[1843] V. Tatischeff and J. P. Thibaud. Is 6-Li in metal-poor halo stars produced in situ by solar-like flares?, *Astron. Astrophys.*, **469**:265, 2007, astro-ph/0610756.

[1844] J. E. Taylor and A. Babul. The evolution of substructure in galaxy, group and cluster haloes I: Basic dynamics, *Mon. Not. Roy. Astron. Soc.*, **348**:811, 2004, astro-ph/0301612.

[1845] J. E. Taylor and A. Babul. The evolution of substructure in galaxy, group and cluster haloes III: Comparison with simulations, *Mon. Not. Roy. Astron. Soc.*, **364**:535–551, 2005, astro-ph/0410049.

[1846] J. E. Taylor and J. F. Navarro. The phase-space density profiles of cold dark matter halos, *Astrophys. J.*, **563**:483–488, 2001, astro-ph/0104002.

[1847] J. E. Taylor and J. Silk. The clumpiness of cold dark matter: Implications for the annihilation signal, *Mon. Not. Roy. Astron. Soc.*, **339**:505, 2003, astro-ph/0207299.

[1848] I. Tereno, C. Schimd, J.-P. Uzan, M. Kilbinger, F. H. Vincent, and L. Fu. CFHTLS weak-lensing constraints on the neutrino masses, 2008, arXiv:0810.0555.

[1849] A. G. R. Thomas *et al.* The effect of laser focusing conditions on propagation and monoenergetic electron production in laser wakefield accelerators, *Phys. Rev. Lett.*, **98**:095004, 2007, physics/0701186.

[1850] S. A. Thomas, F. B. Abdalla, and J. Weller. Constraining modified gravity and growth with weak lensing, 2008, arXiv:0810.4863.

[1851] S. D. Thomas. Baryons and dark matter from the late decay of a supersymmetric condensate, *Phys. Lett.*, B**356**:256–263, 1995, hep-ph/9506274.

[1852] L. Tian, H. Hoekstra, and H. Zhao. The relation between stellar mass and weak lensing signal around galaxies: Implications for MOND, 2008, arXiv:0810.2826.

[1853] M. Tigner. A possible apparatus for electron clashing-beam experiments, *Nuovo Cim.*, **37**:1228–1231, 1965.

[1854] J. L. Tinker *et al.* Toward a halo mass function for precision cosmology: the limits of universality, 2008, arXiv:0803.2706 [astro-ph].

[1855] O. Tiret and F. Combes. Evolution of spiral galaxies in modified gravity, *Astron. Astrophys.*, **464**:517–528, 2007.

[1856] O. Tiret and F. Combes. Evolution of spiral galaxies in modified gravity. II. Gas dynamics, *Astron. Astrophys.*, **483**:719–726, 2008, arXiv:0803.2631.

[1857] M. Toharia. Higgs–radion mixing with enhanced di-photon signal, 2008, arXiv:0809.5245 [hep-ph].

[1858] E. J. Tollerud, J. S. Bullock, L. E. Strigari, and B. Willman. Hundreds of Milky Way satellites? Luminosity bias in the satellite luminosity function, 2008, arXiv:0806.4381 [astro-ph].

[1859] S. Torii *et al.* High-energy electron observations by PPB-BETS flight in Antarctica, 2008, arXiv:0809.0760 [astro-ph].

[1860] D. R. Tovey. Inclusive SUSY searches and measurements at ATLAS, *Eur. Phys. J. direct*, C4:N4, 2002.

[1861] D. R. Tovey. On measuring the masses of pair-produced semi-invisibly decaying particles at hadron colliders, *JHEP*, **04**:034, 2008, arXiv:0802.2879 [hep-ph].

[1862] D. R. Tovey *et al.* Measurement of scintillation efficiencies and pulse-shapes for nuclear recoils in NaI(Tl) and CaF-2(Eu) at low energies for dark matter experiments, *Phys. Lett.*, B**433**:150–155, 1998.

[1863] C. Trachternach, W. J. G. de Blok, F. Walter, E. Brinks, and R. C. Kennicutt, Jr. Dynamical centers and non-circular motions in THINGS galaxies: Implications for dark matter halos, 2008, arXiv:0810.2116 [astro-ph].

[1864] S. Tremaine and J. E. Gunn. Dynamical role of light neutral leptons in cosmology, *Phys. Rev. Lett.*, **42**:407–410, 1979.

[1865] S. Trippe, S. Gillessen, O. E. Gerhard, *et al.* Kinematics of the old stellar population at the Galactic Center, 2008, arXiv:0810.1040.

[1866] R. Trotta. The isocurvature fraction after WMAP 3-year data, *Mon. Not. Roy. Astron. Soc. Lett.*, **375**:L26–L30, 2007, astro-ph/0608116.

[1867] R. Trotta, R. R. de Austri, and L. Roszkowski. Direct dark matter detection around the corner? Prospects in the constrained MSSM, *J. Phys. Conf. Ser.*, **60**:259–263, 2007.

[1868] R. Trotta, R. R. de Austri, and L. Roszkowski. Prospects for direct dark matter detection in the constrained MSSM, *New Astron. Rev.*, **51**:316–320, 2007, astro-ph/0609126.

[1869] R. Trotta, F. Feroz, M. P. Hobson, L. Roszkowski, and R. Ruiz de Austri. The impact of priors and observables on parameter inferences in the Constrained MSSM, 2008, arXiv:0809.3792 [hep-ph].

[1870] C. Tsallis. Possible generalization of Boltzmann–Gibbs statistics, *J. Stat. Phys.*, **52**:479–487, 1988.

[1871] D. Tucker-Smith and N. Weiner. Inelastic dark matter, *Phys. Rev.*, **D64**:043502, 2001, hep-ph/0101138.

[1872] D. Tucker-Smith and N. Weiner. The status of inelastic dark matter, *Phys. Rev.*, **D72**:063509, 2005, hep-ph/0402065.

[1873] M. S. Turner. Quantitative analysis of the thermal damping of coherent axion oscillations, *Phys. Rev.*, **D32**:843, 1985.

[1874] M. S. Turner. Axions from SN 1987a, *Phys. Rev. Lett.*, **60**:1797, 1988.

[1875] M. S. Turner. Windows on the Axion, *Phys. Rep.*, **197**:67–97, 1990.

[1876] M. S. Turner and F. Wilczek. Inflationary axion cosmology, *Phys. Rev. Lett.*, **66**:5–8, 1991.

[1877] C. Tyler. Particle dark matter constraints from the Draco dwarf galaxy, *Phys. Rev.*, **D66**:023509, 2002, astro-ph/0203242.

[1878] J. A. Tyson, R. A. Wenk, and F. Valdes. Detection of systematic gravitational lens galaxy image alignments - Mapping dark matter in galaxy clusters, *Astrophys. J.*, **349**:L1–L4, 1990.

[1879] P. Ullio and L. Bergstrom. Neutralino annihilation into a photon and a Z boson, *Phys. Rev.*, **D57**:1962–1971, 1998, hep-ph/9707333.

[1880] P. Ullio, L. Bergstrom, J. Edsjö, and C. G. Lacey. Cosmological dark matter annihilations into gamma-rays: A closer look, *Phys. Rev.*, **D66**:123502, 2002, astro-ph/0207125.

[1881] P. Ullio and M. Kamionkowski. Velocity distributions and annual-modulation signatures of weakly interacting massive particles, *JHEP*, **03**:049, 2001, hep-ph/0006183.

[1882] P. Ullio, M. Kamionkowski, and P. Vogel. Spin dependent WIMPs in DAMA?, *JHEP*, **07**:044, 2001, hep-ph/0010036.

[1883] P. Ullio, H. S. Zhao, and M. Kamionkowski. A dark-matter spike at the Galactic center?, *Phys. Rev.*, D**64**:043504, 2001, astro-ph/0101481.

[1884] W. G. Unruh and R. M. Wald. On damping mechanisms for coherent oscillations of axions, *Phys. Rev.*, D**32**:831, 1985.

[1885] M. Urban *et al.* Searching for TeV dark matter by atmospheric Cherenkov techniques, *Phys. Lett.*, B**293**:149–156, 1992, hep-ph/9208255.

[1886] C. Vafa and E. Witten. Parity conservation in QCD, *Phys. Rev. Lett.*, **53**:535, 1984.

[1887] O. Valenzuela, G. Rhee, A. Klypin, *et al.* Is there evidence for flat cores in the halos of dwarf galaxies? The case of NGC 3109 and NGC 6822, *Astrophys. J.*, **657**:773–789, 2007.

[1888] J. Valiviita and V. Muhonen. Correlated adiabatic and isocurvature CMB fluctuations in the wake of WMAP, *Phys. Rev. Lett.*, **91**:131302, 2003, astro-ph/0304175.

[1889] G. B. van Albada. Evolution of clusters of galaxies under gravitational forces, *Astron. J.*, **66**:590, 1961.

[1890] K. Van Bibber, N. R. Dagdeviren, S. E. Koonin, A. Kerman, and H. N. Nelson. Proposed experiment to produce and detect light pseudoscalars, *Phys. Rev. Lett.*, **59**:759–762, 1987.

[1891] K. van Bibber, P. M. McIntyre, D. E. Morris, and G. G. Raffelt. A practical laboratory detector for solar axions, *Phys. Rev.*, D**39**:2089, 1989.

[1892] F. C. van den Bosch, T. Abel, R. A. C. Croft, L. Hernquist, and S. D. M. White. The angular momentum of gas in proto-galaxies I. Implications for the formation of disk galaxies, *Astrophys. J.*, **576**:21–35, 2002, astro-ph/0201095.

[1893] L. Van Waerbeke and Y. Mellier. Gravitational lensing by large scale structures: A review, 2003, arXiv:astro-ph/0305089.

[1894] L. Van Waerbeke, Y. Mellier, T. Erben, *et al.* Detection of correlated galaxy ellipticities from CFHT data: first evidence for gravitational lensing by large-scale structures, *Astron. Astrophys.*, **358**:30–44, 2000.

[1895] L. Van Waerbeke, Y. Mellier, M. Radovich, *et al.* Cosmic shear statistics and cosmology, *Astron. Astrophys.*, **374**:757–769, 2001.

[1896] E. Vasiliev. Dark matter annihilation near a black hole: Plateau versus weak cusp, *Phys. Rev. D*, **76**:103532, 2007, 0707.3334.

[1897] E. Vasiliev and M. Zelnikov. Dark matter dynamics in the galactic center, *Phys. Rev. D*, **78**:083506, 2008, arXiv:0803.0002.

[1898] J. D. Vergados. Modulation effect with realistic velocity dispersion of supersymmetric dark matter, *Phys. Rev. Lett.*, **83**:3597–3600, 1999.

[1899] J. D. Vergados, S. H. Hansen, and O. Host. The impact of going beyond the Maxwell distribution in direct dark matter detection rates, *Phys. Rev.*, D**77**:023509, 2008, arXiv:0711.4895 [astro-ph].

[1900] J. D. Vergados and D. Owen. New velocity distribution for cold dark matter in the context of the Eddington theory, *Astrophys. J.*, **589**: 17–28, 2003, astro-ph/0203293.

[1901] M. Viel *et al.* How cold is cold dark matter? Small scales constraints from the flux power spectrum of the high-redshift Lyman-alpha forest, *Phys. Rev. Lett.*, **100**:041304, 2008, arXiv:0709.0131 [astro-ph].

[1902] M. Viel, J. Lesgourgues, M. G. Haehnelt, S. Matarrese, and A. Riotto. Constraining warm dark matter candidates including sterile neutrinos and light gravitinos with WMAP and the Lyman- alpha forest, *Phys. Rev.*, D**71**:063534, 2005, astro-ph/0501562.

[1903] M. Viel, J. Lesgourgues, M. G. Haehnelt, S. Matarrese, and A. Riotto. Can sterile neutrinos be ruled out as warm dark matter candidates?, *Phys. Rev. Lett.*, **97**:071301, 2006, astro-ph/0605706.

[1904] A. Vilenkin and A. E. Everett. Cosmic strings and domain walls in models with Goldstone and pseudoGoldstone bosons, *Phys. Rev. Lett.*, **48**:1867–1870, 1982.

[1905] A. Vilenkin and L. H. Ford. Gravitational effects upon cosmological phase transitions, *Phys. Rev.*, D**26**:1231, 1982.

[1906] A. Vilenkin and T. Vachaspati. Radiation of Goldstone bosons from cosmic strings, *Phys. Rev.*, D**35**:1138, 1987.

[1907] F. Vissani. Do experiments suggest a hierarchy problem?, *Phys. Rev.*, D**57**:7027–7030, 1998, hep-ph/9709409.

[1908] M. Vogelsberger *et al.* Phase-space structure in the local dark matter distribution and its signature in direct detection experiments, 2008, arXiv:0812.0362 [astro-ph].

[1909] M. Vogelsberger, S. D. M. White, A. Helmi, and V. Springel. The fine-grained phase-space structure of Cold Dark Matter halos, 2007, arXiv:0711.1105 [astro-ph].

[1910] M. Volonteri, P. Madau, and F. Haardt. The formation of galaxy stellar cores by the hierarchical merging of supermassive black holes, *Astrophys. J.*, **593**:661–666, 2003.

[1911] M. I. Vysotsky, A. D. Dolgov, and Ya. B. Zeldovich. Cosmological limits on the masses of neutral leptons, *JETP Lett.*, **26**:188, 1977.

[1912] R. V. Wagoner, W. A. Fowler, and F. Hoyle. On the synthesis of elements at very high temperatures, *Astrophys. J.*, **148**:3–49, 1967.

[1913] R. M. Wald. *Quantum Field Theory in Curved Space-time and Black Hole Thermodynamics*, Chicago University Press, 1994.

[1914] M. G. Walker, M. Mateo, and E. W. Olszewski. Systemic proper motions of Milky Way satellites from stellar redshifts: the Carina, Fornax, Sculptor and Sextans dwarf spheroidals, 2008, arXiv:0810.1511 [astro-ph].

[1915] M. G. Walker, M. Mateo, E. W. Olszewski, *et al.* Velocity dispersion profiles of seven dwarf spheroidal galaxies, *Astrophys. J*, **667**:L53–L56, 2007, arXiv:0708.0010.

[1916] D. Walsh, R. F. Carswell, and R. J. Weymann. 0957 + 561 A, B – Twin quasistellar objects or gravitational lens, *Nature*, **279**:381–384, 1979.

[1917] S. Walsh, B. Willman, and H. Jerjen. The invisibles: A detection algorithm to trace the faintest Milky Way satellites, 2008, arXiv:0807.3345 [astro-ph].

[1918] J. Wambsganss. Microlensing search for dark matter at all mass scales. In Y. Mellier and G. Meylan, eds., *Gravitational Lensing Impact on Cosmology*, volume 225 of *IAU Symposium*, 321–332, 2005.

[1919] C. J. Wang, X. X. Martoff, and E. Kaczanowicz. Thin film geau bolometers for particle detection, *Physica B: Cond. Mat.*, **194–196**:11–12, 1994.

[1920] F. Wang and J. M. Yang. SuperWIMP dark matter scenario in light of WMAP, *Eur. Phys. J.*, **C38**:129–133, 2004, hep-ph/0405186.

[1921] L.-T. Wang and I. Yavin. Spin measurements in cascade decays at the LHC, *JHEP*, **04**:032, 2007, hep-ph/0605296.

[1922] S. Wang, L. Hui, M. May, and Z. Haiman. Is modified gravity required by observations? An empirical consistency test of dark energy models, *Phys. Rev. D*, **76**:063503, 2007, arXiv:0705.0165.

[1923] G. Weiglein *et al.* Physics interplay of the LHC and the ILC, *Phys. Rep.*, **426**:47–358, 2006, hep-ph/0410364.

[1924] M. D. Weinberg and N. Katz. The bar–halo interaction–I. From fundamental dynamics to revised N-body Requirements, *Mon. Not. Roy. Astron. Soc.*, **375**:425–459, 2007, astro-ph/0508166.

[1925] S. Weinberg. A model of leptons, *Phys. Rev. Lett.*, **19**:1264–1266, 1967.

[1926] S. Weinberg. A new light boson?, *Phys. Rev. Lett.*, **40**:223–226, 1978.

[1927] S. Weinberg. Baryon and lepton nonconserving processes, *Phys. Rev. Lett.*, **43**:1566–1570, 1979.

[1928] S. Weinberg. Cosmological constraints on the scale of supersymmetry breaking, *Phys. Rev. Lett.*, **48**:1303, 1982.

[1929] M. White and C. S. Kochanek. Constraints on the long-range properties of gravity from weak gravitational lensing, *Astrophys. J.*, **560**:539–543, 2001.

[1930] S. D. M. White. The dynamics of rich clusters of galaxies, *Mon. Not. Roy. Astron. Soc.*, **177**:717–733, 1976.

[1931] S. D. M. White and M. J. Rees. Core condensation in heavy halos – A two-stage theory for galaxy formation and clustering, *Mon. Not. Roy. Astron. Soc.*, **183**:341–358, 1978.

[1932] L. M. Widrow. Distribution functions for cuspy dark matter density profiles, *Astrophys. J. Suppl. Series*, **131**:39–46, 2000.

[1933] L. M. Widrow, B. Pym, and J. Dubinski. Dynamical blueprints for galaxies, 2008, arXiv:0801.3414 [astro-ph].

[1934] F. Wilczek. Problem of strong p and t invariance in the presence of instantons, *Phys. Rev. Lett.*, **40**:279–282, 1978.

[1935] F. Wilczek and A. Zee. Operator analysis of nucleon decay, *Phys. Rev. Lett.*, **43**:1571–1573, 1979.

[1936] B. Willman, M. R. Blanton, A. A. West, *et al.* A new Milky Way companion: Unusual globular cluster or extreme dwarf satellite?, *Astron. J.*, **129**:2692–2700, 2005.

[1937] B. Willman *et al.* A new Milky Way companion: Unusual globular cluster or extreme dwarf satellite?, 2004, astro-ph/0410416.

[1938] B. Willman *et al.* A new Milky Way dwarf galaxy in Ursa Major, *Astrophys. J.*, **626**:L85–L88, 2005, astro-ph/0503552.

[1939] C. Winkelmann *et al.* MACHe3, a prototype for non-baryonic dark matter search: keV event detection and multicell correlation, 2005, astro-ph/0504629.

[1940] E. Witten. Mass hierarchies in supersymmetric theories, *Phys. Lett.*, **B105**:267, 1981.

[1941] E. Witten. Anti-de Sitter space and holography, *Adv. Theor. Math. Phys.*, **2**:253–291, 1998, hep-th/9802150.

[1942] D. M. Wittman, J. A. Tyson, D. Kirkman, I. Dell'Antonio, and G. Bernstein. Detection of weak gravitational lensing distortions of distant galaxies by cosmic dark matter at large scales, *Nature*, **405**:143–148, 2000.

[1943] R. Wojtak *et al.* The distribution function of dark matter in massive haloes, 2008, arXiv:0802.0429 [astro-ph].

[1944] R. Wojtak, E. L. Lokas, S. Gottloeber, and G. A. Mamon. Velocity moments of dark matter haloes, *EAS Publ. Ser.*, **20**:301–302, 2006, astro-ph/0508639.

[1945] J. Wudka. Natural and model-independent conditions for evading the limits on the scale of new physics, 2003, hep-ph/0307339.

[1946] W. Wuensch *et al.* Results of a laboratory search for cosmic axions and other weakly coupled light particles, *Phys. Rev.*, D**40**:3153, 1989.

[1947] C. E. Yaguna. Sterile neutrino production in models with low reheating temperatures, *JHEP*, **06**:002, 2007, arXiv:0706.0178 [hep-ph].

[1948] C. E. Yaguna. Gamma rays from the annihilation of singlet scalar dark matter, 2008, arXiv:0810.4267 [hep-ph].

[1949] M. Yamaguchi, M. Kawasaki, and J. Yokoyama. Evolution of axionic strings and spectrum of axions radiated from them, *Phys. Rev. Lett.*, **82**:4578–4581, 1999, hep-ph/9811311.

[1950] T. Yanagida. Horizontal gauge symmetry and masses of neutrinos, *Prog. Theor. Phys.*, **64**:1103, 1980.

[1951] B. Yanny, H. J. Newberg, E. K. Grebel *et al.* A low-latitude halo stream around the Milky Way, *Astrophys. J*, **588**:824–841, 2003.

[1952] T. Yetkin and M. Spiropulu. Inclusive SUSY searches using missing energy plus multijets in p p collisions at $\sqrt{s} = 14$ TeV with CMS, *Acta Phys. Polon.*, B**38**:661–669, 2007.

[1953] S.-C. Yoon, F. Iocco, and S. Akiyama. Evolution of the first stars with dark matter burning, 2008, arXiv:0806.2662 [astro-ph].

[1954] P. Young. Numerical models of star clusters with a central black hole. I. Adiabatic models, *Astrophys. J.*, **242**:1232–1237, 1980.

[1955] G. Zaharijas and D. Hooper. Challenges in detecting gamma-rays from dark matter annihilations in the galactic center, *Phys. Rev.*, D**73**:103501, 2006, astro-ph/0603540.

[1956] A. F. Zakharov, A. A. Nucita, F. de Paolis, and G. Ingrosso. Apoastron shift constraints on dark matter distribution at the Galactic Center, *Phys. Rev.*, D**76**:062001, 2007, arXiv:0707.4423.

[1957] E. Zavattini *et al.* Experimental observation of optical rotation generated in vacuum by a magnetic field, *Phys. Rev. Lett.*, **96**:110406, 2006, hep-ex/0507107.

[1958] E. Zavattini *et al.* New PVLAS results and limits on magnetically induced optical rotation and ellipticity in vacuum, *Phys. Rev.*, D**77**:032006, 2008, arXiv:0706.3419 [hep-ex].

[1959] Ya. B. Zeldovich. *Adv. Astron. Astrophys.*, **3**:241, 1965.

[1960] M. Zemp *et al.* The graininess of dark matter haloes, *Mon. Not. Roy. Astron. Soc.*, in press.

[1961] M. Zemp, J. Stadel, B. Moore, and C. M. Carollo. An optimum time-stepping scheme for N-body simulations, *Mon. Not. Roy. Astron. Soc.*, **376**:273–286, 2007, astro-ph/0606589.

[1962] A. R. Zentner and J. S. Bullock. Halo substructure and the power spectrum, *Astrophys. J.*, **598**:49, 2003, astro-ph/0304292.

[1963] J. Zhang *et al.* Discriminate different scenarios to account for the PAMELA and ATIC data by synchrotron and IC radiation, 2008, arXiv:0812.0522 [astro-ph].

[1964] L. Zhang and G. Sigl. Dark matter signatures in the anisotropic radio sky, *JCAP*, **0809**:027, 2008, arXiv:0807.3429 [astro-ph].

[1965] P. Zhang, M. Liguori, R. Bean, and S. Dodelson. Probing gravity at cosmological scales by measurements which test the relationship between gravitational lensing and matter overdensity, *Phys. Rev. Lett.*, **99**:141302, 2007, arXiv:0704.1932.

[1966] H. Zhao, D. J. Bacon, A. N. Taylor, and K. Horne. Testing Bekenstein's relativistic modified Newtonian dynamics with lensing data, *Mon. Not. Roy. Astron. Soc.*, **368**:171–186, 2006.

[1967] H. Zhao, J. E. Taylor, J. Silk, and D. Hooper. The first dark microhalos, *Astrophys. J.*, **654**:697, 2007, astro-ph/0508215.

[1968] H. S. Zhao. Constraining TeVeS gravity as effective dark matter and dark energy, *Int. J. Mod. Phys. D*, **16**:2055–2063, 2008, arXiv:astro-ph/0610056.

[1969] H. S. Zhao and B. Famaey. Refining the MOND interpolating function and TeVeS Lagrangian, *Astrophys. J.*, **638**:L9–L12, 2006.

[1970] A. R. Zhitnitsky. On possible suppression of the axion hadron interactions. (In Russian), *Sov. J. Nucl. Phys.*, **31**:260, 1980.

[1971] K. Zioutas and M. Davenport. Running CAST 2008–2010, 2007, CERN-SPSC-2007-021.

[1972] K. Zioutas *et al.* First results from the CERN Axion Solar Telescope (CAST), *Phys. Rev. Lett.*, **94**:121301, 2005, hep-ex/0411033.

[1973] T. G. Zlosnik, P. G. Ferreira, and G. D. Starkman. Vector-tensor nature of Bekenstein's relativistic theory of modified gravity, *Phys. Rev.*, D**74**:044037, 2006.

[1974] T. G. Zlosnik, P. G. Ferreira, and G. D. Starkman. Modifying gravity with the aether: An alternative to dark matter, *Phys. Rev.*, D**75**:044017, 2007.

[1975] D. B. Zucker *et al.* A curious Milky Way satellite in Ursa Major, *Astrophys. J.*, **650**:L41–L44, 2006, astro-ph/0606633.

[1976] D. B. Zucker *et al.* A new Milky Way dwarf satellite in Canes Venatici, *Astrophys. J.*, **643**:L103–L106, 2006, astro-ph/0604354.

[1977] F. Zwicky. Spectral displacement of extra galactic nebulae, *Helv. Phys. Acta*, **6**:110–127, 1933.

Index

Printed in the United States
By Bookmasters